Discrete Mathematics and Applications

Second Edition

TEXTBOOKS in MATHEMATICS

Series Editor: Al Boggess

PUBLISHED TITLES CONTINUED

Discrete Mathematics and Applications

Second Edition

Kevin Ferland

CRC Press
Taylor & Francis Group
Boca Raton London New York

CRC Press is an imprint of the
Taylor & Francis Group, an **informa** business

A CHAPMAN & HALL BOOK

CRC Press
Taylor & Francis Group
6000 Broken Sound Parkway NW, Suite 300
Boca Raton, FL 33487-2742

© 2017 by Taylor & Francis Group, LLC
CRC Press is an imprint of Taylor & Francis Group, an Informa business

No claim to original U.S. Government works

Printed on acid-free paper
Version Date: 20161109

International Standard Book Number-13: 978-1-4987-3065-5 (Hardback)

Visit the Taylor & Francis Web site at
http://www.taylorandfrancis.com

and the CRC Press Web site at
http://www.crcpress.com

Printed and bound by CPI Group (UK) Ltd, Croydon, CR0 4YY

To Sarah and Ethan

Contents

Preface

This book is intended for a one- or two-semester course in discrete mathematics. Such a course is typically taken by mathematics, mathematics education, and computer science majors, usually in their sophomore year. Calculus is not a prerequisite to use this book. Consequently, even freshmen with sufficient maturity can use it. Additionally, the second half of the book can be used for an introductory course in combinatorics and graph theory. The basic organization of the book can be seen in "Contents."

For those designing a one-semester course that covers the core material, the following sections accommodate this approach and make up half of the sections of this book.

Sections	Topic
1.1 through 1.5	Logic and Sets
2.1 through 2.5	Basic Proof Writing
3.1, 3.2	Divisibility, Primes, and Integer Division
4.1 through 4.4	Sequences, Summations, and Induction
5.1, 5.3, 5.4	Basic Relations and Functions
6.1 through 6.3	Fundamentals of Counting
8.1 through 8.4	Fundamentals of Graph Theory
10.1	Trees

A two-semester course may most naturally be set to follow Part I and then Part II, each in order. However, there is sufficient flexibility that other organizational choices may be made as well. The dependence among the chapters of this book is roughly reflected by the following figure.*

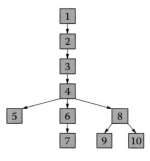

---

* Note that portions of Chapter 5 are needed in the later chapters. For example, the discussion of bijections in Section 5.4 is needed in Section 8.4 when graph isomorphisms are introduced, and the discussion of logarithms is needed in Section 10.4 when growth rates of functions are compared.

The central topic of Part I of the book is learning how to write proofs. We believe that basic set theory is the best area in which to first learn how to write proofs. The connection between properties of sets and the rules of logic (e.g., between De Morgan's Laws for sets and De Morgan's Laws for logic) is so strong that many of the proofs seem to write themselves. This should help students as they learn how to write proofs. After this introduction, we move to number theory, which serves both as a source of interesting applications and as the second setting in which to write proofs. This move also sets up the introduction to proofs by induction, presented in Chapter 4. That introduction is spread over three sections to mitigate the difficulties with induction proofs that many students experience.

In Part II, the emphasis shifts to computations and problem solving. However, we still call on our ability to prove things and to think logically. For example, to conclude that a graph has connectivity 3, it is necessary to do two things. First, a disconnecting set of size 3 must be presented. Second, an argument must be given showing that the removal of fewer than three vertices will not disconnect the graph. The focus of Part II is on combinatorics and graph theory.

The Introduction is a single section that should take approximately one class to cover. Some users may even skip it or assign it as reading. However, it is included to emphasize the link between Parts I and II.

A conscious effort has been made to give suggestive names to the theorems in this book whenever reasonable, with the common names used whenever possible (e.g., the Rational Roots Theorem and the Binomial Theorem). Consequently, theorems may be referred to by name rather than by an unenlightening number. The names should remind the reader of the content of the theorem and hence reduce the amount of page flipping. The index gives page references when desired.

The more than 3600 exercises throughout this book have been purposefully structured. Students are provided a rich supply of straightforward exercises before they encounter those that stretch their thinking. Several exercises that are very much like the examples in the text are included. These provide the students with ample opportunity to sharpen their teeth before moving on to bigger prey. Beyond the exercises that reinforce the central concepts, there are many that explore further both the theory and applications of these concepts. Exercises that may be particularly challenging are marked with an asterisk (*).

Solutions are provided in the back of the book for the odd-numbered exercises from each section and all of the exercises from the review sections. Consequently, a balance is maintained between the odd and even exercises. That is, each odd-numbered exercise is generally followed by a similar even-numbered exercise. Instructors wanting to assign homework without solution references can assign the even problems. Students can use the odd exercises to help them with the even ones. Note that an answer listed in the back may be more brief than that required by the exercise. In particular, exercises demanding proofs are often answered in the back with sketches. Students are expected to flesh out the answer given to obtain the answer requested.

Two appendices containing background material are also provided. In Appendix A, we list the fundamental properties of real numbers and integers that are assumed throughout the book. Appendix B describes the pseudocode in which algorithms are presented throughout the book.

Strengths of *Discrete Mathematics*

- Range, depth, and quantity of exercises and applications, including additional problems that provide extra challenge, give instructors flexible assignment options.

- Mathematics is written with rigor and precision. Definitions and theorems are consistently and clearly highlighted.

- Solid foundation in learning how to write proofs.

- Engaging writing style makes the presentation especially appealing to students.

- Hundreds of worked-out examples feature a wide variety of applications to illustrate concepts and enhance understanding.

Improvements from the first edition

- Hundreds of new exercises and updated exercises with improved organization

- Updated and expanded exposition throughout, including improved clarity in a number of proofs

- Further treatment of encryption calculations

- Reorganization of introduction to graphs and matrices

Supplements

The following ancillaries for this book can be found at https://www.crcpress.com/9781498730655.

Online instructor resource

Complete solutions are provided to all exercises in the text. A test bank with over 400 test items is also provided.

Additionally, lecture notes and teaching tips are given for each section of the text.

Student solutions guide

Solutions to odd-numbered section exercises and *all* the review exercises are provided. The guide also includes a valuable chapter review and summary of key definitions.

Acknowledgments

I have many thanks to give for the help that I have received at various stages in the development of this book. I am grateful to Rachel Hykel, the textbook representative for Houghton Mifflin, who enabled me to present my idea for a new discrete mathematics book in 2002. I thank Richard Stratton and Bob Ross for giving me the chance to move forward with this second edition of the book.

I thank Bloomsburg University for giving me the opportunity to write and use this book as it has developed. I am also grateful to my many students who used early versions and the first edition of this book. They contributed immensely to reducing the number of errors in the current edition.

In addition to the many reviewers who provided valuable feedback at all stages of my writing, I am grateful to several of my colleagues. Those who made a particular effort to help me with this project include Bill Calhoun, Frank Dangello, Keith Ferland, Jaiwant Mulik, Reza Noubary, Jim Pomfret, and Dan Younger. Additionally, I am forever grateful to my late PhD advisor, L. Gaunce Lewis, for teaching me how to write mathematics.

Finally, I thank my family for all of their love and support. They empower me throughout all of my pursuits.

Kevin Ferland

Introduction: Representing Numbers

Throughout this book we encounter **binary sequences**, whose terms have only two possible values. In Part I, we consider sequences of **Boolean variables**; the terms of such sequences can take on either the value True or the value False (briefly T or F). For example, in Part II, sequences of coin flips are considered. There, each coin flip results in either Heads or Tails (briefly H or T). Both of these types of sequences have the same structure as a sequence whose terms are either 1 or 0. Such sequences are used to store data in computers, since a 1 can be represented by the presence of an electrical signal and a 0 by its absence. In that setting, each term is called a **bit**, which is short for *binary digit*.

The primary aim of this introduction is to study sequences of 1's and 0's and learn how each nonnegative integer can be represented by such a sequence. The key to understanding how integers can be so represented is to recall how numbers are ordinarily written in the decimal system.

I.1 Decimal (base ten)

When we see numbers written like 3527 and 60409, we know their values because we understand them to be represented in a familiar form. The **decimal** system, or base-ten place system, for numbers is taught in elementary school. For example, in the number 3527, we learn that 7 is the units digit, 2 is the tens digit, 5 is the hundreds digit, and 3 is the thousands digit. That is,

$$3527 = (3)1000 + (5)100 + (2)10 + (7)1$$
$$= (3)10^3 + (5)10^2 + (2)10^1 + (7)10^0.$$

Similarly,

$$60409 = (6)10^4 + (0)10^3 + (4)10^2 + (0)10^1 + (9)10^0.$$

In the sums above, the exponent on the base ten starts at 0 for the rightmost term and increases by 1 as we move from right to left. In base ten, only the digits $0, 1, \ldots, 9$ are used. The place system then allows any nonnegative integer value to be represented. The reason why base ten is used is simply because we have ten fingers. However, the method of this numbering system can be used for any base $s \geq 2$, where s is an integer.

I.2 Base s

As we will be able to prove in Section 4.5, the system used to represent numbers in base ten can be generalized to represent numbers in any integer base $s \geq 2$. To write integers in base s, digits representing $0, 1, \ldots, s - 1$ are needed. If $s \leq 10$, then the familiar symbols are used (and $s, s + 1, \ldots, 9$ are discarded). If $s > 10$, then symbols for the values $10, 11, \ldots, s - 1$ need to be chosen. For example, for $s = 16$ (the **hexadecimal** system), the convention is to use the digits $0, 1, \ldots, 9, a, b, c, d, e, f$. That is, ten is represented by a, eleven is represented by b, and so on.

In base s, a nonnegative integer n is represented in the form

$$n = a_k a_{k-1} \cdots a_1 a_0 \quad \text{(in base } s\text{)},$$

where the digits $a_k, a_{k-1}, \ldots, a_1, a_0$ represent elements from the set $\{0, 1, \ldots, s - 1\}$. The corresponding value of n is determined by the equation

$$n = a_k s^k + a_{k-1} s^{k-1} + \cdots + a_1 s^1 + a_0 s^0.$$

Our primary focus here is on the base $s = 2$. That is, we focus on **binary numbers**. However, some other related and important bases are also explored.

I.3 Binary (base two)

For binary numbers, 0 and 1 are the only digits used to represent nonnegative integers. We start by practicing the skill of reading binary numbers.

Example I.1.   Determine the decimal value of the following binary numbers:

(a) 110101.

$$(1)2^5 + (1)2^4 + (0)2^3 + (1)2^2 + (0)2^1 + (1)2^0$$
$$= 32 + 16 + 0 + 4 + 0 + 1 = 53$$

$$\text{(in base ten)}$$

TABLE I.1
All 3-digit binary numbers

Binary numbers	Base-ten values
000	0
001	1
010	2
011	3
100	4
101	5
110	6
111	7

(b) 01101110.

$$(0)2^7 + (1)2^6 + (1)2^5 + (0)2^4 + (1)2^3 + (1)2^2 + (1)2^1$$
$$+(0)2^0 = 0 + 64 + 32 + 0 + 8 + 4 + 2 + 0 = 110$$

(in base ten)

Notice in Example I.1(b) that the base-ten result is 110. When working with multiple bases, it is important to specify the base in which a number is represented. Otherwise, a number like 110 could be mistakenly interpreted as binary. Note also in Example I.1(b) that the first digit 0 does not contribute to the value of the integer. Only the digits starting from the leftmost nonzero digit contribute to the result. However, sometimes padding a number with zeros to the left to obtain a certain fixed length is convenient. Table I.1 lists the values of all the 3-digit binary numbers. There are $8 = 2^3$ binary numbers with 3 digits. In general, there are 2^n binary numbers with n digits.

I.4 Connecting Parts I and II

The numerical values of binary sequences provide a natural way to order them. For Boolean sequences, identifying T with 1 and F with 0 gives each sequence a numerical value. Similarly, a sequence of coin flips can be given a numerical value by identifying H with 1 and T with 0. Consequently, this provides a systematic way to list all Boolean sequences and all sequences of coin flips of some fixed length n. Under our identifications, Table I.2 lists the possible sequences of length 3 (in increasing order). In Part I of this book, we study problems related to the second column of Table I.2. In Part II, we study problems related to the third column of Table I.2. The basic notion of binary numbers provides a link between these two considerations.

TABLE I.2
Binary sequences of length 3

Binary numbers	Boolean values	Coin flips
000	FFF	TTT
001	FFT	TTH
010	FTF	THT
011	FTT	THH
100	TFF	HTT
101	TFT	HTH
110	TTF	HHT
111	TTT	HHH

I.5   Expressing numbers in alternative bases

So far, we have learned how to convert from binary to decimal, but we should also be aware of how to go in the other direction. For this, there is a simple algorithm that involves integer division and the computation of remainders. After we develop sufficient proof-writing skills, in the exercises in Section 4.5, we will be asked to prove the validity of the method we now introduce in Example I.2.

Example I.2.   Determine the binary representations of the following numbers given in decimal:

(a) 49.

To start, we perform a sequence of divisions by 2. In each step, the resulting quotient is the number to be divided by 2 in the subsequent step.

$49/2$	$=$	24	remainder	1
$24/2$	$=$	12	remainder	0
$12/2$	$=$	6	remainder	0
$6/2$	$=$	3	remainder	0
$3/2$	$=$	1	remainder	1
$1/2$	$=$	0	remainder	1

The sequence of divisions is terminated when the quotient 0 is obtained. The binary representation is then given by listing the remainders in reverse order. In this case, the binary representation of 49 is

110001.

Note that $(1)2^5 + (1)2^4 + (0)2^3 + (0)2^2 + (0)2^1 + (1)2^0 = 2^5 + 2^4 + 2^0 = 49$.

(b) 72.

72/2	=	36	remainder	0
36/2	=	18	remainder	0
18/2	=	9	remainder	0
9/2	=	4	remainder	1
4/2	=	2	remainder	0
2/2	=	1	remainder	0
1/2	=	0	remainder	1

The binary representation of 72 is

$$1001000.$$

Note that $(1)2^6 + (0)2^5 + (0)2^4 + (1)2^3 + (0)2^2 + (0)2^1 + (0)2^0 = 2^6 + 2^3 = 72$.

Any nonnegative integer can be written in binary by following the procedure from Example I.2. As we shall further prove in Section 4.5, for any integer $s \geq 2$, the base-s representation of a number is obtained by a sequence of divisions by s. In the general case, the remainders are values from $0, 1, \ldots, s - 1$, which are appropriate for base-s digits. After the quotient 0 is obtained, the base-s representation is given by listing the sequence of remainders in reverse order.

I.6   Octal (base eight) and hexadecimal (base sixteen)

Two number systems related to binary are of particular importance in computer science. The **octal** system represents numbers in base eight, and the **hexadecimal** system uses base sixteen. By using the appropriate value of the base s, we can determine the values of numbers expressed using these systems just as in any other.

Example I.3.   The decimal value of the hexadecimal number $3ce0$ is found to be 15584.

$$(3)16^3 + (12)16^2 + (14)16^1 + (0)16^0 =$$

$$12288 + 3072 + 224 + 0 = 15584$$

(in base ten)

Recall that c represents 12 and e represents 14 in hexadecimal.

To convert from decimal to these number systems, we use the general procedure described after Example I.2.

Example I.4.   Writing the decimal number 93 in octal is accomplished by a sequence of divisions by 8.

93/8	=	11	remainder	5
11/8	=	1	remainder	3
1/8	=	0	remainder	1

So 135 is the octal representation of the decimal number 93.

The utility of the octal and hexadecimal systems and their connection to the binary system derives from the fact that their bases are powers of two. This makes it particularly easy to convert between binary and either of these systems. For example, since $16 = 2^4$, the hexadecimal digits of a number are obtained by grouping the binary digits into blocks of size 4, starting from the right.

Example I.5.   The binary number 1110110101 is written in hexadecimal as $3b5$.

$$\underbrace{0011}_{3} \ \underbrace{0011}_{b} \ \underbrace{0011}_{5}$$

The leftmost block is padded to its left with zeros to give it size 4. Since 1011 is the binary representation of 11, the hexadecimal digit for the middle block is b.

The hexadecimal system represents numbers using fewer digits than the binary system. Since a human can more easily parse $3b5$ than 1110110101, when binary data in a computer is printed out for human interpretation, it is often converted to a more readable system such as hexadecimal.

Exercises

In Exercises 1 through 10, convert the given binary numbers to decimal.

1. 1010

2. 1001

3. 10111

4. 11010

5. 101110

6. 110010

7. 1001011

8. 1011011

9. 10101011

10. 11110000

In Exercises 11 through 20, write in binary the numbers given in decimal.

11. 59   12. 73

13. 84   14. 95

15. 117   16. 230

17. 304   18. 500

19. 1024   20. 4096

21. List all of the possible outcomes from a sequence of four coin flips.

22. List all of the possible assignments to a sequence of four Boolean variables.

In Exercises 23 through 28, convert the numbers given in octal to decimal.

23. 163   24. 274

25. 3217   26. 2164

27. 40510   28. 50072

In Exercises 29 through 32, convert the numbers given in hexadecimal to decimal.

29. $dc9$ 30. $2ab$

31. $5a7e$ 32. $b63d$

33. 32   34. $f16$

In Exercises 35 through 38, write in octal the numbers given in decimal.

35. 59   36. 84

37. 117   38. 230

39. 140   40. 5

In Exercises 41 through 46, write in hexadecimal the numbers given in decimal.

41. 59   42. 84

43. 117   44. 230

45. 44252   46. 43962

In Exercises 47 through 50, write the given binary number in (a) octal and (b) hexadecimal.

47. 1100110011.   48. 11100011001.

49. 101100110001011.   50. 11011000001100111.

51. Write the octal number 47 in binary.

52. Write the octal number 162 in binary.

53. Write the hexadecimal number *acc* in binary.

54. Write the hexadecimal number *fda* in binary.

Questions for thought

55. What number is expressed in binary by a one followed by *m* zeros?

56. What number is expressed in binary by *m* consecutive ones?

57. Describe the binary expression of 4^n.

58.* How is multiplication by 8 accomplished with a binary number?

59.* How can a binary number be tested for divisibility by 4?

60.* How can an alternating sum of the digits of a binary number be used to test for divisibility by 3? For example, 1011 has alternating sum $-1 + 0 - 1 + 1 = -1$, and 10101 has alternating sum $1 - 0 + 1 - 0 + 1 = 3$.

61. What number is expressed in octal by *m* consecutive sevens?

62. What number is expressed in octal by *m* consecutive ones?

63.* How can the sum of the digits of an octal number be used to test divisibility by 7?

64.* How can an alternating sum of the digits of an octal number be used to test divisibility by 9?

65.* How would you convert from octal to hexadecimal without using decimal?

66.* If a number written in binary reads the same frontward and backward, will this symmetry remain when it is written in hexadecimal? Explain.

Part I

Proofs

1 Logic and Sets

In this chapter, we lay the conceptual and notational foundation for much of the book, especially Part I. Mathematical reasoning and arguments are based on the rules of logic. Time must be spent learning those rules and eliminating misconceptions that do not follow them. For example, we need to be familiar with the mathematical meaning of the word *or*, in contrast with *exclusive or*, which is sometimes intended in common language.

The notation that needs to be introduced is primarily set notation. This is first presented somewhat informally. It is considered in a more precise way after logic and quantifiers are formally established. However, we do not venture beyond a somewhat naïve approach to set theory.

Throughout the chapter, we consider some applications. Digital circuit design is our featured application of logic. A discussion of software implementation of sets is also highlighted.

1.1 Statement forms and logical equivalences

Our goal in Part I is to learn how to prove statements. However, before we can undertake this, we must study statements and their structure. Consequently, much of our work here involves truth tables and logic identities. As a closely related application, we further consider digital circuits, since their underlying structure is the same as that of statements.

Definition 1.1. A **statement** (or **proposition**) is a sentence that is either true or false, but not both.

It is easy to construct examples of statements. Of course, our primary focus will be on those of a mathematical nature.

Example 1.1. The following sentences are statements:

(a) $2 + 2 = 4$.

(b) $2 + 2 \neq 4$.

(c) $\sqrt{4} = 2$ and $\sqrt{5} > 2$.

(d) The sine function is periodic and 2π is an integer.

(e) $10^2 > 2^{10}$ or $2^{10} > 10^2$.

(f) If $e > 2$, then $e^2 > 4$.

(g) $\sqrt{-1}$ is a real number.

In Example 1.1, it turns out that sentences (a), (c), (e), and (f) are true statements, while (b), (d), and (g) are false. In some cases, such as (a), the truth or falsehood might be obvious. However, in other cases, such as (f), we see the need to understand various forms that statements might have, such as if-then statements. A study of these forms is taken up immediately after the following example showing that not every sentence is a statement.

Example 1.2. The following sentences are not statements:

(a) What is the sum $2 + 2$?

(b) Evaluate the sum $2 + 2$.

(c) This sentence is false.

Sentence (a) is a question, and (b) is a command that can either be obeyed or not, but neither can be true or false. The point is that, in particular, a statement must be a declarative sentence. Sentence (c) warrants a bit more thought, since it is a declarative sentence, so the problem must be in assigning it a value of true or false. If (c) were true, then what it says would be false. If (c) were false, then what it says would be true. Hence, (c) can be neither true nor false.

1.1.1   Statement forms

We are interested in studying statements. However, insight is gained by stepping back and examining their forms. In Example 1.1, both (c) and (d) are statements of the form "p and q", where p and q represent statements. In (c),

$$p = \text{``}\sqrt{4} = 2.\text{''}$$

$$q = \text{``}\sqrt{5} > 2.\text{''}$$

TABLE 1.1
Basic statement forms

Form	Translation
$\neg p$	**not** p (negation of p)
$p \wedge q$	p **and** q
$p \vee q$	p **or** q
$p \to q$	**if** p **then** q (p implies q)

and (c) is true because both p and q are true. In (d),

$$p = \text{"The sine function is periodic."}$$

$$q = \text{"}2\pi \text{ is an integer."}$$

and (d) is false because it is not the case that both p and q are true; in this instance, q is not true. Although statements (c) and (d) are different, their forms are the same. Formally, we say that (c) and (d) both have the **statement form**

$$p \wedge q.$$

Here, the word *and* is denoted by the symbol $\wedge$, and p and q are **statement variables**. The symbols and names for the operations used in the most basic statement forms are introduced in Table 1.1.

General statement forms, or **logical expressions**, can subsequently be built from these, as we shall see. The *definitions* of the basic logical operations $\neg, \wedge, \vee$, and $\to$ are expressed by presenting truth tables for the statement forms named in Table 1.1. A **truth table** for a logical expression is a table that displays how the truth or falsehood of the statement variables involved in the expression affect the truth or falsehood of the expression. The truth table defining the **negation** operator $\neg$ is given in Table 1.2. There we see that the negation operator simply converts each truth value to its opposite. The truth tables for the **and, or**, and **if-then** operators, respectively, $\wedge, \vee$, and $\to$, are combined in Table 1.3. We see in Table 1.3 that $p \wedge q$ is true precisely when both p and q are true. Notice that $p \vee q$ is true if p is true or if q is true, and that it is also true if both p and q are true. In fact, it suffices to say that $p \vee q$ is false precisely

TABLE 1.2
Truth table **defining** $\neg$

p	$\neg p$
F	T
T	F

TABLE 1.3
Truth table **defining** $\wedge$, $\vee$, and $\rightarrow$

p	q	$p \wedge q$	$p \vee q$	$p \rightarrow q$
F	F	F	F	T
F	T	F	T	T
T	F	F	T	F
T	T	T	T	T

when both p and q are false. The if-then, or **conditional**, statement form $p \rightarrow q$ is meant to reflect whether the truth of p implies (or forces) the truth of q. Consequently, $p \rightarrow q$ is false if and only if p is true and q is false. Notice that $p \rightarrow q$ is true whenever p is false. In that case, we say that $p \rightarrow q$ is **vacuously true**. Only in the other case, when p is true, does the truth of q matter.

Note in Tables 1.2 and 1.3 that if F and T are replaced by 0 and 1, respectively, then the portion of the tables corresponding to the input variables represents binary numbers listed in increasing order. This ordering is not required in truth tables but is used throughout this book for consistency. Variables, such as statement variables, that can only take on one of two possible values (e.g., T or F) are called **Boolean variables** and are named after the English mathematician George Boole (1815–1864), who pioneered their use.

In logical expressions involving more than one operation, the order of operations must be understood. The order of **precedence** of the basic operations listed from highest to lowest is

$$\neg,$$

$$\wedge, \vee,$$

$$\rightarrow .$$

The operations $\wedge$ and $\vee$ are listed on the same line because they are considered to be equal in precedence. Of course, any desired order of operations can be forced by using parentheses. For example, $\neg p \wedge q$ should be understood to mean $(\neg p) \wedge q$ and not $\neg(p \wedge q)$.

Remark 1.1.   In programming languages such as C++, it is commonly specified that $\wedge$, which is denoted && in C++, has higher precedence than $\vee$, which is denoted || in C++. Consequently, p && q || r would be interpreted as $(p \wedge q) \vee r$. However, in applications such as *Microsoft Excel*, it is impossible to enter an expression of the form $p \wedge q \vee r$ without explicitly entering it either as $(p \wedge q) \vee r$ or $p \wedge (q \vee r)$. The syntax used in *Excel* is discussed in Example 1.4. That the order of operations matters is seen in Example 1.8.

Example 1.3. Make a truth table for the statement form $p \wedge q \to r$.

Solution. The order of precedence dictates that $p \wedge q$ is performed first, and the operation $\to$ is performed last. In fact, in making a truth table, it is often helpful to include intermediate columns reflecting intermediate steps in the computation.

p	q	r	$p \wedge q$	$p \wedge q \to r$
F	F	F	F	T
F	F	T	F	T
F	T	F	F	T
F	T	T	F	T
T	F	F	F	T
T	F	T	F	T
T	T	F	T	F
T	T	T	T	T

$\square$

Spreadsheet software, such as *Microsoft Excel*, can be used to generate truth tables. In *Excel*, T and F need to be written out as TRUE and FALSE, respectively. The basic operations $\neg$, $\wedge$, and $\vee$ are accomplished using the logical functions NOT, AND, and OR, respectively. A way to accomplish $\to$ is discussed after Example 1.10.

Example 1.4. Make a truth table for the statement form $\neg(p \vee q)$.

Solution. Here, we demonstrate how a truth table could be set up using *Excel*.

	A	B	C	D
1	p	q	$p \vee q$	$\neg (p \vee q)$
2	FALSE	FALSE	FALSE	TRUE
3	FALSE	TRUE	TRUE	FALSE
4	TRUE	FALSE	TRUE	FALSE
5	TRUE	TRUE	TRUE	FALSE

For example, the values in cells C2 and D2 are obtained by the formulas C2 = OR(A2, B2) and D2 = NOT(C2), respectively. The rest of columns C and D are easily filled in using copy and paste. $\square$

TABLE 1.4
Truth table defining $\oplus$ and $\leftrightarrow$

p	q	$p \oplus q$	$p \leftrightarrow q$
F	F	F	T
F	T	T	F
T	F	T	F
T	T	F	T

There are other operations that are used for convenience but are defined in terms of the basic operations.

Definition 1.2.

(a) The **exclusive or** operation $\oplus$ is defined by

$$p \oplus q = (p \vee q) \wedge \neg(p \wedge q).$$

(b) The **if and only if** operation $\leftrightarrow$ is defined by

$$p \leftrightarrow q = (p \rightarrow q) \wedge (q \rightarrow p).$$

Note that **iff** is also used to denote $\leftrightarrow$.

The truth tables for $\oplus$ and $\leftrightarrow$ are combined in Table 1.4. Notice that $p \oplus q$ is different from $p \vee q$ in that $p \oplus q$ is false when both p and q are true. Thus, $p \oplus q$ is true when exactly one of p or q is true. Although *exclusive or* is sometimes what is intended in English when *or* is used, as in "Do you want hamburgers or pizza for dinner?," we must carefully distinguish between $\vee$ and $\oplus$ here. The truth of $p \leftrightarrow q$ holds precisely when p and q have the same truth values.

Definition 1.3.

(a) A **tautology** is a statement form that is always true. We denote a tautology by $\underline{t}$.

(b) A **contradiction** is a statement form that is always false. We denote a contradiction by $\underline{f}$.

A statement whose form is a tautology or contradiction is also said to be a tautology or contradiction, respectively.

Example 1.5.

(a) $p \vee \neg p$ is a tautology.

Solution. In the truth table

p	$\neg p$	$p \vee \neg p$
F	T	T
T	F	T

all of the entries in the column for $p \vee \neg p$ are T, as in a column for $\underline{t}$. ☐

(b) $p \wedge \neg p$ is a contradiction.

Solution. In the truth table

p	$\neg p$	$p \wedge \neg p$
F	T	F
T	F	F

all of the entries in the column for $p \wedge \neg p$ are F, as in a column for $\underline{f}$. ☐

1.1.2 Logical equivalences

Just like we have identities in algebra, such as $a + b = b + a$, there are identities among logical expressions. They are based on an equivalence between statement forms, which we now define.

Definition 1.4. Two statement forms p and q are **logically equivalent**, written $p \equiv q$, if and only if the statement form $p \leftrightarrow q$ is a tautology. We write $p \not\equiv q$ when p and q are not logically equivalent.

Example 1.6. Verify that $\neg(p \rightarrow q) \equiv p \wedge \neg q$.

Solution.

p	q	$\neg q$	$p \rightarrow q$	$\neg(p \rightarrow q)$	$p \wedge \neg q$	$(\neg(p \rightarrow q)) \leftrightarrow (p \wedge \neg q)$
F	F	T	T	F	F	T
F	T	F	T	F	F	T
T	F	T	F	T	T	T
T	T	F	T	F	F	T

Since $(\neg(p \rightarrow q)) \leftrightarrow (p \wedge \neg q)$ is a tautology, we conclude that

$$\neg(p \rightarrow q) \equiv p \wedge \neg q.$$

It would also suffice to confirm that $\neg(p \rightarrow q)$ and $p \wedge \neg q$ have the same truth tables.   □

Example 1.6 tells us how to negate an if-then statement (an and statement is obtained).

Example 1.7. The negation of

"If Alyssa is using tax preparation software, then her tax returns will be accurate."

is

"Alyssa is using tax preparation software, and her tax returns will not be accurate."

Example 1.8 shows that the statement form $p \wedge q \vee r$ is ambiguous. Parentheses are necessary to specify an intended order of operations.

Example 1.8.   Verify that $p \wedge (q \vee r) \not\equiv (p \wedge q) \vee r$.

Solution. Consider the following truth table.

p	q	r	$q \vee r$	$p \wedge q$	$p \wedge (q \vee r)$	$(p \wedge q) \vee r$	$(p \wedge (q \vee r)) \leftrightarrow ((p \wedge q) \vee r)$
F	F	F	F	F	F	F	T
F	F	T	T	F	F	T	F
F	T	F	T	F	F	F	T
F	T	T	T	F	F	T	F
T	F	F	F	F	F	F	T
T	F	T	T	F	T	T	T
T	T	F	T	T	T	T	T
T	T	T	T	T	T	T	T

Since $(p \wedge (q \vee r)) \leftrightarrow ((p \wedge q) \vee r)$ is not a tautology, we conclude that

$$p \wedge (q \vee r) \not\equiv (p \wedge q) \vee r.$$

Of course, we can also see that $p \wedge (q \vee r)$ and $(p \wedge q) \vee r$ have different truth tables.   □

As mentioned in Examples 1.6 and 1.8, two statement forms can be seen to be, or not be, logically equivalent by comparing their truth tables.

Example 1.9. Verify that $p \oplus q \equiv (p \wedge \neg q) \vee (\neg p \wedge q)$.

Solution.

p	q	$\neg p$	$\neg q$	$p \wedge \neg q$	$\neg p \wedge q$	$p \oplus q$	$(p \wedge \neg q) \vee (\neg p \wedge q)$
F	F	T	T	F	F	F	F
F	T	T	F	F	T	T	T
T	F	F	T	T	F	T	T
T	T	F	F	F	F	F	F

Since $p \oplus q$ and $(p \wedge \neg q) \vee (\neg p \wedge q)$ have the same truth tables, we conclude that

$$p \oplus q \equiv (p \wedge \neg q) \vee (\neg p \wedge q). \qquad \square$$

Example 1.10. Verify that $p \rightarrow q \equiv \neg p \vee q$.

Solution.

p	q	$\neg p$	$p \rightarrow q$	$\neg p \vee q$
F	F	T	T	T
F	T	T	T	T
T	F	F	F	F
T	T	F	T	T

Since $p \rightarrow q$ and $\neg p \vee q$ have the same truth tables, we conclude that

$$p \rightarrow q \equiv \neg p \vee q. \qquad \square$$

Example 1.10 gives an alternative expression for an if-then statement form. Since *Excel* and other spreadsheet software do not have built-in logical functions for $\rightarrow$, this alternative expression is useful. Namely, $p \rightarrow q$ can be entered as OR(NOT(p), q).

The if-then statement form $p \rightarrow q$ is present frequently in statements we study. Consequently, it is important to understand some relatives of a given if-then statement form.

Definition 1.5. Given the statement form $p \rightarrow q$,

(a) Its **converse** is $q \rightarrow p$.

(b) Its **contrapositive** is $\neg q \rightarrow \neg p$.

(c) Its **inverse** is $\neg p \rightarrow \neg q$.

The converse of an if-then statement is important in that $p \leftrightarrow q$ combines $p \to q$ and its converse $q \to p$. In Example 1.11, we see that $p \to q$ is logically equivalent to its contrapositive $\neg q \to \neg p$. In fact, this equivalence turns out to be a very useful tool, which we exploit in Section 2.4. The inverse is notable in that it is the contrapositive of the converse.

Example 1.11.   Verify that an if-then statement is logically equivalent to its contrapositive but not to its converse and inverse.

Solution.

p	q	$\neg p$	$\neg q$	$p \to q$	$q \to p$	$\neg q \to \neg p$	$\neg p \to \neg q$
F	F	T	T	T	T	T	T
F	T	T	F	T	F	T	F
T	F	F	T	F	T	F	T
T	T	F	F	T	T	T	T

In the truth table, the column for $p \to q$ is the same as that for $\neg q \to \neg p$ and different from the columns for $q \to p$ and $\neg p \to \neg q$. Hence,

$$p \to q \not\equiv q \to p, \quad p \to q \equiv \neg q \to \neg p, \quad \text{and}$$
$$p \to q \not\equiv \neg p \to \neg q. \qquad \square$$

Example 1.12.   Suppose a computer scientist has written a program to solve a complex mathematical problem, and she guarantees the following statement to be true:

"If a solution exists, then the program terminates."

(a) Its converse is "If the program terminates, then a solution exists."

(b) Its contrapositive is "If the program does not terminate, then a solution does not exist."

(c) Its inverse is "If a solution does not exist, then the program does not terminate."

If we take the programmer's guarantee to be true, then its contrapositive must also be true. However, its converse and inverse need not be true. For example, the relevant program might be designed to terminate after examining finitely many possible solutions and finding none.

Logical equivalences are useful for manipulating and simplifying logical expressions. Theorems 1.1 and 1.2 list some of the basic logical equivalences that can be used to construct others.

Theorem 1.1.   (Basic Logical Equivalences).
Let $p, q,$ and r be statement variables. Then, the following logical equivalences hold:

(a) $\qquad\qquad\qquad \neg\neg p \equiv p.$ $\qquad\qquad$ Double Negative

(b) $\qquad\quad (p \wedge q) \wedge r \equiv p \wedge (q \wedge r).$ $\qquad$ Associativity

$\qquad\qquad\quad (p \vee q) \vee r \equiv p \vee (q \vee r).$

(c) $\qquad\qquad\qquad p \wedge q \equiv q \wedge p.$ $\qquad\qquad$ Commutativity

$\qquad\qquad\qquad p \vee q \equiv q \vee p.$

(d) $\qquad\quad p \wedge (q \vee r) \equiv (p \wedge q) \vee (p \wedge r).$ $\quad$ Distributivity

$\qquad\quad p \vee (q \wedge r) \equiv (p \vee q) \wedge (p \vee r).$

(e) $\qquad\qquad \neg(p \wedge q) \equiv \neg p \vee \neg q.$ $\qquad$ De Morgan's Laws

$\qquad\qquad \neg(p \vee q) \equiv \neg p \wedge \neg q.$

(f) $\quad$ *If $p \to q,$ then* $\quad [p \wedge q \equiv p].$ $\qquad$ Absorption Rules

$\quad$ *If $p \to q,$ then* $\quad [p \vee q \equiv q].$

Theorem 1.2.   (Interactions with Tautologies and Contradictions).
Let p be a statement variable. Then, the following logical equivalences hold:

(a) $\qquad \neg t \quad \equiv \quad f.$

$\qquad\quad \neg f \quad \equiv \quad t.$

(b) $\qquad p \wedge t \quad \equiv \quad p.$

$\qquad p \vee t \quad \equiv \quad t.$

(c) $\qquad p \wedge f \quad \equiv \quad f.$

$\qquad p \vee f \quad \equiv \quad p.$

(d) $\qquad p \wedge \neg p \quad \equiv \quad f.$

$\qquad p \vee \neg p \quad \equiv \quad t.$

(e) $\qquad t \to p \quad \equiv \quad p.$

$\qquad p \to t \quad \equiv \quad t.$

(f) $\qquad p \to f \quad \equiv \quad \neg p.$

$\qquad f \to p \quad \equiv \quad t.$

A **theorem** is a statement that has been verified to be true. Verification of the identities in Theorems 1.1 and 1.2 is left to the exercises. Beyond verification, however, it is important to internalize those identities. Only by having some understanding of the identities can one remember them and be able to use them efficiently and in complicated situations. The first four identities in Theorem 1.1 are easy to accept if one thinks of the obvious algebraic analogs involving the arithmetic operations $-$, $\cdot$, and $+$. For example, the first distributivity rule is analogous to the arithmetic distributivity rule

$$a \cdot (b + c) = (a \cdot b) + (a \cdot c). \tag{1.1}$$

Moreover, the obvious generalizations of (1.1) to more than the two summands b and c also have an analog in logical equivalences. One shortcoming of this analogy is that the second distributivity rule in Theorem 1.1 does not have an arithmetic analog.

The associativity rules imply that expressions like $p \wedge q \wedge r$ and $p \vee q \vee r$ are unambiguous, whereas we already saw in Example 1.8 that $p \wedge q \vee r$ is ambiguous. From the associativity rules, it follows that general n-fold $\wedge$ and $\vee$ statement forms

$$p_1 \wedge p_2 \wedge p_3 \wedge \cdots \wedge p_n \quad \text{and} \quad p_1 \vee p_2 \vee p_3 \vee \cdots \vee p_n$$

may be computed by associating the terms (with parentheses) to reflect any desired order of operations. For example, the association

$$p_1 \wedge (p_2 \wedge (p_3 \wedge \cdots \wedge p_n))$$

shows one such choice in which the computation is performed right to left.

De Morgan's Laws were discovered by the English mathematician Augustus De Morgan (1806–1871) and are the rules that tell us how to negate $\wedge$ and $\vee$ statement forms. Thinking about when such statement forms are false should make De Morgan's Laws seem more natural. De Morgan's Laws, like each of parts (b) through (e) of Theorem 1.1, have obvious analogs in more variables.

The Absorption Rules in Theorem 1.1 are perhaps the most subtle. We will explore their utility in Example 1.15 below. The identities in Theorem 1.2, on the other hand, should be easier to verify and internalize. In the next few examples, the identities from Theorems 1.1 and 1.2 as well as others are used to verify more identities.

Example 1.13.   Verify that $(p \wedge q \wedge r) \vee (p \wedge q \wedge \neg r) \equiv p \wedge q$.

Solution.

$$
\begin{array}{lll}
(p \wedge q \wedge r) \vee (p \wedge q \wedge \neg r) & \equiv ((p \wedge q) \wedge r) \vee ((p \wedge q) \wedge \neg r) & \text{Associativity} \\
& \equiv (p \wedge q) \wedge (r \vee \neg r) & \text{Distributivity} \\
& \equiv (p \wedge q) \wedge \underline{t} & \text{Theorem 1.2(d)} \\
& \equiv p \wedge q & \text{Theorem 1.2(b)} \\
& & \square
\end{array}
$$

Example 1.14.   Verify that $p \wedge q \to \neg r \vee \neg s \equiv r \wedge s \to \neg p \vee \neg q$.

Solution.

$$
\begin{aligned}
p \wedge q \to \neg r \vee \neg s &\equiv (p \wedge q) \to (\neg r \vee \neg s) && \text{Precedence} \\
&\equiv \neg(\neg r \vee \neg s) \to \neg(p \wedge q) && \text{Example 1.11} \\
&\equiv (\neg\neg r \wedge \neg\neg s) \to (\neg p \vee \neg q) && \text{De Morgan's Law} \\
&\equiv (r \wedge s) \to (\neg p \vee \neg q) && \text{Double Negative} \\
&\equiv r \wedge s \to \neg p \vee \neg q && \text{Precedence}
\end{aligned}
$$

$\square$

Example 1.15.   (Basic Absorption Rules).
Verify the following logical equivalences:

(a) $p \wedge (p \vee r) \equiv p$.

(b) $(q \wedge r) \vee q \equiv q$.

Solution. (a) We use the first Absorption Rule with $q = p \vee r$. It is easy to verify that $p \to p \vee r$ is a tautology. From Theorem 1.2(e), it follows that the instance of the Absorption Rule

$$(p \to p \vee r) \to (p \wedge (p \vee r) \equiv p)$$

is equivalent to $p \wedge (p \vee r) \equiv p$, the desired result.

(b) We use the second Absorption Rule with $p = q \wedge r$. Since it is easy to verify that $q \wedge r \to q \equiv t$, again Theorem 1.2(e) reduces the instance of the Absorption Rule

$$(q \wedge r \to q) \to ((q \wedge r) \vee q \equiv q)$$

to the desired result. $\square$

1.1.3 Digital circuits

The basic logic we have considered can be applied to computer hardware design. A **digital circuit** is a piece of electronic hardware that takes input from electrical signals $P_1, P_2, \ldots, P_n$ and produces output signals $S_1, S_2, \ldots, S_m$ determined by those inputs. Figure 1.1 depicts the general structure of a digital circuit, which might represent a computer chip. Each signal P_i and S_j stores a **bit** of data, which may be turned either *on* or *off* and can therefore be represented by either a 1 or a 0, respectively. To simplify our study, we shall only consider digital circuits with a single output signal S.

FIGURE 1.1
The structure of a digital circuit.

TABLE 1.5
An input–output table

Input			Output
P	Q	R	S
0	0	0	0
0	0	1	1
0	1	0	0
0	1	1	1
1	0	0	0
1	0	1	0
1	1	0	1
1	1	1	1

The action of a particular digital circuit, say, with input signals P, Q, R and output signal S, may be recorded in an **input–output table** specifying exactly how the inputs determine the output. Such a table is shown in Table 1.5 and has an obvious connection to a truth table if we replace 1 by T and 0 by F. With this strong similarity to logic, it should not be surprising that complex digital circuits can be built from very basic circuits, called **gates**, which correspond to basic logical operations. The basic gates that we shall consider are the **Inverter** (or **NOT gate**), the **AND gate**, and the **OR gate**. These are displayed and defined in Table 1.6 and correspond to the basic logical operations $\neg$, $\wedge$, and $\vee$, respectively. The basic gates can be used to build a circuit such as the one displayed in Figure 1.2. Note that wires may be split to feed signals to more than one gate. Such a splitting is marked in our diagrams with the symbol •. Other crossings without such a marking reflect no interaction of the wires and may be unavoidable when drawing more complicated circuits.

We are now ready to analyze the inner workings of a particular digital circuit built from basic gates.

TABLE 1.6
Basic gates

Gate	Inverter	AND		OR	
Symbol	P—$\triangleright$NOT$\circ$— S	P —[AND]— S, Q —		P —[OR]— S, Q —	
Input–output table	$P \mid S$ $0 \mid 1$ $1 \mid 0$	$P \mid Q \parallel S$ $0 \mid 0 \parallel 0$ $0 \mid 1 \parallel 0$ $1 \mid 0 \parallel 0$ $1 \mid 1 \parallel 1$		$P \mid Q \parallel S$ $0 \mid 0 \parallel 0$ $0 \mid 1 \parallel 1$ $1 \mid 0 \parallel 1$ $1 \mid 1 \parallel 1$	

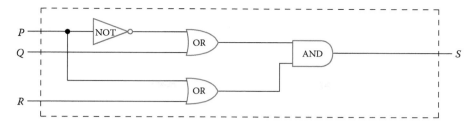

FIGURE 1.2
The inside of a digital circuit.

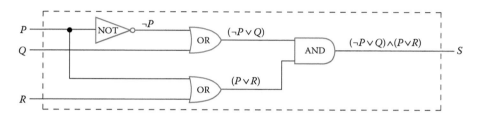

FIGURE 1.3
Tracing the digital circuit from Figure 1.2.

Example 1.16.   Trace the circuit pictured in Figure 1.2 to determine an expression for the output in terms of the input, and make an input–output table.

Solution. The steps of our trace are shown in Figure 1.3. At the output of each gate, we list the logical value of the wire obtained by applying the operation of the gate. Eventually, we discover the expression for S. It is

$$(\neg P \vee Q) \wedge (P \vee R) = S.$$

The input–output table for this expression, and thus for the circuit, is the one shown in Table 1.5.   □

We see in Example 1.16 that a digital circuit corresponds to a logical expression specifying the output in terms of the input. Consequently, given such an expression, we ought to be able to design a circuit that realizes it.

Example 1.17.   Draw a digital circuit that realizes the expression

$$(P \wedge \neg Q \wedge R) \vee (Q \wedge \neg R) = S.$$

Solution. An appropriate circuit is displayed in Figure 1.4.   □

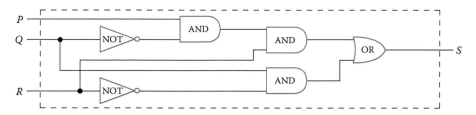

FIGURE 1.4
A circuit realizing $(P \wedge \neg Q \wedge R) \vee (Q \wedge \neg R) = S$.

Exercises

In Exercises 1 through 6, determine if the given sentence is a statement. If so, then determine whether it is true or false.

1. -1 is an integer.

2. $1 + 2 + 3 = 5$.

3. Is e positive?

4. $5^2 < 2^5$.

5. If $\pi > 0$, then compute $\sqrt{\pi}$.

6. The negation of this sentence is false.

In Exercises 7 through 12, make truth tables for the given statement forms.

7. $p \vee \neg q$.

8. $\neg p \wedge (q \vee \neg r)$.

9. $\neg p \rightarrow (q \wedge r)$.

10. $(p \vee \neg q) \rightarrow r$.

11. $(p \rightarrow q) \vee r$.

12. $p \vee (\neg q \rightarrow r)$.

In Exercises 13 through 16, use spreadsheet software, such as *Microsoft Excel*, to generate truth tables for the two given statement forms. In what rows do they differ? These can be done by hand, as well.

13. $p \vee \neg q$ and $p \rightarrow \neg q$.

14. $\neg p \wedge q$ and $\neg(p \wedge q)$.

15. $p \rightarrow (q \vee r)$ and $p \rightarrow (q \vee \neg r)$.

16. $\neg p \wedge (q \vee r)$ and $\neg p \wedge (q \vee \neg r)$.

17. Show that $p \rightarrow p \vee q$ is a tautology.

18. Show that $p \wedge (q \vee \neg p) \wedge \neg q$ is a contradiction.

In Exercises 19 through 36, verify the stated logical equivalences.

19. (a) Double Negative (Theorem 1.1(a)): $\neg\neg p \equiv p$.

    (b) Theorem 1.2(a): $\neg \underline{t} \equiv \underline{f}$ and $\neg \underline{f} \equiv \underline{t}$.

20. (a) Theorem 1.2(b): $p \wedge \underline{t} \equiv p$ and $p \vee \underline{t} \equiv \underline{t}$.

    (b) Theorem 1.2(c): $p \wedge \underline{f} \equiv \underline{f}$ and $p \vee \underline{f} \equiv p$.

21. Theorem 1.2(e): $\underline{t} \rightarrow p \equiv p$ and $p \rightarrow \underline{t} \equiv \underline{t}$.

22. Theorem 1.2(e): $p \rightarrow \underline{f} \equiv \neg p$ and $\underline{f} \rightarrow p \equiv \underline{t}$.

23. $p \vee q \equiv \neg(\neg p \wedge \neg q)$.   24. $p \wedge q \equiv \neg(\neg p \vee \neg q)$.

25. $p \rightarrow p \vee q \equiv \underline{t}$.   26. $p \wedge q \rightarrow p \equiv \underline{t}$.

27. Associativity of $\wedge$ and $\vee$: $(p \wedge q) \wedge r \equiv p \wedge (q \wedge r)$ and $(p \vee q) \vee r \equiv p \vee (q \vee r)$.

28. Commutativity of $\wedge$ and $\vee$: $p \wedge q \equiv q \wedge p$ and $p \vee q \equiv q \vee p$.

29. Associativity of $\oplus$: $(p \oplus q) \oplus r \equiv p \oplus (q \oplus r)$.

30. Commutativity of $\oplus$: $p \oplus q \equiv q \oplus p$.

31. Distributivity with $\wedge$ and $\vee$: $p \wedge (q \vee r) \equiv (p \wedge q) \vee (p \wedge r)$ and $p \vee (q \wedge r) \equiv (p \vee q) \wedge (p \vee r)$.

32. Distributivity with $\wedge$ and $\oplus$: $p \wedge (q \oplus r) \equiv (p \wedge q) \oplus (p \wedge r)$. Also see Exercise 39.

33. De Morgan's Laws: $\neg(p \wedge q) \equiv \neg p \vee \neg q$ and $\neg(p \vee q) \equiv \neg p \wedge \neg q$.

34. Absorption Rules: $(p \rightarrow q) \rightarrow (p \wedge q \leftrightarrow p)$ and $(p \rightarrow q) \rightarrow (p \vee q \leftrightarrow q)$ are tautologies.

35. $\neg(p \oplus q) \equiv p \leftrightarrow q$.

36. The inverse of an if-then statement is equivalent to its converse. That is, $(\neg p \rightarrow \neg q) \equiv q \rightarrow p$.

In Exercises 37 through 42, determine whether the given statement forms are logically equivalent. Justify your answers.

37. $(p \rightarrow q) \rightarrow r$ and $p \rightarrow (q \rightarrow r)$.

38. $(p \leftrightarrow q) \leftrightarrow r$ and $p \leftrightarrow (q \leftrightarrow r)$.

39. $p \oplus (q \wedge r)$ and $(p \oplus q) \wedge (p \oplus r)$.

40. $\neg(p \oplus q)$ and $\neg p \wedge \neg q$.

41. $p \oplus q$ and $\neg p \oplus \neg q$.

42. $\neg(p \oplus q)$ and $\neg p \oplus q$.

In Exercises 43 through 50, for the given if-then statement (form), find and simplify its (a) converse, (b) contrapositive, (c) inverse, and (d) negation.

43. $p \rightarrow \neg q$.

44. $\neg p \rightarrow \neg q$.

45. $p \wedge \neg q \rightarrow r$.

46. $p \rightarrow q \vee \neg r$.

47. If Ted's average is less than 60, then Ted has a failing grade.

48. If Ilia can afford the car, then Ilia is buying the car.

49. If George feels well, then George is going to a movie or going dancing.

50. If Anna is failing history and psychology, then Anna is not graduating.

In Exercises 51 through 56, negate the given statement.

51. $p \vee \neg q$.

52. $\neg p \wedge \neg q$.

53. $\neg p \wedge (q \vee \neg r)$.

54. $p \vee (\neg q \wedge r)$.

55. Helen's average is at least 90, and Helen is getting an A.

56. Raphael is not smiling, but Raphael is bluffing.

57. Verify the logical equivalence
$p \wedge q \rightarrow r \equiv \neg p \vee (q \rightarrow r)$ by

    (a) Making a truth table.

    (b)*Using the result from Example 1.10 and other identities.

58. Verify the logical equivalence
$(p \rightarrow q) \rightarrow (p \wedge r) \equiv p \wedge (q \rightarrow r)$ by

    (a) Making a truth table.

    (b)*Using the result from Example 1.10 and other identities.

In Exercises 59 through 66, verify the stated logical equivalences, not by making a truth table, but by using already established identities.

59. $p \wedge (q \vee r \vee s) \equiv$
$(p \wedge q) \vee (p \wedge r) \vee$
$(p \wedge s)$.

60. $\neg(p \vee q \vee r) \equiv$
$\neg p \wedge \neg q \wedge \neg r$.

61. $(p \wedge q \wedge \neg r) \vee$

    $(p \wedge \neg q \wedge r) \equiv$

    $p \wedge (q \oplus r)$.

62. $(p \wedge q \wedge \neg r) \vee$
$(p \wedge \neg q \wedge \neg r) \equiv$
$p \wedge \neg r$.

63. $p \wedge (\neg(q \wedge r)) \equiv$
$(p \wedge \neg q) \vee (p \wedge \neg r)$.

64. $\neg(p \vee (q \wedge r)) \equiv$
$\neg(p \vee q) \vee \neg(p \vee r)$.

65.* $(p \wedge q) \vee (p \wedge q \wedge r) \equiv$
$p \wedge q$.

66.* $(p \vee q) \wedge (p \vee q \vee r) \equiv$
$p \vee q$.

67.* Show that the operation $\neg$, together with any one of the operations $\wedge$, $\vee$, and $\rightarrow$, can be used to generate the other two. Hint: Example 1.10 accomplishes part of this.

68.* Can the operation $\vee$ be generated from the operations $\oplus$, $\neg$, and $\wedge$? Explain.

In Exercises 69 through 72: (a) Trace the pictured circuit to determine an expression for the output in terms of the input. (b) Make an input–output table. (c) Could the same input–output table be accomplished by a circuit using fewer basic gates? Explain.

69.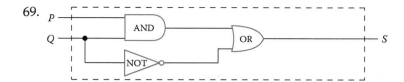

70.

71.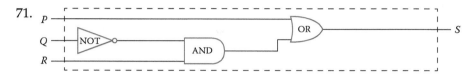

72.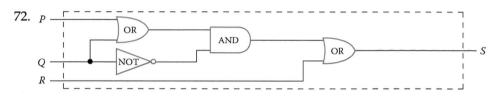

In Exercises 73 through 78, draw a circuit that realizes the given expressions.

73. $\neg(P \wedge Q) \vee R = S$.   74. $\neg(P \vee (Q \wedge R)) = S$.

75. $(\neg P \vee Q) \wedge \neg R = S$.   76. $(P \wedge Q \wedge R) \vee (\neg P \wedge \neg Q) = S$.

77. $P \oplus Q = S$. Accomplish this by

    (a) Using the defining formula for $\oplus$ in Definition 1.2.

    (b) Using the characterization of $\oplus$ given in Example 1.9.

    Which method uses fewer gates?

78. (a) $P \rightarrow Q = S$ by using the characterization of $\rightarrow$ given in Example 1.10.

    (b) $P \leftrightarrow Q = S$ by using the idea from part (a).

1.2   Set notation

Since our interest is in studying mathematical statements, the language of sets is needed. This section therefore gives an informal introduction to set theory. Our main goal here is to gain familiarity with some standard notation and concepts. A more formal consideration of sets is given in Section 1.4.

A **set** is a collection of objects that are referred to as the **elements** of the set. If A is a set and x is an element of A, then we write $x \in A$. If x is not an element of A, then we write $x \notin A$. Two important sets with which we work a great deal are $\mathbb{Z}$, the set of integers, and $\mathbb{R}$, the set of real numbers. We have

$$\ldots, \; -2 \in \mathbb{Z}, \; -1 \in \mathbb{Z}, \; 0 \in \mathbb{Z}, \; 1 \in \mathbb{Z}, \; 2 \in \mathbb{Z}, \; \ldots,$$

but $\sqrt{2} \notin \mathbb{Z}$. Of course, $\sqrt{2} \in \mathbb{R}$, but $\sqrt{-1} \notin \mathbb{R}$. The set of rational numbers $\mathbb{Q}$ (and the fact that $\sqrt{2} \notin \mathbb{Q}$) is considered in Chapter 3.

The most basic way to express a set is in **list notation**, in which the elements of the set are listed between braces. For example, the set S of integers from 1 to 5 can be denoted by

$$S = \{1, 2, 3, 4, 5\}.$$

The set T consisting of the real numbers $-1, 0$, and 1 is given by

$$T = \{-1, 0, 1\}.$$

A more powerful means of expressing sets is provided by **set builder notation** that takes the form

$$\{x \; : \; p(x)\}, \tag{1.2}$$

in which $p(x)$ is sentence involving the variable x. The notation in (1.2) expresses the set of all x such that $p(x)$ is a true statement. That is, an element x earns membership in the set if and only if $p(x)$ holds for that x. For example, the set S of integers from 1 to 5 can be expressed in set builder notation as

$$S = \{n \; : \; n \in \mathbb{Z} \text{ and } 1 \leq n \leq 5\}.$$

The set T of real numbers that are their own cubes can be expressed as

$$T = \{x \; : \; x \in \mathbb{R} \text{ and } x^3 = x\}.$$

The two different expressions for the sets S and T given above raise the issue of characterizing when two sets should be considered the same.

Definition 1.6. (Set Equality (Informal Version)).
Two sets A and B are said to be **equal**, written $A = B$, if and only if A and B contain exactly the same elements.

A set is completely determined by the elements it contains. The two expressions for the set S given above clearly describe the same set. In the case of the two expressions for T, a straightforward algebraic computation verifies that

$$\{x \ : \ x \in \mathbb{R} \text{ and } x^3 = x\} = \{-1, 0, 1\}.$$

When listing the elements of a set between curly braces, it is important to understand that neither order nor repetition matters. For example,

$$\{2, 4, 1, 3, 5\} = \{1, 2, 3, 4, 5\} \quad \text{and}$$
$$\{-1, 0, -1, 1, 0\} = \{-1, 0, 1\}.$$

That equalities like these hold is a consequence of the definition of set equality.

Many of the sets we consider will contain only real numbers. There are some important relatives of $\mathbb{Z}$ and $\mathbb{R}$ for which there is some standard notation.

Definition 1.7.   We define the following sets:

Natural numbers	$\mathbb{N}$	$=$	$\{n \ : \ n \in \mathbb{Z} \text{ and } n \geq 0\}.$
Positive integers	$\mathbb{Z}^+$	$=$	$\{n \ : \ n \in \mathbb{Z} \text{ and } n > 0\}.$
Negative integers	$\mathbb{Z}^-$	$=$	$\{n \ : \ n \in \mathbb{Z} \text{ and } n < 0\}.$
Positive real numbers	$\mathbb{R}^+$	$=$	$\{x \ : \ x \in \mathbb{R} \text{ and } x > 0\}.$
Negative real numbers	$\mathbb{R}^-$	$=$	$\{x \ : \ x \in \mathbb{R} \text{ and } x < 0\}.$

An appropriate way to express the relationship between a set like $\mathbb{N}$ and the set $\mathbb{Z}$ is to say that $\mathbb{N}$ is a subset of $\mathbb{Z}$.

Definition 1.8.   (Subsets (Informal Version)).
Let A and B be sets.

(a) We say that A is a **subset** of B, denoted $A \subseteq B$, if and only if every element of A is also an element of B.

(b) When it is not the case that $A \subseteq B$, we write $A \nsubseteq B$.

(c) If $A \subseteq B$ and B contains at least one element that A does not, then we say that A is a **proper subset** of B and write $A \subset B$.

Example 1.18.   We have the following subset relations:

(a) $\mathbb{Z}^+ \subseteq \mathbb{Z}$.

(b) $\mathbb{Z}^+ \subset \mathbb{Z}$, since $-1 \in \mathbb{Z}$ and $-1 \notin \mathbb{Z}^+$.

(c) $\mathbb{Z} \subseteq \mathbb{R}$.

(d) $\mathbb{R} \subseteq \mathbb{R}$.

(e) $\{1,3,5\} \subseteq \{1,2,3,4,5\}$.

(f) $\{2\} \subset \{2,4\}$, since $4 \in \{2,4\}$ and $4 \notin \{2\}$.

(g) $\{2,5,\mathbb{Z},5,2,\mathbb{R}\} \subseteq \{\mathbb{Z},2,3,5,\mathbb{R}\}$.
Notice that the set on the left is $\{2,5,\mathbb{Z},\mathbb{R}\}$.

(h) $\{\{1\},\{1,2\},\{3\}\} \subseteq \{\{1\},\{2\},\{3\},\{1,2\}\}$.
The elements $\{1\}$, $\{1,2\}$, and $\{3\}$ of the set on the left are also elements of the set on the right.

It is important to notice the distinction between the relations $\in$ and $\subseteq$. This can be especially subtle since sets themselves can be elements of other sets, as we saw in parts (g) and (h) of Example 1.18.

Example 1.19.   We have the following relations:

(a) $1 \in \{1,2,3\}$.

(b) $\{1\} \notin \{1,2,3\}$ but $\{1\} \subseteq \{1,2,3\}$.

(c) $\{1\} \in \{\{1\},\{2\},3\}$ but $\{1\} \not\subseteq \{\{1\},\{2\},3\}$ since $1 \notin \{\{1\},\{2\},3\}$.

(d) $\{\{2\},3\} \subseteq \{\{1\},\{2\},3\}$.

(e) $2 \notin \{\{1\},\{2\},3\}$ but $\{2\} \in \{\{1\},\{2\},3\}$ and $\{\{2\}\} \subseteq \{\{1\},\{2\},3\}$.

(f) $1 \in \mathbb{Z}$ but $1 \neq \mathbb{Z}$ and $1 \not\subseteq \mathbb{Z}$.

(g) $1 \in \{1\}$ but $1 \neq \{1\}$ and $1 \not\subseteq \{1\}$.

Parts (a) and (b) of Example 1.19 illuminate the difference between the number 1 and the set $\{1\}$ containing just the number 1. In general, a set of the form $\{x\}$ is called **singleton** x and should not be confused with the element x alone.

Intervals are examples of subsets of $\mathbb{R}$ that are important enough to warrant their own notation.

Definition 1.9.   (Interval Notation).
Given real numbers a and b, we define the following **intervals**:

$$
\begin{aligned}
(a,b) &= \{x : a < x < b\}. \\
[a,b] &= \{x : a \leq x \leq b\}. \\
[a,b) &= \{x : a \leq x < b\}. \\
(a,b] &= \{x : a < x \leq b\}. \\
(a,\infty) &= \{x : a < x\}. \\
[a,\infty) &= \{x : a \leq x\}. \\
(-\infty,b) &= \{x : x < b\}. \\
(-\infty,b] &= \{x : x \leq b\}.
\end{aligned}
$$

The set notation used in Definition 1.9 is actually not complete. For example, the set description of the interval (a, b) should more precisely be

$$\{x \, : \, x \in \mathbb{R} \text{ and } a < x < b\}.$$

This description makes clear what is implicitly assumed in Definition 1.9 that x represents a real number. In Definition 1.9, it is assumed that we are working in the context of the set of real numbers, although that is not explicitly stated.

Implicitly assuming a certain context is common in mathematical notation. When we work with a collection of sets, it is often the case that all of those sets are subsets of some bigger set called the **universal set** (or **universe of consideration**) for that particular context. Throughout this book we use $\mathcal{U}$ to represent the universal set for any particular context. For example, in Definition 1.9, it is understood that $\mathcal{U} = \mathbb{R}$. The intervals are understood to be subsets of $\mathbb{R}$ and that need not be repeatedly stated. Similarly, when we work with sets like

$$\{1, 2, 3\}, \quad \{-4, -2, 0, 2, 4\}, \quad \mathbb{N}, \quad \text{or} \quad \mathbb{Z}^-,$$

it may be understood that $\mathcal{U} = \mathbb{Z}$.

Examples of sets such as the intervals $(0, 0)$ and $[1, 0]$ bring up the issue of handling a set that does not contain any elements.

Definition 1.10.   The **empty set**, denoted $\emptyset$, is the unique set that contains no elements.

Any set that does contain at least one element is said to be **nonempty**. The fact that there is only one set with no elements (and hence $\emptyset$ is well defined) is a consequence of the definition of set equality and is proved in Section 2.4. Also proved there is the fact that the empty set is a subset of every other set.

Theorem 1.3.   ($\emptyset$ Is Smallest).
Given any set A, we have $\emptyset \subseteq A$.

One of the basic properties of a set is the number of elements in it.

Definition 1.11.   (Cardinality (Informal Version)).
The **cardinality** of a set A, denoted $|A|$, is the number of elements in A.

Cardinality is introduced formally in Section 5.6. Our informal understanding suffices until then.

Example 1.20.   We compute the following cardinalities:

(a) $|\{1, 2, 3\}| = 3$.

(b) $|\{0, 2, 4, 6, 8\}| = 5$.

(c) $|\{2, 3, 2, 4, 5, 3\}| = 4$,
since repeated elements only count once and we are therefore counting the number of elements in the set $\{2, 3, 4, 5\}$.

(d) $|\{n \ : \ n \in \mathbb{Z} \text{ and } 3 \leq n \leq 13\}| = 11$,
since there are 11 elements in the set $\{3, 4, 5, 6, 7, 8, 9, 10, 11, 12, 13\}$.

(e) $|\{x \ : \ x \in \mathbb{R} \text{ and } x^2 - 3x - 1 = 0\}| = 2$,
since the set consists of the two roots $\frac{3+\sqrt{13}}{2}$ and $\frac{3-\sqrt{13}}{2}$.

(f) $|\emptyset| = 0$. Moreover, the definition of $\emptyset$ implies that

$$|A| = 0 \text{ if and only if } A = \emptyset.$$

(g) $|\{\emptyset\}| = 1$,
since $\emptyset$ is the only element.

(h) $|\{\{\emptyset\}\}| = 1$,
since $\{\emptyset\}$ is the only element.

In determining the cardinality of a set, the first issue to address is whether or not there is a finite number of elements.

Definition 1.12.   (Finiteness (Informal Version)).
A set A is said to be **finite** if and only if $|A|$ is a natural number. A set that is not finite is said to be **infinite**.

Example 1.21.   Determining finiteness.

(a) All of the sets in Example 1.20 are finite.

(b) $\mathbb{N}$ and $\mathbb{Z}$ are infinite.

(c) $\mathbb{R}$ is infinite.

(d) If $a < b$, then (a, b) is infinite.

A more formal consideration of the notions of cardinality, finite, and infinite is given in Section 5.6.

1.2.1   Paradoxes

A presentation of sets that is too informal can lead to some undesired inconsistencies. In 1903, the English mathematician

Bertrand Russell (1872–1970) came up with a paradox showing that one cannot be too cavalier with set builder notation $\{x : p(x)\}$. He presented such an example that cannot possibly be a set.

Example 1.22. (Russell's Paradox).
Let

$$P = \{S : S \text{ is a set and } S \notin S\}. \qquad (1.3)$$

Can P be a set?

Solution. Assume that P is a set. Under the assumption that P is a set, it makes sense to explore whether or not $P \in P$.

First, consider the possibility that $P \notin P$. In that case, we have the conditions

$$P \text{ is a set and } P \notin P,$$

which earn P membership in P as described by (1.3). Hence, $P \in P$. Of course, this is inconsistent with our supposition that $P \notin P$.

Second, consider the possibility that $P \in P$. In that case, P fails to satisfy the conditions to be a member of P as described by (1.3). Hence, $P \notin P$. Again, this is inconsistent with our supposition that $P \in P$.

Under the assumption that P is a set, the question of whether or not $P \in P$ cannot be answered. Hence, P cannot be a set. □

A version of Russell's Paradox that involves the same type of paradox, yet is more accessible to the layperson, is the Librarian's Dilemma.

Example 1.23. (Librarian's Dilemma).
A librarian is organizing those books in the library that are indexes of books in the library. After noticing that some of those books list themselves and others do not, the librarian decides to start an index book listing all of those index books that do not list themselves. Should this new index book list itself?

Solution. If the new index book lists itself, then it is no longer an index book that lists only those index books that *do not* list themselves. If the new book does not list itself, then it is no longer an index book that lists *all* of those index books that do not list themselves (since it has excluded itself). The question of whether or not the book should list itself cannot be answered.   □

Subsequent to Russell's Paradox, mathematicians eliminated such problems from set theory by restricting conditions of the form $p(x)$ that can be used in the expression of a set $\{x : p(x)\}$ and by giving a careful axiomatic development of set theory. That is, they specified a list of basic assumptions, called **axioms**, from which all propositions in set theory are required to follow. This book does not concern itself with the intricacies involved in those issues, since the expressions and sets that we will consider shall steer far from the potential problem areas of naïve set theory. However, we mention one axiom that saves us most of the time.

Axiom (The Subset Axiom).
Any subset of a set is also a set.

The Subset Axiom protects us from difficulties because the sets that we consider are subsets of some already well-defined universal set. In particular, most of the sets that we consider are subsets of $\mathcal{U} = \mathbb{R}$.

1.2.2   Software implementation of sets

Sets are fundamental objects in mathematics, and, as we have noted, an underlying feature of a set is the lack of a fixed ordering of its elements. Objects whose elements do have a specified order, such as ordered pairs, must be industriously built upon the foundations of set theory. Formally, an object as simple as the ordered pair (x, y) is defined by the more complicated set construction $\{\{x\}, \{x, y\}\}$. Although, in this book, we do not delve into the formal development of objects as intuitively simple as ordered lists, it is interesting to note a contrast here with computer science.

In a computer, the most natural container for a collection of objects is one that retains an ordering of those objects, as in an array or list. This is true for the same reason that when we write down a set in list notation, we must write down the elements in some order; elements are entered and stored in a computer in some order. If we want a data structure in a computer that behaves like a set, then some effort must be made to strip away that order. We close this section by considering some examples of how and how well this might be addressed in computer software.

Mathematica® is computer software that, in particular, can be used as a symbolic calculator. Although it does not have a data type that acts exactly like a set, it does have the type *list*, which can mimic a set in some ways. Lists are entered using set-like notation.

```
In[1]:= A = {4,2,2,1}
```

```
Out[1]= {4,2,2,1}
```

However, lists can have repeated elements, and the order in which elements are entered is retained as part of the structure. Using the equality test p==, we can see this distinction.

```
In[3]:= A == {4,2,1}
```

```
Out[3]= False
```

```
In[4]:= A == {1,2,2,4}
```

```
Out[4]= False
```

Nonetheless, we can test for the membership of an element in a list as we would for a set.

```
In[5]:= MemberQ[A,2]
```

```
Out[5]= True
```

The size of a list does count element repetitions.

```
In[6]:= Length[A]
```

```
Out[6]= 4
```

The empty list is entered as { } and has 0 elements.

The programming language C++ has a container class called a *set*, which has many of the features one would want. Since the syntax used and overhead required are rather complicated, we shall only mention some available functions and not present detailed coding. For example, the Boolean functions `equal()` and `includes()` test whether two input *sets* are related by $=$ and $\subseteq$, respectively. Emphasizing the structure of a set, the function `insert()` adds a given element to a given *set* but changes nothing if that element is already in the *set*. Similarly, `erase()` deletes an element, if possible. The number of elements in a set is returned by the function `size()`. Difficulties with repeated elements and orderings of elements are not present. However, specifically, a *set* is an example of what is called a sorted associative container class in C++. Thus, there is some ordering that is buried in the internal implementation of a *set*; each *set* has an associated comparison function (like less than) and a construct called an iterator that is used to traverse the elements of the *set*. Therefore, only abstractly is a *set* truly like a set.

Maple is mathematical software similar to *Mathematica*, but it handles sets much more seamlessly. Sets are represented using curly braces.

```
> A := {4,2,2,1} ;
```

$$\{1,2,4\}$$

Note that sets do not retain repeated elements, and *Maple* automatically selects a preferred ordering in which to list the elements. The ordering is dealt with internally and is not a concern of the user. In contrast, lists in *Maple* are represented using square brackets and retain repeats and ordering. Thus, tests for set equality behave as we would like.

```
> evalb(A = {4,2,1}) ;
```

$$\text{true}$$

The same is true for tests of membership.

```
> member(2,A) ;
```

$$\text{true}$$

Also, cardinality is counted correctly.

```
> nops(A) ;
```

$$3$$

The empty set is entered as { }.

Using *Mathematica* and *Maple* to explore various set relations is taken up in the exercises.

Exercises

In Exercises 1 through 8, determine whether the stated equality is true or false.

1. $\{1,2,3\} = \{3,2,1\}$.
2. $\{2,3,4\} = \{4,3,2,1\}$.
3. $\{2,2,2,2\} = \{2,2\}$.
4. $\{1,2,3,2\} = \{1,2,3\}$.
5. $\{x : x \in \mathbb{Z} \text{ and } x^2 < 4\} = \{x : x \in \mathbb{Z} \text{ and } x < 2\}$.
6. $\{x : x \in \mathbb{Z} \text{ and } 0 \leq x \leq \pi\} = \{x : x \in \mathbb{N} \text{ and } -4 < x < 4\}$.
7. $\{x : x \in \mathbb{R} \text{ and } x^3 + x^2 - x - 1 = 0\} = \{x : x \in \mathbb{R} \text{ and } x^3 - x^2 - x + 1 = 0\}$.
8. $\{x : x \in \mathbb{R} \text{ and } x^2 - x - 6 = 0\} = \{x : x \in \mathbb{R} \text{ and } x^2 + x - 6 = 0\}$.

In Exercises 9 through 16, express the given sets in set notation. In each case, use either list notation or set builder notation, whichever is more convenient.

9. The set consisting of the elements $2, 4, 6$.

10. The set consisting of the elements $0, -1, -4, -9$.

11. The set consisting of the elements $\{1\}, \{4\}$.

12. The set consisting of the elements $1, 2, \{3\}$.

13. The set of real solutions to the equation $x^3 - 4x^2 + 5x - 6 = 0$.

14. The set of integers greater than or equal to 100.

15. The set of integers less than -10.

16. The set of real numbers x such that $x^4 - x^2 - 1$ is negative.

In Exercises 17 through 24, express the given sets in interval notation.

17. $\mathbb{R}^+$.

18. $\mathbb{R}^-$.

19. $\{0\}$.

20. $\{1\}$.

21. The set of real numbers x such that $x > 1$.

22. The set of real numbers x such that $3 \le x < 7$.

23. The set of real numbers x such that $-1 < x$ and $1 > x$.

24. The set of real numbers x such that $x < \pi$.

In Exercises 25 through 32, determine whether the stated relations are true or false.

25. $\sqrt{2} \in \mathbb{R}$.

26. $\mathbb{N} \in \mathbb{Z}$.

27. $\{1\} \in \mathbb{Z}$.

28. $0 \subseteq \mathbb{Z}$.

29. $\{2\} \subseteq \{1, 2, 3\}$.

30. $1 \in \{\{1\}, \{2\}, \{3\}\}$.

31. $\emptyset \in \{\emptyset\}$.

32. $\emptyset \subseteq \{\emptyset\}$.

In Exercises 33 through 38, list which of the relations $\in, \subset, \subseteq, =$ can replace the symbol $\lhd$.

33. $\{1\} \lhd \{1, 2\}$.

34. $\{6, 7, 8\} \lhd \{8, 7, 6\}$.

35. $2 \lhd \{1, 2\}$.

36. $\{9\} \lhd \{9, \{9, 10\}\}$.

37.* $\{3\} \lhd \{3, \{3\}, \{\{3\}\}\}$.

38.* $\{4, 5\} \lhd \{\{4\}, \{5\}\}$.

In Exercises 39 through 46, determine whether the given set is finite or infinite. If it is finite, then find its cardinality.

39. $A = \{1, 3, 5, 7, 9\}$.

40. $C = [5, 6]$.

41. $B = \{x : x \in \mathbb{R} \text{ and } x^4 = 16\}$.

42. $D = \{x : x \in \mathbb{R} \text{ and } x^2 > 2\}$.

43. $E = \{n \ : \ n \in \mathbb{N} \text{ and } n \leq 8\}.$

44. $F = \{n \ : \ n \in \mathbb{Z}^- \text{ and } n \leq \frac{1}{2}\}.$

45.* $G = \{\emptyset, \{\emptyset\}\}.$

46.* $H = \{\{\{\emptyset\}\}\}.$

In Exercises 47 and 48, discover a paradox by trying to answer the included question.

47.* **Barber Puzzle.** A male town barber shaves each man and only those men in the town who do not shave themselves. Does the barber shave himself?

48.* **A Television Producer's Nightmare.** Some television shows advertise themselves and others do not. A television producer decides to create a show advertising all shows that do not advertise themselves. Should this show advertise itself?

In Exercises 49 through 52, use *Mathematica* to perform the given tasks.

49. Experiment with == for testing list equality. Specifically, what happens when $\{1,2,\{2,1\}\}$ and $\{1,2,\{1,2\}\}$ are compared?

50. Experiment with MemberQ for testing list membership. Specifically, determine whether $\{\}$ is a member of $\{\{\}\}$?

51. Experiment with Length for counting the size of a list. Specifically, find the size of $\{\{\},\{\}\}$.

52. How can we test whether or not a given list A is empty?

In Exercises 53 through 56, use *Maple* to perform the given tasks.

53. Experiment with evalb for testing set equality. Specifically, what happens when $\{1,2,\{2,1\}\}$ and $\{1,2,\{1,2\}\}$ are compared?

54. Experiment with member for testing set membership. Specifically, determine whether $\{\}$ is a member of $\{\{\}\}$?

55. Experiment with nops for counting the size of a set. Specifically, find the size of $\{\{\},\{\}\}$.

56. How can we test whether or not a given set A is empty?

1.3 Quantifiers

The truth or falsehood of the expression

$$x > 2 \text{ and } x^2 \text{ is an integer} \tag{1.4}$$

depends on the value of x. If x is replaced by $\sqrt{7}$, then (1.4) becomes true. If x is replaced by π, then (1.4) is false. This section concerns expressions of the form $p(x)$, like (1.4), that depend on a **free** (or **unbound**) variable x from some universe of consideration $\mathcal{U}$. Besides replacing x by a specific value, another way to form a statement from $p(x)$ is with the addition of a quantifier. Specifically, we introduce the **quantifiers** $\forall$ and $\exists$.

Definition 1.13.   (Universal Statements).
The statement

$$\forall\, x \in \mathcal{U},\ p(x)$$

is defined to be true if and only if for every value of $x \in \mathcal{U}$, the statement $p(x)$ holds. Consequently, it is false if and only if there is some $x \in \mathcal{U}$ for which $p(x)$ does not hold. The quantifier $\forall$ is read as "for every," "for all," or "for any."

Example 1.24.   Write the following universal statements efficiently using quantifiers and standard notation:

(a) Every real number is greater than 2 and its square is an integer.
Solution.

$$\forall\, x \in \mathbb{R},\ x > 2 \text{ and } x^2 \in \mathbb{Z}.$$

(b) The square of every real number is nonnegative.
Solution.

$$\forall\, x \in \mathbb{R},\ x^2 \geq 0.$$

(c) For every real number x, if x is negative, then $|x| = -x$.
Solution.

$$\forall\, x \in \mathbb{R}, \text{ if } x < 0, \text{ then } |x| = -x.$$

(d) The square of every odd integer is even.
Solution.

$$\forall\, n \in \mathbb{Z}, \text{ if } n \text{ is odd, then } n^2 \text{ is even.}$$

(e) For all real numbers x with $2 \leq x \leq 5$, we have $4 \leq x^2 \leq 25$.
Solution.

$$\forall\, x \in [2,5],\ x^2 \in [4,25].$$

(f) For all positive real numbers x, if $x < -3$, then $x^3 = -1000$.
Solution.

$$\forall\, x \in \mathbb{R}^+,\ \text{ if } x < -3, \text{ then } x^3 = -1000.$$

Note in some cases that a statement can be simplified by restricting the universe of consideration. For example, the statement in part (c) of Example 1.24 can be written as

$$\forall \, x \in \mathbb{R}^-, \; |x| = -x.$$

Similarly, part (d) could be written as

$$\forall \text{ odd integers } n, \; n^2 \text{ is even.}$$

The statements in (b), (c), (e), and (f) are all true, while the statements in (a) and (d) are false. For (d), the statement fails when $n = 1$, for example. That is, the statement

$$\text{If 1 is odd, then } 1^2 \text{ is even}$$

is false. For (f), we say that such a universal if-then statement is **vacuously true** since its hypothesis (here "$x < -3$") is false for every x in the universal set (here "$\mathbb{R}^+$").

Definition 1.14.   (Existential Statements).
The statement

$$\exists \, x \in \mathcal{U} \text{ such that } p(x)$$

is defined to be true if and only if there exists some $x \in \mathcal{U}$ such that the statement $p(x)$ holds. Consequently, it is false if and only if for every $x \in \mathcal{U}$, $p(x)$ does not hold. The quantifier $\exists$ is read as "there exists," "there is," or "there are."

Example 1.25.   Write the following existential statements efficiently using quantifiers and standard notation:

(a) There exists a real number whose square is 2.
Solution.

$$\exists \, x \in \mathbb{R} \text{ such that } x^2 = 2.$$

(b) There is a real zero for the polynomial $x^3 - 2x^2 + x - 2$.
Solution.

$$\exists \, x \in \mathbb{R} \text{ such that } x^3 - 2x^2 + x - 2 = 0.$$

(c) The equation $5x^2 = 40$ has an integer solution.
Solution.

$$\exists \, x \in \mathbb{Z} \text{ such that } 5x^2 = 40.$$

In Example 1.25, statements (a) and (b) are true. For (b), $x = 2$ provides the desired solution. On the other hand, statement (c) is false, since the only solutions, namely, $\pm\sqrt{8}$, are not integers.

In the statements of the forms seen in Definitions 1.13 and 1.14, we say that the variable x in the expression $p(x)$ has been **bound** by its respective quantifier. A statement may involve multiple quantifiers and, consequently, multiple bound variables.

Example 1.26.   Write the following statements efficiently using quantifiers and standard notation:

(a) For every real number x and every integer n, the product nx is a real number.
Solution.

$$\forall\, x \in \mathbb{R}, \forall\, n \in \mathbb{Z}, nx \in \mathbb{R}.$$

(b) The sum of any two real numbers is a real number.
Solution.

$$\forall\, x \in \mathbb{R}, \forall\, y \in \mathbb{R}, x + y \in \mathbb{R}.$$

When two $\forall$ quantifiers have the same domain of consideration, as is the case here, the shorthand

$$\forall\, x, y \in \mathbb{R}, x + y \in \mathbb{R}$$

can be used.

(c) The quotient of any two integers is an integer if the denominator is nonzero.
Solution.

$$\forall\, m, n \in \mathbb{Z}, \text{ if } m \neq 0, \text{ then } \frac{n}{m} \in \mathbb{Z}.$$

(d) There exist two distinct integers with the same cube.
Solution.

$$\exists\, m, n \in \mathbb{Z} \text{ such that } m \neq n \text{ and } m^3 = n^3.$$

(e) Every positive real number has a positive square root.
Solution.

$$\forall\, x \in \mathbb{R}^+, \exists\, y \in \mathbb{R}^+ \text{ such that } y^2 = x.$$

(f) Every polynomial with real coefficients that has odd degree must have a real zero.

Solution.

$$\forall \text{ polynomials } f \text{ with real coefficients,}$$

$$\text{if } f \text{ has odd degree, then}$$

$$\exists\, x \in \mathbb{R} \text{ such that } f(x) = 0.$$

Statements (a), (b), (e), and (f) are true, while statements (c) and (d) are false. For (c), if $m = 2$ and $n = 1$, then $\frac{1}{2} \notin \mathbb{Z}$. For (d), taking cube roots shows that the equality $m^3 = n^3$ forces $m = n$. Statement (f) is a consequence of the Intermediate Value Theorem from calculus.

Remark 1.2.   (Order of Quantifiers).
In statement (e) of Example 1.26, the order of the quantifiers is important. If that order is switched, then the statement

$$\exists\, y \in \mathbb{R}^+ \text{ such that } \forall\, x \in \mathbb{R}^+, y^2 = x$$

is obtained. This says,

> "There is a positive real number whose square equals every positive real number."

and is clearly false. In general, the statements

$$\forall\, x, \exists\, y \text{ such that } p(x, y)$$

and

$$\exists\, y \text{ such that } \forall\, x, p(x, y)$$

are not logically equivalent. The order of the quantifiers matters. However, if a statement starts with a sequence of quantifiers such that all are $\forall$ or all are $\exists$, then order does not matter. It is this last observation that allows us to use shorthand notation like $\forall\, x, y \in \mathcal{U}$ and $\exists\, x, y \in \mathcal{U}$.

Real functions. As an application of quantifiers, we use quantified statements to carefully define some properties of and operations on real functions familiar from precalculus. Somewhat informally, a **real function** is a rule that assigns to each real number (given as input) a unique real number (as output). Functions are considered more generally and more formally in Chapter 5. Until Chapter 5, all real functions are considered to be defined on the entire set of real numbers. Examples illustrating the various properties in Definition 1.15 are shown in Figure 1.5.

Definition 1.15.   (Properties of Real Functions).
A real function f is said to be

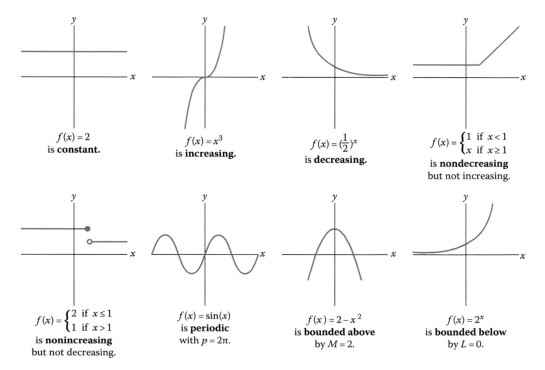

FIGURE 1.5
Illustrating properties of real functions.

(a) **Constant** if

$$\exists\, c \in \mathbb{R} \text{ such that } \forall\, x \in \mathbb{R}, f(x) = c.$$

(b) **Increasing** if

$$\forall\, x, y \in \mathbb{R}, \text{ if } x < y, \text{ then } f(x) < f(y).$$

(c) **Decreasing** if

$$\forall\, x, y \in \mathbb{R}, \text{ if } x < y, \text{ then } f(x) > f(y).$$

(d) **Nondecreasing** if

$$\forall\, x, y \in \mathbb{R}, \text{ if } x \le y, \text{ then } f(x) \le f(y).$$

(e) **Nonincreasing** if

$$\forall\, x, y \in \mathbb{R}, \text{ if } x \le y, \text{ then } f(x) \ge f(y).$$

(f) **Periodic** if

$$\exists\, p \in \mathbb{R}^{+} \text{ such that } \forall\, x \in \mathbb{R}, f(x + p) = f(x).$$

(g) **Bounded above** if

$$\exists\, M \in \mathbb{R} \text{ such that } \forall\, x \in \mathbb{R}, f(x) \le M.$$

(h) **Bounded below** if

$$\exists\, L \in \mathbb{R} \text{ such that } \forall\, x \in \mathbb{R}, f(x) \ge L.$$

We say that f is **bounded** if f is both bounded above and bounded below.

An alternative characterization of constant functions is presented in the exercises.

Definition 1.16.   (Operations on Real Functions).
Given a real number c and real functions f and g, we define

(a) The **constant multiple** cf by

$$\forall\, x \in \mathbb{R}, \quad (cf)(x) = c \cdot f(x).$$

(b) The **product** $f \cdot g$ by

$$\forall\, x \in \mathbb{R}, \quad (f \cdot g)(x) = f(x) \cdot g(x).$$

(c) The **sum** $f + g$ by

$$\forall\, x \in \mathbb{R}, \quad (f + g)(x) = f(x) + g(x).$$

(d) The **composite** $f \circ g$ by

$$\forall\, x \in \mathbb{R}, \quad (f \circ g)(x) = f(g(x)).$$

Exponents can be used to denote the product of a function with itself. Thus, f^2 denotes the product $f \cdot f$. For example, $\sin^2(x)$ denotes $\sin(x)\sin(x)$. The difference $f - g$ is defined to be $f + (-g)$. The function $f + c$ denotes the sum of f and the function whose constant value is c.

1.3.1 Negating quantified statements

There is an important connection between the quantifiers $\forall$ and $\exists$ that arises when quantified statements are negated.

Proposition 1.4.   (Negating $\forall$ and $\exists$).

(a) $\neg\big[\forall\, x \in \mathcal{U}, p(x)\big] \;\equiv\; \exists\, x \in \mathcal{U} \text{ such that } \neg p(x)$

(b) $\neg\big[\exists\, x \in \mathcal{U} \text{ such that } p(x)\big] \;\equiv\; \forall\, x \in \mathcal{U}, \neg p(x)$

Proposition 1.4 follows from the definitions of the quantifiers $\forall$ and $\exists$. The result should also be intuitively reasonable.

If it is not true that everything goes right, then there is something that goes wrong. If it is not true that something goes right, then everything goes wrong.

Example 1.27. Negate the following statements:

(a) The square of every real number is nonnegative.
Solution.

$$\neg \left[\forall\, x \in \mathbb{R}, x^2 \geq 0 \right] \equiv \exists\, x \in \mathbb{R} \text{ such that } x^2 < 0.$$

There is a real number whose square is negative.

(b) There exists an integer whose square is 2.
Solution.

$$\neg \left[\exists\, n \in \mathbb{Z} \text{ such that } n^2 = 2 \right]$$
$$\equiv\ \forall\, n \in \mathbb{Z}, n^2 \neq 2.$$

Every integer has its square not equal to 2.

In (a), the given statement is true, so its negation is false. In (b), the given statement is false, so its negation is true.

Statements involving multiple quantifiers can be negated as well. In fact, Proposition 1.4 can be used to do so.

Example 1.28. Negate the following statements:

(a) The product of any real number and any integer is a real number.
Solution.

$$\neg\, [\forall\, x \in \mathbb{R}, \forall\, n \in \mathbb{Z}, nx \in \mathbb{R}]$$
$$\equiv \exists\, x \in \mathbb{R} \text{ such that } \neg\, [\forall\, n \in \mathbb{Z}, nx \in \mathbb{R}]$$
$$\equiv \exists\, x \in \mathbb{R} \text{ such that } \exists\, n \in \mathbb{Z} \text{ such that } nx \notin \mathbb{R}.$$

Even though the domains of consideration for the consecutive $\exists$ quantifiers are different, this statement may be written more simply as

$$\exists\, x \in \mathbb{R} \text{ and } n \in \mathbb{Z} \text{ such that } nx \notin \mathbb{R}.$$

There is a real number and an integer whose product is not a real number.

(b) $\forall\, x \in \mathbb{R}^+, \exists\, y \in \mathbb{R}^+$ such that $y^2 = x$.
Solution.

$$\neg\left[\forall\, x \in \mathbb{R}^+, \exists\, y \in \mathbb{R}^+ \text{ such that } y^2 = x\right]$$

$$\equiv \exists\, x \in \mathbb{R}^+ \text{ such that}$$

$$\neg\left[\exists\, y \in \mathbb{R}^+ \text{ such that } y^2 = x\right]$$

$$\equiv \exists\, x \in \mathbb{R}^+ \text{ such that } \forall\, y \in \mathbb{R}^+, y^2 \neq x.$$

(c) $\exists\, n \in \mathbb{Z}^+$ such that $\forall\, m \in \mathbb{Z}, m^n > 0$.
Solution.

$$\neg\left[\exists\, n \in \mathbb{Z}^+ \text{ such that } \forall\, m \in \mathbb{Z}, m^n > 0\right]$$

$$\equiv \forall\, n \in \mathbb{Z}^+, \neg\left[\forall\, m \in \mathbb{Z}, m^n > 0\right]$$

$$\equiv \forall\, n \in \mathbb{Z}^+, \exists\, m \in \mathbb{Z} \text{ such that } m^n \leq 0.$$

(d) Every word in this book is illegible.
Solution.

There is a word in this book that is legible.

Statements (a) and (b) are true. Statement (c) is false, as can be seen by considering $m = 0$. Statement (d) is false, since it could be read.

The process of negating quantified statements can be described simply as toggling the quantifiers and negating the subsequent sentence.

Appendix A lists some basic properties of the sets of integers and real numbers. Those properties are expressed there using quantifiers and other standard notation that has been encountered thus far. Consequently, the reader is now encouraged to study the properties listed in Appendix A. Those properties are used freely throughout the book.

Exercises

In Exercises 1 through 10, write the following statements efficiently using quantifiers and standard notation. Note that each statement is true.

1. For every real number x, $x^2 + 1$ is positive.

2. There is an integer n such that $2^n = 1024$.

3. There is an integer whose reciprocal is also an integer.

4. Every integer power of 2 is a real number.

5. There is a natural number n such that for every real number x, x^n is nonnegative.

6. For any real number x, there is a real number y such that $x + y = 0$.

7. There is a real number x such that for every real number y with $2 \le y \le 3$, we have $1 \le xy < 2$.

8. For every integer $n < 0$, there is an integer m such that $mn > 0$.

9. There are real numbers x and y such that $x + y \in \mathbb{Z}$ and $xy \notin \mathbb{Z}$.

10. The product of any two natural numbers is a natural number.

11. Express the fact that the exponential function $f(x) = e^x$ is increasing in a precise statement using quantifiers.

12. Express the fact that the function $f(x) = 2^{-x}$ is decreasing in a precise statement using quantifiers.

In Exercises 13 through 20, negate the given statements. Also, determine whether the original statement or its negation is true.

13. $\forall\, x \in [-2, 2]$, $x^3 \in [0, 8]$.

14. $\exists\, n \in \mathbb{Z}^-$ such that $5n + 2 > 1$.

15. $\forall\, x \in \mathbb{R}^+$, if $x^2 > 4$, then $x > 2$.

16. $\forall\, x \in (-\infty, -4)$, if $\sqrt[3]{x} > 0$, then $x = -5$.

17. $\exists\, n \in \mathbb{Z}$ such that $\forall\, m \in \mathbb{Z}$, $nm < 1$.

18. $\forall\, x \in \mathbb{R}, \exists\, y \in \mathbb{R}$ such that $x^y \in \mathbb{Z}$.

19. $\forall\, m, n \in \mathbb{Z}$, $m + n \in \mathbb{Z}$.

20. $\exists\, m, n \in \mathbb{Z}$ such that $9m - 7n = 1$.

For Exercises 21 through 32, negate the given statements.

21. There is an integer whose reciprocal is also an integer.

22. Every integer power of 2 is a real number.

23. For every real number x, $x^2 + 1$ is positive.

24. There is an integer n such that $2^n = 1024$.

25. There is a natural number n such that for every real number x, x^n is nonnegative.

26. For any real number x, there is a real number y such that $x + y = 0$.

27. There is a real number x such that for every real number y with $2 \le y \le 3$, we have $1 \le xy < 2$.

28. For every integer $n < 0$, there is an integer m such that $mn > 0$.

29. There are real numbers x and y such that $x + y \in \mathbb{Z}$ and $xy \notin \mathbb{Z}$.

30. The product of any two natural numbers is a natural number.

31. Every student at the Harvard University is over the age of 17.

32. There is a planet in our solar system that contains intelligent life.

Exercises 33 through 36 state some popular sayings. If one of them is not the truth, then its negation must be the truth. Negate each statement.

33. Every truly great accomplishment is at first impossible.

34. All students take Calculus.

35. There is no such thing as bad publicity.

36. For everything there is a season.

37. The definition of a constant function is given in Definition 1.15.

    (a) Negate that definition.

    (b) An alternative way to characterize constant functions is to say that a real function f is **constant** if
    $$\forall\, x, y \in \mathbb{R}, \ f(x) = f(y).$$
    Negate this characterization.

38. Roots of Functions
    A real function is said to **have a root** (or **have a zero**) if it takes on the value 0 for some input. Express this definition precisely using quantifiers.

    In Exercises 39 through 44, let f denote a real function, and negate the corresponding quantified statements given in Definition 1.15.

39. f is increasing.   40. f is decreasing.

41. f is nondecreasing.   42. f is nonincreasing.

43. f is bounded above.   44. f is bounded below.

45* Determine whether the statement "$\forall\, x \in \mathbb{R}, x^2 < 0 \to 2x > 1$" is true or false, and explain why.

46* Determine whether the statement "$\forall\, x \in \mathbb{R}, x < 3 \to x^2 \geq 0$" is true or false, and explain why.

47. Explain why the statement "$\forall\, x \in \mathcal{U}, p(x)$" is equivalent to an "and" statement when $|\mathcal{U}| = 2$.

48. Explain why the statement "$\exists\, x \in \mathcal{U}$ such that $p(x)$" is equivalent to an "or" statement when $|\mathcal{U}| = 2$.

49* Explain why Proposition 1.4 is equivalent to De Morgan's Laws when $|\mathcal{U}| = 2$.

50* Are there conditions under which the statements "$\forall\, x, \exists\, y$ such that $p(x, y)$" and "$\exists\, y$ such that $\forall\, x, p(x, y)$" would be logically equivalent?

1.4   Set operations and identities

With quantified statements at our disposal, sets can be explored in more depth. To start, the definitions of set equality and subsets can now be given more precisely.

Definition 1.17.   (Set Equality and Subsets).
Given sets A and B whose elements come from some universal set $\mathcal{U}$,

   (a) We say that A **equals** B, written $A = B$, if and only if

$$\forall\, x \in \mathcal{U}, \quad x \in A \leftrightarrow x \in B.$$

   (b) We say that A is a **subset** of B, written $A \subseteq B$, if and only if

$$\forall\, x \in \mathcal{U}, \quad x \in A \rightarrow x \in B.$$

The notations $\nsubseteq$ and $\subset$ are already clearly defined in Definition 1.8.

For a set A in a universal set $\mathcal{U}$, the assertion

$$A = \{x\ :\ x \in \mathcal{U} \text{ and } p(x)\} \tag{1.5}$$

is equivalent to

$$\forall\, x \in \mathcal{U}, \quad x \in A \leftrightarrow p(x). \tag{1.6}$$

We now introduce five basic set operations, one at a time, using the form (1.5) with an understood universal set $\mathcal{U}$. Following this, we summarize their definitions in Definition 1.18 using the form (1.6).

The **complement** of a set A, denoted A^c, is given by

$$A^c = \{x\ :\ x \notin A\}.$$

Example 1.29.

   (a) If $\mathcal{U} = \{1,2,3,4,5\}$, then $\{1,3,5\}^c = \{2,4\}$.
       The elements of $\{1,2,3,4,5\}$ that are not 1, 3, or 5 are 2 and 4.

   (b) If $\mathcal{U} = \{1,2,3,4,5,6,7,8,9,10\}$, then $\{1,3,5\}^c = \{2,4,6,7,8,9,10\}$.

   (c) If $\mathcal{U} = [0,1]$, then $(0,1)^c = \{0,1\}$.
       Only the endpoints of the closed interval $[0,1]$ are not in the open interval $(0,1)$.

   (d) If $\mathcal{U} = \mathbb{Z}$, then $\mathbb{N}^c = \mathbb{Z}^-$.
       The integers that are not nonnegative are negative.

The difference between the values of $\{1, 3, 5\}^c$ in parts (a) and (b) demonstrates the importance of the universal set $\mathcal{U}$ in determining complements.

The **intersection** of two sets A and B, denoted $A \cap B$, is given by

$$A \cap B = \{x \, : \, x \in A \text{ and } x \in B\}.$$

Example 1.30.

(a) $\{1, 3, 4, 7\} \cap \{2, 3, 5\} = \{3\}$.
   Only the number 3 appears in *both* sets on the left.

(b) $[1, 4) \cap [2, 7] = [2, 4)$.
   The intersection of these two intervals from Figure 1.6 is the portion of the real line that is shaded in both cases.

(c) $\mathbb{Z}^+ \cap \mathbb{Z}^- = \emptyset$.
   There is no integer that is both positive and negative. Note that 0 is neither positive nor negative.

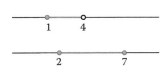

FIGURE 1.6
Diagrams of the intervals $[1, 4)$ and $[2, 7]$.

The **union** of two sets A and B, denoted $A \cup B$, is given by

$$A \cup B = \{x \, : \, x \in A \text{ or } x \in B\}.$$

Example 1.31.

(a) $\{1, 3, 4, 7\} \cup \{2, 3, 5\} = \{1, 3, 4, 7, 2, 5\}$.
   All of the elements on the left are gathered together to form the set on the right. Note that 3 need not be listed twice.

(b) $[1, 4) \cup [2, 7] = [1, 7]$.
   The union of these two intervals from Figure 1.6 is the portion of the real line that is shaded in either case.

(c) $\mathbb{Z}^+ \cup \mathbb{Z}^- = \mathbb{Z} \setminus \{0\}$.
   The set of integers that are either positive or negative excludes only 0.

The **difference** of a set A **minus** a set B, denoted $A \setminus B$, is given by

$$A \setminus B = \{x \, : \, x \in A \text{ and } x \notin B\}.$$

Example 1.32.

(a) $\{1,3,4,7\} \setminus \{2,3,5\} = \{1,4,7\}$.

The elements of the set $\{2,3,5\}$ get removed from the set $\{1,3,4,7\}$. Since 3 is the only element of $\{2,3,5\}$ in $\{1,3,4,7\}$, it is the only element that gets removed from $\{1,3,4,7\}$.

(b) $[1,4) \setminus [2,7] = [1,2)$.

This difference of the two intervals from Figure 1.6 is the portion of the real line that is shaded in the top copy of the real line and *not* in the bottom copy.

(c) $\mathbb{Z}^+ \setminus \mathbb{Z}^- = \mathbb{Z}^+$.

When the set being subtracted has no intersection with the set from which it is subtracted, the result is the same as if nothing were subtracted.

The **symmetric difference** of two sets A and B, denoted $A \triangle B$, is given by

$$A \triangle B = \{x \, : \, x \in A \ \oplus \ x \in B\}.$$

The symmetric difference is so named since

$$A \triangle B = (A \setminus B) \cup (B \setminus A).$$

The expression on the right is symmetric in A and B. That is, switching A and B there yields an expression of the same set.

Example 1.33.

(a) $\{1,3,4,7\} \triangle \{2,3,5\} = \{1,4,7,2,5\}$.

The set on the right consists of the elements that are in one or the other, but not both, of the sets on the left.

(b) $[1,4) \triangle [2,7] = [1,2) \cup [4,7]$.

The symmetric difference of these two intervals from Figure 1.6 is the portion of the real line that is shaded in either the top copy of the real line or the bottom copy, but not both.

(c) $\mathbb{Z}^+ \triangle \mathbb{Z}^- = \mathbb{Z}^+ \cup \mathbb{Z}^- = \mathbb{Z} \setminus \{0\}$.

Definition 1.18.   (Basic Set Operations).
Given sets A and B (subsets of some universal set $\mathcal{U}$):

(a) The **complement** of A, denoted A^c, is defined by

$$\forall \, x \in \mathcal{U}, \quad x \in A^c \leftrightarrow x \notin A \ \text{(i.e., } \neg(x \in A)).$$

(b) The **intersection** of A and B, denoted $A \cap B$, is defined by

$$\forall\, x \in \mathcal{U}, \quad x \in A \cap B \leftrightarrow x \in A \text{ and } x \in B.$$

(c) The **union** of A and B, denoted $A \cup B$, is defined by

$$\forall\, x \in \mathcal{U}, \quad x \in A \cup B \leftrightarrow x \in A \text{ or } x \in B.$$

(d) The **difference** of A **minus** B, denoted $A \setminus B$, is defined by

$$\forall\, x \in \mathcal{U}, \quad x \in A \setminus B \leftrightarrow x \in A \text{ and } x \notin B.$$

(e) The **symmetric difference** of A and B, denoted $A \triangle B$, is defined by

$$\forall\, x \in \mathcal{U}, \quad x \in A \triangle B \leftrightarrow x \in A \,\oplus\, x \in B.$$

Note in Definition 1.18 the correspondence between the set operations

$$^{c}, \cap, \cup, \text{ and } \triangle$$

and the logical operations

$$\neg, \wedge, \vee, \text{ and } \oplus,$$

respectively. Also, in Definition 1.17 there is a correspondence between the set relations

$$= \text{ and } \subseteq$$

and the logical operations

$$\leftrightarrow \text{ and } \rightarrow,$$

respectively. These correspondences are at the heart of the proofs, given in Chapter 2, of properties of these set operations and relations.

A helpful way to visualize the set operations is provided by **Venn diagrams**, in which sets are represented by enclosed regions. In Figure 1.7, Venn diagrams depicting the basic set operations are presented. In each case, the sets A and B involved are represented by circles as labeled, and the relevant portions of the diagram are shaded.

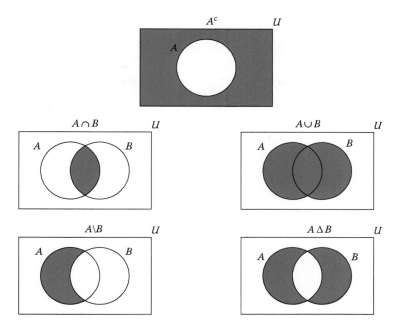

FIGURE 1.7
Venn diagrams for the basic set operations.

Unions like $[1, 2) \cup [4, 7]$ and $\mathbb{Z}^+ \cup \mathbb{Z}^-$ are examples of a special kind of union that we now define.

Definition 1.19.   Given sets A and B:

(a) They are said to be **disjoint** if and only if $A \cap B = \emptyset$.

(b) The union $A \cup B$ is said to be a **disjoint union** if and only if A and B are disjoint.

Example 1.34.

(a) The intervals $[1, 2)$ and $[4, 7]$ are disjoint, since no real number can be in both. Consequently, the union $[1, 2) \cup [4, 7]$ displayed on the right-hand side of Example 1.33(b) is a disjoint union.

(b) The sets $\{1, 3, 4, 7\}$ and $\{2, 3, 5\}$ are not disjoint, since 3 is an element common to both.

Note that $A \cup B$ is a disjoint union if and only if $A \cup B = A \triangle B$. In Example 1.33, we see that $\mathbb{Z}^+ \cup \mathbb{Z}^-$ is a disjoint union and $\mathbb{Z}^+ \cup \mathbb{Z}^- = \mathbb{Z}^+ \triangle \mathbb{Z}^-$.

The basic set operations of Definition 1.18 all yield a set of the same form (in the same universe of consideration) as the sets involved. However, there are also operations that yield sets of different forms.

Definition 1.20. The **product** of two sets A and B, denoted $A \times B$, is given by

$$A \times B = \{(x, y) \, : \, x \in A \text{ and } y \in B\}.$$

That is,

$$\forall \, x \in \mathcal{U} \text{ and } y \in \mathcal{V}, \quad (x, y) \in A \times B \leftrightarrow x \in A \text{ and } y \in B,$$

where A and B are subsets of universes $\mathcal{U}$ and $\mathcal{V}$, respectively. The elements (x, y) of the product $A \times B$ are called **ordered pairs**.

In the case that $A = B$, the product $A \times B$ is denoted A^2. For example, $\mathbb{R}^2$ is the familiar Cartesian plane, in which each element is a point (x, y). When the context is clear, there should not be any confusion between the element (x, y) of a product and the interval (x, y).

Example 1.35. (Products).

(a) $\{2, 5\} \times \{3\} = \{(2, 3), (5, 3)\}$.
The first coordinate of an element (x, y) in this product can be either 2 or 5, but the second coordinate must be 3 in each case.

(b) $\{1, 2\} \times \{2, 4, 6\} =$
$\{(1, 2), (1, 4), (1, 6), (2, 2), (2, 4), (2, 6)\}$.
If the first coordinate is a 1, then the second coordinate can be 2, 4, or 6. Similarly, if the first coordinate is a 2, then the second coordinate can be 2, 4, or 6.

(c) $\{a, b\} \times \{a, b\} = \{(a, a), (a, b), (b, a), (b, b)\}$.
Note that (a, b) is different from (b, a), since the elements are *ordered* pairs.

When A and B are finite,

$$|A \times B| = |A| \cdot |B|. \tag{1.7}$$

The proof of this is left to the exercises in Section 5.6. For now, (1.7) can certainly be seen to be true for the finite products in Example 1.35.

When a product is infinite, it is not possible to list all of its elements. However, since elements of $\mathbb{R}^2$ can be plotted as points, some subsets $\mathbb{R}^2$ can be conveniently drawn in the Cartesian plane.

Example 1.36. Sketch the following subsets of $\mathbb{R}^2$.

(a) $[1,4) \times [2,7]$.
By definition, this is the set $\{(x,y) \ : \ x \in [1,4) \text{ and } y \in [2,7]\} = \{(x,y) \ : \ 1 \le x < 4 \text{ and } 2 \le y \le 7\}$.

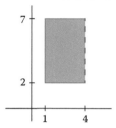

The right-hand edge of the pictured rectangle is excluded since $x = 4$ is not permitted.

(b) $\mathbb{R} \times ((-\infty, -1) \cup (1, \infty))$.
This is $\{(x,y) \ : \ x \in \mathbb{R} \text{ and either } y < -1 \text{ or } 1 < y\}$.

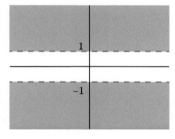

Of course, the lines $y = -1$ and $y = 1$ are excluded from the set.

It is possible to take the product of more than two sets.

Definition 1.21. (General Products).

(a) Given sets $A_1, A_2, \ldots, A_n$, the n-**fold** **product** $A_1 \times A_2 \times \cdots \times A_n$ is given by

$$A_1 \times A_2 \times \cdots \times A_n$$
$$= \{(x_1, x_2, \ldots, x_n) \ : \ x_1 \in A_1, x_2 \in A_2, \ldots, x_n \in A_n\}.$$

(b) The elements $(x_1, x_2, \ldots, x_n)$ of $A_1 \times A_2 \times \cdots \times A_n$ are called **ordered n-tuples**.

(c) The n-fold product $A \times A \times \cdots \times A$ is denoted A^n.

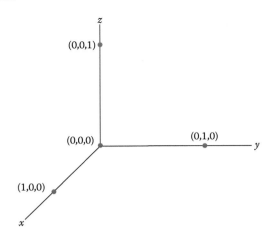

FIGURE 1.8
A picture of R^3.

Example 1.37. The elements of $\mathbb{R}^3$ are the ordered triples (x, y, z) where x, y, and z are real numbers. That is, each point of $\mathbb{R}^3$ is specified by three coordinates. Certainly, the order of these coordinates is important. For example, the point $(1, 0, 0)$ is on the x-axis, the point $(0, 1, 0)$ is on the y-axis, and the point $(0, 0, 1)$ is on the z-axis. See Figure 1.8. Of course, each of these three points happens to be one unit away from the origin $(0, 0, 0)$.

As we have seen in several examples from Section 1.2, sets can be elements of sets.

Definition 1.22. Given a set A, the **power set** of A, denoted $\mathcal{P}(A)$, is the set of subsets of A,

$$\mathcal{P}(A) = \{B \; : \; B \subseteq A\}.$$

That is,

$$\forall\, B, \quad B \in \mathcal{P}(A) \leftrightarrow B \subseteq A.$$

Example 1.38. (Power Sets).

(a) If $A = \{1, 2\}$, then $\mathcal{P}(A) = \{\emptyset, \{1\}, \{2\}, \{1, 2\}\}$.
    By Theorem 1.3, $\emptyset \subseteq A$. Moreover, $\emptyset$ is the only subset of A of cardinality 0. The singletons $\{1\}$

and $\{2\}$ are the subsets of A of cardinality 1. The
set A itself is its only subset of cardinality 2.

(b) If $A = \{2, 4, 6\}$, then
$\mathcal{P}(A) = \{\emptyset, \{2\}, \{4\}, \{6\}, \{2, 4\}, \{2, 6\}, \{4, 6\}, \{2, 4, 6\}\}$.
Although it is not required, the elements on the
right have been ordered according to cardinality.

(c) $\mathcal{P}(\emptyset) = \{\emptyset\}$.
The only subset of $\emptyset$ is $\emptyset$ itself.

(d) $\mathcal{P}(\mathbb{Z})$ is an infinite set, since $\mathbb{Z}$ itself is infinite.
It contains sets such as $\emptyset$, $\{1\}$, $\{2, 4, 6, 8\}$,
$\{1, 3, 5, 7, 9, 11, \ldots\}, \{\ldots, -15, -10, -5, 0, 5, 10, \ldots\}$,
and $\mathbb{Z}$.

The empty set $\emptyset$ and the set A itself are always elements of
$\mathcal{P}(A)$ (although they need not be distinct). The fact that if A is
finite, then

$$|\mathcal{P}(A)| = 2^{|A|} \tag{1.8}$$

motivates the term *power set*. The proof of (1.8), like that of
(1.7), is left for the exercises in Section 5.6.
We now consider some set identities that relate the various operations we have encountered. The similarities of these
identities with the logical equivalences encountered in Section
1.1 should be noted.

Theorem 1.5. (Basic Set Identities).
*Let $A, B,$ and C be sets (in some universe $\mathcal{U}$). Then, the following
identities hold:*

(a) $\qquad\qquad (A^c)^c = A.$ $\qquad\qquad$ Double Complement

(b) $\qquad (A \cap B) \cap C = A \cap (B \cap C).$ $\qquad$ Associativity
$\qquad (A \cup B) \cup C = A \cup (B \cup C).$

(c) $\qquad\qquad A \cap B = B \cap A.$ $\qquad\qquad$ Commutativity
$\qquad\qquad A \cup B = B \cup A.$

(d) $\qquad A \cap (B \cup C) = (A \cap B) \cup (A \cap C).$ Distributivity
$\qquad A \cup (B \cap C) = (A \cup B) \cap (A \cup C).$

(e) $\qquad\quad (A \cap B)^c = A^c \cup B^c.$ $\qquad$ De Morgan's Laws
$\qquad\quad (A \cup B)^c = A^c \cap B^c.$

(f) If $A \subseteq B$, then $A \cap B = A.$ $\qquad$ Absorption Rules
$\quad$ If $A \subseteq B$, then $A \cup B = B.$

Theorem 1.6. (Interactions with $\mathcal{U}$ and $\emptyset$).
*Let A be a set (in some universe $\mathcal{U}$). Then, the following identities
hold:*

(a) $\mathcal{U}^c$ $=$ $\emptyset.$
     $\emptyset^c$ $=$ $\mathcal{U}.$
(b) $A \cap \mathcal{U}$ $=$ $A.$
     $A \cup \mathcal{U}$ $=$ $\mathcal{U}.$
(c) $A \cap \emptyset$ $=$ $\emptyset.$
     $A \cup \emptyset$ $=$ $A.$
(d) $A \cap A^c$ $=$ $\emptyset.$
     $A \cup A^c$ $=$ $\mathcal{U}.$

Identities involving products and power sets are left to Chapter 2, as are the proofs of the identities in Theorems 1.5 and 1.6. Those proofs will come out of the obvious correspondence between these identities and the logical equivalences in Theorems 1.1 and 1.2. For now, familiarity with the logical equivalences should ease intuitive acceptance of the set identities.

From the associativity rules, it follows that general *n*-**fold intersections and unions**

$$A_1 \cap A_2 \cap A_3 \cap \cdots \cap A_n \quad \text{and} \quad A_1 \cup A_2 \cup A_3 \cup \cdots \cup A_n$$

may be computed by associating the sets (with parentheses) to reflect some desired order of operations. For example, the association

$$(((A_1 \cap A_2) \cap A_3) \cap \cdots) \cap A_n$$

shows one such choice in which the computation is performed left to right.

Venn diagrams can be used to help visualize some of the identities in Theorems 1.5 and 1.6. Note that the most general form of a Venn diagram for three sets appears in Figure 1.9.

Example 1.39.   Use Venn diagrams to illustrate the set identity

$$A \cap (B \cup C) = (A \cap B) \cup (A \cap C).$$

Solution.

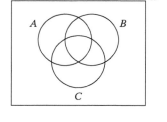

FIGURE 1.9
Venn diagram for three sets.

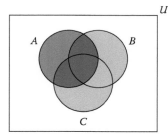 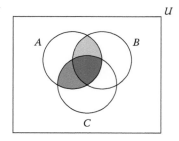

In the left Venn diagram, the set A is shaded medium blue, and $B \cup C$ is shaded light blue. Hence, the set $A \cap (B \cup C)$ is represented by the dark blue shaded portion. In the right Venn diagram, $A \cap B$ is shaded light blue, and $A \cap C$ is shaded medium blue. Hence, the set $(A \cap B) \cup (A \cap C)$ is represented by the total portion that is shaded blue. We see that the regions representing $A \cap (B \cup C)$ and $(A \cap B) \cup (A \cap C)$ are the same. □

Example 1.40.   Use Venn diagrams to illustrate the set identity

$$(A \cap B)^c = A^c \cup B^c.$$

Solution.

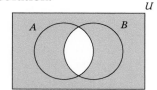 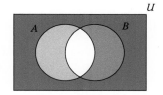

In the left Venn diagram, $(A \cap B)^c$ is shaded light blue. In the right Venn diagram, A^c is shaded medium blue, and B^c is shaded light blue. Hence, the total portion shaded blue represents $A^c \cup B^c$ in the right Venn diagram. This is the same as the portion representing $(A \cap B)^c$ is the left Venn diagram. □

It is important to realize that Examples 1.39 and 1.40 do not comprise proofs of the identities therein. They merely aid intuition. Although we do not prove the identities in Theorems 1.5 and 1.6 until Chapter 2, we use those identities here to obtain more complicated identities. Note that we may allow ourselves also to use obvious generalizations like

$$A \cap (B \cup C \cup D) = (A \cap B) \cup (A \cap C) \cup (A \cap D).$$

Example 1.41.   Verify   the   set   identity $(A \cap B \cap C) \cup (A \cap B \cap C^c) = A \cap B.$

Verification.

$$
\begin{aligned}
(A \cap B \cap C) \cup (A \cap B \cap C^c) &= ((A \cap B) \cap C) \cup ((A \cap B) \cap C^c) && \text{Associativity} \\
&= (A \cap B) \cap (C \cup C^c) && \text{Distributivity} \\
&= (A \cap B) \cap \mathcal{U} && \text{Theorem 1.6} \\
&= A \cap B && \text{Theorem 1.6}
\end{aligned}
$$

□

Example 1.42. Verify the set identity $(A^c \cap (B \cup C))^c = (A \cup B^c) \cap (A \cup C^c)$.

Verification.

$$
\begin{aligned}
(A^c \cap (B \cup C))^c &= (A^c)^c \cup (B \cup C)^c & \text{De Morgan's Law} \\
&= A \cup (B \cup C)^c & \text{Double Complement} \\
&= A \cup (B^c \cap C^c) & \text{De Morgan's Law} \\
&= (A \cup B^c) \cap (A \cup C^c) & \text{Distributivity}
\end{aligned}
$$

$\square$

Example 1.43. Verify the following set identities:

(a) $A \cap (A \cup B) = A$.

(b) $(A \cap B) \cup B = B$.

Verification.

(a) It is easy to see that $A \subseteq A \cup B$ (a formal proof is given in Chapter 2). The Absorption Rule then gives that $A \cap (A \cup B) = A$. To see this, it may be helpful to think of the Absorption Rule as

$$\text{If } S \subseteq T, \text{ then } S \cap T = S.$$

We apply that rule in this case with $S = A$ and $T = A \cup B$.

(b) In this case, we use the fact that $A \cap B \subseteq B$ and the Absorption Rule

$$\text{If } S \subseteq T, \text{ then } S \cup T = T$$

with $S = A \cap B$ and $T = B$. $\square$

1.4.1   Software implementation of set operations

If set-like data structures are needed in computer software, then certainly basic operations like union $\cup$, intersection $\cap$, and difference $\setminus$ are also needed to manipulate these structures. Here, we consider how such operations are addressed in the software examples introduced in Section 1.2.

In *Mathematica,* there do exist operations that behave as one would wish on sets, to some extent.

```
In[1]:= A={4,2,2,1}

Out[1]= {4,2,2,1}

In[2]:= B={5,1,4}
```

Out[2] = {5,1,4}

In[3] := Union[A,B]

Out[3] = {1,2,4,5}

In[4] := Intersection[A,B]

Out[4] = {1,4}

In[5] := Complement[A,B]   ⟵ Note that this is set difference.

Out[5] = {2}

Each of these ignores repetition and order in the input lists and returns a list of distinct elements presented in a standard order. In particular, by applying Union to a single list, we can essentially strip off its structure as a list and obtain its more basic structure as a set.

In C++, the functions

> set_union(), set_intersection(), set_difference(), and set_symmetric_difference()

act on two input sets like ∪, ∩, \, and △, respectively.

In *Maple*, standard operations act as one would expect on sets.

```
> A := {4,2,2,1} ;
```
$$\{1,2,4\}$$

```
> B := {5,1,4} ;
```
$$\{1,4,5\}$$

```
> A union B ;
```
$$\{1,2,4,5\}$$

```
> A intersect B ;
```
$$\{1,4\}$$

```
> A minus B ;
```
$$\{2\}$$

Additionally, the power set of a set can be computed.

```
> powerset({a,b}) ;
```
$$\{\{\}, \{a,b\}, \{b\}, \{a\}\}$$

Using *Mathematica* and *Maple* to explore various set operations is taken up in the exercises.

Exercises

In Exercises 1 through 8, find $A^c, B^c, A \cap B, A \cup B, A \setminus B, B \setminus A$, and $A \bigtriangleup B$ for the given sets.

1. $A = \{1, 2, 3\}, B = \{3, 4\}$, and $\mathcal{U} = \{1, 2, 3, 4\}$.

2. $A = \{a, b, c, d\}, B = \{b, d\}$, and $\mathcal{U} = \{a, b, c, d, e\}$.

3. $A = (-1, 1), B = [0, 1]$, and $\mathcal{U} = \mathbb{R}$.

4. $A = (1, 2], B = [-2, 1]$, and $\mathcal{U} = \mathbb{R}$.

5. $A = \mathbb{N}, B = \mathbb{Z}^+$, and $\mathcal{U} = \mathbb{Z}$.

6. $A = \{0\}, B = \mathbb{Z}^-$, and $\mathcal{U} = \mathbb{Z}$.

7. $A = (0, 3), B = [2, \infty)$, and $\mathcal{U} = \mathbb{R}^+$.

8. $A = (-3, -1], B = [-4, -2)$, and $\mathcal{U} = \mathbb{R}^-$.

9. Are $\mathbb{Z}^-$ and $\mathbb{N}$ disjoint?

10. Are $(-2, 2)$ and $(1, 3)$ disjoint?

11. Is $[2, 4] \cup (3, 5)$ a disjoint union?

12. Is $\{2, 4\} \cup \{3, 5\}$ a disjoint union?

13. Find $\{1, 3\} \times \{2, 4\}$.

14. Find $\{a, b, c\} \times \{1, 2\}$.

15. Find $\{3, 5, 7, 9\} \times \{5\}$.

16. Find $\{a, b\} \times \{a, b, c, d\}$.

In Exercises 17 through 24, sketch the given subsets of $\mathbb{R}^2$.

17. $[2, 4] \times [1, 3]$.

18. $(-1, 1) \times (1, 2)$.

19. $(-1, 1)^2$.

20. $[0, 10]^2$.

21. $\mathbb{R}^+ \times [-1, 1]$.

22. $\mathbb{R}^+ \times \mathbb{R}^-$.

23. $\mathbb{R}^- \times \mathbb{R}^+$.

24. $(0, 1) \times \mathbb{R}^-$.

25. Find $\{1, 3\} \times \{2, 4\} \times \{1, 2\}$.

26. Find $\{a\} \times \{a, b\} \times \{a, b, c\}$.

27. Find $\{a, b\}^3$.

28. What is $\mathbb{Z}^3$?

29. Find $\mathcal{P}(\{a, b, c\})$.

30. Find $\mathcal{P}(\{2\})$.

31. Find $\mathcal{P}(\{1, 2, 3, 4\})$.

32. Find $\mathcal{P}(\{x, y\})$.

33.* Find $|\mathcal{P}(\{n \ : \ n \in \mathbb{Z} \text{ and } 1 \leq n \leq 10\})|$.

34.* Find $|\mathcal{P}(\{n \ : \ n \in \mathbb{Z} \text{ and } -3 \leq n \leq 3\})|$.

35. Give two finite and two infinite elements of $\mathcal{P}(\mathbb{R})$.

36. Give two finite and two infinite elements of $\mathcal{P}(\mathbb{N})$.

For Exercises 37 through 40, determine whether the stated relations are true or false.

37. $1 \in \{1,2\} \times \{1,2\}$.   38. $(1,1) \in \{1,2\} \times \{3,4\}$.

39. $1 \in \mathcal{P}(\{1\})$.   40. $\{1\} \in \mathcal{P}(\{1,2\})$.

For Exercises 41 through 44, write the set identities that correspond to the given logical equivalences.

41. Associativity of $\oplus$: $(p \oplus q) \oplus r \equiv p \oplus (q \oplus r)$.

42. Commutativity of $\oplus$: $p \oplus q \equiv q \oplus p$.

43. Distributivity with $\wedge$ and $\vee$: $p \wedge (q \vee r) \equiv (p \wedge q) \vee (p \wedge r)$ and $p \vee (q \wedge r) \equiv (p \vee q) \wedge (p \vee r)$.

44. Distributivity   with $\wedge$ and $\oplus$: $p \wedge (q \oplus r) \equiv (p \wedge q) \oplus (p \wedge r)$.

In Exercises 45 through 48, use Venn diagrams to illustrate the stated identities.

45. Distributivity: $A \cup (B \cap C) = (A \cup B) \cap (A \cup C)$.

46. De Morgan's Law: $(A \cup B)^c = A^c \cap B^c$.

47. Absorption: If $A \subseteq B$, then $A \cap B = A$ and $A \cup B = B$.

48. Associativity: $(A \cap B) \cap C = A \cap (B \cap C)$ and $(A \cup B) \cup C = A \cup (B \cup C)$.

In Exercises 49 through 56, use Theorems 1.5 and 1.6, as well as obvious generalizations, to verify the stated identities.

49. $A \cap (A^c \cup B \cup C) = (A \cap B) \cup (A \cap C)$.

50. $(A \cup B \cup C)^c = A^c \cap B^c \cap C^c$.

51. $(A \cap B \cap C^c) \cup (A \cap B^c \cap C) = A \cap (B \bigtriangleup C)$.

52. $(A \cap B \cap C^c) \cup (A \cap B^c \cap C^c) = A \cap C^c$.

53. $A \cap ((B \cap C)^c) = (A \cap B^c) \cup (A \cap C^c)$.

54. $(A \cup (B \cap C))^c = (A \cup B)^c \cup (A \cup C)^c$.

55* $(A \cap B) \cup (A \cap B \cap C) = A \cap B$.

56* $(A \cup B) \cap (A \cup B \cup C) = A \cup B$.

In Exercises 57 through 62, use *Mathematica* to perform the given tasks.

57. Define the function

```
In[1] :=  setEq[x_, y_]
       := (Union[x] == Union[y])
```

and experiment with setEq[A,B] for various lists A and B. Does this test for equality as sets? as lists?

58. Define the function

```
In[1] :=  subSet[x_, y_]
       := (Union[x, y] == Union[y])
```

and experiment with `subSet[A,B]` for various lists A and B. Does this test for the subset relation?

59. Define a function `disjoint` that tests whether two given sets are disjoint, and experiment with that function for various lists A and B.

60. Define a function `cardSet` that finds the cardinality of a given list regarded as a set, and experiment with that function for various lists A.

61. Define a function `symmDiff` that finds the symmetric difference of two given lists regarded as sets, and experiment with that function for various lists A and B.

62. Let U = {0,1,2,3,4,5,6,7,8,9}; define the functions

```
In[1]:=  compUn[x_, y_] :=
         Complement[U, Union[x, y]]
In[2]:=  intComp[x_, y_] :=
         Intersection[Complement[U, x],
                      Complement[U, y]]
```

and experiment with `compUn[A,B]` and `intComp[A,B]` for various lists A and B. How are the two results related? Why?

In Exercises 63 through 68, use *Maple* to perform the given tasks.

63. For various sets A and B, experiment with the given expressions, and in each case, determine which of the two operations is computed first.

(a) A intersect B union C

(b) A union B intersect C

64. Compute `powerset({{}})`.

65. Define the functions

```
> subset1 := (x,y) ->
              evalb(x = x intersect y) ;
> subset2 := (x,y) -> member(x,powerset(y)) ;
```

and experiment with `subset1(A,B)` and `subset2(A,B)` for various sets A and B. What do these two functions do? Which of the two ought to be more efficient? Why?

66. Define the function

```
> disjoint := (x,y) ->
               evalb(x intersect y = {}) ;
```

and experiment with `disjoint(A,B)` for various sets A and B. Does it test whether A and B are disjoint sets?

67. Let U = {0,1,2,3,4,5,6,7,8,9}; define a function `compU` that finds the complement of a given subset A of U, and experiment with `compU` for various subsets.

68. Define a function `symmDiff` that finds the symmetric difference of two given sets, and experiment with that function for various sets A and B.

1.5 Valid arguments

In order to prove theorems, the ability to deduce a statement from a sequence of prior statements is needed. This section considers circumstances under which such deductions can be made.

An **argument** consists of a sequence of statements, called **premises**, followed by a final statement, called the **conclusion**. In a formal presentation of an argument, the symbol $\therefore$ is placed before the conclusion and represents the word *therefore*. Example 1.44 gives two examples of arguments from plane geometry.

Example 1.44.

(a) If triangles T_1 and T_2 are congruent, then T_1 and T_2 are similar.
Triangles T_1 and T_2 are congruent.
$\therefore T_1$ and T_2 are similar.

(b) If quadrilateral Q is a rectangle, then both pairs of opposite sides of Q are parallel.
The quadrilateral Q is not a rectangle.
$\therefore Q$ has a pair of nonparallel opposite sides.

In order to analyze the arguments of Example 1.44, we examine their structure. We step back from the specific statements in the arguments and, instead, consider the corresponding statement forms. Of course, we must recognize when the same statement (or its negation) appears in multiple places in an argument. In this way, we obtain the form of a given argument. The argument forms corresponding to the arguments from Example 1.44 are given in Example 1.45.

Example 1.45.

(a) The following is the form of the argument in Example 1.44(a):

$$p \to q$$

$$p$$

$$\therefore \ q.$$

(b) The following is the form of the argument in Example 1.44(b):

$$p \to q$$

$$\neg p$$

$$\therefore \quad \neg q.$$

Definition 1.23.

(a) An **argument form**

$$p_1$$

$$p_2$$

$$\vdots$$

$$p_n$$

$$\therefore \quad r$$

is a sequence of (**premise**) statement forms $p_1, p_2, \ldots, p_n$ followed by a (**conclusion**) statement form r (preceded by the symbol $\therefore$ for "therefore").

(b) The argument form is considered to be **valid** if and only if the statement form $p_1 \wedge p_2 \wedge \cdots \wedge p_n \to r$ is a tautology. Otherwise, it is considered to be **invalid**.

Arguments are analyzed by considering their forms. An argument is said to be **valid** if and only if its corresponding argument form is valid. Of course, checking that a statement of the form $p_1 \wedge p_2 \wedge \cdots \wedge p_n \to r$ is a tautology can be accomplished with a truth table, and it suffices to check that the conclusion r is true in those rows in which each of the premises $p_1, p_2, \ldots, p_n$ is true.

Example 1.46.

(a) Verify that the argument form in Example 1.45(a) is valid (and hence the argument in Example 1.44(a) is valid).

Solution.

p	q	$p \to q$	p	q
F	F	T	F	
F	T	T	F	
T	F	F	T	
T	T	T	T	T

The truth table shows that $(p \to q) \wedge p \to q$ is a tautology. It contains a column for each premise and one of the conclusion. In each row in which both of the premises $p \to q$ and p are true, the conclusion q is also seen to be true. The value of the conclusion q in the other rows need not be checked to now conclude that $(p \to q) \wedge p \to q$ is a tautology, since an if-then statement is always true when its hypothesis is false. Therefore, the argument form in Example 1.45(a) is valid.

(b) Verify that the argument form in Example 1.45(b) is invalid (and hence the argument in Example 1.44(b) is invalid).

Solution.

p	q	$p \to q$	$\neg p$	$\neg q$
F	F	T	T	T
F	T	T	T	F
T	F	F	F	
T	T	T	F	

The second row of the truth table shows that $(p \to q) \wedge \neg p \to \neg q$ is not a tautology. In that row, the hypothesis $(p \to q) \wedge \neg p$ is true, but the conclusion $\neg q$ is false. Therefore, the argument form in Example 1.45(b) is not valid. □

The next example illustrates the fact that the validity of an argument is not characterized solely by the truth of its conclusion.

Example 1.47. Determine if the following argument is valid:

$$2^3 > 0.$$

$$\text{If } 2 > 0, \text{ then } 2^3 > 0.$$

$$\therefore \ 2 > 0.$$

Solution. The argument has the following form:

$$p$$

$$q \to p$$

$$\therefore \ q.$$

The truth table

p	q	p	$q \to p$	q
F	F	F	T	
F	T	F	F	
T	F	T	T	F
T	T	T	T	T

shows that the statement form $p \land (q \to p) \to q$ is not a tautology. Specifically, this is seen in the third row, in which the premises are true but the conclusion is false. Thus, the argument form, and hence the argument, is invalid. □

What can lead to confusion in an example like Example 1.47 is that the conclusion is certainly true under the conditions given in the premises. In fact, in this case the conclusion is true independent of those premises. However, the conclusion does not *follow* from the displayed premises. The important point here is that the validity of an argument is defined in terms of the validity of its form. The argument form is not valid. Therefore, the argument is not considered valid.

Theorem 1.7 lists several valid argument forms. Part (a) included there was shown to be valid in Example 1.46(a). Verification of the validity of some of the other parts is accomplished in subsequent examples. The rest are left to the exercises.

Theorem 1.7.   (Basic Valid Argument Forms).

(a) $p \to q$ Direct Implication
        p
        $\therefore\ q$

(b) $p \to q$ Contrapositive Implication
        $\neg q$
        $\therefore\ \neg p$

(c) $p \to q$ Transitivity of $\to$
        $q \to r$
        $\therefore\ p \to r$

(d) $p \to r$ Two Separate Cases
        $q \to r$
        $p \lor q$
        $\therefore\ r$

(e) $p \lor q$ Eliminating a Possibility
        $\neg p$
        $\therefore\ q$

(f) $p \land q$ In Particular
        $\therefore\ p$

(g) p Obtaining Or
     $\therefore\ p \lor q$

(h) p Obtaining And
     q
     $\therefore\ p \land q$

(i) $p \leftrightarrow q$ Substitution of Equivalent
     p
     $\therefore\ q$

In addition to performing the straightforward verifications of the argument forms in Theorem 1.7, some time should be spent internalizing them. The names attached to each argument form are nonstandard but should aid in this process. Basic arguments of these basic forms appear throughout more complicated arguments. Hence, quick recollection of the basic forms is valuable. Verification of the argument forms may be helpful in the process of remembering them.

Some parts of Theorem 1.7 are in fact consequences of other parts. In particular, this is true of part (i).

Example 1.48. Verify that part (i) of Theorem 1.7 is a valid argument form by using some of the earlier parts of that theorem.

Verification. Starting from the premises as given, we successively apply known logical equivalences and valid argument forms to our current list of statement forms, until the desired conclusion is obtained.

	Statement form	Justification
1.	$p \leftrightarrow q$	Given
2.	p	Given
3.	$(p \to q) \land (q \to p)$	(1), Definition of $\leftrightarrow$
4.	$p \to q$	(3), In Particular
5.	$\therefore\ q$	(2),(4), Direct Implication

$\square$

The argument in the next example is just a special case of part (b) of Theorem 1.7. It is, however, extremely important in its own right.

Example 1.49. (Argument by Contradiction).
Verify that the following argument form is valid:

$$\neg p \to \underline{f}$$

$$\therefore\ p.$$

Verification.

Statement form		Justification
1.	$\neg p \rightarrow f$	Given
2.	t	Definition of t
3.	$\neg f$	(2), Theorem 1.2(a), Substitution of Equivalent
4.	$\neg\neg p$	(1), (3), Contrapositive Implication
5.	$\therefore\ p$	(4), Double Negative, Substitution of Equivalent

Note that the tautology t need not be included as one of the premises, since it is automatically true independent of everything else. □

The basic idea of an argument by contradiction is very simple. If the negation of a statement implies a contradiction, then the statement must, in fact, be true.

Example 1.50. Verify that the following argument is valid:

$$\text{If } \mathbb{R}^+ = \emptyset, \text{ then } 1 \le 0.$$
$$\text{If } \mathbb{R}^+ = \emptyset, \text{ then } -1 \le 0.$$
$$1 > 0 \text{ or } -1 > 0.$$
$$\therefore\ \mathbb{R}^+ \ne \emptyset.$$

Verification. The argument's form

$$p \rightarrow q$$
$$p \rightarrow r$$
$$\neg q \vee \neg r$$
$$\therefore\ \neg p$$

can be seen to be valid.

Statement form		Justification
1.	$p \rightarrow q$	Given
2.	$p \rightarrow r$	Given
3.	$\neg q \vee \neg r$	Given
4.	$\neg q \rightarrow \neg p$	(1), Contrapositive
5.	$\neg r \rightarrow \neg p$	(2), Contrapositive
6.	$\therefore\ \neg p$	(3),(4),(5), Two Separate Cases

Therefore, the argument is valid. □

1.5.1   Arguments involving quantifiers

None of the arguments considered so far involved quantified statements, yet most of the mathematical statements that we prove and use are quantified. Consequently, we need the ability to handle such statements in arguments.

Theorem 1.8.   (Principle of Specification).
If the premises

$$\forall\, x \in \mathcal{U}, \;\; p(x) \qquad \text{and}$$

$$a \in \mathcal{U}$$

hold, then the conclusion

$$p(a)$$

also holds.

Theorem 1.9.   (Principle of Generalization).
From the following steps,

(i) Take an arbitrary element $a \in \mathcal{U}$

(ii) Establish that $p(a)$ holds

The conclusion

$$\forall\, x \in \mathcal{U}, \;\; p(x)$$

is obtained.

The validity of the Principles of Specification and Generalization follows from the definition of the quantified statement

$$\forall\, x \in \mathcal{U}, \;\; p(x),$$

which specifies the conditions under which such a universal statement is true. Note that step (i) in the Principle of Generalization is an *assumption* that is made in order to apply that principle.

It is important to recognize the difference between the two principles. The Principle of Specification starts with a universal statement and concludes with a specific instance. The Principle of Generalization starts with a specific but arbitrarily chosen element a in the universe of consideration $\mathcal{U}$ and concludes with a universal statement. It is critical to understand in this latter case that a represents an arbitrary or generic element of $\mathcal{U}$. That is, a is assumed to have precisely the properties that characterize membership in the set $\mathcal{U}$, no more and no less. Moreover, a should not be taken to be a particular element of $\mathcal{U}$ that can be distinguished from any other element.

All of the arguments in Theorem 1.7 have universal analogs. Examples 1.51 and 1.52 give analogs of part (a), and Example 1.53 gives an analog of part (c).

Example 1.51. Show that the following argument form is valid:

$$\forall\, x \in \mathcal{U},\ p(x) \to q(x)$$

$$a \in \mathcal{U}$$

$$p(a)$$

$$\therefore\ q(a).$$

Verification.

	Statement form	Justification
1.	$\forall\, x \in \mathcal{U}, p(x) \to q(x)$	Given
2.	$a \in \mathcal{U}$	Given
3.	$p(a)$	Given
4.	$p(a) \to q(a)$	(1),(2), Principle of Specification
5.	$\therefore\ q(a)$	(3),(4), Direct Implication

□

Example 1.52. Show that the following argument form is valid:

$$\forall\, x \in \mathcal{U},\ p(x) \to q(x)$$

$$\forall\, x \in \mathcal{U},\ p(x)$$

$$\therefore\ \forall\, x \in \mathcal{U},\ q(x).$$

Verification.

	Statement form	Justification
1.	$\forall\, x \in \mathcal{U}, p(x) \to q(x)$	Given
2.	$\forall\, x \in \mathcal{U}, p(x)$	Given
3.	Let $a \in \mathcal{U}$ be arbitrary.	Assumption
4.	$p(a) \to q(a)$	(1),(3), Principle of Specification
5.	$p(a)$	(2),(3), Principle of Specification
6.	$q(a)$	(4),(5), Direct Implication
7.	$\therefore\ \forall\, x \in \mathcal{U}, q(x)$	(3),(6), Principle of Generalization

□

Example 1.53.   Show that the following argument form is valid:

$$\forall\, x \in \mathcal{U},\; p(x) \to q(x)$$
$$\forall\, x \in \mathcal{U},\; q(x) \to r(x)$$
$$\therefore\; \forall\, x \in \mathcal{U},\; p(x) \to r(x).$$

Verification.

Statement form	Justification
1. $\forall\, x \in \mathcal{U},\, p(x) \to q(x)$	Given
2. $\forall\, x \in \mathcal{U},\, q(x) \to r(x)$	Given
3. Let $a \in \mathcal{U}$ be arbitrary.	Assumption
4. $p(a) \to q(a)$	(1),(3), Principle of Specification
5. $q(a) \to r(a)$	(2),(3), Principle of Specification
6. $p(a) \to r(a)$	(4),(5), Transitivity of $\to$
7. $\therefore\; \forall\, x \in \mathcal{U},\, p(x) \to r(x)$	(3),(6), Principle of Generalization

$\square$

Remark 1.3.   In practice, when the Principle of Generalization is used in an argument, the variable x attached to the universal quantifier is also used to represent the arbitrary element of $\mathcal{U}$, rather than introducing a new variable a. The separate variable a is used here to aid in clarity in the verification of the relevant argument forms. Throughout the rest of the book, a separate variable is not used.

How do we show that an argument form is invalid? If the argument form does not involve quantifiers, then we have seen that a truth table always determines whether or not the argument form is valid. In the case of the argument form in Example 1.45(b)

$$p \to q$$
$$\neg p$$
$$\therefore\; \neg q \tag{1.9}$$

the following line from the truth table in Example 1.46(b)

p	q	$p \to q$	$\neg p$	$\neg q$	
F	T	T	T	F	(1.10)

demonstrates that (1.9) is invalid. Moreover, (1.10) gives an outline for constructing an example of an argument of the form (1.9) in which each of the premises is true but the conclusion is false. For instance, we might take

$$p = \text{``}0 = 1.\text{''}$$

$$q = \text{``}0 = 0.\text{''}$$

The resulting argument

if $0 = 1$, then $0 = 0$.
$0 \neq 1$.
$\therefore \ 0 \neq 0$.

is certainly invalid; its premises are true and its conclusion is false. That is, the invalidity of an argument form can be demonstrated by giving an example of an argument of that form with true premises and a false conclusion. Since truth tables are not an option in argument forms involving quantifiers, this method of showing that an argument form is invalid is useful in the general case.

Example 1.54. Show that the argument form

$$\forall \, x \in \mathcal{U}, \ p(x) \to q(x)$$

$$a \in \mathcal{U}$$

$$p(a)$$

$$\therefore \forall \, x \in \mathcal{U}, \ q(x)$$

is invalid.

Solution. We provide an example.

Let $\mathcal{U} = \mathbb{R}$,

$$p(x) = q(x) = \text{``}x > 0,\text{''} \text{ and}$$

$$a = 1.$$

In the argument

$$\forall \, x \in \mathbb{R}, \ (x > 0) \to (x > 0)$$

$$1 \in \mathbb{R}$$

$$1 > 0$$

$$\therefore \forall \, x \in \mathbb{R}, \ x > 0$$

all of the premises are true, but the conclusion is false. Hence, the given argument form is invalid. □

In general, an argument can be shown to be invalid by providing an example of an argument of the same form each

of whose premises is true but whose conclusion is false. For
example, the argument

$$\forall\, x \in \mathbb{R}, \quad (x > 0) \to (x^3 > 0)$$

$$2 \in \mathbb{R}$$

$$2^3 > 0$$

$$\therefore 2 > 0$$

is invalid, since it has the same form as the obviously invalid
argument

$$\forall\, x \in \mathbb{R}, \quad (x > 0) \to (x^2 > 0)$$

$$-2 \in \mathbb{R}$$

$$(-2)^2 > 0$$

$$\therefore -2 > 0.$$

Exercises

In Exercises 1 through 8, use truth tables to verify the validity
of the argument form.

1. Theorem 1.7(b).
   Contrapositive Implication.

2. Theorem 1.7(c).
   Transitivity of $\to$.

3. Theorem 1.7(d).
   Two Separate Cases.

4. Theorem 1.7(e).
   Eliminating a Possibility.

5. Theorem 1.7(f).
   In Particular.

6. Theorem 1.7(g).
   Obtaining Or.

7. Theorem 1.7(h).
   Obtaining And.

8. Theorem 1.7(i).
   Substitution of Equivalent.

In Exercises 9 through 16, determine whether the argument form is valid or invalid. Justify your answers.

9. $p \to q$

   q

   $\therefore\ p.$

10. $p \to r$

    $q \to r$

    $p \oplus q$

    $\therefore\ r.$

11. $p \lor q$

    $p \to r$

    $\therefore\ q \lor r.$

12. $p \lor q$

    $p \land q \to r$

    $\therefore\ r.$

13. $p \vee q$

p

$\therefore \quad \neg q.$

14. $p \rightarrow q$

$r \rightarrow q$

$\neg(p \wedge r)$

$\therefore \quad \neg q.$

15. $p \vee q$

$p \rightarrow r$

$q \rightarrow s$

$\therefore \quad r \vee s.$

16. $p \rightarrow \neg r$

$r \rightarrow q$

$p \oplus \neg q$

$\therefore \quad \neg r.$

17. (a) Show that the following argument form is valid:

$$p \rightarrow r$$

$$q \rightarrow r$$

$$\therefore \quad p \vee q \rightarrow r.$$

(b) Without using a truth table, use part (a) to show that Theorem 1.7(d), "Two Separate Cases," is valid.

(c)* Without using a truth table, verify the validity of the argument form "Three Separate Cases"

$$p \rightarrow s$$

$$q \rightarrow s$$

$$r \rightarrow s$$

$$p \vee q \vee r$$

$$\therefore \quad s.$$

18. (a) Show that the following argument form is valid:

$$s \vee q$$

$$r \vee \neg q$$

$$\therefore \quad s \vee r.$$

(b)* Use part (a) for $s = \neg p$ together with the logical equivalence $a \rightarrow b \equiv \neg a \vee b$ to prove Theorem 1.7(c).

In Exercises 19 through 22, determine whether the given argument is valid or invalid. Justify your answers.

19. $2^{10} > 10^2$ or $10^2 > 2^{10}$

$\therefore \quad 2^{10} > 10^2.$

20. $3 \in \mathbb{Z}$ and $\pi \notin \mathbb{Z}$

$\therefore \quad 3 \in \mathbb{Z}.$

21. $\mathbb{Z}$ is finite or

   $\mathbb{Z}$ is infinite$\mathbb{Z}$ is infinite

   $\therefore$ $\mathbb{R}$ is infinite.

22. $\mathbb{N}$ is finite or

   $\mathbb{Z}^-$ is finite$\mathbb{Z}^-$ is infinite

   $\therefore$ $\mathbb{N}$ is finite.

In Exercises 23 through 28, use Theorem 1.7 to show that the given argument forms are valid without using a truth table.

23. $p \rightarrow q$

   $q \rightarrow r$

   p

   $\therefore$ r.

24. $p \vee q \rightarrow r$

   p

   $\therefore$ r.

25. $p \rightarrow r$

   $p \wedge q$

   $\therefore$ r.

26. $p \rightarrow r$

   $p \vee q$

   $\neg q \therefore$ r.

27. $p \wedge (q \vee r)$

   $(p \wedge q) \rightarrow s$

   $(p \wedge r) \rightarrow s$

   $\therefore$ s.

28. $p \rightarrow q$

   $p \rightarrow r$

   p

   $\therefore$ $q \wedge r$.

29*. Use part (e) of Theorem 1.7 to prove part (a). Hint: $p \rightarrow q \equiv \neg p \vee q$.

30*. Use part (e) of Theorem 1.7 to prove part (b).

In Exercises 31 through 40, verify that the given argument form is valid.

31. $\forall\, x \in \mathcal{U}, p(x) \rightarrow q(x)$

   $a \in \mathcal{U}$

   $\neg q(a)$

   $\therefore$ $\neg p(a)$.

32. $\forall\, x \in \mathcal{U}, p(x) \vee q(x)$

   $a \in \mathcal{U}$

   $\neg p(a)$

   $\therefore$ $q(a)$.

33. $\forall\, x \in \mathcal{U}, p(x) \rightarrow q(x)$

   $\forall\, x \in \mathcal{U}, \neg q(x)$

   $\therefore$ $\forall\, x \in \mathcal{U}, \neg p(x)$.

34. $\forall\, x \in \mathcal{U}, p(x) \vee q(x)$

   $\forall\, x \in \mathcal{U}, \neg p(x)$

   $\therefore$ $\forall\, x \in \mathcal{U}, q(x)$.

35. $\forall\, x \in \mathcal{U}, p(x)$

$\forall\, x \in \mathcal{U}, q(x)$

$a \in \mathcal{U}$

$\therefore\ p(a) \wedge q(a).$

36. $\forall\, x \in \mathcal{U}, p(x) \to q(x)$

$\forall\, x \in \mathcal{U}, q(x) \to r(x)$

$a \in \mathcal{U}$

$p(a)$

$\therefore\ r(a).$

37. $\forall\, x \in \mathcal{U}, p(x)$

$\forall\, x \in \mathcal{U}, q(x)$

$\therefore\ \forall\, x \in \mathcal{U}, p(x) \wedge q(x).$

38. $\forall\, x \in \mathcal{U}, p(x) \to q(x)$

$\forall\, x \in \mathcal{U}, q(x) \to r(x)$

$\therefore\ \forall\, x \in \mathcal{U},$

$p(x) \to r(x).$

39.* $\forall\, x \in \mathcal{U}, p(x) \vee q(x)$

$a \in \mathcal{U}$

$q(a) \to r(a)$

$\therefore\ p(a) \vee r(a).$

40.* $\forall\, x \in \mathcal{U}, p(x) \wedge \neg q(x)$

$\forall\, x \in \mathcal{U}, q(x) \vee r(x)$

$\therefore\ \forall\, x \in \mathcal{U},$

$\neg p(x) \to r(x).$

In Exercises 41 through 44, verify that the given argument form is invalid.

41. $\forall\, x \in \mathcal{U}, p(x) \vee q(x)$

$\forall\, x \in \mathcal{U}, \neg p(x)$

$\therefore\ \forall\, x \in \mathcal{U}, \neg q(x).$

42. $\forall\, x \in \mathcal{U}, p(x) \vee q(x)$

$\forall\, x \in \mathcal{U}, q(x) \vee r(x)$

$\therefore\ \forall\, x \in \mathcal{U}, p(x) \vee r(x).$

43. $\forall\, x \in \mathcal{U}, p(x) \vee q(x)$

$a \in \mathcal{U}$

$p(a) \wedge q(a)$

$\therefore\ \forall\, x \in \mathcal{U}, p(x) \wedge q(x).$

44. $\forall\, x \in \mathcal{U}, p(x) \to q(x)$

$a \in \mathcal{U}$

$q(a) \to p(a)$

$\therefore\ \forall\, x \in \mathcal{U},$

$q(x) \to p(x).$

In Exercises 45 through 48, determine whether the given argument is valid or invalid. Justify your answers.

45. $\forall\, x \in \mathbb{R},$

$(x > 1) \to (x^2 > 1)$

$4 = 2^2$

$4 > 1$

$\therefore\ 2 > 1.$

46. $\forall\, n \in \mathbb{Z}, (\dfrac{1}{n} \in \mathbb{Z})$

$\to (n = 1 \text{ or } n = -1)$

$\dfrac{1}{2} \notin \mathbb{Z}$

$\therefore\ 2 \neq 1 \text{ and } 2 \neq -1.$

47. $\forall\, n \in \mathbb{Z},$

$(n < 0) \rightarrow (-n > 0)$

$-2 \leq 0$

$\therefore\ 2 \geq 0.$

48. $\forall\, A, B \subseteq \mathbb{R}, (A \subseteq B)$

$\rightarrow (A \cap B = A)$

$\forall\, A, B \subseteq \mathbb{R}, (A \cap B = A)$

$\rightarrow (A \subseteq B)$

$\therefore\ \ \forall\, A, B \subseteq \mathbb{R}, (A \subset B)$

$\rightarrow (A \subseteq B).$

49. State the appropriate Principle of Specification for statements of the form
"$\forall\, x, y \in \mathcal{U}, p(x,y).$"

50. State the appropriate Principle of Generalization to achieve statements of the form
"$\forall\, x, y \in \mathcal{U}, p(x,y).$"

1.6   Review problems

1. Determine whether the sentence "If $0 > 0$, then $1 = 2$" is a statement. If so, is it true?

2. Make a truth table for the statement form $p \rightarrow \neg q$.

3. Verify the logical equivalence $(p \vee q) \equiv (\neg p \wedge q) \vee p$.

4. Verify the logical equivalence $\neg(p \wedge \neg q) \equiv \neg p \vee q$.

5. Verify that $(p \rightarrow q) \vee (q \rightarrow p)$ is a tautology.

6. Are $\neg p \rightarrow q$ and $p \vee q$ logically equivalent? Justify your answer.

7. Are $p \rightarrow (q \vee r)$ and $(p \rightarrow q) \vee r$ logically equivalent? Justify your answer.

8. For the statement form $p \rightarrow p \vee \neg q$, find and simplify its

(a) Converse

(b) Contrapositive

(c) Inverse

(d) Negation

9. Experienced programmers know the truth of the following statement:

> If the program contains a syntax error, then the program does not compile.

Since its contrapositive is an equivalent statement, express its contrapositive.

10. Negate the statement $\neg p \vee (q \wedge \neg r)$.

11. Steve's mom made the following statement:

> "Steve is doing his homework or Steve is not going to the basketball game."

However, Steve does not believe this to be true. Express the negation of his mother's statement.

12. Verify the logical equivalence

$$\neg p \wedge (q \vee \neg r) \equiv (\neg p \wedge q) \vee \neg(p \vee r)$$

by using known logical equivalences.

13. Verify the logical equivalence

$$(p \wedge q \wedge \neg r) \vee (\neg p \wedge q \wedge \neg r) \equiv q \wedge \neg r$$

by using known logical equivalences.

14. Trace the pictured circuit to determine an expression for the output in terms of the input, and make an input–output table.

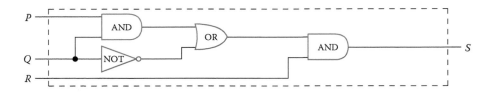

15. Draw a circuit that realizes the expression $\neg((P \vee Q) \wedge R) = S$.

16. Draw a circuit that realizes the expression $(\neg P \wedge Q) \vee (P \wedge R) = S$.

In Exercises 17 through 27, determine whether each of the given relations is true or false.

17. $\{x : x \in \mathbb{R} \text{ and } x^2 < 2\} = \{x : x \in \mathbb{R} \text{ and } x^4 < 4\}$.

18. $\{1, 2, 3, 4\} = \{4, 3, 2, 1\}$.   19. $\{1\} \in \{1, 2, 3\}$.

20. $1 \subseteq \{1\}$.   21. $\{1, 2\} \in \{\{1, 2\}, \{3, 4\}\}$.

22. $[0, 1]$ is finite.   23. $\{2\} \subseteq \{1, 2, 3\}$.

24. $1 \in \{\{1\}, \{2\}\}$.   25. $0 \subset \mathbb{Z}$.

26. $\{1, 2, 1, 2\} \subseteq \{1, 2\}$.   27. $|\{\emptyset\}| = 0$.

In Exercises 28, through 30, express the given sets in set notation.

28. The set of even integers from 4 to 12. (See Definition 3.1 in Section 3.1.)

29. The set of real roots of the polynomial $p(x) = x^5 + x^4 + x^3 + x^2 + x + 1$. See Exercise 38 from Section 1.3.

30. $(-3, -1]$.

31.* Find $|\{x : x \in \mathbb{R} \text{ and } x^2 + x - 1 = 0\}|$.

32. Express the "set" from Russell's Paradox, and explain why it cannot be a set.

In Exercises 33 through 38, write the given statements efficiently using quantifiers and standard notation.

33. Every integer power of 2 is an integer.

34. There is an integer power of 2 that is greater than 1000.

35. The quotient of any real number by any nonzero real number is a real number.

36. For every real number x such that $1 < x \leq 4$, we have $\frac{1}{4} \leq \frac{1}{x} < 1$.

37. The sum of any two integers is an integer.

38. Every real number has a real cube root.

In Exercises 39 through 44, negate the given statements.

39. $\forall\, x \in \mathbb{N}, x^2 \in \mathbb{N}$ and $\frac{1}{2x} \notin \mathbb{N}$.

40. $\exists\, x \in \mathbb{R}$ such that $x^2 - x + 1 = 0$.

41. $\forall\, x \in \mathbb{R}$, if $x^3 < 0$, then $x < 0$.

42. $\forall\, x, y \in \mathbb{R}, (x + y)^2 = x^2 + 2xy + y^2$.

43. $\exists\, n \in \mathbb{Z}$ such that if $n < 0$, then $n^2 - 1 > 0$.

44. $\forall\, x \in \mathbb{R}, \exists\, n \in \mathbb{Z}$ such that $x^n > 0$.

45. The following popular saying may be accepted as wisdom:

> Truth is not always popular, but it is always right.

However, express the negation of this statement.

46. Given that $A = \{1, 2, 3\}$ and $B = \{2, 5\}$, find each of the following:

(a) $A \cap B$ (d) $A \,\triangle\, B$

(b) $A \cup B$ (e) $A \times B$

(c) $B \setminus A$ (f) $\mathcal{P}(B)$

47. If $\mathcal{U} = \mathbb{R}$, then find $(-1, \infty)^{c}$.

48. Find $\mathbb{Z} \cap (-1, 1)$.

49. Find $\{1, 2, 3, 4\} \cup \{2, 4, 6, 8\}$.

50. Find $[3, 5] \setminus [2, 4)$.

51. Find $\{a, b, c, f\} \,\triangle\, \{d, e, f\}$.

52. Are $\mathbb{N}$ and $\mathbb{Z}$ disjoint?

53. Find $\{x, y, z\} \times \{p, q\}$.

54. Sketch $(-1, 1) \times [0, 1]$.

55. Find $\{1\}^2$.

56. Find $\mathcal{P}(\{x, y, z\})$.

In Exercises 57 through 60, use known set identities to verify the given set identities.

57. $(A^c \cap B^c)^c = A \cup B$.

58. $A^c \cap (B \cup C^c) = (A^c \cap B) \cup (A \cup C)^c$.

59. $(A \cap B^c) \cup (A \cap B) = A$.

60. $(A^c \cap B \cap C^c) \cup (A^c \cap (B^c \cup C)) = A^c$.

61. Use truth tables to verify the validity of the given argument form.

$$p \wedge q$$
$$p \rightarrow r$$
$$\therefore \ q \wedge r.$$

62. Use truth tables to verify the validity of the given argument form.

$$p \rightarrow q$$
$$q \rightarrow r$$
$$r \rightarrow p$$
$$\therefore \ p \leftrightarrow r.$$

63. Determine whether the given argument form is valid. Justify your answer.

$$p \vee q$$
$$q \rightarrow r$$
$$\therefore \ p \vee r.$$

64. Determine whether the given argument form is valid. Justify your answer.

$$p \wedge \neg q$$
$$q \vee r$$
$$\therefore \ \neg r \rightarrow p.$$

65. Determine whether the given argument form is valid. Justify your answer.

$$p \rightarrow q$$
$$r \rightarrow q$$
$$q$$
$$\therefore \ p \vee r.$$

66. A geometry student has used the following argument in his homework assignment:

> Rhombus R is a square or a parallelogram.
>
> Rhombus R is a parallelogram.
>
> $\therefore$ Rhombus R is not a square.

Determine whether the argument is valid. Justify your answer.

67. Without using a truth table, verify that the given argument form is valid.

$$p \rightarrow (q \vee r)$$
$$\neg q \wedge \neg r$$
$$\therefore \; \neg p.$$

68. Without using a truth table, verify that the given argument form is valid.

$$\neg r$$
$$p \rightarrow q$$
$$q \rightarrow r$$
$$\therefore \; \neg p.$$

69. Verify that the given argument form is valid.

$$\forall\, x \in \mathcal{U}, \;\; p(x) \wedge q(x)$$
$$\therefore \;\; \forall\, x \in \mathcal{U}, \;\; p(x).$$

70. Verify that the given argument form is valid.

$$\forall\, x \in \mathcal{U}, \;\; p(x) \vee q(x)$$
$$a \in \mathcal{U}$$
$$\neg q(a)$$
$$\therefore \;\; p(a).$$

71.* Verify that the given argument form is valid.

$$\forall\, x \in \mathcal{U}, \;\; p(x) \vee \neg q(x)$$
$$\forall\, x \in \mathcal{U}, \;\; q(x)$$
$$\therefore \;\; \forall\, x \in \mathcal{U}, \;\; p(x).$$

72. Show that the argument

$$\forall\, x \in \mathbb{R}, \;\; x < 1 \text{ or } x > 0$$
$$\forall\, x \in \mathbb{R}, \;\; x > 0 \text{ or } x^2 > -1$$
$$\therefore \;\; \forall\, x \in \mathbb{R}, \;\; x < 1 \text{ or } x^2 > -1.$$

is invalid.

2 Basic Proof Writing

In this chapter, we learn how to prove statements. In addition, emphasis is placed on how to *write* proofs. We want to move away from the rigid two-column (statement–reason) format sometimes used in high school geometry courses. Lists of statements or equations with little or no words connecting them will instead be considered a sketch of a proof and not a proof itself. We move past the formal structure in which the argument forms of Section 1.5 were presented and verified. However, those forms certainly provide the logical foundation on which proofs are built.

Proofs are written in proper English. Correct grammar is necessary. Further, there should be a flow from sentence to sentence that carries the reader through the argument. An analogy can be made comparing a painter to a proof writer. Just like a painter may start with a charcoal sketch on the canvas, a proof writer may start with a sketch of the proof on some scratch paper. For the painter, the desired work of art is not completed until the paint in all of its colors is applied. In a similar fashion, the proof writer succeeds through good writing. Moreover, just like paintings, proofs have styles. With practice, such as that gained in this chapter and subsequent chapters, everyone develops their own proof-writing style.

An additional issue that needs to be addressed before we begin writing proofs regards what facts can be used without proof. In this chapter, the primary context of our proofs will be set theory. We are not allowed to use any properties of sets until we have proven them first. Only the definitions from Chapter 1 are assumed. Additionally, we consider some proofs of facts about the real numbers and real functions. For those, only the facts given in Appendix A may be used without proof.

2.1 Direct demonstration

We start with some simple examples of proofs to illustrate the basic principles. In Example 2.1, a sketch is presented first, and then the proof is provided. This should emphasize the difference between sketches and proofs.

Example 2.1.   Show that the points $(-4, -5)$, $(2, -2)$, and $(8, 1)$ lie on a common line.

Note that *show* is synonymous with *prove*.

Sketch of the proof for Example 2.1.

$$m = \frac{-5 - (-2)}{-4 - 2} = \frac{-5 + 2}{-6} = \frac{-3}{-6} = \frac{1}{2}$$

$$y - (-5) = \tfrac{1}{2}(x - (-4))$$

$$y + 5 = \tfrac{1}{2}(x + 4) = \tfrac{1}{2}x + 2$$

$$y = \tfrac{1}{2}x - 3$$

$$1 = \tfrac{1}{2}(8) - 3$$

Our sketch contains the calculations sufficient to prove our statement. However, it is merely a list of equations, and no explanation is given to carry a reader through the argument. In contrast, our proofs are written in proper English. They contain complete sentences that flow from one step to another, giving appropriate justifications.

Proof for Example 2.1. Let L be the line given by the equation $y = \tfrac{1}{2}x - 3$. Observe that

$$-5 = \tfrac{1}{2}(-4) - 3,$$

$$-2 = \tfrac{1}{2}(2) - 3, \text{ and}$$

$$1 = \tfrac{1}{2}(8) - 3.$$

Therefore, all of the points $(-4, -5)$, $(2, -2)$, and $(8, 1)$ satisfy the equation and thus lie on the common line L.

$\square$

The symbol $\square$ signifies the end of the proof. Some alternatively use Q.E.D., which stands for the Latin phrase *quod erat demonstrandum* and means "which was to be demonstrated."

In some ways, the proof for Example 2.1 turned out to be more efficient than the sketch, since we stripped out some of the extraneous calculations used to discover the equation of our line. The point is that we need to write enough so that the reader sees a clear argument. However, we do not always need to share all of our scratch work. Certainly, no failed attempts at the proof and no false or unnecessary steps should be included in a final proof.

2.1.1 Existential statements

To prove a statement of the form

$$\exists\, x \in \mathcal{U} \text{ such that } p(x),$$

it suffices to present an example of a particular element $x \in \mathcal{U}$ for which $p(x)$ holds.

Example 2.2.   Show: There is a set A such that $A \cap \mathbb{R}^+ = \mathbb{Z}^+$.

Proof. Let $A = \mathbb{Z}$.
Observe that

$$A \cap \mathbb{R}^+ = \mathbb{Z} \cap \mathbb{R}^+ = \mathbb{Z}^+. \qquad\qquad \square$$

For the proof in Example 2.2, $A = \mathbb{Z}^+$ would provide another example for which $A \cap \mathbb{R}^+ = \mathbb{Z}^+$. Of course, there are many others as well. However, a single example suffices. Moreover, examples we might have tried that do not work are certainly not included.

The method of presenting a single example also works, in fact, for any number of existential quantifiers.

Example 2.3. Show: There exist sets A and B such that $|A \cup B| < |A| + |B|$.

Proof. Let $A = \{1,2\}$ and $B = \{2,3\}$.
So, $A \cup B = \{1,2,3\}$.
Observe that

$$|A \cup B| = 3 < 2 + 2 = |A| + |B|. \qquad\qquad \square$$

2.1.2 Counterexamples

If a statement is not true, then we want to disprove it. That is, we want to verify that the statement is false. If the statement has the form

$$\forall\, x \in \mathcal{U}, \;\; p(x), \qquad\qquad (2.1)$$

then it is disproved by presenting an example of a particular element $x \in \mathcal{U}$ for which $p(x)$ does not hold. Such an example is called a **counterexample**. Of course, this technique works since the negation of (2.1) is the existential statement

$$\exists\, x \in \mathcal{U} \text{ such that } \neg p(x).$$

Example 2.4. Disprove: Every interval's complement is not an interval.

Note that we seek an interval whose complement is also an interval.

Counterexample. Let $I = [0, \infty)$.
So, I is an interval.
Observe that

$$I^c = [0, \infty)^c = (-\infty, 0).$$

So, I^c is also an interval.

Hence, there is an interval whose complement is also an interval. $\square$

Sometimes we do not know in advance whether a given statement is true or false. Hence, we must decide whether a proof or disproof is warranted.

Example 2.5.   Prove or Disprove: $\forall x \in \mathbb{R}$, if $x < 2$, then $x^2 < 4$.

Solution. By thinking about values x for which $x \leq -2$, we discover that this statement is not true. Therefore, a counterexample is warranted and comes from picking a particular value x with $x \leq -2$ (whence $x^2 \geq 4$).

Counterexample. Let $x = -3$.
So, $x^2 = 9$.
Observe that $x < 2$ and $x^2 \geq 4$.
That is, for $x = -3$, it is not true that

$$\text{if } x < 2, \text{ then } x^2 < 4. \qquad \square$$

In Example 2.5, it is important to recall that the negation of the if-then statement form $p \rightarrow q$ is the statement form $p \wedge \neg q$.

2.1.3   Universal statements for small universes

If a universe $\mathcal{U}$ is finite and has a very small size, then it may be reasonable to prove a statement of the form

$$\forall x \in \mathcal{U}, \ p(x)$$

by verifying $p(x)$ for each individual element $x \in \mathcal{U}$.

Example 2.6.   Show: $\forall\, x \in \{-1, 0, 1\}, x^3 = x$.

Proof. Observe that

$$(-1)^3 = -1,$$
$$0^3 = 0, \text{ and}$$
$$1^3 = 1. \qquad \square$$

Example 2.7.   Define $A_0 = \emptyset, A_1 = \{\emptyset\}, A_2 = \{\emptyset, \{\emptyset\}\}$,
and $A_3 = \{\emptyset, \{\emptyset\}, \{\emptyset, \{\emptyset\}\}\}$.
Show, for each integer $0 \le i \le 2$, that $A_{i+1} \subseteq \mathcal{P}(A_i)$.

Proof. For $i = 0$, $\mathcal{P}(A_i) = \{\emptyset\}$.
Observe that

$$A_1 = \{\emptyset\} = \mathcal{P}(A_0).$$

In particular, $A_{i+1} \subseteq \mathcal{P}(A_i)$.

For $i = 1$, $\mathcal{P}(A_i) = \{\emptyset, \{\emptyset\}\}$.
Observe that

$$A_2 = \{\emptyset, \{\emptyset\}\} = \mathcal{P}(A_1).$$

In particular, $A_{i+1} \subseteq \mathcal{P}(A_i)$.

For $i = 2$, $\mathcal{P}(A_i) = \{\emptyset, \{\emptyset\}, \{\{\emptyset\}\}, \{\emptyset, \{\emptyset\}\}\}$.
Observe that

$$A_3 = \{\emptyset, \{\emptyset\}, \{\emptyset, \{\emptyset\}\}\} \subseteq \mathcal{P}(A_2).$$

That is, for $i = 0, 1, 2$, we have $A_{i+1} \subseteq \mathcal{P}(A_i)$. $\qquad \square$

The technique used in Examples 2.6 and 2.7 should only be applied when $|\mathcal{U}|$ is small. If that is not the case, then the universal statement should be proved using techniques encountered in the subsequent sections.

Comments on proof writing.   As was done in this section, in much of Part I of this book, we shall start each sentence of our proofs on a new line. This is done to encourage a pause at the end of each statement to reflect on how that fact follows from previous ones. This is in slight conflict with our goal of achieving a natural flow from sentence to sentence in our proofs. However, at this point, learning to write (and read) proofs is an additional focus that motivates this choice. In fact, the proofs and disproofs in this section do not well represent general proof-writing strategies. They are quite brief, in the absence of a need for great explanation and exposition. Better general proof-writing principles are initiated in subsequent sections.

Exercises

1. Show that the points $(-1, -8)$, $(1, -2)$, and $(2, 1)$ lie on a common line.

2. Show that $(-1, 6)$, $(0, 3)$, $(1, 2)$, and $(2, 3)$ lie on a common parabola. Hint: For scratch work, plug three of these points into $y = ax^2 + bx + c$ to obtain three equations determining a, b, and c.

3. Show: There is a set A such that $\{1, 2, 3, 4\} \setminus A = \{1, 3\}$.

4. Show: There is a set A such that $\mathbb{Z} \cup A = \mathbb{R}$.

5. Show: There exist sets A and B such that $A \cup B = A \cap B$.

6. Show: There exist sets A and B such that $|A \setminus B| \neq |A| - |B|$.

7. Show: $\exists\, n \in \mathbb{Z}$ such that $10^n = 0.001$.

8. Show: $\exists\, n \in \mathbb{Z}$ such that $\frac{1}{n} \in \mathbb{Z}$.

9. Show: $\exists\, m, n \in \mathbb{Z}$ such that $9m + 14n = 1$.

10. Show: $\exists\, m, n \in \mathbb{Z}$ such that $3m + 5n = 11$.

11. Show: There exist sets A and B such that $A \setminus B = B \setminus A$.

12. Show: There exist sets A and B such that $A \neq (A \cup B) \setminus B$.

13. Show: $\exists\, x \in \mathbb{Z}$ such that $3x^2 + 8x = 35$.

14. Show: $\exists\, x \in \mathbb{Z}$ such that $x^3 + 2x^2 - 2x + 3 = 0$.

In Exercises 15 through 18, recall that the roots of a function $f(x)$ are the solutions to the equation $f(x) = 0$. Utilize factoring and the quadratic formula.

15. Show: $x^2 - 1$ has two distinct real roots.

16. Show: $x^2 + 1$ has no real roots.

17.* Show: $x^2 - 2x + 5$ has no real roots.

18. Show: $x^3 - 3x^2 + 4$ has only two distinct real roots.

19. If P dollars is invested in a certain account, then the amount A in that account after t years is given by the formula $A = P(1.075)^t$.

    (a) Find the amount A after 10 years for an investment of $6000.

    (b)* Show that there is an investment amount less than $5,000 that will grow to at least $10,000 after 10 years.

20. If a 30-year mortgage for M dollars is obtained from a certain bank, then the monthly payment R is given by the formula

$$R = \frac{.008(1.008^{360})M}{1.008^{360} - 1}.$$

(a) Find the mortgage amount M that can be paid off by monthly payments of \$1000.

(b)* On a budget of no more than \$650 per month, show that it is possible to afford a 30-year mortgage of \$75,000.

21. Show: There is a set A such that $A^2 = A$.

22. Disprove: $\forall\, x \in \mathbb{R}$, $\frac{x^2-4}{x+2} = x - 2$.

23. Disprove: The union of any two intervals is an interval.

24. Disprove: $\forall\, x \in \mathbb{R}$, $x > 2$ and $x^2 \in \mathbb{Z}$.

25. Disprove: $\forall\, x \in \mathbb{R}$, $x^2 > x$.

26. Disprove: $\forall\, x \in \mathbb{R}$, $4x^2 + 4x + 1 > 0$.

27. Prove or disprove: $\forall\, n \in \mathbb{N}$, $n^2 \leq 2^n$.

28. Prove or disprove: For all integers $n \geq 2$, $2^n - 1$ is prime. See Definition 3.3 for a definition of prime.

29. Prove or disprove: There exists $x \in \mathbb{R}$ such that $x < 10$ and $x^2 > 100$.

30. Prove or disprove: For all sets A and B, if $A \times B = \emptyset$, then $A = B = \emptyset$.

31.* Disprove: The sine function is increasing. See Definition 1.15.

32.* Disprove: The cosine function is decreasing. See Definition 1.15.

33. Prove or disprove: Every nonnegative real number is positive.

34. Prove or disprove: $\forall\, x \in \mathbb{R}$, $\sqrt{x^2} = x$.

35. Prove or disprove: For all sets A and B, if $|A| \leq |B|$, then $A \subseteq B$.

36. Prove or disprove: For all sets A and B, $A \cup B = A \,\triangle\, B$.

37. Disprove: For all sets A, B, and C, if $A \neq B$, then $A \cap C \neq B \cap C$.

38. Disprove: For all sets A, B, and C, if $A \setminus C = B \setminus C$, then $A = B$.

39. Show: $\forall\, A \in \{\emptyset, \{1\}, \{1,2\}\}$, $A \cup \{3\} = A \,\triangle\, \{3\}$.

40. Show: For all integers $1 \leq n \leq 3$, $\frac{6}{n} \in \mathbb{Z}$.

41. Prove or disprove: For all integers $2 \leq n \leq 9$, the tens digit and the ones digit of $9n$ sum to 9.

42. Prove or disprove: $\forall\, A \in \{\{1,2\}, \{3,4\}\}$, $|A \times \{3,4\}| = 4$.

43. A divisibility test says that a number is divisible by 11 if and only if the alternating sum of its digits is divisible by 11. Certainly, $11n$ is divisible by 11.
    Show: For all integers $19 \leq n \leq 29$, the alternating sum of

    the hundreds digit $-$ the tens digit $+$ the ones digit

    of $11n$ is a multiple of 11.

44. A divisibility test says that a number is divisible by 3 if and only if the sum of its digits is divisible by 3. Certainly, $3n$ is divisible by 3.
    Show: For all integers $4 \leq n \leq 33$, the tens digit and the ones digit of $3n$ sum to a multiple of 3.

45. Show: For each integer $n \in \{2, 3, 5, 7\}$, $2^n - 1$ is prime. See Definition 3.3 for a definition of prime. A prime of the form $2^n - 1$ is called a **Mersenne prime**. (Compare Exercise 28.)

46. Show: For all integers $5 \leq n \leq 10$, $n^2 < 2^n$. (Compare Exercise 27.)

47.* The **Syracuse Problem**, also known as the $3n + 1$ **problem**, is the famous conjecture in number theory that the function

$$f(n) = \begin{cases} \frac{n}{2} & \text{if } n \text{ is even,} \\ 3n + 1 & \text{if } n \text{ is odd,} \end{cases}$$

when iterated from any starting positive integer n, will return to the value 1. For example, starting from $n = 1$, we see that

$$f(1) = 4, \ f(4) = 2, \ f(2) = 1.$$

We may write $1 \mapsto 4 \mapsto 2 \mapsto 1$. That is, we stop when the function value achieves 1.
Show that this conjecture holds for all $1 \leq n \leq 6$.

48.* Let g be the function given by

$$g(n) = \begin{cases} \frac{n}{2} & \text{if } n \text{ is even,} \\ \frac{3n+1}{2} & \text{if } n \text{ is odd.} \end{cases}$$

Show that for any positive integer $n \leq 8$, starting from n and iterating g, the function values eventually return to 1. (Compare Exercise 47.)

2.2 General demonstration (Part 1)

Often, we want to prove a statement of the form

$$\forall \, x \in \mathcal{U}, \ p(x).$$

Section 2.1 provides a technique for small sets $\mathcal{U}$ that is rarely useful, since $\mathcal{U}$ is usually large, even infinite. More generally, the appropriate technique is to use the Principle of Generalization (Theorem 1.9) from Section 1.5. Since a proof contains finitely many words and can therefore only consider finitely many things explicitly, the ability to work with a generic element $x \in \mathcal{U}$, and to correspondingly prove $p(x)$, is a powerful one. Note that henceforth, we will use the same variable name (here x) that appears in the universal statement to represent the generic element (rather than introducing a new variable a as in Section 1.5).

Example 2.8.   Show: $\forall\, x \in \mathbb{R}$, $x^2 + 1 > 0$.

Proof. Suppose $x \in \mathbb{R}$.
Since the square of any real number is nonnegative (see Appendix A, property 11), we have $x^2 \geq 0$.
Hence,

$$x^2 + 1 \geq 0 + 1 = 1 > 0.$$

We therefore have $x^2 + 1 > 0$.   □

Note that the first sentence of the proof in Example 2.8 establishes that x represents a generic real number. That is, subsequently we can use only general properties of real numbers when working with x. We cannot, for example, assume that x is positive or has some other special property. Equivalently, the first sentence of the proof could be replaced by "Let $x \in \mathbb{R}$." That x is arbitrary is implicitly understood and need not be stated explicitly.

Example 2.9.   Show: $\forall\, n \in \mathbb{Z}^-$, $-n \in \mathbb{Z}^+$.

Proof. Let $n \in \mathbb{Z}^-$.
So $n \in \mathbb{Z}$ and $n < 0$.
Multiplication by -1 gives that $-n > 0$. (See Appendix A, property 11.)
Since $-n \in \mathbb{Z}$ and $-n > 0$, we have $-n \in \mathbb{Z}^+$.   □

2.2.1   If-then statements

An if-then statement of the form

$$\forall\, x \in \mathcal{U}, \;\; p(x) \rightarrow q(x)$$

is a particular example of a universal statement. Hence, it can be proven by working with a generic element $x \in \mathcal{U}$.

However, beyond that, there is a standard method for establishing the resulting if-then statement

$$p(x) \rightarrow q(x). \tag{2.2}$$

We suppose that $p(x)$ is true and then show that $q(x)$ must be true under that assumption. Since (2.2) is vacuously true when $p(x)$ is false, that possibility need not be considered. Very simply, an if-then statement can be verified by supposing its hypothesis holds and then verifying that its conclusion follows.

Example 2.10. Show: $\forall x \in \mathbb{R}$, if $x \in [2,3]$, then $x^2 \in [4,9]$.

Proof. Let $x \in \mathbb{R}$,
and suppose $x \in [2,3]$.
That is, $2 \leq x \leq 3$.
Squaring each term gives $2^2 \leq x^2 \leq 3^2$. (That follows from Appendix A, property 10. See Exercise 11.)
Hence, $4 \leq x^2 \leq 9$.
Therefore, $x^2 \in [4,9]$. □

Example 2.11. Show: For all real functions f, if f is constant, then f^2 is constant.

Solution. Recall from Definition 1.15 that a real function g is **constant** if and only if there exists some $c \in \mathbb{R}$ such that

$$\forall x \in \mathbb{R}, \quad g(x) = c.$$

Also, recall from Definition 1.16 that the function f^2 is defined by

$$\forall x \in \mathbb{R}, \quad f^2(x) = [f(x)]^2.$$

Proof. Let f be a real function,
and suppose that f is constant.
Hence, we have $c \in \mathbb{R}$ such that

$$\forall x \in \mathbb{R}, \quad f(x) = c.$$

Observe that

$$\forall x \in \mathbb{R}, \quad f^2(x) = [f(x)]^2 = c^2.$$

Since $c^2 \in \mathbb{R}$ and $\forall x \in \mathbb{R}$, $f^2(x) = c^2$, it follows that f^2 is constant. □

The comments in Example 2.11 prior to the proof empha-size the importance of definitions in driving most proofs. Before proving any statement, the definition of the terms involved must be clearly understood. In Example 2.11, in order to prove that f^2 is constant, we have to know what f^2 means and, formally, what being constant means. Then, we know precisely what needs to be shown. Note that the first sentence in the proof of Example 2.11 could be expressed more efficiently by the sentence "Suppose f is a constant real function."

Example 2.12.   Show: For all real functions f, if f is con-stant, then f is periodic. That is, constant functions are periodic.

Solution.   Recall from Definition 1.15 that f is periodic if and only if there is a positive real number p such that

$$\forall\, x \in \mathbb{R}, \quad f(x + p) = f(x).$$

Proof. Suppose f is a constant real function. Hence, we have $c \in \mathbb{R}$ such that

$$\forall\, x \in \mathbb{R}, \quad f(x) = c.$$

(Let $p = 1$.) Observe that

$$\forall\, x \in \mathbb{R}, \quad f(x + 1) = c = f(x).$$

That is, f is periodic (with $p = 1$).   □

In the proof of Example 2.12, any positive value p will work to demonstrate that the constant function f is periodic. A single value needed to be chosen for the proof, and we chose $p = 1$.

2.2.2 Subsets

Given two sets S and T in some universe $\mathcal{U}$, Definition 1.17 tells us that the subset relation $S \subseteq T$ is defined by the universal statement

$$\forall\, x \in \mathcal{U}, \ x \in S \ \rightarrow \ x \in T.$$

Hence, verification of the subset relation $S \subseteq T$ is accom-plished by proving the corresponding universal if-then statement. That is, we suppose that we have an element $x \in S$ and show that we must have $x \in T$.

Example 2.13.   Show: For all sets A and B, $A \cap B \subseteq A$.

Proof. Let A and B be sets.
Suppose $x \in A \cap B$.
So $x \in A$ and $x \in B$.
In particular, $x \in A$.
Hence, $x \in A \cap B$ implies $x \in A$.
That is, $A \cap B \subseteq A$.   □

The fact that the argument forms from Section 1.5 provide the logical foundation for our proofs can be seen very clearly in the proof of Example 2.13. The third and fourth sentences of this proof provide an example of the (In Particular) argument form in Theorem 1.7(f). The last two sentences are not completely necessary. They simply state that we have proved what is desired. As we gain more experience, we need not make such closing remarks.

Example 2.14.   (Transitivity of the Subset Relation).
Let A, B, and C be sets.
Show: If $A \subseteq B$ and $B \subseteq C$, then $A \subseteq C$.

Solution. The statement that we wish to prove is the universal statement

For all sets A, B, and C, if $A \subseteq B$ and $B \subseteq C$, then $A \subseteq C$.

However, due to the way it is stated in Example 2.14, we need not state that A, B, and C are generic sets in our proof. We consider that already accomplished by the sentence given prior to what we need to show. Hence, we can dive right into proving the if-then statement.

Proof. Suppose $A \subseteq B$ and $B \subseteq C$.
(Goal: $A \subseteq C$. That is, $\forall\, x,\;\; x \in A \rightarrow x \in C$.)
Suppose $x \in A$.
Since $x \in A$ and $A \subseteq B$, it follows that $x \in B$.
Since $x \in B$ and $B \subseteq C$, it follows that $x \in C$.
Hence, $A \subseteq C$.   □

We included in the proof of Example 2.14 the reminder, in parentheses, of the need to prove the conclusion of the if-then statement and, moreover, what that entails in this case. Such comments can be helpful, especially when first learning to write proofs. The proof in Example 2.14 illustrates the argument form (Transitivity of $\rightarrow$) of Theorem 1.7(c).

The lines in this next proof are numbered for the sole purpose of providing easy reference in the comments that follow. A proper proof would not have its lines numbered.

Example 2.15.   (Intersection Operation Respects Subset Relation).
Let A, B, C, and D be sets.
Show: If $A \subseteq B$ and $C \subseteq D$, then $A \cap C \subseteq B \cap D$.

Proof.

1. Suppose $A \subseteq B$ and $C \subseteq D$.
2. (Goal: $A \cap C \subseteq B \cap D$.)
3. Suppose $x \in A \cap C$.
4. So $x \in A$ and $x \in C$.
5. Since $x \in A$ and $A \subseteq B$, it follows that $x \in B$.
6. Since $x \in C$ and $C \subseteq D$, it follows that $x \in D$.
7. Hence, $x \in B$ and $x \in D$.
8. That is, $x \in B \cap D$.
9. Therefore, $A \cap C \subseteq B \cap D$.   □

Proof-writing strategies. To a large degree, we are on autopilot when we write relatively straightforward proofs. The structure of the asserted statement and the definitions of the terms involved dictate what we do. The preceding proof exemplifies this and provides a good opportunity to point out some of the basic proof-writing principles encountered so far.

Since the statement in Example 2.15 is an if-then statement, we assume its hypothesis in line 1. The second line (though unnecessary) simply reminds us of our aim to verify the conclusion of our if-then statement. Since that goal is a subset relation here, line 3 shoots for it by taking an arbitrary element of the (purportedly) smaller set $A \cap C$. Since line 3 involves a set operation, line 4 accomplishes an unwinding of its definition. In fact, the first four lines of this proof are completely automatic. We are forced out of the gate to naturally take these steps. However, after line 4, our motivation changes.

When our initial momentum subsides, we need to consider where our proof is headed rather than from where it left. We started down the path of proving the subset relation in line 2 by taking an arbitrary element $x \in A \cap C$. To finish this task, we need to get x into $B \cap D$. In fact, looking ahead, this needed assertion appears on line 8. Moreover, immediately following it, line 9 states that we have verified the conclusion of our if-then statement and completes our proof. At this point, lines 1, 2, 3, 4, 8, and 9 have been forced upon us, so getting from line 4 to line 8 is the hole that needs to be filled. Thus, we need to understand what line 8 says. This understanding comes from the unwinding of the definition of the statement in line 8, as stated in line 7. We now realize that after line 4, we need to somehow get the statement in line 7.

We patch the remaining holes by thinking specifically about how to use the statements in lines 1 through 4 to get the statement in line 7. Since up through line 4, we have yet to use the hypothesis of our if-then statement (assumed in line 1), it is not surprising that this hypothesis is the key to filling in the holes here. We have $x \in A$, and we want $x \in B$. So we use the fact that $A \subseteq B$ (from line 1). This is the subject of line 5. We have $x \in C$, and we want $x \in D$. So we use the fact that $C \subseteq D$ (from line 1). This is the subject of line 6. In this example, that suffices to patch the hole.

In summary, there are three main things driving our proof:

- A natural start gets the ball rolling.
- Thinking backward from our destination tells us how to proceed.
- Throughout, the unwinding of definitions provides the details with which and for which we work.

Sometimes, although it was not needed in Example 2.15, appealing to previous results is useful. An additional ingredient of cleverness is always helpful but sometimes unnecessary.

Having discussed proof structure, we should also address style. When writing a proof, one should aim for a flow that leads the reader from one sentence to the next in a smooth and natural way. One thing that can cause writing to be choppy is repeatedly starting sentences in the same way. For example, it is appropriate to start sentences with the word *So*, *Hence*, *Thus*, or *Therefore* when the sentence being expressed follows from an earlier fact. One should resist falling into the habit of always using the same start and, instead, should alternate between different ones. This was done in the proof of Example 2.15.

2.2.3   Set equalities

For the final two examples in this section, we prove some set identities by appealing to Definition 1.17, which tells us that given two sets S and T in some universe $\mathcal{U}$,

$$S = T \quad \text{if and only if} \quad \forall\, x \in \mathcal{U}, (x \in S \leftrightarrow x \in T).$$

An alternative approach for set equalities is presented in the next section.

Example 2.16.   Show: For all sets A, $(A^c)^c = A$.

Proof. Let A be a set,
and let $x \in \mathcal{U}$ (the universal set containing A).
Observe that we have the following string of equivalences:

$$x \in (A^c)^c \leftrightarrow x \notin A^c$$
$$\leftrightarrow \neg(x \in A^c)$$
$$\leftrightarrow \neg(x \notin A)$$
$$\leftrightarrow \neg(\neg(x \in A))$$
$$\leftrightarrow x \in A.$$

That is, $x \in (A^c)^c \leftrightarrow x \in A$. □

Although merely giving a string of logical equivalences like that presented in the proof of Example 2.16 would not constitute a proof by itself, our use of the string there is acceptable. We have incorporated it in a sentence. Displaying it as we did aids in readability. The string by itself would be considered a sketch of the proof.

Example 2.17. Let $A, B,$ and C be sets in some universal set $\mathcal{U}$. Show: $A \cap (B \cup C) = (A \cap B) \cup (A \cap C)$.

Proof. Let $x \in \mathcal{U}$.
From the string of logical equivalences

$$x \in A \cap (B \cup C) \leftrightarrow x \in A \ \wedge \ x \in B \cup C$$
$$\leftrightarrow x \in A \ \wedge \ (x \in B \ \vee \ x \in C)$$
$$\leftrightarrow (x \in A \ \wedge \ x \in B) \ \vee \ (x \in A \ \wedge \ x \in C)$$
$$\leftrightarrow (x \in A \cap B) \ \vee \ (x \in A \cap C)$$
$$\leftrightarrow x \in (A \cap B) \cup (A \cap C),$$

it follows that

$$x \in A \cap (B \cup C) \ \leftrightarrow \ x \in (A \cap B) \cup (A \cap C). \quad \square$$

The proofs in Examples 2.16 and 2.17 make clear the connection between the set identities of Theorem 1.5 and the logical equivalences of Theorem 1.1. When such a strong connection exists, the style used in those examples is appropriate.

Exercises

1. Show: $\forall x \in \mathbb{R}^+, -x \in \mathbb{R}^-$.
2. Show: $\forall x \in [-1, 2), -x \in (-2, 1]$.
3. Show: $\forall x \in \mathbb{R}$, if $x \in (2, 4)$, then $2x \in (4, 8)$.
4. Show: $\forall x \in \mathbb{R}$, if $x \in [1, 2]$, then $(3x - 1) \in [2, 5]$.
5. Prove or disprove: $\forall x \in \mathbb{R}^+, \sqrt{x} < x$.

6. Prove or disprove: $\forall\, x \in \mathbb{R}^+, (x+2)^2 > 4$.

7. Prove or disprove: $\forall\, x \in \mathbb{R}$, if $x < 2$, then $x^2 < 4$.

8. Prove or disprove: $\forall\, x \in \mathbb{R}$, if $x < -4$, then $4 - 3x > 10$.

9. Show: $\forall\, x \in \mathbb{R}$, if $x < -2$, then $x^2 > 4$.

10. Show: $\forall\, x \in \mathbb{R}$, if $x > 2$, then $x^3 > 8$.

11. Show: $\forall\, x, y \in \mathbb{R}$, if $0 < x < y$, then $x^2 < y^2$. Hint: See Appendix A, property 10.

12. Show: $\forall\, x, y \in \mathbb{R}$, if $x, y < -1$, then $xy > 1$.

13. By Ohm's Law, the voltage V (in volts), current I (in amps), and resistance R (in ohms) through a wire are related by the equation $V = IR$. Suppose current is passing through a wire with a voltage of 10 volts.
    Show that if the resistance is greater than 2 ohms, then the current will be less than 5 amps.

14. If $1000 is invested in an account earning 5% annual interest, then the amount A in the account after t years is given by the formula $A = 1000(1.05)^t$.
    Show that if at least 15 years pass, then the amount in the account will be at least $2000.

For Exercises 15 through 26, refer to Definition 1.15.

15. Show: For all real functions f, if f is constant, then $2f$ is constant.

16. Show: For all real functions f, if f is constant, then f is nondecreasing.

17. Show: For all real functions f, if f is periodic, then f^2 is periodic.

18. Show: For all real functions f, if f is periodic, then $-f$ is periodic.

19. Show: For all real functions f and g, if f and g are nondecreasing, then $f + g$ is nondecreasing.

20. Show: For all real functions f and g, if f and g are constant, then $f + g$ is constant.

21. Let f and g be real functions.
    Show: If f and g are constant, then fg is constant.

22. Let f and g be real functions.
    Show: If f and g are bounded above, then $f + g$ is bounded above.

23.* Let f be a real function, and let $c \in \mathbb{R}$.
    Show: If f is periodic, then $f + c$ is periodic.

24.* Let f be a real function, and let $c \in \mathbb{R}$.
    Show: If f is periodic, then cf is periodic.

25. Let f be a real function, and let $c \in \mathbb{R}^+$.
    Show: If f is increasing, then cf is increasing.

26. Let f be a real function, and let $c \in \mathbb{R}$.
    Show: If f is decreasing, then $f + c$ is decreasing.

    For Exercises 27 through 30, we use the following defi-
    nitions from plane geometry. Refer only to these in the
    requested proofs.
    - For each integer $n \geq 3$, an *n-gon* is a union of line seg-
      ments $A_1A_2, A_2A_3, \ldots, A_{n-1}A_n, A_nA_1$ such that no two
      of these segments intersect outside of their endpoints,
      where the endpoints $A_1, A_2, \ldots, A_n$ are distinct points
      in the plane.
    - A *parallelogram* is a 4-gon with each pair of opposite
      sides parallel.
    - A *rectangle* is a parallelogram with all right angles.
    - A *rhombus* is a parallelogram with all sides congruent.
    - A *square* is a rectangle with all sides congruent.

27. Prove that every square is a parallelogram.

28. Prove that every rhombus is a 4-gon.

29. Prove that every rectangle that is also a rhombus is a
    square.

30. Prove that every square is a rhombus.

31. Show: $\mathbb{Z} \cap \mathbb{R}^+ \subseteq \mathbb{N}$.   32. Show: $[2,3] \subseteq (1,4)$.

33. Show: $\mathbb{R}^+ \subseteq (\mathbb{R}^-)^c$.   34. Show: $\mathbb{R}^- \subseteq [0,1]^c$.

    For Exercises 35 through 50, let A, B, and C be arbitrary
    sets.

35. Show: $A \subseteq A \cup B$.   36. Show: $A \cap B \cap C \subseteq A \cap B$.

37. Show: If $A \subseteq A \cap B$, 38. Show: If $A \cup B \subseteq B$,
    then $A \subseteq B$.   then $A \subseteq B$.

39. Show: If $A \subseteq B$, then $A \cap C \subseteq B \cap C$.

40. Show: If $A \subseteq B$ and $A \subseteq C$, then $A \subseteq B \cap C$.

41. Show: $(A \cap B) \cap C =$ 42. Show: $(A \cup B) \cup C =$
    $A \cap (B \cap C)$. $A \cup (B \cup C)$.

43. Show: $A \cap B = B \cap A$.   44. Show: $A \cup B = B \cup A$.

45. Prove or disprove: 46. Prove or disprove:
    $A \times B = B \times A$. $A \triangle B = B \triangle A$.

47. Show: $A \cup (B \cap C) = (A \cup B) \cap (A \cup C)$.

48. Show: $A \cap (B \triangle C) = (A \cap B) \triangle (A \cap C)$. Hint: Use proper-
    ties of $\wedge$ and $\oplus$ from Exercise 32 of Section 1.1.

49. Show: $(A \cap B)^c =$ 50. Show: $(A \cup B)^c =$
    $A^c \cup B^c$. $A^c \cap B^c$.

2.3   General demonstration (Part 2)

We continue the task started in Section 2.2 of learning to write proofs based on the Principle of Generalization.

2.3.1   If and only if statements

A proof of an if and only if statement is often based on the fact that the statement form $p \leftrightarrow q$ is logically equivalent to

$$(p \rightarrow q) \wedge (q \rightarrow p).$$

That is, we can prove a statement of the form $p \leftrightarrow q$ by first proving $p \rightarrow q$ and then proving $q \rightarrow p$.

> **Example 2.18.**   Let A, B, and C be sets.
> Show: $C \subseteq A \cap B$ if and only if $C \subseteq A$ and $C \subseteq B$.
>
> *Proof.* $(\rightarrow)$ Suppose $C \subseteq A \cap B$.
> (Goal: $C \subseteq A$ and $C \subseteq B$.)
> Suppose $x \in C$.
> Since $C \subseteq A \cap B$, it follows that $x \in A \cap B$.
> That is, $x \in A$ and $x \in B$.
> Hence, $C \subseteq A$ and $C \subseteq B$.
>
> $(\leftarrow)$ Suppose $C \subseteq A$ and $C \subseteq B$.
> (Goal: $C \subseteq A \cap B$.)
> Suppose $x \in C$.
> Since $C \subseteq A$, it follows that $x \in A$.
> Since $C \subseteq B$, it follows that $x \in B$.
> Therefore, $x \in A$ and $x \in B$.
> That is, $x \in A \cap B$.
> So $C \subseteq A \cap B$.   □

At the beginning of the proof of Example 2.18, the notation $(\rightarrow)$ denotes the fact that we first aim to prove that the left-hand side of the if and only if statement implies the right-hand side. Similarly, the notation $(\leftarrow)$ denotes the start of the proof that the right-hand side implies the left-hand side. Each of these if-then statements is proved by using the techniques from Section 2.2. Such added notations are certainly not required in a proof, but can be helpful, especially when proof writing is first being learned.

In both parts $(\rightarrow)$ and $(\leftarrow)$ of the proof of Example 2.18, a variable element x is used in the argument. It is important to understand that the x used in the first part $(\rightarrow)$ is not connected to the x used in the second part $(\leftarrow)$. Consequently, in the second part, properties of the x from the first part cannot be referenced. Within each part, x should be thought of as a

local variable. Its properties do not pass globally between the parts, much like variables local to a function used in a computer program. When the first part ends, its variable x dies, and a new one must be created for the second part.

Example 2.19.   Let f be a real function.
Show: f is bounded below if and only if $-f$ is bounded above.

Proof. ($\rightarrow$) Suppose f is bounded below.
So, we have some $L \in \mathbb{R}$ such that

$$\forall\, x \in \mathbb{R}, \quad f(x) \geq L.$$

Multiplication by -1 gives that

$$\forall\, x \in \mathbb{R}, \quad -f(x) \leq -L.$$

Hence, $-L$ provides a bound that demonstrates that $-f$ is bounded above.
   ($\leftarrow$) Suppose $-f$ is bounded above.
   So, we have some $M \in \mathbb{R}$ such that

$$\forall\, x \in \mathbb{R}, \quad -f(x) \leq M.$$

Multiplication by -1 gives that

$$\forall\, x \in \mathbb{R}, \quad f(x) \geq -M.$$

Hence, $-M$ provides a bound that demonstrates that f is bounded below. $\square$

2.3.2   Set equalities revisited

Since set equalities, by definition, correspond to if and only if statements, they can be proved by appealing to a string of if and only if statements, as was done in Section 2.2. However, it is also often useful to prove the corresponding if and only if statement by using the method from Example 2.18. Equivalently, this amounts to using the characterization of the set equality $S = T$ given by

$$S \subseteq T \text{ and } T \subseteq S.$$

That is, proving a set equality can be accomplished by proving the two corresponding subset relations.

Example 2.20.   Let A and B be sets.
Show: If $A \subseteq A \cap B$ and $B \subseteq A \cap B$, then $A = B$.

Proof. Suppose $A \subseteq A \cap B$ and $B \subseteq A \cap B$.
(Goal: $A = B$.)

($\subseteq$) Suppose $x \in A$.
Since $A \subseteq A \cap B$, it follows that $x \in A \cap B$.
That is, $x \in A$ and $x \in B$.
In particular, $x \in B$.
So $A \subseteq B$.

($\supseteq$) Suppose $x \in B$.
Since $B \subseteq A \cap B$, it follows that $x \in A \cap B$.
That is, $x \in A$ and $x \in B$.
In particular, $x \in A$.
So $B \subseteq A$.

Since $A \subseteq B$ and $B \subseteq A$, we have $A = B$.   □

In the proof of Example 2.20, the notation ($\subseteq$) denotes the start of the proof that the set on the left-hand side of the desired set equality ($A = B$) is a subset of the set on the right-hand side. Similarly, the notation ($\supseteq$) denotes the start of the proof that the right-hand side is a subset of the left-hand side. Of course, these notations are not essential.

Example 2.21.   Let A, B, C, and D be sets such that $D \subseteq A \cap B, D \subseteq B \cap C$, and $A \cap C \subseteq D$.
Show: $D = A \cap B \cap C$.

Solution. Note that all of the hypotheses have been stated prior to the desired statement. Hence, we need not restate them in our proof.

Proof. ($\subseteq$) Suppose $x \in D$.
Since $D \subseteq A \cap B$, it follows that $x \in A \cap B$.
That is, $x \in A$ and $x \in B$.
Since $D \subseteq B \cap C$, it also follows that $x \in B \cap C$.
That is, $x \in B$ and $x \in C$.
Altogether, we have $x \in A$ and $x \in B$ and $x \in C$.
So $x \in A \cap B \cap C$.

($\supseteq$) Suppose $x \in A \cap B \cap C$.
So $x \in A$ and $x \in B$ and $x \in C$.
In particular, $x \in A \cap C$.
Since $A \cap C \subseteq D$, it follows that $x \in D$.   □

2.3.3   Sets with particular forms

Some of the set operations that we have encountered yield sets whose elements have a special form. For example, the elements of a product are ordered pairs. We need to take this into account in proofs involving such operations.

Example 2.22.   Let A, B, and C be sets.
Show: If $A \subseteq B$, then $A \times C \subseteq B \times C$.

Solution. On the very surface, the conclusion of this if-then statement is a subset relation. Hence, after supposing that $A \subseteq B$, we should suppose that we have an element in $A \times C$. If we initially name this element with a single variable, say z, then we subsequently have to unwind the fact that z is an element of a product. That is, we have to write $z = (x, y)$ as a pair, where $x \in A$ and $y \in C$. To save this extra writing and to avoid this additional variable z, which is ultimately discarded, it is more efficient to simply start by supposing that (x, y) is an element of $A \times C$. That is, we can immediately write our element in the form appropriate for products.

Proof. Suppose $A \subseteq B$.
(Goal: $A \times C \subseteq B \times C$.)
Suppose $(x, y) \in A \times C$.
So $x \in A$ and $y \in C$.
Since $x \in A$ and $A \subseteq B$, we have $x \in B$.
Thus, $x \in B$ and $y \in C$.
That is, $(x, y) \in B \times C$.
Therefore, $A \times C \subseteq B \times C$. $\square$

Our next example mixes products and complements.

Example 2.23.   Let A and B be sets.
Show: $A^c \times B^c \subseteq (A \times B)^c$.

Proof. Suppose $(x, y) \in A^c \times B^c$.
So, $x \in A^c$ and $y \in B^c$.
In particular, since $x \notin A$, it follows that $(x, y) \notin A \times B$.
(Recall the definition of $A \times B$.)
That is, $(x, y) \in (A \times B)^c$. $\square$

Note in the proof of Example 2.23 that we need only one of the facts, $x \notin A$ or $y \notin B$, to assert that $(x, y) \notin A \times B$. Although we would need both $x \in A$ and $y \in B$ to assert that $(x, y) \in A \times B$, it only takes one coordinate to misbehave to throw the pair (x, y) out of the product $A \times B$.
Power sets warrant their own special considerations.

Example 2.24. Let A and B be sets.
Show: If $A \subseteq B$, then $\mathcal{P}(A) \subseteq \mathcal{P}(B)$.

Solution. Remember that the elements of power sets are themselves sets. The convention is to use capital letters when denoting sets. Hence, we denote the elements of a power set with capital letters, rather than lowercase letters, which are usual for elements. Such a naming convention is not required mathematically, of course. However, it aids in the readability of the proof if the reader can have a sense of an object from its name.

Proof. Suppose $A \subseteq B$.
(Goal: $\mathcal{P}(A) \subseteq \mathcal{P}(B)$.)
Suppose $S \in \mathcal{P}(A)$.
That is, $S \subseteq A$.
Since $S \subseteq A$ and $A \subseteq B$, it follows from the transitivity of $\subseteq$ (proved in Example 2.14) that $S \subseteq B$.
Therefore, $S \in \mathcal{P}(B)$.   □

Exercises

1. Show: $\forall\, x \in \mathbb{R}, \;\; x \in \mathbb{R}^-$ if and only if $-x \in \mathbb{R}^+$.

2. Show: $\forall\, x \in \mathbb{R}, \;\; x \in (-2, 1]$ if and only if $-x \in [-1, 2)$.

3. Show: $\forall\, x \in \mathbb{R}, \;\; x = 2x$ if and only if $x = 0$.

4. Show: $\forall\, x \in \mathbb{R}, \;\; \frac{x+1}{x} = 2$ if and only if $x = 1$.

5. Let $x \in \mathbb{R}$. Show: $x^3 > 0$ if and only if $x > 0$. Hint: See Appendix A, properties 10 and 11.

6. Show: $\forall\, x \in \mathbb{R}^+, \;\; x^2 < x$ if and only if $x < 1$.

7. Show: $\forall\, x \in \mathbb{R}, \;\; 4 - x < 2$ if and only if $x > 2$.

8. Show: $\forall\, x \in \mathbb{R}^+, \;\; x^3 = x$ if and only if $x = 1$.

9. Show: $\forall\, x \in \mathbb{R}, \;\; x^4 - 16 = 0$ if and only if $x^2 - 4 = 0$.

10. Show: $\forall\, x \in \mathbb{R}, \;\; x = \sqrt{6 - x}$ if and only if $x = 2$.

11. Let f be a real function.
    Show: f is constant if and only if $2f$ is constant.

12. Show that the definition of constant functions given in Definition 1.15 and the characterization of constant functions given in Exercise 37(b) from Section 1.3 are equivalent. That is, let f be a real function.
    Show: $\exists\, c \in \mathbb{R}$ such that $\forall\, x \in \mathbb{R}, f(x) = c$ if and only if $\forall\, x, y \in \mathbb{R}, f(x) = f(y)$. Hint: For ($\leftarrow$), consider $c = f(0)$.

13. Let f be a real function.
    Show: f is bounded above if and only if $f + 1$ is bounded above.

14. Let f be a real function.
    Show: f is increasing if and only if $2f$ is increasing.

15*. Let f be a real function.
Show: f is bounded above and below if and only if f^2 is bounded above.

16*. Let f be a nonnegative real function.
Show: f is constant if and only if f^2 is constant.

17. Let f be a real function.
Show: f is periodic if and only if $2f$ is periodic.

18. Let f be a real function.
Show: f is periodic if and only if $f + 1$ is periodic.

19. Assume that a football team can only score touchdowns, worth 7 points each, and field goals, worth 3 points each. Thus, 21 points could be scored either by scoring 3 touchdowns or by scoring 7 field goals. Show that a total of 20 points are scored if and only if the team scores 2 touchdowns and 2 field goals.

20. Suppose that the United States Post Office only sells first class stamps at 39¢ each and postcard stamps at 24¢ each. A certain package requires $3 in postage. Show that $3 can be achieved by stamps if and only if four 39¢ stamps and six 24¢ stamps are used.

21. Show: $\mathbb{R}^+ \cap [-2, 2] = (0, 2]$.

22. Show: $[-2, 2] \setminus \mathbb{R}^+ = [-2, 0]$.

23. Show: $\mathbb{N} \setminus (-1, 1) = \mathbb{Z}^+$.

24. Show: $\mathbb{Z} \cap (-1, 1) = \{0\}$.

25. Show: $[1, 3) \cap [2, 4) = [2, 3)$.

26. Show: $(-\infty, 1] \cap (-1, \infty) = (-1, 1]$.

For Exercises 27 through 46, let A, B, C, D be sets in some universe $\mathcal{U}$.

27. Show: If $A \subseteq B$, then $A \cap B = A$.

28. Show: $A \cap \mathcal{U} = A$.

29. Show: $(A \setminus C) \cap B = (B \setminus C) \cap A$.

30. Show: If $C \subseteq A \cap B$, then $C \cap A = C \cap B$.

31. Show: If $A \cap B = A \cap C$, then $A \cap B \cap C = A \cap B$.

32. Show: If $A \subseteq B, A \subseteq C, A \subseteq D$, and $C \cap D \subseteq A$, then $A = B \cap C \cap D$.

33. Show: $(A \setminus B) \setminus C = A \setminus (B \cup C)$.

34. Show: $A \times (B \setminus C) = (A \times B) \setminus (A \times C)$.

35. Show: If $A \subseteq B$, then $A^2 \subseteq B^2$. Recall that $A^2 = A \times A$.

36. Show: $(A \times \mathcal{U})^c = A^c \times \mathcal{U}$.

37. Show: $(\mathcal{U} \times B) \setminus (A \times B) = A^c \times B$.

38. Show: If $A \subseteq B$ and $C \subseteq D$, then $A \times C \subseteq B \times D$.

39. Show: $A \times (B \cap C) = (A \times B) \cap (A \times C)$.

40. Show: $(B \setminus A) \times (D \setminus C) \subseteq (B \times D) \setminus (A \times C)$.

41.* Assume that $C \neq \emptyset$. So we have some $c \in C$.
    Show: $A \times C = B \times C$ if and only if $A = B$.

42.* Show: $\mathcal{P}(A) = \mathcal{P}(B)$ if and only if $A = B$.

43.* Show: $\mathcal{P}(A^c) \setminus \{\emptyset\} \subseteq \mathcal{P}(A)^c$.

44. Show: If $(C, D) \in \mathcal{P}(A) \times \mathcal{P}(B)$, then $C \times D \in \mathcal{P}(A \times B)$.

45. Show: $\mathcal{P}(A \cap B) \subseteq \mathcal{P}(A) \cap \mathcal{P}(B)$.

46.* Show: $\mathcal{P}(A \setminus B) \setminus \{\emptyset\} \subseteq \mathcal{P}(A) \setminus \mathcal{P}(B)$.

47. Mario is taking two laps in a car around a 1 mile track. He is attempting to achieve a certain average speed over this entire trip.

    (a) Find the average speed for his entire trip if there was an average speed of 40 mph over the first lap and 60 mph over the second lap. Hint: It is not 50 mph.

    (b)* Show that Mario can obtain an average speed of 60 mph over the entire trip if and only if he exceeds an average speed of 30 mph over the first lap.

48. When resistors with resistances R_1 and R_2 are connected in parallel in a circuit, the total resistance R achieved is given by the equation $R = \frac{R_1 R_2}{R_1 + R_2}$.

    (a) If the total resistance is $R = 8$ ohms and we have $R_1 = 10$ ohms, then what must be the value of R_2?

    (b)* Assume that $R_1 = 10$ ohms. Show that a total resistance R is achievable in this circuit if and only if R is less than 10 ohms.

2.4 Indirect arguments

For some statements, a direct approach to a proof is not the best, or may not even be feasible. This section presents some alternatives.

2.4.1 Proofs by contradiction

The basic idea behind the proof technique called **proof by contradiction** is a very simple one. We suppose that the negation of the desired statement holds and show that this leads to a contradiction. The fact that the desired statement follows is a consequence of the argument form (Argument by Contradiction) in Example 1.49 from Section 1.5.

Example 2.25.   Show: $\mathbb{R}^+$ does not have a smallest element.

Proof. (By Contradiction)
Suppose not.
So there is some element $s \in \mathbb{R}^+$ that is the smallest.

However, $\frac{s}{2}$ is a smaller element of $\mathbb{R}^+$ (since $\frac{s}{2} < s$ and $\frac{s}{2} > 0$).
This contradicts the fact that s was supposed to be the smallest element.   □

The first sentence, "Suppose not," in our proof by contradiction is shorthand for supposing the negation of the statement we aim to prove. In this case, it means "Suppose $\mathbb{R}^+$ has a smallest element."

In our proof of Example 2.25, it is important that our choice of $\frac{s}{2}$ is both smaller than s and positive. A choice of $s - 1$ would give an element that is smaller than s but not necessarily positive, and hence not necessarily an element of $\mathbb{R}^+$. A choice of $\sqrt{s}$ would give an element of $\mathbb{R}^+$, but $\sqrt{s}$ is not necessarily smaller than s (definitely not if $s < 1$; e.g., $\sqrt{.01} > .01$).

One of the advantages of a proof by contradiction is that it gives us an additional hypothesis with which to work. The negation of our desired statement generally provides something concrete to pursue. In Example 2.25, forcing a potential smallest element to declare its value was the key to refuting its existence.

Example 2.26.   Prove each of the following:

(a) For all sets A, $A \cap A^c = \emptyset$.

(b) $\mathcal{U}^c = \emptyset$.

(c) $\emptyset^c = \mathcal{U}$.

Proof.

(a) Suppose not.
   So there is some set A for which $A \cap A^c \neq \emptyset$.
   Since $A \cap A^c$ is not empty, there is some element $x \in A \cap A^c$.
   Hence, $x \in A$ and $x \in A^c$.
   That is, $x \in A$ and $x \notin A$.
   This is a contradiction.

(b) Suppose not.
   So $\mathcal{U}^c$ is a nonempty subset of our universal set $\mathcal{U}$.
   Hence, there is an element $x \in \mathcal{U}$ such that $x \in \mathcal{U}^c$.
   That is, $x \in \mathcal{U}$ and $x \notin \mathcal{U}$.
   This is a contradiction.

(c) By the Double Complement Identity (Theorem 1.5(a) and Example 2.16), we know that $\mathcal{U} = (\mathcal{U}^c)^c$.
   From statement (b) above, we get $(\mathcal{U}^c)^c = \emptyset^c$.
   Hence, by substitution, $\mathcal{U} = \emptyset^c$.   □

In Example 2.26, notice that both the contradiction in part (a) (that $x \in A$ and $x \notin A$) and the contradiction in part (b) (that $x \in \mathcal{U}$ and $x \notin \mathcal{U}$) are examples of statements of the form $p \wedge \neg p$. We saw in Example 1.5 that $p \wedge \neg p$ is a contradiction. Although the proof in part (c) is not a proof by contradiction, it does make use of the result in part (b), which was proved by contradiction. Note that an alternative proof of part (b) could have been accomplished by using part (a) in the case that $A = \mathcal{U}$ and the fact that $\mathcal{U}^c = \mathcal{U} \cap \mathcal{U}^c$.

Example 2.27.   Show: $\mathbb{Z}$ is infinite.

Proof. (By Contradiction)
Suppose not.
So $\mathbb{Z}$ is finite.
Let n be the natural number that is the cardinality of $\mathbb{Z}$.
However, there are $n + 1$ distinct integers in the list $1, 2, 3, \ldots, n, n + 1$.
So $\mathbb{Z}$ has more than n elements.
This is a contradiction. $\square$

2.4.2   Proving the contrapositive

We know from Example 1.11 that an if-then statement form $p \rightarrow q$ is logically equivalent to its contrapositive $\neg q \rightarrow \neg p$. Hence, an if-then statement can be proved by instead proving its contrapositive. Since the contrapositive is itself an if-then statement, its proof can use the techniques learned in the earlier sections of this chapter; we can suppose $\neg q$ and establish $\neg p$.

Example 2.28.   Let A be a set. Show: If A^2 is infinite, then A is infinite.

Proof. (Of the Contrapositive)
Suppose that A is not infinite.
That is, A is finite.
Let $n = |A|$, and write $A = \{a_1, a_2, \ldots, a_n\}$.
Observe that $A^2 = A \times A =$

$$\{(a_1, a_1),\ (a_1, a_2),\ \ldots,\ (a_1, a_n),$$
$$(a_2, a_1),\ (a_2, a_2),\ \ldots,\ (a_2, a_n),$$
$$\vdots \qquad\qquad\qquad\qquad \vdots$$
$$(a_n, a_1),\ (a_n, a_2),\ \ldots,\ (a_n, a_n)\}$$

has n^2 elements.
Thus, A^2 is finite (not infinite). $\square$

Our proof in Example 2.28 starts by supposing the negation of the conclusion of the desired if-then statement. It concludes once the negation of the hypothesis is established. That is, we give a direct proof of the contrapositive. In Example 2.28, we therefore gave a direct proof of the statement

For all sets A, if A is finite, then A^2 is finite.

The utility of this choice in that example can be seen by instead trying to prove the desired statement directly. The direct approach immediately runs into an obstacle, since we do not have convenient tools for working with an infinite set A^2.

Example 2.29.   Show: $\forall x \in \mathbb{R}$, if $x^3 \leq 0$, then $x \leq 0$.

Proof. (Of the Contrapositive)
Let $x \in \mathbb{R}$, and suppose $x > 0$ (i.e., not $x \leq 0$).
Since $x > 0$, we have $0 = x \cdot 0 < x \cdot x = x^2$. (See Appendix A, property 10.)
Similarly, we then have $0 = x \cdot 0 < x \cdot x^2 = x^3$.
Hence, $x^3 > 0$ (i.e., not $x^3 \leq 0$).   □

Example 2.29 could alternatively be proved directly. However, such a proof would be more complicated. For example, it could be accomplished by dividing both sides of the inequality $x^3 \leq 0$ by x^2. Of course, the cases in which $x = 0$ and $x \neq 0$ would have to be handled separately. Such proofs involving cases are discussed in the next section. An alternative direct proof using cube roots needs the fact that the cube root function is increasing. The proof of that fact is left to the exercises and itself warrants an indirect argument.

Proving properties of the empty set from Section 1.2. The following example provides a nice illustration of the utility of proving the contrapositive of an if-then statement.

Example 2.30.   (Proving Theorem 1.3).
Let E and A be sets.
Show: If E has no elements, then $E \subseteq A$.

Proof. (Of the Contrapositive)
Suppose $E \not\subseteq A$.
So there is some element x such that $x \in E$ and $x \notin A$.
Since $x \in E$, the set E has an element.   □

In Section 1.2 we encountered examples of sets containing no elements such as $(0, 0)$ and $[1, 0]$. We are now in a position to give the promised proof of the fact that there is a unique set with no elements. This fact enables us to give *the* empty set the name ∅. The proof makes use of the result from Example 2.30.

Theorem 2.1. (∅ is Well-Defined in Definition 1.10).
There is a unique set with no elements, namely, ∅.

Proof. Suppose that two sets A and B each contain no elements.
By Example 2.30, since B has no elements, $B \subseteq A$.
Similarly, since A has no elements, $A \subseteq B$.
Therefore, $A = B$. □

Example 2.30 also implies that for any set A,

$$\emptyset \subseteq A, \quad \text{or equivalently,} \quad \emptyset \in \mathcal{P}(A).$$

Comparing indirect proof techniques for if-then statements. We now have a few options for indirect proofs of an if-then statement. In working with if-then statements, proving the contrapositive is one option. However, a proof by contradiction is also an option, since that technique applies to any statement, if-then or otherwise. Additionally, there is a third option. Since the if-then statements have the form $p \to q$, we could start, as in a direct proof, by supposing p. However, we could then switch to an indirect approach toward our goal of establishing q and apply a proof by contradiction. It turns out that there are strong commonalities among these approaches, as we can see in the following example.

Example 2.31. Let A and B be sets. Show: If $A \subseteq B$, then $A \setminus B = \emptyset$.

Proof.

(a) (Of the Contrapositive)
Suppose $A \setminus B \neq \emptyset$.
Hence, we have some element $x \in A \setminus B$.
That is, $x \in A$ and $x \notin B$.
It follows that $A \nsubseteq B$.

(b) (By Contradiction)
Suppose not.
So $A \subseteq B$ and $A \setminus B \neq \emptyset$.
Since $A \setminus B \neq \emptyset$, we have some element $x \in A \setminus B$.
That is, $x \in A$ and $x \notin B$.
It follows that $A \nsubseteq B$.
This contradicts the fact that $A \subseteq B$.

(c) (Direct Proof with Indirect Proof of Conclusion)
Suppose $A \subseteq B$.

Claim: $A \setminus B = \emptyset$.
Suppose not.
So $A \setminus B \neq \emptyset$.
Since $A \setminus B \neq \emptyset$, we have some element $x \in A \setminus B$.
That is, $x \in A$ and $x \notin B$.
It follows that $A \not\subseteq B$.
This contradicts the fact that $A \subseteq B$ and establishes our claim. □

The three proofs for Example 2.31 can be seen to be very similar. The strongest similarity is between proofs (b) and (c). The first two lines of proof (b) accomplish the same thing as the first four lines of proof (c). The remaining lines are then virtually identical. Therefore, proof (b) is the more efficient of the later two, in this case.

In Example 2.31, there is also much in common between proofs (a) and (b). This stems from the fact that the contradiction $A \not\subseteq B$ obtained in proof (b) is the negation of the hypothesis $A \subseteq B$ of the if-then statement. When that is the case, the proof by contradiction contains the basic argument given in the proof using the contrapositive along with additional structure setting up the proof by contradiction. In that case, proving the contrapositive is more efficient. However, in general, a different contradiction might be obtained in the proof by contradiction. Then, the relative advantages of the two proofs must be compared by other means.

In our final example of this section, we make use of the logical equivalence of an if-then statement with its contrapositive. However, the technique of proving the contrapositive is not used.

Example 2.32. Let A and B be sets in some universal set $\mathcal{U}$. Show: $A \subseteq B$ if and only if $B^c \subseteq A^c$.

Proof. The following sequence of logical equivalences

$$A \subseteq B \leftrightarrow \forall \, x \in \mathcal{U} \; (x \in A \rightarrow x \in B)$$
$$\leftrightarrow \forall \, x \in \mathcal{U} \; (x \notin B \rightarrow x \notin A)$$
$$\leftrightarrow \forall \, x \in \mathcal{U} \; (x \in B^c \rightarrow x \in A^c)$$
$$\leftrightarrow B^c \subseteq A^c$$

yields the desired result. □

Another way in which a contrapositive is useful is in the fact that the statement form $p \leftrightarrow q$ is logically equivalent to $(p \rightarrow q) \wedge (\neg p \rightarrow \neg q)$. Hence, a statement of the form $p \leftrightarrow q$ can be proved by first proving $p \rightarrow q$ and then proving

$\neg p \to \neg q$. That is, the second stage, at which $q \to p$ would typically be proved, can be replaced be proving its contrapositive $\neg p \to \neg q$.

Exercises

1. Show: The interval $(1, 2)$ has no smallest element.

2. Show: $\mathbb{R}$ has no smallest element.

3. Show: $\mathbb{N}$ has no largest element.

4. Show: The interval $(1, 2)$ has no largest element.

5. Show: For any set A, $A \cap \emptyset = \emptyset$.

6. Let A be any set such that $A \subseteq \emptyset$. Show: $A = \emptyset$.

7. Show: The interval $(0, 1]$ is infinite.

8. Show: $\mathbb{Z} \times \mathbb{Z}$ is infinite.

9. Show: $\{(x, y) \; : \; x, y \in \mathbb{R} \text{ and } y = \sqrt{x}\}$ is infinite.

10. Show: $\{k \; : \; k = j^2 \text{ for some } j \in \mathbb{Z}\}$ is infinite.

11. Show: $(1, 0) = \emptyset$.   12. Show: $(0, -1) = \emptyset$.

13. Show: $\mathbb{R}^+ \cap \mathbb{R}^- = \emptyset$.   14. Show: $\mathbb{Z}^- \cap \mathbb{N} = \emptyset$.

15. Tracy Flick is running for class president. Assume that there are a total of n candidates running, where n represents a positive integer. After the votes are tallied, Tracy is only told the fraction of votes that she received.

    (a) Suppose that Tracy received less than $\frac{1}{n}$ of the votes. Show that Tracy cannot have won the election.

    (b) Suppose that Tracy received more than $\frac{1}{n}$ of the votes. Show that Tracy cannot have come in last in the election.

16. **Pigeon Hole Principle**. Suppose that there are n distinct items and m distinct boxes, where n and m are positive integers. We are to place each item into a box, and we are allowed to place more than one item in the same box.

    (a) Suppose that $n > m$. Show that some box must receive more than one item.

    (b) Suppose that $n < m$. Show that some box must be empty.

    Note that the Pigeon Hole Principle is introduced formally in Theorem 5.12 of Section 5.6. There we shall consider the issues above in the context of one-to-one and onto functions. In that setting, we make precise the imprecise notion here of placing items into boxes. However, simple arguments can be made here using basic common sense in the absence of the formal notion of functions.

17. Show: $\forall \, a, b \in \mathbb{R}$, if $(a, b) = \emptyset$, then $b \leq a$.

18. Show: $\forall\, a, b \in \mathbb{R}^+$, if $a^2 \geq b^2$, then $a \geq b$.

For Exercises 19 through 32, let A, B, C be sets in some universe $\mathcal{U}$.

19. Show: If $A^2 = \emptyset$, then $A = \emptyset$.

20. Show: If $A^2 \neq B^2$, then $A \neq B$.

21. Show: If $A^2 \nsubseteq B^2$, then $A \nsubseteq B$.

22. Show: If $\mathcal{P}(A) \nsubseteq \mathcal{P}(B)$, then $A \nsubseteq B$.

23. Show: If $A = \emptyset$, then $A \times B = \emptyset$.

24. Show: If $A^2 \neq \emptyset$, then $A \neq \emptyset$.

25. Show: If $A \nsubseteq B$, then $A \neq B$.

26. Show: If $|\mathcal{P}(A)| > 1$, then $|A| > 0$.

27. Show: If $\mathcal{P}(A) \neq \mathcal{P}(B)$, then $A \neq B$.

28. Show: If $C \nsubseteq A \cup B$, then $C \nsubseteq A$ or $C \nsubseteq B$.

29. Show: If $A \times B$ is infinite, then either A is infinite or B is infinite.

30. Assume that $\mathcal{U}$ is finite. Show: If A^c is finite, then A is infinite.

31. Show: $\forall\, x \in \mathcal{U}$, if $x \in A$ and $x \notin A \cap B$, then $x \notin B$.

32. Show: $\forall\, x \in \mathcal{U}$, if $x \in A \cup B$ and $x \notin B$, then $x \in A$.

33.* **Euclid's first postulate**. For all points P and Q in the plane, if $P \neq Q$, then there is a unique line l that contains P and Q. Prove: Any two distinct lines intersect in at most one point.

34.* Two lines in the plane are said to be *parallel* if they do not intersect. **Euclid's parallel postulate**: For every line l and every point P not on l, there is a unique line m through P and parallel to l.
    Prove: For any three distinct lines k, l, and m, if k is parallel to l and l is parallel to m, then k is parallel to m.

For Exercises 35 through 42, let f be a real function.

35. Assume that f is positive.
    (a) Show: If f^2 is decreasing, then f is decreasing.
    (b) Is it true that if f^2 is increasing, then f is increasing? Explain.

36. (a) Show: If f^2 is not periodic, then f is not periodic.
    (b) Is it true that if f^2 is periodic, then f is periodic? Explain.

37. Show: If $f + 100$ is unbounded above, then f is unbounded above.

38. Show: If f is unbounded below, then $\frac{1}{2}f$ is unbounded below.

39. Show: If f is periodic, then f is not increasing.

40. Show: If f is decreasing, then f is not periodic.

41. Show: If f^2 is not constant, then f is not constant.

42. Show: If f is not increasing, then $2f$ is not increasing.

43. Show: $\forall\, x \in \mathbb{R}$, if $x^2 = 0$, then $x = 0$.

44. Show: $\forall\, x \in \mathbb{R}$, if $x^2 \neq 1$, then $x \neq 1$.

45*. Show: $\forall\, x \in \mathbb{R}$, if $x > 0$, then $\frac{1}{x} > 0$.

46*. Show: $\forall\, x \in \mathbb{R}$, if $\frac{1}{x} < 0$, then $x < 0$.

47*. Show: $\forall\, x, y \in \mathbb{R}$, if $0 < x < y$, then $\frac{1}{y} < \frac{1}{x}$.

48. Show: $\forall\, x, y \in \mathbb{R}$, if $x < 0 < y$, then $\frac{1}{x} < \frac{1}{y}$.

49. Show: $(-\infty, -1) \cap (1, \infty) = \emptyset$.

50. Show that the **cube root** function $f(x) = \sqrt[3]{x}$ is increasing. That is, show: $\forall\, x_1, x_2 \in \mathbb{R}$, if $x_1 < x_2$, then $\sqrt[3]{x_1} < \sqrt[3]{x_2}$. Hint: Use the facts that $\sqrt[3]{x} = y$ if and only if $y^3 = x$ and $\forall\, y_1, y_2 \in \mathbb{R}$, if $y_1 \leq y_2$ then $y_1^3 \leq y_2^3$.

2.5 Splitting into cases

Some proofs naturally split into cases. That is, at some point in the proof, different possibilities need to be considered. Typically, this happens when an "or" statement is encountered. For example, a statement of the form $p \vee q \vee r$ may lead us to consider the different reasons why $p \vee q \vee r$ would be true—if p is true, if q is true, or if r is true.

> **Example 2.33.**   Let A, B, and C be sets.
> Show: If $A \subseteq C$ and $B \subseteq C$, then $A \cup B \subseteq C$.
>
> *Proof.*
> Suppose $A \subseteq C$ and $B \subseteq C$.
> Suppose $x \in A \cup B$.
> So $x \in A$ or $x \in B$.
>
> *Case 1:* $x \in A$.
> Since $A \subseteq C$, it follows that $x \in C$.
>
> *Case 2:* $x \in B$.
> Since $B \subseteq C$, it follows that $x \in C$.
>
> In both cases, we have $x \in C$.
> Hence, $A \cup B \subseteq C$. $\square$

In the proof of Example 2.33, it is the statement just preceding our cases that causes us to split the argument into cases.

For a generic element x, simply knowing that $x \in A$ or $x \in B$ does not determine which one of the two holds. Since the arguments for $x \in A$ and for $x \in B$ require the use of different parts of our hypotheses ($A \subseteq C$ and $B \subseteq C$, respectively), we handle these distinct cases separately. It is important that we consider each of the possible properties for x, of which there are two in this example ($x \in A$ or $x \in B$). It does not matter that there could be some overlap between the cases, if $x \in A \cap B$.

Example 2.34.   Let A, B, and C be sets.
Show: $A \cap (B \cup C) \subseteq (A \cap B) \cup C$.

Proof. Suppose $x \in A \cap (B \cup C)$.
So $x \in A$ and $x \in B \cup C$.
Hence, $x \in B$ or $x \in C$.

Case 1: $x \in B$.
Observe that $x \in A$ and $x \in B$.
That is, $x \in A \cap B$.
Since $A \cap B \subseteq (A \cap B) \cup C$,   it   follows   that $x \in (A \cap B) \cup C$.

Case 2: $x \in C$.
Since $C \subseteq (A \cap B) \cup C$, it follows that $x \in (A \cap B) \cup C$.

In both cases, $x \in (A \cap B) \cup C$.   □

The statements obtained within a case are local to that case and cannot be referenced in other cases. In case 1 of Example 2.34, we get $x \in A \cap B$. This fact cannot be used in case 2, because it may not be true there. However, if a common conclusion (such as $x \in (A \cap B) \cup C$) is reached within *all* of the cases, then that result may be globally referenced subsequent to the cases.

Example 2.35.   Show: $\forall\, x \in \mathbb{R}$, if $|x| > 1$, then $x^2 > 1$.

Solution. The fact that the absolute value function

$$|x| = \begin{cases} x & \text{if } x \geq 0, \\ -x & \text{if } x < 0 \end{cases}$$

is a piecewise function leads us to consider cases. More specifically, it gives that $|x| > 1$ is equivalent to saying that $x > 1$ or $x < -1$.

Proof. Suppose $x \in \mathbb{R}$ and that $|x| > 1$.
So $x > 1$ (when $x \geq 0$) or $-x > 1$ (when $x < 0$).

Case 1: $x > 1$.
Observe that $x^2 > 1^2 = 1$.

Case 2: $-x > 1$.
So, $(-x)^2 > 1^2 = 1$.
Since $(-x)^2 = (-1)^2 x^2 = x^2$, substitution gives that $x^2 > 1$.

In both cases, $x^2 > 1$. □

An alternative proof for Example 2.35 might use the fact that $x^2 = |x|^2$. However, the proof of that fact would warrant the use of cases. That proof is requested in the exercises.

The need for cases was clear in Examples 2.33 through 2.35; it grew out of an "or" statement. Sometimes, however, that need can be a bit more subtle.

Example 2.36.   Let A, B, and C be sets.
Show: $A \setminus C \subseteq (A \setminus B) \cup (B \setminus C)$.

Proof. Suppose $x \in A \setminus C$.
So $x \in A$ and $x \notin C$.

Case 1: $x \in B$.
Since $x \notin C$, we have $x \in B \setminus C$.
Since $B \setminus C \subseteq (A \setminus B) \cup (B \setminus C)$, it follows that $x \in (A \setminus B) \cup (B \setminus C)$.

Case 2: $x \notin B$.
Since $x \in A$, we have $x \in A \setminus B$.
Since $A \setminus B \subseteq (A \setminus B) \cup (B \setminus C)$, it follows that $x \in (A \setminus B) \cup (B \setminus C)$.

In both cases, $x \in (A \setminus B) \cup (B \setminus C)$. □

What drives the split into cases in Example 2.36? Just prior to the split, we do not have the "or" statement

$$x \in B \text{ or } x \notin B.$$

However, that fact is certainly true; it is a tautology. The split in the proof is driven by the proof's ending rather than by its beginning. The aim is to get x in the union $(A \setminus B) \cup (B \setminus C)$. This can happen in two ways. Either we end up with $x \in A \setminus B$ or we end up with $x \in B \setminus C$. What distinguishes these two possibilities is that in the first $x \notin B$ and in the second $x \in B$. As we see in the proof, considering these separate cases leads us to our desired end.

The applicability of cases in the proof of Example 2.36 can also be seen by considering the Venn diagram in Figure 2.1. There, the regions numbered 1 and 2 correspond to cases 1 and 2, respectively, in our argument. We certainly see in our proof that the different regions of $A \setminus C$ warrant different arguments.

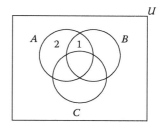

FIGURE 2.1
Two parts of $A \setminus C$.

Example 2.37.   Let $a, b, c \in \mathbb{R}$ with $a \neq 0$.
Show: The number of distinct real roots of the quadratic polynomial

$$ax^2 + bx + c \quad \text{is} \quad \begin{cases} 2 & \text{if } b^2 - 4ac > 0, \\ 1 & \text{if } b^2 - 4ac = 0, \\ 0 & \text{if } b^2 - 4ac < 0. \end{cases}$$

Proof. The roots of the equation

$$ax^2 + bx + c = 0$$

are given by the quadratic formula

$$x = \frac{-b \pm \sqrt{b^2 - 4ac}}{2a}.$$

Case 1: $b^2 - 4ac > 0$.
The roots are

$$\frac{-b + \sqrt{b^2 - 4ac}}{2a} \quad \text{and} \quad \frac{-b - \sqrt{b^2 - 4ac}}{2a}.$$

Moreover, these are distinct. (Check that they are not equal.)

Case 2: $b^2 - 4ac = 0$.
The single root is $\frac{-b}{2a}$.

Case 3: $b^2 - 4ac < 0$.
Since $\sqrt{b^2 - 4ac}$ does not exist as a real number, there are no real roots.   □

In Example 2.37, it is the piecewise conclusion that leads to the use of cases. In that example, unlike the previous ones, there are different conclusions in the different cases. The standard proof of the quadratic formula by the method of completing the square is left to the reader.

An additional way in which cases may be employed is through a strategy called bootstrapping. There, a proof might first consider a case in which some simplifying assumptions have been added to the existing conditions. Then, arguments in subsequent cases, in which those assumptions have been removed, may be streamlined by appealing to the already established result in the simpler case. This is a more advanced strategy than is considered in this section. Bootstrapping, however, is used in examples such as the proofs of the Pigeon Hole Principle (Theorem 5.12) and Brooks' Theorem (Theorem 9.26).

Exercises

Throughout these exercises, let A, B, C, and D denote arbitrary sets in some universal set $\mathcal{U}$.

1. Show: If $A \subseteq B$ and $C \subseteq D$, then $A \cup C \subseteq B \cup D$.

2. Disprove: If $A \subseteq B$ and $C \subseteq D$, then $A \triangle C \subseteq B \triangle D$.

3. Show: If $A \subseteq B$, then $A \cup B = B$.

4. Show: $A \cup \emptyset = A$.

5. Show: $A \cup A^c = \mathcal{U}$.

6. Show: If $A \subseteq B$, then $A \cup C \subseteq B \cup C$.

7. Show: $A \cup \mathcal{U} = \mathcal{U}$.

8. Show: If $A \subseteq B$, then $A \triangle B = B \setminus A$.

9. Show: $A \cup B \subseteq C$ if and only if $A \subseteq C$ and $B \subseteq C$.

10. Show: $(A \cup B) \setminus C = (A \setminus C) \cup (B \setminus C)$.

11. Show: $A \triangle B \subseteq A \cup B$.

12. Show: $A \triangle B \neq \emptyset$ if and only if $A \neq B$.

13. Show: $(A \cup B) \setminus C \subseteq A \cup (B \setminus C)$.

14. Show: $(A \setminus B) \cup (A \setminus C) = A \setminus (B \cap C)$.

15. Show: $A \cup (B \setminus C) = (A \cup B) \setminus (C \setminus A)$.

16. Show: $A \setminus (B \setminus C) = (A \setminus B) \cup (A \cap C)$.

17. Show: $(A \times B)^c = (A^c \times \mathcal{U}) \cup (\mathcal{U} \times B^c)$.

18. Show: $(A \cup B) \times C = (A \times C) \cup (B \times C)$.

19.* Show: $A \cap (B \triangle C) = (A \cap B) \triangle (A \cap C)$.

20.* Show: $A \times (B \triangle C) = (A \times B) \triangle (A \times C)$.

21. Show: $\mathcal{P}(A) \cup \mathcal{P}(B) \subseteq \mathcal{P}(A \cup B)$.

22. Show: If $A \in \mathcal{P}(C)$ and $B \in \mathcal{P}(C)$, then $A \cup B \in \mathcal{P}(C)$.

23. Pictured is a portion of Rome through which the Tiber River flows.

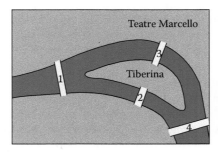

A tourist is interested in the bridges and would like to plan a tour of this part of Rome that starts and ends in the same place and goes over each bridge (*ponte*) exactly once. Show that no such tour exists. Note that it suffices to assume that the starting point is on the north side of the Tiber at the Teatre Marcello and that Ponte Garibaldi is the first bridge to be crossed. From there, consider cases based upon subsequent bridge choices and argue that, in each case, it becomes impossible to complete a desired tour.

24. Notre Dame Cathedral is located on the Ile de la Cité in the portion of Paris pictured below. (Only some of the bridges are included here.)

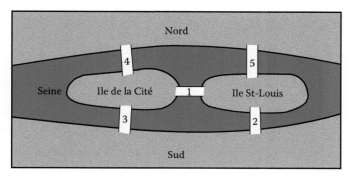

Starting from Notre Dame Cathedral, a tourist would like to take a tour of this portion of Paris that goes over each displayed bridge exactly once and returns to Notre Dame Cathedral. Show that no such tour is possible by considering cases based upon possible bridge choices at each stage of a potential tour.

25. Show: $\forall\, x \in \mathbb{R}, |-x| = \begin{cases} -x & \text{if } x \le 0, \\ x & \text{if } x > 0. \end{cases}$

26. Show: $\forall\, x \in \mathbb{R}, |x+1| = \begin{cases} x+1 & \text{if } x \ge -1, \\ -x-1 & \text{if } x < -1. \end{cases}$

27. Show: $\forall\, x \in \mathbb{R}, |1-2x| = \begin{cases} 1-2x & \text{if } x \le \frac{1}{2}, \\ 2x-1 & \text{if } x > \frac{1}{2}. \end{cases}$

28. Show: $\forall\, x \in \mathbb{R} \setminus \{0\}, \frac{x}{|x|} = \begin{cases} 1 & \text{if } x > 0, \\ -1 & \text{if } x < 0. \end{cases}$

29. Let a be a real number. Show: The number of distinct real roots of the polynomial $(x+3)(x^2 + 2ax + a^2)$ is $\begin{cases} 1 & \text{if } a = 3, \\ 2 & \text{if } a \ne 3. \end{cases}$

30. Let a be a real number. Show: The number of distinct real roots of the polynomial $x^2 - a$ is $\begin{cases} 0 & \text{if } a < 0, \\ 1 & \text{if } a = 0, \\ 2 & \text{if } a > 0. \end{cases}$

31. Show: $\forall\, x \in \mathbb{R}, |x|^2 = x^2$.

32. Show: $\forall\, x \in \mathbb{R}, |(x+2)(x-3)| = \begin{cases} 6+x-x^2 & \text{if } 2 \leq x \leq 3, \\ x^2 - x - 6 & \text{otherwise.} \end{cases}$

33.* Show: $\forall\, x, y \in \mathbb{R}, |x \cdot y| = |x| \cdot |y|$.

34.* Show: $\forall\, x, y \in \mathbb{R}$, if $|x| < y$ then $-y < x < y$.

35. Let $x \in \mathbb{R}$. Show: $x^4 > 1$ if and only if $x < -1$ or $x > 1$.

36. Let $x \in \mathbb{R}$. Show: $x^2 - x - 2 > 0$ if and only if $x < -1$ or $x > 2$.

37. Let $x, y \in \mathbb{R}$. Show: If $x^2 = y^2$, then $x = \pm y$.

38. Let $x, y \in \mathbb{R}$. Show: $x = y$ if and only if $x \leq y$ and $y \leq x$.

39.* **Triangle Inequality.** Show: $\forall\, x, y \in \mathbb{R}, |x + y| \leq |x| + |y|$.

40.* Show: $\forall\, x, y \in \mathbb{R}, ||x| - |y|| \leq |x - y|$.

41.* A regular n-gon is a polygon with n congruent sides and interior angles, where $n \geq 3$. If one wishes to use a fixed regular n-gon as the shape for tiles used to tile the floor in such a way that, for every tile, each edge matches up with the edge of the tile next to it, then the only possible patterns are pictured below.

Prove that these are the only patterns by arguing that only n-gons with $n = 3, 4,$ or 6 will work. Use the fact that the interior angle of a regular n-gon has a measure of $\frac{180(n-2)}{n}$ degrees and that the tiles will join perfectly together around a corner point. First argue that $n \geq 7$ is impossible, and then consider each case $3 \leq n \leq 6$ separately.

42. The **Pythagorean Theorem** says that for a right triangle with side lengths a, b, and c, with c the largest, we have $a^2 + b^2 = c^2$. This well-known result from right triangle trigonometry has many simple proofs, and we shall explore two. Prove the Pythagorean Theorem using the prescribed method.

(a) It suffices to assume that $a \geq b$. In the case $a > b$, we can form the pictured square of area c^2 with the square S inside.

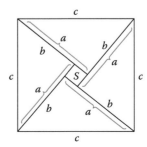

Find the area by adding the areas of the interior pieces, and use algebra to obtain the desired formula. Find a similar picture and argument in the case that $a = b$.

(b) Use the following picture to write a proof that does not require cases.

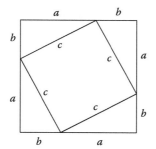

2.6 Review problems

1. Show that $(3,4)$, $(4,-3)$, and $(0,-5)$ all lie on a common circle.

2. Show that 2 and -2 are the only real roots of $x^4 - 2x^2 - 8$.

3. Show: $\exists\, m, n \in \mathbb{Z}$ such that $mn = 100$ and $m + n = 25$.

4. Show: $\exists\, x \in \mathbb{R}$ such that $2^x = x^2$.

5. Show: There exist sets A, B, C such that $A \subset B$ and $A \cup C = B \cup C$.

6. Show: There is a set A such that $|\mathcal{P}(A)| = |A^2|$.

7. Jeremy is taking Calculus, which has three tests, each worth 100 points. On the first test he got an 80, but on the second test he only got a 60. He would like to obtain an average of 75. Show that there is a possible test score on the third test for Jeremy such that he can achieve an average of 75.

8. Disprove: For all real functions f, if f^2 is constant, then f is constant.

9. Disprove: For all sets A, B, C, if $A \subseteq B \cup C$, then $A \subseteq B$ and $A \subseteq C$.

10. Disprove: For   all   sets A, B, C, $A \triangle (B \cap C) = (A \triangle B) \cap (A \triangle C)$.

11. Show: $\forall\, n \in \{-1, 0, 1\}$, $n^4 = n^2$.

12. Show: $\forall\, A \in \{\{1\}, \{2\}\}$, $|A^3| = 1$.

13. Show: $\mathbb{Z} \times \mathbb{N} \neq \mathbb{N} \times \mathbb{Z}$.

14. Show: $\forall\, n \in \mathbb{Z}^+, n^2 \geq n$. Hint: If $n \in \mathbb{Z}^+$, then $n \geq 1$.

15. Show: $\forall\, x \in \mathbb{R}$, if $x \in [2, 4]$, then $x^2 \in [4, 16]$.

16. Show: For all real functions f, if f is constant, then $f + 1$ is constant.

17. Show: For all real functions f and g, if f is periodic and g is constant, then $f + g$ is periodic.

18. Show: For all real functions f and g, if f is bounded above and g is bounded below, then $f - g$ is bounded above.

19. Jennifer is taking statistics, which has three tests, each worth 100 points. She would like to achieve a test average over 80. However, she is very uncomfortable with the material on the first test. Show that if Jennifer gets at most a 40 on the first test, then she can achieve an average of at most 80.

20. Show: For all sets A, B, and C, if $A \subseteq C$, then $A \cap B \subseteq C$.

21. Show: For all sets A and B, $A \setminus B \subseteq A$.

22. Prove or disprove: For all sets A and B, $(A \setminus B)^c = A^c \cup B$.

23. Show: For all sets A and B, if $A \subset B$, then $B \setminus A \neq \emptyset$.

24. Show: $\forall\, x \in \mathbb{R}, \ x \in [1, 2]$ if and only if $2x \in [2, 4]$.

25. Show: $\forall\, x \in \mathbb{R}, \ 3x - 2 \in (1, 4)$ if   and   only   if $5 - 2x \in (1, 3)$.

26. Show: $\forall\, x, y \in \mathbb{R}^+, \ x^2 = y^2$ if and only if $x = y$.

27. Show: $\forall\, x, y \in \mathbb{R}, \ x = y$ if and only if $x \leq y$ and $y \leq x$.

28. Let f be a real function. Show: f is constant if and only if f is nondecreasing and nonincreasing.

29. Let f be a nonnegative real function. Show: f is periodic if and only if f^2 is periodic.

30. Let A and B be sets. Show: $A^2 = B^2$ if and only if $A = B$.

31* Let A, B, and C be sets. Show: $A \setminus B \subseteq C$ if and only if $A \subseteq B \cup C$.

32. Mr. Chipman is very strict with grades. He will only give a student a grade of 100 for the marking period if that student's average is exactly 100; he does not round up. Each test is worth 100 points, and Kevin is a very good student. Show that, no matter what positive number of tests Mr. Chipman gives, Kevin can achieve an average of 100 if and only if Kevin gets a grade of 100 on every test.

33. Show: For all sets A, B, C, $(A \cap B) \setminus C = (A \setminus C) \cap (B \setminus C)$.

34. Show: For   all   sets A, B, C, $A \times (B \cup C) = (A \times B) \cup (A \times C)$.

35. Show: For all sets A and B, $\mathcal{P}(A) \cup \mathcal{P}(B) \subseteq \mathcal{P}(A \cup B)$.

36. Show: $\mathbb{R}^-$ has no largest element.

37. Show: The interval $(-1, 1)$ has no smallest element.

38. Show: The real function $f(x) = x$ is not periodic.

39. Show: The real function $f(x) = x^2$ is not bounded.

40. Let $x \in \mathbb{R}$. Show: If $x^5 \geq 0$, then $x \geq 0$.

41. Show: For all sets A, $A \times \emptyset = \emptyset$.

42. Show: For all sets A and B, if $A = \emptyset$, then $A \cap B = \emptyset$.

43. Show: For all sets A and B, if $A \times B = \emptyset$, then $A = \emptyset$ or $B = \emptyset$.

44. Show: For all sets A, B, C, if $A \times B \neq A \times C$, then $B \neq C$.

45. Show: For all sets A, if $\mathcal{P}(A)$ is infinite, then A is infinite. Hint: If A is finite, then $|\mathcal{P}(A)| = 2^{|A|}$.

46. Let f be a real function. Show: If f^2 is not constant, then f is not constant.

47. Erik is struggling and just wants to pass Discrete Mathematics. His grade is based entirely on the average of the four tests, each worth 100 points. A minimum grade of 60 is required to pass. Show that if Erik is to pass the class, then at least one of his test grades must be 60 or higher.

48. Show: For all sets A, B, C, $(A \cap B \cap C)^c = A^c \cup B^c \cup C^c$.

49. Show: For all sets A, B, C, $(A \setminus C) \cup (B \setminus C) = (A \cup B) \setminus C$.

50. Show: For all sets A, B, C, $(A \bigtriangleup B) \cap (A \bigtriangleup C) \subseteq A \bigtriangleup (B \cap C)$. See Exercise 10.

51. Show: For all sets A and B, if $A = \emptyset$ or $B = \emptyset$, then $A \times B = \emptyset$.
Note that the converse appears in Exercise 43.

52. Show: $\forall\, x, y \in \mathbb{R}$, $xy > 0$ if and only if $x, y > 0$ or $x, y < 0$.

53. Show: $\forall\, x \in \mathbb{R}$, $|x^2| = x^2$.

54. Show: $\forall\, x \in \mathbb{R}, |x^2 - 1| = \begin{cases} 1 - x^2 & \text{if } -1 < x < 1, \\ x^2 - 1 & \text{otherwise.} \end{cases}$

55. Show: $\forall\, x \in \mathbb{R}$, $\sqrt{x^2} = |x|$.

56. Let $x \in \mathbb{R}$. Show: $x^2 - 6x + 8 > 0$ if and only if $x < 2$ or $x > 4$.

57. Let $x \in \mathbb{R}$. Show: $x = \frac{1}{x}$ if and only if $x = \pm 1$.

3 Elementary Number Theory

Quite simply, number theory is the study of numbers, where the primary focus is on the integers. Our studies also yield results that can be used to determine whether a real number is rational or irrational.

Besides learning the basics of number theory, we aim to gain more experience with the proof techniques introduced in Chapter 2. However, instead of working in the context of set theory, we move to the setting of number theory. Despite this change of topic, the basic strategies of our proofs remain the same. For example, if-then statements pertaining to numbers are proved by the same methods as those pertaining to sets. Only the terminology and definitions involved are different.

A main theme throughout this chapter is that the definitions and the basic methods of proof drive the majority of proofs we encounter. There is a degree to which we are on autopilot when writing many proofs. This is because the definitions and the structure of the statements essentially force us to do the right things. Our change of subject should help illuminate this point.

In addition to learning some of the basics in the beautiful area of number theory, we shall see some interesting applications. The check digits appended to the ends of identification numbers are discussed. More generally, we consider error-correcting codes, which enable messages to be read despite the possible introduction of errors during transmission. To send secret messages, we consider linear ciphers. Also, the powerful public key encryption method known as RSA encryption is introduced.

3.1 Divisors

When working with an integer, it is natural to try to factor the integer and thereby write it as a product. For this reason, numbers that cannot be factored themselves, such as 2, are particularly important. To initiate our study of integers and their divisors, we start with the consideration of whether or not an integer is a multiple of 2.

3.1.1 Parity

One of the most basic properties of an integer is whether it is even or odd—its **parity**. Of course, before we can prove anything, we need a formal characterization of even and odd.

Definition 3.1.   An integer n is said to be

(a) **even** if $n = 2k$ for some integer k and

(b) **odd** if $n = 2k + 1$ for some integer k.

The fact that every integer can be written in exactly one of the two forms $2k$ or $2k + 1$ is a consequence of the Division Algorithm proved in Section 3.2. That is, our definition of odd numbers gives that all integers that are not even must be odd. We use Definition 3.1 to prove properties of even and odd integers.

Example 3.1.   Show that for every pair of odd integers, the product is odd.

Proof. Let m and n be arbitrary odd integers. So $m = 2k + 1$ and $n = 2l + 1$ for some $k, l \in \mathbb{Z}$. Hence,

$$mn = (2k + 1)(2l + 1) = 4kl + 2k + 2l + 1$$
$$= 2(2kl + k + l) + 1.$$

Since $2kl + k + l$ is an integer, the form displayed on the right-hand side above shows that mn is odd.   □

The proof in Example 3.1 is basically forced on us. After establishing the odd integers m and n in the first sentence, the second sentence simply expresses m and n according to the definition of odd. The next sentence is merely a calculation of the product mn. The grouping in the expression

$$2(2kl + k + l) + 1$$

is again driven by the definition of odd and our consequent desire to express mn as twice an integer, plus one. Our proof closes with the simple observation that this has been accomplished.

The next example is one in which the contrapositive of an if-then statement is the more natural statement to prove.

Example 3.2.   Let $n \in \mathbb{Z}$. Show: If n^2 is odd, then n is odd.

Proof. Suppose n is not odd. That is, n is even.

So $n = 2k$ for some $k \in \mathbb{Z}$.
Hence, $n^2 = 4k^2 = 2(2k^2)$.
Since $2k^2 \in \mathbb{Z}$, n^2 is even.
That is, n^2 is not odd. □

An attempt to prove Example 3.2 directly quickly runs into a dead end when taking the square root of $n^2 = 2k + 1$ is attempted. This is why the contrapositive was a good choice. The analog of Example 3.2 for even integers is left to the exercises.

3.1.2 Divisibility

An integer being even is simply a special case of one integer being divisible by another.

Definition 3.2. Given integers n and d, we say that d **divides** n, written $d \mid n$, if $n = dk$ for some integer k. In this case, we also say that n is **divisible** by d, that n is a **multiple** of d, that d is a **divisor** of n, and that d is a **factor** of n. When n is not divisible by d, we write $d \nmid n$.

Notice that n is even if and only if $2 \mid n$.

Example 3.3. Let a and m be integers with $m \geq 1$. Show: $a \mid a^m$.

Proof. Observe that $a^m = a \cdot a^{m-1}$ and, since $m - 1 \geq 0$, $a^{m-1} \in \mathbb{Z}$. □

From the algebra of real numbers (Appendix A, property 8), we know that if $a < b$ and $b < c$, then $a < c$. That property is called the transitivity of the less than relation. We have an analogous property of divisibility among integers.

Example 3.4. (Transitivity of the Divides Relation).
Let a, b, and c be integers. Show: If $a \mid b$ and $b \mid c$, then $a \mid c$.

Proof. Suppose $a \mid b$ and $b \mid c$.
So $b = ak$ and $c = bl$ for some $k, l \in \mathbb{Z}$.
Observe that $c = bl = akl = a(kl)$.
Since $kl \in \mathbb{Z}$, we have established that $a \mid c$. □

The proof in Example 3.4 is simply an example of a direct proof of an if-then statement. The details are just the unwinding of the notation from Definition 3.2.

It should come as no surprise that the divisors of a positive integer cannot exceed that integer. However, this assertion warrants a proof.

Theorem 3.1.   *Let $a, b \in \mathbb{Z}$ with $b > 0$. If $a \mid b$, then $a \leq b$.*

Proof. Suppose $a \mid b$.
So $b = ak$ for some $k \in \mathbb{Z}$.

Case 1: $a \leq 0$.
We have $a \leq 0 < b$, and the conclusion $a \leq b$ is immediate.

Case 2: $a > 0$.
Since $ak = b > 0$, it must be that $k > 0$ too. (Otherwise, $ak < 0$.)
Since k is an integer, $1 \leq k$.
Multiplication by (the positive value) a gives that $a = a \cdot 1 \leq a \cdot k = b$.

In both cases, we conclude that $a \leq b$.   □

In Theorem 3.1, the assumption that $b > 0$ is necessary. For example, if $b = -6$, then each of the divisors $a \in \{-3, -2, -1, 1, 2, 3, 6\}$ satisfies $a \mid b$ and $a > b$. Only if an integer b is positive will none of its divisors be larger.

3.1.3 Primes

The most important divisors of an integer are those that cannot be factored any further.

Definition 3.3.   An integer p is said to be **prime** if $p > 1$ and the only positive divisors of p are 1 and p.

Definition 3.4.   An integer $n > 1$ that is not prime is said to be **composite**.

The first ten primes are
$$2, 3, 5, 7, 11, 13, 17, 19, 23, \text{ and } 29.$$
The first ten composites are
$$4 = 2 \cdot 2, \quad 6 = 2 \cdot 3, \quad 8 = 2 \cdot 4, \quad 9 = 3 \cdot 3,$$
$$10 = 2 \cdot 5, \quad 12 = 2 \cdot 6,$$
$$14 = 2 \cdot 7, \quad 15 = 3 \cdot 5, \quad 16 = 2 \cdot 8,$$
$$\text{and} \quad 18 = 3 \cdot 6.$$

Notice that 1 is the only positive integer that can be called neither prime nor composite. We do not want 1 to be a prime, since we want each integer $n > 1$ to have a unique factorization as a product of primes (Theorem 4.9). If 1 were prime, then

$$18 = 1 \cdot 2 \cdot 3 \cdot 3, \quad 18 = 1 \cdot 1 \cdot 2 \cdot 3 \cdot 3, \quad \text{and}$$

$$18 = 1 \cdot 1 \cdot 1 \cdot 2 \cdot 3 \cdot 3$$

would all be different factorizations of 18. We want just the one factorization $18 = 2 \cdot 3 \cdot 3$. Considering 1 to be composite is also not desirable, since 1 is its only positive divisor. Consequently, 2 is the smallest prime, and 4 is the smallest composite.

Example 3.5.   Show that 2 is the only even prime.

Proof. Let p be an even prime.
So $p = 2k$ for some $k \in \mathbb{Z}$.
In fact, since $p > 0$ and $2 > 0$, it must be that $k > 0$. (Otherwise, $2k < 0$.)
Since p is prime, and k is a positive divisor of p, we must have either $k = 1$ or $k = p$.
Since $p = 2p$ is impossible, it has to be that $k = 1$.
Therefore, $p = 2k = 2$.   □

The term *composite* is defined as the negation of the term *prime*. However, it is often useful to have a direct characterization of composites. An integer $n > 1$ is **composite** if and only if

$$\exists\, r, s \in \mathbb{Z} \text{ such that } r > 1, s > 1, \text{ and } rs = n. \qquad (3.1)$$

A proof that this characterization is equivalent to the definition is requested in the exercises. It is the characterization of composites in (3.1) this is most frequently applied in our proofs.

3.1.4 GCD's

An integer c is said to be a **common divisor** of the integers m and n if $c \mid m$ and $c \mid n$. That is, c is a divisor of both m and n. For example, 10 is a divisor of both 40 and 60. Also, 20 is a common divisor of 40 and 60 and is the largest one.

Definition 3.5.   Given integers m and n not both zero, their **greatest common divisor**, denoted $\gcd(m, n)$, is the unique integer d such that

(i) $d > 0$,

(ii) $d \mid m$ and $d \mid n$, and

(iii) $\forall\, c \in \mathbb{Z}^+$, if $c \mid m$ and $c \mid n$, then $c \leq d$.

A proof of the uniqueness assertion in Definition 3.5 is left to the exercises in the next section. Condition (iii) could equivalently be stated for all integers c, not just the positive ones. However, the nonpositive common divisors of m and n are obviously less than or equal to the positive integer d.

Notice that $\gcd(0,0)$ is undefined, since every positive integer is a divisor of 0. Hence, there cannot be a greatest one. At the other extreme, it is possible for two integers to have no positive divisors in common other than 1, which divides all integers.

Definition 3.6. Two integers m and n are said to be **relatively prime** if $\gcd(m,n) = 1$.

Example 3.6.

(a) $\gcd(18, 30) = 6$.
Certainly, $6 > 0$, $6 \mid 18$, and $6 \mid 30$. Also, any element c in the set $\{1,2,3,6\}$ of positive common divisors of 18 and 30 satisfies $c \leq 6$.

(b) Observe that 14 and 9 are relatively prime, since $\gcd(14,9) = 1$.

If one can factor each of two positive integers m and n as a product of powers of primes, then the greatest common divisor can quickly be determined from those factorizations. The proof of the validity of the method introduced in the following example is left to the exercises in Section 4.5, where factorizations are discussed in more depth.

Example 3.7.

(a) $\gcd(81989600,231254595000) = 2928200$.
We have the factorizations

$$81989600 = 2^5 \cdot 5^2 \cdot 7 \cdot 11^4 \quad \text{and}$$

$$231254595000 = 2^3 \cdot 3^5 \cdot 5^4 \cdot 11^4 \cdot 13.$$

We see that 2, 5, and 11 are the prime factors common to both 81989600 and 231254595000. It follows that

$$\gcd(81989600, 231254595000)$$

$$= 2^3 \cdot 5^2 \cdot 11^4 = 2928200.$$

Here the exponents on the primes 2, 5, and 11 are chosen to be the smaller of the two exponents in the factorizations of 81989600 and 231254595000.

(b) $\gcd(26159679,644436) = 3159.$
From the factorizations

$$26159679 = 3^5 \cdot 7^2 \cdot 13^3 \quad \text{and}$$

$$644436 = 2^2 \cdot 3^6 \cdot 13 \cdot 17,$$

we get that

$$\gcd(26159679, 644436) = 3^5 \cdot 13 = 3159.$$

The shortcoming of the method in Example 3.7 is that it is extremely difficult to factor large numbers. What we would prefer is an algorithm that can do our calculations quickly. In general, an **algorithm** is a finite list of simple instructions leading to a desired result. Specifically, Euclid's Algorithm for computing greatest common divisors avoids the difficulty of factoring integers and is presented in Section 3.3. The functions GCD and igcd in *Mathematica* and *Maple*, respectively, interleave Euclid's Algorithm with other methods to calculate greatest common divisors quickly. In C++, however, there is no predefined function for computing greatest common divisors. Hence, some algorithm such as Euclid's would need to be programmed.

Before learning Euclid's Algorithm in Section 3.3, we should further familiarize ourselves here with the basic properties of greatest common divisors. The computation of the greatest common divisor in the following example is driven entirely by Definition 3.5.

Example 3.8. Given any positive integer k, show: $\gcd(k,0) = k$.

Proof. (We verify the conditions in Definition 3.5 for $m = k, n = 0$, and $d = k$.)
Observe that k is positive and that $k \mid k$ and $k \mid 0$.
Hence, conditions (i) and (ii) are satisfied.
If $c \in \mathbb{Z}^+$ and $c \mid k$ and $c \mid 0$, then, in particular, $c \mid k$.
By Theorem 3.1, we have $c \leq k$. Thus, condition (iii) is satisfied.
We conclude that $k = \gcd(k,0)$. $\square$

It follows from Definition 3.5 that $\gcd(m,n)$ does not depend on the sign of m and n. That is, $\gcd(m,n) = \gcd(\pm m, \pm n)$.

Hence, $\gcd(m,n) = \gcd(|m|,|n|)$. (Verification is left for the exercises.) These observations, together with the result in Example 3.8, give the following result.

Theorem 3.2.   *Given any integer $k \neq 0$, $\gcd(k,0) = |k|$.*

The proof in the next example obviously breaks into cases. In each case, the computation of the greatest common divisor simply follows the definition.

Example 3.9.   Let p and n be integers with p prime.
Show: $\gcd(p,n) = \begin{cases} p & \text{if } p \mid n, \\ 1 & \text{if } p \nmid n. \end{cases}$

Proof.
Case 1: $p \mid n$
Since p is prime, p is positive.
Certainly $p \mid p$ and $p \mid n$.
So suppose that c is a positive integer such that $c \mid p$ and $c \mid n$.
Since $c \mid p$, it follows from Theorem 3.1 that $c \leq p$.
Therefore, $p = \gcd(p,n)$.

Case 2: $p \nmid n$
It is clear that 1 is positive and that $1 \mid p$ and $1 \mid n$.
So suppose that c is a positive integer such that $c \mid p$ and $c \mid n$.
Since $c \mid p$ and p is prime, it follows that $c = 1$ or $c = p$.
Since $c \mid n$ and $p \nmid n$, in this case it must be that $c = 1$.
In particular, $c \leq 1$.
Therefore, $1 = \gcd(p,n)$.   □

As we shall see in the next section, it is useful to know that two consecutive integers can never have a factor in common except ± 1.

Lemma 3.3.   Let n be any integer. Then n and $n + 1$ are relatively prime.

Proof. We shall show that $\gcd(n, n + 1) = 1$.
Certainly, $1 > 0$, $1 \mid n$, and $1 \mid (n + 1)$.
So suppose $c \in \mathbb{Z}^+$ and $c \mid n$ and $c \mid (n + 1)$.
So $n = ck$ and $n + 1 = cl$ for some $k, l \in \mathbb{Z}$.
Hence, $1 = (n + 1) - n = cl - ck = c(l - k)$.
That is, $c \mid 1$, and therefore, $c \leq 1$.
It follows that $\gcd(n, n + 1) = 1$.   □

A **lemma** is simply a theorem. However, the term *lemma* is typically used when the theorem is one whose primary use is to help prove subsequent theorems. The previous lemma is important in our proof that the list of all primes goes on forever (Theorem 3.6).

Exercises

1. Show that for every even integer and every odd integer, the product is even.

2. Show that for every even integer and every odd integer, the sum is odd.

3. Let $n \in \mathbb{Z}$. Show: If n^2 is even, then n is even.

4. Let $n \in \mathbb{Z}$. Show: If n^3 is odd, then n is odd.

5. For every odd integer n, $\frac{n+1}{2} \in \mathbb{Z}$.

6. For every even integer n, $n - 1$ is odd.

7. Show: For every even integer n, $(-1)^n = 1$. Hint: Use the Laws of Exponents.

8. Show: For every odd integer n, $(-1)^n = -1$.

9. The power button on a TV remote control turns the TV on if it is off and off if it is on. If the TV is on, then will the TV be on after the power button is pressed 50 times? How does this relate to Exercise 7?

10. Juliet is pulling off the petals of a flower one at a time. When the first petal is pulled off, she says "He loves me." When the second petal is pulled off, she says "He loves me not." And she continues to alternate between these two states of mind until no petals remain. What will her conclusion be if the flower has 17 petals? How does this relate to Exercise 8?

11. Let $a \in \mathbb{Z}$. Show: $a \mid 0$.

12. Let $a \in \mathbb{Z}$. Show: If $0 \mid a$, then $a = 0$.

13. Let $a \in \mathbb{Z}$. Show: If $a \mid 1$, then $a = \pm 1$.

14. Let $a, b \in \mathbb{Z}$. Show: $a \mid ab$.

15. Let $a \in \mathbb{Z}$. (a) Show: $(a - 1) \mid (a^2 - 1)$.
(b) Assume that $a \geq 2$, and consider an $a \times a$ grid of unit squares, such as that shown in Figure 3.1 for $a = 5$. Let R be the region obtained by removing the unit square from the top-right corner of this grid. Observe that the area of R is $a^2 - 1$, and prove the assertion in part (a) by splitting R into rectangles, each of which has a dimension of size $a - 1$.

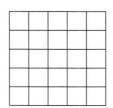

FIGURE 3.1
Grid of unit squares.

16. Let $a \in \mathbb{Z}$. (a) Show: $a \mid ((a + 2)^2 - 4)$.
(b) Assume that $a \geq 1$, and consider an $(a + 2) \times (a + 2)$ grid of unit squares, such as that shown in Figure 3.1 for $a = 3$. Prove the assertion in part (a) by considering the

region obtained by removing the unit square from each corner.

17. Let $a, b \in \mathbb{Z}$. Show: If $a \mid b$ and $b \mid a$, then $a = \pm b$.

18. Let $a, b \in \mathbb{Z}$. Show: $a \mid b$ if and only if $a \mid -b$.

19. Let $n \in \mathbb{Z}$. Show: If n is even, then $4 \mid n^2$.

20. Let $n \in \mathbb{Z}$. Show: If $3 \mid n$, then $9 \mid n^2$.

21. A mother purchased a bag containing 420 m&m's to split among the children attending a birthday party. If there are 8 children attending the party, can this bag be divided evenly? What if one of the children is unable to attend?

22. A large group planning a trip to a baseball game wishes to purchase tickets so that they can sit together in one section. Each row of the section seats 15 people. If there are 275 people in the group, will they completely fill the rows required to contain them? What is the smallest number of extra people they could invite to completely fill the rows they need?

23. Let $a, b, c \in \mathbb{Z}$. Show: If $a \mid b$ and $a \mid c$, then $a \mid (b - c)$.

24. Let $a, b, c \in \mathbb{Z}$. Show: If $a \mid b$ and $a \mid (b + c)$, then $a \mid c$. Hint: $c = (b + c) - b$.

25. For $a, b, c \in \mathbb{Z}$, is it possible that $a \mid (b + c)$ but $a \nmid b$ and $a \nmid c$?

26. For $a, b \in \mathbb{Z}$, is it possible that $a \mid b$ and $a > b$?

27. List the first 20 primes.

28. List the first 20 composites.

29. Show that 3 is the only prime divisible by 3.

30. Show that 5 is the only prime divisible by 5.

31.* Show: For any integer p with $p > 1$,
p is prime if and only if $\forall\, r, s \in \mathbb{Z}$, if $r > 1$ and $s > 1$, then $rs \neq p$.

32.* Show: For any integer p with $p > 1$,
p is prime if and only if the only divisors of p are ± 1 and $\pm p$.

33.* Prove that the characterization of composite integers given in (3.1) is equivalent to that given by Definition 3.4. That is, show: For any integer n with $n > 1$, n is composite if and only if $\exists\, r, s \in \mathbb{Z}$ such that $r > 1$, $s > 1$, and $rs = n$.

34.* Show: Any integer $n > 1$ is composite if and only if the only n has divisors other than ± 1 and $\pm n$.

35.* Show that if an integer $n > 1$ is composite, then n has a factor larger than 1 and less than or equal to $\sqrt{n}$.

36.* Show that if an integer $n > 1$ is a product of three (not necessarily distinct) integers, each of which is greater than 1, then n has a factor larger than 1 and less than or equal to $\sqrt[3]{n}$.

37. The Greek mathematician Eratosthenes of Cyrene (276–194 B.C.E.) used the following method, now known as the **Sieve of Eratosthenes**, to find all primes up to a given integer n provided that all primes $p_1, \ldots, p_k$ less than or equal to $\sqrt{n}$ are known: List the integers $2, \ldots, n$, and, with the exceptions of $p_1, \ldots, p_k$ themselves, cross off all multiples of $p_1, \ldots, p_k$. (See Exercise 35.)

    Use the Sieve of Eratosthenes to find all primes less than or equal to 121. This will yield the first 30 primes (and the first 90 composites), and it requires previous knowledge only of the first 5 primes.

38. A method similar to the Sieve of Eratosthenes in Exercise 37 can be used to determine which integers can be written in the form abc with $a, b, c > 1$. Explain how it follows from Exercise 36 that each integer from 1 to 64 that can be written as a product of three integers greater than 1 must be divisible by 2 and/or 3. Use this fact to find them.

In Exercises 39 through 44, find the greatest common divisors by factoring.

39. Find gcd(56, 42).   40. Find gcd(−360, 270).

41. Find gcd(−108, −90).   42. Find gcd(810, 24500).

43. Find gcd(2475, −780).   44. Find gcd(−1144, 3146).

45. Suppose n beads are to be arranged on a circular loop, and every kth bead must be the same color. That is, for each bead b, both beads k positions away from b must be the same color as b.

    (a) The case in which $n = 10$ and $k = 4$ is reflected in Figure 3.2, where a coloring has been started. What is the maximum number of different colors that could be used in this case?

    (b) For a cycle of 20 beads and a skipping number 8, what is the maximum number of different colors that could be used?

    (c) In general, for a cycle of n beads and a skipping number k, what is the maximum number of different colors that could be used? Try several examples to see a pattern.

46. Some of the squares in an $m \times n$ grid of unit squares are to be numbered. We start with the bottom-left corner. From

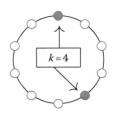

FIGURE 3.2
Cycle of $n = 10$ beads with skipping number $k = 4$.

				6					12		
	⋰		5					11			
16		4					10				
	3					9					15
	2				8					14	
1					7				13		

FIGURE 3.3
Numbering squares on a grid.

then on, we shall next number the square over and up one from the previous numbered square. If this runs us off the top of the grid, then we resume on the square in the next column of the bottom row. If we run off the right of the grid, then we resume on the square in the next row of the left column.

(a) The case in which $m = 15$ and $n = 6$ is displayed in Figure 3.3, where the numbering has also been started. What number gets assigned to the top-right corner?

(b) For a 20×12 grid, what number gets assigned to the top-right corner?

(c) In general, for an $m \times n$ grid, what number gets assigned to the top-right corner?

For Exercises 47 through 52, let m and n be integers that are not both zero.

47. Show that there can only be one value d satisfying the three conditions in Definition 3.5.

48. Let $d = \gcd(m,n)$ and note that $\frac{m}{d}, \frac{n}{d} \in \mathbb{Z}$. Show: $\gcd(\frac{m}{d}, \frac{n}{d}) = 1$.

49. Show: $\gcd(-m,n) = \gcd(m,n)$.

50. Show: $\gcd(n,m) = \gcd(m,n)$.

51. Use Exercises 49 and 50 to show: $\gcd(m, -n) = \gcd(m,n)$.

52. Use Exercises 49, 50, and 51 to show: $\gcd(m,n) = \gcd(|m|, |n|)$.

53. Show that every pair of distinct primes is relatively prime.

54. Let $m \in \mathbb{Z}+$ and $k \in \mathbb{Z} \setminus \{0\}$. Show: $\gcd(m, mk) = m$.

55. Let $n \in \mathbb{Z}$. Show n and $1 - 2n$ are relatively prime. Hint: $1 = (1 - 2n) + 2(n)$.

56. Let $n \in \mathbb{Z}$. Show that $3n + 1$ and $2n + 1$ are relatively prime. Hint: Find $a, b \in \mathbb{Z}$ such that $a(3n + 1) + b(2n + 1) = 1$.

Definition 3.7. Given nonzero integers m and n, their **least common multiple**, denoted $\operatorname{lcm}(m, n)$, is the unique integer l such that

(i) $l > 0$,

(ii) $m \mid l$ and $n \mid l$, and

(iii) $\forall k \in \mathbb{Z}^+$, if $m \mid k$ and $n \mid k$, then $l \le k$.

In Exercises 57 through 60, use the fact, to be proved in Section 4.5, that

$$\forall\, m, n \in \mathbb{Z} \setminus \{0\},\quad \operatorname{lcm}(m, n) = \frac{|mn|}{\gcd(m, n)}. \qquad (3.2)$$

57. Find $\operatorname{lcm}(56, 42)$.   58. Find $\operatorname{lcm}(-36, 27)$.

59. Find $\operatorname{lcm}(-108, -90)$.   60. Find $\operatorname{lcm}(81, 245)$.

61. For two positive integers a and b, we have many cans of height a inches and many of height b inches that we would like to store on some shelves. We insist that (1) the shelves be evenly spaced, (2) only cans of the same height be stacked on the same shelf, and (3) each shelf, when completely filled by stacks of cans, leaves no empty space above the top cans and below the next shelf. For example, if $a = 4$ and $b = 6$, then leaving 12 inches between the shelves will accomplish this. We can stack the 4 inch cans three high, and we can stack the 6 inch cans two high.

   (a) If we have cans of heights 4 and 16 inches, then what is the smallest possible spacing between the shelves that will satisfy our conditions?

   (b) If we have cans of heights 6 and 15 inches, then what is the smallest possible spacing between the shelves that will satisfy our conditions?

   (c) In general, in terms of a and b, what is the smallest possible spacing between the shelves that will satisfy our conditions?

62. Given integers a, b, c, d with b and d nonzero, to compute the fraction sum $\frac{a}{b} + \frac{c}{d}$, we need to find a common denominator.

   (a) What is the smallest possible common denominator we could use in computing $\frac{3}{10} + \frac{2}{25}$?

   (b) In general, in terms of b and d, what is the smallest possible common denominator we could use?

   (c) If $\frac{a}{b}$ and $\frac{c}{d}$ are in lowest terms (i.e., $\gcd(a, b) = \gcd(c, d) = 1$) and we use the smallest possible common denominator to add them, will that always be the smallest possible denominator for the sum?

In Exercises 63 and 64, use Definition 3.7 and not (3.2).

63. Show: $\forall\, m, n \in \mathbb{Z} \setminus \{0\}$, $\mathrm{lcm}(n, m) = \mathrm{lcm}(m, n)$.

64. Show: $\forall\, m, n \in \mathbb{Z} \setminus \{0\}$, $\mathrm{lcm}(m, -n) = \mathrm{lcm}(m, n)$.

3.2   Well-ordering, division, and codes

Since, in this book, we have not axiomatically constructed the set of integers, we have allowed ourselves to accept and use without proof the basic properties listed in Appendix A. For example, this includes the fact that sums and products of integers are integers. In this section, we explore some consequences of a less elementary but extremely important property, which we also assume without proof.

Theorem 3.4.   (Well-Ordering Principle for the Integers). *Each nonempty subset of the nonnegative integers has a smallest element.*

When relatively simple examples are considered, the Well-Ordering Principle may not seem helpful.

Example 3.10.   Let $S = \{s\ :\ s = 15x + 25y$ where $x, y \in \mathbb{Z}$ and $15x + 25y > 0\}$. Obviously, S is a nonempty subset of the nonnegative integers. Hence, the Well-Ordering Principle guarantees that S has a smallest element. However, it is not hard to see directly in this case that the smallest element of S is 5. First, $5 \in S$, since $5 = 15(2) + 25(-1)$. Second, any element $s \in S$ can be expressed as $s = 15x + 25y = 5(3x + 5y)$, a multiple of 5. Since $0 \notin S$ and none of $1, 2, 3, 4$ are divisible by 5, no element smaller than 5 is in S.

The power of the Well-Ordering Principle shows itself in perhaps subtle ways. For example, it enables us to prove that primes are factors of every integer from 2 on up.

Theorem 3.5.   (Existence of Prime Divisors). *Every integer greater than 1 has a prime divisor.*

Proof. Suppose not. Let S be the set of integers greater than 1 that do not have a prime divisor. Since our assumption is that S is nonempty, the Well-Ordering Principle gives us a smallest element $n \in S$. It cannot be the case that n is prime, since n divides itself. Thus, n must be composite. We can therefore write $n = rs$ for some integers r and s such that $1 < r < n$ and $1 < s < n$. (See Exercise 33 of Section 3.1.) In particular, $r > 1$ and r is smaller than n, the smallest element of S. Hence, $r \notin S$. That is, r must be divisible by some prime p. Say $r = pt$, where $t \in \mathbb{Z}$. Therefore, $n = rs = pts$. That is, n is divisible by the prime p. This is a contradiction. □

The proof of Theorem 3.5 additionally illustrates the power of a proof by contradiction. That power is also highlighted in the following proof using Theorem 3.5.

Theorem 3.6.   *There are infinitely many primes.*

Proof. Suppose not. Since there must therefore be only finitely many primes, we can give a complete list $p_1, p_2, \ldots, p_m$ of all of the primes. Form the product $n = p_1 \cdot p_2 \cdots p_m$, and consider the integer $n + 1$. Theorem 3.5 tells us that $n + 1$ must have some prime divisor. Hence, $n + 1$ must be divisible by some prime on our (supposedly complete) list, say p_k. However, n is also divisible by p_k. So n and $n + 1$ have the prime factor p_k in common. This is a contradiction, since Lemma 3.3 tells us that $\gcd(n, n + 1) = 1$. □

3.2.1   Computer searches for large primes

Despite the fact that there is no end to the list of primes, there is an end to the list of *known* primes. As of January 2016, the largest known prime is $2^{74207281} - 1$. However, the determination of all primes up to that value is not yet complete. The prime $2^{74207281} - 1$ has 22,338,618 decimal digits and is an example from a special class of prime numbers named after the French monk Marin Mersenne (1588–1648), who was a connoisseur of mathematics.

A **Mersenne prime** is a prime number of the form $2^n - 1$, for some integer n. The first five Mersenne primes are the numbers $2^n - 1$, when $n = 2, 3, 5, 7, 13$. As we shall see in the exercises, the exponent n must itself be prime for $2^n - 1$ to be prime. However, since $2^{11} - 1 = 23 \cdot 89$ is composite, that condition is not sufficient. The *Great Internet Mersenne Prime Search* (GIMPS) is a worldwide effort to find larger and larger Mersenne primes. One can enlist one's personal computer in this pursuit by downloading software available from the GIMPS website http://www.mersenne.org. An award of $150,000 is being offered by the Electronic Frontier Foundation for the first prime number found to have at least 100 million decimal digits.

3.2.2   Integer division

For another useful application of the Well-Ordering Principle, we revisit the integer division taught in elementary school. For example, the problem of $124 \div 9$

$$
\begin{array}{r}
1\quad 3 \\
9\ \overline{) 1\quad 2\quad 4} \\
-\underline{9} \\
3\quad 4 \\
-\underline{2\quad 7} \\
7
\end{array}
$$

yields the response of 13 remainder 7. This is formalized in the Division Algorithm.

Theorem 3.7.   (Division Algorithm).
Given any integer n and any positive integer d, there exist unique integers q and r such that

(i) $n = dq + r$, and

(ii) $0 \leq r < d$.

Definition 3.8.   In Theorem 3.7, we say that q is the **quotient** and r is the **remainder** upon division of n by d. We also write

$$q = n \text{ div } d \quad \text{and} \quad r = n \text{ mod } d.$$

The operators div and mod occur in some form in most programming languages. In *Mathematica*, the values q and r from the Division Algorithm are returned by the functions `Quotient[n,d]` and `Mod[n,d]`, respectively. In C++, `n/d` and `n%` are used, and in *Maple*, `iquo(n,d)` and `irem(n,d)` can be used.[*] However, when n is negative, the behavior of these operators does not always agree with div and mod, respectively. In that case, `n%` and `irem(n,d)` may be negative. Thus, one must thoroughly understand their precise definitions when using them for computations.

Before proving Theorem 3.7, we first consider examples.

Example 3.11.

(a) If $n = 124$ and $d = 9$, then $q = 13$ and $r = 7$.
   That is, $124 = 9(13) + 7$.
   So 124 div 9 = 13 and 124 mod 9 = 7.

(b) If $n = 60$ and $d = 5$, then $q = 12$ and $r = 0$.
   That is, $60 = 5(12) + 0$.
   So 60 div 5 = 12 and 60 mod 5 = 0.

---

[*] In *Maple*, the function modp and the operator mod can also be used.

(c) If $n = -53$ and $d = 7$, then $q = -8$ and $r = 3$.
   That is, $-53 = 7(-8) + 3$.
   So -53 div $7 = -8$ and -53 mod $7 = 3$.

Proof of Theorem 3.7. Both the existence and the uniqueness of q and r must be shown. Existence is handled first.
   Let $S = \{s : s = n - dk$ where $k \in \mathbb{Z}$ and $n - dk \geq 0\}$. We can argue that S is nonempty. If $n \geq 0$, then $n - d \cdot 0 = n$ is in S. If $n < 0$, then $n - dn = (d-1)(-n)$ is in S. Since S is nonempty, the Well-Ordering Principle provides us with a smallest element r. Thus, $r = n - dq$ for some $q \in \mathbb{Z}$. We just need to show that q and r have the desired properties. Certainly, $n = dq + r$ and $r \geq 0$. Hence, it remains to show that $r < d$. However, if $r \geq d$, then $n - d(q+1) = n - dq - d = r - d \geq 0$. This would make $r - d$ a smaller element of S than r, which is impossible because r is the smallest element of S. Since the assumption that $r \geq d$ leads to a contradiction, we conclude that $0 \leq r < d$.
   To prove uniqueness suppose $n = dq_1 + r_1 = dq_2 + r_2$, $0 \leq r_1 < d$, and $0 \leq r_2 < d$. That is, suppose that the pairs (q_1, r_1) and (q_2, r_2) are both results of the division of n by d. Without loss of generality, say $r_2 \geq r_1$. That is, we can choose to name our pairs so that (q_2, r_2) does not have a smaller second coordinate than (q_1, r_1). Since $d(q_1 - q_2) = dq_1 - dq_2 = r_2 - r_1$, we have that $d \mid (r_2 - r_1)$. If $r_2 \neq r_1$, then $r_2 - r_1 > 0$. Theorem 3.1 then tells us that $d \leq r_2 - r_1 \leq r_2$, which contradicts the fact that $r_2 < d$. Hence, it must be that $r_2 = r_1$. This forces $dq_2 = dq_1$, and therefore $q_2 = q_1$ as well. $\square$

By the Division Algorithm, the only possible remainders upon division of an integer n by a positive integer d are $0, 1, \ldots$, and $d - 1$. In particular, for $d = 2$, the only possible remainders are 0 and 1. Hence, Definition 3.1 is now seen to say that an integer is odd if and only if it is not divisible by 2.
   In general, the Division Algorithm provides a way of proving statements involving an integer d not dividing an integer n. Such a property is characterized by the fact that n mod d is nonzero.

Example 3.12.   Show: For any integer n, $n^2 - 2$ is not divisible by 5.

Proof. Let n be any integer.
By the Division Algorithm, n must take one of the forms

$$5k, 5k + 1, 5k + 2, 5k + 3, \text{ or } 5k + 4,$$

for some integer k. That is, dividing n by 5 leaves a remainder of $0, 1, 2, 3,$ or 4. Here, k is being used to represent the quotient.

We consider each possible form for n.

Case 0: $n = 5k$ for some $k \in \mathbb{Z}$.
Observe that

$$
\begin{aligned}
n^2 - 2 &= 25k^2 - 2 \\
&= 25k^2 - 5 + 3 \\
&= 5(5k^2 - 1) + 3.
\end{aligned}
$$

That is, $(n^2 - 2) \bmod 5 = 3$ and not 0.

Case 1: $n = 5k + 1$ for some $k \in \mathbb{Z}$.
Observe that

$$
\begin{aligned}
n^2 - 2 &= 25k^2 + 10k + 1 - 2 \\
&= 25k^2 + 10k - 5 + 4 \\
&= 5(5k^2 + 2k - 1) + 4.
\end{aligned}
$$

That is, $(n^2 - 2) \bmod 5 = 4$ and not 0.

Case 2: $n = 5k + 2$ for some $k \in \mathbb{Z}$.
Observe that

$$
\begin{aligned}
n^2 - 2 &= 25k^2 + 20k + 4 - 2 \\
&= 25k^2 + 20k + 2 \\
&= 5(5k^2 + 4k) + 2.
\end{aligned}
$$

That is, $(n^2 - 2) \bmod 5 = 2$ and not 0.

Case 3: $n = 5k + 3$ for some $k \in \mathbb{Z}$.
Observe that

$$
\begin{aligned}
n^2 - 2 &= 25k^2 + 30k + 9 - 2 \\
&= 25k^2 + 30k + 5 + 2 \\
&= 5(5k^2 + 6k + 1) + 2.
\end{aligned}
$$

That is, $(n^2 - 2) \bmod 5 = 2$ and not 0.

Case 4: $n = 5k + 4$ for some $k \in \mathbb{Z}$.
Observe that

$$
\begin{aligned}
n^2 - 2 &= 25k^2 + 40k + 16 - 2 \\
&= 25k^2 + 40k + 10 + 4 \\
&= 5(5k^2 + 8k + 2) + 4.
\end{aligned}
$$

That is, $(n^2 - 2) \bmod 5 = 4$ and not 0.

In each case, we see that $5 \nmid (n^2 - 2)$.   □

The approach taken in Example 3.12 can also be used to show that an expression is not divisible by 2.

Example 3.13.   Show: $\forall\, n \in \mathbb{Z},\ n^2 + n - 1$ is odd.

Proof. Let $n \in \mathbb{Z}$.
By the Division Algorithm, n must take either the form $2k$ or $2k + 1$, for some $k \in \mathbb{Z}$. (That is, n is either even or odd.)

Case 0: $n = 2k$ for some $k \in \mathbb{Z}$.
Observe that

$$n^2 + n - 1 = 4k^2 + 2k - 1$$
$$= 4k^2 + 2k - 2 + 1$$
$$= 2(2k^2 + k - 1) + 1.$$

Thus, $(n^2 + n - 1) \bmod 2 = 1$ and not 0. (That is, $n^2 + n - 1$ is odd.)

Case 1: $n = 2k + 1$ for some $k \in \mathbb{Z}$.
Observe that

$$n^2 + n - 1 = 4k^2 + 4k + 1 + 2k + 1 - 1$$
$$= 4k^2 + 6k + 1$$
$$= 2(2k^2 + 3k) + 1.$$

Thus, $(n^2 + n + 1) \bmod 2 = 1$ and not 0. (That is, $n^2 + n + 1$ is odd.)

In both cases, $n^2 + n - 1$ is odd.   □

3.2.3 Rounding numbers

We consider two ways of discarding the decimal part of an arbitrary real number.

Definition 3.9.   Let x be any real number.

(a) The **floor** of x, denoted $\lfloor x \rfloor$, is the largest integer n such that $n \leq x$.

(b) The **ceiling** of x, denoted $\lceil x \rceil$, is the smallest integer n such that $x \leq n$.

In *Mathematica*, the functions in Definition 3.9 are denoted `Floor` and `Ceiling`. In C++ and *Maple*, they are `floor` and `ceil`.

Example 3.14.

(a) $\lfloor 3.2 \rfloor = 3$. (f) $\lceil 3.2 \rceil = 4$.
(b) $\lfloor 5.8 \rfloor = 5$. (g) $\lceil 5.8 \rceil = 6$.
(c) $\lfloor -4.9 \rfloor = -5$. (h) $\lceil -4.9 \rceil = -4$.
(d) $\lfloor -6.3 \rfloor = -7$. (i) $\lceil -6.3 \rceil = -6$.
(e) $\lfloor 8 \rfloor = 8 = \lceil 8 \rceil$. (j) $\lceil e \rceil = 3$, since $e \approx 2.718$.

The characterizations of floor and ceiling given in the following theorem provide a checklist of three items to verify in computations. A proof of Theorem 3.8 is left for the exercises.

Theorem 3.8. *Let x be any real number.*

(a) $\lfloor x \rfloor$ *is the unique value n such that*

  (i) $n \in \mathbb{Z}$,

  (ii) $n \leq x$, *and*

  (iii) $x < n + 1$.

(b) $\lceil x \rceil$ *is the unique value n such that*

  (i) $n \in \mathbb{Z}$,

  (ii) $x \leq n$, *and*

  (iii) $n - 1 < x$.

We apply Theorem 3.8 to prove the following formula for $\lfloor \frac{n}{2} \rfloor$.

Theorem 3.9. $\forall\, n \in \mathbb{Z}$, $\lfloor \frac{n}{2} \rfloor = \begin{cases} \frac{n}{2} & \text{if } n \text{ is even,} \\ \frac{n-1}{2} & \text{if } n \text{ is odd.} \end{cases}$

Proof. Let $n \in \mathbb{Z}$.

Case 1: n is even.
So $n = 2k$ for some $k \in \mathbb{Z}$.
Note that $k = \frac{n}{2}$.
We then have $k \in \mathbb{Z}$, $k \leq \frac{n}{2}$ (actually $=$), and $\frac{n}{2} < k + 1$.
In this case, it follows from Theorem 3.8 that $\lfloor \frac{n}{2} \rfloor = k = \frac{n}{2}$.

Case 2: n is odd.
So $n = 2k + 1$ for some $k \in \mathbb{Z}$.

Note that $k = \frac{n-1}{2}$.
We then have $k \in \mathbb{Z}$, $k \leq \frac{n}{2}$, and $\frac{n}{2} < k + 1$. (Check these.)
In this case, it follows from Theorem 3.8 that $\lfloor \frac{n}{2} \rfloor = k = \frac{n-1}{2}$. $\qquad\qquad\square$

As should be expected, there is an analogous formula for $\lceil \frac{n}{2} \rceil$.

Theorem 3.10. $\forall\, n \in \mathbb{Z}$, $\lceil \frac{n}{2} \rceil = \begin{cases} \frac{n}{2} & \text{if } n \text{ is even,} \\ \frac{n+1}{2} & \text{if } n \text{ is odd.} \end{cases}$

The results in Theorems 3.9 and 3.10 can be combined to yield the following formula.

Corollary 3.11. $\forall\, n \in \mathbb{Z}$, $\lfloor \frac{n}{2} \rfloor + \lceil \frac{n}{2} \rceil = n$.

A **corollary** is a theorem. However, the term *corollary* is typically used when that theorem follows relatively easily from a previous theorem or theorems. The proofs of Theorem 3.10 and Corollary 3.11 are left for the exercises.

It is tempting to overestimate how well-behaved the floor and ceiling functions are. The following example should serve to dispel some misconceptions.

Example 3.15. Disprove: $\forall\, x, y \in \mathbb{R}$, $\lfloor x + y \rfloor = \lfloor x \rfloor + \lfloor y \rfloor$.

Counterexample. Let $x = 1.5$ and $y = 2.5$.
So $x + y = 4$.
Observe that $\lfloor x + y \rfloor = 4$.
and $\lfloor x \rfloor + \lfloor y \rfloor = 1 + 2 = 3$.
Thus, $\lfloor x + y \rfloor \neq \lfloor x \rfloor + \lfloor y \rfloor$. $\qquad\qquad\square$

There is a beautiful connection between the floor function and the $\div$ operation.

Theorem 3.12. *Given any integer n and positive integer d, $\lfloor \frac{n}{d} \rfloor = n$ div d.*

Proof. Let $n, d \in \mathbb{Z}$ with $d > 0$.
By the Division Algorithm, we can write $n = dq + r$,
where $q = n$ div d and $0 \leq r < d$.
Observe that $q \in \mathbb{Z}$, $q = \frac{n}{d} - \frac{r}{d} \leq \frac{n}{d}$, and $q + 1 = \frac{n}{d} + \frac{d-r}{d} > \frac{n}{d}$.
It follows from Theorem 3.8 that $\lfloor \frac{n}{d} \rfloor = q = n$ div d. $\qquad\square$

3.2.4 Applications of mod

Identification numbers are typically appended with a **check digit** that is determined using the remainder function mod.

Example 3.16. (Check Digit Formulas).

(a) A **Universal Product Code** (UPC) number

$$d_1 \quad d_2 \ d_3 \ d_4 \ d_5 \ d_6 \quad d_7 \ d_8 \ d_9 \ d_{10} \ d_{11} \quad d_{12}$$

is attached to each product available for sale. The last digit d_{12} is the check digit and is determined from the first eleven identifying digits by the requirement that

$$[3(d_1 + d_3 + d_5 + d_7 + d_9 + d_{11}) + d_2 + d_4 + d_6$$
$$+ d_8 + d_{10} + d_{12}] \bmod 10 = 0.$$

For example, the UPC number for a bottle of Gatorade is

$$0\ 52000\ 33831\ 7$$

and indeed

$$[3(0 + 2 + 0 + 3 + 8 + 1) + 5 + 0 + 0$$
$$+ 3 + 3 + 7] = 60 \text{ and } 60 \bmod 10 = 0.$$

(b) An **International Standard Book Number** (ISBN-10)

$$d_1 \ - \ d_2 \ d_3 \ d_4 \ d_5 \ d_6 \ d_7 \ d_8 \ d_9 \ - \ d_{10}$$

is attached to each book published. The check digit d_{10} is determined by the requirement that

$$[10d_1 + 9d_2 + 8d_3 + 7d_4 + 6d_5 + 5d_6 + 4d_7$$
$$+ 3d_8 + 2d_9 + d_{10}] \bmod 11 = 0.$$

If the value of d_{10} must be ten, then X is used for that digit. For example, the ISBN for *Curious George* by H. A. Rey is

$$0\text{-}39515023\text{-}X$$

and indeed

$$[10 \cdot 0 + 9 \cdot 3 + 8 \cdot 9 + 7 \cdot 5 + 6 \cdot 1 + 5 \cdot 5$$
$$+ 4 \cdot 0 + 3 \cdot 2 + 2 \cdot 3 + 10] = 187$$
$$\text{and } 187 \bmod 11 = 0.$$

Check digits can be used to detect whether an identification number is read or entered incorrectly.

Example 3.17.   (Error Detection with Check Digits).

(a) If the UPC number for Gatorade is mistakenly scanned as

$$0\ 25000\ 33831\ 7,$$

then the calculation

$$[3(0 + 5 + 0 + 3 + 8 + 1)$$
$$+ 2 + 0 + 0 + 3 + 3 + 7] = 65$$

shows that an error has occurred, since $65 \bmod 10 = 5 \neq 0$.

(b) If the ISBN for *Curios George* is mistakenly typed as

$$0\text{-}36515023\text{-}X,$$

then the calculation

$$[10 \cdot 0 + 9 \cdot 3 + 8 \cdot 6 + 7 \cdot 5 + 6 \cdot 1 + 5 \cdot 5$$
$$+ 4 \cdot 0 + 3 \cdot 2 + 2 \cdot 3 + 10] = 163$$

shows that an error has occurred, since $163 \bmod 11 = 9 \neq 0$.

Having a check digit does not guarantee the detection of all errors, but most typical errors are commonly caught. Some simple errors in identification numbers that might not be caught are explored in the exercises.

Another application in which additional check digits help to catch errors is through **binary linear codes**. These are schemes used to transmit messages when errors may be introduced, and they not only detect errors but also correct some errors. Such systems are employed by NASA to mitigate the errors introduced in sending messages over the vastness of space and by CD burners to circumvent errors caused by scratches and other imperfections.

The binary linear codes we consider in this book are constructed as follows. Given a **message**, encoded as a binary string $b_1 b_2 \ldots b_k$, a binary linear code specifies a **code word** $b_1 b_2 \ldots b_k b_{k+1} \ldots b_n$ by appending extra digits, called **parity check digits**. Each parity check digit is determined by the sum of some of the binary digits in the message.

Example 3.18.   (Constructing a Binary Linear Code). Suppose a message $b_1 b_2 b_3$ is converted to a code word $b_1 b_2 b_3 b_4 b_5 b_6$ by computing the parity check digits according to the following formulas, using sums of the message digits.

$$b_4 = (b_1 + b_2) \bmod 2$$
$$b_5 = (b_1 + b_3) \bmod 2$$
$$b_6 = (b_1 + b_2 + b_3) \bmod 2$$

Thus, the code words determined by the eight possible three-digit binary messages are displayed in Table 3.1.

The **weight** w of a binary linear code is the minimum number of ones that appear in a nonzero code word. For example, the code displayed in Table 3.1 has weight $w = 3$. It turns out that if a code word $c_1 c_2 \ldots c_n$ is received with no more than $\lfloor \frac{w-1}{2} \rfloor$ incorrect digits, then the intended message $b_1 b_2 \ldots b_k$ corresponding to that code word can be recovered by the method of **nearest neighbor decoding**. Very simply, we use the first k digits of the code word $b_1 b_2 \ldots b_n$ differing from $c_1 c_2 \ldots c_n$ in the least number of digits. It is the assumption that no more than $\lfloor \frac{w-1}{2} \rfloor$ errors exist that guarantees the uniqueness of this nearest neighbor $b_1 b_2 \ldots b_n$ to $c_1 c_2 \ldots c_n$.

Example 3.19. Suppose the binary linear code from Example 3.18 is being used, and the code word 010110 is received. Since this is not a true code word from Table 3.1, some error has been introduced. Assuming a very low likelihood of an error, it is not unreasonable to expect that the number of errors introduced is at most $\lfloor \frac{3-1}{2} \rfloor = 1$. The method of nearest neighbor decoding then dictates that we decode this using the code word 011110, which corresponds to the message 011. Notice, in this case, that the message did not turn out to be the first three digits of the received erroneous code word.

TABLE 3.1
A binary linear code for 3-digit messages

Message	Code word
000	000000
001	001011
010	010101
011	011110
100	100111
101	101100
110	110010
111	111001

TABLE 3.2
Converting letters to numbers

	A	B	C	D	E	F	G	H
0	1	2	3	4	5	6	7	8

	I	J	K	L	M	N	O	P	Q
	9	10	11	12	13	14	15	16	17

	R	S	T	U	V	W	X	Y	Z
	18	19	20	21	22	23	24	25	26

Binary linear codes are not designed to encrypt messages for clandestine transmission. The objective of encryption is handled instead by a **cipher**. Suppose, for example, that we wish to send a secret text message using the correspondence of letters to numbers shown in Table 3.2, in which a space is assigned the numerical value 0. Very simply, we seek to scramble the numbers 0–26, so that a message consisting of those characters can be converted to a scrambled message that cannot be read easily. Perhaps the simplest way to assign each number x to a new number y, and thus permute our numbers, is with a **shift cipher**. Such a cipher has the form

$$y = (x + b) \bmod n, \tag{3.3}$$

for some choice of an integer b. In our case, $n = 27$, since we are permuting the numbers 0–26.

Example 3.20.   Using $n = 27$, $b = -1$, Equation (3.3), and the conversions in Table 3.2, the message

IBM IS UP

is encrypted as

HALZHRZTO.

For example, the letter I is represented by the value 9,

$$(9 + (-1)) \bmod 27 = 8,$$

and the value 8 represents the letter H.
    To decrypt a message, the formula

$$x = (y - b) \bmod n \tag{3.4}$$

is used.

Example 3.21. Using $n = 27$, $b = -1$, Equation (3.4), and the conversions in Table 3.2, the encrypted message

RDKKZHSZMNV

is decrypted as

SELL IT NOW.

For example, the letter R is represented by the value 18,

$$(18 - (-1)) \bmod 27 = 19,$$

and 19 represents the letter S.

Generalizations of shift ciphers are considered in Section 3.5.

Exercises

In Exercises 1 through 4, determine the smallest element of the given set S.

1. $S = \{s \ : \ s = 11 - 4n \text{ where } n \in \mathbb{Z} \text{ and } 11 - 4n > 0\}$.

2. $S = \{s \ : \ s = 3n + 20 \text{ where } n \in \mathbb{Z} \text{ and } 3n + 20 > 0\}$.

3. $S = \{s \ : \ s = 12x + 20y \text{ where } x, y \in \mathbb{Z} \text{ and } 12x + 20y > 0\}$.

4. $S = \{s \ : \ s = 15x + 21y \text{ where } x, y \in \mathbb{Z} \text{ and } 15x + 21y > 0\}$.
   Use a calculator or mathematical software to perform the verifications in Exercises 5 and 6. In *Mathematica*, the functions FactorInteger and PrimeQ can be used, and in *Maple*, the functions ifactor and isprime can be used. However, Mersenne (c. 1600) knew the results in these exercises long before the invention of calculators.

5. Verify that $2^n - 1$ is a Mersenne prime for $n = 17$ and 19, but not for $n = 23, 29$. Note that verifying that $2^{17} - 1$ is prime by hand (not recommended) could be done (by Exercise 35 in Section 3.1) by testing for divisibility by the primes from 2 to 359. Also, successively testing for prime divisors of $2^{23} - 1$ could lead to a factorization of $2^{23} - 1$, but that would be rather tedious by hand.

6. Verify that $2^{31} - 1$ is a Mersenne prime, but $2^{37} - 1$ is not. Note that these numbers have 10 and 12 digits, respectively. Hence, verification by hand is not recommended.

7. Show that there is a prime with at least 100 million digits.

8. Show that there is a composite with at least 100 million digits.

9. Use the fact that $\forall\, x \in \mathbb{R}, \forall\, n \in \mathbb{Z}^+$,

$$x^n - 1 = (x-1)(x^{n-1} + x^{n-2} + \cdots + x + 1),$$

to show the following:

(a) For all $b, n \in \mathbb{Z}^+$ with $b \geq 3$ and $n \geq 2$, $b^n - 1$ is composite.

(b) For all $n \in \mathbb{Z}^+$, if $2^n - 1$ is prime, then n is prime. Hint: Use an indirect argument.

10. Show: $\forall\, n \in \mathbb{Z}^+$, there exist n consecutive composites. Hint: Consider $(n+1)! + 2, (n+1)! + 3, \ldots$. Factorials are defined in Section 4.1.

11. Find the quotient and remainder resulting from the integer division problems.

(a) $127 \div 12$.   (b) $216 \div 15$.

12. Find the quotient and remainder resulting from the integer division problems.

(a) $207 \div 8$.   (b) $162 \div 11$.

13. Verify the conclusions of the Division Algorithm (Theorem 3.7) for each of the following:

(a) $n = 45$ and $d = 7$.   (b) $n = -37$ and $d = 4$.

14. Verify the conclusions of the Division Algorithm (Theorem 3.7) for each of the following:

(a) $n = 17$ and $d = 3$.   (b) $n = -53$ and $d = 10$.

15. Compute 73 div 10 and 73 mod 10.

16. Compute 104 div 6 and 104 mod 6.

17. Compute each of the following:

(a) 67 div 13 and 67 mod 13.   (b) -67 div 13 and -67 mod 13.

18. Compute each of the following:

(a) 85 div 12 and 85 mod 12.   (b) -85 div 12 and -85 mod 12.

19. A group of 165 people is planning a trip to a concert. They wish to all sit in one section, each row of which contains 18 seats. How many entirely full rows do they need? How many additional seats in a row that will not be full are needed? Answer these questions using div and mod.

20. A father wishes to split a bag of 375 gum drops as evenly as possible among 18 children, until fewer than 18 gum drops remain. How many gum drops will each child get? How many will be left over for the father to eat? Answer these questions using div and mod.

21.* Given $n, s \in \mathbb{Z}^+$ with $s \geq 2$, the following algorithm (which uses div and mod) gives the base-s representation $a_k a_{k-1} \cdots a_0$ for n.

> Let $k = -1$.
> While $n > 0$,
> > \begin
> > Let $k = k + 1$.
> > Let $a_k = n$ mod s.
> > Let $n = n$ div s.
> > \end.
> Return $a_k a_{k-1} \cdots a_0$.

Follow the algorithm for the given values of n and s.

(a) $n = 39$ and $s = 2$.   (b) $n = 87$ and $s = 8$.

Write a program that implements the algorithm. Prompt the user for n and s, and store $a_k a_{k-1} \cdots a_0$ in an array.

22.* Given an integer $s \geq 2$ and base-s representations $a_k a_{k-1} \cdots a_0$ and $b_k b_{k-1} \cdots b_0$ for two nonnegative integers (both have a given length k), the following algorithm (which uses div and mod) gives the base-s representation $c_{k+1} c_k c_{k-1} \cdots c_0$ for their sum.

> Let $a_{k+1} = b_{k+1} = 0$.
> Let carry $= 0$.
> For $i = 0$ to $k + 1$,
> > \begin
> > Let $c_i = (\text{carry} + a_i + b_i)$ mod s.
> > Let carry $= (\text{carry} + a_i + b_i)$ div s.
> > \end.
> Return $c_{k+1} c_k c_{k-1} \cdots c_0$.

Follow the algorithm for the given s and integers represented in base s.

(a) In base $s = 2$, add 10101 and 01101.

(b) In base $s = 8$, add 4316 and 2570.

Write a program that implements the algorithm. Prompt the user for s and the base-s representations of the summands, and store each base-s representation in an array.

23. Let $n \in \mathbb{Z}^+$. Determine n^2 div n and n^2 mod n.

24. Let $n \in \mathbb{Z}^+$. Determine $6n$ div 3 and $6n$ mod 3.

25. Why does the Well-Ordering Principle not work for any subset of the integers (instead of just the nonnegative integers)?

26. Why is $d = 0$ not allowed in the Division Algorithm?

27. **Generalized Well-Ordering Principle for the integers.** Show: Given a fixed integer a, any nonempty subset of the set of integers greater than or equal to a must have a smallest element. Hint: Starting with a set S consider the set $S - a = \{s - a : s \in S\}$.

28. **Generalized Maximum Principle.** Show: Given a fixed integer b, any nonempty subset of the integers less than or equal to b must have a largest element. Hint: Starting with a set S consider the set $b - S = \{b - s : s \in S\}$.

29. Show: $\forall\, n \in \mathbb{Z}$, $n^3 - n + 2$ is not divisible by 6.

30. Show: $\forall\, n \in \mathbb{Z}$, $n^3 + 3n^2 + 2n + 1$ is not divisible by 6.

31. Show: $\forall\, n \in \mathbb{Z}$, if n is not divisible by 3, then $n^2 \bmod 3 = 1$.

32. Show: $\forall\, n \in \mathbb{Z}$, if n is odd, then $n^2 \bmod 8 = 1$. Hint: Consider $n = 4k + 1$ and $n = 4k + 3$.

33. Show: $\forall\, n \in \mathbb{Z}$, if n is not divisible by 5, then $n^4 - 1$ is divisible by 5.

34. Show: $\forall\, n \in \mathbb{Z}$, if n is not divisible by 7, then $n^6 - 1$ is divisible by 7.

35. Find each of the following:

   (a) $\lfloor 4.4 \rfloor$.   (b) $\lfloor -4.4 \rfloor$.   (c) $\lceil 8.6 \rceil$.   (d) $\lceil -8.6 \rceil$.

36. Find each of the following:

   (a) $\lfloor 3.7 \rfloor$.   (b) $\lfloor -3.7 \rfloor$.   (c) $\lceil 6.2 \rceil$.   (d) $\lceil -6.2 \rceil$.

37. Find each of the following:

   (a) $\lceil -5 \rceil$.   (b) $\lfloor \pi \rfloor$.   (c) $\lfloor \frac{17}{3} \rfloor$.   (d) $\lceil 4e \rceil$.

38. Find each of the following:

   (a) $\lfloor 9 \rfloor$.   (b) $\lceil \sqrt{2} \rceil$.   (c) $\lceil -\frac{22}{7} \rceil$.   (d) $\lfloor (3.2)^2 \rfloor$.

39. A company needs to ship c cans of soup from the cannery to a distribution center. The boxes it has each hold a maximum of m cans. To be efficient, the company wants to use as few boxes as possible.

   (a) What is the minimum number of boxes needed to contain $c = 500$ cans if each box will hold up to $m = 32$ cans?

   (b) In general, what is the minimum number of boxes needed to contain c cans if each box will hold up to m cans?

40. A trucking company has r refrigerators to deliver to a warehouse. Each truck can hold a maximum of m refrigerators, and the company will not send a truck unless it is full.

(a) What is the maximum number of full trucks that can be sent to the warehouse if there are $r = 250$ refrigerators and each truck can hold a maximum of $m = 36$ refrigerators?

(b) In general, what is the maximum number of full trucks that can be sent to the warehouse if there are r refrigerators and each truck can hold a maximum of m refrigerators?

41. Show: $\forall\, x \in \mathbb{R}$ and $n \in \mathbb{Z}$, $\lfloor x + n \rfloor = \lfloor x \rfloor + n$.

42. Show: $\forall\, x \in \mathbb{R}$ and $n \in \mathbb{Z}$, $\lceil x + n \rceil = \lceil x \rceil + n$.

43. Prove Theorem 3.10.

44. Prove Corollary 3.11.

45. Show: $\forall\, n \in \mathbb{Z}$, $\left\lfloor \frac{n}{3} \right\rfloor = \begin{cases} \frac{n}{3} & \text{if } 3 \mid n, \\ \frac{n-1}{3} & \text{if } n \bmod 3 = 1, \\ \frac{n-2}{3} & \text{if } n \bmod 3 = 2. \end{cases}$

46. Show: $\forall\, n \in \mathbb{Z}$, $2\left\lfloor \frac{n}{4} \right\rfloor = \begin{cases} \left\lfloor \frac{n}{2} \right\rfloor & \text{if } n \bmod 4 = 0 \text{ or } 1, \\ \left\lfloor \frac{n}{2} \right\rfloor - 1 & \text{if } n \bmod 4 = 2 \text{ or } 3. \end{cases}$

47. Observe that $\lfloor 2.1 \rfloor + \lfloor 3.2 \rfloor = \lfloor 5.3 \rfloor$ and $\lfloor 4.6 \rfloor + \lfloor 6.7 \rfloor < \lfloor 11.3 \rfloor$.
In general, prove that $\forall\, x, y \in \mathbb{R}$, $\lfloor x \rfloor + \lfloor y \rfloor \leq \lfloor x + y \rfloor$.

48. Observe that $\lceil 3 \rceil + \lceil 4 \rceil = \lceil 7 \rceil$ and $\lceil 2.8 \rceil + \lceil 4.7 \rceil = \lceil 7.5 \rceil$. However, disprove: $\forall\, x, y \in \mathbb{R}$, $\lceil x \rceil + \lceil y \rceil = \lceil x + y \rceil$.

49. Prove or disprove: $\forall\, x \in \mathbb{R}$, $\lfloor 2x \rfloor = 2\lfloor x \rfloor$.

50. Prove or disprove: $\forall\, x \in \mathbb{R}$, $\lfloor \frac{x}{2} \rfloor = \frac{1}{2}\lfloor x \rfloor$.

51. Show: $\forall\, x \in \mathbb{R}$, $\lfloor x \rfloor = \lceil x \rceil$ if and only if $x \in \mathbb{Z}$.

52. Show: $\forall\, x \in \mathbb{R} \setminus \mathbb{Z}$, $\lceil x \rceil = \lfloor x \rfloor + 1$. What is different for $x \in \mathbb{Z}$?

53. Show: $\forall\, x \in \mathbb{R}$, $\lfloor \lfloor x \rfloor \rfloor = \lfloor x \rfloor$.

54. Show: $\forall\, x \in \mathbb{R}$,

(a) $\lfloor -x \rfloor = -\lceil x \rceil$.

(b) $\lceil -x \rceil = -\lfloor x \rfloor$.

55. Show: $\forall\, n \in \mathbb{Z}$, $\left\lfloor \frac{n+1}{2} \right\rfloor = \left\lceil \frac{n}{2} \right\rceil$.

56. Show: $\forall\, n \in \mathbb{Z}$, $\left\lfloor \frac{n}{2} \right\rfloor + 1 = \left\lceil \frac{n+1}{2} \right\rceil$.
For each $x \in \mathbb{R}$, define the integer **round**(x) according to the following rule for rounding. Write $x = n + f$, where $n \in \mathbb{Z}$ and $0 \leq f < 1$, and define
$$\text{round}(x) = \begin{cases} n & \text{if } 0 \leq f < \frac{1}{2}, \\ n + 1 & \text{if } \frac{1}{2} \leq f < 1. \end{cases}$$

57. Express round(x) in terms of the floor function. Since programming languages are more likely to provide a

flooring function than a rounding function, this ability is useful.

58. Express round(x) in terms of the ceiling function. Note that Exercise 54 deals with expressing the ceiling in terms of the floor.

59*. Show: There are no integers in $(0, 1)$. Hint: By the Well-Ordering Principle, there is a smallest positive integer. If it is not 1, then consider its square.

60*. Use the Archimedean Principle from Appendix A to show: $\forall\, x \in \mathbb{R}, \exists\, n \in \mathbb{Z}$ such that $n < x$. Hint: Consider $-x$.

61*. Prove the uniqueness of the value $\lfloor x \rfloor$ specified in Theorem 3.8. Hint: Suppose $n_1, n_2 \in \mathbb{Z}$ with $n_1 \leq x < n_1 + 1$ and $n_2 \leq x < n_2 + 1$. Through subtraction, discover that $-1 < n_1 - n_2 < 1$, and appeal to Exercise 59.

62*. Prove the existence of the value $\lfloor x \rfloor$ specified in Theorem 3.8. Hint: Use the Archimedean Principle from Appendix A and its analog in Exercise 60 to get $a, b \in \mathbb{Z}$ with $a < x < b$. Apply the Generalized Maximum Principle from Exercise 28 to the set $S = \{s\ :\ s \leq x\}$.

63*. Show: $\forall\, n \in \mathbb{Z}^+, \exists\, a \in \mathbb{N}$ and $b \in \mathbb{Z}^+$ with b odd such that $n = 2^a b$. Hint: Apply the Well-Ordering Principle to the set $S = \{s\ :\ n = 2^r s$ where $r \in \mathbb{N}$ and $s \in \mathbb{Z}^+\}$.

64*. Let $a, b \in \mathbb{Z}$ with $b > 0$. Without using the Archimedean Principle from Appendix A, prove: $\exists\, n \in \mathbb{Z}$ such that $n > \frac{a}{b}$. Hint: Apply the Well-Ordering Principle to the set $S = \{s\ :\ s = a - bn \geq 0$ and $n \in \mathbb{Z}^+\}$.

65*. Show that the integer d described in Definition 3.5 does indeed exist. Hint: Apply the Generalized Maximum Principle from Exercise 28.

66*. Show that the integer d described in Definition 3.5 is indeed unique.

67. The manufacturer and the particular product determine the first 11 digits of a UPC number. The value of the check digit is then forced. Compute the missing check digit # in the UPC number for Colgate Toothpaste, 0 35000 74126 #.

68. A cashier is unable to read one of the digits in the UPC number for a bag of Tostitos Tortilla Chips, 0 2#400 00932 4. However, since the check digit is present, that missing digit is determined by the remaining eleven digits. Find the value of that missing digit #.

69. Suppose a cashier mistakenly enters the UPC number for Corn Chex, 0 16000 81160 7, as 0 61000 81160 7. This is a common type of error, called a transposition error, in

which two adjacent digits (16) have been reversed (to 61). Since the register is programmed to verify that the check digit is consistent with the first 11 digits, will the error be detected?

70. Since a UPC number is typically scanned as a bar code, one might expect that scanning the bar code backward would lead to a scanning error. For example, determine if an error would be detected if the UPC number for Gerber Baby Food Carrots, 0 15000 00411 8, is erroneously entered as 8 11400 00015 0. In fact, the error of reversing all of the digits is not likely to occur, since the way in which the first six digits is encoded as bars is different from that for the last six digits. Consequently, scanners can tell the difference between frontward and backward.

71. At a used book store, the cashier notices that one of the digits in the ISBN for *To Kill a Mockingbird* by Harper Lee, 0-446#1078-6, is unreadable. However, from the nine readable digits, she can determine the value of the unreadable digit #. Determine that digit.

72. The publisher, author, and title of a book determine the first nine digits of its ISBN. The value of the check digit is then forced. Compute the missing check digit # in the ISBN for *Fahrenheit 451* by Ray Bradbury, 0-34534296-#.

73. During the holiday rush, as busy cashier accidentally enters the ISBN for *Hop on Pop* by Dr. Seuss, 0-39480029-X, as 0-39480079-6. Since the register is programmed to verify the validity of check digits, will the error be detected? (Note that entering two incorrect digits, as considered here, is an unlikely mistake.)

74. Suppose that a transposition error is made, and the ISBN for *Harry Potter and the Sorcerer's Stone* by J. K. Rowling, 0-59035340-3, is erroneously entered as 0-95035340-3. That is, the two adjacent digits 59 are reversed to 95. Will this error be detected?

75.* Show that if two adjacent digits in a UPC number are switched, then the error will be detected if and only if the digits do not differ by 5. (See Exercise 69.) That is, all but some very special transposition errors will be detected.

76.* Show that if two adjacent digits in an ISBN are switched, then the error will always be detected. (See Exercise 74.) That is, 100% of transposition errors will be detected.

77. A hospital database needs to encode the gender and blood type of each of its patients. First, such data are converted to a four-digit binary message $b_1b_2b_3b_4$, according to Table 3.3. Consider the binary linear code that converts

TABLE 3.3

Encoding gender and blood type

Gender, Blood	Message	Gender, Blood	Message
Female, O^-	0000	Male, O^-	1000
Female, O^+	0001	Male, O^+	1001
Female, B^-	0010	Male, B^-	1010
Female, B^+	0011	Male, B^+	1011
Female, A^-	0100	Male, A^-	1100
Female, A^+	0101	Male, A^+	1101
Female, AB^-	0110	Male, AB^-	1110
Female, AB^+	0111	Male, AB^+	1111

each four-digit message $b_1b_2b_3b_4$ into a seven-digit code word $b_1b_2b_3b_4b_5b_6b_7$ according to the following formulas, using sums of the message digits.

$$b_5 = (b_1 + b_2) \bmod 2$$

$$b_6 = (b_2 + b_3) \bmod 2$$

$$b_7 = (b_3 + b_4) \bmod 2$$

(a) To what code word is "Female, AB^-" converted?

(b) Make the table for this entire code.

(c) What is the weight of this code?

(d) To what gender and blood type should the code word 0101111 be decoded?

(e) Give an example of a seven-digit word that is not a code word from the table and that has more than one code word tied for nearest to it. Such a word cannot be decoded by the nearest neighbor method.

78. The army needs to communicate directions to its troops in the field over communication links that sometimes induce errors. Each compass direction is associated with a four-digit binary message $b_1b_2b_3b_4$, according to Figure 3.4. Consider the binary linear code that converts each four-digit message into a seven-digit code word $b_1b_2b_3b_4b_5b_6b_7$ according to the following formulas, using sums of the message digits.

$$b_5 = (b_1 + b_2 + b_3) \bmod 2$$

$$b_6 = (b_1 + b_2 + b_4) \bmod 2$$

$$b_7 = (b_1 + b_3 + b_4) \bmod 2$$

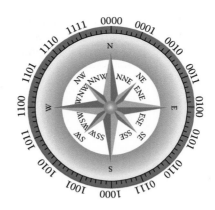

FIGURE 3.4
Direction encoding.

(a) Determine the code word for the south-west direction SW.

(b) Make the table for this entire code.

(c) What is the weight of this code?

(d) Decode the code word 1110100 to a compass direction.

(e) Use the nearest neighbor method to decode the intended compass direction, if the string 0100101 is received.

In Exercises 79 through 82, use the letter-to-number conversions from Table 3.2 (repeated below) and a shift cipher with $n = 27$ and the specified encrypting shift value b. Using software such as *Mathematica* or *Maple* (to evaluate functions on lists of input) might also be wise.

A	B	C	D	E	F	G	H
0	1	2	3	4	5	6	7	8

I	J	K	L	M	N	O	P	Q
9	10	11	12	13	14	15	16	17

R	S	T	U	V	W	X	Y	Z
18	19	20	21	22	23	24	25	26

79. A certain college uses a shift cipher with $b = 8$ to encrypt student course registration data. Determine how the course "DISCRETE MATH" would be encrypted at this college.

80. To protect against hackers, a certain Internet service provider encrypts its customers' passwords using a shift cipher with $b = 3$. Determine how the customer password "BINARY CODE" would be encrypted.

81. In December 2001, Sam Waksal, the CEO of the company ImClone, could have sent the encrypted message

"GT OXAR CBT" to his friend and ImClone stock-holder Martha Stewart. Assuming that Waksal used a shift cipher with $b = 15$, decrypt that message. Note that there are two spaces before the letter O in that message.

82. In 1948, the *Chicago Tribune* could have used a shift cipher to receive encrypted reports from its reporters in the field. Assuming that a reporter was using a shift cipher with $b = 2$, decrypt the message "FGYG BYKPU" that this reporter sent.

3.3   Euclid's Algorithm and Lemma

One of the greatest early contributors to number theory (and mathematics in general) was the Greek mathematician Euclid of Alexandria (fl. 300 B.C.E.). Although his famous work *Elements* is perhaps better known for its contribution to geometry, it also includes important theorems in number theory. For example, it contains the clever argument used in our proof that there are infinitely many primes (Theorem 3.6). This section focuses on two important results in Book VII of Euclid's *Elements*. The first is an algorithm that is useful for finding greatest common divisors and more. The second is a technical lemma that is fundamental to results on factoring.

Before we discuss Euclid's Algorithm for finding gcd's, we should be aware of a very useful form in which greatest common divisors can be expressed. Note that the form on the right-hand side of (3.4) is called a **linear combination** of m and n.

Theorem 3.13. (Expressing   the   GCD   as   a   Linear Combination).
Given integers m and n not both zero, there exist integers x and y such that

$$\gcd(m, n) = mx + ny. \tag{3.4}$$

Proof. Let $S = \{t : t = mu + nv \text{ for some } u, v \in \mathbb{Z} \text{ with } mu + nv > 0\}$.

Since $m \cdot m + n \cdot n > 0$, S is nonempty. The Well-Ordering Principle gives us a smallest element d of S. So, $d = mx + ny$ for some $x, y \in \mathbb{Z}$, and $d > 0$. Our aim is to show that $d = \gcd(m, n)$. We first show that $d \mid m$.

By the Division Algorithm, there are integers q and r such that $m = dq + r$ and $0 \leq r < d$. Observe that

$$r = m - dq = m - (mx + ny)q = m(1 - qx) + n(-qy).$$

If it were the case that $r > 0$, then $r \in S$ and r is smaller than d, the smallest element of S. Hence, it must be that $r = 0$. That is, $m = dq$ and $d \mid m$. A similar argument shows that $d \mid n$.

It remains to show that d is the *greatest* among common divisors. So suppose that c is an integer such that $c \mid m$ and $c \mid n$. Hence, $m = ac$ and $n = bc$ for some $a, b \in \mathbb{Z}$. It follows that

$$d = mx + ny = acx + bcy = (ax + by)c.$$

So $c \mid d$. From Theorem 3.1 it follows that $c \leq d$. Therefore, $\gcd(m, n) = d = mx + ny$.   □

The proof of the following straightforward consequence of Theorem 3.13 is left to the exercises.

Corollary 3.14.   Two integers m and n are relatively prime if and only if there exist integers x and y such that $mx + ny = 1$.

The results in Theorem 3.13 and Corollary 3.14 are illustrated in the following example.

Example 3.22.

(a) Observe that $\gcd(18, 30) = 6$ and $6 = 18(-3) + 30(2)$.
    So the gcd of 18 and 30 can be written as a linear combination of 18 and 30.

(b) Observe that 14 and 9 are relatively prime and $14(2) + 9(-3) = 1$.
    So there is a linear combination of the relatively prime integers 14 and 9 that gives 1.

It is straightforward to verify the linear combinations presented in Example 3.22, but how can we find them?

3.3.1 Euclid's Algorithm

Recall that an algorithm is a finite list of simple instructions leading to a desired result. There does exist a method, known as Euclid's Algorithm, for finding not only greatest common divisors but also the integers x and y promised in Theorem 3.13. However, to see how this algorithm works, we initially focus our attention on the computation of the greatest common divisor.

First observe that since $\gcd(m, n) = \gcd(|m|, |n|)$, it suffices to have a method for computing $\gcd(m, n)$ when m and n are both nonnegative. The critical result exploited by Euclid is an application of the Division Algorithm.

Theorem 3.15.   (GCD Reduction).
Let n and m be integers such that $n \geq m > 0$. Write $n = mq + r$ where $q, r \in \mathbb{Z}$ with $0 \leq r < m$. Then,

$$\gcd(n, m) = \gcd(m, r).$$

Proof. The fact that $\gcd(n, m) = \gcd(m, r)$ is an immediate consequence of the fact, which we now prove, that the pairs (n, m) and (m, r) have the same set of common divisors.

($\rightarrow$) Suppose that c is a common divisor of n and m. That is, $n = ck$ and $m = cl$ for some integers k and l. Observe that

$$r = n - mq = ck - clq = c(k - lq).$$

Thus, $c \mid r$ as well as $c \mid m$. That is, c is a common divisor of m and r.

($\leftarrow$) Suppose that c is a common divisor of m and r. That is, $m = ck$ and $r = cl$ for some integers k and l. Observe that

$$n = mq + r = ckq + cl = c(kq + l).$$

Thus, $c \mid n$ as well as $c \mid m$. That is, c is a common divisor of n and m. $\square$

The point of Theorem 3.15 is that it allows us to exchange a calculation of the gcd of the pair (n, m) for a calculation of the gcd of the smaller pair (m, r). By a sequence of such reductions, we can hope to reduce a general problem to a much smaller problem for which the solution is easy. In particular, these reductions will eventually result in a pair in which one of the integers is 0. At that point, the result from Example 3.8 completes the calculation.

If k is a positive integer, then $\gcd(k, 0) = k$.

We illustrate Euclid's Algorithm in some examples.

Example 3.23.   Compute $\gcd(48, 18)$.

Solution. We repeatedly apply Theorem 3.15.

$$
\begin{aligned}
\gcd(48, 18) &= \gcd(18, 12) && \text{since } 48 = (18)2 + 12 \\
&= \gcd(12, 6) && \text{since } 18 = (12)1 + 6 \\
&= \gcd(6, 0) && \text{since } 12 = (6)2 + 0 \\
&= 6 && \text{by Example 3.8.}
\end{aligned}
$$

$\square$

Algorithm 3.1   Euclid's Algorithm for finding $\gcd(n, m)$

Let $n, m \in \mathbb{Z}^+$ with $n \geq m$.

Algorithm.
>   While $m > 0$,
>>   `\begin`
>>   Let $r = n \bmod m$.
>>   Let $n = m$.
>>   Let $m = r$.
>>   `\end`.
>   Return n.

The basic procedure of Euclid's Algorithm is shown in Algorithm 3.1.[*] The French mathematician Gabriel Lamé (1795–1870) proved that it requires at most $5m$ steps to compute $\gcd(n, m)$, when $n \geq m > 0$. A further consequence of Euclid's Algorithm is that it enables us to find integer coefficients with which we can write $\gcd(n, m)$ as a linear combination of n and m, as we illustrate in the next two examples.

Example 3.24.   Find $\gcd(630, 96)$ and write it in the form $630x + 96y$ for $x, y \in \mathbb{Z}$.

Solution.   The initial steps are the same as those used in Example 3.23.

$$
\begin{aligned}
\gcd(630, 96) &= \gcd(96, 54) && \text{since } 630 = (96)6 + 54 \\
&= \gcd(54, 42) && \text{since } 96 = (54)1 + 42 \\
&= \gcd(42, 12) && \text{since } 54 = (42)1 + 12 \\
&= \gcd(12, 6) && \text{since } 42 = (12)3 + 6 \\
&= \gcd(6, 0) && \text{since } 12 = (6)2 + 0 \\
&= 6 && \text{by Example 3.8.}
\end{aligned}
$$

With some additional work, these calculations can be used to find x and y. Notice that the greatest common divisor 6 is the last nonzero remainder occurring in our calculations. In each row with a nonzero remainder, we can solve for that remainder as follows.

$$54 = 630 - (96)6$$

$$42 = 96 - (54)1$$

$$12 = 54 - (42)1$$

$$6 = 42 - (12)3.$$

---

[*] For an explanation of the pseudocode used to present algorithms throughout this book, see Appendix B.

Substituting backward from the last row we then get

$6 = 42 - (12)3$

$\quad = 42 - (54 - 1(42))3 = (54)(-3) + (42)(4)$

$\quad = (54)(-3) + (96 - 1(54))(4) = (96)(4) + (54)(-7)$

$\quad = (96)(4) + (630 - 6(96))(-7) = (630)(-7) + (96)(46).$

That is, $\gcd(630, 96) = 6 = 630x + 96y$ for $x = -7$ and $y = 46$. $\qquad\square$

Example 3.25.   Find $\gcd(102, 30)$ and write it in the form $102x + 30y$ for $x, y \in \mathbb{Z}$.

Solution.

$$\begin{aligned}
\gcd(102, 30) &= \gcd(30, 12) &&\text{since } 102 = (30)3 + 12, \text{ so } 12 = 102 - (30)3 \\
&= \gcd(12, 6) &&\text{since } 30 = (12)2 + 6, \quad \text{ so } 6 = 30 - (12)2 \\
&= \gcd(6, 0) &&\text{since } 12 = (6)2 + 0 \\
&= 6 &&\text{by Example 3.8.}
\end{aligned}$$

Therefore,

$$6 = 30 - (12)2 = 30 - (102 - 3(30))2$$

$$= (102)(-2) + (30)(7).$$

That is, $\gcd(102, 30) = 6 = 102x + 30y$ for $x = -2$ and $y = 7$. $\qquad\square$

The procedure used in Examples 3.24 and 3.25 to find d, x, and y so that $d = \gcd(m, n) = mx + ny$ is called the **Extended Euclidean Algorithm**. *Mathematica* and *Maple* provide the functions `ExtendedGCD` and `igcdex`, respectively, for performing this task.

3.3.2   Lemmas on factoring

We close this section with some important results on factoring that are needed in subsequent sections. They all stem from a fundamental result known as Euclid's Lemma. The common theme is a determination of conditions under which an integer dividing a product implies that the integer divides one of the factors in the product. For example, $6 \mid (9 \cdot 10)$, but $6 \nmid 9$ and $6 \nmid 10$. Whereas, $5 \mid (9 \cdot 10)$, and, although $5 \nmid 9$, it is true that $5 \mid 10$.

Theorem 3.16.   (Euclid's Lemma).
Let $m, n,$ and c be integers.
If $c \mid mn$ and $\gcd(c, m) = 1$, then $c \mid n$.

Proof. Suppose $c \mid mn$ and $\gcd(c,m) = 1$. So $mn = ck$ for some $k \in \mathbb{Z}$. By Corollary 3.14, there are $x, y \in \mathbb{Z}$ such that $cx + my = 1$. Observe that

$$n = (cx + my)n = cnx + mny = cnx + cky = c(nx + ky).$$

Therefore, $c \mid n$. $\square$

Corollary 3.17.   Let m, n, and p be integers with p prime. If $p \mid mn$, then $p \mid m$ or $p \mid n$.

The proof of Corollary 3.17 is left to the exercises. The next Corollary naturally lends itself to a proof by induction. After induction is introduced in Section 4.3, a proof will be requested in those exercises.

Corollary 3.18.   Let $m_1, m_2, \ldots, m_n$, and p be integers with p prime. If $p \mid (m_1 \cdot m_2 \cdots m_n)$, then $p \mid m_i$ for some $1 \leq i \leq n$.

Corollary 3.19.   Let m, n, and p be integers with $n > 0$ and p prime. If $p \mid m^n$, then $p \mid m$.

In the proof in our final example, we apply Corollary 3.19. An alternative proof is outlined in the exercises in Section 4.5.

Example 3.26.   Let a, b, and n be integers with $n > 0$. Show: If $\gcd(a,b) = 1$, then $\gcd(a, b^n) = 1$.

Proof. Suppose $\gcd(a,b) = 1$. Let $d = \gcd(a, b^n)$. Suppose toward a contradiction that $d > 1$. By Theorem 3.5, we have a prime p such that $p \mid d$. So $p \mid a$ and $p \mid b^n$. By Corollary 3.19, $p \mid b$. So $p \mid \gcd(a,b)$. That is, $p \mid 1$, which is impossible since p is prime. Hence, $d = 1$. $\square$

Exercises

In Exercises 1 through 4, through trial and error find integers x and y that satisfy the given equation.

1. $\gcd(65, 25) = 65x + 25y$.

2. $15x + 11y = 1$.

3. $\gcd(39, 9) = 39x + 9y$.

4. $27x + 8y = 1$.

In Exercises 5 through 20, use Euclid's Algorithm.

5. Find $\gcd(60, 24)$.   6. Find $\gcd(16, 14)$.

7. Find $\gcd(110, 44)$.   8. Find $\gcd(1452, 252)$.

9. Find $\gcd(296, 112)$.   10. Find $\gcd(204, 30)$.

11. Find $\gcd(63, 25)$ and keep a running table of the values of n and m in Algorithm 3.1.

12. Find $\gcd(70, 27)$ and keep a running table of the values of n and m in Algorithm 3.1.

13. Find $\gcd(14, 8)$ and write it in the form $14x + 8y$ for some $x, y \in \mathbb{Z}$.

14. Find $\gcd(56, 42)$ and write it in the form $56x + 42y$ for some $x, y \in \mathbb{Z}$.

15. Find $\gcd(50, 35)$ and write it in the form $50x + 35y$ for some $x, y \in \mathbb{Z}$.

16. Find $\gcd(108, 45)$ and write it in the form $108x + 45y$ for some $x, y \in \mathbb{Z}$.

17. Find $\gcd(81, 60)$ and write it in the form $81x + 60y$ for some $x, y \in \mathbb{Z}$.

18. Find $\gcd(259, 77)$ and write it in the form $259x + 77y$ for some $x, y \in \mathbb{Z}$.

19. Find $x, y \in \mathbb{Z}$ such that $55x + 12y = 1$.

20. Find $x, y \in \mathbb{Z}$ such that $45x + 14y = 1$.

21. Are the integers x and y in Theorem 3.13 unique? Why?

22. Prove Corollary 3.14. Remember that there are two directions to prove.

23.* Since $\gcd(10, 6) = 2$, Theorem 3.13 guarantees that there are $x, y \in \mathbb{Z}$ such that $10x + 6y = 2$. In fact,

$$10(-1) + 6(2) = 2, \ 10(5) + 6(-8) = 2,$$

$$10(11) + 6(-18) = 2, \ \ldots .$$

Since, for any $k \in \mathbb{Z}$, we have $10(-1 + 6k) + 6(2 - 10k) = 2$, observe that there are infinitely many choices for the pair x, y in this case. In general, let m and n be any pair of nonzero integers. Show that there are infinitely many pairs of integers x and y such that $mx + ny = \gcd(m, n)$.

24.* In Example 3.24, we saw that $\gcd(630, 96) = 6$ and, moreover, that $6 = 630x + 96y$ for $x = -7$ and $y = 46$. Observe that these integers guaranteed by Theorem 3.13 are relatively prime in this example. That is, $\gcd(-7,46) = 1$. In general, for any pair of nonzero integers m and n, prove that the integers x and y in Theorem 3.13 are relatively prime.

25. In Corollary 3.17, if p is not required to be prime, is the statement still true? Why?

26. Prove Corollary 3.17.

27. Prove Corollary 3.19 by using Corollary 3.18.

28. Prove the converse of the statement in Example 3.26.

29.* Observe that, in general, $\gcd(a,b) \neq \gcd(a^m, b^n)$, since, for example, $\gcd(42,18) = 6$ and $\gcd(42^2, 18^3) = 36$. However, let a and b be integers not both zero, let m and n be positive integers, and show that $\gcd(a,b)$ and $\gcd(a^m, b^n)$ have the same prime divisors.

30.* Observe that $2^4 \mid (40 \cdot 6)$, and although $2^4 \nmid 40$ and $2^4 \nmid 6$, we do have $2^3 \mid 40$ and $2^1 \mid 6$ (note $2^3 2^1 = 2^4$). In general, let $m,n,p,k \in \mathbb{Z}$ with p prime and $k \geq 0$. Suppose $p^k \mid mn$. Show: There are $a,b \in \mathbb{N}$ such that $a + b = k$, $p^a \mid m$, and $p^b \mid n$. Hint: Use Exercise 28 of Section 3.2 to pick a as large as possible so that $p^a \mid m$. Apply Euclid's Lemma to the fact that $p^{k-a} \mid (\frac{m}{p^a} n)$.

31. Let m and n be integers with $m \neq 0$. Show: $\gcd(m,n) = \gcd(m, n - m)$.

32. Explain why a corollary of the proof of Theorem 3.13 is the fact that for integers m and n not both zero,

$$\gcd(m,n) = \min\{t : t = mu + nv \text{ for some}$$

$$u,v \in \mathbb{Z} \text{ with } mu + nv > 0\}.$$

Use the fact in Exercise 32 to prove the statements in Exercises 33 through 35.

33. Show: For all $m,n \in \mathbb{Z}$ not both zero, $\gcd(m,n) = \gcd(n,m)$.

34. Show: For all $m,n \in \mathbb{Z}$ not both zero, $\gcd(m,-n) = \gcd(m,n)$.

35. Show: For all $k \in \mathbb{Z}^+$, $\gcd(k,0) = k$.

36. Prove Theorem 3.15.

37. Given $m,n \in \mathbb{Z}^+$, use Theorem 3.13 to show: $\forall c \in \mathbb{Z}$, if $c \mid m$ and $c \mid n$, then $c \mid \gcd(m,n)$.

38. Let $d = \gcd(m,n)$. Show: $\exists a,b \in \mathbb{Z}$ with $\gcd(a,b) = 1$ such that $m = ad$ and $n = bd$.

39. Show that for any integer n, the integers $5n + 3$ and $7n + 4$ are relatively prime.

40. Show that for any integer n, the integers $5 - 2n$ and $n - 3$ are relatively prime.

41. A matrix is a rectangular array of real numbers. The **determinant** of a 2 by 2 matrix $\begin{bmatrix} a & b \\ c & d \end{bmatrix}$ is defined to be the value $ad - bc$. The absolute value of this determinant turns out to be the area of the parallelogram in the Cartesian plane with vertices $(0,0)$, (a,b), (c,d), and $(a+c, b+d)$.

   (a) Confirm that the parallelogram with vertices $(0,0)$, $(3,5)$, $(1,2)$, and $(4,7)$ has area 1.

   (b) Show that if a 2 by 2 matrix with integer entries has determinant 1, then, in each row, the two integers must be relatively prime. The same will also be true for the columns.

   (c) Is the converse true?

42. If a vending machine only dispenses stamps worth 6¢ and 15¢, then the monetary amounts $1, \ldots, 5$, $7, \ldots, 11, 13, 14, 16, 17, 19, 20, \ldots$ cannot be achieved with stamps from this machine. In general, suppose a vending machine dispenses stamps worth a¢ and b¢.

   (a) Show that if a and b are not relatively prime, then there will be infinitely many monetary amounts that cannot be achieved with stamps from this machine.

   (b) If stamps worth 4¢ were also available, then what monetary amounts would be unachievable?

3.4 Rational and irrational numbers

3.4.1 Rational numbers

Although the primary focus of number theory is on integers, some of its considerations are relevant to a study of fractions of integers. By expanding our universe of consideration, we also gain further opportunities to practice the proof techniques we have learned.

Definition 3.10. A real number r is said to be **rational** if $r = \frac{a}{b}$ for some integers a and b with $b \neq 0$.

The set of rational numbers is denoted by $\mathbb{Q}$. Very simply, rational numbers are real numbers that *can* be expressed as a fraction of two integers. In fact, every integer is a rational number.

Theorem 3.20. $\mathbb{Z} \subseteq \mathbb{Q}$.

Proof. Suppose n is an arbitrary integer.
Observe that $n = \frac{n}{1}$ and that $n, 1 \in \mathbb{Z}$ with $1 \neq 0$.
Thus, $n \in \mathbb{Q}$. $\qquad\qquad\square$

Example 3.27.   (Recognizing Rational Numbers).

(a) $\frac{5}{7}$ is rational, since $5, 7 \in \mathbb{Z}$ with $7 \neq 0$.
Similarly, $\frac{-2}{3}, \frac{13}{5}$, and $\frac{22}{-9}$ are seen to be rational.

(b) $14\frac{1}{2}$ is rational, since it equals $\frac{29}{2}$.

(c) Real numbers with finite decimal expansions are rational.
For example, $5.604 = 5 + \frac{604}{1000} = \frac{5604}{1000} = \frac{1401}{250}$.

(d) Real numbers with repeating decimal expansions are rational.
For example, $27.\overline{531}$ (which means $27.531531\ldots$) is rational.
Let $x = 27.531531\ldots$.
So, $1000x = 27531.531531\ldots$.
Hence, $999x = 1000x - x = 27504$.
Therefore, $x = \frac{27504}{999} = \frac{3056}{111}$ is rational.

(e) $0.3\overline{5826}$ is rational.
Let $x = 0.3\overline{5826} = 0.35826826\ldots$.
So, $100000x = 35826.\overline{826}$,
and $100x = 35.\overline{826}$.
Hence, $99900x = 100000x - 100x = 35826 - 35 = 35791$.
Therefore, $x = \frac{35791}{99900}$ is rational.

The first eleven properties of $\mathbb{Z}$ and $\mathbb{R}$ listed in Appendix A also hold for $\mathbb{Q}$. In most cases, this is a simple consequence of the fact that rational numbers are special kinds of real numbers. However, some properties do need to be verified separately for $\mathbb{Q}$. These are gathered in the following theorem.

Theorem 3.21.   (Field Properties of $\mathbb{Q}$).
Let $r, s \in \mathbb{Q}$. Then

(a) $0, 1 \in \mathbb{Q}$.

(b) $r + s \in \mathbb{Q}$.

(c) $-s \in \mathbb{Q}$.

(d) $rs \in \mathbb{Q}$.

(e) *if $s \neq 0$, then $\frac{1}{s} \in \mathbb{Q}$.*

Part (a) of Theorem 3.21 follows from Theorem 3.20. We prove parts (b) and (e) here and leave (c) and (d) for the exercises. Each proof is a straightforward illustration of our strategy of letting the definitions of the terms involved dictate our actions.

Proof of Theorem 3.21(b) Since $r, s \in \mathbb{Q}$, we can write $r = \frac{a}{b}$ and $s = \frac{c}{d}$ for some $a, b, c, d \in \mathbb{Z}$ with $b \neq 0$ and $d \neq 0$. Observe that

$$r + s = \frac{a}{b} + \frac{c}{d} = \frac{ad}{bd} + \frac{bc}{bd} = \frac{ad + bc}{bd}.$$

Since $ad + bc$ and bd are integers with $bd \neq 0$, we see that $r + s$ is rational. $\square$

Proof of Theorem 3.21(e) Suppose $s \neq 0$.
Since $s \in \mathbb{Q}$, we can write $s = \frac{a}{b}$ for some $a, b \in \mathbb{Z}$ with $b \neq 0$.
Since $s \neq 0$, it follows that $a \neq 0$.
Observe that (because $\frac{a}{b} \frac{b}{a} = \frac{ab}{ab} = 1$)

$$\frac{1}{s} = \frac{1}{\frac{a}{b}} = \frac{b}{a}.$$

Since b and a are integers with $a \neq 0$, we see that $\frac{1}{s}$ is rational. $\square$

When working with rational numbers, it is often most convenient to express them in lowest terms. That is, we want to cancel out any common factors in the numerator and denominator. For example, $\frac{12}{18} = \frac{2}{3}$ and $\frac{35}{-14} = \frac{-5}{2}$. In general, $\frac{n}{m}$ is completely reduced by canceling $\gcd(m, n)$ from n and m. Consequently, every rational number can be uniquely expressed in lowest terms (with a positive denominator). The following theorem makes an important contribution in the next section, when we first prove the existence of a real number that is not rational.

Theorem 3.22. (Expressing Rational Numbers in Lowest Terms).
*Given $r \in \mathbb{Q}$, there exist unique $a, b \in \mathbb{Z}$ such that $b > 0$, $\gcd(a, b) = 1$, and $r = \frac{a}{b}$. That is, $\frac{a}{b}$ expresses r in **lowest terms**.*

Proof. Write $r = \frac{u}{v}$ for $u, v \in \mathbb{Z}$ with $v \neq 0$. Let

$$S = \{m : m \in \mathbb{Z}, m > 0, \text{ and } r = \frac{n}{m} \text{ for some } n \in \mathbb{Z}\}.$$

If $v > 0$, then $v \in S$. If $v < 0$, then $-v \in S$, since $r = \frac{-u}{-v}$. Hence, S is nonempty. By the Well-Ordering Principle, S has a smallest element b. Thus, $b \in \mathbb{Z}, b > 0$, and $r = \frac{a}{b}$ for some $a \in \mathbb{Z}$.

Let $d = \gcd(a, b)$. So $a = dk$ and $b = dl$ for some $k, l \in \mathbb{Z}$. Since $b, d > 0$, it must be that $l > 0$. Since $\frac{a}{b} = \frac{dk}{dl} = \frac{k}{l}$, it follows that $l \in S$. Since $l \mid b$ and $b > 0$, Theorem 3.1 tells us that $l \leq b$. Since $l \in S$ and b is the smallest element of S, it must be that $l = b$. Hence, $1 = d = \gcd(a, b)$.

To show uniqueness, suppose $\frac{a}{b} = \frac{e}{f}$, where $e, f \in \mathbb{Z}$ with $f > 0$ and $\gcd(e, f) = 1$. Thus $af = be$, so $f \mid be$. Since $\gcd(e, f) = 1$, Euclid's lemma tells us that $f \mid b$. Theorem 3.1 then tells us that $f \leq b$. Similarly, we get $b \leq f$, so $f = b$. It follows that $e = a$ as well. $\square$

3.4.1.1 Decimal form of rational numbers

We often express real numbers in **decimal form**, which is a natural extension of the base-ten representation of integers, as discussed in Introduction. For example,

$$427.3528 = 4 \cdot 10^2 + 2 \cdot 10^1 + 7 \cdot 10^0 + 3 \cdot 10^{-1}$$

$$+ 5 \cdot 10^{-2} + 2 \cdot 10^{-3} + 8 \cdot 10^{-4}.$$

More generally, for $b_i, a_j \in \{0, 1, \ldots, 9\}$, we have

$$b_m b_{m-1} \cdots b_0 . a_1 a_2 \cdots a_n = b_m 10^m + b_{m-1} 10^{m-1}$$

$$+ \cdots + b_0 + \frac{a_1}{10} + \frac{a_2}{100} + \cdots + \frac{a_n}{10^n}.$$

We call $b_m b_{m-1} \cdots b_0$ the **integer part** and $0.a_1 a_2 \cdots a_n$ the **decimal part**. In fact, the decimal part could contain infinitely many digits:

$$0.a_1 a_2 \cdots a_n a_{n+1} \cdots .$$

In that case, it specifies the value

$$\frac{a_1}{10} + \frac{a_2}{100} + \cdots + \frac{a_n}{10^n} + \frac{a_{n+1}}{10^{n+1}} + \cdots,$$

which is an infinite sum that can be proven to lie in the interval $(0, 1]$. Repeating infinite decimal parts

$$0.a_1 a_2 \cdots a_m \overline{a_{m+1} \cdots a_n},$$

as well as a method for computing their values were encountered in Example 3.27. Moreover, from the results seen there, we observe that any real number that can be written in decimal form with either a finite or repeating decimal part represents a rational number. In fact, the converse is also true.

Theorem 3.23.   (When Decimals Are Rational).
A real number written in decimal form represents a rational number if and only if the decimal part is either finite or repeating. Moreover, a rational number $r = \frac{a}{b}$ written in lowest terms has a finite decimal expansion if and only if 2 and/or 5 are the only prime divisors of b. Otherwise, the decimal part of r repeats.

Sketch of proof. Since we saw in Example 3.27 how to write numbers with finite or repeating decimals as rational numbers, we focus only on showing that rational numbers can be written in decimal form with a decimal part that is either finite or repeating. Moreover, it suffices to consider rational numbers $\frac{a}{b}$ with $a, b \in \mathbb{Z}$ and $0 < a < b$. The process of writing a rational number in decimal form is illustrated in the following three examples. After that, the characterization of the divisors of b will be addressed.

Example 3.28.   (A Finite Decimal Expansion).

$$\frac{63}{160} = 0.39375$$

```
                .  3  9  3  7  5
1  6  0 )    6  3. 0
        −    4  8  0
             1  5  0  0                   remainder 150
        −    1  4  4  0
                6  0  0                   remainder 60
           −    4  8  0
                1  2  0  0                remainder 120
           −    1  1  2  0
                   8  0  0                remainder 80
              −    8  0  0
                      0                   remainder 0 ← END
```

The decimal expansion ends when a remainder of 0 is encountered.

Example 3.29.   (A Repeating Decimal Expansion).

$$\frac{389}{3700} = 0.10\overline{513}$$

```
                   .  1  0  5  1  3
3  7  0  0 )    3  8  9. 0
          −     3  7  0  0
                1  9  0  0                   remainder 190
          −              0
                1  9  0  0  0               remainder 1900
             −  1  8  5  0  0
                   5  0  0  0               remainder 500
                −  3  7  0  0
                   1  3  0  0  0            remainder 1300
                −  1  1  1  0  0
                   1  9  0  0  0            remainder 1900
```

The repeating portion is recognized as soon as a remainder is encountered for a second time.

In general, when dividing a by b, either a remainder of 0 will be encountered or some remainder will occur twice. This happens because each remainder is an integer in the interval $[0, b-1]$. Since there are only finitely many such integers, if a remainder of 0 is not encountered, then some remainder must get repeated after at most $b-1$ steps in the division. The following example illustrates the worst possible case.

Example 3.30. (A Fraction $\frac{a}{b}$ with $b-1$ Repeating Digits in Its Decimal Expansion).

$$\frac{2}{7} = 0.\overline{285714}$$

All of the nonzero remainders $1, 2, 3, 4, 5, 6$ are encountered before one gets repeated. Note that the numerator 2, which is being divided by 7, should be considered the first nonzero remainder.

For the final part of the statement in Theorem 3.23, we must show that our rational number $r = \frac{a}{b}$ in lowest terms has a finite decimal expansion if and only if 2 and/or 5 are the only prime divisors of b. If r has a finite decimal expansion

$$r = 0.a_1 a_2 \cdots a_n,$$

then

$$r = \frac{a_1}{10} + \frac{a_2}{100} + \cdots + \frac{a_n}{10^n} = \frac{a_1 a_2 \cdots a_n}{10^n} = \frac{a_1 a_2 \cdots a_n}{2^n 5^n}.$$

Note that $a_1 a_2 \cdots a_n$ is the base-ten representation of the integer $10^n r$. Reducing r to lowest terms would still leave only powers of 2 and/or 5 in the denominator.

Conversely, suppose that $r = \frac{a}{2^s 5^t}$ for some integers $s, t \geq 0$. If $s \geq t$, then

$$r = \frac{5^{s-t}a}{2^s 5^s} = \frac{5^{s-t}a}{10^s}.$$

If $s < t$, then

$$r = \frac{2^{t-s}a}{2^t 5^t} = \frac{2^{t-s}a}{10^t}.$$

Certainly, any fraction of the form $\frac{c}{10^n}$ for some positive integers c and n with $c < 10^n$ has a finite decimal expansion with at most n digits. For example, $\frac{372}{10^5} = 0.00372$.

3.4.2   Irrational numbers

Definition 3.11.   A real number that is not rational is said to be **irrational**.

Since irrational numbers are defined indirectly, statements about irrational numbers tend to be proved by indirect arguments. We are now ready for our first proof that a number is irrational.

Although the fact that $\sqrt{2}$ is irrational may not faze us in modern times, to those who followed the Greek mathematician Pythagoras (c. 580–500 B.C.E.), the discovery of this fact came as quite a shock. The Pythagoreans thought of numbers geometrically in terms of lengths of line segments. They believed that the ratio of the lengths of any two line segments could be expressed as a ratio of two integers. However, $\sqrt{2}$, which can be seen by the Pythagorean Theorem to measure the ratio of the length of the diagonal of a square to the length of its side, cannot be so expressed. The Pythagoreans were so disturbed by this affront to their beliefs that they initially sought to keep it a secret. However, word got out. A modern proof is presented here.

Theorem 3.24. $\sqrt{2}$ *is irrational.*

Proof. Suppose not.
So $\sqrt{2}$ is rational.
That is, $\sqrt{2} = \frac{a}{b}$ for some $a, b \in \mathbb{Z}$ with $b \neq 0$.
Moreover, by Theorem 3.22, we may choose it so that $\frac{a}{b}$ is in lowest terms.
In particular, $\gcd(a, b) = 1$.
We have $a = \sqrt{2}b$.
So $a^2 = 2b^2$.
That is, a^2 is even. (In other words, a^2 is divisible by 2.)

By Exercise 3 of Section 3.1, a must be even. (That is, a is divisible by 2.)

Therefore, $a = 2k$ for some $k \in \mathbb{Z}$.

It follows that $2b^2 = a^2 = 4k^2$.

So $b^2 = 2k^2$.

Since b^2 is even (divisible by 2), so is b (divisible by 2).

The fact that both a and b are divisible by 2 contradicts the fact that $\frac{a}{b}$ was supposed to be in lowest terms.

That is, $\gcd(a, b) \geq 2$ is impossible.   □

Knowing that $\sqrt{2}$ is irrational enables us to more easily show that some other numbers involving $\sqrt{2}$ are also irrational.

Example 3.31.   Show that $\frac{7+\sqrt{2}}{5}$ is irrational.

Proof. Suppose not.

So $r = \frac{7+\sqrt{2}}{5}$ is rational.

From basic algebra, we see that $\sqrt{2} = 5r - 7$.

However, $5r - 7 \in \mathbb{Q}$ and $\sqrt{2} \notin \mathbb{Q}$.

This is a contradiction.   □

Notice in Example 3.31 how we took advantage of Theorem 3.24. We did not have to work through the details of an argument like that given in the proof of Theorem 3.24. In fact, the proof in Example 3.31 did not even require that the rational number r be expressed as a fraction.

For our next example of an irrational number, we turn to **logarithms**. Given real numbers b, x, and y with $b > 1$ and $x > 0$, recall that $\log_b x = y$ is equivalent to $b^y = x$. This description suffices for our purposes here, and we shall consider logarithms in more depth in Section 5.4.

Example 3.32.   Show that $\log_3 5$ is irrational.

Proof. Suppose not.

So $\log_3 5$ is rational.

That is, $\log_3 5 = \frac{a}{b}$ for some $a, b \in \mathbb{Z}$ with $b \neq 0$.

Moreover, (since both a and b could be multiplied by -1 if necessary) we may assume that $b > 0$.

Since $\log_3 5 > 0$, it follows that $a > 0$.

By the definition of logarithms, $3^{\frac{a}{b}} = 5$.

Raising both sides of this equation to the power b gives $3^a = 5^b$.

Since $a > 0$, we see that $3 \mid 5^b$.

Corollary 3.19 then tells us that $3 \mid 5$.

This is a contradiction.   □

A **root** or **zero** of a real function $f(x)$ is a value r such that $f(r) = 0$. The following result about *rational* roots turns out to be a useful tool for showing that numbers are *irrational*.

Theorem 3.25.   (Rational Roots Theorem).
Let $n \in \mathbb{Z}^+$ and let

$$f(x) = c_n x^n + c_{n-1} x^{n-1} + \cdots + c_1 x + c_0$$

be a polynomial with integer coefficients $c_n, c_{n-1}, \ldots, c_1, c_0$ such that $c_n \neq 0$. If r is a rational root of f (that is, $r \in \mathbb{Q}$ and $f(r) = 0$), then $r = \frac{a}{b}$ for some $a, b \in \mathbb{Z}$ such that $a \mid c_0$ and $b \mid c_n$.

Proof. Suppose $f(r) = 0$ for some rational number r. By Theorem 3.22, we can choose $a, b \in \mathbb{Z}$ with $b > 0$ and $\gcd(a, b) = 1$ so that $r = \frac{a}{b}$ is written in lowest terms. The equation

$$c_n \frac{a^n}{b^n} + c_{n-1} \frac{a^{n-1}}{b^{n-1}} + \cdots + c_1 \frac{a}{b} + c_0 = 0$$

can be written as

$$c_n a^n + c_{n-1} b a^{n-1} + \cdots + c_1 b^{n-1} a + c_0 b^n = 0.$$

Since

$$c_0 b^n = -a(c_n a^{n-1} + c_{n-1} b a^{n-2} + \cdots + c_1 b^{n-1}),$$

it follows that $a \mid c_0 b^n$. Since $\gcd(a, b) = 1$, the result in Example 3.26 tells us that $\gcd(a, b^n) = 1$. Euclid's lemma then tells us that $a \mid c_0$. Similarly, since

$$c_n a^n = -b(c_{n-1} a^{n-1} + \cdots + c_1 b^{n-2} a + c_0 b^{n-1}),$$

it follows that $b \mid c_n a^n$ and therefore $b \mid c_n$. $\square$

The statement of the Rational Roots Theorem is perhaps best digested in an example.

Example 3.33.   Consider the polynomial

$$f(x) = 4x^4 + 3x^3 - 9x^2 - 6x + 2,$$

which clearly has integer coefficients. The divisors of the constant term 2 are

$$1, -1, 2, \text{ and } -2. \tag{3.6}$$

The divisors of the leading coefficient 4 are

$$1, -1, 2, -2, 4, \text{ and } -4. \tag{3.7}$$

The Rational Roots Theorem says that any rational root of f must be the value of a fraction whose numerator comes from list (3.6) and whose denominator comes from list (3.7).

$$\frac{1}{1}, \ \frac{1}{-1}, \ \frac{1}{2}, \ \frac{1}{-2}, \ \frac{1}{4}, \ \frac{1}{-4},$$

$$\frac{-1}{1}, \ \frac{-1}{-1}, \ \frac{-1}{2}, \ \frac{-1}{-2}, \ \frac{-1}{4}, \ \frac{-1}{-4},$$

$$\frac{2}{1}, \ \frac{2}{-1}, \ \frac{2}{2}, \ \frac{2}{-2}, \ \frac{2}{4}, \ \frac{2}{-4},$$

$$\frac{-2}{1}, \ \frac{-2}{-1}, \ \frac{-2}{2}, \ \frac{-2}{-2}, \ \frac{-2}{4}, \ \frac{-2}{-4}$$

Stripping out the obvious redundancies, the list of possible rational roots for f is

$$1, \ -1, \ \frac{1}{2}, \ -\frac{1}{2}, \ \frac{1}{4}, \ -\frac{1}{4}, \ 2, \ -2.$$

By simply plugging the numbers from this list into f, we can determine all of the rational roots for f.

$$f(1) = -6 \neq 0$$

$$f(-1) = 0$$

$$f\left(\frac{1}{2}\right) = -\frac{21}{8} \neq 0$$

$$f\left(-\frac{1}{2}\right) = \frac{21}{8} \neq 0$$

$$f\left(\frac{1}{4}\right) = 0$$

$$f\left(-\frac{1}{4}\right) = \frac{93}{32} \neq 0$$

$$f(2) = 42 \neq 0$$

$$f(-2) = 18 \neq 0$$

We see that -1 and $\frac{1}{4}$ are the only rational roots of f. Consequently, $(x+1)$ and $(4x-1)$ must be factors of $f(x)$. Therefore, long division could be used to discover that

$$f(x) = (x+1)(4x-1)(x^2 - 2).$$

In this case, the final term $x^2 - 2$ is easily factored as $(x - \sqrt{2})(x + \sqrt{2})$. From this, it follows that a complete factorization of $f(x)$ is given by

$$f(x) = (x + 1)(4x - 1)(x - \sqrt{2})(x + \sqrt{2}).$$

Hence, the roots of f are

$$-1, \quad \frac{1}{4}, \quad \sqrt{2}, \quad \text{and } -\sqrt{2}.$$

Since $\sqrt{2}$ and $-\sqrt{2}$ are not on the list of rational roots, they must be irrational roots. Of course, in this case, we already knew that $\sqrt{2}$ and $-\sqrt{2}$ are irrational.

We can now justify our inclusion of a theorem on rational roots in this section on irrational numbers. In Example 3.31, we showed that $r = \frac{7+\sqrt{2}}{5}$ is irrational. Basic algebra gave us the critical result that $\sqrt{2} = 5r - 7$. If we further, square both sides of this equation, then we get $2 = 25r^2 - 70r + 49$. Hence, $25r^2 - 70r + 47 = 0$. That is, r is a root of the polynomial $f(x) = 25x^2 - 70x + 47$. Since the Rational Roots Theorem tells us that $\pm 1, \pm 47, \pm \frac{1}{5}, \pm \frac{47}{5}, \pm \frac{1}{25}$, and $\pm \frac{47}{25}$ are the only possible rational roots of f, a simple check that none of those rational numbers are roots of f shows that r must be irrational. Moreover, the other root $\frac{7-\sqrt{2}}{5}$ of f must also be irrational. This general line of argument can be exploited in many examples.

Example 3.34.   Show that $\sqrt{2} + \sqrt{3}$ is irrational.

First, we do some scratch work.
Let $r = \sqrt{2} + \sqrt{3}$.
So $r^2 = 2 + 2\sqrt{6} + 3 = 5 + 2\sqrt{6}$.
Hence, $r^2 - 5 = 2\sqrt{6}$.
So $r^4 - 10r^2 + 25 = (r^2 - 5)^2 = (2\sqrt{6})^2 = 24$.
Therefore, $r^4 - 10r^2 + 1 = 0$.
This equation now inspires our choice of a function to consider in our proof.

Proof. Let $f(x) = x^4 - 10x^2 + 1$.
By the Rational Roots Theorem, the only possible rational roots of f are ± 1.
Of course, $\sqrt{2} + \sqrt{3}$ is neither 1 nor -1, since $\sqrt{2} + \sqrt{3} > 1 + 1 > 1 > -1$.
Since $f(\sqrt{2} + \sqrt{3}) = 0$, it follows that $\sqrt{2} + \sqrt{3}$ must be irrational. $\square$

The proof technique used for Theorem 3.24 depends on the fact that 2 is prime. Consequently, that technique does not lend itself to a proof that $\sqrt{6}$ is irrational. Instead, the approach taken in Example 3.34 works quite well.

Example 3.35.   Show that $\sqrt{6}$ is irrational.

Proof. Certainly, $\sqrt{6}$ is a root of $f(x) = x^2 - 6$.
By the Rational Roots Theorem, the only possible rational roots of f are $\pm 1, \pm 2, \pm 3, \pm 6$.
Since none of these are roots of f, $\sqrt{6}$ must be irrational.   □

The numbers that lend themselves to the type of argument in Examples 3.34 and 3.35 form a special class.

Definition 3.12.   A real number r is said to be **algebraic** if r is a root of a polynomial with integer coefficients.

A proof of the fact that rational numbers are algebraic is left to the exercises. However, not all algebraic numbers are rational. In Example 3.34, we showed that $\sqrt{2} + \sqrt{3}$ is algebraic but not rational.

Definition 3.13.   Real numbers that are not algebraic are said to be **transcendental**.

Since rational numbers are algebraic, all of the transcendental numbers must be irrational. Although verification is beyond the scope of this book, the numbers π and e are transcendental and hence irrational.

Exercises

In Exercises 1 through 12, show that the specified real number is rational.

1. $5\frac{1}{2}$

2. $7\frac{2}{3}$

3. $-13\frac{2}{5}$

4. $-25\frac{1}{4}$

5. 5.821

6. -3.72

7. $3.\overline{14}$

8. $-2.\overline{718}$

9. $-4.3\overline{21}$

10. $17.83\overline{2}$

11. $12.75\overline{8}$

12. $59.17\overline{36}$

For Exercises 13 through 18, let $r, s \in \mathbb{Q}$ and let $n \in \mathbb{Z}$. Use only the definitions of the terms involved, and not previous results, in each requested proof.

13. Show: $nr \in \mathbb{Q}$.

14. Show: If $n \neq 0$, then
$\frac{r}{n} \in \mathbb{Q}$.

15. Show: (a) $-s \in \mathbb{Q}$.
(b) $r - s \in \mathbb{Q}$.

16. Show: (a) $rs \in \mathbb{Q}$.
(b) If $s \neq 0$, then $\frac{r}{s} \in \mathbb{Q}$.

17. Show: If $n \geq 0$, then $r^n \in \mathbb{Q}$.

18. Show: If $n < 0$ and $r \neq 0$, then $r^n \in \mathbb{Q}$.

In Exercises 19 through 26, express the given rational number in lowest terms.

19. $\frac{65}{39}$

20. $\frac{21}{147}$

21. $\frac{-513}{72}$

22. $\frac{144}{-36}$

23. 3.14

24. 0.045

25. -0.378

26. -17.832

In Exercises 27 through 32, write the given rational number in decimal form, without using a calculator.

27. $\frac{12}{25}$

28. $\frac{9}{150}$

29. $\frac{3}{7}$

30. $\frac{4}{11}$

31. $\frac{8}{15}$

32. $\frac{24}{35}$

33. Let a, b be nonzero integers. Is it possible for $\frac{a}{b}$ to have a finite decimal expansion if b has prime factors other than 2 and 5? Note that Theorem 3.23 deals with rational numbers expressed in lowest terms. Think about what changes if this restriction is lifted.

34. Observe that $\frac{4}{7}$ and $\frac{5}{7}$ are both in lowest terms. In general, let a, b be nonzero integers. If $\frac{a}{b}$ is in lowest terms, must $\frac{a+1}{b}$ also be in lowest terms? What about when $b = 7$, in particular?

35. Observe that $\frac{2}{7}$ and $\frac{5}{11}$ are in lowest terms, and the computation

$$\frac{2(11) + 5(7)}{7 \cdot 11} = \frac{57}{77}$$

gives their sum in lowest terms. In general, let a, b, c, d be nonzero integers. If $\frac{a}{b}$ and $\frac{c}{d}$ are in lowest terms, does $\frac{ad+bc}{bd}$ necessarily express the sum $\frac{a}{b} + \frac{c}{d}$ in lowest terms?

36. Observe that $\frac{6}{11}$ and $\frac{32}{35}$ are in lowest terms, and the computation

$$\frac{6 \cdot 32}{11 \cdot 35} = \frac{192}{385}$$

gives their product in lowest terms. In general, let a, b, c, d be nonzero integers. If $\frac{a}{b}$ and $\frac{c}{d}$ are in lowest terms, does $\frac{ac}{bd}$ necessarily express the product $\frac{a}{b} \cdot \frac{c}{d}$ in lowest terms?

37. Observe that $\frac{5}{6}$, $\frac{5}{14}$, and $\frac{5}{23}$ are in lowest terms. If p and n are integers with p prime, then under what conditions is $\frac{p}{n}$ in lowest terms?

38. Show: If $n \in \mathbb{Z}^+$, then $\frac{n-1}{n}$ is in lowest terms.

39. If we represent numbers in binary form (e.g., $\frac{25}{4}$ is represented as 110.01), then does the result in Theorem 3.23 change? If so, how?

40. If we represent numbers in base 12 (e.g., $\frac{17}{6}$ is represented as 2.a), then does the result in Theorem 3.23 change? If so, how?

41. Observe that $\frac{4}{3}$, $\frac{7}{5}$, and $\frac{10}{7}$ are in lowest terms. In general, let $k \in \mathbb{Z}^+$. Show that $\frac{3k+1}{2k+1}$ is in lowest terms.

42. Observe that the rational numbers $\frac{5}{6}$ and $\frac{7}{15}$ can alternatively be written with the same denominator, namely as $\frac{25}{30}$ and $\frac{14}{30}$, respectively. Let $r_1, r_2 \in \mathbb{Q}$. Show: There exist $a_1, a_2, b \in \mathbb{Z}$ with $b > 0$ such that $r_1 = \frac{a_1}{b}$ and $r_2 = \frac{a_2}{b}$.

43. The Least Upper Bound Property (Appendix A, property 13) does not hold when restricted to only the rational numbers. Consider the set $A = \{r \in \mathbb{Q} : r^2 < 2\}$. What is lub($A$)? What condition is not met?

44. Use the Density of the Rationals (Appendix A, property 15) to show that $\forall \, x, y \in \mathbb{R}$, if $x < y$, then $\exists \, n \in \mathbb{Z}$ such that $(nx, ny) \cap \mathbb{Z} \neq \emptyset$.

In Exercises 45 through 52, mimic the proof of Theorem 3.24. Hint: To play the role played there by Exercise 3 of Section 3.1 use Corollary 3.19.

45. Show that $\sqrt{3}$ is irrational.

46. Show that $\sqrt{5}$ is irrational.

47. Show that $\sqrt{13}$ is irrational.

48. Let p be any prime. Show that $\sqrt{p}$ is irrational.

49. Show that $\sqrt[3]{2}$ is irrational.

50. Show that $\sqrt[3]{5}$ is irrational.

51. Show that $\sqrt[3]{7}$ is irrational.

52. Show that $\sqrt[4]{2}$ is irrational.

53. If a population is growing continuously and doubling every year, then the function $p(t)$ that measures the size of the population after t years from some start time has the form $p(t) = a2^t$, where a is the population at time 0.

Observe, for example, that $p(6) = 2p(5)$. It turns out that the tripling time for the population is $\log_2 3$. That is, $\forall\, t \geq 0$, $p(t + \log_2 3) = 3p(t)$. (This can be confirmed by using (5.5) from Section 5.4.) Show that the tripling time $\log_2 3$ is irrational.

54. Suppose that the amount of a radioactive substance in a rock at time t is given by the function $r(t) = a\left(\frac{1}{5}\right)^t$, for some constant $a = r(0)$. The half-life of the radioactive substance is defined to be the length of time it takes the amount of radioactive substance to decrease by 50%. It turns out that the half-life in this case will be $\log_5 2$. That is, $\forall\, t \geq 0$, $r(t + \log_5 2) = \frac{1}{2}r(t)$. (This can be confirmed by using (5.5) from Section 5.4.) Show that the half-life $\log_5 2$ is irrational.

55. Show that $\log_3 7$ is irrational.

56. Show that $\log_7 5$ is irrational.

57.* Is $\dfrac{\sqrt{2}+\sqrt{6}}{\sqrt{2+\sqrt{3}}}$ irrational? Why?

58.* Is $\sqrt{13 + \sqrt{2} + \dfrac{7}{3+\sqrt{2}}}$ irrational? Why?

59. Is the sum of irrational numbers always irrational? Explain.

60. Is the negative of an irrational number always irrational? Explain.

In Exercises 61 through 66, mimic the proof of Example 3.31. Refer to the results in Exercises 45 through 52.

61. If the ratio of the length to the width of a rectangle is $\frac{1+\sqrt{5}}{2}$ and a square is cut off of one end of that rectangle, then the remaining rectangle will also have $\frac{1+\sqrt{5}}{2}$ as the ratio of its length to its width. The number $\frac{1+\sqrt{5}}{2}$ is called the **golden ratio**. Show that $\frac{1+\sqrt{5}}{2}$ is irrational.

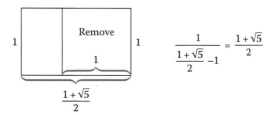

62. The formula $s(n) = \left(\frac{2+\sqrt{3}}{3}\right)^n + \left(\frac{2-\sqrt{3}}{3}\right)^n$ generates a rational number for all positive integers n. Compute the values $s(1)$, $s(2)$, and $s(3)$, and observe that they are rational. However, show that $\frac{2+\sqrt{3}}{3}$ is irrational. Also, use this together with the computation of $s(1)$ to show that $\frac{2-\sqrt{3}}{3}$ is irrational.

63. A regular dodecahedron is a solid that has twelve congruent regular pentagons for its faces. The volume V of a regular dodecahedron with side length s is given by $V = \frac{15+7\sqrt{5}}{4}s^3$. Show that $\frac{15+7\sqrt{5}}{4}$ is irrational.

64. A regular icosahedron is a solid that has twenty congruent equilateral triangles for its faces. The volume V of a regular icosahedron with side length s is given by $V = \frac{5(3+\sqrt{5})}{12}s^3$. Show that $\frac{5(3+\sqrt{5})}{12}$ is irrational.

65. Show that $\frac{7-\sqrt{2}}{3+\sqrt{2}}$ is irrational.

66. Show that $3 - \sqrt[4]{2}$ is irrational. Hint: Use the result from Exercise 52.

In Exercises 67 through 70, use the Rational Roots Theorem to find all of the rational roots of the given polynomial.

67. $6x^4 - 7x^3 + 3x^2 - 7x - 3$.

68. $x^6 - 5x^4 + 2x^2 - 10$.

69. $x^4 - x^3 + 5x^2 - 6x - 6$.

70. $10x^4 - x^3 - 22x^2 + 2x + 4$.

In Exercises 71 through 78, use the Rational Roots Theorem.

71. Show that $\sqrt{10}$ is irrational.

72. Show that $\sqrt{35}$ is irrational.

73. Using the identity $\cos(A - B) = \cos(A)\cos(B) + \sin(A)\sin(B)$ for $A = 45°$ and $B = 30°$, we get

$$\cos(15°) = \cos(45°)\cos(30°) + \sin(45°)\sin(30°)$$

$$= \frac{\sqrt{2}}{2} \cdot \frac{\sqrt{3}}{2} + \frac{\sqrt{2}}{2} \cdot \frac{1}{2} = \frac{\sqrt{6}+\sqrt{2}}{4}.$$

Show that $\sqrt{6} + \sqrt{2}$ is irrational. Conclude that $\cos(15°)$ is irrational.

74. Show that $\sqrt{3} + \sqrt{5}$ is irrational. Keep Exercise 58 in mind.

75. Using the identity $\tan^2(\theta) = \frac{1-\cos(2\theta)}{1+\cos(2\theta)}$ for $\theta = 22.5°$, we get

$$\tan(22.5°) = \sqrt{\frac{1-\cos(45°)}{1+\cos(45°)}} = \sqrt{\frac{1-\frac{\sqrt{2}}{2}}{1+\frac{\sqrt{2}}{2}}} = \sqrt{3-2\sqrt{2}}.$$

Show that $\sqrt{3-2\sqrt{2}}$ is irrational. Conclude that $\tan(22.5°)$ is irrational.

76. Using the identity $\cos^2(\theta) = \frac{1+\cos(2\theta)}{2}$ for $\theta = 15°$, we get

$$\cos(15°) = \sqrt{\frac{1+\cos(30°)}{2}} = \sqrt{\frac{1+\frac{\sqrt{3}}{2}}{2}} = \frac{\sqrt{2+\sqrt{3}}}{2}.$$

Show that $\sqrt{2+\sqrt{3}}$ is irrational. Conclude that $\cos(15°)$ is irrational, and consider Exercise 57 in light of this result and Exercise 73.

77. Show that $\frac{\sqrt[4]{3}}{\sqrt{2}}$ is irrational.

78. Show that $\frac{\sqrt[3]{2}}{\sqrt[3]{5}}$ is irrational.

79. In 1768, the French mathematician Johann Heinrich Lambert (1728–1777) proved that π is irrational. Use the fact that π is irrational to show that $\frac{\pi+1}{2}$ is irrational.

80. In 1737, Euler proved that the Euler number e is irrational. Use the fact that e is irrational to show that $\frac{e-1}{2}$ is irrational.

81. Let $x \in \mathbb{R}$. Show: If x is irrational, then $\sqrt{x}$ is irrational.

82. Show that for a monic polynomial (a polynomial with 1 as the coefficient of the highest power of the variable), all of the rational roots are integers.

83. Show that rational numbers are algebraic.

84.* Is $\sqrt{1+\sqrt{1+\sqrt{1+\cdots}}}$ algebraic? Why?

85.* Show that the roots of a polynomial with rational coefficients are algebraic.

86.* Show: $\forall\, a \in \mathbb{R}$, if a is algebraic, then $\sqrt{a}$ is algebraic.

87. Are all irrational numbers transcendental? Why?

88. Is the sum of two transcendental numbers necessarily transcendental? Why?

89.* In 1873, the French mathematician Charles Hermite (1822–1901) proved that the Euler number e is transcendental. Given that e is transcendental, show that $2e$ must be transcendental.

90.* In 1882, the German mathematician Ferdinand von Lindemann (1852–1939) proved that π is transcendental. Given that π is transcendental, show that $\frac{\pi}{6}$ must be transcendental.

3.5 Modular arithmetic and encryption

Many things are numbered cyclically. For example, four hours after 10 o'clock, it is 2 o'clock, not 14 o'clock. Seven hours before 3 o'clock, it is 8 o'clock. Twenty-eight hours after 1 o'clock, it is 5 o'clock. In specifying time of day, we equate $10 + 4$ with 2, we equate $3 - 7$ with 8, and we equate $1 + 28$ with 5. These equivalences hold because the differences $(10 + 4) - 2$, $(3 - 7) - 8$, and $(1 + 28) - 5$, respectively, are divisible by 12. In the same way, two dates fall on the same day of the week if and only if the number of days by which they differ is divisible by 7. These types of calculations are sometimes called clock arithmetic. The more formal term is *modular arithmetic*.

Definition 3.14. Given integers a, b, and n with $n > 1$, we say that a is **congruent** to b **modulo** n, written $a \equiv b \pmod{n}$, if $n \mid (a - b)$.

Example 3.36. (Some Congruences).

$$14 \equiv 2 \ (\mathrm{mod}\ 12), \ \text{since} \ 12 \mid (14 - 2).$$

$$-4 \equiv 8 \ (\mathrm{mod}\ 12), \ \text{since} \ 12 \mid (-4 - 8).$$

$$34 \equiv 6 \ (\mathrm{mod}\ 7), \ \text{since} \ 7 \mid (34 - 6).$$

$$25 \equiv 0 \ (\mathrm{mod}\ 5), \ \text{since} \ 5 \mid (25 - 0).$$

Formal properties of congruence are listed in the following two theorems. Their proofs are straightforward applications of the definitions, and almost all are left for the exercises. Very simply, congruence $\equiv$ has many properties similar to those of equality $=$. In fact, the properties of $\equiv$ listed in the following theorem constitute those of an equivalence relation, as we shall study in Section 5.2.

Theorem 3.26. (Congruence is an Equivalence Relation).
Let a, b, and n be integers with $n > 1$.

(a) $a \equiv a \pmod{n}$.

(b) *If $a \equiv b \pmod{n}$, then $b \equiv a \pmod{n}$.*

(c) *If* $a \equiv b \pmod{n}$ *and* $b \equiv c \pmod{n}$,   *then* $a \equiv c \pmod{n}$.

Theorem 3.27. (Arithmetic Properties of Congruence).
Let a_1, a_2, b_1, b_2, and n be integers with $n > 1$.
If $a_1 \equiv a_2 \pmod{n}$ and $b_1 \equiv b_2 \pmod{n}$, then

(a) $a_1 + b_1 \equiv a_2 + b_2 \pmod{n}$, *and*

(b) $a_1 b_1 \equiv a_2 b_2 \pmod{n}$.

Proof of Theorem 3.27(b) Suppose $a_1 \equiv a_2 \pmod{n}$ and $b_1 \equiv b_2 \pmod{n}$. Therefore, $n \mid (a_1 - a_2)$ and $n \mid (b_1 - b_2)$. That is, $a_1 - a_2 = nk$ and $b_1 - b_2 = nl$ for some $k, l \in \mathbb{Z}$. Observe that

$$a_1 b_1 - a_2 b_2 = a_1 b_1 - a_1 b_2 + a_1 b_2 - a_2 b_2$$
$$= a_1 (b_1 - b_2) + b_2 (a_1 - a_2)$$
$$= a_1 nl + b_2 nk = n(a_1 l - b_2 k).$$

Therefore, $n \mid (a_1 b_1 - a_2 b_2)$, and it follows that $a_1 b_1 \equiv a_2 b_2 \pmod{n}$. $\qquad\square$

Example 3.37. Note that $4 \equiv -6 \pmod{10}$ and $22 \equiv 2 \pmod{10}$. As promised by Theorem 3.27, $4 + 22 \equiv -6 + 2 \pmod{10}$ and $4 \cdot 22 \equiv -6 \cdot 2 \pmod{10}$.

A proof of part (a) of the following straightforward consequence of Theorem 3.27 is left for the exercises at the end of this section. A proof of part (b) is left for the exercises in Section 4.3, after induction is available.

Corollary 3.28. Let m, a_1, a_2, and n be integers with $n > 1$. If $a_1 \equiv a_2 \pmod{n}$, then

(a) $ma_1 \equiv ma_2 \pmod{n}$, and

(b) if $m \geq 0$, then $a_1^m \equiv a_2^m \pmod{n}$.

Example 3.38. Note that $3 \equiv -2 \pmod{5}$. As promised by Corollary 3.28, we also have $4(3) \equiv 4(-2) \pmod{5}$ and $3^4 \equiv (-2)^4 \pmod{5}$.

The notation **mod** has now been used in two ways in this chapter. In fact, there is an important relationship that mitigates any potential confusion. For example, note that $25 \bmod 7 = 4$ and $4 \equiv 25 \pmod{7}$.

Lemma 3.29. Given integers n and d with $d > 1$,

$$n \bmod d \equiv n \pmod{d},$$

and, moreover, $n \bmod d$ is the unique integer in $\{0, 1, \ldots, d-1\}$ congruent to n modulo d.

Proof. Let $r = n \bmod d$. That is, r is the remainder upon division of n by d guaranteed by the Division Algorithm. By letting $q = n \operatorname{div} d$, we have $n = dq + r$ and $0 \le r < d$. So $d(-q) = r - n$. Since $d \mid (r - n)$, in the notation of congruence, we have $r \equiv n \pmod{d}$. That is, $n \bmod d \equiv n \pmod{d}$.

To establish our uniqueness assertion, suppose $r' \in \{0, 1, \ldots, d - 1\}$ and $r' \equiv n \pmod{d}$. Since $d \mid (r' - n)$, there is some $q' \in \mathbb{Z}$ such that $d(-q') = r' - n$. That is, $n = dq' + r'$. Since $0 \le r' < d$, the uniqueness assertion in the Division Algorithm tells us that $r' = r = n \bmod d$. $\square$

We can use Lemma 3.29 together with Theorem 3.27 to simplify computations of the form $n \bmod d$.

Example 3.39.   Use Theorem 3.27 to help with the following calculations:

(a) Compute $(5162387 + 83645) \bmod 10$.
Note that the value of a positive integer mod 10 is its ones digit.
Since $5162387 \equiv 7 \pmod{10}$ and $83645 \equiv 5 \pmod{10}$, we get
$5162387 + 83645 \equiv 5 + 7 \equiv 12 \equiv 2 \pmod{10}$.
That is, we can conclude that $(5162387 + 83645) \bmod 10 = 2$ without having to compute $5162387 + 83645 = 5246032$.

(b) Compute $(34861 + 571028) \bmod 5$.
Note that it is easy to determine the value of an integer mod 5 based on its ones digit.
Since $34861 \equiv 1 \pmod{5}$ and $571028 \equiv 8 \equiv 3 \pmod{5}$, we get
$34861 + 571028 \equiv 1 + 3 \equiv 4 \pmod{5}$.   That is, we can conclude that $(34861 + 571028) \bmod 5 = 4$ without having to compute $34861 + 571028 = 605889$.

(c) Compute $3^{35} \bmod 10$.
Note that $3^4 \equiv 81 \equiv 1 \pmod{10}$.
So $3^{32} = (3^4)^8 \equiv 1^8 \equiv 1 \pmod{10}$.
Hence, $3^{35} \equiv 3^{32}3^3 \equiv 1(27) \equiv 27 \equiv 7 \pmod{10}$.
We conclude that $3^{35} \bmod 10 = 7$, and we did not need to compute the value $3^{35} = 50031545098999707$.

(d) Compute $2^{13} \bmod 209$.
Note that $2^8 \equiv 256 \equiv 47 \pmod{209}$
and $2^5 \equiv 32 \pmod{209}$.

Hence, $2^{13} \equiv 2^8 2^5 \equiv 47(32) \equiv 1504 \equiv$
41 (mod 209).
We conclude that 2^{13} mod 209 = 41.

(e) Compute 41^7 mod 209.
Note that $41^2 \equiv 1681 \equiv 9$ (mod 209).
Hence, $41^6 \equiv (41^2)^3 \equiv 9^3 \equiv 729 \equiv$
102 (mod 209).
Therefore, $41^7 \equiv 41^6(41) \equiv 102(41) \equiv 4182 \equiv$
2 (mod 209).
We conclude that 41^7 mod 209 = 2.

In Example 3.39, part (c) is made easy by the observation that 3^4 mod 10 = 1, and part (e) becomes easier after it is observed that 41^2 mod 209 = 9. When doing calculations modulo n, being fortunate to find a small power a^k that simplifies the computation of a larger power a^m is certainly helpful. However, we cannot always count on being so lucky. After we appreciate an application of such computations in Example 3.43, a systematic means of exploiting small powers in an ultimate calculation of a larger one is introduced.

The notion of congruences also enables us to make some divisibility arguments. In Example 3.12, we showed that for any integer n, $n^2 - 2$ is not divisible by 5. Here, we can give a new proof by noting that $5 \mid (n^2 - 2)$ is equivalent to $n^2 - 2 \equiv 0$ (mod 5), and that is further equivalent to $n^2 \equiv 2$ (mod 5).

Example 3.40.   Show: For   any   integer n,
$n^2 \not\equiv 2$ (mod 5).

Proof. It suffices to consider $n \equiv r$ (mod 5) for each of $r = 0, 1, 2, 3,$ and 4.
If $n \equiv 0$ (mod 5), then $n^2 \equiv 0$ (mod 5).
If $n \equiv 1$ (mod 5), then $n^2 \equiv 1$ (mod 5).
If $n \equiv 2$ (mod 5), then $n^2 \equiv 4$ (mod 5).
If $n \equiv 3$ (mod 5), then $n^2 \equiv 9 \equiv 4$ (mod 5).
If $n \equiv 4$ (mod 5), then $n^2 \equiv 16 \equiv 1$ (mod 5).
In each case, $n^2 \not\equiv 2$ (mod 5). $\square$

Note how much cleaner the proof in Example 3.40 is than that in Example 3.12. Working with congruences can be convenient.

Example 3.41. Let n be any integer. Show that
$n^3 \equiv n$ (mod 6).

Proof. It suffices to consider $n \equiv r \pmod 6$ for each of $r = 0, 1, 2, 3, 4$ and 5.

If $n \equiv 0 \pmod 6$, then $n^3 \equiv 0 \pmod 6$.
If $n \equiv 1 \pmod 6$, then $n^3 \equiv 1 \pmod 6$.
If $n \equiv 2 \pmod 6$, then $n^3 \equiv 8 \equiv 2 \pmod 6$.
If $n \equiv 3 \pmod 6$, then $n^3 \equiv 27 \equiv 3 \pmod 6$.
If $n \equiv 4 \pmod 6$, then $n^3 \equiv 64 \equiv 4 \pmod 6$.
If $n \equiv 5 \pmod 6$, then $n^3 \equiv 125 \equiv 5 \pmod 6$.
In each case, $n^3 \equiv n \pmod 6$. $\square$

In the equation $n^3 \equiv n \pmod 6$, it may be tempting to cancel an n from both sides. However, $n^2 \equiv 1 \pmod 6$ does not follow in general (for instance, consider $n = 2$). Cancellation works only under certain conditions.

Lemma 3.30.   (Modular Cancellation Rule).
Let a, b_1, b_2, and n be integers with $n > 1$. Suppose that $ab_1 \equiv ab_2 \pmod n$ and $\gcd(a, n) = 1$. Then, $b_1 \equiv b_2 \pmod n$.

Proof. Since $n \mid a(b_1 - b_2)$ and $\gcd(n, a) = 1$, it follows from Euclid's lemma that $n \mid (b_1 - b_2)$. That is, $b_1 \equiv b_2 \pmod n$. $\square$

Canceling a from both sides of the congruence $ab_1 \equiv ab_2 \pmod n$ is effectively dividing both sides by a. From another point of view, we are multiplying both sides by an inverse for a.

Definition 3.15.   Given $a, n \in \mathbb{Z}$ with $n > 1$, a **multiplicative inverse of a modulo n** is an integer c such that $ac \equiv 1 \pmod n$.

Lemma 3.31.   Given $n \in \mathbb{Z}$ with $n > 1$, an integer a has a multiplicative inverse modulo n if and only if $\gcd(a, n) = 1$.

Proof. Observe that the following statements are equivalent:
(1) $\exists\, c \in \mathbb{Z}$ such that $ac \equiv 1 \pmod n$.
(2) $\exists\, c, d \in \mathbb{Z}$ such that $n(-d) = ac - 1$.
(3) $\exists\, c, d \in \mathbb{Z}$ such that $ac + nd = 1$. $\square$

Remark 3.1.   From the proof of Lemma 3.31, it follows that Euclid's Algorithm can be applied to find multiplicative inverses modulo n. If $\gcd(a, n) = 1$, then the Extended Euclidean Algorithm can be used to find integers c and d such that $ac + nd = 1$. We see in the proof of Lemma 3.31 that c is a multiplicative inverse of a modulo n.

TABLE 3.4
Converting letters to numbers

	A	B	C	D	E	F	G	H
0	1	2	3	4	5	6	7	8

I	J	K	L	M	N	O	P	Q
9	10	11	12	13	14	15	16	17

R	S	T	U	V	W	X	Y	Z
18	19	20	21	22	23	24	25	26

3.5.1   Linear ciphers

Lemma 3.31 can be applied to give a generalization of the shift cipher introduced in Section 3.2. As in Section 3.2, to be concrete, suppose $n = 27$ and that we wish to assign each number x from 0 to 26 (representing letters as shown in Table 3.4) to a new number y in that range. A **linear cipher** has the form

$$y = (ax + b) \bmod n, \tag{3.8}$$

for some choice of integers a and b. Moreover, in order for the assignment of x to y in Equation (3.8) to be useful, it must be the case that x can be subsequently recovered from y. As is seen in the exercises, this occurs precisely when a has a multiplicative inverse c modulo n. In this case, the deciphering equation corresponding to Equation (3.8) is

$$x = c(y - b) \bmod n, \tag{3.9}$$

where $ac \equiv 1 \pmod{n}$.

Example 3.42.   Using $n = 27$, $a = 4$, $b = 2$, Equation (3.8), and the conversions of letters to numbers in Table 3.2, the message

JUMP

is encrypted to the message

OE L.

For example, the letter P is represented by the value 16,

$$(4 \cdot 16 + 2) \bmod 27 = 66 \bmod 27 = 12,$$

and 12 represents the letter L.

In the other direction, using $c = 7$ (since $4 \cdot 7 \equiv 1 \pmod{27}$) and Equation (3.9), the encrypted message

GHMBGKCG

is decrypted to the message

HOW HIGH.

For example, the letter K is represented by the value 11,

$$7(11 - 2) \bmod 27 = 63 \bmod 27 = 9,$$

and 9 represents the letter I.

3.5.2 RSA encryption

Computations like those in Example 3.39 are critical to a means of encryption known as **RSA encryption**, developed by Ronald Rivest, Adi Shamir, and Leonard Adleman in 1977. The method requires that an integer n be known to the sender and receiver of a secret message. Also, the sender must have a particular integer a, and the receiver must have a particular integer c, where n, a, and c have certain properties. Specifically, n is chosen by the receiver to be the product of two (large) primes p and q. Also, the receiver picks an integer a that has a multiplicative inverse c modulo $m = \mathrm{lcm}(p - 1, q - 1)$. The sender then takes a message x (which is assumed to be in the form of a number[*]) and computes the encrypted message

$$y = x^a \bmod n. \tag{3.10}$$

The encrypted message y can then be sent to the receiver, who decrypts the message by computing $y^c \bmod n$. That is,

$$x = y^c \bmod n, \tag{3.11}$$

as we shall prove in the exercises.

Example 3.43.   An ATM is set up to receive transactions from many customers of a bank. However, no customer should be able to intercept another customer's transaction. The ATM therefore uses an RSA encryption scheme with $p = 11$ and $q = 19$, so $n = 209$. Since $m = \mathrm{lcm}(10, 18) = 90$ and $\gcd(13, 90) = 1$, the ATM chooses $a = 13$ and consequently decrypts transactions with the inverse of a modulo m, namely, $c = 7$. Only the values n and a are stored on all ATM cards to enable

---

[*] In this case, since n can be quite large, a number may represent much more than merely a single character of a message.

them to use Equation (3.10) to encrypt transactions (messages). The value c needed in Equation (3.11) for decrypting received messages is kept secret by the bank.

(a) An ATM card encrypts the message $x = 2$, which represents a 'Balance Inquiry' transaction. The details presented in Example 3.39(d) show that

$$y = 2^{13} \bmod 209 = 41.$$

The encrypted message 41 is sent to the bank.

(b) The bank can decrypt $y = 41$. The details presented in Example 3.39(e) show that

$$x = 41^7 \bmod 209 = 2.$$

The decrypted message 2 is thus received and interpreted as a 'Balance Inquiry'.

Since factoring n as the product pq is extremely difficult in general,[*] the sender's knowledge of n and a is not sufficient to determine the exponent c needed for decrypting. Consequently, multiple senders can encrypt messages by the same method, while none of them can decrypt another sender's message. A system having this property is called a **public key encryption** scheme. Such schemes have applications in wireless communications, Internet commerce, and security.

A combination of the techniques in Example 3.39 and the ability to represent natural numbers in binary gives us a straightforward procedure for performing the calculations central to RSA encryption.

Example 3.44.   Compute $7^{53} \bmod 143$.

Solution. The binary representation of 53 is 110101. That is, $53 = 32 + 16 + 4 + 1$. Thus, $7^{53} = 7^{32} 7^{16} 7^4 7^1$, and the powers on the right-hand side can be obtained by calculating a sequence of squares, whose values we only need modulo 143.

$$7^2 \equiv 49 \ (\mathrm{mod} \ 143),$$

$$7^4 \equiv 49^2 \equiv 2401 \equiv 113 \ (\mathrm{mod} \ 143),$$

$$7^8 \equiv 113^2 \equiv 12769 \equiv 42 \ (\mathrm{mod} \ 143),$$

$$7^{16} \equiv 42^2 \equiv 1764 \equiv 48 \ (\mathrm{mod} \ 143), \text{ and}$$

$$7^{32} \equiv 48^2 \equiv 2304 \equiv 16 \ (\mathrm{mod} \ 143).$$

---

[*] There are nearly two quintillion 20 digit primes, for example.

Therefore,

$$7^{53} \equiv 16 \cdot 48 \cdot 113 \cdot 7 \equiv 768 \cdot 113 \cdot 7 \equiv 53 \cdot 113 \cdot 7$$

$$\equiv 5989 \cdot 7 \equiv 126 \cdot 7 \equiv 882 \equiv 24 \pmod{143}.$$

That is, $7^{53} \bmod 143 = 24$.   □

Example 3.45. Consider the RSA encryption scheme with $p = 11$ and $q = 13$ (so $n = 143$ and $m = \operatorname{lcm}(10, 12) = 60$) in which we use $a = 53$ to encrypt messages and $c = 17$ to decrypt messages (since $ac \equiv 1 \pmod{60}$). The result of Example 3.44 shows that the message $x = 7$ encrypts to $y = 24$. Use the method there to decrypt $y = 24$.

Solution. Since 10001 is the binary representation of 17, we have $24^{17} = 24^{16}24^{1}$. Observe that

$$24^2 \equiv 576 \equiv 4 \pmod{143},$$

$$24^4 \equiv 4^2 \equiv 16 \pmod{143},$$

$$24^8 \equiv 16^2 \equiv 256 \equiv -30 \pmod{143}, \text{ and}$$

$$24^{16} \equiv (-30)^2 \equiv 900 \equiv 42 \pmod{143}.$$

Hence, $24^{17} \equiv 42 \cdot 24 \equiv 1008 \equiv 7 \pmod{143}$. This establishes that $24^{17} \bmod 143 = 7$ and decrypts $y = 24$ to the message $x = 7$, as we would expect.   □

3.5.3 Primality conditions

The French mathematician Pierre de Fermat (1601–1665) is perhaps best known for what is now called Fermat's Last Theorem. It states that for every integer $n \geq 3$, there are no nonzero integers x, y, and z satisfying the equation $x^n + y^n = z^n$. This was proved in 1994 by the English mathematician Andrew Wiles (1953–). Our study of congruences is aided by a smaller but extremely useful result about primes that we owe to Fermat.

Theorem 3.32. (Fermat's Little Theorem).
If p is a prime, $a \in \mathbb{Z}$, and $p \nmid a$, then $a^{p-1} \equiv 1 \pmod{p}$.

Proof. Since no pair of integers from the list

$$0, 1, 2, \ldots, p - 1 \tag{3.12}$$

is congruent modulo p (Exercise 45), it follows from the Modular Cancellation Rule that no pair of integers from the list

$$a \cdot 0, a \cdot 1, a \cdot 2, \ldots, a(p - 1) \tag{3.13}$$

is congruent modulo p (Exercise 46). Consequently, there must be a way to pair off the entries in (3.12) with the entries in (3.13) so that each pair is congruent modulo p. Certainly, 0 pairs with $a \cdot 0$.

Since the list $1, 2, \ldots, p - 1$ gets paired, in some order, with the list $a \cdot 1, a \cdot 2, \ldots, a(p - 1)$, it follows from Theorem 3.27(b) (generalized to $p - 1$ factors) that the products of these lists are congruent modulo p. That is, $(p - 1)! \equiv a^{p-1}(p - 1)! \pmod{p}$. Since p is prime, $\gcd(p, (p - 1)!) = 1$. The Modular Cancellation Rule then tells us that $1 \equiv a^{p-1} \pmod{p}$. $\square$

Example 3.46.   Compute $7^{2222} \bmod 13$.

Solution. Fermat's Little Theorem tells us that $7^{12} \equiv 1 \pmod{13}$. Thus,

$$7^{2222} \equiv (7^{12})^{185} 7^2 \equiv 1^{185} 49 \equiv 10 \pmod{13}.$$

We conclude that $7^{2222} \bmod 13 = 10$. $\square$

The proofs of the following corollaries to Fermat's Little Theorem are left for the exercises.

Corollary 3.33.   If p is prime, then, for all $a \in \mathbb{Z}$, $a^p \equiv a \pmod{p}$.

Corollary 3.34.   Let $n \in \mathbb{Z}$ with $n > 1$. If there exists $a \in \mathbb{Z}$ such that $a^n \not\equiv a \pmod{n}$, then n is not prime.

Corollary 3.34 can be used to test whether an integer n is prime.

Example 3.47.   Consider $n = 221$. Observe that

$$3^5 \equiv 243 \equiv 22 \pmod{221} \quad \text{and}$$

$$22^4 \equiv 234256 \equiv -4 \pmod{221}.$$

Thus,

$$3^{220} \equiv (3^5)^{44} \equiv 22^{44} \equiv (22^4)^{11}$$

$$\equiv (-4)^{11} \equiv -4194304 \equiv 55 \pmod{221}.$$

Since $3^{221} \equiv 165 \not\equiv 3 \pmod{221}$, we see from Corollary 3.34 that 221 is not prime. (Of course, $221 = 13 \cdot 17$.)

If an integer n satisfies $a^n \equiv a$ (mod n) for a particular value of a, then that is not enough to conclude that n is prime. For example, $n = 91 = 7 \cdot 13$ is certainly not prime despite the fact that $3^{91} \equiv 3$ (mod 91). A more useful computation in this case is the observation that $2^{91} \equiv 37 \not\equiv 2$ (mod 91).

3.5.4   The additive group of integers modulo n

We close this section with some notions that are of further interest to us in Chapter 5 and Section 7.4, where formal introductions to equivalence relations and groups are given. Proofs of our assertions here are left for the exercises. Let n be an integer such that $n > 1$. Given any integer a, let $[a]_n$ denote the set

$$[a]_n = \{k \; : \; k \in \mathbb{Z} \text{ and } k \equiv a \text{ (mod } n)\}.$$

We refer to $[a]_n$ as the **equivalence class of a modulo n**. We shall prove in the exercises that for all $a, b \in \mathbb{Z}$,

$$[a]_n = [b]_n \quad \text{if and only if} \quad a \equiv b \text{ (mod } n).$$

In fact we usually represent the set $[a]_n$ as $[b]_n$, where b is the unique integer such that $0 \le b \le n - 1$ and $a \equiv b$ (mod n). That is, $b = a \bmod n$.

Example 3.48.   Consider the case in which $n = 15$.

(a) $[25]_{15} = [10]_{15}$

(b) $[73]_{15} = [13]_{15}$

(c) $[-36]_{15} = [9]_{15}$

(d) $[8]_{15} = [8]_{15}$

(e) $[15]_{15} = [0]_{15}$

Let $\mathbb{Z}_n$ denote the set of equivalence classes of integers modulo n. That is,

$$\mathbb{Z}_n = \{[a]_n \; : \; a \in \mathbb{Z}\}.$$

More simply, we have

$$\mathbb{Z}_n = \{[0]_n, [1]_n, \dots, [n-1]_n\}.$$

We define arithmetic operations on subsets S and T of the integers by

$$S + T = \{s + t \; : \; s \in S \text{ and } t \in T\},$$
$$-T = \{-t \; : \; t \in T\}, \text{ and}$$
$$S - T = S + (-T).$$

It is straightforward to verify that, given integers a and b, we have

$$[a]_n + [b]_n = [a + b]_n,$$

$$-[b]_n = [-b]_n, \text{ and}$$

$$[a]_n - [b]_n = [a - b]_n.$$

With $[0]_n$ playing the role of the additive identity and $[-a]_n$ playing the role of the additive inverse of $[a]_n$, the set $\mathbb{Z}_n$ under addition (and subtraction) satisfies arithmetic properties familiar from the integers. In fact, $\mathbb{Z}_n$ forms what is called a group, as we shall study in Section 7.4.

Theorem 3.35. ($\mathbb{Z}_n$ Forms a Group Under $+$).
Let $n > 1$ be an integer, and let $a, b, c \in \mathbb{Z}$. Then

(a) $([a]_n + [b]_n) + [c]_n = [a]_n + ([b]_n + [c]_n)$,   Associativity
(b) $[0]_n + [a]_n = [a]_n$,   Identity
(c) $[-a]_n + [a]_n = [0]_n$.   Inverse

Exercises

For Exercises 1 through 4, determine whether the given assertion is true or false.

1. $55 \equiv 15 \pmod{10}$.   2. $38 \equiv 8 \pmod{8}$.

3. $-7 \equiv 21 \pmod{6}$.   4. $10 \equiv -8 \pmod{3}$.

5. The stock market crash on October 19, 1987, is referred to as Black Monday. Consequently, it is easy to remember that October 19, 1987, was a Monday. Use this to determine the day of the week on January 8, 1987. That was the first day that the Dow Jones Industrial Average ever closed at over 2000 points.

6. The Rev. Martin Luther King Jr. was shot on Thursday April 4, 1968. In that same year, Robert F. Kennedy was shot on June 6. Use the fact that MLK was shot on a Thursday to determine the day of the week that RFK was shot.

7. In 1986, Susan Butcher, only the second woman to win the Iditarod Dog-Sled Race, won her first of three consecutive (and four overall) Iditarods in just over 279 hours. If the race started at 6 a.m., at what time of day did she finish?

8. In October of 1997, the *Cassini* spacecraft was launched from Earth to take detailed pictures of Jupiter. If the trip takes approximately 81 months, then in what month (of 2004) did the Cassini spacecraft reach Jupiter?

9. Prove Theorem 3.26.

10. (a) Prove Theorem 3.27(a).
    (b) Prove Corollary 3.28(a).

    In Exercises 11 through 14, give proofs using only the definitions of the terms involved and basic properties of integer arithmetic. Do not use other results.

11. Let a_1, a_2, and n be integers with $n > 1$.
    Show: If $a_1 \equiv a_2 \pmod{n}$, then $-a_1 \equiv -a_2 \pmod{n}$.

12. Let a_1, a_2, b_1, b_2, and n be integers with $n > 1$.
    Show: If $a_1 \equiv a_2 \pmod{n}$ and $b_1 \equiv b_2 \pmod{n}$, then $a_1 - b_1 \equiv a_2 - b_2 \pmod{n}$.

13. Let a, b and n be integers with $n > 1$.
    Show: If $a \equiv b \pmod{n}$ then $a^2 \equiv b^2 \pmod{n}$.
    This is the special case of Corollary 3.28(b) in which $m = 2$. A general proof can be given after mathematical induction is introduced in Chapter 4.

14. Let a, b, c and n be integers with $n > 1$.
    Show: If $a \equiv b \pmod{n}$, then $a + c \equiv b + c \pmod{n}$.

15. Let $d, n_1, n_2 \in \mathbb{Z}$ with $d > 1$. Show: $n_1 \bmod d = n_2 \bmod d$ if and only if $n_1 \equiv n_2 \pmod{d}$.

16. Let $d, n_1, n_2 \in \mathbb{Z}$ with $d > 1$.
    Prove or disprove: $(n_1 \bmod d)(n_2 \bmod d) = n_1 n_2 \bmod d$.

17. Compute $(8763 + 536) \bmod 25$, without using the value of $8763 + 536$.

18. Compute $(517 + 229) \bmod 3$, without using the value of $517 + 229$.

19. Compute $(10^{5379} + 14) \bmod 10$.

20. Compute $(2^{5379} + 5) \bmod 4$.

21. Compute $25^{17} \bmod 19$. Hint: $25^2 \equiv -2 \pmod{19}$.

22. Compute $150^{1000} \bmod 13$. Hint: $150 \equiv 7 \pmod{13}$ and $7^6 \equiv -1 \pmod{13}$.

23. Compute $20^{50} \bmod 3$.

24. Compute $17^{100} \bmod 5$.

25. Compute $13^{200} \bmod 7$.   26. Compute $30^{421} \bmod 11$.

27. Show: $\forall n \in \mathbb{Z}, n^3 - n - 1 \equiv 2 \pmod{3}$.

28. Show: $\forall n \in \mathbb{Z}, n^3 - 3n^2 - 4n \equiv 0 \pmod{6}$.

29. Let $n \in \mathbb{Z}$. Show: If $n \not\equiv 0 \pmod{7}$, then $n^3 \equiv \pm 1 \pmod{7}$.

30. Let $n \in \mathbb{Z}$. Show: $\begin{cases} n^4 \equiv 0 \pmod{5} & \text{if } n \equiv 0 \pmod{5}, \\ n^4 \equiv 1 \pmod{5} & \text{if } n \not\equiv 0 \pmod{5}. \end{cases}$

31. Let n be an odd integer.

    (a) Show: $n^2 \equiv 1 \pmod 8$.   (b) Show: $n^3 \equiv n \pmod 8$.

32. Let $n \in \mathbb{Z}$. Show: $\begin{cases} n^3 \equiv 0 \pmod 4 & \text{if } n \text{ is even,} \\ n^3 \equiv 1 \pmod 4 & \text{if } n \equiv 1 \pmod 4, \\ n^3 \equiv 3 \pmod 4 & \text{if } n \equiv 3 \pmod 4. \end{cases}$

33. Let $n, r \in \mathbb{N}$. Show: If $n \equiv r \pmod 3$, then $2^n \equiv 2^r \pmod 7$.

34. Let $n, r \in \mathbb{Z}$.
    Show: If $n \equiv r \pmod 4$, then $3^n \equiv 3^r \pmod{10}$.
    Further, determine the values of $3^r \bmod 10$ for $r = 0, 1, 2, 3$.

35. Let a, b, m, and n be integers with $m > 1$ and $n > 1$. Suppose that $a \equiv b \pmod m$ and $a \equiv -b \pmod n$.
    Show: $a^2 \equiv b^2 \pmod{mn}$.

36.* Let a, b, m, and n be integers with $m > 1$ and $n > 1$. Suppose that $a \equiv b \pmod m$ and $a \equiv b \pmod n$.

    (a) Must it follow that $a \equiv b \pmod{mn}$? Explain.

    (b) What if m and n are distinct primes?

In Exercises 37 through 40, use the Extended Euclidean Algorithm to find the requested multiplicative inverse. See Remark 3.1.

37. A multiplicative inverse of 12 modulo 55.

38. A multiplicative inverse of 14 modulo 45.

39. A multiplicative inverse of 18 modulo 25.

40. A multiplicative inverse of 33 modulo 74.

In Exercises 41 through 44, use the letter-to-number conversions in Table 3.4 (repeated below) and a **linear cipher** with $n = 27$ and the specified encrypting values a and b. Using software such as *Mathematica* or *Maple* (to evaluate functions on lists of input) might also be wise.

A	B	C	D	E	F	G	H
0	1	2	3	4	5	6	7	8

I	J	K	L	M	N	O	P	Q
9	10	11	12	13	14	15	16	17

R	S	T	U	V	W	X	Y	Z
18	19	20	21	22	23	24	25	26

41. A certain university uses a linear cipher with $a = 2$ and $b = 8$ to encrypt student course registration data. Determine how the course "DISCRETE MATH" would be encrypted at this college.

42. To protect against hackers, a certain Internet service provider encrypts its customers' passwords using a linear

cipher with $a = 5$ and $b = 3$. Determine how the customer password "BINARY CODE" would be encrypted.

43. In 2007, Jack Nicholson might have used the encrypted message "QPVOZ" to submit his vote for the Academy Award for Best Adapted Screenplay. Assuming that Jack used a linear cipher with $a = 2$ and $b = 13$, decrypt that message.

44. In December of 1941, Japanese commander Admiral Isoroku Yamamoto might have sent the encrypted message "KIIKY DUAC" (here translated into English) to his fleet. Assuming that Admiral Yamamoto used a linear cipher with $a = 7$ and $b = 4$, decrypt that message.

45. Let $n \in \mathbb{Z}$ with $n > 1$. Show that no pair of distinct integers from the list $0, 1, 2, \ldots, n - 1$ is congruent modulo n.

46. Let $a, n \in \mathbb{Z}$ with $n > 1$ and $\gcd(a, n) = 1$. Show that no pair of distinct integers from the list $a \cdot 0, a \cdot 1, a \cdot 2, \ldots, a(n - 1)$ is congruent modulo n.

47.* Given $a, n \in \mathbb{Z}$ with $n > 1$ and $\gcd(a, n) = 1$, show that there is a unique value $c \in \{0, 1, \ldots, n - 1\}$ that is the multiplicative inverse of a modulo n. This uniqueness result expands upon the existence result in Lemma 3.31.

48.* Given $a, b, n \in \mathbb{Z}$ with $n > 1$, show that for each $y \in \{0, 1, \ldots, n - 1\}$, there is a unique solution $x \in \{0, 1, \ldots, n - 1\}$ to the equation $y = (ax + b) \bmod n$ if and only if $\gcd(a, n) = 1$.

In Exercises 49 through 52, use binary expansion as in Example 3.44 to do the calculations.

49. $13^{29} \bmod 77$.

50. $11^9 \bmod 65$.

51. $31^{17} \bmod 87$.

52. $13^{19} \bmod 62$.

53. The BooksForCheap Web site uses the RSA encryption method to encrypt customer credit card orders. Consequently, even if a criminal intercepts a customer's order, the criminal will not be able to decrypt that customer's

TABLE 3.5
Messages representing credit cards

Message (x)	Credit card
8	American Express
15	Diner's Club
23	Discover
30	MasterCard
38	Sears
49	Visa

credit card information. Table 3.5 shows the numerical codes (messages) used by BooksForCheap to represent the credit cards it accepts. Assume that BooksForCheap uses the RSA encryption method with $p = 5$, $q = 11$, $a = 7$, and $c = 3$.

(a) Determine the numerical value to which "American Express" encrypts.

(b) Determine the numerical value to which "Visa" encrypts.

(c) Suppose BooksForCheap receives an order with the encrypted credit card value $y = 12$. Decrypt this value to determine the type of credit card used by the customer.

(d) Suppose BooksForCheap receives an order with the encrypted credit card value $y = 35$. Decrypt this value to determine the type of credit card used by the customer.

54. The cable company TvByWire uses the RSA encryption method to receive pay-per-view movie orders from its customers. This prevents nonpaying customers from receiving the movies for free. Table 3.6 shows the numerical codes (messages) used by TvByWire to represent the movies it has available. Assume that TvByWire uses the RSA encryption method with $p = 13$, $q = 17$, $a = 5$, and $c = 29$.

(a) Determine the numerical value to which *Final Analysis* encrypts.

(b) Determine the numerical value to which *The Matrix* encrypts.

(c) Suppose TvByWire receives an order with the encrypted movie value $y = 2$. Decrypt this value to determine the movie requested by the customer.

(d) Suppose TvByWire receives an order with the encrypted movie value $y = 11$. Decrypt this value to determine the movie requested by the customer.

TABLE 3.6
Messages representing movies

Message (x)	Movie
7	A Beautiful Mind
9	Final Analysis
20	Good Will Hunting
32	It's My Turn
33	The Matrix
41	Seven
45	Stand and Deliver

55. Compute 10^{1000} mod 17.

56. Compute 3^{400} mod 7.

57. Prove Corollary 3.33.

58. Prove Corollary 3.34.

59. Use Corollary 3.34 to show that 253 is not prime.

60. Use Corollary 3.34 to show that 209 is not prime.

61.* Prove **Wilson's Theorem**: If p is prime, then $(p-1)! \equiv -1 \pmod{p}$. Hint: In the product $(p-1)!$, pair off each integer in $\{2, \ldots, p-2\}$ with its multiplicative inverse modulo p. Factorials are defined in Section 4.1.

62.* Prove the converse of Wilson's Theorem: If $n \in \mathbb{Z}$ with $n > 1$ and $(n-1)! \equiv -1 \pmod{n}$, then n is prime. Hint: If not, then find a common divisor of $(n-1)!$ and $(n-1)! + 1$.

63.* Recall from Example 3.16 in Section 3.2 that an ISBN $d_1 d_2 d_3 d_4 d_5 d_6 d_7 d_8 d_9 d_{10}$ satisfies $[10d_1 + 9d_2 + 8d_3 + 7d_4 + 6d_5 + 5d_6 + 4d_7 + 3d_8 + 2d_9 + d_{10}] \bmod 11 = 0$.
Show that this holds if and only if

$$[d_1 + 2d_2 + 3d_3 + 4d_4 + 5d_5 + 6d_6$$

$$+ 7d_7 + 8d_8 + 9d_9 + 10d_{10}] \bmod 11 = 0.$$

64.* **RSA decryption**. Given integers a and c and primes p and q, let $n = pq$ and $m = \text{lcm}(p-1, q-1)$. Show that if $ac \equiv 1 \pmod{m}$ and $y = x^a \bmod n$ for some $x \in \{0, 1, \ldots, n-1\}$, then $x = y^c \bmod n$. Hint: Use Fermat's Little Theorem.

65. Find an integer x such that $x^2 \equiv -1 \pmod{10}$. Observe that $9! \not\equiv -1 \pmod{10}$, and compare Exercise 61.

66. Find an integer x such that $3x \equiv 5 \pmod{7}$.

67. Find integers x and y such that $x \not\equiv 0 \pmod{10}$ and $y \not\equiv 0 \pmod{10}$ but $xy \equiv 0 \pmod{10}$. This shows that the set of integers modulo 10 does not have the zero multiplication property (Appendix A, property 6) as $\mathbb{R}$ does.

68. Find integers x and y such that $x \not\equiv \pm y \pmod{8}$ but $x^2 \equiv y^2 \pmod{8}$. This shows that the set of integers modulo 8 cannot have uniquely defined square roots (Appendix A, property 16) as $\mathbb{R}^+$ does.

In Exercises 69 through 74, for the relevant n, express the given set in the form $[b]_n$, where $0 \le b \le n-1$.

69. $[8]_3$.

70. $[52]_{10}$.

71. $[10]_4 + [5]_4$.

72. $[12]_5 + [17]_5$.

73. $[18]_{10} + [217]_{10} +$
    $[3146]_{10}.$

74. $[18]_{12} + [217]_{12} +$
    $[3146]_{12}.$

In Exercises 75 through 82, let $a, b, c, n \in \mathbb{Z}$ with $n > 1$.

75. Show: $[a]_n = [b]_n$ if and only if $a \equiv b \pmod{n}$.

76. Show: If $[a]_n \cap [b]_n \neq \emptyset$, then $[a]_n = [b]_n$.

77. Show: $[a]_n + [b]_n = [a + b]_n$. Consequently, $[a]_n + [b]_n = [b]_n + [a]_n$.

78. Show: $-[b]_n = [-b]_n$.

79. Show: $[a]_n - [b]_n = [a - b]_n$.

80. Show: $([a]_n + [b]_n) + [c]_n = [a]_n + ([b]_n + [c]_n)$.

81. Show: $[0]_n + [a]_n = [a]_n$.

82. Show: $[-a]_n + [a]_n = [0]_n$.

Divisibility tests. For Exercises 83 through 86, let $a_k a_{k-1} \cdots a_0$ be the base-ten representation of an integer n. That is,

$$n = 10^k a_k + 10^{k-1} a_{k-1} + \cdots + 10^0 a_0.$$

83. Show: $n \equiv a_0 \pmod{5}$. This proves that the last digit alone determines whether or not an integer is divisible by 5.

84. Let m be the integer whose base-ten representation is $a_1 a_0$. Show: $n \equiv m \pmod{4}$. This proves that the last two digits determine whether or not an integer is divisible by 4.

85.* Let $m = a_k + a_{k-1} + \cdots + a_0$.
    Show: $n \equiv m \pmod{9}$. This proves that an integer is divisible by 9 if and only if the sum of its digits is divisible by 9. Prove that this is also true if 9 is replaced by 3.

86.* Let $m = (-1)^k a_k + (-1)^{k-1} a_{k-1} + \cdots + (-1)^0 a_0$.
    Show: $n \equiv m \pmod{11}$. This proves that an integer is divisible by 11 if and only if the alternating sum of its digits is divisible by 11.

3.6   Review problems

1. Show that the product of any two even integers is even.

2. Show: $\forall\, n \in \mathbb{Z}$, if $4 \nmid n^2$ then n is odd.

3. In a game of bridge, all 52 cards in a standard deck are dealt to four players so that each gets the same number of cards. Suppose we wish to invent a game like bridge, but for six players. Can a deck of 52 cards be split evenly among six players? Why?

4. Let $a, b \in \mathbb{Z}$. Show: $a \mid ((a + b)^3 - b^3)$.

5. Let $a, b \in \mathbb{Z}$. Show: $a \mid b$ if and only if $-a \mid b$.

6. Is 91 prime? Why? Is 1 prime? Why?

7. Are 14 and 33 relatively prime? Why?

8. Let a and n be integers. Show: If $a \mid n$ and $a \mid (n + 2)$, then $a \mid 2$.

9. Find $\gcd(1001, 455)$ by factoring.

10. Let $n \in \mathbb{Z}$ with $n \neq 0$. Show: $\gcd(n, -n) = |n|$. Note that
$$|n| = \begin{cases} n & \text{if } n \geq 0, \\ -n & \text{if } n < 0. \end{cases}$$

11. Let a and b be integers not both zero, and let m and n be positive integers.

    (a) Find specific values for a, b, m, n with $m, n > 1$ and $\gcd(a, b) = \gcd(a^m, b^n)$.

    (b) In general, show that $\gcd(a, b) \leq \gcd(a^m, b^n)$.

    (c)* What additional conditions on a, b, m, n are forced by the strict inequality $\gcd(a, b) < \gcd(a^m, b^n)$?

12. Find $\text{lcm}(120, 84)$.

13. Observe that $\text{lcm}(6^9, 6^{15}) = 6^{15}$.
    Show: $\forall\, m, j, k \in \mathbb{N}$ with $m > 0$, that $\text{lcm}(m^j, m^k) = m^{\max\{j,k\}}$.

14. Find the smallest element of $\{s : s = 10x + 25y$ where $x, y \in \mathbb{Z}$ and $10x + 25y > 0\}$.

15. Recall that a Mersenne prime is a prime of the form $2^p - 1$ for some prime p. Decide whether the following statement is true or false. Since 113 is prime, so is $2^{113} - 1$. Hint: See the GIMPS website.

16. Find the quotient and remainder resulting from the integer division problem $101 \div 8$.

17. Compute

    (a) 43 div 7 and 43 mod 7.

    (b) -51 div 9 and -51 mod 9.

18. Suppose two standard decks of cards (52 cards in each) are shuffled together and dealt to 6 players as follows. For each round of the deal, if there are at least 6 cards left in the pile, then each player is given one more card. When fewer than 6 cards remain, the dealing stops. How many cards will remain when the dealing stops? How many cards will each player have? Compare Exercise 3.

19. Show: $\forall\, n \in \mathbb{Z}$, $(n^3 - n) \bmod 3 = 0$.

20. Let $n \in \mathbb{Z}$. Show: If n is odd, then $4 \nmid n^2$. Note that this is the converse of Exercise 2.

21. Find (a) $\lfloor 6.6 \rfloor$. (b) $\lfloor -6.6 \rfloor$.

22. Find (a) $\lceil 5.4 \rceil$. (b) $\lceil -5.4 \rceil$.

23. Show: $\forall \, n \in \mathbb{Z}$, if $4 \mid n$, then $\lfloor \frac{n+2}{4} \rfloor = \frac{n}{4}$.

24. Show: $\forall \, x \in \mathbb{R}$, $\lceil \lceil x \rceil \rceil = \lceil x \rceil$.

25. One of the digits in the ISBN of a book purchased at a yard sale is smudged. If the ISBN is 0-8218-#461-4, what is the value of the smudged digit #? If two digits were smudged, could they be determined by the remaining digits?

26. Two high school girls have agreed to send secret messages using the letter-to-number conversions " " = 0, A = 1, ..., Z = 26 and a shift cipher with $n = 27$ and encrypting shift value $b = 4$. Thus, if the teacher intercepts a message, he will not be able to read it. How would one of these girls encrypt the message "I LOVE TOM"?

27.* A shift cipher for a given n and b will contain cycles

$$x_1 \text{ encrypts to } x_2 \text{ encrypts to}$$

$$\cdots \text{ encrypts to } x_k \text{ encrypts to } x_1.$$

For example, when $n = 27$ (with letter-to-number conversions " " = 0, A = 1, ..., Z = 26) and $b = 6$, we have the cycle

$$A \to G \to M \to S \to Y \to D \to J$$
$$\to P \to V \to A$$

of length 9. In general, in terms of n and b, what is the length k of these cycles? Try various examples, and discover a pattern.

28. Through trial and error, find $x, y \in \mathbb{Z}$ such that $35x + 49y = 7$.

29. Prove or disprove: $\exists \, x, y \in \mathbb{Z}$ such that $85x + 39y = 1$.

30. Use Euclid's Algorithm to find $\gcd(110, 88)$.

31. Use Euclid's Algorithm to find $\gcd(810, 245)$.

32. Use the Extended Euclidean Algorithm to find integers x and y such that $\gcd(81, 45) = 81x + 45y$.

33. Use the Extended Euclidean Algorithm to find integers x and y such that $77x + 24y = 1$.

34. Observe that $5 \mid 4^3 10^2$ and, in particular, $5 \mid 10$. Let $a, b, m, n \in \mathbb{Z}$ with $m, n > 0$. Show: If $5 \mid a^m b^n$, then $5 \mid a$ or $5 \mid b$.

35. In Exercises 3 and 18, we considered dealing decks of cards to six players. Show that n decks will split evenly among six players if and only if n is divisible by 3.

36. Let $a, b \in \mathbb{Z}$. Is it true that if $10 \mid ab$, then $10 \mid a$ or $10 \mid b$? Justify your answer.

37. Suppose you wish to pay for everything using only quarters and dimes.

    (a) Show that there are infinitely many monetary amounts (dollars and cents) that cannot be achieved in this way.

    (b) What multiples of 5 cents cannot be achieved in this way?

    (c) If we also had coins worth two cents available, then what values could we not achieve?

38. Show that $6\frac{3}{4}$ is rational.

39. Show that $1.\overline{414}$ is rational. Is this number bigger or smaller than $\sqrt{2}$? Justify your answer without using a calculator.

40. Show that $1.6\overline{25}$ is rational.

41. Show: $\forall\, r \in \mathbb{R}$, if $r \in \mathbb{Q}$, then $\frac{3r}{4} \in \mathbb{Q}$.

42. Let $r \in \mathbb{Q}$. Show: r^2 is rational.

43. Observe that both $\frac{14}{15}$ and $(\frac{14}{15})^2 = \frac{196}{225}$ are in lowest terms. Suppose $r = \frac{a}{b}$ is a rational number written in lowest terms. Show that $r^2 = \frac{a^2}{b^2}$ is also in lowest terms.

44. Without using a calculator, write $\frac{5}{11}$ in decimal form. Show your work.

45. Show that $\sqrt{7}$ is irrational by mimicking the proof that $\sqrt{2}$ is irrational.

46. Use the result from Exercise 45 to show that $\frac{5+\sqrt{7}}{3}$ is irrational.

47. Show that $\log_3 11$ is irrational.

48. In 1737, Euler proved that e and e^2 are irrational. In 1768, Lambert proved that all nonzero integer powers of e are irrational. Given that every nonzero integer power of e is irrational, do the following:

    (a) Show that $\frac{e^2-4}{3}$ is irrational.

    (b) Show that $\ln 2$ is irrational. Note that $\ln 2$ is defined to be $\log_e 2$.

49. Show that $\sqrt{3 + \sqrt{2}}$ is irrational.

50. Show that $\frac{\sqrt[3]{2}}{\sqrt{5}}$ is irrational.

51. Using the trigonometric identity $(2\cos^2(\theta) - 1)^2 = \cos^2(2\theta) = \frac{1+\cos(4\theta)}{2}$ for $\theta = 11.25°$, observe that we get $\cos(11.25°) = \frac{1}{2}\sqrt{2 + \sqrt{2 + \sqrt{2}}}$. Use this to prove that $\cos(11.25°)$ is irrational.

52. Is $\sqrt{6 + 4\sqrt{2}} + \sqrt{6 - 4\sqrt{2}}$ irrational? Why?

53. Are the roots of $\frac{5}{4}x^2 - \frac{2}{3}x + 1$ algebraic? Why?

54. February 1, 2004, the New England Patriots won Super Bowl XXXVIII. That same year, on October 27, the Boston Red Sox won the World Series and finally broke the curse of the Bambino. Since the Super Bowl occurred on a Sunday, determine the day of the week that the Red Sox won the 2004 World Series.

55. Let a, b, c and n be integers with $n > 1$.
    Show: If $a \equiv b \pmod{n}$, then
    $ac \equiv bc \pmod{n}$.

56. Compute (a) 11^{10} mod 9. (b) 23^{4321} mod 12.

57. Let $n \in \mathbb{Z}$. Show: If n is odd, then $n^2 \equiv 1 \pmod{4}$.

58. Let $n \in \mathbb{Z}$. Show: If $n \not\equiv 0 \pmod{3}$, then $n^2 \equiv 1 \pmod{3}$.

59. To enter a high security area, a CIA operative needs a three-character code to key into the lock. Since this code must be kept secret, the letter-to-number conversions " " = 0, A = 1, ..., Z = 26 and a linear cipher with $n = 27$, $a = 4$, and $b = 1$ are used to encrypt the code, before sending it to the operative. If the encrypted code is "SWE", what code opens the lock?

60. Use the Extended Euclidean Algorithm to find the multiplicative inverse of 11 modulo 50.

61. Use binary expansion and repeated squaring to compute 19^{49} mod 391.

62. All of the CIA operatives in the field send messages to CIA headquarters using the RSA encryption method for $p = 7$, $q = 13$, and $a = 17$. The numerical equivalents of some standard messages are shown in Table 3.7. Since only CIA headquarters knows the value $c = 5$ needed to decrypt the messages, no operative can decrypt another operative's message.

TABLE 3.7
CIA message codes

Message (x)	Translation
2	I'm in position.
3	Target is destroyed.
60	The package has been delivered.
61	The package has been received.
99	I've been caught.

   (a) Determine the numerical value to which "I'm in position" encrypts.

   (b) Suppose CIA headquarters receives the encrypted value 3 from one of its operatives. Decrypt this value to determine the message that was sent.

63. Compute 9^{5432} mod 11.

In Exercises 64 through 66, for the relevant n, express the given set in the form $[b]_n$, where $0 \leq b \leq n - 1$.

64. $[7]_5$.

65. $[8]_3 + [2]_3$.

66. $[17]_7 - [208]_7 + [1343]_7$.

67. Let $a, n \in \mathbb{Z}$ with $n > 1$. Show that $[a]_n$ is not empty.

4 Indexed by Integers

We continue the study begun in Chapter 3 of results and proofs in number theory. Our main goal here is to introduce the powerful proof technique known as mathematical induction. Since induction is used to prove statements that are indexed by the integers, sequences are discussed first. Recursion is also introduced, and we explore sequences expressed recursively.

After some results have been presented that can be proved by induction, we then proceed to do so. In the exercises, we explore extensions of two-variable identities to n variables, and we consider applications to annuities and mortgages. Summation formulas provide another major source of examples.

After the more powerful form of induction known as strong induction is introduced, we prove some important theorems in number theory, highlighted by the Fundamental Theorem of Arithmetic. In the last section, the Binomial Theorem is presented.

4.1 Sequences, indexing, and recursion

A basic acquaintance with sequences is required before induction can be presented. Familiarity with some standard notation is also required.

4.1.1 Factorials and binomial coefficients

For all integers $n \geq 1$, the notation $n!$, read **n factorial**, is defined to be the product

$$n! = n(n-1)(n-2)\cdots 2 \cdot 1.$$

Additionally, we define

$$0! = 1.$$

It is easy to compute small examples, such as

$$0! = 1, \quad 1! = 1, \quad 2! = 2, \quad 3! = 6, \quad 4! = 24,$$
$$5! = 120, \quad 6! = 720, \ldots, \tag{4.1}$$

but the numbers get large rather quickly.

Given integers n and k with $0 \le k \le n$, the **binomial coefficient** $\binom{n}{k}$ is defined as

$$\binom{n}{k} = \frac{n!}{k!(n-k)!}.$$

The reason for calling $\binom{n}{k}$ a binomial coefficient is seen in Section 4.6, where we see values of the form $\binom{n}{k}$ serving as coefficients in the Binomial Theorem. The notation $\binom{n}{k}$ is read "n choose k." In Chapter 6, we learn that $\binom{n}{k}$ counts the number of ways to choose k objects from a set of n distinct objects. For our purposes here, simple examples of binomial coefficients can be computed explicitly from the defining formula.

Example 4.1.   (Computations of Binomial Coefficients).

$$\binom{6}{4} = \frac{6!}{4!2!} = \frac{6 \cdot 5 \cdot 4!}{4! \cdot 2 \cdot 1} = 15.$$

$$\binom{6}{3} = \frac{6!}{3!3!} = \frac{6 \cdot 5 \cdot 4}{3 \cdot 2 \cdot 1} = 20.$$

$$\binom{n}{0} = \frac{n!}{0!n!} = 1.$$

$$\binom{n}{1} = \frac{n!}{1!(n-1)!} = n.$$

$$\binom{n}{2} = \frac{n!}{2!(n-2)!} = \frac{n(n-1)}{2}.$$

$$\binom{n}{n-4} = \frac{n!}{(n-4)!4!} = \frac{n(n-1)(n-2)(n-3)}{24}.$$

The computation $\binom{n}{0} = 1$ motivates our choice that $0! = 1$. Factorials and binomial coefficients are used heavily in the counting problems of Chapter 6. It is there that the true value of reading $\binom{n}{k}$ as "n choose k" can be appreciated.

Although C++ has no predefined functions for factorials and binomial coefficients, *Mathematica* provides `n!` (or `Factorial[n]`) and `Binomial[n,k]`, and *Maple* similarly provides `n!` (or `factorial(n)`) and `binomial(n,k)`.

4.1.2   Sequences

A **sequence** is simply an ordered list of real numbers. All sequences have an initial term, but only finite sequences have a final term. Our primary focus will be on infinite sequences. We have already encountered the sequence of factorials in (4.1). Other examples are plentiful.

$$1, 2, 3, 4, 5, 6, \ldots \qquad (4.2)$$

$$7, 10, 13, 16, 19, 22, \ldots \qquad (4.3)$$

$$1, 2, 4, 8, 16, 32, 64, \ldots \qquad (4.4)$$

$$6e, 18e, 54e, 162e, 486e, 1458e, \ldots \qquad (4.5)$$

$$5, -9, 13, -17, 21, -25, \ldots \qquad (4.6)$$

$$10, 15, 21, 28, 36, 45, \ldots \qquad (4.7)$$

$$-2, 11, 0, 23, -5, 4, \ldots \qquad (4.8)$$

Sequence (4.8) is meant to emphasize the fact that there need not be an apparent pattern. However, we work primarily with sequences for which there is a visible pattern, and consequently, the terms in the sequence can be given by a formula.

We express a sequence in the form $\{s_n\}_{n \geq a}$ and mean that our sequence consists of the terms

$$s_a, \ s_{a+1}, \ s_{a+2}, \ s_{a+3}, \ s_{a+4}, \ s_{a+5}, \ldots,$$

where $a \in \mathbb{Z}$. That is, the sequence has been **indexed** by the integers

$$a, a + 1, a + 2, a + 3, a + 4, a + 5, \ldots.$$

Usually, $a = 0$ or 1, but that need not be the case. If the indexing is clear in context, then a sequence may be displayed as $\{s_n\}$, or just a formula for s_n may be given. Sequences (4.1) through (4.7) can each be expressed by formulas.

$$(4.1) \qquad \forall n \geq 0, \ s_n = n!.$$

$$(4.2) \qquad \forall n \geq 1, \ s_n = n.$$

$$(4.3) \qquad \forall n \geq 0, \ s_n = 7 + 3n.$$

$$(4.4) \qquad \forall n \geq 0, \ s_n = 2^n.$$

$$(4.5) \qquad \forall n \geq 1, \ s_n = 2e \cdot 3^n.$$

$$(4.6) \qquad \forall n \geq 0, \ s_n = (-1)^n (5 + 4n).$$

$$(4.7) \qquad \forall n \geq 5, \ s_n = \binom{n}{2}.$$

Sequence (4.8) has no obvious formula. Note that the alternating signs in sequence (4.6) are obtained from the factor $(-1)^n$ in its formula. The simple formula given in sequence (4.7) shows one reason why starting the indexing at $a = 0$ or 1 may not always be the best choice. The alternative formula, $\binom{n+5}{2}$, for $n \geq 0$, is not as concise.

Infinite sequences can be handled in software implementations as special cases of functions.[*] In *Mathematica*, sequence (4.5) can be entered as

```
In[1]:= s[n_] := 2*E*3^n.
```

The underline symbol after the n on the left-hand side denotes that n is the variable when it occurs on the right-hand side. In C++, sequence (4.6) can be programmed as the function

```
int s(int n)
    {
     return (int) pow(-1,n)*(5+4*n);
    }.
```

The occurrences of the type identifier int simply announce that the object following it shall be an integer. In *Maple*, sequence (4.7) can be entered as

```
> s := n -> binomial(n,2) ; .
```

The symbol -> dictates that each input index n shall yield the output value that follows it. In each case, a desired term in the sequence can be computed by evaluating the function s at the appropriate index. Note that the initial index (here, 1, 0, or 5, respectively) is not specified in these definitions, and hence the user needs to take care in each call of s that only an appropriate index is used.

Formulas for sequences (4.1) through (4.7) were presented without explaining how to find them. Being able to find a formula for a sequence is a valuable skill that comes from seeing examples and discerning patterns. Some exercises are devoted to practicing that skill after some more examples are seen.

4.1.3   Two special kinds of sequences

A sequence in which each successive term is obtained by adding a fixed number c to the previous term is called an **arithmetic sequence**. Such a sequence, with given first term s_0, is generated by the formula

$$\forall n \geq 0, \ s_n = s_0 + cn.$$

Sequences (4.2) and (4.3) are examples of arithmetic sequences.

A sequence in which each successive term is obtained by multiplying the previous term by a fixed number r is called a **geometric sequence**. Such a sequence, with given first term s_0, is generated by the formula

$$\forall n \geq 0, \ s_n = s_0 r^n.$$

Sequences (4.4) and (4.5) are examples of geometric sequences.

---

[*] We address functions in general in Chapter 5.

4.1.4 Reindexing a sequence

One of the issues that arises in giving a formula for a sequence is how to index it. Sometimes the indexing in which an original formulation is written may not be the most convenient, and reindexing may be desired. Reindexing a sequence is a simple matter of substituting the index variable.

Example 4.2. The sequence given by

$$\forall n \geq 3, \; s_n = 5 + 2^n$$

has terms

$$s_3 = 13, \; s_4 = 21, \; s_5 = 37, \; s_6 = 69, \ldots.$$

If we prefer to have this sequence indexed starting at 1, then we accomplish this with the variable substitution $m = n - 2$. A new formula t_m for the sequence is obtained by observing that $n = m + 2$. That is,

$$\forall m \geq 1, \; t_m = 5 + 2^{m+2} = 5 + 4 \cdot 2^m.$$

The terms

$$t_1 = 13, \; t_2 = 21, \; t_3 = 37, \; t_4 = 69, \ldots$$

are the same as before. Only the indexing has changed. Since there is nothing special about the choice of variable used for indexing, we may also write

$$\forall n \geq 1, \; t_n = 5 + 4 \cdot 2^n.$$

Example 4.3. Reindex the sequence given by $\forall n \geq 5, \; s_n = \binom{n-3}{2}$, so that the indexing starts at 0.

Solution. Let $m = n - 5$. So $n = m + 5$, and

$$\forall m \geq 0, \; t_m = \binom{m + 5 - 3}{2} = \binom{m + 2}{2}.$$

Equivalently, $\forall n \geq 0, \; t_n = \binom{n+2}{2}$. $\qquad\qquad\square$

4.1.5 Recursion

A **closed formula** for a sequence $\{s_n\}_{n \geq a}$ is a formula that expresses the value of s_n in terms of the index n. That is, it allows s_n to be computed directly from n. Each of the formulas that we have seen so far for sequences has been a closed formula. In contrast, a **recursive formula** is a formula in which s_n is expressed not only in terms of n but also in terms of some of the earlier values $s_a, \ldots, s_{n-1}$ in the sequence. More generally, a recursive function is a function whose value for a given input is expressed in terms of its values for smaller inputs.

A recursive formula can provide a more convenient and more natural way to express some sequences. For example, reconsider the sequence of factorials in (4.1). We can express this recursively by

$$(4.1) \qquad s_0 = 1 \text{ and } \forall n \geq 1, \ s_n = ns_{n-1}.$$

The equation $s_n = ns_{n-1}$, which is defined for all integers $n \geq 1$, is called the **recurrence relation**. This equation expresses the value of s_n in terms of earlier values in the sequence. It may refer to n directly, as is done in this case, and could refer to more than one earlier value. The assignment $s_0 = 1$ is called the **initial condition** and specifies the value of the first term in the sequence. In some cases, more than one initial term may be included, when additional initial conditions are warranted.

Since the recurrence relation expresses s_n in terms of earlier values, and those earlier values may be defined in terms of yet earlier values, and so on, the initial conditions play the important role of terminating what would otherwise be a non-terminating cycle of self-reference by a sequence. We can see this nicely by tracing how our recursive formula for sequence (4.1) enables us to compute the value of s_3.

$$
\begin{array}{ll}
s_3 = 3s_2, & \text{(What is } s_2?) \\
\quad s_2 = 2s_1, & \text{(What is } s_1?) \\
\qquad s_1 = 1s_0, & \text{(What is } s_0?) \\
\qquad\quad s_0 = 1, & \text{(The initial condition.)} \\
\qquad s_1 = 1, & \text{(We get } s_1 \text{ from } s_0.) \\
\quad s_2 = 2, & \text{(We get } s_2 \text{ from } s_1.) \\
s_3 = 6. & \text{(We get } s_3 \text{ from } s_2.)
\end{array}
$$

Of course, 6 is the expected value for 3!.

Sequences (4.2) through (4.7) can also be expressed recursively.

(4.2) $\quad s_1 = 1$ and $\forall n \geq 2, \ s_n = s_{n-1} + 1.$

(4.3) $\quad s_0 = 7$ and $\forall n \geq 1, \ s_n = s_{n-1} + 3.$

(4.4) $\quad s_0 = 1$ and $\forall n \geq 1, \ s_n = 2s_{n-1}.$

(4.5) $\quad s_1 = 6e$ and $\forall n \geq 2, \ s_n = 3s_{n-1}.$

(4.6) $\quad s_0 = 5$ and $\forall n \geq 1, \ s_n = -s_{n-1}\left(1 + \dfrac{4}{|s_{n-1}|}\right).$

(4.7) $\quad s_5 = 10$ and $\forall n \geq 6, \ s_n = s_{n-1} + n - 1.$

Another way of expressing sequence (4.6) is

$$s_0 = 5, s_1 = -9, \text{ and } \forall n \geq 2, \quad s_n = \begin{cases} s_{n-2} + 8 & \text{if } n \text{ is even,} \\ s_{n-2} - 8 & \text{if } n \text{ is odd.} \end{cases}$$

This second expression for sequence (4.6) expresses values in the sequence in terms of the second most recent value.

Example 4.4. (Recursion in Arithmetic and Geometric Sequences).
   (a) An arithmetic sequence is very naturally expressed recursively by specifying s_0 and a constant c for which

$$\forall n \geq 1, \quad s_n = s_{n-1} + c.$$

   (b) A geometric sequence is also naturally expressed recursively by specifying s_0 and a constant r for which

$$\forall n \geq 1, \quad s_n = rs_{n-1}.$$

Example 4.5. The following recurrence relation involves more than one previous value in the sequence:

$$s_0 = 1, s_1 = 2, \text{ and } \forall n \geq 2, \quad s_n = s_{n-1} + 2s_{n-2}. \qquad (4.9)$$

Note that the earliest term on the right-hand side of the recurrence relation is two terms prior to the term on the left-hand side, and there are two initial conditions. In general, the minimum requirement of initial conditions is determined by the right-hand side of the recurrence relation in this way. The terms in sequence (4.9) start out

$$1, \ 2, \ 4, \ 8, \ 16, \dots.$$

In fact, sequence (4.9) is the same as sequence (4.4).

Example 4.6. The sequence expressed recursively by

$$s_0 = 1 \text{ and } \forall n \geq 1, \quad s_n = 2s_{n-1}$$

might alternatively be expressed recursively by

$$s_0 = 1 \text{ and } \forall n \geq 0, \quad s_{n+1} = 2s_n.$$

To see this formally, rewrite the second expression (using m as the index) as

$$s_0 = 1 \text{ and } \forall m \geq 0, \ s_{m+1} = 2s_m.$$

Set $n = m + 1$, so $m = n - 1$, and the inequality $m \geq 0$ becomes the inequality $n \geq 1$. Substitution now yields the first expression

$$s_0 = 1 \text{ and } \forall n \geq 1, \ s_n = 2s_{n-1}.$$

Of course, it is also easy to see this equivalence informally. Both expressions say that after the initial value, each subsequent value in the sequence is twice the previous value.

Example 4.7. Given the sequence expressed recursively by

$$s_1 = 1, s_2 = 3, \text{ and } \forall n \geq 3, \ s_n = 2s_{n-2} + s_{n-1},$$

and given an index $k \geq 2$, find an expression for s_{k+1} in terms of earlier values in the sequence.

Solution.
Let $n = k + 1$. Since $k \geq 2$, we have $n \geq 3$. So the formula

$$s_n = 2s_{n-2} + s_{n-1}$$

may be applied. This gives

$$s_{k+1} = 2s_{k+1-2} + s_{k+1-1}.$$

That is,

$$s_{k+1} = 2s_{k-1} + s_k. \qquad \square$$

4.1.6   Software implementation of recursion

Since recursion can be a powerful programming tool, mathematical software and programming languages generally support it.

In *Mathematica*, sequence (4.3) can be defined recursively as follows:

```
In[1]:= s[n_] := s[n-1] + 3

In[2]:= s[0] := 7
```

Both the recurrence relation and the initial conditions must be specified.

In C++, we can recursively define a function `factorial` (which we have previously noted is not predefined) to produce the values in sequence (4.1) as follows:

```
int factorial(int n)
    {
      if (n > 0) return n*factorial(n-1);
      else return 1;
    }
```

The input is assumed to be a natural number. If it is positive, then the function calls itself with smaller input. Otherwise, the value $0! = 1$ is returned.

In *Maple*, sequence (4.4) can be defined recursively as follows:

```
> s := n -> 2*s(n-1) ;
```

```
> s(1) := 1 ;
```

Both the recurrence relation and the initial conditions must be specified.

Exercises

Each of the following exercises can be completed by hand. However, those requesting computations might also be done with the aid of a symbolic calculator such as *Mathematica* or *Maple*. The results can then be compared with hand computations.

1. Compute 10!.

2. Compute 12!.

3. Compute $\binom{7}{5}$.

4. Compute $\binom{8}{3}$.

5. Compute $\binom{9}{4}$.

6. Compute $\binom{10}{4}$.

7. Use the defining formula for binomial coefficients to verify, for all integers $0 \le k \le n$, that $\binom{n}{k} = \binom{n}{n-k}$.

8. Use the defining formula for binomial coefficients to verify, for all integers $1 \le k \le n$, that $\binom{n}{k-1} + \binom{n}{k} = \binom{n+1}{k}$.

In Exercises 9 through 12, determine whether the stated formula is true or false.

9. $\forall n \in \mathbb{N}, \ (n^2)! = (n!)^2$.

10. $\forall n \in \mathbb{N}, \ (n+1)! = (n+1)n!$.

11. $\forall$ even $n \in \mathbb{N}$, $(\frac{n}{2})! = \frac{n!}{2}$.

12. $\forall k \le n \in \mathbb{N}$, $\binom{n}{k} = \binom{2n}{2k}$.

In Exercises 13 through 18, find the first four terms of each sequence.

13. $\forall n \geq 0, \; s_n = 4 - 2n.$ 14. $\forall n \geq 1, \; s_n = 10^{n-3}.$

15. $\forall n \geq 3, \; s_n = \frac{n!}{n-2}.$ 16. $\forall n \geq 0, \; s_n = 6 - 3 \cdot 2^n.$

17. $\forall n \geq 2, \; s_n = 3 + 2n.$ 18. $\forall n \geq 6, \; s_n = \binom{n}{3}.$

19. What is the 15th term in the sequence of prime numbers? Recall that the first is 2, not 1.

20. What is the 15th term in the sequence of composite numbers? The first is 4.

In Exercises 21 through 28, find a closed formula for each sequence.

21. $2, 4, 6, 8, 10, \ldots$ 22. $1, -4, 9, -16, 25, \ldots$

23. $3, 6, 12, 24, 48, \ldots$ 24. $\frac{\pi}{2}, \frac{3\pi}{2}, \frac{5\pi}{2}, \frac{7\pi}{2}, \frac{9\pi}{2}, \ldots$

25. $1, -3, 5, -7, 9, \ldots$ 26. $5\sqrt{2}, 10\sqrt{2}, 20\sqrt{2}, 40\sqrt{2}, 80\sqrt{2}, \ldots$

27. $1, \frac{1}{2}, \frac{1}{3}, \frac{1}{4}, \frac{1}{5}, \ldots$ 28.* $1, 4, 10, 20, 35, 56, \ldots$
Hint: Consider binomial coefficients.

29. **Compound interest.** An initial investment of $s_0 = P$ dollars is made into an account earning periodic interest rate i. At the end of each period, interest is added to the previous balance.

(a) If $P = \$6000$ and $i = 0.03$, then what is the amount in the account after 1 period? after 2 periods?

(b) If $P = \$1000$, then express, in terms of an unspecified interest rate i, the amount in the account after 2 periods. Is twice as much interest earned when $i = 0.02$ versus when $i = 0.01$? Note that the interest earned is the total amount minus $1000.

(c) In general, if an initial investment of P dollars is made, then to what future value s_n will the balance grow after n periods? What kind of sequence is $\{s_n\}$?

30. **Resource management.** A storehouse is stocked with $s_0 = a$ items. Each month, c items are sold, and b new items are bought.

(a) If $a = 100$, $c = 40$, and $b = 25$, then how many items are in the storehouse after 2 months?

(b) If $a = 5000$, then express, in terms of general c and b, the number of items in the storehouse after 3 months.

(c) In general, after n months, how many items s_n are in the storehouse? What kind of sequence is $\{s_n\}$?

In Exercises 31 through 36, reindex the sequence so that the indexing starts at 0.

31. $\forall n \geq 1,\ s_n = 10^{n-3}$.

32. $\forall n \geq 3,\ s_n = \frac{n!}{n-2}$.

33. $\forall n \geq 2,\ s_n = 3 + 2n$.

34. $\forall n \geq 6,\ s_n = \binom{n}{3}$.

35. $\forall n \geq 2,\ s_n = (-1)^n \frac{n-2}{n}$.

36. $\forall n \geq 1,\ s_n = \frac{(-1)^n}{(n-1)!}$.

In Exercises 37 through 42, find the first four terms of each sequence.

37. $s_1 = 4$ and $\forall n \geq 2,\ s_n = 3s_{n-1} - 2$.

38. $s_0 = \frac{1}{3}$ and $\forall n \geq 1,\ s_n = 1 - s_{n-1}$.

39. $s_2 = 5$ and $\forall n \geq 3,\ s_n = s_{n-1} - 2$.

40. $s_0 = \frac{\pi}{4}$ and $\forall n \geq 1,\ s_n = s_{n-1} + \pi$.

41.* Let $s_1 = -1$ and $\forall n \geq 2,\ s_n = 5s_{n-1} + 1$. Observe that this sequence is decreasing. That is, $\forall n \geq 1,\ s_n > s_{n+1}$. However, varying the initial condition alone can change this. Find the smallest possible value for s_1 such that $\{s_n\}$ is not decreasing.

42.* Let $s_1 = 2$ and $\forall n \geq 2,\ s_n = s_{n-1}^2 - 1$. Observe that this sequence is increasing. That is, $\forall n \geq 1,\ s_n < s_{n+1}$. However, varying the initial condition alone can change this. Find the largest possible value for s_1 such that $\{s_n\}$ is not increasing.

In Exercises 43 through 50, find a recursive formula for each sequence.

43. $2, 4, 6, 8, 10, \ldots$

44.* $1, -4, 9, -16, 25, \ldots$

45. $3, 6, 12, 24, 48, \ldots$

46. $\frac{\pi}{2}, \frac{3\pi}{2}, \frac{5\pi}{2}, \frac{7\pi}{2}, \frac{9\pi}{2}, \ldots$

47.* $1, -3, 5, -7, 9, \ldots$

48. $5\sqrt{2}, 10\sqrt{2}, 20\sqrt{2}, 40\sqrt{2}, 80\sqrt{2}, \ldots$

49.* $1, \frac{1}{2}, \frac{1}{3}, \frac{1}{4}, \frac{1}{5}, \ldots$

50. $1, 4, 10, 20, 35, 56, \ldots$

51. **Annuity.** An annuity is an investment into which regular deposits are made. For example, retirement annuities often receive automatic deposits taken from an employee's regular paychecks. In an ordinary annuity earning periodic interest rate i, a deposit of D dollars is made at the end of each month, after interest is added to the previous balance. Assuming an initial balance of zero and letting s_n be the balance after n months, we have $s_0 = 0,\ s_1 = D,\ s_2 = D(1 + i) + D,\ \ldots$.

(a) If $D = 1000$ and $i = 0.05$, then find s_2 and s_3.

(b) If $s_{10} = \$1200.61$, $i = 0.04$, and $D = \$100$, then find s_{11}.

(c) In general, find a recursive formula for the balance s_n after n months.

Exercises in subsequent sections will establish the closed formula $s_n = D\frac{(1+i)^n - 1}{i}$.

52. **Mortgage**. A mortgage is a loan that is paid off in periodic (typically monthly) installments, while interest is also charged each period. A mortgage that has annual interest rate r is understood to have monthly interest rate $i = \frac{r}{12}$. A mortgage of M dollars at monthly interest rate i has monthly (rent) payments of R dollars. At the end of each month, interest is added to the previous balance, and then the payment of R dollars is subtracted from that result. Letting s_n be the balance due after n months, we have $s_0 = M$, $s_1 = M(1 + i) - R, \ldots$.

(a) If $M = \$10,000$, $i = 0.02$, and $R = \$94.56$, then find s_2 and s_3.

(b) If $s_{20} = \$79,495.98$, $i = 0.01$, and $R = \$822.89$, then find s_{21}.

(c) In general, find a recursive formula for the balance s_n. Exercises in subsequent sections will establish the closed formula $s_n = M(1 + i)^n - R\frac{(1+i)^n - 1}{i}$.

In Exercises 53 through 58, given that $k \geq 1$, find an expression for s_{k+1} in terms of prior values in the sequence.

53. $s_1 = 4$ and $\forall n \geq 2$, $s_n = 3s_{n-1} - 2$.

54. $s_0 = \frac{1}{3}$ and $\forall n \geq 1$, $s_n = 1 - s_{n-1}$.

55. $s_1 = 5$ and $\forall n \geq 2$, $s_n = s_{n-1} - 2$.

56. $s_0 = \frac{\pi}{4}$ and $\forall n \geq 1$, $s_n = 3s_{n-1} + \pi$.

57. $s_0 = 1, s_1 = -1$, and $\forall n \geq 2$, $s_n = 5s_{n-2} - 3s_{n-1}$. First find s_2 and s_3.

58. $s_1 = 2, s_2 = 1$, and $\forall n \geq 2$, $s_n = (s_{n-1})^2 - (s_{n-2})^2$. First find s_3 and s_4.

In Exercises 59 through 62, find a recursive formula that expresses s_n in terms of prior values in the sequence.

59. $s_1 = 5$ and $\forall n \geq 1$, $s_{n+1} = s_n - 2$.

60. $s_0 = \frac{\pi}{4}$ and $\forall n \geq 0$, $s_{n+1} = 3s_n + \pi$.

61. $s_0 = 1, s_1 = -1$, and $\forall n \geq 0$, $s_{n+2} = 5s_n - 3s_{n+1}$.

62. $s_1 = 2, s_2 = 1$, and $\forall n \geq 1$, $s_{n+2} = (s_{n+1})^2 - (s_n)^2$.

In Exercises 63 and 64, use software, such as *Mathematica*, *Maple*, or C++.

63. (a) Write a recursive program that implements the function

```
AppRt2(n)
    \begin
        if (n > 0),
            return   1 + 1/(1+AppRt2(n-1)).
        else
            return   1.
    \end.
```

where n is understood to be a natural number.

   (b) Use the program to find `AppRt2(10)` to five decimal places.

   (c) The sequence `AppRt2(n)` gives a better and better approximation for $\sqrt{2}$ as n gets larger. What is the first decimal place in which `AppRt2(15)` and $\sqrt{2}$ differ?

64. (a) Write a recursive program that implements the function

```
Gold(n)
    \begin
        if (n > 0),
            return   Sqrt(1+Gold(n-1)).
        else
            return   1.
    \end.
```

where n is understood to be a natural number.

   (b) Use the program to find `Gold(11)` to seven decimal places.

   (c) The sequence `Gold(n)` gives a better and better approximation for the golden ratio $\phi = \frac{1+\sqrt{5}}{2}$ as n gets larger. What is the first decimal place in which `Gold(20)` and ϕ differ?

4.2 Sigma notation

Given a sequence $\{s_n\}$, one may be interested in a sum of several of its terms:

$$S = s_a + s_{a+1} + s_{a+2} + \cdots + s_{b-1} + s_b, \qquad (4.10)$$

where $a, b \in \mathbb{Z}$. The notation used to represent the sum in Equation (4.10) is

$$S = \sum_{i=a}^{b} s_i. \qquad (4.11)$$

Due to the use of the Greek letter Σ, the right-hand side of Equation (4.11) is said to be written in **sigma notation**. This notation is also known as **summation notation**. The notation on the right-hand side of Equation (4.11) is read as the sum, as i goes from a to b, of s_i. However, the index does not have to be denoted by i. If $b < a$, then the sum in (4.11) is assigned the value 0. Infinite sums (when $b = \infty$ or $a = -\infty$) are called **series**. They are studied extensively in calculus and will not be considered here, where we focus on finite sums.

Example 4.8.   Compute $\displaystyle\sum_{i=0}^{5} i!$.

Solution.

$$\sum_{i=0}^{5} i! = 0! + 1! + 2! + 3! + 4! + 5!$$

$$= 1 + 1 + 2 + 6 + 24 + 120$$

$$= 154. \qquad\qquad \square$$

Given a sequence $\{s_n\}$ and indices $a, b \in \mathbb{Z}$, the computation of the sum $\sum_{i=a}^{b} s_i$ can be thought of as the implementation of a for loop.

> Let $S = 0$.
> For $i = a$ to b,
>   Let $S = S + s_i$.
> Return S.

Since C++ has no predefined function for computing sums, an algorithm like this could be used to program one. Assuming that s is our sequence and a and b are integers, *Mathematica* uses

```
In[1]:= Sum[s[i], {i, a, b}]
```

and *Maple* uses

```
> sum(s(i), i=a..b) ;
```

to compute the sum.

Although software can be used to do many summation computations, there are limits to the sizes it can handle. Having a deeper understanding of sigma notation is thus warranted. Since sigma notation represents summation, it has some formal properties that are immediate consequences of the associative and distributive properties of addition.

Theorem 4.1.   *Let $a, b \in \mathbb{Z}$, let $\{s_n\}$ and $\{t_n\}$ be sequences, and let $c \in \mathbb{R}$.*

(a) $\displaystyle\sum_{i=a}^{b}(s_i \pm t_i) = \sum_{i=a}^{b}s_i \pm \sum_{i=a}^{b}t_i.$

(b) $\displaystyle\sum_{i=a}^{b}cs_i = c\sum_{i=a}^{b}s_i.$

Sigma notation provides an efficient and clear way to represent long sums.

Example 4.9.   (Writing Fixed Sums in Sigma Notation).

$$15 + 16 + 17 + 18 + \cdots + 84 = \sum_{i=15}^{84}i.$$

$$1 + 2 + 4 + 8 + \cdots + 1024 = \sum_{i=0}^{10}2^i.$$

Although concrete sums like those in Example 4.9 do arise, the real power of sigma notation comes from its utility in handling a variable number of terms.

Example 4.10.   (Writing General Sums in Sigma Notation).

$$1 + 2 + 3 + 4 + \cdots + n = \sum_{i=1}^{n}i.$$

$$1 + 4 + 9 + 16 + \cdots + n^2 = \sum_{i=1}^{n}i^2.$$

$$1 + 2 + 4 + 8 + \cdots + 2^n = \sum_{i=0}^{n}2^i.$$

$$\binom{n}{0} + \binom{n}{1} + \binom{n}{2} + \binom{n}{3} + \cdots + \binom{n}{n} = \sum_{i=0}^{n}\binom{n}{i}.$$

$$7\cdot 2^3 + 7\cdot 2^4 + 7\cdot 2^5 + 7\cdot 2^6 + \cdots + 7\cdot 2^{n-1} = \sum_{i=3}^{n-1}7\cdot 2^i.$$

$$1 - \frac{1}{2} + \frac{1}{6} - \frac{1}{24} + \cdots + (-1)^{n+1}\frac{1}{n!} = \sum_{i=1}^{n}\frac{(-1)^{i+1}}{i!}.$$

Most of the summations in Example 4.10 have formulas that make them easy to compute. For example, the first three summations there can be computed using the following two theorems.

Theorem 4.2.   *Let* $n \in \mathbb{Z}$ *with* $n \geq 1$. *Then*

(a) $\displaystyle\sum_{i=1}^{n} 1 = n.$

(b) $\displaystyle\sum_{i=1}^{n} i = \frac{n(n+1)}{2}.$

(c) $\displaystyle\sum_{i=1}^{n} i^2 = \frac{n(n+1)(2n+1)}{6}.$

(d) $\displaystyle\sum_{i=1}^{n} i^3 = \left[\frac{n(n+1)}{2}\right]^2.$

The formulas in Theorem 4.2 tell us how to sum certain powers of the index variable. Formulas for all sums of the form $1^m + 2^m + 3^m + \cdots + n^m = \sum_{i=1}^{n} i^m$ are discussed at the end of this section in Theorem 4.4. The next theorem handles the sum $1 + r + r^2 + \cdots + r^n$ of the terms in a finite geometric sequence.

Theorem 4.3.   *Let* $r \in \mathbb{R}$ *with* $r \neq 1$ *and* $n \in \mathbb{Z}$ *with* $n \geq 0$. *Then*

$$\sum_{i=0}^{n} r^i = \frac{r^{n+1} - 1}{r - 1}.$$

Theorems 4.2 and 4.3 can easily be proved by induction. Hence, their proofs are postponed until Section 4.4. Here, we instead focus on computing sums using Theorems 4.2 and 4.3 together with the properties in Theorem 4.1.

Example 4.11.   Compute $1 + 2 + 3 + 4 + \cdots + 100$.

Solution. By Theorem 4.2(b), with $n = 100$, we see that

$$\sum_{i=1}^{100} i = \frac{100(101)}{2} = 5050. \qquad \square$$

The computation in Example 4.11 is the subject of a well-known story about the German mathematician Carl Friedrich Gauss (1777–1855). His elementary school teacher, in an attempt to give Gauss' class a time-consuming calculation, asked them to compute the sum of the first 100 positive integers. Only moments later, young Gauss had obtained the solution through the observation that the sums of the first and last, the second and second to last, and so on, all equaled 101.

$$1+2+3+\cdots+98+99+100$$

Since there are $\frac{100}{2}$ such pairwise sums, the total is immediately seen to be $\frac{100}{2}(101) = 5050$. In fact, this calculation is essentially the one used in Example 4.11. Moreover, Gauss' technique motivates a clever proof of Theorem 4.2(b), outlined in Exercise 45.

Example 4.12.   Compute $\displaystyle\sum_{i=1}^{n}(6i^2 - 4i + 5)$.

Solution.

$$\sum_{i=1}^{n}(6i^2 - 4i + 5) = \sum_{i=1}^{n}6i^2 - \sum_{i=1}^{n}4i + \sum_{i=1}^{n}5$$

$$= 6\sum_{i=1}^{n}i^2 - 4\sum_{i=1}^{n}i + 5\sum_{i=1}^{n}1$$

$$= 6\left[\frac{n(n+1)(2n+1)}{6}\right]$$

$$- 4\left[\frac{n(n+1)}{2}\right] + 5n$$

$$= 2n^3 + n^2 + 4n.$$

The first two equalities follow from Theorem 4.1(a). The third equality follows from Theorem 4.2, and the last is basic algebra. □

Example 4.13.   Compute $\displaystyle\sum_{i=0}^{n-1}3\cdot 2^i$.

Solution.

$$\sum_{i=0}^{n-1}3\cdot 2^i = 3\sum_{i=0}^{n-1}2^i = 3\left[\frac{2^{n-1+1}-1}{2-1}\right] = 3(2^n - 1).$$

Since Theorem 4.1(b) enables us to factor out the constant 3, Theorem 4.3 takes care of the summation. □

Theorems 4.2 and 4.3 are directly applicable only for computing summations like those in Examples 4.12 and 4.13, where the indexing starts at $i = 1$ or $i = 0$, respectively. For summations in which this is not the case, reindexing is useful.

Example 4.14. Compute $15 + 16 + 17 + 18 + \cdots + 84$.

Solution. This sum is most naturally written in sigma notation as $\sum_{i=15}^{84} i$. Theorem 4.2 can be used to compute such a sum, but first we need to reindex it so that the indexing starts at 1.

Let $j = i - 14$. So $i = j + 14$. Since $j = 1$ when $i = 15$, and $j = 70$ when $i = 84$, we have

$$\sum_{i=15}^{84} i = \sum_{j=1}^{70} (j + 14) = \sum_{j=1}^{70} j + \sum_{j=1}^{70} 14$$

$$= \frac{70(71)}{2} + 70(14) = 3465.$$

Alternatively, we could compute $\sum_{i=15}^{84} i$ via

$$\sum_{i=15}^{84} i = \sum_{i=1}^{84} i - \sum_{i=1}^{14} i = \frac{84(85)}{2} - \frac{14(15)}{2} = 3465. \qquad \square$$

That is, from the sum $1 + 2 + 3 + \cdots + 84$, we subtract the partial sum $1 + 2 + 3 + \cdots + 14$ of the terms that are excluded from our desired sum.

Example 4.15. Compute $\displaystyle\sum_{i=5}^{n} 3^i$.

Solution. To enable us to apply Theorem 4.3, we reindex this sum so that the indexing starts at 0. Let $j = i - 5$. So $i = j + 5$. Since $j = 0$ when $i = 5$, and $j = n - 5$ when $i = n$, we have

$$\sum_{i=5}^{n} 3^i = \sum_{j=0}^{n-5} 3^{j+5} = \sum_{j=0}^{n-5} 3^5 3^j$$

$$= 3^5 \sum_{j=0}^{n-5} 3^j = 3^5 \left[\frac{3^{n-5+1} - 1}{3 - 1} \right] = \frac{3^5(3^{n-4} - 1)}{2}.$$

Alternatively,

$$\sum_{i=5}^{n} 3^i = \sum_{i=0}^{n} 3^i - \sum_{i=0}^{4} 3^i = \frac{3^{n+1}-1}{3-1} - \frac{3^{4+1}-1}{3-1}$$

$$= \frac{3^{n+1}-1-3^5+1}{2} = \frac{3^5(3^{n-4}-1)}{2}. \qquad \square$$

4.2.1 Product notation

Just like summation notation represents sums, there is **product notation** for representing products. In this notation, the product

$$P = s_a \cdot s_{a+1} \cdot s_{a+2} \cdot \cdots \cdot s_b$$

is represented by

$$P = \prod_{i=a}^{b} s_i.$$

When $b < a$, it is assigned the value 1. For example, factorials can be represented in product notation by

$$\forall n \geq 0, \quad n! = \prod_{i=1}^{n} i.$$

Products can be computed in *Mathematica* and *Maple* with the functions `Product` and `product`, respectively.

Example 4.16. Let $n \geq 1$. We represent a factorization of $x^{2^n} - y^{2^n}$ in product notation.

Solution. Factoring results from repeatedly taking advantage of the difference of two squares. For example,

$$x^4 - y^4 = (x^2 - y^2)(x^2 + y^2) = (x-y)(x+y)(x^2+y^2).$$

In general,

$$x^{2^n} - y^{2^n} = (x^{2^{n-1}} - y^{2^{n-1}})(x^{2^{n-1}} + y^{2^{n-1}})$$

$$= (x^{2^{n-2}} - y^{2^{n-2}})(x^{2^{n-2}} + y^{2^{n-2}})(x^{2^{n-1}} + y^{2^{n-1}})$$

$$= \cdots$$

$$= (x-y) \prod_{i=0}^{n-1} (x^{2^i} + y^{2^i}).$$

A careful proof of this fact (without the $\cdots$) can be given by induction and is left for the exercises in Section 4.4. $\square$

4.2.2 General summation formulas

Summation formulas like those in Theorem 4.2, but for $\sum_{i=1}^{n} i^m$ when $m \geq 4$, can be generated from the following recursive relationship.

Theorem 4.4.   *Let $m \geq 1$ and $n \geq 1$ be integers. Then*

$$\sum_{i=1}^{n} i^m = \frac{(n+1)((n+1)^m - 1) - \sum_{j=1}^{m-1}\left[\binom{m+1}{j}\sum_{i=1}^{n} i^j\right]}{m+1}.$$

A proof of Theorem 4.4 is left for the exercises in Section 4.6 after the Binomial Theorem is presented. Here, we demonstrate the utility of Theorem 4.4 by finding $\sum_{i=1}^{n} i^4$.

Example 4.17.   Find a formula for $\displaystyle\sum_{i=1}^{n} i^4$.

Solution.

$$\sum_{i=1}^{n} i^4 = \frac{(n+1)((n+1)^4 - 1) - \left[\binom{5}{1}\sum_{i=1}^{n} i + \binom{5}{2}\sum_{i=1}^{n} i^2 + \binom{5}{3}\sum_{i=1}^{n} i^3\right]}{5}$$

$$= \frac{(n+1)((n+1)^4 - 1) - 5\frac{n(n+1)}{2} - 10\frac{n(n+1)(2n+1)}{6} - 10\left[\frac{n(n+1)}{2}\right]^2}{5}$$

$$= \frac{n(n+1)(2n+1)(3n^2 + 3n - 1)}{30}.$$

The first equality follows from Theorem 4.4 with $m = 4$. The second results from applications of Theorem 4.2, and the last is basic algebra. We conclude that

$$\sum_{i=1}^{n} i^4 = \frac{n(n+1)(2n+1)(3n^2 + 3n - 1)}{30}. \qquad \square$$

Exercises

Each of the following exercises can be completed by hand. However, those requesting computations might also be done with the aid of a symbolic calculator such as *Mathematica* or *Maple*. The results should then be compared with hand computations.

1. Compute $\displaystyle\sum_{i=0}^{4} \frac{1}{i!}$.

   As n gets larger and larger, $\sum_{i=0}^{n} \frac{1}{i!}$ becomes a better and better approximation for the Euler number $e \approx 2.718$.

2. Compute $\displaystyle\sum_{i=2}^{6} \binom{i}{2}$.

3. Use a known summation formula to compute $\displaystyle\sum_{i=1}^{25} i^2$.

4. Use a known summation formula to compute $\displaystyle\sum_{i=1}^{15} i^3$.

   In Exercises 5 through 16, (a) write the given sum in sigma notation and (b) compute the sum using known summation formulas.

5. $1 + 8 + 27 + 64 + \cdots$ $+ 1000$.

6. $1 + 2 + 3 + 4 + \cdots$ $+ 1000$.

7. The sum of the powers of 2 from 1 to 1024.

8. The sum of the powers of 3 from 1 to 3^{10}.

9. $-2 + 4 - 8 + 16 - \cdots$ $- 512$.

10. The sum of the odd integers from 5 to 1001.

11. $12 + 27 + 48 + 75 + \cdots$ $+ 3n^2$.

12. $1 + 8 + 27 + 64 + \cdots$ $+ n^3$.

13. The sum of the powers of 4 from 4 to 4^n.

14. $6 + 18 + 54 + 162 + \cdots$ $+ 2 \cdot 3^n$.

15. $9 - 27 + 81 - 243 + \cdots$ $+ (-3)^n$.
    This is the alternating sum of powers of 3 from 9 to 3^n or, equivalently, the sum of powers of -3 from 9 to $(-3)^n$.

16. $2 - 4 + 6 - 8 + \cdots$ $+ (-1)^{n+1} 2n$.
    This is the alternating sum of the even integers from 2 to $2n$.

17. **Annuity.** In an ordinary annuity with periodic interest rate i and deposit D, the balance s_n at the end of each period can be shown to satisfy

    $$s_n = D(1+i)^{n-1} + D(1+i)^{n-2} + \cdots + D(1+i) + D. \tag{4.12}$$

    (a) Use the recursive definition of s_n given in Exercise 51 from Section 4.1 to obtain expressions for s_2, s_3, s_4, and verify that they can be written in the form of (4.12).

    (b) With $i = 0.01$ and $D = 100$, use (4.12) to compute s_4.

(c) Use (4.12) and known summation formulas to verify the closed formula for s_n asserted at the end of Exercise 51 from Section 4.1.

(d) If $i = 0.01$ and \$10,000 is desired after 12 periods, then what should be the amount D of each deposit?

(e) In general, at periodic interest rate i, if F dollars is desired after N periods, then what should be the amount D of each deposit?

18. **Mortgage.** For a mortgage of M dollars with periodic interest rate i and payment R, the balance s_n due at the end of each month can be shown to satisfy

$$s_n = M(1 + i)^n - R(1 + i)^{n-1} - \cdots - R(1 + i) - R. \quad (4.13)$$

(a) Use the recursive definition of s_n given in Exercise 52 from Section 4.1 to obtain expressions for s_1, s_2, s_3, and verify that they can be written in the form of (4.13).

(b) With $M = $ \$80,000, $i = 0.01$, and $R = $ \$822.89, use (4.13) to find s_4.

(c) Use (4.13) and known summation formulas to verify the closed formula for s_n asserted at the end of Exercise 52 from Section 4.1.

(d) A mortgage of \$80,000 at interest rate $i = 0.01$ is desired. If the term of the mortgage is 480 months (40 years), then what should be the amount of each payment R so that the balance due is exactly zero after month 480?

(e) A mortgage of M dollars at interest rate i is desired. In general, if the term of the mortgage is N months, then what should be the amount of each payment R so that the balance due is exactly zero after month N?

In Exercises 19 through 30, compute the sums.

19. $\displaystyle\sum_{i=1}^{n}(4i^3 - 6i - 1)$.

20. $\displaystyle\sum_{i=1}^{n}(12i^2 + 2i - 3)$.

21. $\displaystyle\sum_{i=1}^{n}(i - 1)^2$.

22. $\displaystyle\sum_{i=1}^{n}(i + 1)^3$.

23. $\displaystyle\sum_{i=0}^{n}\left(\frac{1}{3}\right)^i$.

24. The sum of the powers of 5 from 1 to 5^n.

25. The sum of the powers of 2 from 1024 to 2^{100}.

26. $\displaystyle\sum_{i=100}^{400} 3^i$.

27. $\displaystyle\sum_{i=2}^{64} 5 \cdot 4^i.$ 28. $\displaystyle\sum_{i=1}^{40} (-2) \cdot 3^i.$

29*. $\displaystyle\sum_{i=1}^{100} \frac{4}{3^i}.$ 30*. $\displaystyle\sum_{i=2}^{50} \frac{3}{2^i}.$

In Exercises 31 through 38, verify the stated formulas.

31. $\displaystyle\sum_{i=1}^{n} (4i - 3) = n(2n - 1).$ 32. $\displaystyle\sum_{i=1}^{n} (3i - 2) = \frac{n(3n - 1)}{2}.$

33. $\displaystyle\sum_{i=1}^{n} (3i^2 - i) = n^2(n + 1).$ 34. $\displaystyle\sum_{i=1}^{n} (4i^3 - 2i) =$
$n(n + 1)(n^2 + n - 1).$

35. $\displaystyle\sum_{i=1}^{2n} i = n(2n + 1).$ 36. The sum of the first $2n$ cubes (i.e., from 1 to $(2n)^3$) is $n^2(2n + 1)^2$.

37*. The sum of the odd integers from 3 to $2n + 1$ is $n(n + 2)$.   38. $\displaystyle\sum_{i=1}^{n} 2^{2i+1} = \frac{2}{3}(4^{n+1} - 1).$
Hint: $2^{2i+1} = 2 \cdot 4^i.$

39. Express k^n in product notation.   40. Express $\binom{n}{k}$ by using product notation.

41. Find a simple formula for $\displaystyle\prod_{i=1}^{n} 2^i.$ 42. Find a simple formula for $\displaystyle\prod_{i=1}^{n} i^2.$

43*. Given distinct $x_1, x_2, x_3 \in \mathbb{R}$, a polynomial $L(x)$ through $(x_1, y_1), (x_2, y_2), (x_3, y_3)$ is given by

$$L(x) = y_1 \frac{(x - x_2)(x - x_3)}{(x_1 - x_2)(x_1 - x_3)} + y_2 \frac{(x - x_1)(x - x_3)}{(x_2 - x_1)(x_2 - x_3)}$$
$$+ y_3 \frac{(x - x_1)(x - x_2)}{(x_3 - x_1)(x_3 - x_2)}.$$

Use summation and product notation to write a formula for the generalization of this to a polynomial through $(x_1, y_1), \ldots, (x_n, y_n)$, where $x_1, \ldots, x_n$ are distinct. It is called the **Lagrange polynomial** and is named after the French mathematician Joseph-Louis Lagrange (1736–1813), though it was not first discovered by him.

44. Observe the pattern

$$\sin(2\theta) = 2\sin\theta\cos\theta,$$
$$\sin(4\theta) = 4\sin\theta\cos 2\theta,$$
$$\sin(8\theta) = 8\sin\theta\cos\theta\cos 2\theta\cos 4\theta,\ldots,$$

and use product notation to write the general formula for $\sin(2^n\theta)$ for $n \geq 1$.

45. Use the following outline to verify Theorem 4.2(b).
Let $S = 1 + 2 + 3 + \cdots + (n-2) + (n-1) + n$. So we can also write $S = n + (n-1) + (n-2) + \cdots + 3 + 2 + 1$. Adding these two equations together, by first observing that the sums of corresponding terms (column sums) are all the same, we obtain an expression for $2S$ that is easily used to get S.

46. Use the following outline to verify Theorem 4.3.
Let $S = 1 + r + r^2 + \cdots + r^{n-1} + r^n$. So $rS = r + r^2 + r^3 + \cdots + r^n + r^{n+1}$. Subtracting the first equation from the second, and taking advantage of the resulting cancellations, we obtain an equation that can be used to get S.

47.* Use Theorem 4.4 and the result in Example 4.17 to find a formula for $\displaystyle\sum_{i=1}^{n} i^5$.

48.* Use Theorem 4.4 and the result of Exercise 47 to find a formula for $\displaystyle\sum_{i=1}^{n} i^6$.

49. A spiral pattern is created by pasting together squares, each with side length half that of the previous. The initial square is a unit square, and the first four iterations of this process are pictured.

   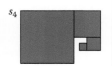

(a) Draw s_6.

(b) Which is the first s_n to have area greater than 1.33?

(c) In general, express the area a_n of s_n as a sum.

(d) Use Theorem 4.3 and the expression from part (c) to get a closed formula for a_n. Note that, as n gets larger, a_n approaches $\frac{4}{3}$.

50. Suppose we have an unlimited supply of rectangular blocks, each 2 feet long and weighing 1 pound. For each n, by stacking n blocks in the pictured stair-step pattern s_n, we get the greatest possible extension of the blocks

beyond the edge of a table, while having them balance on their own.

That is, at the nth stage, the stack s_{n-1} is placed atop the nth block, and the nth block extends $\frac{1}{n}$ feet beyond the edge of the table.

(a) Draw s_5.

(b) Which is the first stack s_n for which the topmost block extends at least 3 feet beyond the edge of the table?

(c) Express the distance d_n that the topmost block of s_n extends beyond the edge of the table as a sum. As a consequence of Exercise 34 from Section 4.4, for any desired distance from the edge of the table, we can form a stack that obtains that distance.

4.3   Mathematical induction: An introduction

Often, we want to prove statements of the form

$$\forall \text{ integers } n \geq a, \ P(n). \tag{4.14}$$

Here, a is some fixed integer (usually $a = 0$ or 1), and $P(n)$ is a statement that depends on n. Two examples to keep in mind are

$$\forall \, n \geq 2, \ n^2 > n. \tag{4.15}$$

$$\forall \, n \geq 0, \ 2^n \geq n + 1. \tag{4.16}$$

Note that when it is clear in context that we are working with integers n, we may omit mention of this in our statements. Also, statements of the form (4.14) can be thought of as sequences of statements. For example, (4.15) is the sequence

$$2^2 > 2, \ 3^2 > 3, \ 4^2 > 4, \ 5^2 > 5, \ 6^2 > 6, \ldots$$

Sometimes a statement of the form (4.14) can be proved directly. That is, it may be possible to prove $P(n)$ independently of n. For example, in statement (4.15), the restriction $n \geq 2$ is equivalent to $n > 1$. Multiplying both sides of this by n gives the inequality $n^2 > n$. Thus, a proof of statement (4.15) is immediate. However, many statements of the form (4.14), such as statement (4.16), are not proved so easily.

For statement (4.16), it is certainly easy to check a bunch of cases.

$$\text{For } n = 0, \ 2^0 = 1 \geq 0 + 1.$$

$$\text{For } n = 1, \ 2^1 = 2 \geq 1 + 1.$$

$$\text{For } n = 2, \ 2^2 = 4 \geq 2 + 1.$$

$$\text{For } n = 3, \ 2^3 = 8 \geq 3 + 1.$$

Moreover, we have the sense that this checking will continue to work for as many cases as we wish to consider. However, since we can only explicitly check only a finite number of cases, such a technique does not yield a proof of statement (4.16). This is where we need a more powerful tool like mathematical induction.

Here is an outline for a proof by induction.

Outline 4.1.   (Proof by Mathematical Induction).
To show: $\forall \, n \geq a, \ P(n)$.

Proof by induction

1. Base cases:
    Show: $P(a), \dots, P(b)$ are true.

2. Inductive step:
    Show: $\forall \, k \geq b$, if $P(k)$ is true, then $P(k + 1)$ is true.
    That is,

    (a) Suppose $k \geq b$ and that $P(k)$ is true.

    (b) Show: $P(k + 1)$ is true.

In Outline 4.1, although steps 2a and 2b are merely the expected steps in a standard proof of the if-then statement listed in step 2, step 2a does deserve special mention. It is called the **inductive hypothesis**. The power behind a proof by induction comes from the inductive step, where it is shown that the truth of $P(k)$ implies the truth of $P(k + 1)$. Very simply, this establishes that the truth of later cases follows from the truth of the early cases. In fact, in many examples only a single base case needs to be verified, and $b = a$. In Example 4.21, we will see one reason why $b > a$ is sometimes used.

Before discussing why induction works, it may be helpful to see it applied to a particular example. In the following proof of statement (4.16), note how the general structure of Outline 4.1 is followed.

Example 4.18.   Show: $\forall \, n \geq 0, \ 2^n \geq n + 1$.

Proof. (By Induction)
Base case: $(n = 0)$

It is straightforward to see that $2^0 = 1 \geq 0 + 1$.
Inductive step:
Suppose $k \geq 0$ and that $2^k \geq k + 1$.
(Goal: $2^{k+1} \geq (k+1) + 1$.)
Observe that

$$2^{k+1} = 2(2^k)$$

$$\geq 2(k+1) \qquad \text{(By the inductive hypothesis)}$$

$$= (k+1) + (k+1)$$

$$\geq (k+1) + 1.$$

That is, $2^{k+1} \geq (k+1) + 1$. $\qquad\qquad\qquad$ □

In the displayed string of equalities and inequalities, the left-hand side of the first line is not meant to be copied on each row of the left-hand side. Instead, the right-hand side of each line should be thought of as the left-hand side of its subsequent line. Very simply, the string should be read as if it were all on a single line. The convention of displaying it on several lines is intended to improve readability. This style is repeated throughout the book.

Although unnecessary, the parenthesized comment expressing our goal in Example 4.18 is a good idea. In more complicated examples, having our goal written out explicitly can greatly help our construction of a proof. Notice that, in the process of achieving our goal, the inequality

$$2(2^k) \geq 2(k+1)$$

is used. That is, we used the inductive hypothesis (times 2 in this example). Using the truth of $P(k)$ in the proof of $P(k+1)$ is a feature of every proof by induction. The final inequality

$$k + (k+1) + 1 \geq (k+1) + 1$$

follows from the assumption that $k \geq 0$, since dropping the first term, k, from the sum $k + (k+1) + 1$ will certainly not cause an increase in magnitude.

Why does a proof by induction work? We answer this question in two ways. An intuitive explanation is given first and then a formal one. The intuitive explanation comes in the form of an analogy.

Consider the way in which dominoes are lined up with the intent of knocking them all down with one push (see Figure 4.1). For the purpose of our analogy, we assume that we have infinitely many dominoes lined up, and they are numbered starting at 1. We equate knocking over the nth domino with proving $P(n)$. Consequently, the base case in our proof by induction corresponds to knocking down the first domino. The inductive step corresponds to the fact that

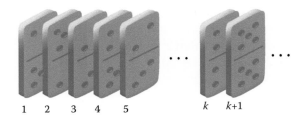

FIGURE 4.1
Dominoes.

if the kth domino falls down, then the $(k+1)$st domino will also fall down. Basically, the inductive step means that consecutive dominoes have been set up close enough together. Given this, once the first domino is knocked down, they will all fall in succession. That is, $P(n)$ is proved for all $n \geq 0$.

The analogy brings out the importance of both the base case and the inductive step. Without the base case (i.e. if we cannot knock down the first domino), not all of the dominoes will fall, no matter how close together they are. Without the inductive step (i.e. if the dominoes are not placed close enough together), even though we knock down the first domino, the dominoes will not all fall down.

The analogy to dominoes is helpful to satisfy our intuition, but there must be a formal reason why induction works. The formal reason comes out of the well-ordering of the integers. For convenience, we state here the straightforward generalization of the Well-Ordering Principle that we need.

Theorem 4.5. (Generalized Well-Ordering Principle).
Given a fixed integer a, each nonempty subset of $\{n : n \in \mathbb{Z} \text{ and } n \geq a\}$ has a smallest element.

We use Theorem 4.5 to prove the Principle of Mathematical Induction.

Theorem 4.6. (Principle of Mathematical Induction).
Let $a \leq b$ be integers, and let $P(n)$ be an expression that depends on the free integer variable n. If

(i) *$P(a), \ldots, P(b)$ hold, and*

(ii) *$\forall\, k \geq b$, if $P(k)$ holds, then $P(k+1)$ holds,*

then the statement

$\forall\, n \geq a, \;\; P(n) \quad$ *holds.*

Proof. Assume conditions (i) and (ii) in the hypotheses of the theorem. Suppose it is not true that $P(n)$ holds $\forall\, n \geq a$.

Let S be the set of those integers $n \geq a$ for which $P(n)$ does not hold. By our assumptions, S is nonempty. Hence, by the Generalized Well-Ordering Principle, S has a smallest element, say s. Since $P(a), P(a + 1), \ldots, P(b)$ all hold, it must be that $s > b$. Therefore, $s - 1 \geq b$. Since $s - 1 \notin S$, it follows that $P(s - 1)$ holds. However, for $k = s - 1$, by condition (ii), since $P(k)$ holds, $P(k + 1)$ must also hold. That is, $P(s)$ holds. This contradicts the fact that $s \in S$. □

Now that induction has been shown to be a valid proof technique, some more practice using it is warranted.

Example 4.19.   Show: $\forall\, n \geq 4,\ n^2 \geq 3n + 4$.

Proof. (By Induction)
Base case: $(n = 4)$
Note that $4^2 = 16 \geq 3(4) + 4$.
(In fact, we have equality there.)
Inductive step:
Suppose $k \geq 4$ and that $k^2 \geq 3k + 4$.
(Goal: $(k + 1)^2 \geq 3(k + 1) + 4 = 3k + 7$.)
Observe that

$$(k + 1)^2 = k^2 + (2k + 1)$$
$$\geq (3k + 4) + (2k + 1) \quad \text{(By the inductive hypothesis)}$$
$$= 3k + (2k + 5)$$
$$\geq 3k + 7.$$

That is, $(k + 1)^2 \geq 3(k + 1) + 4$. □

In the string of equalities and inequalities in Example 4.19, the equalities all follow from basic algebra. The first inequality comes from the inductive hypothesis. It follows by adding $2k + 1$ to both sides of the inequality

$$k^2 \geq 3k + 4.$$

The final inequality holds since $k \geq 4$ and it is easy to see that

$$\forall\, k \geq 1,\ 2k + 5 \geq 7.$$

Example 4.20.   Show: $\forall\, n \geq 0,\ 4 \mid (5^n - 1)$.

Proof. (By Induction)
Base case: $(n = 0)$
Here, $5^0 - 1 = 0$ and $4 \mid 0$.

Inductive step:
Suppose $k \geq 0$ and that $4 \mid (5^k - 1)$.
So $5^k - 1 = 4c$ for some $c \in \mathbb{Z}$.
(Goal: $4 \mid (5^{k+1} - 1)$.)
Observe that

$$5^{k+1} - 1 = 5^k(5) - 1$$
$$= 5^k(4 + 1) - 1$$
$$= 4 \cdot 5^k + 5^k - 1$$
$$= 4 \cdot 5^k + 4c \qquad \text{(By the inductive hypothesis)}$$
$$= 4(5^k + c).$$

Therefore, $4 \mid (5^{k+1} - 1)$. □

A simpler argument than that given in Example 4.20 can be made using the ideas in Section 3.5, since $4 \mid (5^n - 1)$ is equivalent to $5^n \equiv 1 \pmod 4$. However, we are practicing induction here.

In the next example, we see that sometimes more than one base case can be helpful.

Example 4.21.   Show: $\forall \, n \geq 0, \; 2^n \geq n^2 - 1$.

Proof. (By Induction)
Base cases: $(n = 0, 1, 2, 3)$
Observe that $2^0 \geq 0^2 - 1$, $2^1 \geq 1^2 - 1$, $2^2 \geq 2^2 - 1$, and $2^3 \geq 3^2 - 1$.
Inductive step:
Suppose $k \geq 3$ and that $2^k \geq k^2 - 1$.
(Goal: $2^{k+1} \geq (k+1)^2 - 1 = k^2 + 2k$.)
Observe that

$$2^{k+1} = 2(2^k)$$
$$\geq 2(k^2 - 1) \qquad \text{(By the inductive hypothesis)}$$
$$= k^2 + (k^2 - 2)$$
$$\geq k^2 + 2k \text{ (since } \forall \, k \geq 3, \; k^2 - 2 \geq 2k, \text{ by Exercise 2)}$$
$$= (k+1)^2 - 1.$$

That is, $2^{k+1} \geq (k+1)^2 - 1$. □

Note that in our proof, the fact that $k \geq 3$ is used to get

$$k^2 - 2 \geq 2k \;\; \text{and hence} \;\; k^2 + (k^2 - 2) \geq k^2 + (2k).$$

However, this does not mean that extra base cases are necessary. It is certainly true that

$$\forall\, k \geq 0, \text{ if } 2^k \geq k^2 - 1, \text{ then } 2^{k+1} \geq (k+1)^2 - 1.$$

However, in the particular argument that we chose, it was convenient that $k \geq 3$.

Induction and recursion. When a sequence is defined recursively, we can often figure out the closed formula for that sequence by listing several terms in the sequence. Induction can then be used to prove that a proposed closed formula is correct. The skill of guessing a closed formula for a sequence from a listing of several terms was practiced in the exercises in Section 4.1. Here, we practice the skill of proving that such a guess is correct.

Example 4.22.   Sequence (4.7) from Section 4.1 was expressed recursively by the formula

$$s_5 = 10 \quad \text{and } \forall\, n \geq 6, \quad s_n = s_{n-1} + n - 1. \qquad (4.17)$$

Show: $\forall\, n \geq 5,\ s_n = \binom{n}{2}$.

Remark 4.1.   The closed formula $\binom{n}{2}$ was also asserted in Section 4.1. However, at that time, no proof was given that both sequences are indeed the same. Here, we are taking (4.17) as the definition of the sequence.

Proof for Example 4.22. (By Induction)
Base case: ($n = 5$)
Observe that $\binom{5}{2} = 10$, and 10 is the defined value for s_5.
Inductive step:
Suppose $k \geq 5$ and that $s_k = \binom{k}{2}$.
(Goal: $s_{k+1} = \binom{k+1}{2}$.)
From (4.17), it follows that

$$s_{k+1} = s_k + (k+1) - 1$$

$$= \binom{k}{2} + k \qquad \text{(By the inductive hypothesis)}$$

$$= \frac{k(k-1)}{2} + k$$

$$= \frac{k^2 - k + 2k}{2}$$

$$= \frac{(k+1)k}{2}$$

$$= \binom{k+1}{2}.$$

That is, $s_{k+1} = \binom{k+1}{2}$. $\qquad\qquad\qquad \square$

Exercises

In Exercises 1 through 20, use induction to prove the given statements.

1. Show: $\forall\, n \geq 3,\ n^2 + 1 \geq 3n$.

2. Show: $\forall\, n \geq 3,\ n^2 - 2 \geq 2n$.

3. Show: $\forall\, n \geq 3,\ n^2 \geq 2n + 1$.

4. Show: $\forall\, n \geq 3,\ 2n^3 \geq 3n^2 + 3n + 1$.

5. Show: $\forall\, n \geq 4,\ 2^n \geq n^2$.

6. Show: $\forall\, n \geq 0,\ 3^n \geq n^3$. Hint: Consider four base cases.

7. Show: $\forall\, n \geq 4,\ n! \geq n^2$.

8. Show: $\forall\, n \geq 6,\ n! > 2n^3$.

9. Show: $\forall\, n \geq 4,\ n! > 2^n$.

10. Show: $\forall\, n \geq 7,\ n! > 3^n$.

11. Show: $\forall\, n \geq 0,\ 3 \mid (4^n - 1)$.

12. Show: $\forall\, n \geq 0,\ 5 \mid (6^n - 1)$.

13. Show: $\forall\, n \geq 0,\ 4 \mid (6^n - 2^n)$.

14. Show: $\forall\, n \geq 0,\ 5 \mid (9^n - 4^n)$.

15.* Show: $\forall\, n \geq 0,\ 6 \mid (n^3 - n)$.

16.* Show: $\forall\, n \geq 0,\ 6 \mid (n^3 + 5n)$.

17. Let $\{s_n\}$ be the sequence defined by

$$s_1 = 4 \text{ and } \forall\, n \geq 2,$$

$$s_n = 3s_{n-1} - 2.$$

Show: $\forall\, n \geq 1,\ s_n = 3^n + 1$.

18. Let $\{s_n\}$ be the sequence defined by

$$s_0 = \frac{\pi}{4} \text{ and } \forall\, n \geq 1,$$

$$s_n = s_{n-1} + \pi.$$

Show: $\forall\, n \geq 0,\ s_n = \frac{4n+1}{4}\pi$.

19. Let $\{s_n\}$ be the sequence defined by

$$s_2 = 5 \text{ and } \forall\, n \geq 3,$$

$$s_n = s_{n-1} - 2.$$

Show: $\forall\, n \geq 2,\ s_n = 9 - 2n$.

20.* Let $\{s_n\}$ be the sequence defined by

$$s_0 = \frac{1}{3} \text{ and } \forall\, n \geq 1,$$

$$s_n = 1 - s_{n-1}.$$

Show: $\forall\, n \geq 0,\ s_n = \begin{cases} \frac{1}{3} & \text{if } n \text{ is even,} \\ \frac{2}{3} & \text{if } n \text{ is odd.} \end{cases}$

21. **Annuity.** The sequence given recursively by

$$s_0 = 0 \quad \text{and} \quad \forall\, n \geq 1,\ s_n = (1+i)s_{n-1} + D$$

gives the value after n periods of an annuity with periodic interest rate i and deposit D. See Exercise 51 in Section 4.1.

(a) For $n = 0, 1, 2$, determine s_n and confirm that it agrees with the corresponding value of the expression $D\frac{(1+i)^n - 1}{i}$.

(b) Show that the sequence whose nth term is $D\frac{(1+i)^n - 1}{i}$ satisfies the recurrence relation above for all $n \geq 1$.

(c) Show: $\forall\, n \geq 0,\ s_n = D\frac{(1+i)^n - 1}{i}$. (By Induction)

(d) Use the closed formula in part (c) to determine the value after 24 months of an annuity with a monthly interest rate of 0.75% (i.e., $i = 0.0075$) and a monthly deposit of \$200.

22. **Mortgage.** The sequence given recursively by

$$s_0 = M \quad \text{and} \quad \forall\, n \geq 1,\ s_n = (1+i)s_{n-1} - R$$

gives the balance due after n periods of a mortgage for M dollars with periodic interest rate i and payment R. See Exercise 52 in Section 4.1.

(a) For $n = 0, 1, 2, 3$, determine s_n and confirm that it agrees with the corresponding value of the expression $M(1+i)^n - R\frac{(1+i)^n - 1}{i}$.

(b) Show that the sequence whose nth term is given by $M(1+i)^n - R\frac{(1+i)^n - 1}{i}$ satisfies the recurrence relation above for all $n \geq 1$.

(c) Show: $\forall\, n \geq 0,\ s_n = M(1+i)^n - R\frac{(1+i)^n - 1}{i}$. (By Induction)

(d) Use the closed formula in part (c) to determine the balance due after 24 months on a 30-year \$60,000 mortgage with a monthly interest rate of 0.75% (i.e., $i = 0.0075$) and a monthly payment of \$482.77.

23. **Generalized distributivity for sets.** We generalize (the $n = 2$ case of) Theorem 1.5(d) from Section 1.4. Let A be a set.

(a) Show: For all sets B_1, B_2, B_3 that
$A \cap (B_1 \cup B_2 \cup B_3) = (A \cap B_1) \cup (A \cap B_2) \cup (A \cap B_3)$,
by associating $B_1 \cup B_2 \cup B_3 = (B_1 \cup B_2) \cup B_3$.

(b) Show: $\forall\, n \geq 1$, for all sets $B_1, B_2, \ldots, B_n$,
$A \cap (B_1 \cup B_2 \cup \cdots \cup B_n) = (A \cap B_1) \cup (A \cap B_2) \cup \cdots \cup (A \cap B_n)$.

(c) Show: $\forall\, n \geq 1$, for all sets $B_1, B_2, \ldots, B_n$,
$A \cup (B_1 \cap B_2 \cap \cdots \cap B_n) = (A \cup B_1) \cap (A \cup B_2) \cap \cdots \cap (A \cup B_n)$.

Hint: For the inductive step, use Associativity and Distributivity ($n = 2$).

24. **Generalized distributivity for logic.** We generalize (the $n = 2$ case of) Theorem 1.1(d) from Section 1.1. Let p be a statement form.

(a) Show: For all statement forms q_1, q_2, q_3 that
$p \wedge (q_1 \vee q_2 \vee q_3) \equiv (p \wedge q_1) \vee (p \wedge q_2) \vee (p \wedge q_3)$,
by associating $q_1 \vee q_2 \vee q_3 \equiv (q_1 \vee q_2) \vee q_3$.

(b) Show: $\forall\, n \geq 1$, for all statement forms $q_1, q_2, \ldots, q_n$,
$p \wedge (q_1 \vee q_2 \vee \cdots \vee q_n) \equiv (p \wedge q_1) \vee (p \wedge q_2) \vee \cdots \vee (p \wedge q_n)$.

(c) Show: $\forall\, n \geq 1$, for all statement forms $q_1, q_2, \ldots, q_n$,
$p \vee (q_1 \wedge q_2 \wedge \cdots \wedge q_n) \equiv (p \vee q_1) \wedge (p \vee q_2) \wedge \cdots \wedge (p \vee q_n)$.

Hint: For the inductive step, use Associativity and Distributivity ($n = 2$).

25. **Generalized De Morgan's Laws for logic.** We generalize (the $n = 2$ case of) Theorem 1.1(e) from Section 1.1.

(a) Show: For all statement forms p_1, p_2, p_3 that
$\neg(p_1 \vee p_2 \vee p_3) \equiv \neg p_1 \wedge \neg p_2 \wedge \neg p_3$,
by associating $p_1 \vee p_2 \vee p_3 \equiv (p_1 \vee p_2) \vee p_3$.

(b) Show: $\forall\, n \geq 1$, for all statement forms $p_1, p_2, \ldots, p_n$,
$\neg(p_1 \vee p_2 \vee \cdots \vee p_n) \equiv \neg p_1 \wedge \neg p_2 \wedge \cdots \wedge \neg p_n$.

(c) Show: $\forall\, n \geq 1$, for all statement forms $p_1, p_2, \ldots, p_n$,
$\neg(p_1 \wedge p_2 \wedge \cdots \wedge p_n) \equiv \neg p_1 \vee \neg p_2 \vee \cdots \vee \neg p_n$.

Hint: For the inductive step, use Associativity and De Morgan's Law.

26. **Generalized De Morgan's Laws for sets.** We generalize (the $n = 2$ case of) Theorem 1.5(e) from Section 1.4.

(a) Show: For all sets A_1, A_2, A_3 that
$(A_1 \cup A_2 \cup A_3)^c = A_1{}^c \cap A_2{}^c \cap A_3{}^c$,
by associating $A_1 \cup A_2 \cup A_3 = (A_1 \cup A_2) \cup A_3$.

(b) Show: $\forall\, n \geq 1$, for all sets $A_1, A_2, \ldots, A_n$,
$(A_1 \cup A_2 \cup \cdots \cup A_n)^c = A_1{}^c \cap A_2{}^c \cap \cdots \cap A_n{}^c$.

(c) Show: $\forall\, n \geq 1$, for all sets $A_1, A_2, \ldots, A_n$,
$(A_1 \cap A_2 \cap \cdots \cap A_n)^c = A_1{}^c \cup A_2{}^c \cup \cdots \cup A_n{}^c$.

Hint: For the inductive step, use Associativity and De Morgan's Law.

27. Prove the following generalization of Theorem 3.27(a).
Let $n \in \mathbb{Z}$ with $n > 1$.
Show: $\forall\, m \geq 1,\ \forall\, a_1, a_2, \ldots, a_m, b_1, b_2, \ldots, b_m \in \mathbb{Z}$,
if $a_1 \equiv b_1 \pmod{n}, a_2 \equiv b_2 \pmod{n}, \ldots, a_m \equiv b_m \pmod{n}$,
then

(a) $\sum_{i=1}^{m} a_i \equiv \sum_{i=1}^{m} b_i \pmod{n}$.

(b) $\prod_{i=1}^{m} a_i \equiv \prod_{i=1}^{m} b_i \pmod{n}$.

Hint: Use induction on m.

28. Prove Corollary 3.28(b). Hint: Use Theorem 3.27 for the base case and to help with the inductive step.

29. Show that any nonempty finite subset of $\mathbb{R}$ has a maximum element. That is, for any finite set S such that $S \subset \mathbb{R}$, there is an element $m \in S$ such that $m = \max(S)$. Hint: If $S = \{s_1, \ldots, s_k, s_{k+1}\}$, then $\max(S) = \max(\max(s_1, \ldots, s_k), s_{k+1})$. Use induction on $|S|$.

30. Prove Corollary 3.18. Hint: For the base case, use Corollary 3.17. For the inductive step, treat the $(k+1)$-fold product $m_1 \cdot m_2 \cdots m_{k+1}$ as a twofold product $(m_1 \cdot m_2 \cdots m_k) \cdot m_{k+1}$ and apply Corollary 3.17.

A matrix is a rectangular array of real numbers. In Exercises 31 through 35, we focus on 2 by 2 matrices (i.e. matrices with 2 rows and 2 columns) and use the standard definition of matrix multiplication and exponentiation.

$$\begin{bmatrix} a_1 & b_1 \\ c_1 & d_1 \end{bmatrix} \begin{bmatrix} a_2 & b_2 \\ c_2 & d_2 \end{bmatrix} = \begin{bmatrix} a_1 a_2 + b_1 c_2 & a_1 b_2 + b_1 d_2 \\ c_1 a_2 + d_1 c_2 & c_1 b_2 + d_1 d_2 \end{bmatrix}$$

For any matrix A, we define $A^1 = A$ and, $\forall\, n \geq 2,\ A^n = A \cdot A^{n-1}$.

31. Show: $\forall\, n \geq 1$,

$$\begin{bmatrix} 1 & 1 \\ 0 & 1 \end{bmatrix}^n = \begin{bmatrix} 1 & n \\ 0 & 1 \end{bmatrix}.$$

32. Show: $\forall\, n \geq 1$,

$$\begin{bmatrix} -1 & 1 \\ 1 & -1 \end{bmatrix}^n = (-2)^{n-1} \begin{bmatrix} -1 & 1 \\ 1 & -1 \end{bmatrix}.$$

33. Show: $\forall\, n \geq 1$,

$$\begin{bmatrix} 1 & 1 \\ 0 & 2 \end{bmatrix}^n = \begin{bmatrix} 1 & 2^n - 1 \\ 0 & 2^n \end{bmatrix}.$$

34. Show: $\forall\, n \geq 1$,

$$\begin{bmatrix} 1 & 1 \\ 0 & -2 \end{bmatrix}^n = \begin{bmatrix} 1 & \frac{1 - (-2)^n}{3} \\ 0 & (-2)^n \end{bmatrix}.$$

For Exercises 35 through 37, use the trigonometric identities

$$\sin(\theta_1 + \theta_2) = \sin\theta_1 \cos\theta_2 + \cos\theta_1 \sin\theta_2$$

$$\cos(\theta_1 + \theta_2) = \cos\theta_1 \cos\theta_2 - \sin\theta_1 \sin\theta_2$$

35.* Let $\theta \in \mathbb{R}$. Show: $\forall\, n \geq 1$,

$$\begin{bmatrix} \cos\theta & \sin\theta \\ -\sin\theta & \cos\theta \end{bmatrix}^n = \begin{bmatrix} \cos n\theta & \sin n\theta \\ -\sin n\theta & \cos n\theta \end{bmatrix}.$$

36.* Using $i = \sqrt{-1}$, multiplication and exponentiation of complex numbers are defined by the formulas
$(a_1 + ib_1) \cdot (a_2 + ib_2) = (a_1 a_2 - b_1 b_2) + i(a_1 b_2 + b_1 a_2)$,
$(a + ib)^1 = a + ib$, and,
$\forall\, n \geq 2,\ (a + ib)^n = (a + ib) \cdot (a + ib)^{n-1}$.

(a) Observe that $(\cos\frac{\pi}{3}, \sin\frac{\pi}{3}) = (\frac{1}{2}, \frac{\sqrt{3}}{2})$ and compute
$(\frac{1}{2} + i\frac{\sqrt{3}}{2}) \cdot (\frac{1}{2} + i\frac{\sqrt{3}}{2})$. How is the answer related to $\frac{2\pi}{3}$?

(b) Let $\theta \in \mathbb{R}$. Prove **De Moivre's Theorem**. That is, show:
$\forall\, n \geq 1, (\cos\theta + i\sin\theta)^n = \cos n\theta + i\sin n\theta$.

37.* Let $\theta \in \mathbb{R}$.

(a) Use the first identity above with $\theta = \theta_1 = \theta_2$ to obtain the identity for $\sin 2\theta$, known as the double-angle identity for the sine function.

(b) Use the double-angle identity to prove that $\sin 4\theta = 4\sin\theta \cos\theta \cos 2\theta$.

(c) Show: $\forall\, n \geq 1,\ \sin 2^n\theta = 2^n \sin\theta \prod_{i=0}^{n-1} \cos 2^i\theta$.

See Exercise 46 from Section 4.2.

38. Let $x > -1$. Prove **Bernoulli's Inequality**. That is, show: $\forall\, n \geq 0,\ (1 + x)^n \geq 1 + nx$.

39. Give an example of a statement of the form (4.14) for which the inductive step holds but the base case does not.

40. Give an example of a statement of the form (4.14) for which the base case holds but the inductive step does not.

41. Show that if $\forall\, n \geq a,\ P(n)$, then $\forall\, k \geq a,\ P(k) \rightarrow P(k+1)$.

42. Show that if $\forall\, n \geq a,\ P(n)$, then $P(a)$. What principle is this?

43. Show: $\forall\, n \geq 0,\ 8^n \equiv 1 \pmod 7$.

44. Show: $\forall\, n \geq 0,\ 9^n \equiv 4^n \pmod 5$.

45. The **Catalan numbers** are defined by $C_0 = 1$ and $\forall\, n \geq 1,\ C_n = \frac{2(2n-1)}{n+1} C_{n-1}$.

(a) List the Catalan numbers C_0, C_1, C_2, C_3, C_4.

(b)* Show: $\forall\, n \geq 0, C_n = \frac{1}{n+1}\binom{2n}{n}$.

(c) C_n counts the number of ways of parenthesizing a product consisting of $n + 1$ factors (e.g., abc can be written as $(ab)c$ or $a(bc)$). Verify that this is true for $n = 0, 1, 2, 3, 4$.

46. In the outline for a proof by induction, replace the inductive step with
$\forall\, k > b$, if $P(k - 1)$ is true, then $P(k)$ is true.
Explain why this new format for induction provides another valid technique.

4.4   Induction and summations

One class of statements whose proofs are often done by induction consists of those that involve summations. This section focuses on this type and also provides more practice with inductive proofs. The promised proofs of the summation formulas in Theorems 4.2 and 4.3 are also provided.

In the first example, we prove part (c) from Theorem 4.2. The proofs of the remaining parts are left for the exercises.

Example 4.23.   Show: $\forall\, n \geq 1$, $\displaystyle\sum_{i=1}^{n} i^2 = \frac{n(n + 1)(2n + 1)}{6}$.

Proof. (By Induction)
Base case: ($n = 1$)
It is easy to see that $\sum_{i=1}^{1} i^2 = 1 = \frac{1(2)(3)}{6}$.
Inductive step:
Suppose $k \geq 1$ and that $\sum_{i=1}^{k} i^2 = \frac{k(k+1)(2k+1)}{6}$.
(Goal: $\sum_{i=1}^{k+1} i^2 = \frac{(k+1)((k+1)+1)(2(k+1)+1)}{6} = \frac{(k+1)(k+2)(2k+3)}{6}$.)
Observe that

$$\sum_{i=1}^{k+1} i^2 = \left(\sum_{i=1}^{k} i^2 \right) + (k + 1)^2$$

$$= \frac{k(k + 1)(2k + 1)}{6}$$

$$+ (k + 1)^2 \qquad \text{(By the inductive hypothesis)}$$

$$= \frac{k(k + 1)(2k + 1) + 6(k + 1)^2}{6}$$

$$= \frac{(k + 1)[k(2k + 1) + 6(k + 1)]}{6}$$

$$= \frac{(k + 1)(k + 2)(2k + 3)}{6}$$

$$= \frac{(k + 1)((k + 1) + 1)(2(k + 1) + 1)}{6}. \qquad \square$$

In Example 4.23, our first step toward achieving our goal is to split the sum $\sum_{i=1}^{k+1} i^2$ into $\sum_{i=1}^{k} i^2$ plus $(k+1)^2$. That is, the sum from the inductive hypothesis is split off, and the remaining term (which could be terms in general) is added on separately. This enables the inductive hypothesis to be used in the subsequent step. The rest is basic algebra. This splitting of the sum is a feature common to inductive proofs involving summations.

In our second example, we prove Theorem 4.3.

Example 4.24. Let $r \in \mathbb{R}$ with $r \neq 1$. Show: $\forall\, n \geq 0$,
$$\sum_{i=0}^{n} r^i = \frac{r^{n+1} - 1}{r - 1}.$$

Proof. (By Induction)
Base case: $(n = 0)$
It is straightforward to see that $\sum_{i=0}^{0} r^i = 1 = \frac{r^1 - 1}{r - 1}$.
Inductive step:
Suppose $k \geq 0$ and that $\sum_{i=0}^{k} r^i = \frac{r^{k+1} - 1}{r - 1}$.
(Goal: $\sum_{i=0}^{k+1} r^i = \frac{r^{k+2} - 1}{r - 1}$.)
Observe that

$$\sum_{i=0}^{k+1} r^i = \left(\sum_{i=0}^{k} r^i \right) + r^{k+1}$$

$$= \frac{r^{k+1} - 1}{r - 1} + r^{k+1} \quad \text{(By the inductive hypothesis)}$$

$$= \frac{r^{k+1} - 1 + r^{k+1}(r - 1)}{r - 1}$$

$$= \frac{r^{k+1} - 1 + r^{k+2} - r^{k+1}}{r - 1}$$

$$= \frac{r^{k+2} - 1}{r - 1}. \qquad \square$$

In both examples so far, we took advantage of the sum in the inductive hypothesis by splitting off a single term from the sum in our goal. In the next example, more than one term gets split off.

Example 4.25. Show: $\forall\, n \geq 1$, $\displaystyle\sum_{i=1}^{3n} i(i - 1) = n(9n^2 - 1)$.

Proof. (By Induction)
Base case: $(n = 1)$
A straightforward computation shows that
$\sum_{i=1}^{3} i(i - 1) = 1(0) + 2(1) + 3(2) = 8 = 1(9 \cdot 1^2 - 1).$

Inductive step:

Suppose $k \geq 1$ and that $\sum_{i=1}^{3k} i(i-1) = k(9k^2 - 1)$.

(Goal: $\sum_{i=1}^{3(k+1)} i(i-1) = (k+1)(9(k+1)^2 - 1) = (k+1)(9k^2 + 18k + 8)$.)

Observe that

$$\sum_{i=1}^{3(k+1)} i(i-1) = \sum_{i=1}^{3k+3} i(i-1)$$

$$= \left(\sum_{i=1}^{3k} i(i-1) \right)$$

$$+ \left(\sum_{i=3k+1}^{3k+3} i(i-1) \right) \qquad \text{(Split off 3 terms)}$$

$$= \left(\sum_{i=1}^{3k} i(i-1) \right)$$

$$+ [(3k+1)(3k) + (3k+2)(3k+1)$$

$$+ (3k+3)(3k+2)]$$

$$= k(9k^2 - 1)$$

$$+ (9k^2 + 3k + 9k^2 + 9k + 2 + 9k^2 + 15k + 6)$$

$$= 9k^3 + 27k^2 + 26k + 8$$

$$= (k+1)(9k^2 + 18k + 8)$$

$$= (k+1)(9(k+1)^2 - 1). \qquad \square$$

Our last example is a beautiful result about the sum of binomial coefficients. Its proof uses a more involved splitting of the sum in the inductive step. It uses the identity

$$\binom{k+1}{i} = \binom{k}{i-1} + \binom{k}{i}, \quad \text{for } 1 \leq i \leq k \qquad (4.18)$$

known as **Pascal's identity**. The proof of Pascal's identity is immediate from the formulas for the binomial coefficients. Pascal's identity is discussed further in Section 4.6.

Example 4.26.   Show: $\forall\, n \geq 0, \ \sum_{i=0}^{n} \binom{n}{i} = 2^n$.

Proof. (By Induction)

Base case: $(n = 0)$

It is straightforward to see that $\sum_{i=0}^{0} \binom{0}{i} = 1 = 2^0$.

Inductive step:

Suppose $k \geq 0$ and that $\sum_{i=0}^{k} \binom{k}{i} = 2^k$.

(Goal: $\sum_{i=0}^{k+1} \binom{k+1}{i} = 2^{k+1}$.)

Observe that

$$\sum_{i=0}^{k+1} \binom{k+1}{i} = \binom{k+1}{0} + \sum_{i=1}^{k} \binom{k+1}{i}$$

$$+ \binom{k+1}{k+1} \quad \text{(Split off first and last)}$$

$$= 1 + \sum_{i=1}^{k} \binom{k+1}{i} + 1$$

$$= 1 + \sum_{i=1}^{k} \left[\binom{k}{i-1} + \binom{k}{i} \right] + 1$$

$$= 1 + \sum_{i=1}^{k} \binom{k}{i-1} + \sum_{i=1}^{k} \binom{k}{i} + 1$$

$$= 1 + \sum_{j=0}^{k-1} \binom{k}{j} + \sum_{i=1}^{k} \binom{k}{i} + 1$$

$$= \binom{k}{k} + \sum_{j=0}^{k-1} \binom{k}{j} + \sum_{i=1}^{k} \binom{k}{i} + \binom{k}{0}$$

$$= \sum_{j=0}^{k} \binom{k}{j} + \sum_{i=0}^{k} \binom{k}{i}$$

$$= 2^k + 2^k \quad \text{(By the inductive hypothesis)}$$

$$= 2 \cdot 2^k$$

$$= 2^{k+1}. \qquad \square$$

A much more elegant proof of Example 4.26 can be given after the Binomial Theorem is covered in Section 4.6. That proof is presented in Example 4.31.

Exercises

1. Use induction to prove:

$$\forall\ n \geq 1, \ \sum_{i=1}^{n} 0 = \underbrace{0 + 0 + \cdots + 0}_{n \text{ times}} = 0.$$

2. Use induction to prove Theorem 4.2(a):

$$\forall\ n \geq 1, \sum_{i=1}^{n} 1 = \underbrace{1 + 1 + \cdots + 1}_{n \text{ times}} = n.$$

3. In Section 4.2, we saw that Gauss used the summation formula

$$\forall n \geq 1, \ \sum_{i=1}^{n} i = \frac{n(n+1)}{2}$$

from Theorem 4.2(b) to compute the sum of the first n positive integers (in the case that $n = 100$).

(a) Prove that result here by using induction.

(b) What is the formula for the sum of the first n positive even integers?

(c) When n is even, what is the formula for the sum of the positive even integers up to and including n?

(d) When n is odd, what is the formula for the sum of the positive even integers less than n?

4. Suppose we wish to build a pyramid-like structure by placing a cube with side length 1 inch, on top of a cube of side length 2 inches, on top a cube of side length 3 inches, and so forth. If we use a total of n cubes, then the volume of this structure is given by the formula

$$\forall\, n \geq 1, \quad \sum_{i=1}^{n} i^3 = \left[\frac{n(n+1)}{2}\right]^2$$

from Theorem 4.2(d).

(a) Prove this result here by using induction.

(b) How many cubes are required to make the volume of the structure exceed the volume of a single cube with side length 10 inches?

(c) How many cubes are required to make the structure at least 6 feet tall?

In Exercises 5 through 24, do not use Theorems 4.2 and 4.3. Instead, give a proof by induction.

5. Show: $\forall\, n \geq 1,\ \displaystyle\sum_{i=1}^{n}(3i^2 - i) = n^2(n+1)$.

6. Show: $\forall\, n \geq 1,\ \displaystyle\sum_{i=1}^{n}(4i^3 - 2i) = n(n+1)(n^2 + n - 1)$.

7. Show: $\forall\ n \geq 1,\ \displaystyle\sum_{i=1}^{n}(2i)^3 = 2n^2(n+1)^2$.

8. Show: $\forall\ n \geq 1,\ \displaystyle\sum_{i=1}^{n}(6i)^2 = 6n(n+1)(2n+1)$.

9. Show: $\forall\, n \geq 1,\ 1 + 5 + 9 + \cdots + (4n - 3) = n(2n - 1)$.

10. Show: $\forall\, n \geq 1,\ 1 + 4 + 7 + \cdots + (3n - 2) = \dfrac{n(3n - 1)}{2}$.

11. Show: $\forall\, n \geq 1,\ 3 + 5 + 7 + \cdots + (2n + 1) = n(n + 2)$.

12. Consider the sequence of binary numbers

$$10, 1010, 101010, \ldots$$

whose digits alternate between 1 and 0. If we start our indexing at 0, then the value of the nth number in this sequence is given by

$$2^{2n+1} + \cdots + 2^5 + 2^3 + 2.$$

Show: $\forall\, n \geq 0,\ 2 + 2^3 + 2^5 + \cdots + 2^{2n+1} = \frac{2}{3}(4^{n+1} - 1).$

13. A binary tournament, such as the type used for the NCAA basketball tournament held each March, involves a number of teams t that is a power of 2, say $t = 2^{n+1}$. In the first round, the t teams are paired off into 2^n games, and only the winners advance to the second round. There are then 2^{n-1} games in the second round, 2^{n-2} games in the third round, and so forth, until the one game in the final round determines the champion. The total number of games played in this tournament is thus $2^n + 2^{n-1} + 2^{n-2} + \cdots + 1$. In the case that $t = 64 = 2^{5+1}$, the number of games played is $32 + 16 + 8 + 4 + 2 + 1 = 63$. In general, show: $\forall\, n \geq 0,\ 1 + 2 + 4 + 8 + \cdots + 2^n = 2^{n+1} - 1.$

14. Show: $\forall\, n \geq 0,\ 1 + 3 + 9 + 27 + \cdots + 3^n = \frac{1}{2}(3^{n+1} - 1).$

15. Show: $\forall\, n \geq 2,\ \displaystyle\sum_{i=2}^{n} i2^i = (n-1)2^{n+1}.$

16. Show: $\forall\, n \geq 1,\ \displaystyle\sum_{i=1}^{n} i3^i = \frac{3}{4}[(2n-1)3^n + 1].$

17. Show: $\forall\, n \geq 1,\ \displaystyle\sum_{i=1}^{n} i^2 2^i = (n^2 - 2n + 3)2^{n+1} - 6.$

18. Show: $\forall\, n \geq 1,\ \displaystyle\sum_{i=1}^{n} i^2 3^i = \frac{3}{2}[(n^2 - n + 1)3^n - 1].$

19. Show: $\forall\, n \geq 1,\ \displaystyle\sum_{i=1}^{n} (i \cdot i!) = (n+1)! - 1.$

20. A landscaper wants to have rows of bricks emanating from the backdoor of a house in the following pattern:

By adding the number of bricks in each row, we see that the total number of bricks needed to make n rows in this arrangement is $\sum_{i=1}^{n} (2i - 1)$.

Show: $\forall\, n \geq 1,\ \displaystyle\sum_{i=1}^{n} (2i - 1) = n^2.$

21. Show: $\forall\, n \geq 1,\ \displaystyle\sum_{i=1}^{2n} i = n(2n + 1).$

22. Show: $\forall\, n \geq 1$, $\displaystyle\sum_{i=1}^{2n} i^3 = n^2(2n+1)^2$.

23. Show: $\forall\, n \geq 1$, $\displaystyle\sum_{i=1}^{n} \frac{1}{i(i+1)} = \frac{n}{n+1}$.

24. Show: $\forall\, n \geq 1$, $\displaystyle\sum_{i=1}^{n} \frac{2}{i(i+2)} = \frac{3}{2} - \frac{2n+3}{(n+1)(n+2)}$.

25. If a fair coin is tossed i times, then the probability that the first occurrence of "heads" is on the ith toss is $\frac{1}{2^i}$. This is true because, on each of the first $i-1$ tosses, there is a probability of $\frac{1}{2}$ that the coin will be "tails," and on the ith toss, there is a probability of $\frac{1}{2}$ that the coin will be "heads." Consequently, if the coin is tossed n times, then the probability that there is some occurrence of heads is $\sum_{i=1}^{n} \frac{1}{2^i}$.

   Show: $\forall\, n \geq 1$, $\displaystyle\sum_{i=1}^{n} \frac{1}{2^i} = 1 - \frac{1}{2^n}$.

26. Show: $\forall\, n \geq 0$, $\displaystyle\sum_{i=0}^{n} (3^{i+1} - 3^i) = 3^{n+1} - 1$.

27. Show: $\forall\, n \geq 1$, $\displaystyle\prod_{i=1}^{n} \frac{i}{i+2} = \frac{2}{(n+1)(n+2)}$.

28. Show: $\forall\, n \geq 1$, $\displaystyle\prod_{i=1}^{n} \frac{i}{i+1} = \frac{1}{n+1}$.

29. Let $r \in \mathbb{R}$. Show: $\forall\, n \geq 1$, $\displaystyle\prod_{i=1}^{n} r^{2i} = r^{n(n+1)}$.

30. Let $m \in \mathbb{Z}$. Show: $\forall\, n \geq 1$, $\displaystyle\prod_{i=1}^{n} i^m = (n!)^m$.

31. Show: $\forall\, n \geq 0$, $x^{2^n} - y^{2^n} = (x-y)\displaystyle\prod_{i=0}^{n-1} (x^{2^i} + y^{2^i})$.

32. Show: $\forall\, n \geq 1$, $\displaystyle\prod_{i=1}^{n} (4i-2) = \frac{(2n)!}{n!}$.

33. Let $\{s_n\}_{n \geq 0}$ be a sequence of nonnegative real numbers such that for all $n \geq 0$, $s_{n+1} \geq 2s_n$. Show:
    $$\forall\, n \geq 0, \quad s_{n+1} \geq \sum_{i=0}^{n} s_i.$$

34.* In Exercise 50 of Section 4.2, we saw that N 2-foot blocks can be stacked to extend $\sum_{i=1}^{N} \frac{1}{i}$ feet off the end of a table. By varying the value of N, we shall see here that we can exceed any desired extension off the end of the table.

(a) Show: $\forall\, n \geq 0,\ \displaystyle\sum_{i=1}^{2^n} \frac{1}{i} \geq \frac{n+2}{2}$.

(b) Suppose we wish to extend the blocks greater than n feet off the end of the table. Use the result in part (a) with $N = 2^{2n}$ to show that we can achieve this.

35.* Show: $\forall\, n \geq 1,\ \displaystyle\sum_{i=1}^{n} \frac{1}{i^2} \geq \frac{3}{2} - \frac{1}{n+1}$.

36. Show: $\forall\, n \geq 1,\ \displaystyle\sum_{i=1}^{n} \frac{i}{2^i} = 2 - \frac{n+2}{2^n}$.

37.* Show: $\forall\, n \geq 1,\ \displaystyle\sum_{i=1}^{n} \frac{1}{i^2} \leq 2 - \frac{1}{n}$.

38.* Show: $\forall\, n \geq 1,\ \displaystyle\sum_{i=1}^{n} \frac{1}{i^3} \leq \frac{1}{2}\left(3 - \frac{1}{n^2}\right)$.

39. Prove Theorem 4.1(a): $\forall\, a, s_i, t_i,$
$$\forall\, b \geq a,\ \sum_{i=a}^{b} (s_i \pm t_i) = \sum_{i=a}^{b} s_i \pm \sum_{i=a}^{b} t_i.$$

40. Prove Theorem 4.1(b): $\forall\, a, c, s_i,$
$$\forall\, b \geq a,\ \sum_{i=a}^{b} c s_i = c \sum_{i=a}^{b} s_i.$$

Remark 4.2.   From Exercises 35 and 37, it follows that $\dfrac{3}{2} \leq \displaystyle\sum_{i=1}^{\infty} \frac{1}{i^2} \leq 2$. In fact, $\displaystyle\sum_{i=1}^{\infty} \frac{1}{i^2} = \frac{\pi^2}{6} \approx 1.645$.

4.5   Strong induction

Sections 4.3 and 4.4 show the power of mathematical induction for proving statements of the form

$$\forall \text{ integers } n \geq a,\ P(n).$$

However, there are also many such examples whose proofs are more naturally done by a stronger form of induction.

Outline 4.2.   (Proof by Strong Induction).
To show: $\forall\, n \geq a,\ P(n)$.

Proof by strong induction

1. Base cases:
    Show: $P(a), \ldots, P(b)$ are true.

2. Inductive step:
   Show: $\forall\, k \geq b$, if $P(a), \ldots, P(k)$ are true, then $P(k+1)$ is true. That is,

   (a) Suppose $k \geq b$ and that $P(i)$ is true for all $a \leq i \leq k$.

   (b) Show: $P(k+1)$ is true.

The difference between regular induction and strong induction is in their inductive steps and, more specifically, in their inductive hypotheses. In strong induction, we assume not only that $P(k)$ is true but further that all of $P(a), \ldots, P(k)$ are true. That is, we carry with us a stronger inductive hypothesis, as we aim to show that $P(k+1)$ is true. Our interest in doing this can be appreciated after seeing a specific example of strong induction in action. After that, a formal proof of why it works is provided. Our example is a second proof of Theorem 3.5 from Section 3.2.

Theorem 4.7.   *Every integer greater than 1 has a prime divisor.*

Proof. First note that the statement in the theorem has the form

$$\forall \text{ integers } n \geq 2,\ \exists p \text{ a prime such that } p \mid n.$$

Of course, different integers n may have different prime divisors p. With this in mind, our proof proceeds *by strong induction*.

Base case: $(n = 2)$
Certainly, 2 is already prime and divides itself. That is, $2 \mid 2$.
Inductive step:
Suppose $k \geq 2$ and that each integer i with $2 \leq i \leq k$ has a prime divisor.
(Goal: $k + 1$ has a prime divisor.)
Our proof naturally breaks into two cases.
Case 1: $k + 1$ is prime.
Obviously, $k + 1$ divides itself.
Case 2: $k + 1$ is composite.
We can express $k + 1$ as a product $k + 1 = rs$, where $2 \leq r \leq k$ and $2 \leq s \leq k$.
Thus, in particular, r is an integer for which the inductive hypothesis applies.
Consequently, there exists a prime p that divides r.
That is, $r = pt$ for some integer t.
It follows that $k + 1 = rs = pts$.
Therefore, $k + 1$ is divisible by the prime p.   □

The key to our proof in case 2 is our ability to call on a prime factor of a potentially much smaller number r to also be a prime factor of the bigger number $k + 1$. We cannot count on a prime factor of k to also be a prime factor of $k + 1$, as would be needed in a proof by regular induction. In fact, $k + 1$ and k never share any factors other than ± 1. That is, it is important that our induction hypothesis is

Each integer i, with $2 \leq i \leq k$, has a prime divisor

and not merely

k has a prime divisor.

Strong induction is a good tool for this example.

That strong induction works can be seen, in fact, as a consequence of regular induction. An alternative proof using the Well-Ordering Principle is suggested in the exercises.

Theorem 4.8.   (Principle of Strong Induction).
Let $a \leq b$ be integers, and let $P(n)$ be an expression that depends on the free integer variable n. If

(i) $P(a), \ldots, P(b)$ hold, and

(ii) $\forall \, k \geq b$, if $P(i)$ holds for each $a \leq i \leq k$, then $P(k + 1)$ holds,

then the statement

$\forall \, n \geq a, \;\; P(n) \qquad$ *holds.*

Proof. Assume conditions (i) and (ii) in the hypotheses of the theorem. For each $n \geq a$, let $Q(n)$ be the statement that

$P(i)$ holds for each $a \leq i \leq n$.

Since $Q(n)$ implies $P(n)$, it is sufficient to show that $Q(n)$ holds $\forall \, n \geq a$. We accomplish this with a proof by regular induction.

Base cases: $(n = a, \ldots, b)$

The truth of $Q(a), \ldots, Q(b)$ follows from the truth of $P(a), \ldots, P(b)$.

Inductive step:

Suppose $k \geq a$ and that $Q(k)$ holds. That is, $P(i)$ holds for each $a \leq i \leq k$. From (ii), it follows that $P(k + 1)$ holds. By definition, $Q(k + 1)$ also holds. □

4.5.1   Standard factorization

Given an integer $n > 1$, not only can a single prime divisor be found, as asserted in Theorem 3.5, but we further know from practice that n can be completely factored into a product of primes.

$$21 = 3 \cdot 7$$

$$440 = 2^3 \cdot 5 \cdot 11$$

$$29 = 29$$

$$27061671 = 3 \cdot 7^4 \cdot 13 \cdot 17^2$$

The convention is to list the prime factors in increasing order and to use exponents to denote repeated use of the same prime.

Definition 4.1.   An expression of an integer $n > 1$ as a product of the form

$$n = p_1^{e_1} \cdot p_2^{e_2} \cdot \ \cdots \ \cdot p_m^{e_m},$$

where m is a positive integer, $p_1 < p_2 < \cdots < p_m$ are primes, and $e_1, e_2, \ldots, e_m$ are positive integers, is referred to as a **standard factorization** of n.

The following theorem is also known as the **Unique Factorization Theorem**, which is suggestive of what it asserts.

Theorem 4.9.   (Fundamental Theorem of Arithmetic). *Every integer greater than 1 has a unique standard factorization.*

Proof. Existence and uniqueness are proved separately, but each *by strong induction*.
Existence is handled first.
Base case: ($n = 2$)
Certainly, $2 = 2^1$ is a standard factorization.
Inductive step:
Suppose $k \geq 2$ and that each integer i with $2 \leq i \leq k$ has a standard factorization.
(Goal: $k + 1$ has a standard factorization.)
Our proof naturally breaks into two cases.
Case 1: $k + 1$ is prime.
Here, $k + 1 = (k + 1)^1$ is already a standard factorization.
Case 2: $k + 1$ is composite.
We can express $k + 1$ as a product $k + 1 = rs$, where $2 \leq r \leq k$ and $2 \leq s \leq k$. By the inductive hypothesis, r and s have standard factorizations. By appropriately grouping the primes in the product rs, we obtain a standard factorization for $k + 1$.

Now we handle uniqueness.
Base case: ($n = 2$)
Since 2 is prime, 2 has the unique standard factorization $2 = 2^1$. (Note that any other product of powers of primes is greater than 2.)
Inductive step:
Suppose $k \geq 2$ and that each integer i with $2 \leq i \leq k$ has a unique standard factorization.
(Goal: $k + 1$ has a unique standard factorization.)
Suppose $k + 1$ has two standard factorizations

$$k + 1 = p_1^{e_1} \cdots p_m^{e_m} = q_1^{d_1} \cdots q_l^{d_l}.$$

Since $p_1 \mid (k+1)$, Corollary 3.18 tells us that $p_1 \mid q_i^{d_i}$ for some i. Moreover, Corollary 3.19 tells us that $p_1 \mid q_i$. Since p_1 and q_i are prime, it must be that $p_1 = q_i$. Since the primes in a standard factorization are listed in increasing order, we have $q_1 \leq q_i = p_1$. By a symmetric argument (reversing the roles of the p's and the q's), we see that $p_1 \leq q_1$. Thus, $p_1 = q_1$. The integer $\frac{k+1}{p_1}$ now has the two standard factorizations

$$\frac{k+1}{p_1} = p_1^{e_1-1} \cdots p_m^{e_m} = q_1^{d_1-1} \cdots q_l^{d_l}$$

By the inductive hypothesis, those standard factorizations must be the same. Therefore, the two standard factorizations for $k+1$ must be the same. $\square$

4.5.2 The Fibonacci numbers

The sequence of numbers

$$1, 1, 2, 3, 5, 8, 13, 21, 34, 55, \ldots$$

is known as the **Fibonacci sequence**. It is named after the Italian mathematician Leonardo of Pisa (known as Fibonacci) who lived in the late twelfth and early thirteenth centuries. After the first two terms, each term in the sequence is obtained as the sum of the previous two. If we denote the Fibonacci sequence by $\{F_n\}_{n\geq 0}$, then

$$F_0 = 1, F_1 = 1, \text{ and } \forall n \geq 2, \ F_n = F_{n-2} + F_{n-1}.$$

That is, the sequence is defined recursively.

It is a beautiful fact that the Fibonacci sequence can also be expressed with a closed formula (that happens to be somewhat ugly).

Example 4.27. Show that the Fibonacci sequence can be expressed by the formula

$$\forall n \geq 0, \ F_n = \frac{1}{\sqrt{5}} \left[\left(\frac{1+\sqrt{5}}{2} \right)^{n+1} - \left(\frac{1-\sqrt{5}}{2} \right)^{n+1} \right].$$

This is known as **Binet's formula**, after the French mathematician Jacques Philippe Marie Binet (1786–1856), although it was known much earlier.

Proof. Since the recursive formula for the nth Fibonacci number defines the value F_n as a function of the two previous Fibonacci numbers, we consider two base cases. Subsequently, we may appeal to the recursive formula.

Base cases: $(n = 0, 1)$
It is straightforward to check that

$$\frac{1}{\sqrt{5}} \left[\left(\frac{1+\sqrt{5}}{2} \right)^1 - \left(\frac{1-\sqrt{5}}{2} \right)^1 \right] = 1, \text{ and}$$

$$\frac{1}{\sqrt{5}} \left[\left(\frac{1+\sqrt{5}}{2} \right)^2 - \left(\frac{1-\sqrt{5}}{2} \right)^2 \right] = 1.$$

Inductive step:
Suppose $k \geq 1$ and that, for each $0 \leq i \leq k$,

$$F_i = \frac{1}{\sqrt{5}} \left[\left(\frac{1+\sqrt{5}}{2} \right)^{i+1} - \left(\frac{1-\sqrt{5}}{2} \right)^{i+1} \right].$$

(Goal: $F_{k+1} = \frac{1}{\sqrt{5}} \left[\left(\frac{1+\sqrt{5}}{2} \right)^{k+2} - \left(\frac{1-\sqrt{5}}{2} \right)^{k+2} \right]$.)

Notice that $k + 1 \geq 2$ and that both $k - 1$ and k lie in the interval $[0, k]$.

Observe that

$$F_{k+1} = F_{k-1} + F_k$$

$$= \frac{1}{\sqrt{5}} \left[\left(\frac{1+\sqrt{5}}{2} \right)^k - \left(\frac{1-\sqrt{5}}{2} \right)^k \right]$$

$$+ \frac{1}{\sqrt{5}} \left[\left(\frac{1+\sqrt{5}}{2} \right)^{k+1} - \left(\frac{1-\sqrt{5}}{2} \right)^{k+1} \right]$$

$$= \frac{1}{\sqrt{5}} \left[\left(\frac{1+\sqrt{5}}{2} \right)^k + \left(\frac{1+\sqrt{5}}{2} \right)^{k+1} \right.$$

$$\left. - \left(\frac{1-\sqrt{5}}{2} \right)^k - \left(\frac{1-\sqrt{5}}{2} \right)^{k+1} \right]$$

$$= \frac{1}{\sqrt{5}} \left[\left(\frac{1+\sqrt{5}}{2} \right)^k \left(1 + \frac{1+\sqrt{5}}{2} \right) \right.$$

$$\left. - \left(\frac{1-\sqrt{5}}{2} \right)^k \left(1 + \frac{1-\sqrt{5}}{2} \right) \right]$$

$$= \frac{1}{\sqrt{5}} \left[\left(\frac{1+\sqrt{5}}{2} \right)^k \left(\frac{3+\sqrt{5}}{2} \right) \right.$$

$$-\left(\frac{1-\sqrt{5}}{2}\right)^k\left(\frac{3-\sqrt{5}}{2}\right)\Bigg]$$

$$=\frac{1}{\sqrt{5}}\left[\left(\frac{1+\sqrt{5}}{2}\right)^k\left(\frac{1+\sqrt{5}}{2}\right)^2\right.$$

$$\left.-\left(\frac{1-\sqrt{5}}{2}\right)^k\left(\frac{1-\sqrt{5}}{2}\right)^2\right]$$

$$=\frac{1}{\sqrt{5}}\left[\left(\frac{1+\sqrt{5}}{2}\right)^{k+2}-\left(\frac{1-\sqrt{5}}{2}\right)^{k+2}\right].\qquad\square$$

For our final example, we move to the world of sports. What is and what is not possible for the score of a football game?

Example 4.28.   Assume that, on any one possession, a football team can only score either 3 points (for a field goal) or 7 points (for a touchdown).

Show that it is then mathematically possible (assuming that there are no time constraints) for a football team to score any number of points from 12 on up.

Remark 4.3.   Note that $1, 2, 4, 5, 8,$ and 11 points are not possible in the scheme in Example 4.28. Of course, $3, 6, 7, 9,$ and 10 are possible numbers of points.

Proof of Example 4.28 (By Strong Induction)
Base cases: ($n = 12, 13, 14$)
We see that
12 points can be accomplished by 4 field goals,
13 points can be accomplished by 2 field goals and 1 touchdown, and
14 points can be accomplished by 2 touchdowns.

Inductive step:
Suppose $k \geq 14$ and that, for each $12 \leq i \leq k$, it is possible to attain i points through field goals and touchdowns.
(Goal: Show that $k + 1$ points can be accomplished through field goals and touchdowns.)
Consider the possibility of scoring $(k + 1) - 3$ points (i.e. $k - 2$ points). Since $k \geq 14$, we know that $12 \leq k - 2 \leq k$.
Hence, by our inductive hypothesis, $k - 2$ points can be accomplished by some number f of field goals and some number t of touchdowns.

Consequently, scoring one more field goal would yield $k+1$ points.

That is, $k+1$ points can be accomplished by $f+1$ field goals and t touchdowns. $\quad\square$

Why does the proof in Example 4.28 consider three base cases? The reason can be seen in the inductive step. Three base cases are considered since a field goal is worth 3 points. Consequently, if a team can score 3 fewer than some desired number of points, then they need only score an additional field goal to get the desired total. Alternatively, we could have considered seven base cases and made similar use of a touchdown. However, using the smaller point value was certainly more efficient.

Formally, Example 4.28 is the statement that

$$\forall\, n \geq 12, \ \exists\, f, t \in \mathbb{N} \text{ such that } 3f + 7t = n.$$

However, the connection with football perhaps made it more digestible for some.

Exercises

1. Let $\{s_n\}$ be the sequence defined by

   $$s_0 = 0, s_1 = 1, \text{ and } \forall\, n \geq 2, \ s_n = 3s_{n-1} - 2s_{n-2}.$$

   Show: $\forall\, n \geq 0, \ s_n = 2^n - 1$.

2. Let $\{s_n\}$ be the sequence defined by

   $$s_0 = 2, s_1 = 6, \text{ and } \forall\, n \geq 2, \ s_n = 6s_{n-1} - 8s_{n-2}.$$

   Show: $\forall\, n \geq 0, \ s_n = 2^n(2^n + 1)$.

3. Let $\{s_n\}$ be the sequence defined by

   $$s_0 = 2, s_1 = 11, \text{ and } \forall\, n \geq 2, \ s_n = 11s_{n-1} - 28s_{n-2}.$$

   Show: $\forall\, n \geq 0, \ s_n = 4^n + 7^n$.

4. Let $\{s_n\}$ be the sequence defined by

   $$s_0 = 6, s_1 = 8, \text{ and } \forall\, n \geq 2, \ s_n = 4s_{n-1} - 3s_{n-2}.$$

   Show: $\forall\, n \geq 0, \ s_n = 5 + 3^n$.

5. Let $\{s_n\}$ be the sequence defined by

   $$s_0 = 1, s_1 = 2, \text{ and } \forall\, n \geq 2, \ s_n = 4s_{n-2}.$$

   Show: $\forall\, n \geq 0, \ s_n = 2^n$.

6. Let $\{s_n\}$ be the sequence defined by

   $$s_0 = 0, s_1 = 3, \text{ and } \forall\, n \geq 2, \ s_n = s_{n-2} + 6.$$

   Show: $\forall\, n \geq 0, \ s_n = 3n$.

254 DISCRETE MATHEMATICS AND APPLICATIONS

7. Let $\{s_n\}$ be the sequence defined by

$$s_0 = -1, s_1 = 0, s_2 = 12, \text{ and } \forall\, n \geq 3,$$
$$s_n = 10s_{n-1} - 31s_{n-2} + 30s_{n-3}.$$

Show: $\forall\, n \geq 0$, $s_n = 5^n - 3^n - 2^n$.

8. Let $\{s_n\}$ be the sequence defined by

$$s_0 = 2, s_1 = 5, s_2 = 5, \text{ and } \forall\, n \geq 3,$$
$$s_n = 14s_{n-1} - 59s_{n-2} + 70s_{n-3}.$$

Show: $\forall\, n \geq 0$, $s_n = 2^n + 2 \cdot 5^n - 7^n$.

9. Let $\{s_n\}$ be the sequence defined by

$$s_0 = 1, s_1 = 3, \text{ and } \forall\, n \geq 2, \quad s_n = 3s_{n-2} - 2s_{n-1}.$$

Show: $\forall\, n \geq 0$, s_n is odd.

10. Let $\{s_n\}$ be the sequence defined by

$$s_0 = 0, s_1 = 2, s_2 = 6, \text{ and } \forall\, n \geq 3,$$
$$s_n = s_{n-3} - s_{n-2} + s_{n-1}.$$

(a) Show: $\forall\, n \geq 0$, s_n is even.

(b) Find s_{6543}. (Hint: Find a pattern.)

(c) State and prove the pattern used in part (b).

(d) How many base cases were used in part (c)? versus part (a)?

11. Let $\{s_n\}$ be the sequence defined by

$$s_0 = -1, s_1 = 0, \text{ and } \forall\, n \geq 2, \quad s_n = -6s_{n-2} + 5s_{n-1}.$$

(a) Find s_2, s_3, s_4.

(b) Show: $\forall\, n \geq 0$, $s_n = 2 \cdot 3^n - 3 \cdot 2^n$.

(c) Show: $\{s_n\}$ is increasing. Hint: Consider $s_{n+1} - s_n$.

12. The **Lucas sequence** $\{L_n\}_{n \geq 1}$ is a sequence very similar to the Fibonacci sequence, but with different initial terms. It is defined by

$$L_1 = 1, L_2 = 3, \text{ and } \forall\, n \geq 3, \quad L_n = L_{n-2} + L_{n-1}.$$

(a) Find $L_1, \ldots, L_6$.

(b) Why would $L_1 = 1, L_2 = 2$ not be an interesting choice for the initial terms of the Lucas sequence?

(c) Show: $\forall\, n \geq 1$, $L_n = \left(\frac{1+\sqrt{5}}{2}\right)^n + \left(\frac{1-\sqrt{5}}{2}\right)^n$.

13. The following is a recursive function that encodes a given array (of specified length) of single-digit positive integers into a string of digits:

> Function **Encode**$(n, [a_1, \ldots, a_n])$
>> \begin
>>> If $n = 0$, then Print 0.
>>> If $n = 1$, then Print a_1.
>>> Otherwise,
>>>> \begin
>>>> Encode$(n - 1, [a_2, \ldots, a_n])$,
>>>> Encode$(n - 2, [a_2, \ldots, a_{n-1}])$,
>>>> Encode$(n - 1, [a_1, \ldots, a_{n-1}])$.
>>>> \end.
>> \end.

If s_n denotes the length of the resulting string when an array of length n is input, then

$$s_0 = 1, s_1 = 1, \quad \text{and} \quad \forall\, n \geq 2, \quad s_n = 2s_{n-1} + s_{n-2}.$$

(a) Find s_2, s_3, s_4.

(b) Show: $\forall\, n \geq 0$, $s_n = \frac{1}{2}((1 + \sqrt{2})^n + (1 - \sqrt{2})^n)$.

(c) Find the string resulting from the 3-digit input array $[5, 2, 7]$.

14. Let $\{s_n\}$ be the sequence defined by $s_0 = 1$ and $\forall\, n \geq 1$,
$s_n = 3 - s_{n-1}$. Show: $\forall\, n \geq 0, s_n = \begin{cases} 2 & \text{if } n \text{ is even,} \\ 1 & \text{if } n \text{ is odd.} \end{cases}$
Is strong induction needed for this?

15. Show: $\forall\, n \geq 0$, $\sum_{i=0}^{n}(-1)^i = \begin{cases} 1 & \text{if } n \text{ is even,} \\ 0 & \text{if } n \text{ is odd.} \end{cases}$

16. Let $\{s_n\}$ be the sequence defined by

$$s_0 = 3, s_1 = 3 + \sqrt{2}, \quad \text{and} \quad \forall\, n \geq 2, \quad s_n = 2s_{n-1} + s_{n-2}.$$

Show: $\forall\, n \geq 0, s_n = 2(1 + \sqrt{2})^n + (1 - \sqrt{2})^n$.

17. (a) Show that any dollar amount of $4 or greater can be achieved with $2 bills and $5 bills. That is, only $1 and $3 are unachievable.

(b) Would using $4 bills instead of $5 bills increase or decrease the number of dollar amounts than cannot be achieved? Explain.

18. A vending machine dispenses only 5¢ stamps and 12¢ stamps.

(a) Show that any monetary value of 44¢ or greater can be obtained from stamps from this machine.

(b) Using this machine to get stamps, what is the largest amount of extra postage (beyond the value required) that one would ever have to put on an item to be mailed?

19. A hardware store is promoting sales of cement blocks of heights 4 and 9 inches to college students. The suggestion is that such blocks can be stacked to form the legs of a table with a sheet of plywood 1 inch thick used as the top. The claim is that a table of any (integer) height in inches of 25 or greater can be achieved by this method. Prove that the store's claim is true. Do not forget about the plywood.

20. Show that all monetary values except $1¢, 2¢, 4¢$, and $7¢$ can be obtained from $3¢$ and $5¢$ stamps. Compare Exercise 18.

21.* (a) Show that any monetary value of $20¢$ or greater that is a multiple of $5¢$ can be achieved with dimes and quarters.

   (b) If exactly four pennies are available as well, then what monetary values are unachievable?

22.* Show that any even number of points greater than 2 can be achieved with items worth 4 points and 6 points.

For Exercises 23 through 28, find the standard factorization of the given integer.

23. 1260.   24. 1210.

25. 3549.   26. 70125.

27. 12!.   28. $\binom{18}{3}$.

29. Show: $\forall \, n \geq 2$, there is a prime p such that $p^2 \mid n^2$

   (a) By mimicking the proof of Theorem 3.5 and *not* using the Fundamental Theorem of Arithmetic

   (b) By using the Fundamental Theorem of Arithmetic

30. Show: $\forall \, n \geq 2$, there is a prime p such that $p^3 \mid n^3$

   (a) By mimicking the proof of Theorem 3.5 and *not* using the Fundamental Theorem of Arithmetic

   (b) By using the Fundamental Theorem of Arithmetic

31.* Let $p_1, p_2, \ldots, p_m$ be primes and $e_1, e_2, \ldots, e_m, f_1, f_2, \ldots, f_m$ be natural numbers. Show that $\gcd(p_1^{e_1} p_2^{e_2} \cdots p_m^{e_m}, p_1^{f_1} p_2^{f_2} \cdots p_m^{f_m}) = p_1^{\min\{e_1, f_1\}} p_2^{\min\{e_2, f_2\}} \cdots p_m^{\min\{e_m, f_m\}}$.

32.* Let m and n be integers, not both zero. Show that $\text{lcm}(m, n) = \frac{|mn|}{\gcd(m,n)}$.

33. Show that $\forall \, a, b \in \mathbb{Z}^+$, $\gcd(a^2, b^2) = (\gcd(a, b))^2$.

34. Show: $\forall \, a, b, d \in \mathbb{Z}^+$, $d = \gcd(a, b)$ if and only if there exist relatively prime integers x and y such that $a = dx$ and $b = dy$.

35. (a) **Unique binary representations of nonnegative integers.** Show: Each integer $n > 0$ has a unique representation in the form

$$n = b_m \cdot 2^m + b_{m-1} \cdot 2^{m-1} + \cdots + b_1 \cdot 2 + b_0,$$

where $m \geq 0, b_m = 1$, and $b_i = 0$ or 1 for each $0 \leq i \leq m - 1$. That is, the string $b_m b_{m-1} \cdots b_1 b_0$ is the binary representation of n.

(b) **Unique base-s representations of nonnegative integers.** Show: Each integer $n > 0$ has a unique representation in the form

$$n = a_m \cdot s^m + a_{m-1} \cdot s^{m-1} + \cdots + a_1 \cdot s + a_0,$$

where $m \geq 0, a_m \neq 0$, and $a_i \in \{0, 1, \ldots, s - 1\}$ for each $0 \leq i \leq m$. That is, the string $a_m a_{m-1} \cdots a_1 a_0$ is the base-s representation of n.

36. Given an integer n and a base s, show that the following algorithm returns the base-s representation $a_m a_{m-1} \cdots a_1 a_0$ of n. That is, $n = a_m s^m + a_{m-1} s^{m-1} + \cdots + a_1 s + a_0$.

Let $m = 0$.
While $n > 0$,
    \begin
    Let $a_m = n \bmod s$.
    Let $n = n \operatorname{div} s$.
    Let $m = m + 1$.
    \end.
Return $a_m a_{m-1} \cdots a_1 a_0$.

37. Observe that $n(n + 1)$ is a factor of the numerator in each of the formulas

$$\sum_{i=1}^{n} i = \frac{n(n + 1)}{2}, \sum_{i=1}^{n} i^2 = \frac{n(n + 1)(2n + 1)}{6}, \text{ and}$$

$$\sum_{i=1}^{n} i^3 = \left[\frac{n(n + 1)}{2} \right]^2.$$

Show that, for all integers $m \geq 1$ and $n \geq 1$, the product $n(n + 1)$ is a factor of the numerator in the formula for $\sum_{i=1}^{n} i^m$. Hint: Use Theorem 4.4 and strong induction on m.

In Exercises 38 through 44, we explore sum of the beautiful identities involving the Fibonacci and Lucas numbers.

38. Show: $\forall\, n \geq 0$, $\displaystyle\sum_{i=0}^{n} F_i^2 = F_n F_{n+1}$. That is, the sum of the squares of any initial string of the Fibonacci sequence can be easily obtained.

39. Show: $\forall\, n \geq 0$, $\displaystyle\sum_{i=0}^{n} F_i = F_{n+2} - 1$. That is, the sum of any initial string of the Fibonacci sequence can be easily obtained.

40. Show: $\forall\, n \geq 0$, $F_n = \displaystyle\sum_{i=0}^{n \text{ div } 2} \binom{n-i}{i}$. Hint: Consider odd and even n separately.

41. Show: $\forall\, n \geq 2$, $\begin{bmatrix} 1 & 1 \\ 1 & 0 \end{bmatrix}^n = \begin{bmatrix} F_n & F_{n-1} \\ F_{n-1} & F_{n-2} \end{bmatrix}$. The relevant matrix multiplication is defined prior to Exercise 31 of Section 4.3.

42. Show: $\forall\, n \geq 2$, $L_n = F_n + F_{n-2}$.
    See Exercise 7 for the definition of L_n.

43. Show that $\forall\, n \geq 0$, F_n and F_{n+1} are relatively prime.

44. Show that $\forall\, n \geq 1$, F_n and L_{n+1} have the same parity.
    See Exercise 7 for the definition of L_n.

45. Prove the Principle of Strong Induction (Theorem 4.8) by mimicking the proof of the Principle of Induction (Theorem 4.6) that uses the Well-Ordering Principle.

46. Prove the existence result in the Fundamental Theorem of Arithmetic (Theorem 4.9) by using the Well-Ordering Principle and Theorem 3.5.

47. **Rugby.** A rugby team may score a try, which is worth 5 points, and only after doing so, they may score a conversion goal, which is worth 2 points. Assuming that these are the only scoring possibilities, show that 23 points is not achievable but that any number of points from 24 on up is achievable. That is, show that $\forall\, n \geq 24, \exists\, t, c \in \mathbb{N}$ with $t \geq c$ such that $5t + 2c = n$. Is there a certain number of base cases that must be considered?

48. **Frobenius coin exchange problem.** Given relatively prime positive integers a and b (thought of as the monetary values of two different coins), $(a - 1)(b - 1) - 1$ is the largest value that cannot be achieved as a linear combination $ax + by$ for natural numbers x and y. For example, in Exercise 20, we have $a = 3$ and $b = 5$, and we see that 7 is the largest unachievable value. Observe that this formula also agrees with the result in Exercise 18 that 43 is the largest unachievable value when $a = 5$ and $b = 12$. In general, do the following:

(a) Show that $(a-1)(b-1)-1$ cannot be achieved.

(b) Show that $(a-1)(b-1)$ can be achieved.

4.6 The Binomial Theorem

The Binomial Theorem is a result that tells us how to expand an expression of the form

$$(a+b)^n \quad \text{for some integer } n \geq 0.$$

In fact, it expresses the pattern that starts to show itself for the first few values of n.

$$
\begin{aligned}
(a+b)^0 &= 1 \\
(a+b)^1 &= a+b \\
(a+b)^2 &= a^2 + 2ab + b^2 \\
(a+b)^3 &= a^3 + 3a^2b + 3ab^2 + b^3 \\
(a+b)^4 &= a^4 + 4a^3b + 6a^2b^2 + 4ab^3 + b^4 \\
&\vdots
\end{aligned} \tag{4.19}
$$

One feature to observe in the developing pattern is that in each row, the powers of a decrease from left to right, while the powers of b increase. Another feature is how the coefficients change.

$$
\begin{array}{ccccccccc}
 & & & & 1 & & & & \\
 & & & 1 & & 1 & & & \\
 & & 1 & & 2 & & 1 & & \\
 & 1 & & 3 & & 3 & & 1 & \\
1 & & 4 & & 6 & & 4 & & 1 \\
 & & & & \vdots & & & &
\end{array}
$$

This infinite array of integers is known as **Pascal's triangle**. Its beauty comes from the way in which each subsequent row can be determined from the previous. To express this, let $c_{n,k}$ denote the kth entry in the nth row. Here, both n and k start counting from 0.

$$
\begin{array}{ccccccccc}
 & & & & c_{0,0} & & & & \\
 & & & c_{1,0} & & c_{1,1} & & & \\
 & & c_{2,0} & & c_{2,1} & & c_{2,2} & & \\
 & c_{3,0} & & c_{3,1} & & c_{3,2} & & c_{3,3} & \\
c_{4,0} & & c_{4,1} & & c_{4,2} & & c_{4,3} & & c_{4,4} \\
 & & & & \vdots & & & &
\end{array}
$$

The edges of the triangle display the identity

$$\forall\, n \geq 0, \quad c_{n,0} = c_{n,n} = 1.$$

The internal entries of each row are determined from the previous row by Pascal's identity

$$\forall n \geq 2 \text{ and } 1 \leq k \leq n-1, \quad c_{n,k} = c_{n-1,k-1} + c_{n-1,k}. \quad (4.20)$$

For example, the entries in the 5th row of Pascal's triangle are the following values of $c_{n,k}$:

$$c_{5,0} \quad c_{5,1} \quad c_{5,2} \quad c_{5,3} \quad c_{5,4} \quad c_{5,5}$$

The internal entry $c_{5,3}$ is determined by (4.20) as

$$c_{5,3} = c_{4,2} + c_{4,3} = 4 + 6 = 10.$$

In a similar way, each internal entry in the 5th row can be determined from entries in the in the 4th row. Of course, the initial and final entries are defined to be 1. Consequently, the 5th row of Pascal's triangle is seen to be

$$1 \quad 5 \quad 10 \quad 10 \quad 5 \quad 1.$$

The disadvantage of this method is that if the 100th row is desired, then rows 0 through 99 would first have to be computed. Fortunately, in practice, all that extra work is not necessary. This is a consequence of the fact that the entries in Pascal's triangle turn out to be the binomial coefficients. That is, $c_{n,k} = \binom{n}{k}$.

$$\binom{0}{0}$$
$$\binom{1}{0} \quad \binom{1}{1}$$
$$\binom{2}{0} \quad \binom{2}{1} \quad \binom{2}{2}$$
$$\binom{3}{0} \quad \binom{3}{1} \quad \binom{3}{2} \quad \binom{3}{3}$$
$$\binom{4}{0} \quad \binom{4}{1} \quad \binom{4}{2} \quad \binom{4}{3} \quad \binom{4}{4}$$
$$\vdots$$

This equality is certainly apparent in the first few rows, if we compute the values $\binom{n}{k}$. However, a careful proof comes from the identities

$$\forall n \geq 0, \quad \binom{n}{0} = \binom{n}{n} = 1 \quad (4.21)$$

$$\forall n \geq 2 \text{ and } 1 \leq k \leq n-1, \quad \binom{n}{k} = \binom{n-1}{k-1} + \binom{n-1}{k}. \quad (4.22)$$

Equation (4.22) is **Pascal's identity**, and we also encountered it in Equation (4.18) of Section 4.4. Of course, Equation (4.22) agrees with Equation (4.20).

We now have all of the pieces that go into the Binomial Theorem. In fact, the binomial coefficients are so named because of their role as coefficients in the Binomial Theorem.

Theorem 4.10. (The Binomial Theorem). *Let* $a, b \in \mathbb{R}$ *and* $n \in \mathbb{N}$. *Then,*[*]

$$(a + b)^n = \sum_{i=0}^{n} \binom{n}{i} a^{n-i} b^i.$$

Proof. More than enough base cases, namely, $n = 0, 1, 2, 3, 4$, were confirmed in (4.19). Hence, we move right to the inductive step.

Suppose $k \geq 0$ and that $(a + b)^k = \sum_{i=0}^{k} \binom{k}{i} a^{k-i} b^i$.
(Goal: $(a + b)^{k+1} = \sum_{j=0}^{k+1} \binom{k+1}{j} a^{k+1-j} b^j$.)
Observe that

$$(a + b)^{k+1} = (a + b)(a + b)^k$$

$$= (a + b) \sum_{i=0}^{k} \binom{k}{i} a^{k-i} b^i$$

(by the inductive hypothesis)

$$= \sum_{i=0}^{k} \binom{k}{i} a^{k+1-i} b^i + \sum_{i=0}^{k} \binom{k}{i} a^{k-i} b^{i+1}$$

$$= \sum_{j=0}^{k} \binom{k}{j} a^{k+1-j} b^j + \sum_{j=1}^{k+1} \binom{k}{j-1} a^{k+1-j} b^j$$

$$= \left(a^{k+1} + \sum_{j=1}^{k} \binom{k}{j} a^{k+1-j} b^j \right)$$

$$+ \left(\sum_{j=1}^{k} \binom{k}{j-1} a^{k+1-j} b^j + b^{k+1} \right)$$

$$= a^{k+1} + \left(\sum_{j=1}^{k} \left[\binom{k}{j} + \binom{k}{j-1} \right] a^{k+1-j} b^j \right)$$

$$+ b^{k+1}$$

$$= a^{k+1} + \left(\sum_{j=1}^{k} \binom{k+1}{j} a^{k+1-j} b^j \right) + b^{k+1}$$

$$= \sum_{j=0}^{k+1} \binom{k+1}{j} a^{k+1-j} b^j. \qquad \square$$

Pascal's identity is used in the second to last equality of the inductive step. In that case, the form of Pascal's identity

---

[*] Any occurrence of 0^0 here is taken to have the value 1.

expressed in (4.18) from Section 4.4 may be the more convenient reference. Notice that the terms a^{k+1} and b^{k+1} need to be split off so that Pascal's identity can be applied for the indices in the remaining sums. This is analogous to the fact that the 1's on the edges of Pascal's triangle need to be specified separately from the description of the internal numbers.

One obvious use of the Binomial Theorem is in performing algebraic expansions.

Example 4.29. Use the Binomial Theorem to expand each of the following:

(a) Expand $(x + y)^6$.

Solution. We use $a = x$, $b = y$, and $n = 6$ in Theorem 4.10.

$$
\begin{aligned}
(x+y)^6 &= \sum_{i=0}^{6} \binom{6}{i} x^{6-i} y^i \\
&= \binom{6}{0} x^6 + \binom{6}{1} x^5 y^1 \\
&\quad + \binom{6}{2} x^4 y^2 + \binom{6}{3} x^3 y^3 \\
&\quad + \binom{6}{4} x^2 y^4 + \binom{6}{5} x^1 y^5 + \binom{6}{6} y^6 \\
&= x^6 + 6x^5 y^1 + 15x^4 y^2 + 20x^3 y^3 \\
&\quad + 15x^2 y^4 + 6x^1 y^5 + y^6. \qquad \square
\end{aligned}
$$

(b) Expand $(2x + 3y)^5$.

Solution. We use $a = 2x$, $b = 3y$, and $n = 5$ in Theorem 4.10.

$$
\begin{aligned}
(2x+3y)^5 &= \sum_{i=0}^{5} \binom{5}{i} (2x)^{5-i} (3y)^i \\
&= \binom{5}{0} (2x)^5 + \binom{5}{1} (2x)^4 (3y) \\
&\quad + \binom{5}{2} (2x)^3 (3y)^2 + \binom{5}{3} (2x)^2 (3y)^3 \\
&\quad + \binom{5}{4} (2x)(3y)^4 + \binom{5}{5} (3y)^5 \\
&= 2^5 x^5 + 5 \cdot 2^4 \cdot 3 x^4 y + 10 \cdot 2^3 \cdot 3^2 x^3 y^2 \\
&\quad + 10 \cdot 2^2 \cdot 3^3 x^2 y^3 + 5 \cdot 2 \cdot 3^4 x y^4 + 3^5 y^5
\end{aligned}
$$

$$= 32x^5 + 240x^4y + 720x^3y^2$$
$$+ 1080x^2y^3 + 810xy^4 + 243y^5. \qquad \square$$

(c) Expand $(x + 1)^n$.

Solution. We use $a = x$ and $b = 1$ in Theorem 4.10.

$$(x + 1)^n = \sum_{i=0}^{n} \binom{n}{i} x^{n-i} 1^i = \sum_{i=0}^{n} \binom{n}{i} x^{n-i}$$

$$= x^n + nx^{n-1} + \binom{n}{2} x^{n-2}$$

$$+ \cdots + nx + 1. \qquad \square$$

(d) Expand $(x - y)^n$.

Solution. We use $a = x$ and $b = -y$ in Theorem 4.10.

$$(x - y)^n = (x + (-y))^n = \sum_{i=0}^{n} \binom{n}{i} x^{n-i}(-y)^i$$

$$= \sum_{i=0}^{n} (-1)^i \binom{n}{i} x^{n-i} y^i$$

$$= x^n - nx^{n-1}y + \binom{n}{2} x^{n-2}y^2$$

$$- \binom{n}{3} x^{n-3}y^3 + \cdots$$

$$+ (-1)^{n-1} nxy^{n-1} + (-1)^n y^n.$$

Notice that the first term has a positive coefficient, and then the signs alternate. Of course, the sign of the last term depends on the parity of n. $\qquad \square$

Another way in which the Binomial Theorem may be used is to find, with little effort, a particular coefficient in a polynomial that has not been expanded.

Example 4.30. Find the coefficient of x^{40} in $(1 + 2x)^{50}$.

Solution. By the Binomial Theorem,

$$(1 + 2x)^{50} = \sum_{i=0}^{50} \binom{50}{i} 1^{50-i}(2x)^i = \sum_{i=0}^{50} \binom{50}{i} 2^i x^i.$$

Consequently, the coefficient of x^{40} (i.e. when $i = 40$) is

$$\binom{50}{40}2^{40}. \qquad \square$$

The ability to quickly extract a particular coefficient from a factored polynomial without having to expand the polynomial entirely is of great use in Section 7.3, where polynomials are used to solve counting problems. The functions Expand and Coefficient in *Mathematica*, and the functions expand and coeff in *Maple*, may also be of some help.

For our final applications, we see how the Binomial Theorem can be used to give quick proofs of some beautiful identities.

Example 4.31.   Verify each of the following identities:

(a) $\forall\, n \geq 0, \ \displaystyle\sum_{i=0}^{n} \binom{n}{i} = 2^n.$

Proof. By Theorem 4.10,

$$2^n = (1+1)^n = \sum_{i=0}^{n} \binom{n}{i}1^{n-i}1^i = \sum_{i=0}^{n} \binom{n}{i}. \qquad \square$$

(b) $\forall\, n \geq 0, \ \displaystyle\sum_{i=0}^{n} \binom{n}{i}2^i = 3^n.$

Proof. By Theorem 4.10,

$$3^n = (1+2)^n = \sum_{i=0}^{n} \binom{n}{i}1^{n-i}2^i = \sum_{i=0}^{n} \binom{n}{i}2^i. \qquad \square$$

(c) $\forall\, n \geq 1, \ \displaystyle\sum_{i=0}^{n}(-1)^i\binom{n}{i} = 0.$

Proof. By Theorem 4.10,

$$0 = (1-1)^n = \sum_{i=0}^{n}\binom{n}{i}1^{n-i}(-1)^i = \sum_{i=0}^{n}(-1)^i\binom{n}{i}. \qquad \square$$

Exercises

Each of the following exercises can be completed by hand. Those requesting computations might also be done with the aid of a symbolic calculator such as *Mathematica* or *Maple*. However, the results should then be compared with hand computations.

1. **Hexagon identity.** Let $n, k \in \mathbb{Z}$ with $1 \leq k \leq n - 1$.
   Show: $\binom{n-1}{k-1}\binom{n}{k+1}\binom{n+1}{k} = \binom{n-1}{k}\binom{n}{k-1}\binom{n+1}{k+1}$.
   These binomial coefficients form a hexagon in Pascal's triangle.

2. **Second-order Pascal's identity.** Let $n, k \in \mathbb{Z}$ with $2 \leq k \leq n - 2$. Show: $\binom{n-2}{k-2} + 2 \cdot \binom{n-2}{k-1} + \binom{n-2}{k} = \binom{n}{k}$.
   What shape in Pascal's triangle is formed by the binomial coefficients in this identity?

In Exercises 3 through 22, use the Binomial Theorem to expand the given expressions, and perform any other requested computations.

3. $(x + y)^5$.

4. $(x + y)^7$.

5. $(3x + y)^6$.

6. $(x + 2y)^5$.

7. $(2x - y)^5$.

8. $(x - 3y)^6$.

9. $(x - 1)^n$.

10. $(x + 2)^n$.

11. (a) Expand $(x + \frac{1}{2})^n$.

    (b) What is the coefficient of x^5?

12. (a) Expand $(x - \frac{1}{3})^n$.

    (b) What is the coefficient of x^6?

13. $(x^2 + y^2)^4$.

14. $(x^2 + y^2)^5$.

15. $(3x^2 + y^3)^5$.

16. $(x^3 + 2y^2)^4$.

17. (a) Expand $(x^2 + 1)^n$.

    (b) What is the coefficient of x^8?

18. (a) Expand $(x^2 - y^2)^n$.

    (b) What is the coefficient of y^{10}?

19. $(n - 1)^n$.

20. $(n + 1)^n$.

21.* The expression $p(n) = (1 + \frac{1}{n})^n$ gives a better and better approximation for the **Euler number** $e \approx 2.718$, as n gets larger. (a) Compute $p(100)$, $p(1000)$, and $p(10000)$, and observe how the approximation improves. (b) Give an expression for the expansion of $p(n)$.

22.* The expression $q(n) = (1 - \frac{1}{n})^n$ gives a better and better approximation for the reciprocal of the Euler number $\frac{1}{e} \approx 0.3679$, as n gets larger. (a) Compute $q(100)$, $q(200)$, and $q(300)$, and observe how the approximation improves. (b) Give an expression for the expansion of $q(n)$.

23. Find the coefficient of $x^{60}y^{40}$ in $(3x + 2y)^{100}$.

24. Find the coefficient of $x^{493}y^7$ in $(x - 2y)^{500}$.

25. Find the coefficient of $x^{20}y^{1170}$ in $(2x^2 + y^3)^{400}$.

26. Find the coefficient of $x^{840}y^{40}$ in $(x^3 - 2y^2)^{300}$.

27. Find the coefficient of $x^{50}y^{50}$ in $(x^2 + y^2)^{30}$.

28. Find the coefficient of $x^{35}y^{15}$ in $(x^2 + y^2)^{50}$.

29. Let $a, b \in \mathbb{N}$. A game consists of tossing a coin. If heads is obtained, then a dollars is won. If tails is obtained, then b dollars is lost. If the coin is tossed 10 times, then the probability of winning exactly w dollars equals $\frac{1}{2^{10}}$ times the coefficient of x^w in $(x^a + x^{-b})^{10}$. Determine the probability of winning \$4, in each of the following cases:
(a) $a = 1, b = -1$. (b) $a = 2, b = -2$. (c) $a = 3, b = -1$.

30. Let $a, b \in \mathbb{N}$. A vending machine dispenses stamps worth 3¢ and 7¢. A man has inserted a dollar into the machine, and 12 times he randomly presses either the button to receive a 3¢ stamp or the button to receive a 7¢ stamp. If at the end he retrieves his change and the 12 stamps that he purchased, then the probability that he ends up with s¢ in stamps is $\frac{1}{2^{12}}$ times the coefficient of x^s in $(x^3 + x^7)^{12}$. Determine the probability that he ends up with s¢ in stamps in each of the following cases:
(a) $s = 68$¢. (b) $s = 60$¢.

In Exercises 31 through 42, use the Binomial Theorem to verify the given identities.

31. $\sum_{i=0}^{n} \binom{n}{i}8^i = 9^n$.

32. $\sum_{i=0}^{n} \binom{n}{i}2^{n-i}3^i = 5^n$.

33. Since base-twelve numbers use the "digits" $0, \ldots, 9, a, b$, it is natural to classify such numbers according to how many of their digits are "letters" (a or b). Among all base-twelve numbers of length n, for each $0 \le i \le n$, there are $\binom{n}{i}10^{n-i}2^i$ that contain exactly i letters. Summing over all $0 \le i \le n$, we see that there are $\sum_{i=0}^{n} \binom{n}{i}10^{n-i}2^i$ possible base-twelve numbers of length n. A much easier formula is given by 12^n. Verify that $\sum_{i=0}^{n} \binom{n}{i}10^{n-i}2^i = 12^n$.

34. A deoxyribonucleic acid (DNA) strand contains a sequence of nitrogen bases, each chosen from one of cytosine (C), guanine (G), thymine (T), or adenine (A). Scientists studying a particular class of genes are categorizing strands according to the number of occurrences of the nitrogen base T. Among stands of length n, for each $0 \le i \le n$, there are $\binom{n}{i}3^{n-i}$ strands containing exactly i occurrences of T. Summing over all $0 \le i \le n$, we see that there are $\sum_{i=0}^{n} \binom{n}{i}3^{n-i}$ possible DNA sequences of length n. A much easier formula is given by 4^n. Verify that $\sum_{i=0}^{n} \binom{n}{i}3^{n-i} = 4^n$.

35. $\sum_{i=0}^{n} \binom{n}{i}5^{n-i}3^i = 2^{3n}$.

36. $\sum_{i=0}^{n} \binom{n}{i}2^{2i}3^{n-i} = 7^n$.

37. $\sum_{i=0}^{n}(-1)^i\binom{n}{i}3^{n-i}=2^n.$ 38. $\sum_{i=0}^{n}(-1)^i\binom{n}{i}2^{n-i}3^i=(-1)^n.$

39. (a) Use the fact that $6=2+4$ to prove $\sum_{i=0}^{n}\binom{n}{i}2^{n+i}=6^n.$

(b) This identity is 2^n times what simpler identity?

40. (a) Use the fact that $-2=2-4$ to prove $\sum_{i=0}^{n}(-1)^i\binom{n}{i}2^{n+i}$
$=(-2)^n.$

(b) This identity is 2^n times what simpler identity?

41. $\sum_{i=0}^{n}(-1)^i\binom{n}{i}\left(\frac{1}{3}\right)^i$ 42. $\sum_{i=0}^{n}\binom{n}{i}\left(\frac{1}{2}\right)^i=\left(\frac{3}{2}\right)^n.$
$=\left(\frac{2}{3}\right)^n.$

43.* Let a and b be integers that are not both zero, and let n be a positive integer.
Show: If $\gcd(a,b)=1$, then $\gcd(a,b^n)=1$. Hint: If we can write $by=1-ax$, then how can we write b^ny^n?

44.* Prove Theorem 4.4. Hint: Follow these steps.

1. Observe that

$$(i+1)^{m+1}-i^{m+1}$$

$$=1+\left[\sum_{j=1}^{m-1}\binom{m+1}{j}i^j\right]+(m+1)i^m. \qquad (4.23)$$

2. Sum both sides of Equation (4.23) from $i=1$ to n, and observe that the left-hand side gives $-1+(n+1)^{m+1}$ while the right-hand side gives

$$n+\left[\sum_{i=1}^{n}\left(\sum_{j=1}^{m-1}\binom{m+1}{j}i^j\right)\right]+(m+1)\sum_{i=1}^{n}i^m.$$

3. Solve this resulting equation for $\sum_{i=1}^{n}i^m$.

4.7 Review problems

1. Find the first five terms in the sequence $\{s_n\}$ given by $\forall n\geq 3, s_n=2^n\binom{n}{3}.$

2. Find the first five terms in the sequence $\{s_n\}$ given by $s_1=3$, and $\forall n\geq 2, s_n=(s_{n-1})^2-4.$

3. Find a closed formula for the sequence $\frac{1}{2},\frac{2}{4},\frac{3}{8},\frac{4}{16},\frac{5}{32},\ldots.$

4. Find a recursive formula for the sequence $-6,6,18,30,42,\ldots.$

5. Find a closed formula for the sequence $-2, 6, -18, 54,$ $-162, \ldots$.

6. **Compound interest.** In an account earning annual interest rate r, for which interest is compounded m times per year (e.g., if interest is compounded monthly, then $m = 12$), the periodic interest rate is $\frac{r}{m}$.

   (a) Assume that $r = 0.06$, $m = 12$, and an initial investment of \$500 is made. To what will that investment grow after 2 months? after 2 years?

   (b) In general, if an initial investment of $s_0 = P$ dollars is made, then to what future value s_t will the balance grow after t years?

7. Reindex the sequence in Exercise 1 so that the indexing starts at 0.

8. Given an index $k \geq 1$, find an expression for s_{k+1} from the sequence in Exercise 2.

9. Write $12 + 24 + 48 + 96 + \cdots + 3 \cdot 2^{n-1}$ in summation notation.

10. Compute $1 + 2 + 3 + 4 + \cdots + 500$.

11. Compute $1 + 4 + 16 + 64 + \cdots + 4^{10}$.

12. Ultramarathoner Dean Karnazes will run 1 mile on the first day of spring, 2 miles on each of the next 2 days, 3 miles on each of the subsequent 3 days, and so on, until completing 20 miles on each of 20 consecutive days.

    (a) For how many days will he run?

    (b) How many total miles will he run?

13. Compute $\displaystyle\sum_{i=1}^{100} (3i^2 - i + 2)$.

14. Compute $\displaystyle\sum_{i=1}^{n} (i - 2)^2$.

15. Compute $\displaystyle\sum_{i=0}^{n} (-2)^i$.

16. Compute $\displaystyle\prod_{i=0}^{4} (2i + 1)$.

17. **Mortgage.** A mortgage of M dollars with a 6% annual interest rate and a \$500 monthly payment has, after t years a balance s_t, given by

$$s_t = M(1.005)^{12t} - 500(1.005)^{12t-1}$$

$$- \cdots - 500(1.005) - 500.$$

(a) Assume the $M = \$80,000$. Then what is the balance due after 10 years? after 20 years?

(b) What amount M can be paid off exactly at the end of 30 years?

18. Consider the sum $\displaystyle\sum_{i=5}^{203} \frac{2}{3^i}$.

(a) Reindex the sum so that the indexing starts at 0.

(b) Compute the sum.

19. Show: $\forall\, n \geq 9,\ n! > 4^n$.

20. Show: $\forall\, n \geq 6,\ n^2 > 4(n+2)$.

21.* Show: $\forall\, n \geq 0,\ 3^n \geq n^2 + 1$.

22. Show: $\forall\, n \geq 0,\ 3 \mid (n^3 - 4n + 6)$.

23. Show: $\forall\, n \geq 0,\ 6 \mid (7^n - 1)$.

24. Show: $\forall\, n \geq 0,\ 3 \mid (5^n - 2^n)$.

25. **Annuity.** In an annuity earning a 6% annual interest rate and receiving a \$300 monthly payment, the balance at the end of each month is increased by 0.5% just prior to adding the \$300 payment. That is, the balance s_n after n months satisfies the recurrence relation

$$s_0 = 0 \text{ and } \forall\, n \geq 1,\ s_n = 1.005 s_{n-1} + 300.$$

(a) Find s_1, s_2, s_3.

(b) Show: $\forall\, n \geq 0,\ s_n = 60000(1.005^n - 1)$.

26. Let $a, x_1, x_2, \ldots$ be real numbers.
Show: $\forall\, n \geq 2,\ a(x_1 + \cdots + x_n) = ax_1 + \cdots ax_n$. Hint: Use the Distributive Law for the base case, and additionally use Associativity for the inductive step.

27. Show that any nonempty finite subset of $\mathbb{R}$ has a minimum element. That is, for any finite set S such that $S \subset \mathbb{R}$, there is an element $m \in S$ such that $m = \min(S)$. Hint: Use induction on $|S|$.

28. Show: $\displaystyle\forall\, n \geq 1,\ \sum_{i=1}^{n}\left(\frac{1}{i} - \frac{1}{i+1}\right) = 1 - \frac{1}{n+1}$.

29. Show: $\forall\, n \geq 1,\ 4 + 7 + 10 + \cdots + (3n+1) = \frac{n}{2}(3n+5)$.

30. Show: $\displaystyle\forall\, n \geq 1,\ \sum_{i=1}^{n} 3^i = \frac{3}{2}(3^n - 1)$.

31. Show: $\displaystyle\forall\, n \geq 0,\ \sum_{i=0}^{n}(i+1)2^i = n2^{n+1} + 1$.

32. Show: $\displaystyle\forall\, n \geq 1,\ \sum_{i=1}^{n}(3i^2 + 5i) = n(n+1)(n+3)$.

33. Show: $\forall\, n \geq 1,\ \sum_{i=1}^{n} i4^i = \frac{4}{9}[4^n(3n-1)+1].$

34. Let $b, a_1, a_2, a_3, \ldots$ be real numbers with $b > 0$.
    Show: $\forall\, n \geq 1,\ b^{\sum_{i=1}^{n} a_i} = \prod_{i=1}^{n} b^{a_i}.$

35. Let $\{s_n\}$ be the sequence defined by

    $$s_0 = 6, s_1 = 3, \quad \text{and} \quad \forall\, n \geq 2,\ s_n = 2s_{n-2} + s_{n-1}.$$

    Show: $\forall\, n \geq 0,\ 3 \mid s_n.$

36. Let $\{s_n\}$ be the sequence defined by

    $$s_0 = 7, s_1 = 32, \quad \text{and} \quad \forall\, n \geq 2,\ s_n = -20s_{n-2} + 9s_{n-1}.$$

    Show: $\forall\, n \geq 0,\ s_n = 4 \cdot 5^n + 3 \cdot 4^n.$

37. Let $\{s_n\}$ be the sequence defined by

    $$s_0 = 5, s_1 = 16, \quad \text{and} \quad \forall\, n \geq 2,\ s_n = 6s_{n-1} - 8s_{n-2}.$$

    Show: $\forall\, n \geq 0,\ s_n = 2^{n+1} + 3 \cdot 2^{2n}.$

38. If a vending machine only dispenses 3¢ and 8¢ stamps, then any monetary value of 14¢ or greater can be obtained from this machine.

    (a) How could 20¢ be obtained?

    (b) Prove the general result.

39. Display the standard factorization for 1001.

40. Display the standard factorization for 78408.

41. Display the standard factorization for $\binom{50}{9}$.

42. Recall that the Lucas sequence $\{L_n\}_{n \geq 1}$, as defined in Exercise 12 of Section 4.5, is $1, 3, 4, 7, 11, \ldots$.

    Show $\forall\, n \geq 1,\ \sum_{i=1}^{n} L_i = L_{n+2} - 3.$

43. Let $\{s_n\}$ be the sequence defined by

    $$s_0 = 0, s_1 = 6, \quad \text{and} \quad \forall\, n \geq 2,\ s_n = 3(4s_{n-1} - 9s_{n-2}).$$

    Show: $\forall\, n \geq 0,\ s_n = 3^n(3^n - 1).$

44. Use the Binomial Theorem to expand $(x + y)^4$.

45. Use the Binomial Theorem to expand $(3x - 4y)^8$.

46. Use the Binomial Theorem to expand $(x^2 - y^2)^5$.

47. Find the coefficient of x^{10} in $(x - 2)^{100}$.

48. Find the coefficient of $x^{25}y^{75}$ in $(3x - y)^{100}$.

49. Find the coefficient of $x^{30}y^{40}$ in $(7x + 2y)^{80}$.

50. You are on a team of 6 that is completing a large job by splitting it into smaller tasks and having an individual team member volunteer for each task. From your point of view, what is important is how many tasks you end up doing. If there are a total of n tasks, then for each $0 \leq i \leq n$, the number of possible task assignments in which you are responsible for i tasks is $\binom{n}{i} 5^i$. Summing over all $0 \leq i \leq n$, we see that there are $\sum_{i=0}^{n} \binom{n}{i} 5^i$ different possible task assignments for the job. A much easier formula is given by 6^n. Use the Binomial Theorem to verify that $\forall\, n \geq 0,\ \sum_{i=0}^{n} \binom{n}{i} 5^i = 6^n$.

51. Show: $\forall\, n \geq 0,\ \displaystyle\sum_{i=0}^{n} \binom{n}{i} 2^{2i} = 5^n.$

52. Show: $\forall\, n \geq 0,\ \displaystyle\sum_{i=0}^{n} (-1)^i \binom{n}{i} 3^{n-i} 4^i = (-1)^n.$

5

Relations

Relations provide the means of comparing mathematical objects, such as sets or numbers. In some cases, we seek to decide whether one object is bigger or smaller than another in terms of some measure of size. In other cases, we care only whether two objects ought to be considered the same relative to some property. Relations that keep track of some ordering are called partial order relations, and those that keep track of some notion of sameness are called equivalence relations.

Functions are another special kind of relation. They assign values from one set to another. Via input values and output values, a function provides a link between two sets. Among the important kinds of functions we study are one-to-one functions, onto functions, and bijective functions. As we shall see, bijections provide the strongest communication between a pair of sets and enable us to pair off the elements in one set with another. By pairing off elements, we can decide whether two sets have the same number of elements. Thus, we formally study the cardinality of sets.

Throughout this chapter, the general notions associated with relations are applied to relational databases. Beyond the simple observation that databases provide examples of relations, the implementations of natural database queries can be understood through general constructs for relations. The utility of hash functions and tables is also addressed.

5.1 General relations

The previous chapters have utilized notations such as $x \in A$, $x < y$, and $a \mid b$. In these cases, the symbols $\in$, $<$, and $\mid$ express a relationship between the two items they separate. Between members of a family, there exist relationships, such as parent and child. Within data collections, certain pieces of data are related to others. Here, we take up a more formal study of relationships.

Definition 5.1. Given sets X and Y, a **relation from X to Y** is a subset R of $X \times Y$. When $(x, y) \in R$, we say that x **is related to y** (by R) and write $x \mathrel{R} y$. Similarly, $(x, y) \notin R$ is denoted by $x \mathrel{\not R} y$.

The relations described in Definition 5.1 are sometimes called binary relations, since $X \times Y$ is a product of two sets. More generally, n-ary relations are subsets of n-fold products. Since, in this book, we focus on binary relations, they are simply referred to as relations. General n-ary relations are encountered here only in the context of databases, as we shall see in Table 5.2 on page 277.

Example 5.1. (The "Is an Advisee of" Relation).
Let X be the set of students at a certain college, and let Y be its faculty. Define the relation R from X to Y by,

$$x \, R \, y \quad \text{if and only if} \quad x \text{ is an advisee of } y.$$

Suppose we have

$$
\begin{aligned}
R = \{ \ &(\text{Megan Johnson, Dr. Gauss}) \\
&(\text{Richard Kelley, Dr. Dijkstra}) \\
&(\text{Charles Murphy, Dr. Gauss}) \\
&(\text{Susan Brower, Dr. Euler}) \\
&(\text{Martha Lang, Dr. Dijkstra}) \\
&(\text{Richard Kelley, Dr. Gauss}) \ \}.
\end{aligned}
$$

For example, Charles Murphy is an advisee of Dr. Gauss, since

$$(\text{Charles Murphy, Dr. Gauss}) \in R.$$

However, Nate Cobb is not an advisee of Dr. Gauss, since

$$(\text{Nate Cobb, Dr. Gauss}) \notin R.$$

Here, Nate Cobb happens to be a student with no advisor. At another extreme, a faculty member may have several advisees. For example, Dr. Gauss has three advisees. Also, a student with multiple majors may have an advisor for each major. For example, Richard Kelley has two advisors.

We shall also explore several examples of relations of a purely mathematical nature. For our next example, recall that a zero of a real function[*] f is a value a such that $f(a) = 0$.

Example 5.2. (The "Is a Zero of" Relation).
Let Y be the set of real functions. Define the relation R from $\mathbb{R}$ to Y by

$$a \, R \, f \quad \text{if and only if} \quad f(a) = 0.$$

---

[*] Our informal understanding of real functions from Section 1.3 suffices here, before functions are defined more formally and generally in Section 5.3.

That is, $a\,R\,f$ means a is a zero of f. For example, suppose $f(x) = x^2 - 1$. Then $-1\,R\,f$, since $f(-1) = 0$. However, $0\,\not\!R\,f$, since $f(0) \neq 0$. Note that $f\,\not\!R\,-1$, since order is important here and $(f, -1) \notin \mathbb{R} \times Y$.

Example 5.3.   (The "Is an Element of" Relation $\in$).
Let $\mathcal{U}$ be some universal set and $Y \subseteq \mathcal{P}(\mathcal{U})$. To be concrete, consider $\mathcal{U} = \mathbb{Z}$ and $Y = \mathcal{P}(\mathbb{Z})$. Let

$$R = \{(x, A) \,:\, x \text{ is an element of } A\} \subseteq \mathcal{U} \times Y.$$

That is,

$$x\,R\,A \quad \text{if and only if} \quad x \in A.$$

So, R is the relation $\in$. Consequently,

$$x\,\not\!R\,A \quad \text{if and only if} \quad x \notin A.$$

For example, if $\mathcal{U} = \mathbb{Z}$ and $Y = \mathcal{P}(\mathbb{Z})$, then $0\,R\,\mathbb{N}$ since $0 \in \mathbb{N}$. However, $0\,\not\!R\,\mathbb{Z}^+$, since $0 \notin \mathbb{Z}^+$.

Example 5.4.   (The "Is Less Than" Relation $<$).
Define the relation R from $\mathbb{R}$ to $\mathbb{R}$ by

$$x\,R\,y \quad \text{if and only if} \quad x \text{ is less than } y.$$

That is, R is the relation $<$. Formally, the symbol $<$ represents a subset of $\mathbb{R} \times \mathbb{R}$, as just described. For example, $2\,R\,4$ since $2 < 4$. That is, $(2, 4) \in R$. However, $4\,\not\!R\,2$, since $4 \not< 2$.

In Example 5.4, we see an example of a relation from a set X to itself.

Definition 5.2.   A **relation on a set** X is a relation from X to itself.

Example 5.5.   (The "Is the Father of" Relation).
Let X be the set of living male members of a certain family, and define the relation R on X by

$$x\,R\,y \quad \text{if and only if} \quad x \text{ is the father of } y.$$

Suppose that X is the set of men listed in Figure 5.1 and that each arrow there points from a father to his son. For example,

$$\text{Nicholas } R \text{ Mark}$$

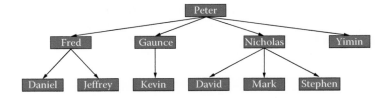

FIGURE 5.1
A family tree.

since Nicholas is the father of Mark. However,

$$\text{Mark } \not{R} \text{ Nicholas}$$

since Mark is not the father of Nicholas.

Example 5.6. (The "Divides" Relation |).
Define the relation R on $\mathbb{Z}$ by

$$a \, R \, b \quad \text{if and only if} \quad a \mid b.$$

For example, $3 \, R \, 6$ since $3 \mid 6$. However, $6 \, \not{R} \, 3$ since $6 \nmid 3$.

Further emphasizing the fact that a relation goes *from* one set *to* another is the notion of an inverse, which reverses that direction.

Definition 5.3.   The **inverse** of a given relation R from a set X to a set Y is the relation $R^{-1} = \{(y, x) : y \in Y, x \in X, \text{ and } (x, y) \in R\}$ from Y to X. That is, $\forall \, y \in Y, x \in X,$

$$y \, R^{-1} \, x \quad \text{if and only if} \quad x \, R \, y.$$

Example 5.7.   The inverse of the "is an advisee of" relation from Example 5.1 is the "is an advisor for" relation R^{-1} from the faculty Y to the students X given by

$$y \, R^{-1} \, x \quad \text{if and only if} \quad x \text{ is an advisee of } y$$

$$\text{if and only if} \quad y \text{ is an advisor for } x.$$

For example,

$$\text{Dr. Dijkstra } R^{-1} \text{ Martha Lang,}$$

since Martha Lang is an advisee of Dr. Dijkstra.

Example 5.8.   The inverse of the "divides" relation R on $\mathbb{Z}$ is the "is a multiple of" relation R^{-1} on $\mathbb{Z}$ given by

$$b \, R^{-1} \, a \quad \text{if and only if} \quad a \mid b.$$

The symbol $\ni$ is sometimes used to represent the inverse of $\in$; the notation $A \ni x$ is read as A "contains" x and is taken to mean $x \in A$.

5.1.1 Databases

A **relational database** is a collection of **tables**. Each table is organized by its **columns**, or **fields**, and in a fixed column, each entry comes from a specific set of potential data. The **rows** of a table are also called its **records**, and a table may have some positive number of rows or no rows at all. Generally, the columns of a table represent characteristics of the objects recorded in its rows. For example, if the objects of interest are advisee–advisor pairings at a college, then one column would contain the advisees and the other the advisors. Data are entered into a database by adding a row to some table in the database, and data are deleted by removing a row.

The Registrar's Office at a college might keep a database containing tables like those in Tables 5.1 and 5.2. Very simply, each table is a relation, and its columns represent the sets involved in the relation. For example, Table 5.1 is the "is an advisee of" relation from Example 5.1. Table 5.2 is a relation that specifies class schedules. For each course, that table lists who is teaching the course, the name of the course, and when and where it meets. This "class schedule" relation is a 4-ary relation, since it lists the elements in a subset of the fourfold product

Faculty $\times$ Courses $\times$ Meeting Times $\times$ Classrooms.

TABLE 5.1
Advisee assignments

Student	Faculty member
Megan Johnson	Dr. Gauss
Richard Kelley	Dr. Dijkstra
Charles Murphy	Dr. Gauss
Susan Brower	Dr. Euler
Martha Lang	Dr. Dijkstra
Richard Kelley	Dr. Gauss

TABLE 5.2
Class schedules

Faculty member	Course	Meeting time	Room
Dr. Gauss	Discrete Math	MWF 8-9	Kingsbury 314
Dr. Dijkstra	Algorithms	MWF 10-11	Carnegie 217
Dr. Gauss	Number Theory	TuTh 8-9:30	Kingsbury 146
Dr. Euler	Number Theory	MWF 8-9	Carnegie 217

5.1.2 Representing relations

Since a relation is a set, it might be represented simply by listing its elements, as in list notation. In relational databases, this approach is adopted by presenting each element of a relation as a row in a table. Here, we consider other means of representing a relation that can be more efficient or more digestible.

Arrow diagrams for finite relations. A relation R from a finite set X to a finite set Y can be displayed pictorially. An **arrow diagram** for R is constructed by representing the sets X and Y in disjoint regions and drawing an arrow from an element $x \in X$ to an element $y \in Y$ if and only if $x \, R \, y$.

> **Example 5.9.** (A Restricted "Is a Zero of" Relation).
> Let R be the "is a zero of" relation from the set $X = \{-2, -1, 0, 1, 2\}$ to the set $Y = \{x^2 - 1, x - 2, x^2 + x - 2, x^2 - 3\}$. That is, $a \, R \, f$ if and only if $f(a) = 0$. An arrow diagram for R is shown in Figure 5.2.

If R is a relation from a finite set X to itself, then, in a more compact way than an arrow diagram, we can represent R with a directed graph. In fact, the family tree displayed in Figure 5.1 is a directed graph for the "is the father of" relation from Example 5.5. In general, given a relation R on a finite set X, a **directed graph**, or **digraph**, for R is obtained by displaying the elements of X and drawing an arrow from an element x to an element y if and only if $x \, R \, y$.

Directed graphs are introduced in more depth in Section 8.6. Here, we simply make superficial use of digraphs as a means of representing relations.

> **Example 5.10.** (A Restricted "Divides" Relation).
> Let R be the "divides" relation on the set $X = \{1, 2, 3, 4, 6, 12\}$. That is, $a \, R \, b$ if and only if $a \mid b$.

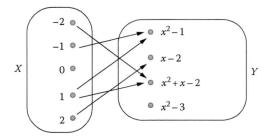

FIGURE 5.2
An arrow diagram for the "is a zero of" relation.

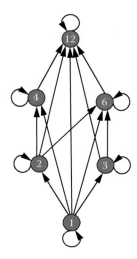

FIGURE 5.3
A digraph for the "divides" relation |.

A digraph for R is shown in Figure 5.3. For example, there is an arrow from 3 to 6, since 3 | 6. However, there is no arrow from 3 to 2, since $3 \nmid 2$. Note that each loop represents the fact that an element is related to itself in this relation.

Representing finite relations with matrices. A relation R from a finite set X to a finite set Y may be represented by a zero-one matrix. That is, we use a rectangular array of numbers, each of whose entries is either 0 or 1. The rows of this matrix are labeled by the elements of X and the columns by the elements of Y. For each $x \in X$ and $y \in Y$, the entry in row x and column y is assigned the value 1 if $x \, R \, y$ and 0 if $x \, \cancel{R} \, y$.

Example 5.11.   Let $X = \{1, 2, 3, 4\}$, $Y = \{\emptyset, \{1, 2\}, \{2, 3\}\}$, and R be the "is an element of" relation from X to Y, defined as in Example 5.3. The matrix

$$
\begin{array}{c}
 \\ 1 \\ 2 \\ 3 \\ 4
\end{array}
\begin{array}{ccc}
\emptyset & \{1,2\} & \{2,3\} \\
\left[\begin{array}{ccc}
0 & 1 & 0 \\
0 & 1 & 1 \\
0 & 0 & 1 \\
0 & 0 & 0
\end{array}\right]
\end{array}
$$

represents R. For example, the entry in row 2 and column $\{1, 2\}$ is a 1, since $2 \in \{1, 2\}$. However, the entry in row 3 and column $\{1, 2\}$ is a 0, since $3 \notin \{1, 2\}$.

Example 5.12. Let $X = \{1,2,3,4\}$ and R be the "is less than" relation on X, defined as in Example 5.4. The matrix

$$
\begin{array}{c}
\\
1\\
2\\
3\\
4
\end{array}
\begin{array}{c}
1\ 2\ 3\ 4\\
\left[\begin{array}{cccc}
0 & 1 & 1 & 1\\
0 & 0 & 1 & 1\\
0 & 0 & 0 & 1\\
0 & 0 & 0 & 0
\end{array}\right]
\end{array}
$$

represents R.

If a relation R is represented by a matrix A, then its inverse relation R^{-1} is represented by the **transpose** matrix A^T, whose columns are the rows of A (and whose rows are the columns of A). That is, the entry in row y and column x of A^T is the entry from row x and column y of A. For example, the matrix

$$
\begin{array}{c}
\\
\varnothing\\
\{1,2\}\\
\{2,3\}
\end{array}
\begin{array}{c}
1\ 2\ 3\ 4\\
\left[\begin{array}{cccc}
0 & 0 & 0 & 0\\
0 & 0 & 1 & 1\\
0 & 1 & 1 & 0
\end{array}\right]
\end{array}
$$

represents R^{-1}, the inverse of the relation R from Example 5.11.

Graphing relations on $\mathbb{R}$. Although digraphs and matrices work well for representing finite relations, they do not work well for infinite relations. For infinite relations, we consider a different method that works in the particular case of a relation on the set of real numbers $\mathbb{R}$, which is a common case in our studies. Since a relation on $\mathbb{R}$ is a subset of $\mathbb{R} \times \mathbb{R}$, it may be drawn in the Cartesian plane.

Example 5.13. Let R be the "is less than" relation on $X = \mathbb{R}$ from Example 5.4. In the drawing of the graph of R, all points (x,y) for which $x < y$ are shaded.

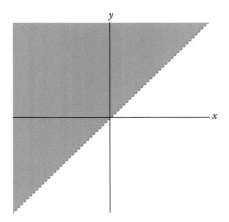

Example 5.14.   Let R be the relation on $X = \mathbb{R}$ defined by

$$x \, R \, y \quad \text{if and only if} \quad x = y^2.$$

The drawing of the graph of R is a parabola that opens to the right.

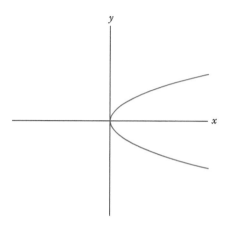

For any relation R on $\mathbb{R}$, the graph of R^{-1} is the reflection of the graph of R about the line $y = x$. This follows since $\forall \, y, x \in \mathbb{R}$,

$$(y, x) \in R^{-1} \quad \text{if and only if} \quad (x, y) \in R,$$

and the point (y, x) is the reflection of the point (x, y).

Example 5.15. Let R be the inverse of the relation $x = y^2$ on $X = \mathbb{R}$ from Example 5.14. The drawing of the graph of R is the reflection of the solid curve in the picture on the left and is shown on the right.

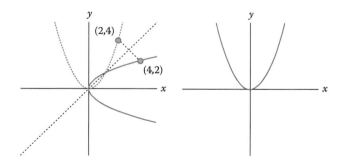

Of course, R is the relation $y = x^2$.

Example 5.16. The inverse of $<$ on $X = \mathbb{R}$ is $>$. The drawing of the graph of the relation $>$ is the reflection of the picture from Example 5.13.

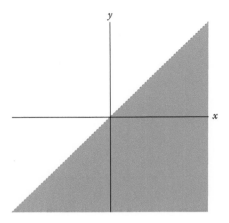

5.1.3   Properties of relations on sets

We close this section with a consideration of four basic properties that a relation on a set might have. These properties play a prominent role in Section 5.2.

Definition 5.4.   A relation R on a set X is said to be

(a) **Reflexive** if $\forall\, x \in X$,

$$x\, R\, x.$$

(b) **Symmetric** if $\forall\, x, y \in X$,

$$x\, R\, y \;\rightarrow\; y\, R\, x.$$

(c) **Antisymmetric** if $\forall\, x, y \in X$,

$$x\, R\, y \;\text{and}\; y\, R\, x \;\rightarrow\; x = y.$$

(d) **Transitive** if $\forall\, x, y, z \in X$,

$$x\, R\, y \;\text{and}\; y\, R\, z \;\rightarrow\; x\, R\, z.$$

Very simply, a relation on a set is reflexive if each element is related to itself. It is symmetric if the order in which elements are related is interchangeable. If it is antisymmetric, then two distinct elements can be related in at most one order. Transitivity says that what one might write as a triple $x\, R\, y\, R\, z$ can have its middle element dropped to obtain $x\, R\, z$. Formally, a relation is checked for each of these properties by appealing to Definition 5.4.

Example 5.17.   (The "Intersects" Relation).
Let X be the set of nonempty subsets of $\mathbb{R}$, and let R be the relation on X given by

$$A \, R \, B \quad \text{if and only if} \quad A \cap B \neq \emptyset.$$

For each of the properties reflexive, symmetric, antisymmetric, and transitive, prove or disprove that R has that property.

Solution.
Reflexive: Let A be a nonempty subset of $\mathbb{R}$. Since $A \cap A = A \neq \emptyset$, it follows that $A \, R \, A$.
Symmetric: Let A and B be nonempty subsets of $\mathbb{R}$. Suppose $A \, R \, B$. That is, $A \cap B \neq \emptyset$. By the commutativity of $\cap$, we have $B \cap A = A \cap B \neq \emptyset$. That is, $B \, R \, A$.
Not antisymmetric: Let $A = \{1, 2\}$ and $B = \{2, 3\}$. So $A \, R \, B$ and $B \, R \, A$. However, $A \neq B$.
Not transitive: Let $A = \{1, 2\}$, $B = \{2, 3\}$, and $C = \{3, 4\}$. So, $A \, R \, B$ and $B \, R \, C$. However, $A \not\!R \, C$. $\square$

Example 5.18.   (The "Divides" Relation on $\mathbb{Z}^+$).
Let R be the relation on $\mathbb{Z}^+$ given by

$$a \, R \, b \quad \text{if and only if} \quad a \mid b.$$

For each of the properties reflexive, symmetric, antisymmetric, and transitive, prove or disprove that R has that property.

Solution.
Reflexive: Let $a \in \mathbb{Z}^+$. Since $a = a \cdot 1$, we see that $a \mid a$. That is, $a \, R \, a$.
Not symmetric: Let $a = 2$ and $b = 4$. So $a \, R \, b$. However, $b \not\!R \, a$.
Antisymmetric: Let $a, b \in \mathbb{Z}^+$. Suppose $a \mid b$ and $b \mid a$. By Exercise 17 of Section 3.1, we have $a = \pm b$. Since $a, b > 0$, it must be that $a = b$.
Transitive: Let $a, b, c \in \mathbb{Z}^+$. Suppose $a \mid b$ and $b \mid c$. By Example 3.4 of Section 3.1, we have $a \mid c$. $\square$

With a minor additional assumption, properties (b) and (d) in Definition 5.4 imply property (a).

Theorem 5.1.   *Let R be any relation on a set X. Suppose R is symmetric and transitive and every $x \in R$ has some $y \in R$ to which it is related (i.e. $\forall \, x \in R, \exists \, y \in R$ such that $x \, R \, y$). Then R is reflexive.*

Proof. Let $x \in R$ be arbitrary. Our assumptions give us some $y \in R$ such that $x \, R \, y$. It follows from symmetry that $y \, R \, x$. Since both $x \, R \, y$ and $y \, R \, x$, transitivity implies $x \, R \, x$. Hence, R is reflexive.   □

Exercises

In Exercises 1 through 8, decide whether the given statements are true or false.

1. Let X be the set of primes, let $Y = \mathbb{Z}$, and define the relation R from X to Y by $a \, R \, b$ if and only if $a \mid b$.

   (a) 5 is related to 1.   (c) 2 is related to 7.

   (b) 3 is related to 6.

2. Let $X = \mathbb{R}$, let $Y = \mathcal{P}(\mathbb{Z})$, and define the relation R from X to Y by $x \, R \, A$ if and only if $x \in A$.

   (a) 0 is related to $\{-1, 0, 1\}$.   (c) $\{2\}$ is related to 2.

   (b) 1 is related to $\{0, \frac{1}{2}, 1\}$.

3. The "is a subset of" relation.
   Given the universal set $\mathcal{U} = \mathbb{R}$, let $X = \mathcal{P}(\mathcal{U})$, and define the relation R on X by $A \, R \, B$ if and only if $A \subseteq B$.

   (a) $\emptyset$ is related to $\mathbb{Z}$.   (c) $\{1, 2\}$ is related to $\mathbb{R}^+$.

   (b) 0 is related to $\emptyset$.

4. Let X be the set of circles in $\mathbb{R}^2$ with positive radius, and define the relation R on X by $C_1 \, R \, C_2$ if and only if C_1 is tangent to C_2.

   (a) $(x + 1)^2 + y^2 = 1$ is related to $(x - 1)^2 + y^2 = 1$.

   (b) $x^2 + y^2 = 2$ is related to $(x - 1)^2 + y^2 = 1$.

   (c) $x^2 + y^2 = 1$ is related to $x^2 + y^2 = 4$.

5. The "is a superset of" relation $\supseteq$ is the inverse of $\subseteq$.

6. The "is equal to" relation $=$ on $\mathbb{R}$ is its own inverse.

7. The "is perpendicular to" relation $\perp$ on the set of lines in $\mathbb{R}^2$ is the inverse of the "is parallel to" relation $\parallel$.

8. The "is congruent to" relation $\cong$ on the set of triangles in $\mathbb{R}^2$ is the inverse of the "is not congruent to" relation $\ncong$. Recall that congruent triangles have the same side lengths.

9. A map of South America is shown in Figure 5.4. Consider the "shares a border with" relation on the set of countries in South America. That is, one country "shares a border with" another if and only if they meet at more than just isolated points.

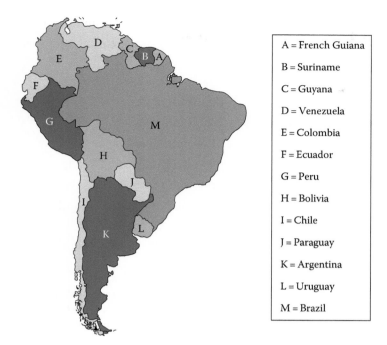

FIGURE 5.4
South America.

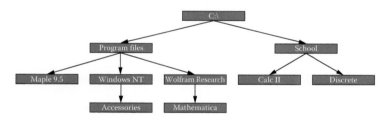

FIGURE 5.5
Folders in a directory system.

    (a) Does Chile share a border with Paraguay?

    (b) List the countries with which Venezuela shares a border.

10. A hierarchy of folders on a C:\ drive is displayed in Figure 5.5. Each arrow points from a folder to an immediate subfolder and therefore corresponds to an element of the "immediate subfolder" relation on the set of folders on this disk.

    (a) Is Accessories an immediate subfolder of Program Files?

(b) List the immediate subfolders of Program Files.

In Exercises 11 through 18, find the inverse of the specified relation.

11. The "is the father of" relation from Example 5.5.

12. The "is less than" relation from Example 5.4.

13. The "is a subset of" relation from Exercise 3.

14. The "is tangent to" relation from Exercise 4.

15. The "is perpendicular to" relation $\perp$ from Exercise 7.

16. The "is a zero of" relation from Example 5.2.

17. The relation R on $\mathbb{R}$ defined by $x\,R\,y$ if and only if $x^2 + y^2 = 1$.

18. The relation R on $\mathbb{R}$ defined by $x\,R\,y$ if and only if $x = y + 1$.

19. SoftJobs is a company capable of producing a variety of software packages, ranging from data management software to video games. SoftJobs itself uses a database to keep track of which programmers are working on which projects for which clients. Two tables from that database are displayed in Figure 5.6.

    (a) For what client is the project Skate Rats being produced?

    (b) Is Rx Tracker contracted out to GameCo?

    (c) Is Charles Murphy assigned to work on NBA Dunkfest?

20. HardSell is a mail-order hardware store. It needs to keep track of both the customers to which it sells items and the vendors from which it purchases items. Two tables from the database used by HardSell are displayed in Figure 5.7.

    (a) Which customer(s) ordered a wrench?

    (b) Were pliers ordered by Susan Brower?

    (c) Will Maxtool supply a wrench?

In Exercises 21 through 24, draw an arrow diagram for the specified relation.

Project	Client
NBA Dunkfest	GameCo
Rx Tracker	MediComp
Skate Rats	GameCo

Contracts

Programmer	Project
Martha Lang	NBA Dunkfest
Megan Johnson	Rx Tracker
Charles Murphy	NBA Dunkfest
Martha Lang	Skate Rats
Charles Murphy	Skate Rats

Work assignments

FIGURE 5.6
Database for SoftJobs.

Part	Customer
Hammer	Richard Kelley
Wrench	Susan Brower
Pliers	Susan Brower
Wrench	Charles Murphy

Orders

Vendor	Part
Stanley	Hammer
Maxtool	Hammer
Stanley	Wrench
Maxtool	Pliers

Suppliers

FIGURE 5.7
Database for HardSell.

21. The "is an element of" relation $\in$ from $\{0, 1, 2\}$ to $\mathcal{P}(\{0, 1, 2\})$.

22. The "is tangent to" relation from the set of lines

$$\{y = 1, y = -1, y = -3, x = 3\}$$

to the set of circles $\{x^2 + y^2 = 1, x^2 + (y + 1)^2 = 4\}$.

23. The relation in the "Contracts" table in Exercise 19 from projects to clients.

24. The relation in the "Orders" table in Exercise 20 from parts to customers.

In Exercises 25 through 28, draw a digraph for the specified relation.

25. The "is less than" relation $<$ on $\{2, 4, 6, 8\}$.

26. The "divides" relation $|$ on $\{2, 5, 6, 10, 15\}$.

27. The "is a subset of" relation $\subseteq$ on $\{\{1\}, \{2\}, \{1, 2\}, \{2, 3\}, \{1, 2, 3\}\}$.

28. The "is perpendicular to" relation $\perp$ on $\{y = x, y = -x, y = x + 1, y = x - 1\}$.

In Exercises 29 through 32, find the matrix representing the specified relation.

29. The "divides" relation from the set $\{0, 1, 2\}$ to the set $\{0, 3, 6, 9\}$.

30. The relation $x = y^2$ in Example 5.14 on the set $\{-4, -2, -1, 1, 2, 4\}$.

31. The "is a subset of" relation on $X = \{\emptyset, \{1\}, \{2\}, \{1, 2\}\}$.

32. The "is tangent to" relation in Exercise 4 on the set

$$\{C_{(-1,-1)}, C_{(-1,1)}, C_{(1,-1)}, C_{(1,1)}\},$$

where $C_{(h,k)}$ is the circle in $\mathbb{R}^2$ of radius 1 with center (h, k).

In Exercises 33 through 36,

(a) Draw the graph of the specified relation on $\mathbb{R}$.

(b) Draw the graph of its inverse.

33. $y > x^2$.

34. $y = x + 1$.

35. $x = y^2 + 1$.

36. $x < y^2$.

In Exercises 37 through 54, a set X and a relation R on X are given. For each of the properties reflexive, symmetric, antisymmetric, and transitive, determine whether the given relation has that property, and justify your answer.

37. $X = \mathbb{R}$ and $x \, R \, y$ if and only if $\frac{1}{x} = y$.

38. $X = \mathbb{R}$ and $x \, R \, y$ if and only if $x \leq y$.

39. X is the set of lines in the Cartesian plane $\mathbb{R}^2$ and

$$l_1 \, R \, l_2 \quad \text{if and only if} \quad l_1 \text{ intersects } l_2.$$

40. $X = \mathbb{R}$ and $x \, R \, y$ if and only if $\frac{1}{x} = \frac{1}{y} \in \mathbb{R}$.

41. $X = \mathbb{R}$ and $x \, R \, y$ if and only if $\sqrt{x} = \sqrt{y} \in \mathbb{R}$.

42. $X = \mathbb{R}$ and $x \, R \, y$ if and only if $x \neq y$.

43. $X = \mathcal{P}(\mathcal{U})$ for some nonempty universal set $\mathcal{U}$ and $A \, R \, B$ if and only if $A \subset B$.

44. X is the set of circles in $\mathbb{R}^2$ and

$$C_1 \, R \, C_2 \quad \text{if and only if} \quad \text{center}(C_1) = \text{center}(C_2).$$

45. $X = \mathbb{R}$ and $x \, R \, y$ if and only if $x + 1 = y$.

46. $X = \mathbb{R}$ and $x \, R \, y$ if and only if $x - 1 < y$.

47. $X = \mathbb{Z}$ and $a \, R \, b$ if and only if a and b have the same prime factors.

48. $X = \mathbb{R}$ and $x \, R \, y$ if and only if $2x = y$.

49. $X = \mathbb{R}$ and $x \, R \, y$ if and only if $x \leq |y|$.

50. $X = \mathbb{R}$ and $x \, R \, y$ if and only if $x < y$.

51. $X = \mathcal{P}(\mathbb{R})$ and $A \, R \, B$ if and only if $A \subseteq B \cup \mathbb{Z}$.

52. $X = \{n \, : \, n \in \mathbb{Z} \text{ and } n \geq 2\}$ and $a \, R \, b$ if and only if $\gcd(a, b) > 1$.

53. The "is a subset of" relation in Exercise 3.

54. The relation in Exercise 4.

55. Consider "shares a border with" relations on regions of a map such as that for the countries of South America in Exercise 9. Such relations will have exactly one of the properties reflexive, symmetric, antisymmetric, or transitive. Which one? Explain.

56. Consider "immediate subfolder" relations on the folders on a disk such as that for the folders on the C:\ drive in Exercise 10. Such relations in general will have exactly one of the properties reflexive, symmetric, antisymmetric, or transitive. Which one? Explain.

For Exercises 57 through 60, given a set X, let $\Delta = \{(x, x) :$ $x \in X\} \subseteq X^2$. We call Δ the **diagonal**.

57. Show that a relation R on a set X is symmetric if and only if $R = R^{-1}$. When X is finite, observe that the representing matrix A is symmetric (i.e. $A^T = A$). When $X = \mathbb{R}$, observe that the graph of R is symmetric about the line $y = x$.

58. Show that a relation R on a set X is reflexive if and only if $\Delta \subseteq R$. When X is finite, observe that the representing matrix A contains a 1 in each entry of its main diagonal. When $X = \mathbb{R}$, observe that the graph of R contains the line $y = x$.

59.* Show that a relation R on a set X is antisymmetric if and only if $R \cap R^{-1} \subseteq \Delta$. When X is finite, observe that the representing matrix A cannot contain a 1 in both of any two distinct symmetric positions (i, j) and (j, i). When $X = \mathbb{R}$, observe that the graphs of R and R^{-1} intersect only in the line $y = x$.

60.* Show that a relation R on a set X is both symmetric and antisymmetric if and only if $R \subseteq \Delta$. When X is finite, observe that the representing matrix A can only contain 1's on the diagonal. When $X = \mathbb{R}$, observe that the graph of R is contained in the line $y = x$.

5.2 Special relations on sets

Relations that have most, but not all, of the properties in Definition 5.4 (reflexive, symmetric, antisymmetric, transitive) are particularly important. As we shall see, certain such relations provide a good means of comparing elements in a set. For example, the "is greater than" relation $>$ has properties that render it a good means of "ordering" real numbers, and the congruence relation $\equiv$ modulo some fixed integer $n > 1$ has properties that allow us to "equate" integers (modulo n). In this section, we introduce partial order relations and equivalence relations.

5.2.1 Order relations

Relations like $\leq, \geq, <, >, \subseteq, \supseteq, \subset$, and $\supset$ establish an ordering between objects. However, we should clarify what we mean by an ordering.

Definition 5.5. A **partial order relation** or **partial ordering** on a set X is a relation on X that is reflexive, antisymmetric, and transitive. In this case, we say that X is a **partially ordered set** or **poset** (under R). Given a partial order relation R on a set X and two elements $x, y \in X$, we say that x is **comparable** to y

(under R) if and only if $x \mathrel{R} y$ or $y \mathrel{R} x$. Otherwise, x and y are said to be **incomparable**.

Since $<, >, \subset$, and $\supset$ are not reflexive, they are not partial orderings as specified in Definition 5.5. The allowance of equality is a necessary condition. A prototypical example of a partial ordering is the "is a subset of" relation $\subseteq$ on $\mathcal{P}(\mathcal{U})$, for some fixed universal set $\mathcal{U}$.

Example 5.19. Let $\mathcal{U}$ be a set. Show that $\subseteq$ is a partial order relation on $\mathcal{P}(\mathcal{U})$, the set of all subsets of $\mathcal{U}$.

Proof. Let A, B, and C represent arbitrary elements of $\mathcal{P}(\mathcal{U})$. That is, A, B, and C are subsets of $\mathcal{U}$.
Reflexive: Since $A \subseteq A$, we see that $\subseteq$ is reflexive.
Antisymmetric: Suppose $A \subseteq B$ and $B \subseteq A$. By our characterization of set equalities given in Section 2.3, it follows that $A = B$.
Transitive: It was shown in Example 2.14 of Section 2.2 that if $A \subseteq B$ and $B \subseteq C$, then $A \subseteq C$. $\qquad\square$

The adjective *partial* is used in the term partial ordering because, in general, not every pair of elements will be comparable.

Example 5.20. Consider the subset relation $\subseteq$ on $\mathcal{P}(\{1,2,3\})$.

(a) $\{1,3\}$ is comparable to $\{1\}$,
     since $\{1\} \subseteq \{1,3\}$.

(b) $\{1,3\}$ is incomparable to $\{2\}$,
     since $\{1,3\} \not\subseteq \{2\}$ and $\{2\} \not\subseteq \{1,3\}$.

Hasse diagrams. Just as a digraph provides a useful way to represent an arbitrary relation on a set, a **Hasse diagram** provides a useful way to represent a partial order relation R on a set X. Hasse diagrams are named after the German mathematician Helmut Hasse (1898–1979) and are relatives of digraphs in the sense that one can be constructed from a digraph as follows:

1. Construct a digraph with the elements of X arranged so that all of the arrows point upward.

2. Delete all loops, since these follow immediately from reflexivity.

3. Delete any arrows that follow from transitivity. That is, if x, y, and z are distinct elements of X with $x \mathrel{R} y$ and $y \mathrel{R} z$, then delete the arrow reflecting $x \mathrel{R} z$.

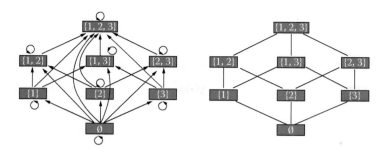

FIGURE 5.8
Constructing the Hasse diagram for $\mathcal{P}(\{1, 2, 3\})$.

4. Delete the direction indicators from the arrows, since the relative heights of the elements indicate the appropriate direction.

Since a Hasse diagram is obtained from a digraph by deleting several of its features, a Hasse diagram provides a more efficient means of representing a partial order relation.

The left-hand side of Figure 5.8 shows a digraph for the relation $\subseteq$ on $\mathcal{P}(\{1, 2, 3\})$, and the right-hand side shows its Hasse diagram. The Hasse diagram is obviously less cluttered than the digraph. Yet, by reversing the process used to construct the Hasse diagram, all of the details in the corresponding digraph can be recovered.

Example 5.21. In Example 5.18 from Section 5.1, we saw that the "divides" relation on $\mathbb{Z}^+$ is reflexive, antisymmetric, and transitive. Therefore, it is a partial order relation on $\mathbb{Z}^+$. The more general observation that, given any subset X of $\mathbb{Z}^+$, the "divides" relation is a partial order relation on X, is left for the exercises. The Hasse diagram in the case that $X = \{1, 2, 3, 4, 6, 12\}$, the set of divisors of 12, is displayed on the left-hand side of Figure 5.9. The corresponding digraph was given in Figure 5.3 in Example 5.10.

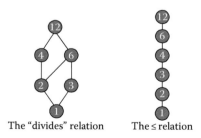

The "divides" relation   The $\leq$ relation

FIGURE 5.9
Hasse diagrams for relations on $\{1, 2, 3, 4, 6, 12\}$.

Example 5.22. It is left for the exercises to verify that, given any subset X of $\mathbb{R}$, $\leq$ is a partial ordering on X. The Hasse diagram in the case that $X = \{1, 2, 3, 4, 6, 12\}$ is displayed on the right-hand side of Figure 5.9.

The striking difference between the two diagrams in Figure 5.9 is that the diagram on the right appears naturally as a line and that on the left does not. This is a reflection of the fact that, in a sense, the ordering on the left is only "partial," whereas that on the right is "total."

Definition 5.6. A partial order relation R on a set X is called a **total order relation** or **linear order relation** if every pair of elements is comparable. In this case, we say that X is a **totally ordered set** or **linearly ordered set** (under R).

Given any $X \subseteq \mathbb{R}$, the relation $\leq$ on X is a total ordering; any two real numbers can always be compared with $\leq$ (or $\geq$), by the Trichotomy Law (Appendix A, property 7).

Lexicographic order. Very simply, lexicographic order is a generalization of alphabetical order. It provides a practical way to order and store words or word-like objects in a computer, given a fixed character order.

Given a character set C, a **word** over C is a string $x_1 x_2 \cdots x_n$, where $n \in \mathbb{Z}^+$ and $x_1, x_2, \ldots, x_n \in C$. When C is partially ordered by some relation $\preceq$, we construct a corresponding partial ordering, which we shall denote $\trianglelefteq$, called the **lexicographic ordering**, on the set of words $W = \{x_1 x_2 \cdots x_n : n \in \mathbb{Z}^+, x_1, x_2, \ldots, x_n \in C\}$. Given two words $\vec{x} = x_1 x_2 \cdots x_m$ and $\vec{y} = y_1 y_2 \cdots y_n$, let k be the largest index such that $x_1 x_2 \cdots x_k = y_1 y_2 \cdots y_k$, and define $\vec{x} \triangleleft \vec{y}$ if either $k = m < n$ (so $\vec{x}$ is a proper subword of $\vec{y}$) or $k < m, n$ and $x_{k+1} \preceq y_{k+1}$ (note that $x_{k+1} \neq y_{k+1}$). The lexicographic order $\trianglelefteq$ on W is thus defined by taking $\vec{x} \trianglelefteq \vec{y}$ precisely if $\vec{x} \triangleleft \vec{y}$ or $\vec{x} = \vec{y}$.

Example 5.23. (Dictionary Order).
The best-known example of a lexicographic order is alphabetical order, as used in a dictionary, based upon the standard ordering of the letters a, b, ..., z. For example,

$$\text{algebra} \trianglelefteq \text{algebraic}$$

and

$$\text{combination} \trianglelefteq \text{combinatorics}.$$

Example 5.24. (Set Powers and Job Priority).
Given a partially ordered set X and a positive integer
n, we have a corresponding lexicographic order on the
power X^n of X. That is, we can compare the elements
of X^n, which are ordered n-tuples, by regarding them as
words of length n.

Suppose X consists of tasks listed in order of priority
from highest to lowest $t_1, t_2, \ldots, t_m$. Hence, X^3 consists of
ordered task lists of length 3, which we regard as jobs. By
using lexicographic order, we can assign a corresponding priority to these jobs. Here, we use $\succeq$ for the relation
between two tasks (e.g. $t_1 \succeq t_2$) and $\unrhd$ for the relation
between two jobs. For instance,

$$(t_2, t_4, t_1) \unrhd (t_3, t_1, t_2).$$

This might reflect a situation in which students are
registering for classes and are allowed to list their top
three desired classes in order of preference. If the relative demand for the classes is reflected by the relation
$\succeq$ (i.e., t_1 is the class of highest demand), then the registrar might schedule the students using the order given
by $\unrhd$. In that way, the students that have the strongest
preference for the high-demand classes will be sched-
uled first.

5.2.2   Equivalence relations

Relations like $=$ and $\equiv$ suggest equality or equivalences. What
properties are important here?

Definition 5.7.   An **equivalence relation** on a set X is a relation on X that is reflexive, symmetric, and transitive. Given an
equivalence relation R on a set X and two elements $x, y \in X$,
we say that x is **equivalent** to y (under R) if and only if $x \, R \, y$.

In our first example, we "equate" two points in the xy-plane
if their y-coordinates round down to the same integer.

Example 5.25.   Let R be the relation on $\mathbb{R}^2$ defined by

$$(x_1, y_1) \, R \, (x_2, y_2) \quad \text{if and only if} \quad \lfloor y_1 \rfloor = \lfloor y_2 \rfloor.$$

For example, $(7, \sqrt{3}) \, R \, (\frac{1}{5}, \sqrt{2})$, since $\lfloor \sqrt{3} \rfloor = \lfloor \sqrt{2} \rfloor$, and
$(0, \pi) \, \cancel{R} \, (0, 2)$, since $\lfloor \pi \rfloor \neq \lfloor 2 \rfloor$. Show that R is an equivalence relation.

Proof. Let $(x_1, y_1), (x_2, y_2)$, and (x_3, y_3) represent arbitrary
elements of X.

Reflexive: Since $\lfloor y_1 \rfloor = \lfloor y_1 \rfloor$, we automatically have $(x_1, y_1) \, R \, (x_1, y_1)$.

Symmetric: Suppose $(x_1, y_1) \, R \, (x_2, y_2)$. That is, $\lfloor y_1 \rfloor = \lfloor y_2 \rfloor$. Since the symmetry of equality gives $\lfloor y_2 \rfloor = \lfloor y_1 \rfloor$, we have $(x_2, y_2) \, R \, (x_1, y_1)$.

Transitive: Suppose $(x_1, y_1) \, R \, (x_2, y_2)$ and $(x_2, y_2) \, R \, (x_3, y_3)$. That is, $\lfloor y_1 \rfloor = \lfloor y_2 \rfloor$ and $\lfloor y_2 \rfloor = \lfloor y_3 \rfloor$. From the transitivity of equality, it follows that $\lfloor y_1 \rfloor = \lfloor y_3 \rfloor$. Therefore, $(x_1, y_1) \, R \, (x_3, y_3)$. $\qquad\qquad\square$

To understand a particular equivalence relation, it is helpful to group together equivalent elements.

Definition 5.8. Given an equivalence relation R on a set X, the **equivalence class** (under R) of an element $x \in X$ is the set $\{y \, : \, y \in X \text{ and } y \, R \, x\}$ of all elements of X that are equivalent to x. It is denoted by $[x]_R$, and the subscript is dropped from our notation if the relation R is understood in context. A **representative** of an equivalence class $[x]$ is an element $y \in [x]$. That is, it is an element y such that $y \, R \, x$.

Given an equivalence relation R on a set X and an element $x \in X$, it is always the case that x is equivalent to itself. That is, the reflexivity of R gives $x \, R \, x$. Consequently, x can serve as a representative for $[x]$. Moreover, since $x \in [x]$, an equivalence class $[x]$ is never empty.

Example 5.26. Let R be the equivalence relation on $\mathbb{R}^2$ from Example 5.25. For each point $(x_1, y_1) \in \mathbb{R}^2$, we have

$$[(x_1, y_1)] = \{(x, y) \, : \, x, y \in \mathbb{R} \text{ and } \lfloor y \rfloor = \lfloor y_1 \rfloor\}.$$

Letting n denote the integer $\lfloor y_1 \rfloor$, we have

$$\lfloor y \rfloor = n \quad \text{if and only if} \quad n \leq y < n + 1.$$

Hence,

$$
\begin{aligned}
[(x_1, y_1)] &= \{(x, y) \, : \, x, y \in \mathbb{R} \text{ and } n \leq y < n + 1\} \\
&= \mathbb{R} \times [n, n + 1),
\end{aligned}
$$

the horizontal strip shown in Figure 5.10. Any of its points, such as $(0, n)$, $(1, n)$, or $(-\sqrt{2}, n + \frac{1}{\pi})$, can serve as representatives for this equivalence class.

The equivalence classes of equivalent elements are always the same.

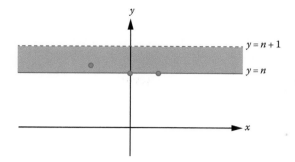

FIGURE 5.10
An equivalence class for the relation in Example 5.25.

Lemma 5.2. Let X be any set and R be any equivalence relation on X. For all $x, y \in X$,

$$y \; R \; x \quad \text{if and only if} \quad [y] = [x].$$

That is, y is equivalent to x if and only if the equivalence class of y is the same as that of x.

Proof. Let x and y be arbitrary elements of X.
  ($\rightarrow$) Suppose $y \; R \; x$.
  To show that $[y] \subseteq [x]$, suppose $z \in [y]$. That is, $z \; R \; y$. Since $z \; R \; y$ and $y \; R \; x$, it follows from transitivity that $z \; R \; x$. Hence, $z \in [x]$. We have thus shown that $[y] \subseteq [x]$.
  To show that $[x] \subseteq [y]$, suppose $z \in [x]$. That is, $z \; R \; x$. Since $y \; R \; x$, symmetry of R gives that $x \; R \; y$. Since $z \; R \; x$ and $x \; R \; y$, it follows from transitivity that $z \; R \; y$. Hence, $z \in [y]$. We have thus shown that $[x] \subseteq [y]$.
  It follows that $[y] = [x]$.
  ($\leftarrow$) Suppose $[y] = [x]$. Since $y \in [y] = [x]$, we have $y \; R \; x$. $\square$

Distinct equivalence classes never intersect.

Theorem 5.3. *Let X be any set and R be any equivalence relation on X. For all $x, y \in X$,*

$$[y] \neq [x] \quad \textit{if and only if} \quad [y] \cap [x] = \emptyset.$$

That is, two equivalence classes are distinct if and only if they are disjoint.

Proof. Let x and y be arbitrary elements of X. We prove each implication by considering its contrapositive.
  ($\rightarrow$) Suppose $[y] \cap [x] \neq \emptyset$. So, we have some element $z \in [y] \cap [x]$. That is, $z \in [x]$ and $z \in [y]$. So $z \; R \; x$ and $z \; R \; y$. By symmetry, we have $y \; R \; z$. Since $y \; R \; z$ and $z \; R \; x$, it follows

from transitivity that $y \, R \, x$. That is, $y \in [x]$. Lemma 5.2 then gives that $[y] = [x]$.

($\leftarrow$) If $[y] = [x]$, then certainly $x \in [x] = [y] \cap [x]$. Thus, $[y] \cap [x] \neq \emptyset$. $\qquad \square$

Recall from Section 3.5 that, given an integer $n > 1$, integers a and b are said to be congruent modulo n, written $a \equiv b \pmod{n}$, if and only if $n \mid (a - b)$. That is, for some $k \in \mathbb{Z}$, $a - b = nk$ or, equivalently, $a = b + nk$. A prototypical example of an equivalence relation is congruence modulo n.

Example 5.27. (Congruence Modulo n).
Fix an integer $n > 1$, and define a relation R on $\mathbb{Z}$ by

$$a \, R \, b \quad \text{if and only if} \quad a \equiv b \pmod{n}.$$

Theorem 3.26 tells us that R is an equivalence relation. For any $a \in \mathbb{Z}$,

$$[a] = \{b \; : \; b \in \mathbb{Z} \text{ and } b \equiv a \pmod{n}\}$$
$$= \{\ldots, a - 3n, a - 2n, a - n,$$
$$a, a + n, a + 2n, a + 3n, \ldots\}.$$

If the value n needs to be made clear, then this equivalence class may be denoted $[a]_n$, as it was in Section 3.5. In this context, $[a]_n$ is called the equivalence class of a modulo n. It follows from Lemma 5.2 that

$$a \equiv b \pmod{n} \quad \text{if and only if} \quad [a]_n = [b]_n.$$

Moreover, the set of equivalence classes of the integers modulo n is

$$\mathbb{Z}_n = \{[0], [1], \ldots, [n - 1]\}.$$

(a) Fix $n = 10$.
We have, for example, $37 \in [7]_{10}$, $-16 \in [4]_{10}$, and $[28]_{10} = [8]_{10} = \{\ldots, -22, -12, -2, 8, 18, 28, 38, \ldots\}$.
In the presentation of the integers displayed in Figure 5.11, the equivalence classes are the vertical strips.

(b) Fix $n = 5$.
We have, for example, $37 \in [2]_5$, $-16 \in [4]_5$, and $[28]_5 = [3]_5 = \{\ldots, -7, -2, 3, 8, 13, 18, 23, 28, 33, \ldots\}$.

5.2.3 Partitions

The concept of a partition is very closely tied to that of an equivalence relation. Before we can define a partition, we

⋮	⋮	⋮	⋮	⋮	⋮	⋮	⋮	⋮	⋮
20	21	22	23	24	25	26	27	28	29
10	11	12	13	14	15	16	17	18	19
0	1	2	3	4	5	6	7	8	9
−10	−9	−8	−7	−6	−5	−4	−3	−2	−1
−20	−19	−18	−17	−16	−15	−14	−13	−12	−11
⋮	⋮	⋮	⋮	⋮	⋮	⋮	⋮	⋮	⋮

FIGURE 5.11
Equivalence classes of integers modulo 10.

need to generalize some set notions from Section 1.4. Further generalizations are pursued in Section 5.5.

Definition 5.9.   Let $\mathcal{A}$ be a collection of sets from some universe $\mathcal{U}$.

(a) The **union** of $\mathcal{A}$, denoted $\bigcup_{A \in \mathcal{A}} A$, is the set defined by

$$\forall\, x \in \mathcal{U}, \quad x \in \bigcup_{A \in \mathcal{A}} A \;\leftrightarrow\; x \in A \text{ for some } A \in \mathcal{A}.$$

(b) We say that $\mathcal{A}$ is a collection of **disjoint** sets if

$$\forall\, A, B \in \mathcal{A}, \quad \text{if } A \neq B, \text{ then } A \cap B = \emptyset.$$

Example 5.28.   Let $\mathcal{U} = \mathbb{R}$, and let

$$\mathcal{A} = \{[n, n+1] \,:\, n \in \mathbb{Z} \text{ and } -10 \le n \le 20\}$$
$$= \{[-10, -9], [-9, -8], [-8, -7], \ldots, [20, 21]\}.$$

(a) $\bigcup_{A \in \mathcal{A}} A = [-10, 21]$.
For   example, $-7.5 \in \bigcup_{A \in \mathcal{A}} A$
since $-7.5 \in [-8, -7] \in \mathcal{A}$.   Moreover,   since
$\mathcal{A}$ is a finite collection, we see that

$$\bigcup_{A \in \mathcal{A}} A = [-10, -9] \cup [-9, -8] \cup [-8, -7]$$
$$\cup \cdots \cup [20, 21]$$
$$= [-10, 21].$$

(b) $\mathcal{A}$ is not a collection of disjoint sets.
For example, $[0, 1] \cap [1, 2] = \{1\} \neq \emptyset$.

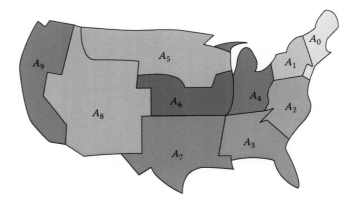

FIGURE 5.12
Partition of the United States based on first digit of ZIP code.

Alternatively, $\{(n, n+1) : n \in \mathbb{Z}$ and $-10 \leq n \leq$ $20\}$ is a collection of disjoint sets. Its union is $(-10, 21) \setminus \{-9, -8, \ldots, 20\}$.

Definition 5.10.   A **partition** of a set X is a collection $\mathcal{A}$ of disjoint nonempty subsets of X whose union is X.

The diagram in Figure 5.12 shows what a partition $\mathcal{A} = \{A_0, A_1, \ldots, A_9\}$ might look like. That particular partition shows the regions into which the continental United States is divided based on the first digit of the ZIP code. That is, A_i is the region in which the first digit of the ZIP code is i. Essentially, a partition of a set X carves it into pieces.

Example 5.29.   Let $X = \{1, 2, 3, 4, 5, 6, 7, 8\}$. In each case, does the specified collection $\mathcal{A} = \{A_1, A_2, A_3\}$ form a partition of X?

(a) $A_1 = \{1, 4\}$, $A_2 = \{2, 3, 5, 7\}$, and $A_3 = \{6, 8\}$.

> **Solution.** Yes. The sets A_1, A_2, and A_3 are disjoint, and their union $A_1 \cup A_2 \cup A_3$ equals X. $\square$

(b) $A_1 = \{1, 3, 7\}$, $A_2 = \{2, 8\}$, and $A_3 = \{5, 6\}$.

> **Solution.** No. Despite the fact that A_1, A_2, and A_3 are disjoint, their union fails to be X, since $4 \notin A_1 \cup A_2 \cup A_3$. $\square$

(c) $A_1 = \{2, 4, 6\}$, $A_2 = \{1, 5, 7\}$, and $A_3 = \{3, 4, 8\}$.

Solution. No. Despite the fact that $A_1 \cup A_2 \cup A_3 = X$, the sets A_1, A_2, and A_3 are not disjoint, since $4 \in A_1 \cap A_3$. □

An equivalence relation can be used to form a partition.

Lemma 5.4. Given an equivalence relation R on a nonempty set X, the collection of equivalence classes

$$\{[x] \ : \ x \in X\}$$

is a partition of X.

Proof. By our observations after Definition 5.8, the equivalence classes are nonempty sets. Theorem 5.3 tells us that distinct equivalence classes are disjoint. That X is the union of its equivalence classes follows from the fact that each element $x \in X$ is indeed in some equivalence class, namely, $[x]$. □

Definition 5.11. Given an equivalence relation R on a nonempty set X, the partition specified in Lemma 5.4 is called the **partition of X corresponding to** R.

Example 5.30. Let $X = \mathbb{Z}$ and R be congruence modulo n. The collection

$$\mathbb{Z}_n = \{[a] \ : \ a \in \mathbb{Z}\} = \{[0], [1], \ldots, [n-2], [n-1]\}$$

forms the partition of the integers corresponding to R. Note that the equivalence classes $\ldots, [-2], [-1], [n]$, $[n+1], \ldots$ are included in the listing above, since $\ldots, [-2] = [n-2]$, $[-1] = [n-1]$, $[n] = [0]$, $[n+1] = [1], \ldots$.

Example 5.31. Let R be the equivalence relation on $\mathbb{R}^2$ from Example 5.25 defined by

$$(x_1, y_1) \ R \ (x_2, y_2) \quad \text{if and only if} \quad \lfloor y_1 \rfloor = \lfloor y_2 \rfloor.$$

Find the partition of $\mathbb{R}^2$ corresponding to R.

Solution. As observed in Example 5.26, each equivalence class is a horizontal strip. For each $n \in \mathbb{Z}$, let

$$A_n = \{(x, y) \ : \ x, y \in \mathbb{R} \text{ and } \lfloor y \rfloor = n\} = \mathbb{R} \times [n, n+1).$$

The collection

$$\mathcal{A} = \{A_n \ : \ n \in \mathbb{Z}\}$$

is the partition of $\mathbb{R}^2$ corresponding to R.

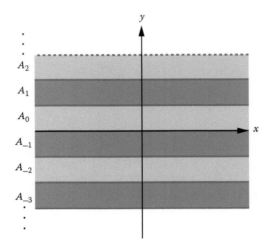

Clearly, any two distinct horizontal strips are disjoint, each strip is nonempty, and each point of $\mathbb{R}^2$ lies on exactly one horizontal strip.   □

Lemma 5.4 describes how to form a partition from an equivalence relation. Conversely, an equivalence relation can be formed from a partition.

Lemma 5.5.   Given a partition $\mathcal{A}$ of a set X, the relation R defined by, $\forall \, x, y \in X$,

$$x \, R \, y \quad \text{if and only if} \quad \exists \, A \in \mathcal{A} \text{ such that } x, y \in A$$

is an equivalence relation on X.

Proof. Let x, y, and z represent arbitrary elements of X.
Reflexive: Since $X = \bigcup_{A \in \mathcal{A}} A$, there must be some $A \in \mathcal{A}$ such that $x \in A$. Since $x, x \in A$, it follows that $x \, R \, x$.
Symmetric: Suppose $x \, R \, y$. That is, $x, y \in A$ for some $A \in \mathcal{A}$. Since $y, x \in A$, we have $y \, R \, x$.
Transitive: Suppose $x \, R \, y$ and $y \, R \, z$. That is, $x, y \in A_1$ for some $A_1 \in \mathcal{A}$, and $y, z \in A_2$ for some $A_2 \in \mathcal{A}$. Since $y \in A_1 \cap A_2$, we must have $A_1 = A_2$. Otherwise, distinct sets A_1 and A_2 would have to be disjoint and could not both contain y. It follows that $x, z \in A_1$, and thus $x \, R \, z$.   □

Definition 5.12. Given a partition $\mathcal{A}$ of a set X, the equivalence relation R specified in Lemma 5.5 is called the **equivalence relation on X corresponding to $\mathcal{A}$.**

Example 5.32. Let $X = \{1, 2, 3, 4, 5, 6, 7, 8\}$, and take the partition from Example 5.29(a) given by $A_1 = \{1, 4\}$, $A_2 = \{2, 3, 5, 7\}$, and $A_3 = \{6, 8\}$. The equivalence relation corresponding to this partition is the relation represented by the matrix

$$
\begin{array}{c}
\\ 1\\ 2\\ 3\\ 4\\ 5\\ 6\\ 7\\ 8
\end{array}
\begin{array}{c}
1\ 2\ 3\ 4\ 5\ 6\ 7\ 8\\
\left[
\begin{array}{cccccccc}
1 & 0 & 0 & 1 & 0 & 0 & 0 & 0\\
0 & 1 & 1 & 0 & 1 & 0 & 1 & 0\\
0 & 1 & 1 & 0 & 1 & 0 & 1 & 0\\
1 & 0 & 0 & 1 & 0 & 0 & 0 & 0\\
0 & 1 & 1 & 0 & 1 & 0 & 1 & 0\\
0 & 0 & 0 & 0 & 0 & 1 & 0 & 1\\
0 & 1 & 1 & 0 & 1 & 0 & 1 & 0\\
0 & 0 & 0 & 0 & 0 & 1 & 0 & 1
\end{array}
\right]
\end{array}.
$$

Example 5.33. Let $\mathcal{A} = \{A_n \ : \ n \in \mathbb{Z}\}$ be the partition of $\mathbb{R}^2$, where $\forall\, n \in \mathbb{Z}$,

$$A_n = \{(x, y) \ : \ x, y \in \mathbb{R} \text{ and } \lfloor y \rfloor = n\}.$$

Find the equivalence relation R on $\mathbb{R}^2$ corresponding to $\mathcal{A}$.

Solution. As specified in Lemma 5.5, R is given by

$$(x_1, y_1)\, R\, (x_2, y_2) \leftrightarrow \exists\, A_n \in \mathcal{A} \text{ such that } (x_1, y_1), (x_2, y_2) \in A_n$$

$$\leftrightarrow \exists\, n \in \mathbb{Z} \text{ such that } \lfloor y_1 \rfloor = n \text{ and } \lfloor y_2 \rfloor = n$$

$$\leftrightarrow \lfloor y_1 \rfloor = \lfloor y_2 \rfloor.$$

That is, R is the equivalence relation from Example 5.25.

$\square$

The fact that the partition $\mathcal{A}$ found in Example 5.31 recovered the equivalence relation R from which it came illustrates a general result.

Theorem 5.6. (Correspondence Between Equivalence Relations and Partitions).

Let X be a set, R be an equivalence relation on X, and $\mathcal{A}$ be a partition of X. Then $\mathcal{A}$ is the partition of X corresponding to R if and only if R is the equivalence relation on X corresponding to $\mathcal{A}$.

The proof of Theorem 5.6 is left for the exercises.

Exercises

1. Let X be a subset of $\mathbb{Z}^+$. Show that the "divides" relation $|$ is a partial order relation on X.

2. Let X be a subset of $\mathbb{R}$. Show that the relation $\leq$ is a partial order relation on X.

3. Let R be a partial order relation on a set X. Show that R^{-1} is also a partial order relation on X. Conclude that $\supseteq$ and $\geq$ are partial order relations.

4. Let R be a partial order relation on a set X such that any two elements of X are comparable. Show that the same is true for R^{-1}. Conclude that $\geq$ is a total order relation.

5. Among the relations in the odd-numbered exercises in Exercises 37 through 54 from Section 5.1, which are partial order relations?

6. Among the relations in the even-numbered exercises in Exercises 37 through 54 from Section 5.1, which are partial order relations?

In Exercises 7 through 12, determine whether the specified relation R on the given set X is a partial order relation. Justify your answer.

7. Define R on $\mathcal{P}(\{1, 2, \ldots, 10\}) \setminus \{\emptyset\}$ by $A \, R \, B$ if and only if $\min(A) \leq \min(B)$ and $\max(A) \leq \max(B)$.

8. Let R be the "is a proper divisor" relation on $\mathbb{Z}^+$. That is, $a \, R \, b$ if and only if $a \mid b$ and $a \neq b$.

9. A college is tracking those students who took a math class both semesters of their first year in college. In the displayed digraph, an arrow points from the class taken first semester to that taken second semester.

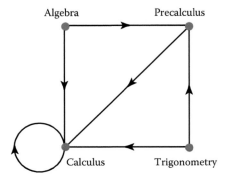

Let R be the relation on the set

{Algebra, Calculus, PreCalculus, Trigonometry}

of first-year math courses reflected by this digraph.

10. Some of the birds on the islands Alki, Balu, Cerf, and Dago have been tagged for a study of their migration habits from year to year. In the displayed digraph, an arrow points from the island inhabited by a tagged bird 1 year to that inhabited by that same bird the next year.

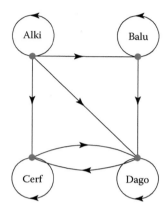

Let R be the relation on the set {Alki, Balu, Cerf, Dago} of islands reflected by this digraph.

11. The relation on {1, 2, 3, 4} whose matrix representation is displayed.

$$
\begin{array}{c}
\ 1\ 2\ 3\ 4 \\
\begin{array}{c} 1 \\ 2 \\ 3 \\ 4 \end{array}
\left[\begin{array}{cccc}
1 & 1 & 1 & 0 \\
0 & 1 & 0 & 0 \\
0 & 1 & 1 & 0 \\
0 & 0 & 0 & 1
\end{array} \right]
\end{array}
$$

12. The relation on {1, 2, 3, 4} whose matrix representation is displayed.

$$
\begin{array}{c}
\ 1\ 2\ 3\ 4 \\
\begin{array}{c} 1 \\ 2 \\ 3 \\ 4 \end{array}
\left[\begin{array}{cccc}
1 & 1 & 1 & 0 \\
0 & 1 & 0 & 0 \\
0 & 1 & 0 & 1 \\
0 & 0 & 0 & 1
\end{array} \right]
\end{array}
$$

In Exercises 13 through 16, make a Hasse diagram for the specified relation on the given set.

13. The "is a subset of" relation $\subseteq$ on $\mathcal{P}(\{a, b\})$.

14. The "divides" relation $\mid$ on {2, 3, 5, 6, 10, 12}.

15. The "divides" relation $\mid$ on the set of positive divisors of 72.

16. The "is a subset of" relation $\subseteq$ on the subsets of {1, 2, 3, 4} containing at least three elements.

TABLE 5.3
Blood types determined by antigens

A	B	Rh	Type
F	F	F	O^-
F	F	T	O^+
F	T	F	B^-
F	T	T	B^+
T	F	F	A^-
T	F	T	A^+
T	T	F	AB^-
T	T	T	AB^+

17. **Blood donors.** A person's blood type is determined by the presence (T) or absence (F) of the antigens A, B, and Rh, as shown in Table 5.3. A person with blood type X can donate blood to a person with blood type Y if and only if Y contains all of the antigens present in X. Let P be the set of the eight possible blood types, and let R be the relation on P such that $X \, R \, Y$ if and only if a person with blood type X can donate blood to a person with blood type Y.

    (a) Can a person with A^+ blood donate to one with A^- blood?

    (b) What types of blood can a person with A^+ blood receive?

    (c) Draw a digraph for R, and observe that R is a partial order relation.

    (d) Make a Hasse diagram for R.

18. The employees in Bob's grocery store are divided into six departments: Baggers, Cashiers, Deli, Produce, Meat, and Management. Although promotions are always being sought, they are restricted according to the matrix in Table 5.4. That is, a promotion from department X to department Y is permitted only if there is a 1 in row X and column Y of the matrix. Let R be the relation on the grocery store departments determined by this matrix.

    (a) Can an employee be promoted from the Deli department to the Produce department?

    (b) From what departments might an employee be promoted to the Meat department?

    (c) Draw a digraph for R, and observe that R is a partial order relation.

    (d) Make a Hasse diagram for R.

TABLE 5.4
Promotion matrix for Bob's grocery store

	Bag	Deli	Cash	Meat	Prod	Manage
Bag	1	1	1	1	1	1
Deli	0	1	0	1	1	1
Cash	0	0	1	0	1	1
Meat	0	0	0	1	0	1
Prod	0	0	0	0	1	1
Manage	0	0	0	0	0	1

19.* Show that a Hasse diagram never contains triangles (three mutually joined elements).

20.* Show that a Hasse diagram for a total order relation on a finite set can always be represented in a vertical line.

21. The login IDs are stored in a computer using a lexicographic ordering based on the character ordering $a < b < \cdots < z < 0 < 1 < \cdots < 9$. This allows for efficient lookup of an ID, when a user attempts to log in. Place the two login ID's b6vitamin and b12vitamin in lexicographic order.

22. The American Standard Code for Information Interchange (ASCII Code) imposes the following ordering on characters

$$0 < 1 < \cdots < 9 < a < b < \cdots < z.$$

The file names on a computer are stored in a list using a lexicographic ordering based upon this character ordering. When a user searches for a file, its name is easily checked against the list of existing files. Determine the proper order for the two file names doggy2 and dog8cat.

23. Using lexicographic order on $\mathbb{Z}^4$, based on the normal ordering $\leq$ on $\mathbb{Z}$, determine the proper order of $(2, -3, 1, 0)$ and $(2, -1, 5, 3)$.

24. Using lexicographic order on $\mathbb{N}^2$, based on the normal ordering $\leq$ on $\mathbb{N}$, determine the proper order of $(3, 1)$ and $(2, 7)$.

25.* Let $n \in \mathbb{Z}^+$, and let X be the set of n-bit binary numbers (with zeros padded at the left). Using $0 < 1$, show that the lexicographic ordering on X is the same as the ordering given by $\leq$.

26.* Show that if $\preceq$ is a total ordering on a character set C, then the lexicographic ordering $\trianglelefteq$ on the set of words W will be a total ordering as well.

In Exercises 27 through 32, show that the specified relation R on the given set X is an equivalence relation.

27. On $X = \mathbb{R}$, define $x \, R \, y$ if and only if $\lfloor x \rfloor = \lfloor y \rfloor$.

28. On $X = \mathbb{R}$, define $x \, R \, y$ if and only if $\cos x = \cos y$ and $\sin x = \sin y$.

29. On $X = \mathbb{N}^2$, define $(m_1, n_1) \, R \, (m_2, n_2)$ if and only if $m_1 - n_1 = m_2 - n_2$.

30. On $X = \mathbb{R}^2$, define $(x_1, y_1) \, R \, (x_2, y_2)$ if and only if $x_1^2 + y_1^2 = x_2^2 + y_2^2$.

31. On $X = \mathbb{R}^2 \setminus \{(0,0)\}$, define $(x_1, y_1) \, R \, (x_2, y_2)$ if and only if $\exists \, c \in \mathbb{R} \setminus \{0\}$ such that $cx_1 = x_2$ and $cy_1 = y_2$. That is, $c(x_1, y_1) = (x_2, y_2)$.

32. On $X = \mathbb{R}^2$, define $(x_1, y_1) \, R \, (x_2, y_2)$ if and only if $y_1 - x_1 = y_2 - x_2$.

33. Show that the relation R on $\mathbb{R}^2$ defined by $(x_1, y_1) \, R \, (x_2, y_2)$ if and only if $x_1 + x_2 = y_1 + y_2$ is not an equivalence relation.

34. Show that the relation R on $\mathbb{R}^2$ defined by $(x_1, y_1) \, R \, (x_2, y_2)$ if and only if $x_1 y_1 = x_2 + y_2$ is not an equivalence relation.

35. Among the relations in the odd-numbered exercises in Exercises 37 through 54 from Section 5.1, which are equivalence relations?

36. Among the relations in the even-numbered exercises in Exercises 37 through 54 from Section 5.1, which are equivalence relations?

37. Let R be an equivalence relation on a set X.
Show: $\forall \, x, y \in X$, $x \in [y]$ if and only if $y \in [x]$.

38. Let R be an equivalence relation on a set X.
Show: $\forall \, x, y, z \in X$, if $x \in [y]$ and $y \in [z]$, then $[x] = [z]$.

39. In Exercise 29, find a representative for $[(m_1, n_1)]$ such that one of the coordinates is zero.

40. In Exercise 30, find a representative for $[(x_1, y_1)]$ that lies on the x-axis.

41. In Exercise 31, find a representative for $[(x_1, y_1)]$ that lies on the unit circle.

42. In Exercise 32, find a representative for $[(x_1, y_1)]$ that lies on the y-axis.

In Exercises 43 through 46, determine whether a partition of the given set X is formed by the specified collection $\{A_1, A_2, \ldots\}$.

43. $X = \{1, 2, 3, 4, 5, 6\}$, $A_1 = \{1, 2\}$, $A_2 = \{3, 4\}$, $A_3 = \{5\}$, and $A_4 = \{6\}$.

44. $X = \mathbb{R}^2$ the Cartesian plane, and A_1, A_2, A_3, A_4 represent the four quadrants. That is,

$$A_1 = \{(x, y) : x > 0 \text{ and } y > 0\},$$
$$A_2 = \{(x, y) : x < 0 \text{ and } y > 0\},$$
$$A_3 = \{(x, y) : x < 0 \text{ and } y < 0\}, \text{ and}$$
$$A_4 = \{(x, y) : x > 0 \text{ and } y < 0\}.$$

45. $X = \mathbb{R}^+$, $A_1 = (0, 1)$, $A_2 = (1, 2)$, and, in general, $A_n = (n - 1, n)$, for all $n \in \mathbb{Z}^+$.

46. $X = \mathbb{Z}$,

$$A_1 = \{m : m = 2k \text{ for some } k \in \mathbb{Z}\},$$
$$A_2 = \{m : m = 3k \text{ for some } k \in \mathbb{Z}\},$$
$$A_3 = \{m : m = 5k \text{ for some } k \in \mathbb{Z}\},$$

and, in general, taking p_n to be the nth prime number,

$$A_n = \{m : m = p_n k \text{ for some } k \in \mathbb{Z}\}.$$

47. The members of a committee—Jones, Halmos, Carlson, McBride, Damon, and Yokus—have been divided into subcommittees. One consists of Damon, Halmos, and McBride. Another consists of Yokus and Jones. And another consists of Carlson and Halmos. Do these subcommittees form a partition of the committee? Explain.

48. Major League Baseball (MLB) is divided into two leagues: the American League (AL) and the National League (NL). The teams in each league are then divided into three divisions: East, Central, and West. Do the MLB divisions (AL East, AL Central, AL West, NL East, NL Central, and NL West) form a partition of the MLB teams? Explain.

In Exercises 49 through 52, show that a partition of the given set X is formed by the specified collection of subsets of X.

49. $X = \mathbb{Z}$, A_1 is the set of odd integers, and A_2 is the set of even integers.

50. $X = \mathbb{R} \times \mathbb{R}$, and, for each $a \in \mathbb{R}$, let A_a be the set of points on the vertical line through $(a, 0)$.

51. Let $\mathbb{Z}^*$ denote the set of nonzero integers and $X = \mathbb{Z} \times \mathbb{Z}^*$. For each $r \in \mathbb{Q}$, let $A_r = \{(a, b) : \frac{a}{b} = r\}$.

52. $X = \mathbb{Z}$, and, for each $i \in \mathbb{N}$, let $A_i = \{i, -i\}$.

In Exercises 53 through 58, find the partition on the given set X that corresponds to the specified equivalence relation R on X.

53. The relation from Exercise 27.

54. The relation from Exercise 28.

55. The relation from Exercise 29.

56. The relation from Exercise 30.

57. The relation from Exercise 31.

58. The relation from Exercise 32.

59. Partition the set of words {apple, eat, peace, car, call, ear} according to the specified equivalence relation.

    (a) "starts with the same letter as"

    (b) "has the same number of letters as"

    (c) "has the same set of vowels as"

60. Partition the set of numbers {20, 500, 5, 176, 80, 605} according to the specified equivalence relation.

    (a) "has the same number of digits as"

    (b) "has the same set of prime divisors as"

    (c) "ends with the same digit as"

    In Exercises 61 through 64, for the given set X, find the equivalence relation R on X that corresponds to the specified partition of X.

61. The partition from Exercise 49.

62. The partition from Exercise 50.

63. The partition from Exercise 51.

64. The partition from Exercise 52.

65. Find a simple equivalence relation that yields the given partition on the set of file names {hw1.doc, hw1.xls, hw2.doc, hw2.xls, hw2.ppt}.

    (a) {hw1.doc, hw2.doc}, {hw1.xls, hw2.xls}, {hw2.ppt}.

    (b) {hw1.doc, hw1.xls}, {hw2.doc, hw2.xls, hw2.ppt}.

66. Find a simple equivalence relation that yields the given partition on the set of ZIP codes {03264, 17815, 01104, 13210, 13244, 90210}.

    (a) {03264, 01104}, {17815, 13210, 13244}, {90210}.

    (b) {03264, 01104, 13244}, {17815}, {13210, 90210}.

67.* Prove Theorem 5.6.

68.* Show that an equivalence relation on a finite set always can be represented by a block diagonal matrix. Such a matrix has the form

$$\begin{bmatrix} J_1 & 0 & 0 & \cdots & 0 \\ 0 & J_2 & 0 & \cdots & 0 \\ 0 & 0 & J_3 & \cdots & 0 \\ \vdots & \vdots & \vdots & \ddots & \vdots \\ 0 & 0 & 0 & \cdots & J_k \end{bmatrix},$$

where each zero represents a square array, all of whose entries are zero, and each J_i represents a square array, all of whose entries are one. For example, $\begin{bmatrix} J_1 & 0 \\ 0 & J_2 \end{bmatrix}$ is the

block form of the matrix $\begin{bmatrix} 1 & 1 & 0 & 0 & 0 \\ 1 & 1 & 0 & 0 & 0 \\ 0 & 0 & 1 & 1 & 1 \\ 0 & 0 & 1 & 1 & 1 \\ 0 & 0 & 1 & 1 & 1 \end{bmatrix}.$

What is important about the ordering of the rows and columns?

69. Refer to Definition 5.9. Let $\mathcal{A}_1$ and $\mathcal{A}_2$ be collections of sets from some universal set $\mathcal{U}$. Show that

$$\left(\bigcup_{A \in \mathcal{A}_1} A \right) \cup \left(\bigcup_{A \in \mathcal{A}_2} A \right) = \bigcup_{A \in \mathcal{A}_1 \cup \mathcal{A}_2} A.$$

70. Refer to Definition 5.9. Let B be a set and $\mathcal{A}$ be a collection of sets from some universal set $\mathcal{U}$. Show that

$$B \cap \left(\bigcup_{A \in \mathcal{A}} A \right) = \bigcup_{A \in \mathcal{A}} (B \cap A).$$

5.3 Basics of functions

Most lower-level mathematics books work with an informal definition of function. Such books say that a function f from a set X to a set Y is a rule that, to each $x \in X$, assigns a *unique* $y \in Y$, denoted $f(x)$. That is, $f(x) = y$. Although this understanding typically proves adequate for the intended audience, it is not precise. It relies on the undefined terms *rule* and *assigns*. However, these holes are filled by our formal understanding of relations from Section 5.1.

Definition 5.13. Given sets X and Y, a **function** f **from** X **to** Y, denoted $f : X \longrightarrow Y$, is a relation from X to Y such that each $x \in X$ is related to a *unique* $y \in Y$. In this context, we write $f(x) = y$ in place of $x f y$ or $(x, y) \in f$. When f is understood, we may also write $x \mapsto y$.

(a) When $f(x) = y$, so $x \mapsto y$, we say that f **maps** the element x to the element y or that y is the **image** of x (under f). We also say that f **maps** the set X to the set Y. In fact, functions are sometimes called maps.

(b) The **domain** of f is the set X. That is, domain$(f) = X$.

(c) The **codomain**, or **target**, of f is the set Y.

(d) The **range**, or **image**, of f is the set

$$\text{range}(f) = \{y \; : \; y \in Y \text{ and } f(x) = y \text{ for some } x \in X\}.$$

Figure 5.13 depicts the structure of a function, and Figure 5.14 shows the principle feature that a relation avoids to satisfy the conditions of a function. A function f cannot have $f(x) = y_1$ and $f(x) = y_2$ with $y_1 \neq y_2$.

Example 5.34.   Let f be the function from $\{0, 1, 2, 3, 4\}$ to $\{1, 2, 3, 4, 5, 6\}$ defined by $f(x) = \binom{4}{x}$. That is,

$$f : \{0, 1, 2, 3, 4\} \longrightarrow \{1, 2, 3, 4, 5, 6\}.$$

(a) For each integer $0 \leq x \leq 4$, the function f maps x to the binomial coefficient $\binom{4}{x}$. For example, the image of 1 is 4, and the image of 2 is 6. That is, $1 \mapsto 4$ and $2 \mapsto 6$.

(b) The domain of f is $\{0, 1, 2, 3, 4\}$.

(c) The codomain of f is $\{1, 2, 3, 4, 5, 6\}$.

(d) The range of f is $\{1, 4, 6\}$. Observe that $\{y \; : \; f(x) = y \text{ for some } x \in \{0, 1, 2, 3, 4\}\}$ is

$$\{f(0), f(1), f(2), f(3), f(4)\} = \{1, 4, 6, 4, 1\}$$

$$= \{1, 4, 6\}.$$

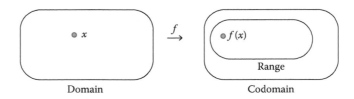

Domain

Codomain

FIGURE 5.13
The structure of a function.

FIGURE 5.14
A relation that is not a function.

Example 5.35.

(a) Define f from $\mathcal{P}(\mathbb{R})$ to $\{0,1\}$ by

$$f(A) = \begin{cases} 1 & \text{if } \sqrt{2} \in A, \\ 0 & \text{if } \sqrt{2} \notin A. \end{cases}$$

This defines a function $f : \mathcal{P}(\mathbb{R}) \longrightarrow \{0,1\}$ that maps each subset A of $\mathbb{R}$ to 1 precisely when A contains the number $\sqrt{2}$.
The range of f is all of $\{0,1\}$. For example, $f(\mathbb{Z}) = 0$ and $f(\{0, \sqrt{2}\}) = 1$. So both 0 and 1 lie in the range.
This function is an example of an **indicator function**. In this case, f indicates whether or not a set contains $\sqrt{2}$.

(b) Define f from $\mathcal{P}(\mathbb{R})$ to $\{0,1\}$ by

$$f(A) = \begin{cases} 1 & \text{if } \sqrt{2} \in A, \\ 0 & \text{if } \sqrt{3} \in A. \end{cases}$$

This formula does not define a function. The problem is that a *unique* value for $f(A)$ is not specified when A contains both $\sqrt{2}$ and $\sqrt{3}$. The formula merely defines a relation. For example, $\{\sqrt{2}, \sqrt{3}\} \, f \, 1$ and $\{\sqrt{2}, \sqrt{3}\} \, f \, 0$.

Well-defined functions. In an example like Example 5.35(b), we say that the *relation* f is not a **well-defined** function. Very simply, this is equivalent to saying that f is not a function. However, the extra adjective *well-defined* is sometimes added for emphasis; a formula that appears on the surface to yield a function might, upon careful analysis, be seen not to do so. In fact, it is inappropriate to use notation like $f(x) = y$ when f is not a function.

The reasons some relations fail to satisfy the conditions of a function are sometimes quite subtle. In Example 5.35(b), the specified formula is ambiguous and therefore does not yield unique function values. More basic problems can occur if, for some x in the proposed domain X, a value $f(x)$ either is not defined or does not land in the proposed codomain Y. There are also ambiguity problems that are particularly plaguing when X is a set of equivalence classes under some equivalence relation. Such a difficulty is illustrated in the next example.

Example 5.36.   Let $d \in \mathbb{Z}$ with $d > 1$. Recall that $\mathbb{Z}_d = \{[0], [1], \ldots, [d-1]\}$ is the set of equivalence classes of integers modulo d.

(a) Define f from $\mathbb{Z}_d$ to $\mathbb{Z}$ by $f([a]) = a$.
Show that f is not a well-defined function.

Solution. The problem is that the representative a for $[a]$ is not used in a valid way to specify a unique function value. For example, since $[0] = [d]$, our function f should have $f([0]) = f([d])$. However, under the specified formula, we have $f([0]) = 0$ and $f([d]) = d$. Since $0 \neq d$, a unique output value has not been specified for the input value $[0] = [d]$. Thus, f is not a well-defined function. $\square$

(b) Define f from $\mathbb{Z}_d$ to $\mathbb{Z}$ by $f([a]) = a \bmod d$.
Show that f is a well-defined function.

Solution. In this case, we shall show that the representative a for $[a]$ is used in a valid way to specify a unique function value. Let $r = a \bmod d$, the proposed value for $f([a])$. So $0 \leq r < d$ and $a = dq + r$, for some $q \in \mathbb{Z}$.
Suppose $[b] = [a]$. Since the formula for f specifies that $f([b]) = b \bmod d$, we must establish that $b \bmod d = r$. Since $b \equiv a \pmod{d}$, we have that $b - a = dk$, for some $k \in \mathbb{Z}$. Consequently,

$$b = a + dk = dq + r + dk = d(q + k) + r.$$

Since $q + k$ is an integer, we see, by the Division Algorithm, that $b \bmod d = r$. That is, $f([b]) = f([a])$, and thus each input value has a unique output value. $\square$

Example 5.37.   Define $f : [-1, 1] \longrightarrow (-4, 4)$ by $x \mapsto 1 - 2x$.

(a) Show that f indeed maps elements from the domain $[-1, 1]$ to the codomain $(-4, 4)$, so f is a well-defined function from $[-1, 1]$ to $(-4, 4)$.

Solution. If $x \in [-1, 1]$, then the inequalities

$$-1 \leq x \leq 1 \qquad \text{(Now, multiply by } -2.)$$
$$-2 \leq -2x \leq 2 \qquad \text{(Now, add 1.)}$$
$$-1 \leq 1 - 2x \leq 3$$

show that $f(x) \in [-1, 3] \subseteq (-4, 4)$. $\square$

(b) Show that the range of f is $[-1,3]$.

> **Solution.** In part (a) we showed that the range is a subset of $[-1,3]$. To establish equality, we must show conversely that, for each $y \in [-1,3]$, there is some $x \in [-1,1]$ such that $x \mapsto y$. So suppose that $y \in [-1,3]$. We seek a value x such that $-1 \le x \le 1$ and $1 - 2x = y$. Since this equation is satisfied by $x = \frac{1-y}{2}$, it suffices to verify that $-1 \le \frac{1-y}{2} \le 1$. The inequalities
>
> $$
> \begin{array}{ll}
> -1 \le y \le 3 & \text{(Now, multiply by } -1.) \\
> -3 \le -y \le 1 & \text{(Now, add 1.)} \\
> -2 \le 1 - y \le 2 & \text{(Now, divide by 2.)} \\
> -1 \le \frac{1-y}{2} \le 1
> \end{array}
> $$
>
> thus establish that $[-1,3]$ is a subset of the range, and, therefore, $[-1,3]$ is the range.   □

5.3.1 Composing functions

In some cases, it makes sense to follow one function by another. That is, we may want to use the output from one function as input to another.

Definition 5.14.   Given   functions $f : X \longrightarrow Y$ and $g : W \longrightarrow Z$ such that the range of f is a subset of the domain of g, their **composite**, denoted $g \circ f$, is the function $g \circ f : X \longrightarrow Z$ defined by the formula

$$
\forall x \in X, \quad (g \circ f)(x) = g(f(x)).
$$

The notation $g \circ f$ is read "g composed with f."

Figure 5.15 depicts the structure of a composition of two functions. In the picture at its center, the oval denotes the range of f and the rectangle denotes the domain of g.

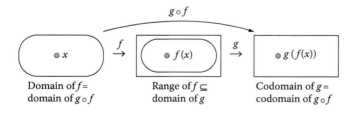

FIGURE 5.15
The structure of a composition of functions.

Example 5.38. Define a function $f : \{0, 1, 2\} \longrightarrow \{0, 1, 2, 3, 4\}$ by $f(x) = 2x$, and define $g : \{0, 1, 2, 3, 4\} \longrightarrow \{1, 2, 3, 4, 5, 6\}$ by $g(x) = \binom{4}{x}$. Then,

$$g \circ f : \{0, 1, 2\} \longrightarrow \{1, 2, 3, 4, 5, 6\}$$

is given by

$$(g \circ f)(x) = g(f(x)) = g(2x) = \binom{4}{2x}, \quad \forall\, x \in \{0, 1, 2\}.$$

Example 5.39. Define a function $f : [-1, 1] \longrightarrow (-4, 4)$ by $f(x) = 1 - 2x$, and define $g : (-2, 3] \longrightarrow [0, \infty)$ by $g(x) = \frac{1}{x+2}$. By Example 5.37(b), the range$(f) = [-1, 3] \subseteq (-2, 3]$. Therefore, the composite $g \circ f$ is defined and maps $[-1, 1]$ to $[0, \infty)$ by the formula

$$(g \circ f)(x) = g(f(x)) = g(1 - 2x) = \frac{1}{(1 - 2x) + 2} = \frac{1}{3 - 2x}.$$

It follows from the definition of a function that two functions $F : X \longrightarrow W$ and $F' : X' \longrightarrow W'$ are equal (i.e., $F = F'$) if and only if $X = X'$, $W = W'$, and $\forall\, x \in X, F(x) = F'(x)$. We use this fact in our proof of the following result.

Theorem 5.7. (Associativity of Function Composition). *Given any functions $f : X \longrightarrow Y, g : Y \longrightarrow Z$, and $h : Z \longrightarrow W$,*

$$(h \circ g) \circ f = h \circ (g \circ f).$$

Proof. By the definition of composition (Definition 5.14), both $(h \circ g) \circ f$ and $h \circ (g \circ f)$ have domain X and codomain W. So we focus on the output values of these two compositions. Let $x \in X$. Observe that

$$((h \circ g) \circ f)(x) = (h \circ g)(f(x)) = h(g(f(x)))$$

and

$$(h \circ (g \circ f))(x) = h((g \circ f)(x)) = h(g(f(x))).$$

Since, $((h \circ g) \circ f)(x) = (h \circ (g \circ f))(x)$ and x is arbitrary, it follows that $(h \circ g) \circ f = h \circ (g \circ f)$.   □

Generalizing Definition 5.14, it is also possible to define the composition of two relations. This is taken up in the exercises, where its applications to databases are also explored.

5.3.2 Focusing on real functions

Real functions were informally introduced in Section 1.3. We now define them more formally, in light of Definition 5.13, as functions that map a subset of $\mathbb{R}$ to a subset of $\mathbb{R}$. Real polynomial functions and exponential functions are important examples of real functions.

Definition 5.15.

(a) A **real polynomial function** is a real function f for which there are $n \in \mathbb{N}$ and $c_n, c_{n-1}, \ldots, c_0 \in \mathbb{R}$ such that

$$\forall\, x \in \mathbb{R}, \quad f(x) = c_n x^n + c_{n-1} x^{n-1} + \cdots + c_1 x + c_0.$$

When $n = 0$, the polynomial $f(x) = c_0$ is a constant function.

(b) An **exponential function**[*] is a real function f for which there is some $b \in \mathbb{R}^+$ such that

$$\forall\, x \in \mathbb{R}, \quad f(x) = b^x.$$

The number b is called its **base**. When b is the Euler number $e \approx 2.718$, the function $f(x) = e^x$ is called the **natural exponential** function.

Examples of polynomials are given by the formulas $3x^2 + 2$ and $\sqrt{2}x^3 - \frac{1}{4}x + \pi$ but not $4x^{\frac{1}{2}} + 1 = 4\sqrt{x} + 1$. Important examples of exponential functions besides e^x include 2^x and $2^{-x} = \frac{1}{2^x} = (\frac{1}{2})^x$.

The **graph of a function** $f : X \longrightarrow Y$ is the set

$$\{(x,y) \,:\, x \in X, y \in Y, \text{ and } f(x) = y\}. \tag{5.1}$$

When f is a real function, a drawing of (5.1) is a valuable tool for understanding properties of f. Basic examples of drawings of graphs of real functions are shown in Figure 5.16.

---

[*] A formal development of general powers b^a is given in Appendix A. However, an informal understanding gained from a precalculus course ought to suffice for our purposes.

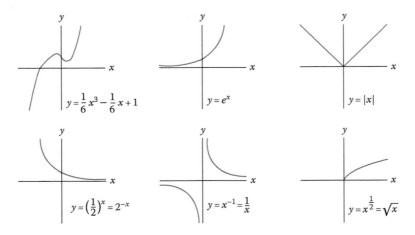

$$y = \frac{1}{6}x^3 - \frac{1}{6}x + 1 \qquad y = e^x \qquad y = |x|$$

$$y = \left(\frac{1}{2}\right)^x = 2^{-x} \qquad y = x^{-1} = \frac{1}{x} \qquad y = x^{\frac{1}{2}} = \sqrt{x}$$

FIGURE 5.16
Examples of real functions.

Example 5.40.   The relation on $\mathbb{R}$ given by $\{(x,y) : x = y^2\}$ can be seen not to be a function, since it contains both $(4,2)$ and $(4,-2)$; there is no unique value y specified as the image of 4.

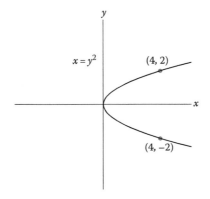

For relations on $\mathbb{R}$, the definition of a function can be recast in a form known as the Vertical Line Test.

Test (Vertical Line Test).
Let $X \subseteq \mathbb{R}$, $Y \subseteq \mathbb{R}$, and let R be a relation from X to Y. Consider all vertical lines in $\mathbb{R}^2$ (of the form $x = a$ for some $a \in \mathbb{R}$).
   If every such vertical line intersects R *at most once*, then R is a function.
   Moreover, each vertical line $x = a$ should intersect R

   (i) *exactly once* if $a \in X$, and

   (ii) *not at all* if $a \notin X$.

The fact that the relation $x = y^2$ from Example 5.40 is not a function can be seen with the Vertical Line Test. The argument given in Example 5.40 shows that the line $x = 4$ intersects the relation $x = y^2$ *more than once*. Much as Venn diagrams give us intuition about set identities, the Vertical Line Test gives us a way of seeing the difference between functions and relations on $\mathbb{R}$. Rigorous arguments, however, must appeal to the definition of a function (as was done in Example 5.40) and not just to a picture.

In a spirit similar to the Vertical Line Test, the range of a real function can be seen by considering horizontal lines.

Test (The Horizontal Line Range Test).
Let $X \subseteq \mathbb{R}$, $Y \subseteq \mathbb{R}$, and $f : X \longrightarrow Y$. Consider all horizontal lines in $\mathbb{R}^2$ (of the form $y = b$ for some $b \in \mathbb{R}$).

The range of f consists of all values b such that the line $y = b$ intersects the graph of f *at least once*.

Throughout this book, if a function is expressed by a formula, and the domain is not specified, it is understood that, within the context of some universe of consideration, the domain is the set of values at which the formula can be evaluated. Similarly, if the codomain is not specified, then it is taken to be the set of images of elements from the domain (i.e., the range).

Example 5.41. For each real function displayed in Figure 5.16, its domain and range is listed in Table 5.5. For example, $\frac{1}{x}$ is defined for all real numbers x except $x = 0$. Consequently, the domain of $\frac{1}{x}$ is $\mathbb{R} \setminus \{0\}$, and there is no point (x, y) on its graph with $x = 0$. To determine the range of $\frac{1}{x}$, let y be an arbitrary real number. Can we find a value x in $\mathbb{R} \setminus \{0\}$ for which $\frac{1}{x} = y$? Since this equation is equivalent to $x = \frac{1}{y}$, such an x can be found if and only if $y \neq 0$. Consequently, the range of $\frac{1}{x}$ is $\mathbb{R} \setminus \{0\}$, and there is no point (x, y) on its graph with $y = 0$.

TABLE 5.5
Domain and range for the functions in Figure 5.16

f(x)	Domain of f	Range of f
$\frac{1}{6}x^3 - \frac{1}{6}x + 1$	$\mathbb{R}$	$\mathbb{R}$
e^x	$\mathbb{R}$	$(0, \infty)$
$|x|$	$\mathbb{R}$	$[0, \infty)$
$(\frac{1}{2})^x$	$\mathbb{R}$	$(0, \infty)$
$\frac{1}{x}$	$\mathbb{R} \setminus \{0\}$	$\mathbb{R} \setminus \{0\}$
$\sqrt{x}$	$[0, \infty)$	$[0, \infty)$

Example 5.42. The **floor function** $f(x) = \lfloor x \rfloor$ is an example of a real function. Recall, for each $x \in \mathbb{R}$, that $\lfloor x \rfloor$ is the unique integer n such that $n \leq x < n + 1$.

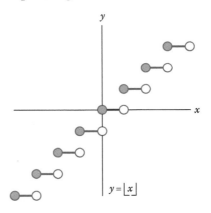

The domain of f is $\mathbb{R}$, and the range of f is $\mathbb{Z}$. Consequently, the floor function may alternatively be treated as a function from $\mathbb{R}$ to $\mathbb{Z}$.

Exercises

1. Is codomain synonymous with range?

2. Does a function map its domain to its codomain?

   In Exercises 3 through 14, determine whether the given f, as "defined," is a function. If it is not well-defined, then explain why.

3. Define f from $\mathbb{Z}$ to $\mathbb{Z}^+$ by
   $$f(n) = 2^{n-1}.$$

4. Define f from $\mathbb{Z}^+$ to $\mathbb{Z}^+$ by
   $$f(n) = 3n^2 - 2.$$

5. Define f from $[0, \infty)$ to $\mathbb{R}$ by
   $$x \mapsto \pm\sqrt{x}.$$

6. Define f from $\mathbb{R}$ to $\mathcal{P}(\mathbb{R})$ by
   (a) $x \mapsto \{x\}$.
   (b) $x \mapsto \{x, -x\}$.

7. Define f from $\mathbb{Z}$ to $\mathbb{Z}$ by
   (a) $f(n) = \frac{n}{2}$.
   (b) $f(n) = 2n$.

8. Define f from $\mathbb{Z}$ to $\mathbb{Q}$ by
   (a) $f(n) = \frac{1}{n^2+1}$.
   (b) $f(n) = \frac{1}{n^2}$.

9. Define f from $\mathbb{Z}$ to $\mathbb{Q}$ by
   $$n \mapsto \frac{3n+1}{2n-1}.$$

10. Define f from $\mathbb{Q}$ to $\mathbb{Q}$ by
    $$r \mapsto \frac{r-1}{3r+2}.$$

11. Define f from $\mathbb{Q}$ to $\mathbb{Q}$ by
    (a) $f(\frac{m}{n}) = n$.
    (b) $f(\frac{m}{n}) = m$.
    (c) $f(\frac{m}{n}) = \begin{cases} \frac{n}{m} & \text{if } m \neq 0, \\ 0 & \text{if } m = 0. \end{cases}$

12. Define f from $\mathbb{Q}$ to $\mathbb{Q}$ by
    (a) $f(\frac{m}{n}) = m - n$.
    (b) $f(\frac{m}{n}) = \frac{m-n}{n}$.
    (c) $f(\frac{m}{n}) = \frac{m+1}{n}$.

13.* Define f from $\mathbb{Z}_5$ to $\mathbb{Z}_{10}$ by

(a) $[a]_5 \mapsto [a]_{10}$.

(b) $[a]_5 \mapsto [2a]_{10}$.

14.* Define f from $\mathbb{Z}_{10}$ to $\mathbb{Z}_5$ by

(a) $[a]_{10} \mapsto [a]_5$.

(b) $[a]_{10} \mapsto [\frac{a}{2}]_5$.

15. Tamera's boss promised that employee salaries would be a function of how hard the employees worked. After a year of putting in far more hours than her coworkers, Tamera found out that she and her coworkers all made the same salary. Did her boss keep his promise? Explain.

16. For the purposes of a large medical study, height is measured in inches, and weight is measured in pounds, each rounded to the nearest integer. Would it be reasonable for this study to conclude that weight is a function of height? Explain.

In Exercises 17 through 26, specify the domain and range of the given function. You need not prove your assertions.

17. The function $f : \{-3, -2, \ldots, 3\} \longrightarrow \{-10, -9, \ldots, 10\}$ defined by $n \mapsto n^2$.

18. The function $f : \{-4, -2, \ldots, 4\} \longrightarrow \{-3, -2, \ldots, 3\}$ defined by $n \mapsto \frac{n}{2}$.

19. The function $f : \{0, 1, \ldots, 4\} \longrightarrow \{0, 1, \ldots, 25\}$ defined by $n \mapsto 2^n$.

20. The function $f : \{0, 1, \ldots, 4\} \longrightarrow \{0, 1, \ldots, 25\}$ defined by $n \mapsto n!$.

21. The function $f : \mathbb{R} \longrightarrow \mathbb{R}$ defined by $x \mapsto x^2 - 1$.

22. The function $f : (\mathbb{R} \setminus \{0\}) \longrightarrow \mathbb{R}$ defined by $f(x) = \frac{1}{x^2}$.

23. $f(x) = \sqrt{x-1}$.

24. $f(x) = \sqrt{x} - 1$.

25. $f(x) = \frac{1}{x+1}$.

26. $f(x) = \frac{1}{x^2+1}$.

In Exercises 27 through 32, prove that the range is the stated set.

27. The function $f : \{0, 1, \ldots, 6\} \longrightarrow \mathbb{Z}$ defined by $n \mapsto (n-2)^2$ has range $\{0, 1, 4, 9, 16\}$.

28. The function $f : \{0, 1, \ldots, 6\} \longrightarrow \mathbb{Z}$ defined by $n \mapsto |4 - n|$ has range $\{0, 1, 2, 3, 4\}$.

29. The function $f : [0, 2] \longrightarrow \mathbb{R}$ defined by $f(x) = 3x - 2$ has range $[-2, 4]$.

30. The function $f : [1, 4) \longrightarrow \mathbb{R}$ defined by $f(x) = 3 - 2x$ has range $(-5, 1]$.

31. The function $f : [0, 2] \longrightarrow \mathbb{R}$ defined by $f(x) = x^2$ has range $[0, 4]$.

32. The function $f : [-8, 0] \longrightarrow \mathbb{R}$ defined by $f(x) = x^{\frac{1}{3}}$ has range $[-2, 0]$.

In Exercises 33 through 38, for the given functions f and g, find $g \circ f$.

33. Let $f : \mathbb{N} \longrightarrow \mathbb{N}$ and $g : \mathbb{N} \longrightarrow \mathbb{N}$ be the functions defined by $f(n) = n!$ and $g(n) = n^2$, respectively.

34. Let $f : \mathbb{N} \longrightarrow \mathbb{Z}^+$ and $g : \mathbb{Z}^+ \longrightarrow \mathbb{Z}^+$ be the functions defined by $f(n) = 2^n$ and $g(n) = \binom{n+1}{2}$, respectively.

35. Let $f(x) = 1 + 3x$ and $g(x) = 1 - 3x$.

36. Let $f(x) = 1 - x^2$ and $g(x) = 1 + x^2$.

37. Let $f : \mathbb{R} \setminus \{0\} \longrightarrow \mathbb{R}^+$ and $g : \mathbb{R}^+ \longrightarrow [0, \infty)$ be the functions defined by $f(x) = \frac{1}{x^2}$ and $g(x) = \sqrt{x}$, respectively.

38. Let $f : \mathbb{R} \longrightarrow \mathbb{R}^+$ and $g : \mathbb{R}^+ \longrightarrow [0, \infty)$ be the functions defined by $f(x) = x^2 + 1$ and $g(x) = x^e$, respectively.

39. Since a bit can take the value 0 or 1, an operation on a bit is a function from $\{0, 1\}$ to itself. Consider only functions $f : \{0, 1\} \longrightarrow \{0, 1\}$.

    (a) Are $x \mapsto 1 - x$ and $x \mapsto 1 - x^2$ the same function? Explain.

    (b) How many different functions are there from $\{0, 1\}$ to itself?

40. Are the functions $f(x) = \frac{x^2 - 4}{x - 2}$ and $g(x) = x + 2$ the same function? Explain.

In Exercises 41 through 48, use the following definition of the **composite of two relations**.

Definition 5.16.   Given relations R from X to Y, and S from W to Z, such that

$$\{y \in Y : \exists x \in X, x \ R \ y\} \subseteq \{w \in W : \exists z \in Z, w \ R \ z\},$$
$$(5.2)$$

their **composite**, $S \circ R$, is the relation from X to Z defined by

$$x \ (S \circ R) \ z \quad \text{if and only if} \quad \exists y \in Y \text{ such that}$$
$$x \ R \ y \text{ and } y \ S \ z.$$

Note that condition (5.2) is satisfied if $Y \subseteq W$.

41. Let R be the "is the father of" relation from Example 5.5 in Section 5.1. Determine $R \circ R$.

42. Let R be the "is a proper divisor of" relation on $\mathbb{Z}^+$. That is, $a\,R\,b$ if and only if $a \mid b$ and $\frac{b}{a} \neq 1$. Determine $R \circ R$.

43. Transitivity. Let R be a relation on a set X.

    (a) Show: R is transitive if and only if $R \circ R \subseteq R$.

    (b) Show: If R is reflexive and transitive, then $R \circ R = R$.

    Conclude that any partial order relation composed with itself is itself. This is true of $\leq$ on $\mathbb{R}$, $\subseteq$ on any collection of sets, and $\mid$ on $\mathbb{Z}^+$.

44. Associativity. Given relations R from X to Y, S from Y to Z, and T from Z to W, prove that $(T \circ S) \circ R = T \circ (S \circ R)$.

45. RapItUp is a music company that distributes CDs and specializes in rap music. For each album that is sells, RapItUp keeps track of both the music company and the artist that produced the album. Figure 5.17 shows two tables from the database for RapItUp.

    (a) Does M.C. Escher have a record deal with Bald Boy Records?

    (b) Compose the "Contracts" table with the "Projects" table to obtain a new "Record Deals" table.

    (c) For which music companies does MandM have record deals?

46. CheapTripTics is a Web-based travel referral service that allows customers to compare the travel packages offered by various travel agents. It keeps track of both the packages offered and the packages purchased in a database. The relevant tables are shown in Figure 5.18.

    (a) Does Tony Nistler have an agreement with the Riding High travel agency?

    (b) Compose the "Agreements" table with the "Purchases" table to obtain a new "Buys From" table.

    (c) From which agents is Steve Terry buying a travel package?

Album	Music company
Slim Pickins	Aristotle Records
Half Full of It	Bald Boy Records
Can't Make This	Aristotle Records
8 Kilometer	Bald Boy Records

Contracts

Artist	Album
MandM	Slim Pickins
Fifty Percent	Half Full of It
MandM	8 Kilometer
M.C. Escher	Can't Make This

Projects

FIGURE 5.17
Database for RapItUp.

Package	Travel agent
Paris Passions	Fly by Night
Rome Ruins	Jump Ship
London Laughs	Fly by Night
New York Nights	Riding High

Agreements

Traveler	Package
Steve Terry	Paris Passions
Steve Terry	London Laughs
Tony Nistler	New York Nights

Purchases

FIGURE 5.18
Database for CheapTripTics.

47. SoftJobs is the software company described in Exercise 19 from Section 5.1. Using the tables they have in their database, SoftJobs now wishes to determine which client each programmer "works for."

   (a) Compose the "Contracts" table with the "Work Assignments" table to obtain a new "Works For" table.

   (b) For what clients is Martha Lang working?

48. HardSell is the mail-order hardware company described in Exercise 20 from Section 5.1. Using the tables they have in their database, SoftJobs now wishes to determine which customer(s) each vendor "sells to."

   (a) Compose the "Orders" table with the "Suppliers" table to obtain a new "Sells To" table.

   (b) To what customers does Stanley sell parts?

In Exercises 49 through 54, a relation is given on $\mathbb{R}$.

(a) Draw the graph of the relation.

(b) Use the Vertical Line Test to decide whether the relation is a function and, if so, to find its domain.

(c) If the relation is a function, then use the Horizontal Line Range Test to find its range.

49. $x^2 + y = 1$.   50. $x^2 + y = 0$.

51. $x^2 + y^2 = 1$.   52. $x^2 + y^2 = 0$.

53. $x^2 - y^2 = 1$.   54. $x^2 - y^2 = 0$.

55. For a relation with a finite domain and finite codomain, state the matrix representation version of the Vertical Line Test.

56. For a function with a finite domain and finite codomain, state the matrix representation version of the Horizontal Line Range Test.

In Exercises 57 through 60, let f, g, and h be functions from $\mathbb{R}$ to $\mathbb{R}$. Recall the definitions of the operations $+$ and $\cdot$ on functions given in Definition 1.16 of Section 1.3.

57. (a) Show: $(f + g) \circ h = f \circ h + g \circ h$.

(b)* Give an example for which $f \circ (g + h) \neq f \circ g + f \circ h$.

58. (a) Show: $(f \cdot g) \circ h = (f \circ h) \cdot (g \circ h)$.

(b)* Give an example for which $f \circ (g \cdot h) \neq (f \circ g) \cdot (f \circ h)$.

59. Let $c \in \mathbb{R}$.

(a) Show: $c(f \circ g) = (cf) \circ g$.

(b)* Give an example for which $f \circ (cg)$ is not the same as the functions in part (a).

60. The sum of two subsets A and B of $\mathbb{R}$ is defined by

$$A + B = \{c \ : \ c = a + b \text{ for some } a \in A \text{ and } b \in B\}.$$

(a) Show: $\text{range}(f + g) \subseteq \text{range}(f) + \text{range}(g)$.

(b)* Give an example for which $\text{range}(f + g) \neq \text{range}(f) + \text{range}(g)$.

61. Is it possible that $f \circ f = f$ besides the function $f(x) = x$ or the function $f(x) = -x$?

62. If $f \circ g = g \circ f$, must it be true that $f = g$?

5.4 Special functions

Definition 5.13 tells us that a function is a relation with the property that each input yields a unique output. In this section, we consider additional important properties a function might have.

Definition 5.17. Given a set X, the **identity function** on X, denoted id_X, is the function $\text{id}_X : X \longrightarrow X$ given by the formula

$$\forall x \in X, \quad \text{id}_X(x) = x.$$

Under the operation of composition, the identity function acts just as the multiplicative identity 1 does under multiplication.

Lemma 5.8.   Let $f : X \longrightarrow Y$ be any function. Then

$$f \circ \text{id}_X = f \quad \text{and} \quad \text{id}_Y \circ f = f.$$

The straightforward proof of Lemma 5.8 is left for the exercises.

5.4.1　One-to-one and onto functions

A function, by definition, cannot send one input value to two or more output values. However, two or more input values may be sent to one common output value. That feature is pictured in the left-hand arrow diagram in Figure 5.19. Also, there may be elements of the codomain that do not occur as output values. This is pictured in the right-hand arrow diagram in Figure 5.19. We now consider functions that avoid one or more of the features represented in Figure 5.19.

Definition 5.18.　Let a function $f : X \longrightarrow Y$ be given.

(a) We say that f is **one-to-one**, or **injective**, if

$$\forall\, x_1, x_2 \in X, \quad \text{if}\ \ f(x_1) = f(x_2)\ \ \text{then}\ \ x_1 = x_2.$$

That is, each element of Y is the image of *at most* one element of X.

(b) We say that f is **onto**, or **surjective**, if range$(f) = Y$. That is, each element of Y is the image of *at least* one element of X.

(c) We say that f is **bijective** if f is both one-to-one and onto. That is, each element of Y is the image of *exactly* one element of X. A bijective function is said to be a **bijection**, or a **one-to-one correspondence**.

Remark 5.1.

(a) We prove that a function $f : X \longrightarrow Y$ is one-to-one by proving the if-then statement in Definition 5.18(a). To prove that f is not one-to-one, we provide an example of $x_1, x_2 \in X$ with $f(x_1) = f(x_2)$ and $x_1 \neq x_2$.

(b) To prove that $f : X \longrightarrow Y$ is onto, since range$(f) \subseteq Y$, it suffices to prove $Y \subseteq$ range(f). We can prove that f is not onto by providing an example of $y \in Y$ such that $y \notin$ range(f).

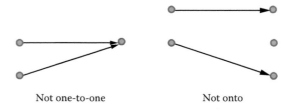

Not one-to-one　　　　　　Not onto

FIGURE 5.19
Functions without a special property.

(c) We prove that $f : X \longrightarrow Y$ is a bijection by both establishing that f is one-to-one and establishing that f is onto. To prove that f is not a bijection, it suffices to show that either f is not one-to-one or f is not onto.

As is reflected in Figure 5.19, an understanding of the properties introduced in Definition 5.18 can be gained by considering an example of a function without those properties.

Example 5.43. The function $f : \{0, 1, 2, 3, 4\} \longrightarrow \{1, 2, 3, 4, 5, 6\}$ defined by $f(x) = \binom{4}{x}$ that we considered in Example 5.34 is

(a) Not one-to-one,
  since $f(1) = 4 = f(3)$. (Of course, $1 \neq 3$.)

(b) Not onto,
  since range$(f) = \{1, 4, 6\} \neq \{1, 2, 3, 4, 5, 6\}$. (For example, $2 \notin$ range(f).)

In our next example, we see functions that do exhibit the properties in Definition 5.18.

Example 5.44. Let d be any positive integer.

(a) Show that the "multiplication by d" function $f : \mathbb{Z} \longrightarrow \mathbb{Z}$ defined by

$$\forall\, n \in \mathbb{Z}, \quad f(n) = dn$$

is one-to-one.

Solution. Suppose that $n_1, n_2 \in \mathbb{Z}$ and $f(n_1) = f(n_2)$. That is, $dn_1 = dn_2$. Canceling d from both sides gives $n_1 = n_2$. Hence,

$$\forall\, n_1, n_2 \in \mathbb{Z}, \quad \text{if } f(n_1) = f(n_2) \text{ then } n_1 = n_2.$$

That is, f is one-to-one. □

(b) Show that the "mod d" function $f : \mathbb{Z} \longrightarrow \{0, 1, \ldots, d - 1\}$ defined by

$$\forall\, n \in \mathbb{Z}, \quad f(n) = n \bmod d$$

is onto and not one-to-one.

Solution. (Onto) Suppose $r \in \{0, 1, \ldots, d - 1\}$. We seek an integer n such that $f(n) = r$. That is, we want $n \bmod d = r$. Pick $n = r$.

Since $0 \le r < d$, by definition of mod, we have $r \bmod d = r$. That is, $f(r) = r$. So, every $r \in \{0, 1, \dots, d-1\}$ is the image of some $n \in \mathbb{Z}$. Consequently, range$(f) = \{0, 1, \dots, d-1\}$, the codomain of f.

(Not one-to-one) Observe that $f(d) = 0 = f(0)$, but $d \ne 0$. $\qquad\qquad\square$

Hash functions. One method used for storing and retrieving information in a computer, such as in a database, is known as **hashing**. It employs a function, called a **hash function**, that determines the storage location of a particular item of data on the basis of some identifying key for that item. For example, a library might choose to keep track of the books that are checked out by using each book's International Standard Book Number (ISBN) for its key. When identification numbers are used for the keys, a common choice for the hashing function is a "mod d" function from Example 5.44(b), for a particular choice of d. For example, the library might use the function

$$h(n) = n \bmod 10000,$$

where n is the first 9 digits of the ISBN

$$d_1 - d_2 d_3 d_4 d_5 d_6 d_7 d_8 d_9 - d_{10},$$

regarded as a base-ten number $d_1 d_2 d_3 d_4 d_5 d_6 d_7 d_8 d_9$. (We are ignoring here the last digit, which we saw in Section 3.2 is a check digit.) Consequently, in this case, when a book with identification key n is checked out, it would be recorded in row $h(n)$ of a table with $d = 10{,}000$ rows, as shown in Table 5.6. We call such a table a **hash table**.

One major issue involved in choosing a hash function is dealing with **collisions**. These arise when two keys n_1 and n_2 hash to the same storage location, that is $h(n_1) = h(n_2)$. This is precisely the phenomenon that is avoided if h is one-to-one. However, storage considerations may not allow such a choice. Alternatively, if an item of data with key n_1 has already been

TABLE 5.6
Hash table for books checked out

Row	ISBN	Title	Borrower
⋮			
5023	0-39515023-X	*Curious George*	Megan Johnson
⋮			
5340	0-59035340-3	*Harry Potter*	Charles Brower
⋮			

entered and $h(n_2) = h(n_1)$, then a common choice is to instead hash the item with key n_2 to the next available location in the hash table. Of course, the size of the codomain of h, which is the number of rows in the hash table, must exceed the number of expected inputs. Otherwise, there would eventually be a collision that could not be resolved.

Another important feature of the hash function is that it should be onto. Otherwise, there is wasted space in the hash table that would never be filled. Of course, there will also be unused space, in general, when the number of stored records is smaller than the table's capacity. Since we can never exceed the capacity of a hash table, finding an appropriate balance between minimizing collisions and conserving space influences the choice of a good hash function.

The value of a hash table comes from the ease with which a lookup of an entry in the table can be implemented. For example, suppose a librarian using Table 5.6 to record checked-out books wants to determine whether the book *Hop on Pop* by Dr. Seuss is checked out. Since the book's ISBN 0-39480029-X hashes to row 29 of the table, if there is no entry in that row, then the book must still be available in the library.

Our next example emphasizes the fact that whether a function is one-to-one or onto is determined not only by its formula but also by its domain and codomain.

Example 5.45. Define the four functions

$$f : \mathbb{R} \longrightarrow \mathbb{R},$$
$$g_1 : \mathbb{R} \longrightarrow \mathbb{R},$$
$$g_2 : \mathbb{R} \longrightarrow [0, \infty), \text{ and}$$
$$g_3 : [0, \infty) \longrightarrow [0, \infty)$$

by the formulas $f(x) = x^3$ and $g_1(x) = g_2(x) = g_3(x) = x^2$.

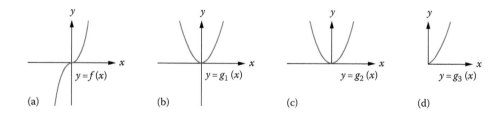

The functions $g_1, g_2,$ and g_3 are distinguished by their domains and codomains.

(a) Show that f is one-to-one.

> **Solution.** Suppose that $x_1, x_2 \in \mathbb{R}$ and $f(x_1) = f(x_2)$. That is, $x_1{}^3 = x_2{}^3$. Taking the cube

root of both sides gives $(x_1{}^3)^{\frac{1}{3}} = (x_2{}^3)^{\frac{1}{3}}$. That is, $x_1 = x_2$. Therefore,

$$\forall\, x_1, x_2 \in \mathbb{R}, \ \text{if } f(x_1) = f(x_2), \text{ then } x_1 = x_2. \ \square$$

(b) Show that g_1 is not one-to-one.

Solution. Observe that $g_1(-1) = g_1(1)$ and $-1 \neq 1$. Consequently,

$$\exists\, x_1, x_2 \in X \text{ such that } g_1(x_1)$$
$$= g_1(x_2) \text{ and } x_1 \neq x_2.$$

Namely, $x_1 = 1$ and $x_2 = -1$. $\hfill\square$

(c) Show that f is onto.

Solution. Suppose $y \in \mathbb{R}$. We seek a value $x \in \mathbb{R}$ such that $f(x) = y$. That is, we want $x^3 = y$. So, pick $x = y^{\frac{1}{3}}$. Hence, $f(x) = f(y^{\frac{1}{3}}) = (y^{\frac{1}{3}})^3 = y$. Since y is arbitrary, every element of the codomain $\mathbb{R}$ is the image of some element from the domain $\mathbb{R}$. That is, range$(f) = Y$, where $Y = \mathbb{R}$ in this case. $\hfill\square$

(d) Show that g_1 is not onto.

Solution. Let $y = -1$. Since $\forall\, x \in \mathbb{R}, \ x^2 \geq 0$, there can be no value $x \in \mathbb{R}$ such that $x^2 = -1$. That is, there is no x in the domain such that $g_1(x) = y$. Hence, $y = -1$ is not in the image of g_1. That is, range$(g_1) \neq Y$, where $Y = \mathbb{R}$ in this case. Namely, $-1 \notin$ range(g_1). $\hfill\square$

(e) Show that g_2 is onto.

Solution. Suppose $y \in [0, \infty)$. Since $y \geq 0$, we may pick $x = \sqrt{y}$. Observe that $g_2(x) = (\sqrt{y})^2 = y$. Hence, $g_2(\mathbb{R}) = [0, \infty)$. $\hfill\square$

(f) Show that f is bijective.

Solution. By parts (a) and (c), f is one-to-one and onto. $\hfill\square$

(g) Show that g_2 is not bijective.

Solution. The arguments from part (b) can also be used to show that g_2 is not one-to-one. Hence, g_2 is not bijective. $\hfill\square$

(h) Show that g_3 is bijective.

> **Solution.** The arguments from part (e) can also be used to show that g_3 is onto. So it remains to show that g_3 is one-to-one. Suppose $g_3(x_1) = g_3(x_2)$. That is, $x_1{}^2 = x_2{}^2$. Since $x_1, x_2 \geq 0$, it follows that
>
> $$x_1 = |x_1| = \sqrt{x_1{}^2} = \sqrt{x_2{}^2} = |x_2| = x_2. \qquad \square$$

The assertions in Example 5.45 can also be seen graphically. For real functions, the definitions of one-to-one and onto can be recast in the form of the following horizontal line tests. The first of these is commonly called *the* Horizontal Line Test.

Test (Horizontal Line Tests).
Let $X \subseteq \mathbb{R}$, $Y \subseteq \mathbb{R}$, and $f : X \longrightarrow Y$. Consider all horizontal lines in $\mathbb{R}^2$ of the form $y = b$ for some $b \in Y$.

(a) If every such horizontal line intersects the graph of f at *most once*, then f is one-to-one.

(b) If every such horizontal line intersects the graph of f at *least once*, then f is onto.

Certain properties held by functions f and g are retained by their composition $g \circ f$.

Theorem 5.9. (Composition Preserves One-to-one and Onto).
Let $f : X \longrightarrow Y$ and $g : Y \longrightarrow Z$ be functions.

(a) *If f and g are one-to-one, then $g \circ f$ is one-to-one.*

(b) *If f and g are onto, then $g \circ f$ is onto.*

(c) *If f and g are bijective, then $g \circ f$ is bijective.*

Proof. For part (a), suppose that f and g are one-to-one. To show that $g \circ f$ is one-to-one, suppose that $x_1, x_2 \in X$ and $(g \circ f)(x_1) = (g \circ f)(x_2)$. Let $y_1 = f(x_1)$ and $y_2 = f(x_2)$. Then

$$g(y_1) = g(f(x_1)) = (g \circ f)(x_1) = (g \circ f)(x_2)$$
$$= g(f(x_2)) = g(y_2).$$

Since g is one-to-one and $g(y_1) = g(y_2)$, it follows that $y_1 = y_2$. That is, $f(x_1) = f(x_2)$. Since f is one-to-one and $f(x_1) = f(x_2)$, it follows that $x_1 = x_2$. We conclude that $g \circ f$ is one-to-one. The proofs for parts (b) and (c) are left for the exercises. $\square$

5.4.2 Inverse functions

Since a function $f : X \longrightarrow Y$ is, in particular, a relation, Definition 5.3 from Section 5.1 defines its inverse relation f^{-1} by, for all $y \in Y$ and $x \in X$,

$$y\, f^{-1}\, x \quad \text{if and only if} \quad x\, f\, y \; (\text{i.e.,}\, f(x) = y). \qquad (5.3)$$

However, f^{-1} is not always a function.

Example 5.46. The function $y = x^2$ has inverse relation $x = y^2$. However, we saw in Example 5.40 that $x = y^2$ is not a function.

To understand the cases in which f^{-1} is a function, and to appreciate the properties f^{-1} then has, we utilize the following notion.

Definition 5.19. Two functions $f : X \longrightarrow Y$ and $g : Y \longrightarrow X$ are said to be **inverses of one another** if

$$g \circ f = \mathrm{id}_X \quad \text{and} \quad f \circ g = \mathrm{id}_Y.$$

Example 5.47.   The   functions $f : \mathbb{R} \longrightarrow \mathbb{R}$ and $g : \mathbb{R} \longrightarrow \mathbb{R}$ defined by $f(x) = x^3$ and $g(x) = x^{\frac{1}{3}}$ are inverses of one another.

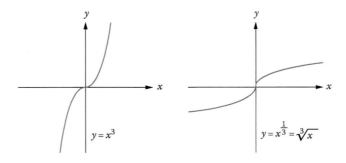

Proof. Observe that, $\forall\, x \in \mathbb{R}$,

$$(g \circ f)(x) = g(f(x)) = g(x^3) = (x^3)^{\frac{1}{3}} = x,$$

and, $\forall\, y \in \mathbb{R}$,

$$(f \circ g)(y) = f(g(y)) = f(y^{\frac{1}{3}}) = (y^{\frac{1}{3}})^3 = y. \qquad \square$$

Of course, rather than using y in the second equation, we could use the same letter, say x, in both equations.

Example 5.48. The functions $f : \mathbb{R} \longrightarrow (-1,1)$ and $g : (-1,1) \longrightarrow \mathbb{R}$ defined by $f(x) = \frac{x}{\sqrt{1+x^2}}$ and $g(x) = \frac{x}{\sqrt{1-x^2}}$ are inverses of one another.

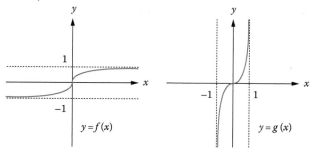

Proof. Observe that, $\forall\, x \in \mathbb{R}$,

$$g(f(x)) = \frac{\frac{x}{\sqrt{1+x^2}}}{\sqrt{1 - \left(\frac{x}{\sqrt{1+x^2}}\right)^2}} = \frac{\frac{x}{\sqrt{1+x^2}}}{\sqrt{\frac{1+x^2}{1+x^2} - \frac{x^2}{1+x^2}}}$$

$$= \frac{\frac{x}{\sqrt{1+x^2}}}{\sqrt{\frac{1}{1+x^2}}} = \frac{\frac{x}{\sqrt{1+x^2}}}{\frac{1}{\sqrt{1+x^2}}} = x,$$

and, $\forall\, x \in (-1,1)$,

$$f(g(x)) = \frac{\frac{x}{\sqrt{1-x^2}}}{\sqrt{1 + \left(\frac{x}{\sqrt{1-x^2}}\right)^2}} = \frac{\frac{x}{\sqrt{1-x^2}}}{\sqrt{\frac{1-x^2}{1-x^2} + \frac{x^2}{1-x^2}}}$$

$$= \frac{\frac{x}{\sqrt{1-x^2}}}{\sqrt{\frac{1}{1-x^2}}} = \frac{\frac{x}{\sqrt{1-x^2}}}{\frac{1}{\sqrt{1-x^2}}} = x.$$ □

In Example 5.47, the fact that $g = f^{-1}$ may be obvious from (5.3). In Example 5.48, that relationship is not so obvious. However, part (b) of the following theorem says that $g = f^{-1}$ must hold when f and g are inverses of one another.

Theorem 5.10. *Let $f : X \longrightarrow Y$ be any function.*

   (a) *If f is a bijection, then*

      (i) *f^{-1} is a function and*

      (ii) *f and f^{-1} are inverses of one another.*

   (b) *If there is a function $g : Y \longrightarrow X$ such that f and g are inverses of one another, then*

      (i) *f is a bijection and*

      (ii) *$g = f^{-1}$.*

Proof.

(a) Suppose f is a bijection.

(i) We need to show that the relation f^{-1} defined by (5.3) is a function. For each $y \in Y$, since f is a bijection, there is exactly one value $x \in X$ such that $f(x) = y$. Hence, there is a *unique* value x specified by (5.3) such that $y f^{-1} x$. So, f^{-1} is indeed a function, and we may write $f^{-1}(y) = x$.

(ii) It must be verified that $f^{-1} \circ f = \mathrm{id}_X$ and $f \circ f^{-1} = \mathrm{id}_Y$. For each $x \in X$, if we let $y = f(x)$, then (5.3) tell us that $f^{-1}(y) = x$. That is, $f^{-1}(f(x)) = x$. Thus, $f^{-1} \circ f = \mathrm{id}_X$. For each $y \in Y$, if we let $x = f^{-1}(y)$, then (5.3) tell us that $f(x) = y$. That is, $f(f^{-1}(y)) = y$. So $f \circ f^{-1} = \mathrm{id}_Y$. Hence, f and f^{-1} are inverses of one another.

(b) Suppose $g : Y \longrightarrow X$ is a function such that f and g are inverses of one another.

(i) To show that f is one-to-one, suppose $x_1, x_2 \in X$ and $f(x_1) = f(x_2)$. Evaluating g on both sides of this equation gives $g(f(x_1)) = g(f(x_2))$. So

$$x_1 = (g \circ f)(x_1) = g(f(x_1)) = g(f(x_2))$$
$$= (g \circ f)(x_2) = x_2.$$

Therefore, f is one-to-one. To show that f is onto, suppose $y \in Y$. Let $x = g(y) \in X$ and observe that

$$f(x) = f(g(y)) = (f \circ g)(y) = y.$$

Since y is the image of x, it follows that f is onto.

(ii) It must be verified that, $\forall\, y \in Y$ and $x \in X$,

$$g(y) = x \;\; (\text{i.e. } y\, g\, x) \quad \text{if and only if} \quad f(x) = y.$$

If $g(y) = x$, then $f(x) = f(g(y)) = (f \circ g)(y) = y$.
If $f(x) = y$, then $g(y) = g(f(x)) = (g \circ f)(x) = x$. □

Theorem 5.10 provides, in part (b), a useful tool for showing that a function is bijective. It also enables us to properly define the inverse of a function as a *function*, when possible.

Definition 5.20.   If a function $f : X \longrightarrow Y$ satisfies the hypotheses of either part (a) or part (b) of Theorem 5.10, then the *function* f^{-1} defined by, $\forall\, x \in X$ and $y \in Y$,

$$f^{-1}(y) = x \quad \text{if and only if} \quad f(x) = y$$

is called the **inverse function** for f.

As we observed in Section 5.1, the graphs of real functions f and f^{-1} are reflections of one another about the line $y = x$. This relationship can be seen, for example, in the graphs in Examples 5.47 and 5.48.

5.4.3 Logarithms

Let b be a real number such that $b > 1$. The exponential function

$$f : \mathbb{R} \longrightarrow \mathbb{R}^+ \quad \text{given by} \quad \forall\, x \in \mathbb{R}, \quad f(x) = b^x$$

can be seen to be a bijection from $\mathbb{R}$ to $\mathbb{R}^+$.

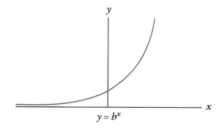

Since f satisfies the hypothesis of Theorem 5.10(a), Definition 5.20 tells us that f has an inverse function $f^{-1} : \mathbb{R}^+ \longrightarrow \mathbb{R}$. In this case, for each $y \in \mathbb{R}^+$, the function value $f^{-1}(y)$ is called the **logarithm base** b of y and is denoted $\log_b(y)$. Moreover, $\forall\, x \in \mathbb{R}$ and $y \in \mathbb{R}^+$,

$$\log_b(y) = x \quad \text{if and only if} \quad b^x = y. \tag{5.4}$$

The graph of $y = \log_b(x)$ is obtained by reflecting the graph of $y = b^x$ about the line $y = x$.

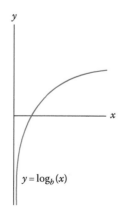

Moreover, since the logarithm base b and the exponential function base b are inverses of one another, we have, $\forall\, x \in \mathbb{R}$ and $y \in \mathbb{R}^+$,

$$\log_b(b^x) = x \quad \text{and} \quad b^{\log_b(y)} = y. \tag{5.5}$$

Definition 5.21. The **natural logarithm** function, which we denote by ln, is the function $\ln : \mathbb{R}^+ \longrightarrow \mathbb{R}$ given by $\forall\, x \in \mathbb{R}^+$, $\ln(x) = \log_e(x)$. That is, the natural logarithm function is the inverse of the natural exponential function $x \mapsto e^x$.

Example 5.49. (Computing Logarithms from Their Definition, Equation 5.4)

(a) $\log_2(8) = 3$,   since $2^3 = 8$.

(b) $\log_{10}(0.001) = -3$,   since $10^{-3} = 0.001$.

(c) $\log_9(3) = \frac{1}{2}$,   since $9^{\frac{1}{2}} = 3$.

(d) $\ln(e^2) = 2$,   since $e^2 = e^2$.

The fundamental properties of logarithms are developed in the exercises.

Exercises

1. Show that $f(x) = x^3 + 8$ is one-to-one.

2. Show that $f(x) = (x + 2)^3$ is one-to-one.

3. Show that the function $f : \mathbb{Z}^- \longrightarrow \mathbb{Z}$ defined by $n \mapsto 1 - n^2$ is one-to-one.

4. Show that the function $f : \mathbb{Z}^+ \longrightarrow \mathbb{Z}$ defined by $n \mapsto n - n^2$ is one-to-one.

5. Show that the function $f : \mathbb{N} \longrightarrow \mathbb{Z}^-$ defined by $n \mapsto n - n^2 - 1$ is not one-to-one.

6. Show that the function $f : \mathbb{Z} \longrightarrow \mathbb{N}$ defined by $n \mapsto n^2 + n$ is not one-to-one.

7. Show that the function $f : \mathbb{R}^- \longrightarrow (1, \infty)$ defined by $x \mapsto x^2 + 1$ is onto.

8. Show that the function $f : \mathbb{R}^- \longrightarrow \mathbb{R}^+$ defined by $x \mapsto (x - 1)^2 - 1$ is onto.

9. Show that the function $f : \mathbb{R} \longrightarrow \mathbb{R}^+$ defined by $x \mapsto x^2 + 1$ is not onto.

10. Show that the function $f : \mathbb{R} \longrightarrow \mathbb{R}$ defined by $x \mapsto x^2 + x$ is not onto.

11* Show that the function $f : \mathbb{R}^+ \longrightarrow \mathbb{R}^+$ defined by $x \mapsto x^2 + x$ is onto.

12. Show that the function $f : \mathbb{R}^+ \longrightarrow \mathbb{R}^-$ defined by $x \mapsto -x^2$ is onto.

13. Define $f : \mathbb{Z} \times \mathbb{Z} \longrightarrow \mathbb{Z}$ by $f(m,n) = mn$.

   (a) Show that f is onto.   (b) Show that f is not one-to-one.

14. Define $f : \mathbb{Z} \longrightarrow \mathbb{Q}$ by $f(n) = \frac{n}{2^n}$.

   (a) Show that f is not one-to-one.   (b) Show that f is not onto.

15. Let f be an arbitrary real function. Show: If f is increasing, then f is one-to-one. See Definition 1.15(b).

16. Give an example of a real function f that is increasing but not onto.

17* Define the multiplication by 2 map $f : \mathbb{Z}_6 \longrightarrow \mathbb{Z}_6$ by $[n] \mapsto [2n]$.

   (a) Show that f is well-defined.

   (b) Show that f is not one-to-one.

   (c) Use Lemma 3.30 to explain why the cancellation argument used in Example 5.44(a) does not work here.

18* Define the multiplication by 5 map $f : \mathbb{Z}_6 \longrightarrow \mathbb{Z}_6$ by $[n] \mapsto [5n]$.

   (a) Show that f is well-defined.

   (b) Show that f is bijective by computing the six function values.

   (c) Use Lemma 3.30 to explain why the cancellation argument used in Example 5.44(a) also works here to show that f is one-to-one.

19. A cash register is using a hash function h to keep track of the items purchased by a customer. For each UPC number

$$d_1 \, d_2 d_3 d_4 d_5 d_6 \, d_7 d_8 d_9 d_{10} d_{11} \, d_{12},$$

it takes n to be the base-ten value of the string $d_1 d_2 d_3 d_4 d_5 d_6 d_7 d_8 d_9 d_{10} d_{11}$ and computes $h(n) = n \bmod 625$.

   (a) To what value does *Corn Chex* 0 16000 81160 7 hash?

(b) Suppose no item hashes to 321. Give two examples of UPC numbers that cannot possibly represent a product purchased by the customer. Refer to Example 3.16 from Section 3.2 to see how to form the check digit.

20. The Internet bookseller BooksForCheap is using a hash function h to keep track of the books ordered by a customer. For each ISBN

$$d_1 - d_2d_3d_4d_5d_6 \ d_7d_8d_9 - d_{10},$$

it takes n to be the base-ten value of the string $d_1d_2d_3d_4d_5d_6d_7d_8d_9$ and computes $h(n) = n \bmod 250$.

(a) To what value does *Harry Potter and the Sorcerer's Stone* by J. K. Rowling, 0-59035340-3, hash?

(b) Suppose no book hashes to 123. Give two examples of ISBNs that cannot possibly represent a book ordered by the customer. Refer to Example 3.16 from Section 3.2 to see how to form the check digit.

21. Let X and Y be sets with $Y \neq \emptyset$. Say, $y_0 \in Y$. The function $p : X \times Y \longrightarrow X$ defined by $p(x, y) = x$ is called a projection. The function $i : X \longrightarrow X \times Y$ defined by $i(x) = (x, y_0)$ is called an **injection**.

(a) Show that p is onto.

(b) Show that i is one-to-one.

(c) Show $p \circ i = \mathrm{id}_X$.

(d) Find $i \circ p$.

22. Let P be the set of all real polynomial functions. Define $f : P \longrightarrow \mathbb{R}$ by $f(p) = p(0)$, and define $g : \mathbb{R} \longrightarrow P$ by taking $g(c)$ to be the constant function with constant value c.

(a) Show f is onto.

(b) Show g is one-to-one.

(c) Find $f \circ g$.

(d) Find $g \circ f$.

23. Let $A \subseteq X$. The function $i : A \longrightarrow X$ defined by $i(a) = a$ is called an **inclusion function**. Show that i is one-to-one.

24.* Let R be an equivalence relation on a set X. Let $Y = \{[x]_R : x \in X\}$, the set of equivalence classes under R. The function $p : X \longrightarrow Y$ defined by $p(x) = [x]_R$ is called a **quotient map**. Show that p is onto.

25. Define $f : \mathcal{P}(\mathbb{R}) \longrightarrow \mathcal{P}([0, 1])$ by $A \mapsto A \cap [0, 1]$. Show f is onto.

26. Define functions $f, g : \mathcal{P}(\mathbb{R}) \longrightarrow \mathcal{P}(\mathbb{R}^2)$ by $f(A) = A \times [0, 1]$ and $g(A) = A^2$. Are f and g onto? Are they one-to-one?

27. Define functions $f, g : \mathcal{P}(\mathbb{R})^2 \longrightarrow \mathcal{P}(\mathbb{R})$ by $f((A, B)) = A \cap B$ and $g((A, B)) = A \cup B$. Are f and g onto? Are they one-to-one?

28. Define functions $f, g : \mathcal{P}(\mathbb{R}) \longrightarrow \mathcal{P}(\mathbb{R}) \times \mathcal{P}(\mathbb{R})$ by $f(A) = (A, A)$ and $g(A) = (A, \emptyset)$. Are f and g onto? Are they one-to-one?

29. A couple has reserved a reception hall for their wedding and wishes to specify a seating chart for the guests.

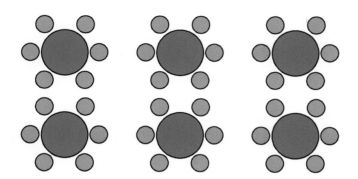

In fact, a seating chart is a function from the set of guests to the set of seats in the reception hall. Which one of the following must this function be: one-to-one, onto, or bijective? Explain.

30. A contestant throws darts at the pictured board until each one of the nine regions contains a dart.

Assuming that the contestant never misses the board, a record of a contestant's throws can be regarded as a function from the set of her or his throws to the set of regions on the dart board. Which one of the following must this function be: one-to-one, onto, or bijective? Explain.

31. Prove Lemma 5.8.

32. Show, for any set X, that id_X is a bijection.

33. Show that the function $f(x) = x^3 + 8$ is a bijection. See Exercise 1.

34. Show that the function $f : \mathbb{R}^+ \longrightarrow \mathbb{R}^-$ defined by $f(x) = -x^2$ is a bijection. See Exercise 12.

35. Show that the function $f : \{-2, -1, 0, 1, 2\} \longrightarrow \{0, 2, 4, 6, 8\}$ defined by $k \mapsto 2k + 4$ is a bijection.

36. Let $n \in \mathbb{Z}^+$. Show that the function

$$f : \{m \; : \; m \in \mathbb{Z} \text{ and } -n \leq m \leq 2n\}$$
$$\longrightarrow \{m \; : \; m \in \mathbb{Z} \text{ and } 0 \leq m \leq 3n\}$$

defined by $k \mapsto 2n - k$ is a bijection.

37. Let $n \in \mathbb{Z}^+$. Show that the function

$$f : \{m \; : \; m \in \mathbb{Z} \text{ and } 1 \leq m \leq 2n + 1\}$$
$$\longrightarrow \{m \; : \; m \in \mathbb{Z} \text{ and } -n \leq m \leq n\}$$

defined by $f(k) = k - 1 - n$ is a bijection.

38.* Let $d \in \mathbb{Z}$ with $d > 1$. Show that the function $f : \{0, 1, \ldots, d - 1\} \longrightarrow \mathbb{Z}_d$ given by $f(r) = [r]_d$ is a bijection.

39.* Let $m, n \in \mathbb{Z}$ with $n > 1$. Define functions $f, g : \mathbb{Z}_n \longrightarrow \mathbb{Z}_n$ by $f([k]) = [mk]$ and $g([k]) = [m + k]$.

   (a) Show that f and g are well-defined.

   (b) Show f is a bijection if and only if $\gcd(m, n) = 1$.

   (c) Under what conditions is g a bijection?

40.* Let $m, n \in \mathbb{Z}^+$ with $n > 1$. Define functions $f : \mathbb{Z}_{mn} \longrightarrow \mathbb{Z}_n$ and $g : \mathbb{Z}_n \longrightarrow \mathbb{Z}_{mn}$ by $f([k]_{mn}) = [k]_n$ and $g([k]_n) = [mk]_{mn}$.

   (a) Show that f and g are well-defined.

   (b) Compute both compositions $f \circ g$ and $g \circ f$.

   (c) Show that g is one-to-one.

   (d) Under what conditions is f one-to-one?

41. For a function with a finite domain and finite codomain, state the matrix representation version of part (a) of the Horizontal Line Test.

42. For a function with a finite domain and finite codomain, state the matrix representation version of part (b) of the Horizontal Line Test.

43. Prove parts (b) and (c) of Theorem 5.9.

44. Let $f : X \longrightarrow X$ be a bijection. Show that $f \circ f \circ f$ is also a bijection.

45. Let X be an arbitrary set and $f : X \longrightarrow X$ an arbitrary function. Show: f is a symmetric relation on X if and only if $f \circ f = \text{id}_X$.

46. Let X be an arbitrary set and $f : X \longrightarrow X$ an arbitrary function. Show: f is a reflexive relation on X if and only if $f = \text{id}_X$.

In Exercises 47 through 50, let $f : X \longrightarrow Y$ and $g : Y \longrightarrow Z$ be functions.

47* Suppose $g \circ f$ is one-to-one.

   (a) Show: f is one-to-one.

   (b) Give an example in which g is not one-to-one.

48* Suppose $g \circ f$ is onto.

   (a) Show: g is onto.

   (b) Give an example in which f is not onto.

49* Suppose $g \circ f$ is a bijection.

   (a) Show: f is one-to-one.

   (b) Show: g is onto.

   (c) Give an example in which neither f nor g is a bijection.

50* Suppose $Z = X$. Give an example in which $g \circ f = \text{id}_X$ but f and g are not inverses of one another.

51. Suppose f and g are real functions.

   (a) Show: If f and g are increasing, then $g \circ f$ is increasing.

   (b) Give an example for which $g \circ f$ is increasing, but neither f nor g is increasing.

52. Suppose f and g are real functions.

   (a) Show: If f or g is constant, then $g \circ f$ is constant.

   (b) Give an example for which $g \circ f$ is constant, but neither f nor g is constant.

In Exercises 53 through 58, show that the given functions f and g are inverses of one another.

53. Define $f, g : \mathbb{R} \longrightarrow \mathbb{R}$ by $f(x) = 2x + 5$ and $g(x) = \frac{x-5}{2}$.

54. $f(x) = \sqrt[3]{x} + 1$ and $g(x) = (x - 1)^3$.

55. $f(x) = 4 - 2x$ and $g(x) = 2 - \frac{1}{2}x$.

56. Define $f, g : \mathbb{R} \longrightarrow \mathbb{R}$ by $f(x) = \frac{x}{3} + 1$ and $g(x) = 3x - 3$.

57. $f : \{2, 3, 4, 5\} \longrightarrow \{1, 3, 6, 10\}$ defined by $f(n) = \binom{n}{2}$ and
$g : \{1, 3, 6, 10\} \longrightarrow \{2, 3, 4, 5\}$ defined by $g(n) = \frac{1 + \sqrt{8n+1}}{2}$.

58. $f : \{1, 2, 3\} \longrightarrow \{2, 4, 8\}$ defined by $f(n) = 2^n$ and
$g : \{2, 4, 8\} \longrightarrow \{1, 2, 3\}$ defined by $g(n) = \lceil \frac{n}{3} \rceil$.

59. Define $f : \mathbb{Q}^+ \longrightarrow \mathbb{Q}^+$ by $r \mapsto \frac{1}{r}$. Show that f is its own inverse.

60. Define $f : \mathbb{Z} \longrightarrow \mathbb{Z}$ by $n \mapsto -n$. Show that f is its own inverse.

61. Find the inverse of $f(x) = 4x^3 + 1$.

62. Find the inverse of $f(x) = \frac{2x+3}{5}$.

63. The listings of phone numbers in a phone book can be regarded as a function from the set of people with phones to the set of possible phone numbers.

Name	Phone number
Blair, Tina	555-3148
Jennings, Robert	555-6301
Tillman, Paul	555-4500
Walsh, Carol	555-3992

Caller ID is a service that inverts this function and allows one to determine the identity of a caller based upon his or her phone number. Specify the inverse listing of the small phone book above, while putting the phone numbers in numerical order. Must the phone listing function be one-to-one? onto?

64. The appearance of terms in a book can be regarded as a relation from the set of pages (each specified by its number) in the book to the set of terms included in the book.

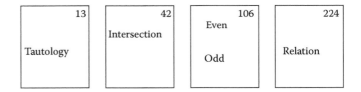

An index of terms in the back of the book is the inverse of this relation and allows one to easily find the page on which a term is defined. Form the index for the portion of the book above, while putting the terms in alphabetical order. Must either relation be a function?

65. Show, for any bijection $f : X \longrightarrow Y$, that $f^{-1} : Y \longrightarrow X$ is a bijection. Hint: Use Theorem 5.10.

66. Show that, in Theorem 5.10(b), the conclusions g is a bijection and $f = g^{-1}$ may be added.

67. Given functions $f_1 : X_1 \longrightarrow Y_1$ and $f_2 : X_2 \longrightarrow Y_2$, define a function, denoted $f_1 \times f_2$, such that $f_1 \times f_2 : X_1 \times X_2 \longrightarrow Y_1 \times Y_2$ and $(f_1 \times f_2)((x_1, x_2)) = (f_1(x_1), f_2(x_2))$.

(a) Show: If f_1 and f_2 are one-to-one, then $f_1 \times f_2$ is one-to-one.

(b) Show: If f_1 and f_2 are onto, then $f_1 \times f_2$ is onto.

(c) Show: If f_1 and f_2 are bijective, then $f_1 \times f_2$ is bijective.

68. Let X_1 and X_2 be disjoint sets. Given functions $f_1 : X_1 \longrightarrow Y$ and $f_2 : X_2 \longrightarrow Y$, define a function $f : X_1 \cup X_2 \longrightarrow Y$ by $f(x) = \begin{cases} f_1(x) & \text{if } x \in X_1 \\ f_2(x) & \text{if } x \in X_2 \end{cases}$.

(a) Prove or disprove: If f_1 and f_2 are one-to-one, then f is one-to-one.

(b) Prove or disprove: If f_1 or f_2 is onto, then f is onto.

(c) Under what conditions would f be well-defined if X_1 and X_2 were not disjoint?

69. Define $f : \mathbb{N} \times \mathbb{N} \longrightarrow \mathbb{N}$ by $f((m,n)) = n + \sum_{i=1}^{m+n} i$.

(a) Show f is a bijection. Hint: Draw a picture, and label each point $(m,n) \in \mathbb{N} \times \mathbb{N}$ with its function value $f((m,n))$.

(b) Show that the map $g : \mathbb{Z}^+ \longrightarrow \mathbb{N}$ defined by $g(n) = n - 1$ is a bijection.

(c) Show that the map $(g \times g) : \mathbb{Z}^+ \times \mathbb{Z}^+ \longrightarrow \mathbb{N} \times \mathbb{N}$ defined by $(g \times g)(m,n) = (g(m), g(n))$ is a bijection.

(d) Show that the map $(g^{-1} \circ f \circ (g \times g)) : \mathbb{Z}^+ \times \mathbb{Z}^+ \longrightarrow \mathbb{Z}^+$ is a bijection.

70. Let $P = \{(a,b) : a \in \mathbb{Z}, b \in \mathbb{Z}^+, \text{ and } \gcd(a,b) = 1\} \subset \mathbb{Z} \times \mathbb{Z}^+$, and define $f : \mathbb{Q} \longrightarrow P$ by $f(r) = (a,b)$, where $\frac{a}{b}$ is the unique expression of r in lowest terms as guaranteed by Theorem 3.22. Show that f is a bijection.

71. Find the following values:

(a) $\log_2(\frac{1}{4})$.

(b) $\log_3(81)$.

(c) $\log_{16}(2)$.

(d) $\ln(\frac{1}{e})$.

72. Find the following values:

(a) $\log_3(\frac{1}{9})$.

(b) $\log_{10}(100)$.

(c) $\log_4(8)$.

(d) $\ln(\sqrt{e})$.

73. If a population is growing continuously and doubling every year, then the function $p(t)$ that measures the size of the population after t years from some start time has the form $p(t) = a2^t$, where a is the population at time 0. Observe, for example, that $p(6) = 2p(5)$. In general, the population indeed doubles from 1 year to the next. Use Equation 5.5 from the end of this section as a tool to determine the tripling time for the population.

74. Suppose that the amount of a radioactive substance in a rock at time t is given by the function $r(t) = a(\frac{1}{5})^t$, for some constant $a = r(0)$. The half-life of the radioactive

substance is defined to be the length of time it takes the amount of radioactive substance to decrease by 50%. Use Equation 5.5 from the end of this section as a tool to determine the half-life of the substance.

75. Let $a \in \mathbb{R}$ and $b, c, y, z \in \mathbb{R}^+$ with $b, c > 1$. Prove the following:

   (a) $\log_b(1) = 0$.

   (b) $\log_b(yz) = \log_b(y) + \log_b(z)$.

   (c) $\log_b(y^a) = a \log_b(y)$.

76. Let $a \in \mathbb{R}$ and $b, c, y, z \in \mathbb{R}^+$ with $b, c > 1$. Prove the following:

   (a) $\log_b(b) = 1$.

   (b) $\log_b(\frac{y}{z}) = \log_b(y) - \log_b(z)$.

   (c) $\log_c(y) = \log_c(b) \cdot \log_b(y)$.

77. Use the Fundamental Theorem of Arithmetic to argue that the exponential function $f : \mathbb{Z}^+ \longrightarrow \mathbb{Z}^+$ given by $f(n) = 2^n$ is one-to-one.

78. Show that the function $f : \mathbb{Z} \longrightarrow \mathbb{R}^2$ defined by $f(n) = (\cos n, \sin n)$ is one-to-one. Note that angles are measured in radians here.

5.5   General set constructions

Our focus in the previous sections on a function's behavior relative to individual values shifts to a consideration of its behavior relative to sets. Such considerations play a prominent role in higher mathematics courses, such as advanced calculus and real analysis. We shall study arbitrary unions and intersections of sets and generalize earlier special cases of these operations, such as finite unions and intersections. Applications to database queries are also discussed.

5.5.1   Images and inverse images

Definition 5.22.   Let a function $f : X \longrightarrow Y$ be given.

   (a) Given a subset S of X, the **image of S under f**, denoted $f(S)$, is the set given by

$$f(S) = \{t \ : \ t \in Y \text{ and } f(s) = t \text{ for some } s \in S\}.$$

   (b) Given a subset T of Y, the **inverse image of T under f**, denoted $f^{-1}(T)$, is the set given by

$$f^{-1}(T) = \{s \ : \ s \in X \text{ and } f(s) = t \text{ for some } t \in T\}.$$

To develop some comfort with the notions in Definition 5.22, we start by revisiting a function we studied earlier, in Example 5.34 of Section 5.3.

Example 5.50. Consider the function $f : \{0, 1, 2, 3, 4\} \longrightarrow \{1, 2, 3, 4, 5, 6\}$ defined by $f(x) = \binom{4}{x}$.

(a) $f(\{1, 2\}) = \{4, 6\}$,
   since $f(1) = 4$ and $f(2) = 6$.

(b) $f(\{0, 1, 4\}) = \{1, 4\}$,
   since $f(0) = f(4) = 1$ and $f(1) = 4$.

(c) $f^{-1}(\{4, 6\}) = \{1, 2, 3\}$,
   since $f(1) = f(3) = 4$, $f(2) = 6$, and for no other value x outside of $\{1, 2, 3\}$ is $f(x) \in \{4, 6\}$.

(d) $f^{-1}(\{4, 5, 6\}) = \{1, 2, 3\}$,
   by part (c) together with the fact that there is no value x for which $f(x) = 5$.

(e) $f^{-1}(\{1, 4, 6\}) = \{0, 1, 2, 3, 4\}$,
   since Example 5.34 tells us that $\{1, 4, 6\}$ is the range of f. That is, the entire domain of f is mapped to the set $\{1, 4, 6\}$.

Remark 5.2. Let a function $f : X \longrightarrow Y$, an element $x \in X$, a subset $S \subseteq X$, and a subset $T \subseteq Y$ be given.

(a) The term *image* has been defined in three contexts. We say

   (i) $f(x)$ is the image of the *element x*,

   (ii) $f(X)$ is the image of the *function f*, and

   (iii) $f(S)$ is the image of the *subset S*.

   The intended type of image being considered should always be clear in context. Of course, all three types are closely related. For example, the image of the function $f : X \longrightarrow Y$ is the image of the subset X.

(b) The image of S is the image of the **restricted function**

   $$f_S : S \longrightarrow Y \quad \text{defined by } \forall\, x \in S, \ f_S(x) = f(x).$$

   That is, $f(S)$ is the range of f_S. Consequently, images of subsets are computed using the techniques introduced in Section 5.3 to compute ranges.

(c) The inverse image operator f^{-1} should not be confused with an inverse function (or relation). The inverse image operator f^{-1} is defined for any function, whereas the inverse function f^{-1} exists only for some functions. If a given function $f : X \longrightarrow Y$ has an inverse function $f^{-1} : Y \longrightarrow X$, then there are two potential interpretations of the notation $f^{-1}(T)$. This might be thought of as the inverse image of T under f or as the image of T under f^{-1}. Fortunately, the result is the same either way. Namely,

$$\{s \ : \ s \in X \text{ and } f(s) = t \text{ for some } t \in T\}$$

$$= \{s \ : \ s \in X \text{ and } s = f^{-1}(t) \text{ for some } t \in T\},$$

since

$$f(s) = t \quad \text{if and only if} \quad s = f^{-1}(t)$$

in this special case.

As an illustration of Remark 5.2(b), consider the function $f : \mathbb{R} \longrightarrow \mathbb{R}$ defined by $x \mapsto 1 - 2x$. The image $f([-1, 1]) = [-1, 3]$ can be seen by the computations in Example 5.37, where the domain of f was restricted to $[-1, 1]$. In our next example, we consider another real function.

Example 5.51.   Define $f : \mathbb{R} \setminus \{0\} \longrightarrow \mathbb{R}$ by $x \mapsto \frac{1}{x^2}$.

(a) Show: $f((2, 3]) = \left[\frac{1}{9}, \frac{1}{4}\right)$.

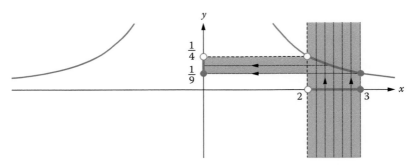

Proof.
($\subseteq$) Suppose $t \in f((2, 3])$. So $t = f(s) = \frac{1}{s^2}$ for some $2 < s \leq 3$.
Since $4 < s^2 \leq 9$, we see that $\frac{1}{9} \leq t < \frac{1}{4}$. That is, $t \in \left[\frac{1}{9}, \frac{1}{4}\right)$.
($\supseteq$) Suppose $t \in \left[\frac{1}{9}, \frac{1}{4}\right)$. Since $\frac{1}{9} \leq t < \frac{1}{4}$, we see that $4 < \frac{1}{t} \leq 9$.
Let $s = \sqrt{\frac{1}{t}}$, and observe that $2 < s \leq 3$. Since

$$f(s) = \frac{1}{\left(\sqrt{\frac{1}{t}}\right)^2} = \frac{1}{\frac{1}{t}} = t,$$

it follows that $t \in f((2, 3])$. $\square$

(b) Show: $f^{-1}\left(\left[\frac{1}{9}, \frac{1}{4}\right)\right) = [-3, 2) \cup (2, 3]$.

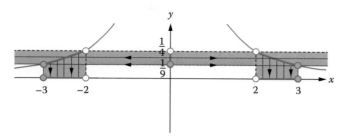

Proof. ($\subseteq$) Suppose $s \in f^{-1}([\frac{1}{9}, \frac{1}{4}))$. So $\frac{1}{s^2} = f(s)$ is in $[\frac{1}{9}, \frac{1}{4})$.

Since $\frac{1}{9} \leq \frac{1}{s^2} < \frac{1}{4}$, we see that $4 < s^2 \leq 9$.

Thus, $-3 \leq s < 2$ or $2 < s \leq 3$.   That   is, $s \in [-3, 2) \cup (2, 3]$.

($\supseteq$) Suppose $s \in [-3, 2) \cup (2, 3]$. So $-3 \leq s < 2$ or $2 < s \leq 3$.

Thus, $4 < s^2 \leq 9$, and we have $\frac{1}{9} \leq \frac{1}{s^2} < \frac{1}{4}$.

Since $f(s) = \frac{1}{s^2}$ is in $[\frac{1}{9}, \frac{1}{4})$, it follows that $s \in f^{-1}([\frac{1}{9}, \frac{1}{4}))$. $\qquad\square$

Example 5.52.   (Database Queries)
A library might keep track of which books are checked out in a table like Table 5.7. In fact, this table can be seen to reflect a function $f : B \longrightarrow P$, where B is the set of books in the library, P is the set of patrons of the library, and $f(b) = p$ precisely when (b, p) is a row of Table 5.7. That is, $f(b) = p$ means that book b has been checked out by patron p. From this point of view, natural questions one might ask about the database can be seen to be either an image or an inverse image for the function f.

(a) Which members have checked out books by John Steinbeck?

TABLE 5.7
Books checked out

Book	Borrower
Grapes of Wrath, by J. Steinbeck	Abe Roth
Hop on Pop, by Dr. Seuss	Megan Johnson
Of Mice and Men, by J. Steinbeck	Carla Torre
Fahrenheit 451, by R. Bradbury	Richard Kelley
Curious George, by H. Rey	Martha Lang
The Martian Chronicles, by R. Bradbury	Carla Torre
Cannery Row, by J. Steinbeck	Abe Roth

Solution. The set of books by John Steinbeck that have been checked out is

$$S = \{Grapes\ of\ Wrath,$$

$$Of\ Mice\ and\ Men, Cannery\ Row\}.$$

What we seek is the image of S under f. Since

$$f(S) = \{Abe\ Roth, Carla\ Torre\},$$

the answer to our question is Abe Roth and Carla Torre. □

(b) Which books have been checked out by Carla Torre?

Solution. We are inquiring about the checked out books for the set of patrons

$$T = \{Carla\ Torre\}.$$

What we seek is the inverse image of T under f. Since

$$f^{-1}(T) = \{Of\ Mice\ and\ Men,$$

$$The\ Martian\ Chronicles\},$$

the answer to our question is *Of Mice and Men* and *The Martian Chronicles*. □

Although Table 5.7 reflects a function, we have seen in Section 5.1 that tables in databases more generally reflect relations. In fact, the notions of image and inverse image in Definition 5.22 naturally generalize to relations, as we shall see in the exercises.

Although images may seem easier to compute than inverse images, it turns out that inverse images are better behaved with respect to basic set operations. That is, the operation of taking an inverse image tends to commute with a set operation more often than does the operation of taking an image. The behavior relative to complements is demonstrated in the next example. The behavior relative to other operations, such as unions and intersections, is explored in the exercises. (See Exercises 31 through 38.)

Example 5.53. Let $f : X \longrightarrow Y$ be an arbitrary function.

(a) Show: For any subset $T \subseteq Y$, that $f^{-1}(T^c) = (f^{-1}(T))^c$.

(b) Give an example of a function f and a subset $S \subseteq X$ such that both $f(S^c) \not\subseteq f(S)^c$ and $f(S)^c \not\subseteq f(S^c)$. Hence, certainly $f(S)^c \neq f(S^c)$.

Solution.

(a) ($\subseteq$) Suppose $x \in f^{-1}(T^c)$. Hence, $f(x) \in T^c$. That is, $f(x) \notin T$. Therefore, $x \notin f^{-1}(T)$. That is, $x \in (f^{-1}(T))^c$.

($\supseteq$) Suppose $x \in (f^{-1}(T))^c$. That is, $x \notin f^{-1}(T)$. Hence, $f(x) \notin T$. Therefore, $f(x) \in T^c$. That is, $x \in f^{-1}(T^c)$.

(b) Define $f : [0,4] \longrightarrow [0,3]$ by

$$f(x) = \begin{cases} x & \text{if } x \in [0,2], \\ x - 2 & \text{if } x \in (2,4], \end{cases}$$

and let $S = [0,2]$. Observe that $f(S)^c = (2,3]$ and $f(S^c) = (0,2]$. Neither is a subset of the other. $\square$

5.5.2 Indexed set operations

The union of a finite list of sets was defined in Section 1.4. The union of a general (possibly infinite) collection of sets was introduced in Section 5.2. Although taking unions of collections of sets was sufficient for our study of partitions, it does not cover all possible types of unions. Since $\mathcal{A}$ represents a *set* of sets in $\bigcup_{A \in \mathcal{A}} A$, each set A in $\mathcal{A}$ occurs exactly once in this union. However, in the finite case (such as, $A \cup A$), a set A can be repeated. In fact, when dealing with an arbitrary union, such as $A \cup B$, we may not know if B is a repeat of A, and we prefer not needing to fuss over such possibilities when making a general argument. To accomplish both degrees of flexibility, arbitrary infinite unions and arbitrary repetitions, we introduce the most general notion of indexed unions. Of course, intersections are handled similarly.

Definition 5.23. Let $\mathcal{U}$ be some fixed universal set. Given a set $\mathcal{I}$ and a function that assigns to each $i \in \mathcal{I}$ a set A_i in $\mathcal{U}$, we say that $\mathcal{I}$ is the **indexing set** for the **indexed collection** $\{A_i\}_{i \in \mathcal{I}}$ of sets.[*]

Example 5.54.   Let $\mathcal{U} = \{2,3,5,7\}$ and $\mathcal{I} = \{1,2,3\}$. Define $A_1 = \{2,5\}$, $A_2 = \{5,7\}$, and $A_3 = \{2,5\}$. The sets A_1, A_2, A_3 form an indexed collection. In this case, $A_1 = A_3$, but the indices of A_1 and A_3, namely, 1 and 3, respectively, are different.

Definition 5.24.   Let $\mathcal{I}$ be the indexing set for an indexed collection of sets $\{A_i\}_{i \in \mathcal{I}}$ from some universe $\mathcal{U}$.

---

[*] Notation similar to that of sequences is used here to emphasize the allowance of repeats, as in sequences.

(a) The **union** of $\{A_i\}_{i\in\mathcal{I}}$, denoted $\bigcup_{i\in\mathcal{I}}A_i$, is the set defined by

$$\forall\, x \in \mathcal{U}, \quad x \in \bigcup_{i\in\mathcal{I}}A_i \;\leftrightarrow\; x \in A_i \text{ for some } i \in \mathcal{I}.$$

(b) The **intersection** of $\{A_i\}_{i\in\mathcal{I}}$, denoted $\bigcap_{i\in\mathcal{I}}A_i$, is the set defined by

$$\forall\, x \in \mathcal{U}, \quad x \in \bigcap_{i\in\mathcal{I}}A_i \;\leftrightarrow\; x \in A_i \text{ for every } i \in \mathcal{I}.$$

The unions of collections of sets in Definition 5.9 are the special cases of the unions in Definition 5.24(a) in which the indexing set and the collection of sets coincide. The repetitions permitted in indexed unions and intersections certainly do not affect the final result, but the flexibility of the notation that we sought is achieved. For other set operations, such as products, the difference is more significant. Those cases are not considered in this book, but will be encountered in higher mathematics.

Example 5.55. Consider the indexed collection of sets from Example 5.54.

(a) $\bigcup_{i\in\{1,2,3\}}A_i = \{2,5,7\}$,
   since $2 \in A_1$, $5 \in A_1$, $7 \in A_2$, and 3 is not in any of A_1, A_2, or A_3. In this case, $\bigcup_{i\in\{1,2,3\}}A_i = A_1 \cup A_2 \cup A_3$ is a finite union.

(b) $\bigcap_{i\in\{1,2,3\}}A_i = \{5\}$,
   since only 5 is in each of A_1, A_2, and A_3.

Example 5.56. (Birth Rates).
For each natural number n, let A_n denote the set of women living in the United States who have given birth to exactly n children. Hence,

$$\bigcup_{n\geq 5} A_n$$

would be the set of women living in the United States who have had at least five children.

Example 5.57. Compute $\bigcup_{r\in\mathbb{R}}(-r^2, r^2)$.

Solution. Here, $\mathcal{U} = \mathbb{R}$, $\mathcal{I} = \mathbb{R}$, and, $\forall\, r \in \mathbb{R}, A_r = (-r^2, r^2)$. Observe that each real number x is in some set

A_r. Namely, since

$$- (|x| + 1)^2 \leq -(|x| + 1) < -|x| \leq x$$
$$\leq |x| < |x| + 1 \leq (|x| + 1)^2,$$

it follows that $x \in (-(|x| + 1)^2, (|x| + 1)^2)$. That is, $x \in A_r$ for $r = |x| + 1$. Therefore,

$$\bigcup_{r \in \mathbb{R}} (-r^2, r^2) = \mathbb{R}.$$

☐

Example 5.58. Show $\bigcap_{r \in \mathbb{Q}^+} (0, r] = \emptyset$.

Proof. Here, $\mathcal{U} = \mathbb{R}$, $\mathcal{I} = \mathbb{Q}^+$, and, $\forall\, r \in \mathbb{Q}^+$, $A_r = (0, r]$.
    Suppose $x \in \mathbb{R}$. We claim that there is some set $(0, r]$ that does not contain x. If $x \leq 0$, then $x \notin (0, 1]$. If $x > 0$, then by the density of the rational numbers (Appendix A, property 15), there is some $r \in \mathbb{Q}^+$ with $0 < r < x$. So $x \notin (0, r]$. Thus, $x \notin \bigcap_{r \in \mathbb{Q}^+} (0, r]$. Since $\bigcap_{r \in \mathbb{Q}^+} (0, r]$ contains no real numbers, it must be empty.   ☐

Example 5.59. (Frequent Travelers).
Let $\mathcal{S}$ be the set of the 50 states in the United States, and for each $S \in \mathcal{S}$, let A_S denote the set of living Americans who have visited state S. Hence,

$$\bigcap_{S \in \mathcal{S}} A_S$$

is the set of living Americans who have visited all 50 states.

Exercises

1. Define $f : \mathbb{N} \longrightarrow \mathbb{N}$ by $n \mapsto n!$. (a) Find $f(\{0, 1, 2, 3\})$. (b) Is $f(\{3\}) = 6$? Explain.

2. Define $f : \mathbb{Z}^+ \longrightarrow \mathbb{Q}$ by $n \mapsto \frac{2^n}{n}$. (a) Find $f(\{1, 2, 3\})$. (b) Is $f(2) = \{2\}$? Explain.

3. For $f(x) = x^4 - 6$, find $f(\{-2, -1, 0, 1\})$.

4. For $f(x) = x^3 - 4x$, find $f(\{0, 1, 2, 3\})$.

5. For $f(x) = 2x + 1$, find $f([-2, 2])$ and prove your result.

6. For $f(x) = x^2$, find $f([-1, 2])$ and prove your result.

7. For $f(x) = x^2 + 1$, find $f([1, 3])$ and prove your result.

8. For $f(x) = x^3$, find $f([-2, 3])$ and prove your result.

9. Given functions $f : X \longrightarrow Y$ and $g : Y \longrightarrow Z$, Definition 5.14 tells us that the domain of $g \circ f$ is the domain of f and that the codomain of $g \circ f$ is the codomain of g. Show that the image under g of the range of f is the range of $g \circ f$. That is, $g(\text{range}(f)) = \text{range}(g \circ f)$.

10. Let $f : X \longrightarrow Y$ and $S \subseteq X$ be given. Show: $f(S) \subseteq \text{range}(f)$.

11. Define $f : \mathbb{N} \longrightarrow \mathbb{N}$ by $n \mapsto n!$. See Exercise 1.
    (a) Find $f^{-1}(\{1, 2, 24\})$. (b) Find $f^{-1}(\{6, 10\})$. (c) Is $f^{-1}(\{120\}) = 5$? Explain.

12. Define $f : \mathbb{Z}^+ \longrightarrow \mathbb{Q}$ by $n \mapsto \frac{2^n}{n}$. See Exercise 2.
    (a) Find $f^{-1}(\{2, 4\})$. (b) Is it correct to say $f^{-1}(\frac{8}{3}) = 3$? Explain.

13. For $f(x) = x^4 - 6$, find $f^{-1}(\{-5, 10\})$. See Exercise 3.

14. For $f(x) = x^3 - 4x$, find $f^{-1}(\{0, 15\})$. See Exercise 4.

15. Define $f : \mathbb{Z} \longrightarrow \{-1, 1\}$ by $f(n) = (-1)^n$. Find $f^{-1}(\{-1\})$ and prove your result.

16. Define $f : \mathbb{Z} \longrightarrow \{-1, 1\}$ by $f(n) = (-1)^{\frac{n(n-1)}{2}}$. Find $f^{-1}(\{1\})$ and prove your result.

17. Define $f : \mathbb{Z}^+ \times \mathbb{Z}^+ \longrightarrow \mathbb{Z}^+$ by $(m, n) \mapsto \gcd(m, n)$. Find $f^{-1}(\{1\})$.

18. Let $d \in \mathbb{Z} \setminus \{0\}$, and define $f : \mathbb{Z} \longrightarrow \mathbb{Q}$ by $n \mapsto \frac{n}{d}$. Find $f^{-1}(\mathbb{Z})$.

19. For $f(x) = x^2$, find $f^{-1}([1, 4])$ and prove your result.

20. For $f(x) = x^3$, find $f^{-1}([-1, 8])$ and prove your result.

21. Define functions $f : \mathbb{R} \longrightarrow \mathbb{R} \times \mathbb{R}$ and $g : \mathbb{R} \times \mathbb{R} \longrightarrow \mathbb{R}$ by $f(x) = (x, x)$ and $g(x, y) = x - y$. Show: $f(\mathbb{R}) = g^{-1}(\{0\})$.

22. Let $d \in \mathbb{Z}^+$. Define functions $f : \mathbb{Z} \longrightarrow \mathbb{Z}$ and $g : \mathbb{Z} \longrightarrow \{0, 1, \ldots, d - 1\}$ by $f(n) = dn$ and $g(n) = n \bmod d$. Show: $f(\mathbb{Z}) = g^{-1}(\{0\})$.

23. Some colleges are not unique in their choice of a mascot. Table 5.8 represents a function in a database that lists the nicknames used by certain colleges. After answering each of the following questions, determine which type of computation is needed for this function, an image or an inverse image.

    (a) Which nicknames are used for the Massachusetts schools? Note that Boston College and Northeastern University are in Massachusetts.

    (b) Which schools use the nickname "Wildcats"?

24. The software company SoftJobs uses a database to keep track of its current projects and clients. In particular, Table 5.9 is a function in this database that assigns to

TABLE 5.8

Mascots

College	Nickname
Boston College	Eagles
Gonzaga University	Bulldogs
Kansas State University	Wildcats
Morehead State University	Eagles
Northeastern University	Huskies
Northwestern University	Wildcats
University of Georgia	Bulldogs
University of New Hampshire	Wildcats
Villanova University	Wildcats
Yale University	Bulldogs

TABLE 5.9

Contracts

Project	Client
NBA Dunkfest	GameCo
Rx Tracker	MediComp
Skate Rats	GameCo
Claim Pro	HealthCorp
Slap Shot 3	PlayBox

each project the client for which it is under contract. After answering each of the following questions, determine which type of computation is needed for this function, an image or an inverse image.

(a) Which projects are being done for at least one of GameCo or HealthCorp?

(b) For which clients is a video game being designed? Note that NBA Dunkfest, Skate Rats, and Slap Shot 3 are games.

For Exercises 25 through 30, use the following analog of Definition 5.22 for a relation R from a set X to a set Y:

(a) Given a subset S of X, the **image of S under R** is

$$R(S) = \{t \; : \; t \in Y \text{ and } s \, R \, t \text{ for some } s \in S\}.$$

(b) Given a subset T of Y, the **inverse image of T under R** is

$$R^{-1}(T) = \{s \; : \; s \in X \text{ and } s \, R \, t \text{ for some } t \in T\}.$$

25. Let R be the "divides" relation from the set P of primes to $\mathbb{Z}$.
    (a) Find $R(\{2\})$. (b) Find $R^{-1}(\{15, 35\})$.

26. Let R be the "is a zero of" relation from $\mathbb{R}$ to the set P of real polynomial functions.

    (a) List four elements of $R(\{1\})$. Can you describe the entire set?

    (b) Find $R^{-1}(\{x^2 - 4x + 3, 2x - 8\})$.

27. Let R be the "is an element of" relation from $\mathbb{Z}$ to $\mathcal{P}(\mathbb{Z})$.

    (a) List three elements of $R(\{0, 1\})$. Can you describe the entire set?

    (b) Find $R^{-1}(\{\{0, 1, 2, 3\}, \{0, 2, 4, 6\}\})$.

28. Let R be the "is less than" relation from $\mathbb{R}$ to $\mathbb{R}$.
    (a) Find $R(\{1, 3, 5\})$. (b) Find $R^{-1}(\{1, 3, 5\})$.

29. HardSell is a mail-order hardware company that uses a database to keep track of customer orders. In particular, Table 5.10 is a relation that assigns, to each part ordered, the customer who ordered it. After answering parts (b) and (c) below, determine which type of computation is needed for this relation, an image or an inverse image.

    (a) Is the relation in Table 5.10 a function? Explain.

    (b) Which parts were ordered by Susan Brower?

    (c) Which customers ordered wrenches?

30. The software company SoftJobs from Exercise 24 also uses its database to keep track of its programmers' work assignments. In particular, Table 5.11 is a relation that assigns, to each programmer, the project on which he or she is working. After answering parts (b) and (c) below,

TABLE 5.10
Hardware orders

Part	Customer
Hammer	Richard Kelley
Wrench	Susan Brower
Pliers	Susan Brower
Wrench	Abe Roth

TABLE 5.11
Programming assignments

Programmer	Project
Martha Lang	NBA Dunkfest
Megan Johnson	Rx Tracker
Abe Roth	NBA Dunkfest
Martha Lang	Skate Rats
Abe Roth	Skate Rats

determine which type of computation is needed for this relation, an image or an inverse image.

(a) Is the relation in Table 5.11 a function? Explain.

(b) On which projects is Martha Lang working?

(c) Which programmers are working on NBA Dunkfest?

In Exercises 31 through 38, let $f : X \longrightarrow Y$ be an arbitrary function, and let $S, S_1, S_2 \subseteq X$ and $T, T_1, T_2 \subseteq Y$ be arbitrary subsets.

31.* (a) Show: If $S_1 \subseteq S_2$, then $f(S_1) \subseteq f(S_2)$.

   (b) Show: If $T_1 \subseteq T_2$, then $f^{-1}(T_1) \subseteq f^{-1}(T_2)$.

32.* (a) Show: $S \subseteq f^{-1}(f(S))$.

   (b) Show: $f(f^{-1}(T)) \subseteq T$.

   (c) Show: If f is one-to-one, then $S = f^{-1}(f(S))$.

   (d) Show: If f is onto, then $f(f^{-1}(T)) = T$.

33. (a) Show: $X = f^{-1}(Y)$.

   (b) Show: $f(X) \subseteq Y$.

   (c) Give an example of a function $f : X \longrightarrow Y$ such that $f(X) \neq Y$.

34. (a) Show: $f(\emptyset) = \emptyset$.

   (b) Show: $f^{-1}(\emptyset) = \emptyset$.

35. (a) Show: $f^{-1}(T_1 \cup T_2) = f^{-1}(T_1) \cup f^{-1}(T_2)$.

   (b) Show: $f^{-1}(T_1 \cap T_2) = f^{-1}(T_1) \cap f^{-1}(T_2)$.

36. (a) Show: $f(S_1 \cup S_2) = f(S_1) \cup f(S_2)$.

   (b) Show: $f(S_1 \cap S_2) \subseteq f(S_1) \cap f(S_2)$.

   (c) Give an example of a function $f : X \longrightarrow Y$ such that $f(S_1 \cap S_2) \neq f(S_1) \cap f(S_2)$.

37.* (a) Show: $f(S \cap f^{-1}(T)) \subseteq f(S) \cap T$.

   (b) Show: $S \cap f^{-1}(T) \subseteq f^{-1}(f(S) \cap T)$.

38.* (a) Show: $f(f^{-1}(f(S))) = f(S)$.

   (b) Show: $f^{-1}(f(f^{-1}(T))) = f^{-1}(T)$.

39.* Let $f : X \longrightarrow Y$ be given, and define $F : \mathcal{P}(X) \longrightarrow \mathcal{P}(Y)$ by $F(A) = f(A)$. Show that

(a) f is one-to-one if and only if F is one-to-one.

(b) f is onto if and only if F is onto.

40.* Let $f_1 : X_1 \longrightarrow Y_1$ and $f_2 : X_2 \longrightarrow Y_2$ be given. The function $f_1 \times f_2 : X_1 \times X_2 \longrightarrow Y_1 \times Y_2$ is defined by

$$(f_1 \times f_2)(x_1, x_2) = (f_1(x_1), f_2(x_2)), \forall\, x_1 \in X_1, x_2 \in X_2.$$

Suppose $A_1 \subseteq X_1, A_2 \subseteq X_2, B_1 \subseteq Y_1$, and $B_2 \subseteq Y_2$.

(a) Prove: $(f_1 \times f_2)(A_1 \times A_2) = f_1(A_1) \times f_2(A_2)$.

(b) Prove   or   disprove: $(f_1 \times f_2)^{-1}(B_1 \times B_2) = f_1^{-1}(B_1) \times f_2^{-1}(B_2)$.

41. Find $\bigcap_{r \in \mathbb{R}} (-r^2, r^2)$.

42. Find $\bigcup_{r \in \mathbb{Q}^+} (0, r)$.

43. Find $\bigcup_{n \in \mathbb{Z}^+} [n, n+2)$.

44. Find $\bigcap_{n \in \mathbb{Z}^+} [n, n+2)$.

45. Find $\bigcap_{n \in \mathbb{Z}^+} [0, n]$.

46. Find $\bigcup_{r \in \mathbb{R}^+} [0, 1 - \frac{1}{r}]$.

In Exercises 47 through 50, express the described set S using either an indexed union or an indexed intersection.

47. At Bloomsburg University, the possible grade point averages (GPAs) range from 0 to 4. For each $x \in [0, 4]$, let A_x be the set of students at Bloomsburg University whose GPA is x. Let S be the set of students whose GPA is greater than or equal to 3 and less than 4, the "B students."

48. Some baseball fans aim to visit as many of the Major League Baseball parks as possible. Let M be the set of Major League Baseball teams, and for each $t \in M$, let A_t be the set of people living in the United States who have attended a game in the home park of team t. The American League East division is

$$E = \{\text{Blue Jays, Orioles, Rays, Red Sox, Yankees}\}.$$

Let S be the set of people living in the United States who have attended a game in each American League East home park.

49. The sentence "The quick brown fox jumps over the lazy dog." is used to test keyboards, since it contains every letter in the alphabet. It is also interesting to think of individual words that use a variety of letters. For each letter α in the alphabet $\mathcal{A} = \{a, b, \ldots, z\}$, let A_α be the set of words in the English language that contain the letter α. Let S be the set of words in the English language that contain all of the vowels, such as the word *hyperstimulation*.

50. Many doctors prescribe cholesterol-lowering medication to their patients whose cholesterol levels exceed 200. For each $n \in \mathbb{N}$, let A_n be the set of living Americans whose

cholesterol level is n. Let S be the set of living Americans whose cholesterol level is at most 200.

In Exercises 51 through 60, X and Y are sets, $B \subseteq X$, and for some indexing set $\mathcal{I}$, $\{A_i\}_{i \in \mathcal{I}}$ is an indexed collection of subsets of X.

51.* Let $f : X \longrightarrow Y$. Show $f \left(\bigcup_{i \in \mathcal{I}} A_i \right) = \bigcup_{i \in \mathcal{I}} f(A_i)$.

52.* Let $f : Y \longrightarrow X$. Show $f^{-1} \left(\bigcup_{i \in \mathcal{I}} A_i \right) = \bigcup_{i \in \mathcal{I}} f^{-1}(A_i)$.

53.* Let $f : Y \longrightarrow X$. Show $f^{-1} \left(\bigcap_{i \in \mathcal{I}} A_i \right) = \bigcap_{i \in \mathcal{I}} f^{-1}(A_i)$.

54.* Let $f : X \longrightarrow Y$. Show $f \left(\bigcap_{i \in \mathcal{I}} A_i \right) \subseteq \bigcap_{i \in \mathcal{I}} f(A_i)$.

55. Suppose $\mathcal{J} \subseteq \mathcal{I}$.

   (a) Show: $\bigcup_{i \in \mathcal{J}} A_i$ $\subseteq \bigcup_{i \in \mathcal{I}} A_i$.

   (b) Show: $\bigcap_{i \in \mathcal{I}} A_i$ $\subseteq \bigcap_{i \in \mathcal{J}} A_i$.

56. Suppose, for all $i \in \mathcal{I}$, that $B_i \subseteq A_i$.

   (a) Show: $\bigcup_{i \in \mathcal{I}} B_i$ $\subseteq \bigcup_{i \in \mathcal{I}} A_i$.

   (b) Show: $\bigcap_{i \in \mathcal{I}} B_i$ $\subseteq \bigcap_{i \in \mathcal{I}} A_i$.

57. **Distributive Laws.**

   (a) Show: $B \cup \bigcap_{i \in \mathcal{I}} A_i$ $= \bigcap_{i \in \mathcal{I}} (B \cup A_i)$.

   (b) Show: $B \cap \bigcup_{i \in \mathcal{I}} A_i$ $= \bigcup_{i \in \mathcal{I}} (B \cap A_i)$.

58. **De Morgan's Laws.**

   (a) Show: $\left(\bigcup_{i \in \mathcal{I}} A_i \right)^c$ $= \bigcap_{i \in \mathcal{I}} A_i{}^c$.

   (b) Show: $\left(\bigcap_{i \in \mathcal{I}} A_i \right)^c$ $= \bigcup_{i \in \mathcal{I}} A_i{}^c$.

59. Show: $\left(\bigcup_{i \in \mathcal{I}} A_i \right) \setminus B = \bigcup_{i \in \mathcal{I}} (A_i \setminus B)$.

60. Show: $B \times \bigcap_{i \in \mathcal{I}} A_i = \bigcap_{i \in \mathcal{I}} (B \times A_i)$.

61.* Given an onto function $f : X \longrightarrow Y$ and a partition $\mathcal{A}$ of Y, must it be true that the collection $\{f^{-1}(A) : A \in \mathcal{A}\}$ is a partition of X? Justify your answer.

62.* Given an onto function $f : X \longrightarrow Y$ and a partition $\mathcal{A}$ of X, must it be true that the collection $\{f(A) : A \in \mathcal{A}\}$ is a partition of Y? Justify your answer.

5.6 Cardinality

The cardinality of a set was introduced informally in Section 1.2 as the number of elements in the set. What was not made precise is exactly how one counts elements. This is settled with the notion of a bijection. If we agree that $\{1, 2, \ldots, n\}$ should have cardinality n, then the same should be true of sets for which there is a one-to-one correspondence with $\{1, 2, \ldots, n\}$.

Definition 5.25.

(a) Two sets A and B are said to have the same **cardinality** if there is a bijection from A to B.

(b) Given $n \in \mathbb{N}$, a set A is said to have **cardinality** n if A has the same cardinality as the set $\{k : k \in \mathbb{Z}$ and $1 \le k \le n\}$.

(c) A set A is said to be **finite**, or to have **finite cardinality**, if A has cardinality n for some $n \ge 0$. Otherwise, A is said to be **infinite**.

The empty set $\emptyset$ has cardinality 0, since

$$\{k : k \in \mathbb{Z} \text{ and } 1 \le k \le 0\} = \emptyset.$$

Example 5.60.   Show that the set $\{-5, -4, \ldots, 4, 5\}$ has cardinality 11.

Proof. Define $f : \{-5, -4, \ldots, 4, 5\} \longrightarrow \{1, 2, \ldots, 10, 11\}$ by $f(k) = k + 6$.

$$
\begin{array}{ccccccccccc}
-5 & -4 & -3 & -2 & -1 & 0 & 1 & 2 & 3 & 4 & 5 \\
\downarrow & \downarrow & \downarrow & \downarrow & \downarrow & \downarrow & \downarrow & \downarrow & \downarrow & \downarrow & \downarrow \\
1 & 2 & 3 & 4 & 5 & 6 & 7 & 8 & 9 & 10 & 11
\end{array}
$$

with f at the left.

It is straightforward to verify that f is a bijection with inverse function f^{-1} given by $f^{-1}(k) = k - 6$.   □

The bijection f in Example 5.60 can be interpreted as counting the elements in the set $\{-5, -4, \ldots, 4, 5\}$. That is, -5 is the first, -4 is the second, and so on. Ultimately, 11 elements are counted. For finite sets, using cardinality to measure size is quite natural. For infinite sets, our intuitions may be challenged as we gain comfort with the issues.

Example 5.61.   Let E be the set of even integers. Show that $\mathbb{Z}$ and E have the same cardinality.

Proof. Define functions $f : \mathbb{Z} \longrightarrow E$ and $g : E \longrightarrow \mathbb{Z}$ by
$f(n) = 2n$

$\mathbb{Z}$	$\cdots$	-3	-2	-1	0	1	2	3	$\cdots$
$\downarrow f$		$\downarrow$	$\downarrow$	$\downarrow$	$\downarrow$	$\downarrow$	$\downarrow$	$\downarrow$	
E	$\cdots$	-6	-4	-2	0	2	4	6	$\cdots$

and $g(n) = \frac{n}{2}$. Note that when n is even, $\frac{n}{2}$ is an integer. So the values taken by g are indeed in $\mathbb{Z}$. Since

$$\forall n \in E, \ (f \circ g)(n) = f(g(n)) = f\left(\frac{n}{2}\right) = 2\left(\frac{n}{2}\right) = n, \text{ and}$$

$$\forall n \in \mathbb{Z}, \ (g \circ f)(n) = g(f(n)) = g(2n) = \frac{2n}{2} = n,$$

we see that f and g are inverses of one another. Therefore, f is a bijection displaying the fact that $\mathbb{Z}$ and E have the same cardinality. $\qquad\square$

The subset relation is one means of comparing sizes of sets. From that perspective, since the set E of even integers is a proper subset of $\mathbb{Z}$, the set E is a "smaller" set. However, that is not the view taken with cardinality. Example 5.61 shows that $\mathbb{Z}$ and E have the same "size" when cardinality is the measure.

Example 5.62. Show that, for any $a, b \in \mathbb{R}$ with $a < b$, the open intervals (a, b) and $(0, 1)$ have the same cardinality.

Proof. Define functions $f : (a, b) \longrightarrow (0, 1)$ and $g : (0, 1) \longrightarrow (a, b)$ by

$$f(x) = \frac{x - a}{b - a} \quad \text{and} \quad g(x) = a + (b - a)x.$$

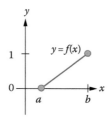

It is straightforward to verify that f and g are inverses of one another. Hence, Theorem 5.10 tells us that f is a bijection from (a, b) to $(0, 1)$. $\qquad\square$

In some contexts, we might want to measure the "size" of an interval (a, b) by its length $b - a$. However, Example 5.62 shows that cardinality is independent of length.

The following theorem says, in particular, that in Definition 5.25(a), the order of the two sets is unimportant when asserting that they have the same cardinality.

Theorem 5.11. (Common Cardinality Is an Equivalence Relation).
Let A, B, and C be any sets in some fixed universal set U.

(a) *A has the same cardinality as itself.*

(b) *If A has the same cardinality as B, then B has the same cardinality as A.*

(c) *If A has the same cardinality as B and B has the same cardinality as C, then A has the same cardinality as C.*

Proof. For part (a), the identity function id_A is a bijection from A to A (see Exercise 32 from Section 5.4). The proofs of parts (b) and (c) are left for the exercises.   □

The bijection in Example 5.48 shows that $\mathbb{R}$ and $(-1, 1)$ have the same cardinality. That result combined with the result in Example 5.62 gives that $\mathbb{R}$ has the same cardinality as any open interval.

In Definition 5.25(b), we want to additionally assert that a single set A cannot be assigned both cardinality m and cardinality n when $m \neq n$. The Pigeon Hole Principle accomplishes that and more.

Theorem 5.12.   (The Pigeon Hole Principle).
If A is any set of cardinality n, and B is any set of cardinality m with $n > m$, then there is no one-to-one function from A to B. That is, any function from A to B must send two distinct elements of A to the same element of B.

Remark 5.3.   Very simply, the Pigeon Hole Principle says that if $n > m$ and n pigeons are placed into m holes, then some hole must receive two or more pigeons.

Proof of Theorem 5.12. It suffices to assume that $A = \{1, 2, \ldots, n\}$ and $B = \{1, 2, \ldots, m\}$. Since $B \subseteq A$, we may define the inclusion function $i : B \longrightarrow A$ given by the formula $i(b) = b$ for each $b \in B$. It is easy to see that i is one-to-one. Given any function $f : A \longrightarrow B$, we can define a new function $\bar{f} : A \longrightarrow A$ by $\bar{f} = i \circ f$. Since $m < n$ and

$$\text{range}(\bar{f}) \subseteq \text{range}(i) = \{1, 2, \ldots, m\} \subset \{1, 2, \ldots, m, \ldots, n\},$$

the element n cannot be in the range of $\bar{f}$. Thus, $\bar{f}$ cannot be onto. If f is one-to-one, then Theorem 5.9(a) tells us that $\bar{f}$ must also be one-to-one. Hence, to show that there is no one-to-one function from A to B, it suffices to show that there is no one-to-one function from A to itself that is not onto. We therefore show that any one-to-one function $\overline{f} : A \longrightarrow A$ must be onto. This is accomplished by induction on n, for $n \geq 1$.

If $n = 1$, then the identity function is the only one-to-one function from $\{1\}$ to itself (in fact, the only function), and it is certainly onto. Since this takes care of the base case, we move on to the inductive step. Suppose $k \geq 1$ and all one-to-one functions $g : \{1, 2, \ldots, k\} \longrightarrow \{1, 2, \ldots, k\}$ are onto. Let $h : \{1, 2, \ldots, k, k + 1\} \longrightarrow \{1, 2, \ldots, k, k + 1\}$ be any one-to-one function.

Case 1: $h(a) \leq k$ for all $1 \leq a \leq k$.

Here, we may define $g : \{1, 2, \ldots, k\} \longrightarrow \{1, 2, \ldots, k\}$ by $g(a) = h(a)$ for all $1 \leq a \leq k$. Since the fact that h is one-to-one forces g to also be one-to-one, it follows from the inductive hypothesis that g must be onto. Consequently, the only possibility is that $h(k + 1) = k + 1$, and thus h is onto.

Case 2: $h(a') = k + 1$ for some $1 \leq a' \leq k$.

Define a new function $h' : \{1, 2, \ldots, k, k + 1\} \longrightarrow \{1, 2, \ldots, k, k + 1\}$ by $h'(a') = h(k + 1), h'(k + 1) = h(a') = k + 1$, and, for all other values of a, $h'(a) = h(a)$. The fact that h is one-to-one forces h' also to be one-to-one. Since h' also satisfies the conditions of Case 1, it must be that h' is onto. Consequently, h must be onto. □

The Pigeon Hole Principle has an important consequence in our study of cardinality.

Corollary 5.13.   Let A be a set with cardinality n, and let $m \in \mathbb{Z}$ with $m \neq n$. Then A does not have cardinality m.

The proof of Corollary 5.13 is left for the exercises. Corollary 5.13 tells us that for each finite set A, there is a unique natural number n such that A has cardinality n. Hence, we may assign to each finite set A the unique number, denoted $|A|$, characterized by

$$|A| = n \quad \text{if and only if} \quad A \text{ has cardinality } n.$$

That is, A has cardinality $|A|$.

It follows from Example 5.60 that $|\{-5, -4, \ldots, 4, 5\}| = 11$. That the set of all positive integers

$$1, 2, 3, 4, 5, 6, 7, 8, 9, 10, 11, 12, \ldots$$

cannot be assigned a finite cardinality n is suggested by the "..." in our listing. However, how can we justify this carefully?

Example 5.63.   Show that $\mathbb{Z}^+$ is infinite.

Proof. Suppose to the contrary that $\mathbb{Z}^+$ is finite. Let $m = |\mathbb{Z}^+|$. By definition, there must be a bijection $f : \mathbb{Z}^+ \longrightarrow \{1, 2, \ldots, m\}$. Let

$$i : \{1, 2, \ldots, m, m+1\} \longrightarrow \mathbb{Z}^+$$

be the inclusion given by $i(k) = k$. Since f and i are one-to-one, Theorem 5.9(a) tells us that the composite

$$f \circ i : \{1, 2, \ldots, m, m+1\} \longrightarrow \{1, 2, \ldots, m\}$$

must be one-to-one. However, this contradicts the Pigeon Hole Principle (with $n = m + 1$). We conclude that $\mathbb{Z}^+$ must be infinite. $\qquad \square$

The proof in Example 5.63 is essentially the same as that used in Example 2.27 to show that $\mathbb{Z}$ is infinite. However, here we utilize our more formal understanding of cardinality and its properties.

We have defined the notion of an infinite set and seen the— perhaps surprising— result that infinite sets such as the set E of even integers and the set $\mathbb{Z}$ of all integers have the same cardinality. In light of this, what may now come as a surprise is the fact that there are indeed different infinite cardinalities. That is, not all infinities are equal.

Definition 5.26.   Let A be any set.

(a) A is said to be **countably infinite** if A has the same cardinality as $\mathbb{Z}^+$.

(b) A is said to be **countable** if A is finite or countably infinite.

(c) A is said to be **uncountable** if A is not countable.

Example 5.64.   Show that $\mathbb{Z}$ is countably infinite.

Proof. Define $f : \mathbb{Z}^+ \longrightarrow \mathbb{Z}$ and $g : \mathbb{Z} \longrightarrow \mathbb{Z}^+$ by

$$f(n) = (-1)^n \left\lfloor \frac{n}{2} \right\rfloor \quad \text{and} \quad g(n) = \begin{cases} 2n & \text{if } n > 0, \\ 2|n| + 1 & \text{if } n \le 0. \end{cases}$$

$$
\begin{array}{ccccccccc}
\mathbb{Z}^+ & 1 & 2 & 3 & 4 & 5 & 6 & 7 & \cdots \\
\downarrow f & \downarrow & \downarrow & \downarrow & \downarrow & \downarrow & \downarrow & \downarrow & \\
\mathbb{Z} & 0 & 1 & -1 & 2 & -2 & 3 & -3 & \cdots
\end{array}
$$

The equations

$$\forall \text{ even } n \in \mathbb{Z}^+, \qquad g(f(n)) = g(\tfrac{n}{2})$$
$$= 2(\tfrac{n}{2}) = n,$$
$$\forall \text{ odd } n \in \mathbb{Z}^+, \qquad g(f(n)) = g(-\tfrac{n-1}{2})$$
$$= 2(\tfrac{n-1}{2}) + 1 = n,$$
$$\forall \text{ positive } n \in \mathbb{Z}, \qquad f(g(n)) = f(2n)$$
$$= \lfloor \tfrac{2n}{2} \rfloor = n,$$
$$\forall \text{ nonpositive } n \in \mathbb{Z}, \ f(g(n)) = f(-2n+1) = -\lfloor \tfrac{-2n+1}{2} \rfloor$$
$$= -(-n) = n$$

show that f is a bijection (with inverse g). $\qquad\qquad \square$

If a set X is countably infinite, then we could write a computer program that prints each of its elements. Namely, a bijection $f : \mathbb{Z}^+ \longrightarrow X$ could be used in a program

```
Let n = 1.
While n > 0,
    \begin
    Print f(n).
    Let n = n + 1.
    \end.
```

that implements an infinite while loop. Although this program will never finish, any particular element for which we wish to wait will eventually be printed, if we wait long enough. In contrast, no such program could be written to list the elements of an uncountable set.

The following theorem shows us that not all infinite sets are countable. We owe the method of proof presented here to the German mathematician Georg Cantor (1845–1918). It is known as the **diagonal argument**.

Theorem 5.14. $\mathbb{R}$ *is uncountable.*

Proof. Since $\mathbb{R}$ has the same cardinality as $(0, 1)$, it suffices to prove that $(0, 1)$ is uncountable. By using the ideas discussed in Section 3.4, each real number $x \in (0, 1)$ can be expressed in decimal form

$$0. \ a_1 \ a_2 \ a_3 \ \cdots \ a_n \ \cdots \qquad\qquad (5.6)$$

Moreover, if we do not allow decimal expressions to end in repeating 9's, then each $x \in (0, 1)$ has a unique representation in form (5.6). For example, $0.4\overline{9} = 0.5$, and we take $0.5\overline{0}$ as the unique representation of this number.

Suppose to the contrary that $(0,1)$ is countable. Thus, we have a bijective function $f : \mathbb{Z}^+ \longrightarrow (0,1)$ that we may use to list all of the elements of $(0,1)$.

A contradiction is obtained by finding a real number x in $(0,1)$ that is not on this list and thus showing that f is not onto. We describe x in decimal form

$$x = 0. \; b_1 \; b_2 \; b_3 \; \cdots \; b_n \; \cdots$$

by defining, for each $i \in \mathbb{Z}^+$,

$$b_i = \begin{cases} 1 & \text{if } a_{i,i} \neq 1, \\ 2 & \text{if } a_{i,i} = 1. \end{cases}$$

For each i, we have chosen $b_i \neq a_{i,i}$, and therefore $x \neq f(i)$. That is, x differs from $f(i)$ in the (boxed) digit on the diagonal above. So x is not on the list. This contradiction shows that $(0,1)$ is uncountable. □

Despite the fact that the density of the rationals (Appendix A, property 15) tells us that every real number has rational numbers arbitrarily close to it, $\mathbb{R}$ and $\mathbb{Q}$ do not have the same cardinality.

Theorem 5.15. $\mathbb{Q}$ *is countably infinite.*

A proof of Theorem 5.15 is laid out in Exercise 51. An alternative proof follows from Exercise 70 in Section 5.4.

Theorem 5.16.   *All intervals containing more than one element (including the interval $(-\infty, \infty) = \mathbb{R}$) have the same cardinality.*

The proof of Theorem 5.16 follows from Theorem 5.11, Examples 5.48 and 5.62, and Exercises 19 through 24.

Exercises

In Exercises 1 through 6, specify the cardinality of the given set.

1. $\{-65, -64, \ldots, 9, 10\}$.

2. $\{1, \frac{1}{2}, \frac{1}{4}, \frac{1}{8}, \ldots, \frac{1}{2048}\}$.

3. $\{\binom{4}{0}, \binom{4}{1}, \binom{4}{2}, \binom{4}{3}, \binom{4}{4}\}$.

4. $\{\frac{2^1}{1}, \frac{2^2}{2}, \frac{2^3}{3}, \frac{2^4}{4}\}$.

5. $\{x : x \in \mathbb{R} \text{ and } x^3 - 4x = 0\}$.

6. $\{x : x \in \mathbb{R} \text{ and } x^3 - 3x^2 + 4x = 0\}$.

7. The software company SoftJobs keeps track of its projects in the following table from its database.

Project	Client
NBA Dunkfest	GameCo
Rx Tracker	MediComp
Skate Rats	GameCo
Claim Pro	HealthCorp
Slap Shot 3	PlayBox

How many clients does SoftJobs have?

8. The mail-order hardware company HardSell keeps track of its of its customer orders in the following table from its database.

Part	Customer
Hammer	Richard Kelley
Wrench	Susan Brower
Pliers	Susan Brower
Wrench	Abe Roth

How many different types of parts have been ordered?

9. Let $n \in \mathbb{N}$. Show that $\{0, 1, \ldots, n\}$ has cardinality $n + 1$.

10. Let $n \in \mathbb{N}$. Show that $\{m : m \text{ is even and } 1 \le m \le 2n\}$ has cardinality n.

11. Let $n \in \mathbb{N}$. Show that $\{n^2, n^2 + 1, \ldots, (n+1)^2\}$ has cardinality $2n + 2$.

12. Let $n \in \mathbb{N}$. Show that $\{-n, -n+1, \ldots, n\}$ has cardinality $2n + 1$.

13. Show that $\mathbb{N}$ has the same cardinality as $\mathbb{Z}^-$.

14. Show that $\mathbb{Z}^+$ has the same cardinality as the set $\{n : n = 2^k \text{ for some } k \in \mathbb{Z}^+\}$.

15. Show that $\mathbb{Z}^+$ has the same cardinality as the set $\{n : n = k^2 \text{ for some } k \in \mathbb{Z}^+\}$.

16. Show that $\mathbb{Z}$ has the same cardinality as the set O of odd integers.

17. Show that the intervals $[3, 8]$ and $[1, 11]$ have the same cardinality.

18. Show that the intervals $[-1, 1)$ and $[0, 8)$ have the same cardinality.

19.* Let $a, b \in \mathbb{R}$ with $a < b$. Prove the following:

   (a) The closed intervals $[a, b]$ and $[0, 1]$ have the same cardinality.

   (b) The right-half-open intervals $[a, b)$ and $[0, 1)$ have the same cardinality.

   (c) The left-half-open intervals $(a, b]$ and $(0, 1]$ have the same cardinality.

20.* Let $a \in \mathbb{R}$. Prove the following:

   (a) The intervals (a, ∞) and $(0, \infty)$ have the same cardinality.

   (b) The intervals $[a, \infty)$ and $[0, \infty)$ have the same cardinality.

   (c) The intervals $(-\infty, a)$ and $(-\infty, 0)$ have the same cardinality.

   (d) The intervals $(-\infty, a]$ and $(-\infty, 0]$ have the same cardinality.

21.* Let $a, b \in \mathbb{R}$ with $a < b$. Show that $[a, b)$ and $(0, 1]$ have the same cardinality. Hint: For $a = -1$ and $b = 0$, use $x \mapsto -x$.

22.* Let $a \in \mathbb{R}$.

   (a) Show that (a, ∞) and $(-\infty, 0)$ have the same cardinality.

   (b) Show that $[a, \infty)$ and $(-\infty, 0]$ have the same cardinality.

23.* Let $a \in \mathbb{R}$.

   (a) Show that (a, b) and $(0, \infty)$ have the same cardinality.

   (b) Show that $[a, b)$ and $[0, \infty)$ have the same cardinality.

   Hint: For $a = 0$ and $b = 1$, restrict the function $g : (-1, 1) \longrightarrow \mathbb{R}$ used in Example 5.48.

24.* (a) Show that $[0, 1]$ and $[0, 1)$ have the same cardinality. Hint: Use a one-to-one correspondence between $\{\frac{1}{n} : n \geq 1\}$ and $\{\frac{1}{n} : n \geq 2\}$, and use the identity for the rest.

   (b) Show that $(0, 1]$ and $(0, 1)$ have the same cardinality by restricting the bijection from part (a).

25.* Let $a, b, c, d \in \mathbb{R}$ with $a < b \leq c < d$. Show that $[a, b) \cup [c, d)$ has the same cardinality as $[0, 1)$.

26. Let $a, b, c \in \mathbb{R}$ with $a < b \leq c$. Show that $[a, c] \setminus \{b\}$ has the same cardinality as $[0, 1)$. Hint: By Theorem 5.11 and Exercise 21, $(b, c]$ and $[b, c)$ have the same cardinality.

27. Prove Theorem 5.11(b). Hint: Use Exercise 65 from Section 5.4.

28. Prove Theorem 5.11(c). Hint: Use Theorem 5.9.

29. Prove Corollary 5.13. Hint: It suffices to show that $\{1, 2, \ldots, n\}$ and $\{1, 2, \ldots, m\}$ cannot have the same cardinality when $0 \leq m < n$.

30. Let A be a set with cardinality n and B a set with cardinality m.

   (a) Show that if $n > m$, then there is no onto function $f : B \longrightarrow A$. Hint: $\forall\, a \in A$ pick an element $g(a)$ of $f^{-1}(\{a\})$.

   (b) Show that if there is an onto function $f : B \longrightarrow A$, then $m \geq n$.

31. Let A be a set with cardinality n and B a set with cardinality m. Show that if there is a bijection $f : A \longrightarrow B$, then $n = m$.

32. Let A be a set with cardinality n and B a set with cardinality m. Show that if there is a one-to-one function $f : B \longrightarrow A$, then $m \leq n$.

33. Suppose that nonnegative integers are stored in a computer in binary using 16 bits. An array of such integers can be regarded as a function mapping each index in the array to the integer value contained at that index. Use the Pigeon Hole Principle to argue that an array of 70,000 nonnegative 16-bit integers cannot contain all distinct values.

34. Before every NBA basketball game, the starters on each team are introduced. Suppose that a coach uses the same 5 starters in every game, but he wants to have them introduced in a different order each game. (a) Can he accomplish this in an 82-game season? (b) Use the Pigeon Hole Principle to argue that the coach cannot achieve his goal if there are 132 games in a season. Hint: There are $n(n - 1)(n - 2) \cdots 2 \cdot 1$ ways to order n items.

35. Let A, B, C, D be sets, and suppose that A and C have the same cardinality and that B and D have the same cardinality. Show that $A \times B$ and $C \times D$ have the same cardinality.

36. Let A and B be disjoint sets in some universal set $\mathcal{U}$, and let C and D be disjoint sets in some universal set $\mathcal{V}$. Suppose that A and C have the same cardinality and that B and D have the same cardinality. Show that $A \cup B$ and $C \cup D$ have the same cardinality.

37.* Let A and B be finite sets. Show: $A \times B = |A| \cdot |B|$. Hint: Consider $|A|$ arbitrary but fixed, and use induction on $|B|$.

38.* Let A be a finite set. Show: $|\mathcal{P}(A)| = 2^{|A|}$. Hint: Use induction on $|A|$.

39. Show that $\mathbb{Z}^+ \times \mathbb{Z}^+$ is countably infinite. Hint: Use Exercise 69 from Section 5.4.

40. Show that the set E^+ of even positive integers is countable and that the set O^+ of odd positive integers is countable.

41.* Show that the product of countable sets is countable. Hint: Use Exercises 35 and 39 to build a map $C_1 \times C_2 \longrightarrow \mathbb{Z}^+ \times \mathbb{Z}^+ \longrightarrow \mathbb{Z}^+$.

42.* Show that the disjoint union of countably infinite sets is countably infinite. Hint: Use Exercises 36 and 40 to build a map $C_1 \cup C_2 \longrightarrow \mathbb{Z}^+ \cup \mathbb{Z}^+ \longrightarrow E^+ \cup O^+ = \mathbb{Z}^+$.

43. Imagine that there is a hotel, called Hotel Infinity, that contains a countably infinite number of rooms. The nice feature of this hotel is that even when it is full, it can always take another guest. For convenience, suppose the rooms are numbered $1, 2, 3, \ldots$ and G is the set of guests. If the hotel is full, then we have a bijection $g : \mathbb{N} \longrightarrow G$. Suppose that a new guest z would also like a room. Find a new bijection $h : \mathbb{N} \longrightarrow G \cup \{z\}$ to accomplish this. Hint: Effectively, we can ask each current guest to move up one room.

44. We have seen in Exercise 19 that whenever $b > a$, the line segments (intervals) $[a, b]$ and $[0, 1]$ have the same cardinality, even though their lengths may be different. Similarly, we can consider two-dimensional shapes. Show that for any positive length L and positive width W, a rectangle with dimensions $L \times W$ will have the same cardinality as the unit square (i.e., with dimensions 1×1), even though their areas may be different. Hint: Refer to Exercise 35 and describe each rectangle as a product of closed intervals.

45.* (a) Show: Any subset of any finite set is finite. Hint: Use induction on the cardinality of the finite set.

(b) Show: Any superset of any infinite set is infinite.

46.* Let A be an infinite set. Show: There is a sequence $\{a_n\}_{n \geq 1}$ such that $\forall\, n \geq 1, a_n \in A \setminus \{a_1, \ldots, a_{n-1}\}$.

(a) Conclude that for each $n \geq 1$, A contains a set of cardinality n.

(b) Conclude that A contains a countable subset.

47.* Let A be an infinite set.

(a) Show: If $A \subseteq \mathbb{Z}^+$, then A is countably infinite. Hint: The Well-Ordering Principle for $\mathbb{Z}^+$ allows one to define a function $f : \mathbb{Z}^+ \longrightarrow A$ by $f(1) = \min(A)$ and,

for $n \geq 1$, $f(n) = \min(A \setminus \{f(1), f(2), \ldots, f(n-1)\})$.
Argue that f is a bijection.

(b) Show: If there is a one-to-one map $g : A \longrightarrow \mathbb{Z}^+$, then A is countably infinite.

48. Let A be an infinite set. Show: If there is an onto map $f : \mathbb{Z}^+ \longrightarrow A$, then A is countably infinite. Hint: For each $a \in A$, pick an element $g(a)$ from $f^{-1}(\{a\})$. Then apply Exercise 47(b) for this function g. A systematic choice can be accomplished using the Well-Ordering Principle for $\mathbb{Z}^+$ by always picking the smallest element of $f^{-1}(\{a\})$.

49. Let $A \subseteq B$. Show: If B is countable, then A is countable. Hint: Use Exercise 47.

50. Show that the union of two countable sets is countable. Hint: $A \cup B = A \cup (B \setminus A)$.

51. Show that $\mathbb{Q}$ is countably infinite. Hint: Consider the function $h : \mathbb{Z} \times \mathbb{Z}^+ \longrightarrow \mathbb{Q}$ defined by $h((m, n)) = \frac{m}{n}$. Specify an onto function $g : \mathbb{Z}^+ \longrightarrow \mathbb{Z} \times \mathbb{Z}^+$, and apply Exercise 48 to the composite $h \circ g$.

52. Show that the subset of $(0, 1)$ consisting of those real numbers whose decimal digits are in $\{0, 1\}$ is uncountable.

53. Show that the countable union of countable sets is countable. That is, if B is countable, and, for each $b \in B$, A_b is countable, then $\cup_{b \in B} A_b$ is countable. Hint: Given bijections $g : \mathbb{Z}^+ \longrightarrow B$ and, for each $b \in B, f_b : \mathbb{Z}^+ \longrightarrow A_b$, define $h : \mathbb{Z}^+ \times \mathbb{Z}^+ \longrightarrow \cup_{b \in B} A_b$ by $h((m, n)) = f_{g(m)}(n)$.

54. (a) Show that the countable product of finite sets need not be countable. That is,

$$\prod_{n \in \mathbb{Z}} A_n = \{(x_1, x_2, x_3, \ldots) : x_i \in A_i \; \forall \, i\}$$

need not be countable even though A_i is finite for all i.

(b) Could it be countable for some choices?

5.7 Review problems

For Exercises 1 through 4, let R be the relation from $\{0, 1, 2, 3, 4\}$ to $\{0, 1, 2\}$ defined by $x \, R \, y$ if and only if $x = 2^y$.

1. (a) Is $2 \, R \, 1$?   (c) Is $1 \, R \, 2$?
   (b) Is $0 \, R \, 0$?

2. Find the matrix that represents R.

3. Find the inverse of R.

4. Find the matrix that represents the inverse of R.

5. The registrar's office at a science and engineering college is keeping track of their students' majors in the table below. Note that a student may have two majors.

Student	Major
Abe Roth	Computer science
Megan Johnson	Mathematics
Richard Kelley	Computer science
Martha Lang	Physics
Abe Roth	Mathematics

(a) Is Martha Lang a mathematics major?

(b) Is computer science a major for Richard Kelley?

(c) What are Abe Roth's majors?

6. Draw an arrow diagram for the relation $\in$ from $\{a, b\}$ to $\mathcal{P}(\{a, b\})$.

7. A game is being played on the grid of points with integer coordinates shown on the left-hand side of Table 5.12. Starting from the origin $(0, 0)$, a player's first move can be one unit in any of the four compass directions N, S, E, or W. However, each subsequent move is restricted by the previous move according to the relation on $\{N, S, E, W\}$ represented by the matrix on the right-hand side of Table 5.12.

(a) Find a sequence of moves taking a player from $(0, 0)$ to $(1, 1)$.

(b) Starting from $(0, 0)$, is it possible to get to $(0, -2)$?

(c) Draw a digraph for the relation on $\{N, S, E, W\}$ represented by the given matrix.

For Exercises 8 through 11, let R be the relation on $\mathbb{R}$ defined by

$$x \, R \, y \quad \text{if and only if} \quad y^2 - x^2 = 1.$$

TABLE 5.12

A random walking game

$$\begin{array}{c}
\begin{array}{cccc} N & S & E & W \end{array} \\
\begin{array}{c} N \\ S \\ E \\ W \end{array}
\begin{bmatrix}
1 & 1 & 0 & 0 \\
1 & 0 & 1 & 1 \\
1 & 0 & 0 & 0 \\
0 & 0 & 0 & 0
\end{bmatrix}
\end{array}$$

8. (a) Is $\sqrt{3}\ R\ \sqrt{2}$?

   (b) Is $0\ R\ -1$?

   (c) Is $1\ R\ 0$?

9. Draw the graph of R.

10. Find the inverse of R.

11. Draw the graph of the inverse of R.

12. Given the binary relation R on $\mathbb{R}$ defined by

$$x\ R\ y \quad \text{if and only if} \quad x^2 + y^2 = 1,$$

for each of the properties reflexive, symmetric, antisymmetric, and transitive, prove or disprove that R has that property.

13. Given the binary relation R on $\mathbb{Z}^+$ defined by

$$a\ R\ b \quad \text{if and only if} \quad a \mid b \text{ and } a \neq b,$$

for each of the properties reflexive, symmetric, antisymmetric, and transitive, prove or disprove that R has that property.

14. Let R_1 and R_2 be partial order relations on a set X, and define a new relation R on X by $x\ R\ y$ if and only if $x\ R_1\ y$ and $x\ R_2\ y$. Show that R is also a partial order relation on X.

15. Given any set X, what is the only relation on X that is both a partial order relation and an equivalence relation?

16. Draw a Hasse diagram for the "divides" relation $\mid$ on the set of positive divisors of 30.

17. Given the standard ordering $\leq$ on $\mathbb{Z}$, what is the lexicographic ordering $\preceq$ for $(6, -1, 2, -4, 7)$ and $(6, -1, -3, 5, 2)$ in $\mathbb{Z}^5$?

18. If a sports league allows ties in its games, then a team's record would consist of wins W, losses L, and ties T. At the end of the season, the teams are ranked from lowest to highest according to the value $\frac{W + \frac{1}{2}T}{W + L + T}$ associated with each team's record (W, L, T). Assume that each team plays 60 games in a season.

   (a) Show that the ranking of the teams from lowest to highest need not be the same as the order determined by the lexicographic ordering on the triples (W, L, T) in $\mathbb{Z}^3$.

   (b) How could we change the way we keep track of records so that these two orderings would agree?

19. Let $X = \mathbb{R}$, and let R be the relation on X defined by

$$x\ R\ y \quad \text{if and only if} \quad x^2 = y^2.$$

Show that R is an equivalence relation.

20. For the equivalence relation R on $\mathbb{R}$ from Exercise 19, for each $x \in \mathbb{R}$,

    (a) Give an explicit description of the set $[x]$ by listing its elements.

    (b) Specify the nonnegative representative for $[x]$.

21. For each $n \in \mathbb{Z}$, let $A_n = (n - 1, n]$. Let $\mathcal{A} = \{A_n : n \in \mathbb{Z}\}$. Show that $\mathcal{A}$ partitions $\mathbb{R}$.

22. For each $x \in \mathbb{R}$, let $A_x = \{x\} \times \mathbb{R}$.

    (a) What does A_x look like in $\mathbb{R}^2$?

    (b) Let $\mathcal{A} = \{A_x : x \in \mathbb{R}\}$. Show that $\mathcal{A}$ partitions $\mathbb{R}^2$.

23. For   each $n \in \mathbb{Z}^+$,   let $A_n = \{r : r \in \mathbb{Q}, nr \in \mathbb{Z}\}$. Does $\{A_n : n \in \mathbb{Z}^+\}$ form a partition of $\mathbb{Q}$? Why?

24. Find the partition corresponding to the equivalence relation R on $\mathbb{R}$ from Exercise 19.

25. Find the equivalence relation R that corresponds to the partition from Exercise 21.

26. Find the equivalence relation R that corresponds to the partition from Exercise 22.

27. Why does the formula

$$\forall\, x \in \mathbb{R}, \quad f(x) = \begin{cases} 2 - x & \text{if } x \leq 1, \\ x + 1 & \text{if } x \geq 1 \end{cases}$$

not define a function?

28. Define f from $\mathbb{Z}$ to $\mathbb{Z}$ by $n \mapsto 2^n$. Does this give a function $f : \mathbb{Z} \longrightarrow \mathbb{Z}$?

29. Define $f : \{2, 3, 4, 5\} \longrightarrow \{1, 2, \ldots, 10\}$ by $n \mapsto \binom{n}{2}$. Find the domain and range of f.

30. Define $f : [-1, 2] \longrightarrow \mathbb{R}$ by $x \mapsto 2 - x$. Show that the range of f is $[0, 3]$.

31. Specify the domain and range of $f(x) = 1 + \frac{1}{x-2}$.

32. Define functions $f : \mathbb{Z}^+ \longrightarrow \mathbb{R}$ and $g : \mathbb{R} \longrightarrow \mathbb{R}$ by $f(n) = \sqrt{n - 1}$ and $g(x) = x^2 - 1$. Find $g \circ f$.

33. BookEmDotCom is an Internet book supplier that uses an extensive database to keep track of its business operations. In particular, it must keep track of both the sales it makes and the suppliers of the items ordered. Two tables from the database for BookEmDotCom are shown in Table 5.13.

    (a) Among the books sold by BookEmDotCom, which ones are published by Book Farm?

    (b) Compose the "Orders" table with the "Publications" table to obtain a new "Sells To" table.

    (c) To what customers is Book Farm selling books?

TABLE 5.13
Database for BookEmDotCom

Book	Customer	Publisher	Book
Of Mice and Cats	Raul Cortez	Book Farm	Of Mice and Cats
Fahrenheit 212	Mary Wright	Authority Pubs	Fahrenheit 212
Raisins of Wrath	David Franklin	Word Factory	War and Peas
War and Peas	Mary Wright	Book Farm	Raisins of Wrath

| Orders | Publications |

34. Let R be a relation from X to Y and S be a relation from Y to Z. Show that the inverse of $S \circ R$ is $R^{-1} \circ S^{-1}$.

35. Define $f : \mathbb{Z} \longrightarrow \mathbb{Z}$ by $f(n) = 3n - 2$. Show that f is one-to-one.

36. Show that $f(x) = x^3 - x$ is not one-to-one.

37. Define $f : \{-3, -1, 2, 3\} \longrightarrow \{0, 3, 8\}$ by $n \mapsto n^2 - 1$. Show that f is onto.

38. Show that $f(x) = x^2 - 10$ is not onto.

39. The town tax collector is using social security numbers to identify each resident's account in the town database. The hashing function h that is being used for social security numbers $d_1 d_2 d_3 \text{-} d_4 d_5 \text{-} d_6 d_7 d_8 d_9$ is defined by taking n to be the base-ten value of $d_6 d_7 d_8 d_9$ and then computing $h(n) = n \bmod 225$.

    (a) To what value does 036-77-5484 hash?

    (b) Give examples of two distinct social security numbers that hash to the same value.

40. Let $n \in \mathbb{Z}^+$, and define $f : \mathbb{Q} \longrightarrow \mathbb{Q}$ by $r \mapsto \frac{r}{n}$. Show that f is a bijection.

41. Define $f : \mathbb{Z} \longrightarrow \mathbb{Z} \times \{0, 1\}$ by

$$f(n) = \begin{cases} (\frac{n}{2}, 0) & \text{if } n \text{ is even,} \\ (\frac{n-1}{2}, 1) & \text{if } n \text{ is odd.} \end{cases}$$

Show that f is a bijection.

42. Suppose $f : X \longrightarrow Y$ is a one-to-one function, and let Y' be the range of f. Define $f' : X \longrightarrow Y'$ by $f'(x) = f(x)$. Show that f' is a bijection.

43. Define the function $g : [0, 2] \longrightarrow [0, 1]$ by

$$g(x) = \begin{cases} x & \text{if } x \in [0, 1], \\ 1 & \text{if } x \in (1, 2]. \end{cases}$$

    (a) Show: For any function $f : [0, 2] \longrightarrow [0, 1]$, that $g \circ f = f$.

    (b) Find a function $f : [0, 2] \longrightarrow [0, 1]$ such that $f \circ g \neq f$.

44. Define functions $f, g : \mathbb{R} \longrightarrow \mathbb{R}$ by $f(x) = \frac{2-x}{3}$ and $g(x) = 2 - 3x$. Show that f and g are inverses of one another.

45. Show: Given any function $f : X \longrightarrow Y$, there is at most one function $g : Y \longrightarrow X$ such that f and g are inverses of one another.

46. Find $\log_5(125)$.

47. Find $\ln(e)$.

48. Define $f : \mathbb{Z} \longrightarrow \mathbb{Z}$ by $f(n) = (-1)^n n^2$.

    (a) Find $f(\{-1, 0, 1, 2\})$.

    (b) Find $f^{-1}(\{-1, 0, 1, 4, 9\})$.

49. For $f(x) = 4 - x^2$,

    (a) Find $f([-1, 2])$.

    (b) Find $f^{-1}([1, 4])$.

50. Consider the function $f : [-1, 1] \longrightarrow (-4, 4)$ defined by $x \mapsto 1 - 2x$. Fill in the missing values.

S	f(S)	T	f⁻¹(T)
$\{1\}$		$\{1\}$	
$[0, 1]$		$[1, 4)$	
$(-1, 0)$		$(-4, -2)$	

51. A table from the database maintained by the registrar's office at a science and engineering college is given in Exercise 5. That table can be used to determine the students in each major.

    (a) What students are majoring in mathematics or physics?

    (b) Which type of computation is needed in part (a), an image or an inverse image?

52. Determine whether the following statements are true or false for all functions $f : X \longrightarrow Y$:

    (a) For all $A, B \subseteq Y$, $f^{-1}(A \cap B) = f^{-1}(A) \cap f^{-1}(B)$.

    (b) For all $A, B \subseteq X$, $f(A \cap B) = f(A) \cap f(B)$.

53.* Let $f : X \longrightarrow Y$ be an arbitrary function, and let $S_1, S_2 \subseteq X$ and $T_1, T_2 \subseteq Y$ be arbitrary subsets. Prove or disprove:

    (a) $f(S_1 \triangle S_2) \subseteq f(S_1) \triangle f(S_2)$.

    (b) $f^{-1}(T_1 \triangle T_2) = f^{-1}(T_1) \triangle f^{-1}(T_2)$.

54. Find $\bigcup_{r \in [0,2]} (r, r+3)$.

55. Find $\bigcap_{k \in \mathbb{Z}^+} \{m : m = nk \text{ for some } n \in \mathbb{Z}\}$.

56. The town tax collector needs to distinguish various income brackets. For each $r \in \mathbb{R}$, let A_r be the set of

Americans whose annual salary in thousands of dollars is at least r but less than $r + 10$. Use either an indexed union or an indexed intersection to express the set S of Americans whose annual salary is at least \$80,000 but less than \$125,000. Could a finite union or intersection be used?

57.* Let B be a set, and, for some nonempty indexing set $\mathcal{I}$, $\{A_i\}_{i \in \mathcal{I}}$ be an indexed collection of sets in some universal set $\mathcal{U}$. Show: $\bigcup_{i \in \mathcal{I}} (B \cup A_i) = B \cup \bigcup_{i \in \mathcal{I}} A_i$.

58. What is the cardinality of $\{4, 5, 6, 5, 4\}$?

59. Show: $|\{-100, -99, \ldots, 199, 200\}| = 301$.

60. Refer to the table from the registrar's database in Exercise 5. How many different majors are there at this science and engineering college?

61. Show that $[-1, 0)$ and $(1, 7]$ have the same cardinality. Give a direct proof based on the definition of cardinality.

62. Show that $\{m : m = 2k \text{ for some } k \in \mathbb{Z}, 0 \leq k \leq n\}$ has cardinality $n + 1$.

63. Show that $\mathbb{Z}$ and $T = \{n : n = 10k \text{ for some } k \in \mathbb{Z}\}$ have the same cardinality.

64. Let A and B be sets. Show that $A \times B$ and $B \times A$ have the same cardinality. Note that A and B need not be finite.

65. Let A, B, C, D be sets, and suppose that A and C have the same cardinality and B and D have the same cardinality. Prove or Disprove: If $A \subseteq B$ and $C \subseteq D$, then $B \setminus A$ and $D \setminus C$ have the same cardinality.

66.* Show that $[1, 2]$ is infinite.

67. Show that $\{n : n = 3k \text{ for some } k \in \mathbb{Z}^+\}$ is countable.

68.* Show that $\mathbb{R}^2$ is uncountable.

Part II

Combinatorics

6 Basic Counting

The basic techniques of counting are introduced. Our interest is in enumerating the number of ways a desired outcome can be achieved given certain constraints. For example, we count the number of possible license plates having a certain fixed number of characters subject to conditions on the characters. Such a computation might be useful for police officers trying to apprehend a suspect in a hit-and-run accident on the basis of a partial license plate given by a witness. By using factorials, permutations, and combinations as our basic building blocks and the multiplication and addition principles as our tools for assembling these blocks, we are able to build up to solving more and more complex enumeration problems.

Our counting skills also enable us to compute probabilities. For example, by counting the number of ways to get a full house in 5-card stud poker, we can determine the likelihood of getting a full house. When gambling is involved, it is valuable to know relative probabilities. With our tools, we will be able to determine the extremely low likelihood of winning state lotteries, for example.

A great benefit of developing our counting skills is that those skills can be applied to many and varied contexts, which nonetheless share a common mathematical structure. For example, tools used to count license plates can also be used to count ways in which workers might be broken into teams. Those same tools also apply to analyses of jury selections and can be used to study the security of passwords and other codes. Several applications of counting techniques are encountered throughout this chapter. Ways in which initial overcounting can be corrected to a desired value are also explored.

6.1 The multiplication principle

The primary technique of counting is to break a complex problem into a sequence of simpler ones. Additionally, one must understand how to combine the answers to the simple problems to obtain the desired answer to the complex one. This section introduces one of the most basic tools of counting and starts with an example.

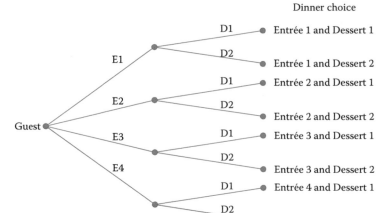

FIGURE 6.1
A guest's dinner choice.

Example 6.1. (Dinner Choices).
A guest at a formal dinner has 4 entrée choices and 2 dessert choices. If a guest's dinner is entirely determined by these two choices, then how many different dinner choices are there?

Solution. The total number of dinner choices is given by the product

$$4 \cdot 2 = 8.$$

The reason can be seen in the tree in Figure 6.1. Each of the 4 branches connected to the root of that tree splits off into 2 more branches. The 8 different ends represent the various possible pairings of an entrée choice with a dessert choice.   □

Example 6.1 motivates our first fundamental principle of counting.

Theorem 6.1. (Multiplication Principle).
Let A be a set of outcomes we wish to count. If there is a set of outcomes A_1, and, for each outcome in A_1, there is a set of outcomes A_2 such that

(i) *each outcome from A can be uniquely characterized by a pair of outcomes, the first from A_1 and the second from its corresponding set A_2, and*

(ii) *for each outcome from A_1, the number $|A_2|$ is the same,* *then*

$$|A| = |A_1| \cdot |A_2|.$$

In Example 6.1, the set A to which the Multiplication Principle applies is the set of dinner choices, A_1 is the set of entrée choices, and (for each entrée choice) A_2 is the set of dessert choices. It is important in that example that the number of dessert choices does not depend on any entrée choice.

Example 6.2.   (Home Security).
If a home security code consists of a sequence of two distinct letters (A to Z), then how many different codes are possible? If a thief could only try 5 possible codes every 30 minutes, then how long would it take the thief to try all possible codes?

Solution.   The first letter can be any of the letters A to Z. Hence, there are 26 choices for the first letter. Once the first letter is chosen, there remain only 25 letters from which to choose the second letter. Since the number 25 is independent of the particular first letter chosen, the Multiplication Principle tells us that there are

$$26 \cdot 25 = 650$$

possible security codes. At a rate of 5 codes per each 30 minutes, it would take the thief

$$\frac{650}{5/30} \frac{\text{codes}}{\text{codes/minute}} = 3900 \text{ minutes} = 65 \text{ hours}$$

to try all possible codes.   □

Note in Example 6.2 that the choice for the second letter does depend on the choice of the first letter. However, the *number* of choices for the second letter does not depend on the choice of the first letter. This is all that is required to apply the Multiplication Principle. In Example 6.1, the fact that the dessert choice did not depend on the entrée choice was a stronger-than-necessary condition (though certainly sufficient) for the Multiplication Principle.

A better appreciation for the Multiplication Principle can be gained by considering an example in which it does not apply.

Example 6.3.   A 2-digit code is constructed using the digits 1, 2, 3. If the second digit is required to be at least as large as the first digit, then how many such codes are possible?

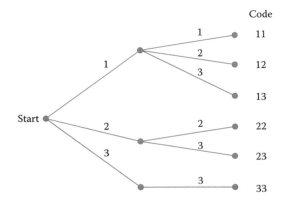

FIGURE 6.2
Constructing a security code.

Solution. There are certainly 3 choices for the first digit. However, the number of choices for the second digit does depend on the first digit. For example, if the first digit is a 1, then there are 3 choices for the second digit. Alternatively, if the first digit is a 2, then there are only 2 choices for the second digit. Therefore, the Multiplication Principle does not apply in the way it did in Example 6.2. Instead, the problem can be analyzed with the tree in Figure 6.2. We see that there are 6 possible codes.   □

Our primary focus here is on problems to which the Multiplication Principle does apply. Of course, the Multiplication Principle as stated in Theorem 6.1 generalizes to problems characterized by a sequence of more than two steps.

Suppose that $n \geq 2$ and each outcome in a set A is uniquely characterized by a sequence of outcomes, one from each of a sequence of sets $A_1, A_2, \ldots, A_n$. In this general case, the requirement is that for each $2 \leq k \leq n$, the number $|A_k|$ must not depend on any of the sets A_i for $1 \leq i \leq k - 1$. Under those conditions, $|A|$ is given by the product

$$|A| = |A_1| \cdot |A_2| \cdots |A_n|.$$

Of course, this **General Multiplication Principle** follows from Theorem 6.1 by an inductive argument.

Example 6.4. (License Plates).
In a state that has just under 13 million cars registered, the Department of Motor Vehicles is considering a new requirement that license plates must consist of 3 letters

(A to Z) followed by 4 digits (0–9). Could all of the cars registered in this state be accommodated by this system?

Solution. There are 7 places on the license plate

$$\underbrace{\quad\ \ \quad\ \ \quad\ \ \quad}_{\text{Letters}}\ \underbrace{\quad\ \ \quad\ \ \quad\ \ \quad}_{\text{Digits}}$$

that need to be filled. For each of the first three slots, there are 26 choices. For each of the last four slots, there are 10 choices. Moreover, the choice for each slot is independent of the choices for the other slots. Hence, there are

$$26 \cdot 26 \cdot 26 \cdot 10 \cdot 10 \cdot 10 \cdot 10 = 26^3 \cdot 10^4 = 175760000$$

different license plates possible. Since this number is greater than 13 million, the system could accommodate all of the registered cars.   □

Example 6.5. How many octal (base-8) numbers are there of length n? Note that the number is allowed to start with zeros.

Solution. Since each of the n digits can take any one of the 8 values 0 through 7, there are

$$\underbrace{8 \cdot 8 \cdots\cdots 8}_{n \text{ terms}} = 8^n$$

octal numbers of length n.   □

Example 6.6. (Rolling Dice).
Three standard (6-sided) dice are rolled. One is red, one is blue, and one is green. If the particular number showing on each colored die is important, then how many different outcomes are possible?

Solution. The outcome of a roll can be thought of as a sequence of length 3 (red, blue, green) in which each entry is one of the numbers 1 through 6. There are

$$6^3 = 216$$

such sequences.   □

We saw in Example 6.3 that not every counting problem is solved by a direct application of the Multiplication Principle. Of course, those that are not do not all require drawing trees. Some counting problems are simply handled by basic principles.

Example 6.7. How many 3-digit (base-10) numbers (with nonzero hundreds digit) are divisible by 7?

Solution. Of the 3-digit numbers

$$100, 101, \ldots, 104, 105, 106, \ldots, 111, 112, 113,$$

$$\ldots, 993, 994, 995, \ldots, 999$$

the numbers

$$105, 112, \ldots, 994$$

are divisible by 7. That is, we need to count the multiples of 7:

$$105 = 7(15), \quad 112 = 7(16), \ldots, 994 = 7(142).$$

Hence, there is a correspondence between the multiples of 7 and the numbers

$$15, 16, \ldots, 142.$$

Therefore, the number of 3-digit positive integers (with nonzero hundreds digit) divisible by 7 equals the number of integers from 15 to 142, inclusive. That is, there are

$$142 - 15 + 1 = 128$$

3-digits multiples of 7. □

Exercises

1. An outfit consists of a skirt and a blouse. If there are 6 skirts and 8 blouses from which to choose, then how many different outfits are possible? We assume that any possible combination can be worn together.

2. A computer system consists of a tower, a monitor, and a printer. If an electronics store has 8 different towers, 4 different monitors, and 6 different printers, then how many different computer systems can be purchased from this store?

3. Baskin-Robbins made itself famous for selling 31 flavors of ice cream. Suppose a Baskin-Robbins ice cream shop has waffle, sugar, and regular cones in sizes kiddie, small, medium, and large. How many different ice cream cones are possible in that shop?

4. Two college students are planning a date that will consist of dinner, a movie, and dancing. They are deciding among Applebee's, Red Lobster, Ruby Tuesdays, TGI Friday's, and the Outback Steakhouse for dinner. There are 8 movies playing, and they are considering Disco Dan's,

The Calypso, and The Night Light for dancing after the movie. How many different dates might they plan?

5. How many sequences of two letters (A, B, C, or D) are in alphabetical order (repeats are allowed)? Why does the Multiplication Principle not apply by considering each of the two letters as an outcome?

6. A certain code is to consist of two digits (0–9). If the first digit is a multiple of three, then the second digit must be even. Otherwise, the second digit must be a multiple of three. How many such codes are possible? Why does the Multiplication Principle not apply by considering each of the two digits as an outcome?

7. A certain game consists of two steps. First, a standard die is rolled, and the number showing is recorded. If the number is even, then a coin is tossed, and its result is recorded. Otherwise, a card is drawn from a standard deck, and the suit is recorded.

   (a) How many different outcomes are there for this game?

   (b) Is the number of outcomes involving a coin the same as the number involving a card? Explain.

8. To remotely retrieve messages from a certain answering machine, a code is entered after the beep. That code, which needs to have been set by the user, must consist of a digit (0, 1, or 2) followed by that number of letters (A to Z).

   (a) How many such codes are possible?

   (b) Since letters are entered into a phone using the scheme shown in Figure 6.3, the number of possible key sequences is smaller than the number of codes

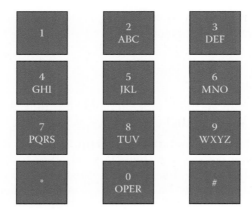

FIGURE 6.3
Phone keypad.

determined in part (a). For example, "1D" and "1F" are both entered as "13" on the keypad. Consequently, how many different key sequences represent all of the possible codes?

9. The Department of Motor Vehicles (DMV) in a certain state is discontinuing vanity plates. Moreover, this DMV is considering requiring license plates to consist of 3 letters (A to Z) followed by 3 digits (0–9), such as

┌─────────┐
│ HLP 911 │
└─────────┘

   (a) How many cars could this system accommodate?

   (b) What if "911" is reserved for emergency vehicles and cannot be used on a civilian plate?

   (c) What if the three-letter sequence "ASS" is also not allowed?

10. A home security code consists of 2 letters (A to Z) followed by 4 digits (0–9). To gain access to the home, the code must be entered in a keypad. A thief has a device that enables him to attempt 1 code every second.

   (a) How many hours would it take him to try all possible codes?

   (b) What if he knew the first two letters were the same?

   (c) What if he knew the first two letters were different?

11. A certain ATM code is required to consist of 4 distinct digits (0–9). If the ATM machine will only allow 3 guesses per day, then how many days would it take to try all possible ATM codes?

12. A computer account requires a password consisting of 6 characters (letters A to Z or digits 0–9). If one password can be tried every 5 seconds, then how long would it take to try all possible passwords?

13. How many license plates consisting of 2 nonrepeating letters (A to Z) followed by 3 even digits (0–8) are possible?

14. How many 6-digit numbers have odd digits in the odd places and even digits in the even places?

15. A certain Internet provider requires that its e-mail addresses consist of 6 characters (letters A to Z or digits 0–9). Further, the first character must be a letter, and the last character must be a digit. How many different e-mail accounts are possible from this provider?

16. How many 5-digit numbers neither start with zero nor end in zero?

17. How many 6-digit numbers (leading zeros are allowed) have the last two digits the same as the first two digits (in the same order)?

18. A palindromic number, or **palindrome**, is a natural number whose digits remain the same if the ordering of the digits is reversed. How many 6-digit numbers (leading zeros are allowed) are palindromes? What if leading zeros are not allowed?

19. A certain license plate is required to consist of 3 distinct letters (A to Z) followed by 4 digits (0–9). How many are possible?

20. A certain license plate is required to consist of 2 letters (A to Z) followed by 5 distinct digits (0–9). How many are possible?

21. A **DNA** (deoxyribonucleic acid) molecule consists of two parallel strands of nitrogen bases. The possible bases are cytosine (C), guanine (G), thymine (T), and adenine (A). However, the two strands are bonded together pairwise and are complementary in the sense that the only possible base pairs are C with G and T with A. Hence,

$$\begin{matrix} \text{G} & \text{T} & \text{C} & \text{A} & \text{C} & \text{A} \\ \text{C} & \text{A} & \text{G} & \text{T} & \text{G} & \text{T} \end{matrix}$$

is a possible segment of length 6 in a DNA molecule, but one with G paired over A is not. How many different segments of length 6 are possible?

22. A **chromosome** is a body that contains DNA in the nucleus of a cell. Variations in chromosomes allow for variations in cell functions. If a chromosome has a DNA molecule (see Exercise 21) with 10^8 base pairs, then how many different chromosomes are mathematically possible?

23. How many hexadecimal (base-16) numbers of length 10 are there? Note that such numbers may start with zeros.

24. How many different sequences of 5 coin flips are possible?

25. Some countries have flags that consist of three vertical stripes, each of different colors. For example, the stripes on the flag of Ireland are green–white–yellow, and those on the flag of Côte d'Ivoire (also known as the Ivory Coast) are yellow–white–green. How many such flags are possible if the color choices are red, white, blue, green, and yellow?

26. A melody consists of three notes (C, D, E, F, G, A, or B). How many are possible if repeats are allowed?

27. A game consists of rolling a standard (6-sided) die until an even number is showing (only that even number is

recorded), flipping a coin, and drawing a card from a standard deck. How many possible outcomes does this game have?

28. A game consists of flipping a coin, rolling a 4-sided die, and drawing a card from a standard deck until a heart is obtained (only the denomination of that heart is recorded). How many possible outcomes does this game have?

29. How many integers are there from 85 to 850?

30. How many integers are there from 17 to 1700?

31. A roadside guard rail is built by placing a post every 10 yards and then attaching a metal barrier connecting them. How many posts are needed to build a 100 yard guard rail?

32. A bucket of paint will paint 1 mile of the yellow lines on the highway. How many buckets are needed to paint from the 236 mile marker to the 287 mile marker on the highway?

33. How many of the integers from 1 to 7000 are divisible by 28?

34. How many of the integers from 1 to 3000 are divisible by 20?

35. How many of the integers from 80 to 8000 are divisible by 7?

36. How many of the integers from 20 to 2000 are divisible by 13?

37. How many of the integers from 200 to 2200 are divisible by 6?

38. How many of the integers from 150 to 1515 are divisible by 8?

39. Find $|\{n \,:\, 17 \le n \le 777 \text{ and } n \text{ is even}\}|$.

40. Find $|\{n \,:\, 30 \le n \le 3003 \text{ and } n \text{ is odd}\}|$.

41.* Determine the number of 8-digit binary numbers that are divisible by 3. Leading zeros are allowed.

42.* Determine the number of 6-digit base-3 numbers that are divisible by 4. Leading zeros are allowed.

Calendar problems. For Exercises 43 through 48, recall that leap years are years that are divisible by 4, with the exception that years divisible by 100 are leap years only if they are also divisible by 400.

43. How many days after 12/7/1941 (Pearl Harbor bombed) was 9/11/2001 (the September 11 terrorist attack)?

44. How many days after 4/14/1865 (Lincoln shot) was 4/4/1968 (Martin Luther King, Jr. shot)?

45. How many days after 12/17/1903 (First manned flight) was 7/20/1969 (First man on moon)?

46. How many days after 3/10/1876 (Bell's first telephone call) was 1/26/1926 (Baird's first public demonstration of television)?

47. How many days after 7/4/1776 (Declaration of Independence) was 1/1/1863 (Emancipation Proclamation signed by Lincoln)?

48. How many days after 9/4/1929 (Stock market crash) was 10/19/1987 (Stock market crash)?

6.2 Permutations and combinations

Many counting problems can be thought of as a selection of some objects from a fixed set of objects. In some cases, the order in which the objects are selected is important, and in others it is not. As we shall see, permutations handle the former case, and combinations handle the latter.

6.2.1 Permutations

A **permutation** of a set of objects is an ordering of those objects. Permutations reflect selections for which an ordering is important.

Example 6.8. (Lining Up).
How many ways are there to put 8 children in a line to get ice cream?

Solution. Of course, the ordering of the children is important, especially to them! There are 8 children from which to pick the child who is first. Then there are 7 children left from which to pick the second child. Then there are 6 left, and so on. Therefore, there are

$$8 \cdot 7 \cdot 6 \cdot 5 \cdot 4 \cdot 3 \cdot 2 \cdot 1 = 8! = 40320$$

ways to line up the children.   □

Example 6.9. (Listing Elements of a Set).
The set $S = \{a, b, c, d, e\}$ is entered into the computer (e.g. using *Maple*) by listing its elements in some order between curly braces, such as

```
> S := {b, a, c, e, d} ;
```

Assuming that no element is listed more than once, in how many ways might this be done?

Solution. There are

$$5 \cdot 4 \cdot 3 \cdot 2 \cdot 1 = 5! = 120$$

ways to enter S. □

The factorials that arose in the previous two examples suggest a general result.

Theorem 6.2.   *The number of ways to put n distinct items in order is n!.*

In many problems, the entire set of objects under consideration need not be placed in order; only a subset may need ordering. Consequently, not every example in which there is an ordering requirement results in a complete factorial.

Example 6.10.   (Student Officers).
How many ways are there to elect a class president, vice president, and treasurer from a class of 30 students? We assume that each student is eligible for any office and that no student may be elected to more than one office.

Solution. Regard the election results as an ordered list of length 3.

President, Vice President, Treasurer

There are 30 ways to fill in the first office, then 29 ways left to fill in the second office, and finally 28 ways left to fill in the last. Hence, there are

$$30 \cdot 29 \cdot 28 = 24360$$

possible election results. □

Example 6.11.   How many different license plates consisting of 6 *distinct* digits (0–9) are possible?

Solution. Any digit can be the first. Then there are 9 digits left from which to choose the second. There are then 8 choices left for the third, and so on. Hence, there are

$$10 \cdot 9 \cdot 8 \cdot 7 \cdot 6 \cdot 5 = 151200$$

6-digit license plates with distinct digits. □

The computations in the previous two examples warrant the following terminology and notation.

$$
\begin{array}{llllllllll}
abc & abd & abe & acd & ace & ade & bcd & bce & bde & cde \\
acb & adb & aeb & adc & aec & aed & bdc & bec & bed & ced \\
bac & bad & bae & cad & cae & dae & cbd & cbe & dbe & dce \\
bca & bda & bea & cda & cea & dea & cdb & ceb & deb & dec \\
cab & dab & eab & dac & eac & ead & dbc & ebc & ebd & ecd \\
cba & dba & eba & dca & eca & eda & dcb & ecb & edb & edc
\end{array}
$$

FIGURE 6.4
Permutations of size 3 from $S = \{a, b, c, d, e\}$.

Definition 6.1. A **permutation** of k objects from a set of size n is an ordered list of k of the n objects. The number of permutations of k objects from n is denoted $P(n, k)$.

Example 6.12. If $S = \{a, b, c, d, e\}$ is the set of $n = 5$ objects under consideration and $k = 3$, then all of the possible permutations of 3 elements from S are listed in Figure 6.4. We see that $P(5, 3) = 60$.

In Example 6.10, we computed $P(30, 3) = 24360$. Example 6.11 showed us that $P(10, 6) = 151200$. The arguments used in those examples can easily be adapted to obtain a general formula for $P(n, k)$.

Theorem 6.3. *Let* $n, k \in \mathbb{Z}$ *with* $0 \leq k \leq n$. *Then*

$$
P(n, k) = n(n - 1) \cdots (n - k + 1) = \frac{n!}{(n - k)!}.
$$

Example 6.13. $P(25, 5) = \frac{25!}{20!} = 25 \cdot 24 \cdot 23 \cdot 22 \cdot 21 = 6375600$.

6.2.2 Combinations

Although some counting problems contain an ordering requirement, there are also situations in which order is not important. For example, we saw in Chapter 1 that the elements in a set need not be listed in any particular order.

Definition 6.2. A **combination** of k elements from a set of size n is a subset of size k.

Example 6.14. If $S = \{a, b, c, d, e\}$ is the set of $n = 5$ objects under consideration and $k = 3$, then all of the possible combinations of 3 elements from S are listed in Figure 6.5. We see that the number of combinations of 3 objects from 5 is equal to 10.

$$\{a,b,c\} \quad \{a,b,d\} \quad \{a,b,e\} \quad \{a,c,d\} \quad \{a,c,e\}$$
$$\{a,d,e\} \quad \{b,c,d\} \quad \{b,c,e\} \quad \{b,d,e\} \quad \{c,d,e\}$$

FIGURE 6.5
The combinations of size 3 from $S = \{a, b, c, d, e\}$.

The following theorem shows that we already have notation that keeps track of numbers of combinations.

Theorem 6.4.   *Let $n, k \in \mathbb{Z}$ with $0 \le k \le n$. Given a set of n distinct elements, the number of subsets of size k is given by the binomial coefficient*

$$\binom{n}{k} = \frac{n!}{k!(n-k)!}.$$

*That is, $\binom{n}{k}$ counts the number of combinations of k elements from a
set of size n.*

In Section 4.6, we saw that the binomial coefficient $\binom{n}{k}$ is
the kth entry in the nth row of Pascal's triangle and gets its
name from its occurrence in the Binomial Theorem as the
coefficient of $a^{n-k}b^k$ in the expansion of $(a+b)^n$. Theorem 6.4
is the reason why $\binom{n}{k}$ is read "n choose k." The quantity $\binom{n}{k}$ is
the number of ways to choose k things out of n. A proof of
Theorem 6.4 is given at the end of this section. Prior to that,
we consider a few basic examples involving combinations.
Many more examples are explored in Section 6.5.

Example 6.15.   (Student Council).
Given a class of size 30, in how many ways can a
student council of 3 students be selected (assuming
that all 3 members of the student council are equal in
power)?

Solution. Since the 3 student council positions are
considered equivalent, there is no order inherent in the
selection. We simply want a subset of 3 students from
the set of 30 students. There are

$$\binom{30}{3} = 4060$$

ways to make this selection.   □

Example 6.16.   How many binary numbers of length n have exactly k ones (where $0 \leq k \leq n$)? Leading zeros are allowed.

Solution. A binary number with exactly k ones can be characterized by specifying which k of the n positions are occupied by ones; by default, the remaining $n - k$ positions must be occupied by zeros. There are

$$\binom{n}{k}$$

ways to choose k of the n positions to be ones.   □

Example 6.17.   (Jury Selection).
When a jury is selected for a court case, the lawyers for both the prosecution and the defense have opportunities to accept or reject candidates from a jury pool. Suppose a jury pool contains 18 men and 16 women. How many possibilities are there for a jury that is composed of 5 men and 7 women?

Solution. We want to select a subset of size 5 from the set of 18 men and a subset of size 7 from the set of 16 women. By the Multiplication Principle, there are

$$\binom{18}{5}\binom{16}{7} = 8568 \cdot 11440 = 98017920$$

such selections.   □

Problems need not involve permutations exclusively or combinations exclusively.

Example 6.18.   How many license plates consisting of 6 digits (0 to 9) with exactly 2 of the same digit are possible?

Solution. These license plates can be characterized by the value and locations of the repeated digit, together with an ordered list of the remaining distinct digits. There are 10 digits from which to choose the repeated one. There are $\binom{6}{2}$ ways to choose the two positions to contain the repeated digit. Then there are $P(9,4)$ ways to fill in the 4 remaining distinct digits. By the Multiplication Principle, there are

$$10 \cdot \binom{6}{2} \cdot P(9,4) = 453600$$

6-digit license plates with exactly 2 digits the same.   □

The solution in Example 6.18 shows that the factors multiplied together need not always be in one-to-one correspondence with the entries in a sequence being considered. The first two factors 10 and $\binom{6}{2}$ do correspond to the placement of two digits into the license plate. However, neither one of them alone corresponds to the placement of any single digit.

Example 6.19. (Postnet Codes).
The U.S. Postal Service encodes ZIP + 4 codes on business reply cards and envelopes using the digit-to-code conversion shown in Table 6.1. In the **Postnet code**, each digit is thus encoded by a sequence of 5 bars, in which exactly 2 are long (and 3 are short). Since there are $\binom{5}{2}$ such sequences and 10 digits, the fact that $\binom{5}{2} = 10$ is at the heart of this system! For example, the Postnet code for Capital One in Richmond, VA 23286-9291 is

$$2 \quad 3 \quad 2 \quad 8 \quad 6 \quad 9 \quad 2 \quad 9 \quad 1 \quad 8$$

A check digit (an 8 in this case) is appended to the ZIP + 4 code so that the sum of all the digits is divisible by 10. Also, long guard bars are added to mark each end of the code.

Identification numbers and check digits were introduced in Section 3.2.

TABLE 6.1
Postnet code conversions

Digit	Code
0	‖|||
1	|||‖
2						
3	||‖|
4						
5						
6						
7					
8					
9						

6.2.3 Permutations vs. combinations

It is important to distinguish between situations that require permutations and those that require combinations. For example, suppose we wish to select 2 cards from a deck of 52 distinct cards. A common incorrect analysis of this problem gives that there are

$$52 \cdot 51 = P(52, 2) = 2652$$

possible 2-card hands. The argument is that there are 52 choices for the first card and then 51 choices left for the second card. However, in that reasoning, it is assumed that the order of the cards is important. For example, the selections $3\heartsuit$, $8\clubsuit$ and $8\clubsuit$, $3\heartsuit$ would be counted as different. If it only matters which 2 cards are obtained, then there are

$$\binom{52}{2} = 1326$$

ways to select 2 cards from the deck. The desired outcome (the *set* of 2 cards) is what is being counted, not the process by which someone might obtain those cards (first one, then another).

6.2.4 The proof of Theorem 6.4

For convenience here, let $C(n, k)$ denote the number of combinations of k elements from a set of size n. Our aim is to show that $C(n, k) = \frac{P(n,k)}{k!}$. We accomplish this by relating the permutations counted by $P(n, k)$ with the combinations counted by $C(n, k)$. Before we give a general proof, it is helpful to think about a particular example.

Consider the case in which $n = 5$, $k = 3$, and the set of 5 elements is $S = \{a, b, c, d, e\}$. The $P(5, 3) = 60$ permutations of 3 elements from S are listed in Figure 6.4, and the $C(5, 3) = 10$ combinations of 3 elements from S are listed in Figure 6.5. There are certainly more permutations than combinations. In fact, there are distinct permutations that correspond to the same combination. For example, all of the permutations in the first column of Figure 6.4 correspond to the combination $\{a, b, c\}$. In general, we can group together permutations that correspond to the same combination. In Figure 6.4, these groups are the columns. Each column lists the $3! = 6$ different orderings of the 3 elements contained therein. Since each group has size 6, the number of groups (combinations) must be $\frac{60}{6} = 10$. That is, $C(5, 3) = \frac{P(5,3)}{3!}$.

Proof of Theorem 6.4.
Let $S = \{a_1, a_2, \ldots, a_n\}$ be a set of size n. A listing of all of the permutations of size k from S has size $P(n, k)$. Each permutation $a_{i_1} a_{i_2} \cdots a_{i_k}$ in this list corresponds to a combination $\{a_{i_1}, a_{i_2}, \ldots, a_{i_k}\}$ of size k. Group together all of the

permutations that correspond to the same combination. The number of groups obtained is the number of combinations $C(n, k)$.

Two permutations $a_{i_1} a_{i_2} \cdots a_{i_k}$ and $a_{j_1} a_{j_2} \cdots a_{j_k}$ correspond to the same combination if and only if $a_{j_1} a_{j_2} \cdots a_{j_k}$ is an ordering of $a_{i_1} a_{i_2} \cdots a_{i_k}$. Since there are $k!$ orderings of $\{a_{i_1}, a_{i_2}, \ldots, a_{i_k}\}$, each group has size $k!$. Hence, the number of groups must be $\frac{P(n,k)}{k!}$. That is, $C(n, k) = \frac{P(n,k)}{k!}$.   □

Exercises

1. In how many ways can a family of 4 line up in a row to pose for a picture?

2. Tyrone Ecke, Seema Khan, and Carol Masters have applied for a programming job at a software company. The personnel director has asked the search committee to rank these candidates, and ties are not allowed. How many different rankings are possible?

3. Anna, Billy, Erica, Glen, Sam, and Tammy will take turns jumping on a trampoline. If each child takes one turn, then how many orderings of these children are possible?

4. A standard deck consists of 52 distinct cards. Prior to any card game, the deck is typically shuffled to place the cards in some unknown order. How many ways are there to order the cards?

5. In how many ways can a movie critic make a list of the top 10 movies (ranked 1–10) out of the 200 in the past year?

6. A neighborhood of 40 families has agreed to have one family volunteer for each night of the week to oversee the neighborhood watch program. In how many ways can the weekly schedule be set?

7. Compute $P(20, 4)$.   8. Compute $P(15, 6)$.

9. The final of the Olympic 100 meter dash results in 3 medalists. If there are 8 finalists, then in how many ways might the gold, silver, and bronze medals be awarded? Assume that there are no ties.

10. The United States has a team of 5 women ready to compete in the 4 by 100 meter relay. In how many ways can a running order of 4 of these women be selected?

11. Compute $\binom{20}{4}$.   12. Compute $\binom{15}{6}$.

13. The witness to the hit-and-run accident was not able to get a clear view of the license plate as the offending vehicle sped off. However, she did notice that it consisted of 6 letters (A to Z) and had exactly three A's. How many possible license plates meet the witness's description?

14. The license plates issued by a certain state have the form of 4 letters (A to Z) followed by 4 digits. A rumor is circulating that plates with two 7's designate that the owner is a secret agent. Find the number of possible plates that consist of 4 letters (A to Z) followed by 4 digits and have exactly two 7's.

15. How many hexadecimal (base-16) numbers of length 10 have exactly 5 twos? Leading zeros are allowed.

16. How many octal (base-8) numbers of length 10 have 5 consecutive twos and no other twos? Leading zeros are allowed.

17. A bag contains 8 gold, 10 silver, and 20 bronze coins, no two of the same size. In how many ways can 2 of each kind be selected if we can distinguish between any two of the 38 coins?

18. A DNA strand consists of a sequence of nitrogen bases. Each is one of cytosine, guanine, thymine, or adenine. How many possible strands of length 12 have 3 of each type of nitrogen base?

19. A bag contains 8 black balls and 6 white balls. How many ways are there to select 3 black and 2 white balls from the bag? We are assuming that the balls are distinguishable beyond their color. Therefore, which 3 black balls are chosen is important. However, the order in which they are chosen is not.

20. A computer gaming company employs 10 programmers, 6 graphics specialists, and 3 creative designers. In how many ways can a team of 4 programmers, 2 graphics specialists, and 1 creative designer be selected to work on a new project?

21. A child riding in a car notices a license plate consisting of 6 distinct letters (A to Z) that happen to be in alphabetical order. How many such license plates are possible?
Hint: The *set* of 6 letters determines the plate.

22. How many 4-digit (base-ten) numbers have strictly decreasing digits?

23. An NBA team consists of 12 players, and exactly 5 play at one time. Before each game, the coach must decide which players will start the game. Determine the number of ways to pick the 5 starters based on the given conditions.

    (a) There is no bias regarding a player's position.

    (b) The team consists of 3 centers, 5 guards, and 4 forwards. To match up against the opponent, the starting 5 must contain 1 center, 2 guards, and 2 forwards.

24. George is on a strict diet that requires him to eat 9 pieces of fruit each day. On a day he is visiting his mother, she

has a fruit basket containing 10 apples, 8 oranges, and 6 bananas. However, each of these 24 pieces of fruit is unique in its own way, relative to ripeness, size, and color differences.

(a) How many ways are there for George to select 9 pieces of fruit from this basket to maintain his diet for the day?

(b) What if he wants to select 4 apples, 2 oranges, and 3 bananas?

25. How many different 8-digit hexadecimal sequences have exactly 2 zeros and 2 ones?

26. A researcher is studying a DNA strand of length 6 that has exactly 1 guanine and 2 adenine nitrogen bases. How many such DNA strands are possible? See Exercise 18 for an explanation of DNA.

27. How many different 10-digit octal sequences have exactly 4 zeros and no other repeats?

28. How many different 8-digit hexadecimal sequences have 4 zeros and 4 ones?

29. A **BINGO** card has 5 rows and 5 columns, and the columns are labeled by the letters in BINGO. The third cell in column N is a free cell (containing no number) on which the player places a chip before the game starts. The first column contains numbers from 1 to 15, the second from 16 to 30, ... , and the last from 61 to 75. In a game of BINGO, letter and number combinations, such as B12 or N37, are called out until some card achieves 5-in-a-row along a row, column, or diagonal.

B	I	N	G	O
1	21	33	54	70
3	23	31	47	65
15	16	★	53	67
13	29	43	55	62
9	19	39	50	71

(a) How many different BINGO cards are possible?

(b) How many different cards could achieve BINGO with the same numbers

(i) If it happens in column B, I, G, or O?
(ii) If it happens in column N?

30. A dart board is a circle divided into 20 equal-sized sectors, numbered 1 through 20 in a certain order. The dart board is always oriented so that the sector containing the number 20 is at the top.

(a) How many different dart boards are possible?

(b) What if we relax the restriction that the number 20 must be at the top? This would change the game, since the topmost and bottommost numbers tend to be the easiest to hit.

Ordered dice. For Exercises 31 through 36, consider the task of rolling a standard (6-sided) die 5 times and recording the sequence of 5 numerical outcomes. In each case, determine the number of possible sequences satisfying the given conditions.

31. No repeats.

32. Increasing order.

33. Exactly 2 sixes.

34. 3 twos and 2 fours.

35. 3 of one number and 2 of another.

36. Sum is 6.

Scrambled dice. For Exercises 37 through 42, consider the task of rolling 4 dice and recording the 4 numbers showing. The dice are considered identical and have no order to them. In each case, determine the number of possible outcomes satisfying the given conditions.

37. A run of four consecutive numbers (such as 2, 3, 4, 5).

38. Three of a kind (and not four of a kind).

39. Four distinct numbers.

40. 2 sixes and 2 other distinct values.

41. A pair and 2 other distinct values.

42. Two pair (and not four of a kind).

Exercises 43 through 46 refer to the **Postnet codes** used by U.S. Postal Service, as described in Example 6.19. See Table 6.1.

43. Use the pictured Postnet code

for INFORMS (Institute for Operations Research and the Management Sciences) in Linthicum, Maryland to determine the ZIP + 4 code for INFORMS.

44. The Postnet code for the MAA (Mathematical Association of America) in Washington, DC, is

Consequently, what is the ZIP + 4 code for the MAA?

45. Suppose we wish to create a Postnet code to encode hexadecimal numbers using a fixed number of long and short bars for each digit.

   (a)* What is the smallest fixed number of bars that will enable us to encode all of the possible hexadecimal digits?

   (b) How many long bars will be used for each digit?

46. In a Postnet code for decimal digits, each digit uses the same number of bars. However, suppose we now allow the number of long bars to vary from digit to digit.

   (a)* What is the smallest fixed number of bars that will enable us to encode all of the possible decimal digits?

   (b) Can this be done using the number of bars from part (a) and only two different numbers of long bars? How?

47. In the 2005 Kentucky Derby, a family betting on the race won the superfecta, which was worth $854,253. That is, they correctly picked the first four horses to finish the race and in the correct order. This was particularly remarkable because the horses that did well were huge long-shots. The strategy they employed was to select 6 of the 20 horses in the race and purchase a 6-horse $1 superfecta box. That is, they bet $1 on each of the possible top-four finishes using only the 6 horses they selected.

   (a) In total, how much did they bet?

   (b) How much would it cost to bet on all possible top-four finishes from the entire 20-horse field?

   (c)* Why might the strategy in part (b) not be a good choice?

48. In dog racing, a quiniela (or quinella) bet is a bet on two dogs to be the first two to cross the finish line. However,

the order in which they cross does not matter. For any number n of dogs, an n-dog \$2 quiniela box is a bet of \$2 each on each possible pair from a set of n dogs.

(a) How much does it cost to purchase a 5-dog \$2 quiniela box?

(b)* If there is a field of 14 dogs in the race, then how much would it cost to guarantee a quiniela win from one of the \$2 bets?

6.3   Addition and subtraction

Besides the Multiplication Principle and its consequences, we need tools to handle problems that naturally break into cases. For such problems, we need to appropriately combine results obtained in those cases. A couple of examples will motivate another useful counting principle.

6.3.1   Disjoint events

Some counting problems are handled by taking the sum of separate counts.

Example 6.20.   How many possible license plates consisting of 6 digits (0–9) have either all digits distinct or all digits the same?

Solution. There are $P(10,6)$ plates with all digits distinct and 10 with all digits the same. Certainly no one plate can have both of these properties. Hence, the total number of license plates under consideration is

$$P(10,6) + 10 = 151210.$$

We add the counts from the two disjoint cases.   □

Example 6.21.   How many possible sequences of 10 coin flips result in exactly 3 or exactly 4 heads?

Solution. A sequence cannot have both exactly 3 and exactly 4 heads. Hence, we add the counts for these distinct possibilities. Therefore, there are

$$\binom{10}{3} + \binom{10}{4} = 330$$

sequences with 3 or 4 heads.   □

Theorem 6.5.   (Addition Principle).
Finite sets A and B are disjoint (equivalently, $|A \cap B| = 0$) if and only if

$$|A \cup B| = |A| + |B|.$$

An inductive argument shows that the Addition Principle generalizes to any finite number of finite disjoint sets, giving the **General Addition Principle**

$$|A_1 \cup \cdots \cup A_n| = |A_1| + \cdots + |A_n| \text{ if and only if}$$

$A_1, \ldots, A_n$ are disjoint.

To fully appreciate the Addition Principle, it may be helpful to consider an example in which it does not apply.

Example 6.22.   Consider subsets of the set $\mathcal{U}$ of 6-digit license plates. Let A contain those with exactly 2 ones, and let B contain those with exactly 3 nines. Hence, $A \cup B$ contains those with 2 ones or 3 nines. However,

$$|A \cup B| \neq |A| + |B|,$$

since A and B are not disjoint. It is possible to have a plate with both 2 ones and 3 nines (e.g., 115999). Since $A \cap B$ is nonempty, the Addition Principle does not apply. It is important to take this into account before blindly invoking the Addition Principle. Theorem 6.7 will tell us how to handle situations in which the sets being considered are not disjoint.

In the applications of the Addition Principle in Examples 6.20 and 6.21, the disjoint cases that we considered were explicitly described in the problem. More typically, we must discover for ourselves a convenient decomposition of a counting problem into disjoint cases.

Example 6.23.   (Jury Selection).
The lawyer for the prosecution in a certain court case wants the jury of 12 to contain more women than men. If the jury pool contains 15 men and 9 women, then how many different possible juries would satisfy this lawyer?

Solution.   To have more women than men on the jury, there must be one of the following possibilities:

5 men and 7 women,
4 men and 8 women, or
3 men and 9 women.

Since there are only 9 women, we cannot have 10 or more on the jury. Since the listed possibilities are disjoint, we obtain

$$\binom{15}{5}\binom{9}{7} + \binom{15}{4}\binom{9}{8} + \binom{15}{3}\binom{9}{9} = 120848$$

possible juries with more women than men.   □

Example 6.24. (Cracking Passwords). There are 2.6 million possible computer passwords consisting of a letter (A to Z) followed by 5 digits (0–9). If such a password is known to contain at least 3 zeros, then to how many possibilities is the password narrowed down?

Solution. There are 26 choices for the beginning letter. In general, if a password is to contain k zeros, then there are $\binom{5}{k}$ choices for their locations and 9 choices (1–9) for each of the $5 - k$ remaining locations. Of course, there must be one of 3, 4, or 5 zeros (exactly), and these are disjoint possibilities. Hence, there are

$$26 \cdot \binom{5}{3} \cdot 9^2 + 26 \cdot \binom{5}{4} \cdot 9 + 26 = 22256$$

possible passwords with at least 3 zeros.   □

6.3.2 Complements

Recall that given a subset A of a universal set $\mathcal{U}$, the complement of A, denoted A^c, is the set of elements in $\mathcal{U}$ that are not in A. There is a simple corollary to the Addition Principle that makes use of set complements and turns out to be a valuable counting tool.

Theorem 6.6. (Complement Principle).
Given a subset A of a finite universal set $\mathcal{U}$,

$$|A| = |\mathcal{U}| - |A^c|.$$

Proof. The Addition Principle applies to the disjoint union $A \cup A^c = \mathcal{U}$ and gives $|A| + |A^c| = |\mathcal{U}|$. Subtracting $|A^c|$ from both sides gives the desired equality.   □

The Complement Principle is useful in problems involving a set whose complement is more natural or easier to consider.

Example 6.25. How many of the integers from 1 to 300 are not divisible by 15?

Solution. Let $\mathcal{U} = \{1, 2, \ldots, 300\}$, and let A be the subset of those integers that are not divisible by 15. It is easy to count the subset A^c of integers that are divisible by 15. We get

$$|A^c| = \frac{300}{15} = 20.$$

The Complement Principle now tells us that there are

$$|A| = |\mathcal{U}| - |A^c| = 300 - 20 = 280$$

integers in $\mathcal{U}$ that are not divisible by 15.   □

Example 6.26.   A **byte** is a binary number consisting of 8 digits. How many bytes have at least 2 zeros?

Solution. There are 2^8 binary sequences of length 8. Of them, 1 has no zeros and 8 have one zero. Since the rest have at least 2 zeros, there are

$$2^8 - (1 + 8) = 247$$

sequences with at least 2 zeros. It would have been tedious to count directly those that have 2 zeros, or 3 zeros, $\ldots$, or 8 zeros.   □

Example 6.27.   A bag contains 8 red, 4 blue, 7 green, and 5 yellow balls. A box is to be filled with 3 balls. How many ways are there to do this so that at least two colors are used? Note that the 24 balls are considered distinguishable.

Solution. The total number of ways to fill the box with 3 balls is

$$\binom{8 + 4 + 7 + 5}{3} = \binom{24}{3} = 2024.$$

The number of monochromatic ways to fill the box is

$$\binom{8}{3} + \binom{4}{3} + \binom{7}{3} + \binom{5}{3} = 105.$$

Hence, there are

$$2024 - 105 = 1919$$

ways to fill the box using at least 2 colors.   □

The utility of the Complement Principle can be appreciated by trying to solve Example 6.27 directly rather than by using complements.

6.3.3   Basic inclusion and exclusion

The Addition Principle does not handle problems in which the relevant sets are not disjoint. In those cases, some subtraction is needed.

Theorem 6.7.   (Basic Inclusion–Exclusion Principle). *Given finite sets A and B,*

$$|A \cup B| = |A| + |B| - |A \cap B|.$$

Proof. Our approach is motivated by the Venn diagram reflecting this general situation.

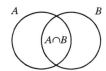

The basic idea is that $A \cap B$ is counted twice in the sum $|A| + |B|$ and must therefore be subtracted off. A careful argument is based on the disjoint unions

$$A = (A \setminus B) \cup (A \cap B), \; B = (B \setminus A) \cup (A \cap B), \; \text{and}$$
$$A \cup B = (A \setminus B) \cup (A \cap B) \cup (B \setminus A).$$

The Addition Principle then gives that

$$|A| = |A \setminus B| + |A \cap B|, \; |B| = |B \setminus A| + |A \cap B|, \; \text{and}$$

$$|A \cup B| = |A \setminus B| + |A \cap B| + |B \setminus A|$$
$$= |A| + |B \setminus A|$$
$$= |A| + |B| - |A \cap B|$$

as asserted.   □

The Addition Principle is simply the special case of the Basic Inclusion–Exclusion Principle in which $|A \cap B| = 0$. The more general result, which further covers cases in which $A \cap B$ is nonempty, enables us to handle problems like the one encountered in Example 6.22.

Example 6.28.   (Getaway Car).
Two witnesses of a bank robbery had different memories of the license plate on the getaway car. Both agreed that the plate consisted of 6 digits. However, one noticed that there were exactly 2 ones, and the other noticed that there were exactly 3 nines on the plate. To be safe, and

realizing that one of the witnesses might be mistaken, the police want to consider both possibilities. How many license plates consisting of 6 digits (0–9) have exactly 2 ones or 3 nines?

Solution. Our universe of consideration is the set of 6-digit license plates. Let A denote the set of plates with exactly 2 ones. Let B denote the set of plates with exactly 3 nines. Hence, $A \cap B$ is the set of plates with both 2 ones and 3 nines, and $A \cup B$ is the set of plates with 2 ones or 3 nines. By the Basic Inclusion–Exclusion Principle,

$$|A \cup B| = |A| + |B| - |A \cap B|$$

$$= \binom{6}{2} \cdot 9^4 + \binom{6}{3} \cdot 9^3 - \binom{6}{2}\binom{4}{3} \cdot 8$$

$$= 113475.$$

Of course, knowing the make, model, or color of the car could further narrow down the possibilities significantly. □

Example 6.29. A standard (6-sided) die is rolled a sequence of 5 times. In how many ways can the sequence of numbers resulting be all even or all multiples of 3?

Solution. Since there are 3 even values (2, 4, and 6), there are 3^5 ways to get all even numbers. Since there are 2 multiples of 3 (3 and 6), there are 2^5 ways to get all multiples of 3. Only the value 6 is both even and a multiple of 3, so there is 1 way to do both (namely, 66,666). Therefore,

$$3^5 + 2^5 - 1 = 274$$

is the desired number of ways. □

Our final example in this section makes use of the following corollary of the Basic Inclusion-Exclusion Principle.

Corollary 6.8. Given subsets A and B of a finite universal set $\mathcal{U}$,

$$|A^c \cap B^c| = |\mathcal{U}| - |A| - |B| + |A \cap B|.$$

Proof. By De Morgan's Law, $A^c \cap B^c = (A \cup B)^c$. By the Complement Principle, $|(A \cup B)^c| = |\mathcal{U}| - |A \cup B|$. Basic Inclusion–Exclusion now gives

$$|A^c \cap B^c| = |\mathcal{U}| - (|A| + |B| - |A \cap B|)$$
$$= |\mathcal{U}| - |A| - |B| + |A \cap B|. \qquad \square$$

Example 6.30.   How many of the integers from 1 to 1000 are relatively prime to 1000?

Solution. Our universe of consideration is $\mathcal{U} = \{1, 2, \ldots, 1000\}$. We want to count those integers that do not have any factors in common with $1000 = 2^3 \cdot 5^3$. Since it suffices to consider only prime factors, we want to know how many integers are divisible neither by 2 nor 5.

Let A be the set of integers that are divisible by 2, and let B be the set of integers that are divisible by 5. So $A^c \cap B^c$ is the set that we wish to count, and Corollary 6.8 applies. We have $|A| = \frac{1000}{2}$, $|B| = \frac{1000}{5}$, and $|A \cap B| = \frac{1000}{10}$. Therefore,

$$|A^c \cap B^c| = 1000 - \frac{1000}{2} - \frac{1000}{5} + \frac{1000}{10}$$
$$= 1000 - 500 - 200 + 100$$
$$= 400$$

is the number of integers in $\mathcal{U}$ relatively prime to 1000. $\qquad \square$

Example 6.30 shows a computation of a particular value of the **Euler phi-function** ϕ. For any positive integer n, the value $\phi(n)$ is defined to be the number of integers from 1 to n that are relatively prime to n. In particular, we see that $\phi(1000) = 400$. Computations of values of ϕ and other similar computations are taken up in the exercises. Further consideration of ϕ is given in Section 7.1.

Exercises

1. A bag contains 10 black and 12 white balls. How many ways are there to select 8 balls so that exactly 5 or 6 of them are black? The 22 balls are considered distinguishable.

2. How many bytes have 2 or 3 zeros? See Example 6.26.

3. Determine the number of possible 6-digit (0–9) license plates that have exactly 4 threes or exactly 4 sixes.

4. Determine the number of possible 6-digit (0–9) license plates that have at least 4 nines.

5. A coin is tossed 10 times. In how many different ways can there be at most 3 heads?

6. Anakin is in fourth grade and has 6 different classes. His mother told him that she will take him to the pod races if

he gets at least 3 A's on his next report card. If each class grade will be one of A, B, C, D, or F, then how many possible report cards could earn Anakin his reward?

7. How many different sequences of 6 coin flips result in at least 4 tails?

8. A standard (6-sided) die is rolled 5 times. How many possible sequences result in at most 2 ones?

9. The defense attorney figures that she will have a good chance of winning the case if the jury contains at least 2 men. The jury pool contains 18 men and 17 women. How many ways are there to select 12 so that this attorney is satisfied?

10. How many different sequences of 10 coin flips result in at least 3 tails?

11. How many different sequences of 5 coin flips result in an even number of heads?

12. How many different bytes have at least 5 zeros? See Example 6.26.

13. How many of the integers from 500 to 1000 are not divisible by 3?

14. How many of the integers from 10 to 999 are not divisible by 9?

15. How many of the integers from 1 to 500 are divisible by 3 or 7?

16. How many of the integers from 500 to 5000 are divisible by 11 or 13?

17. How many of the integers from 200 to 2000 are divisible by 2 or 5?

18. How many of the integers from 1 to 3500 are divisible by 3 or 11?

19. Two children on a long car ride are playing a license plate game. One counts the plates she sees consisting of 8 digits (0–9) with all digits greater than or equal to 4. The other counts the 8-digit plates he sees with all digits less than or equal to 5. After playing for a while, the children complain that there are not many such plates to be found.

    (a) How many possible plates satisfy the girl's restrictions?

    (b) How many satisfy the boy's restrictions?

    (c) Overall, how many possible plates fit into the children's game?

20. A man has 3 dimes and 5 nickels in his pocket. To pay for a newspaper, he will reach into his pocket and retrieve exactly 40¢. Assume that the 8 coins are distinguishable.

    (a) How many ways are there to pay for the paper using all 3 dimes?

(b) How many ways are there to pay using no dimes?

(c) Overall, how many ways are there for him to pay for the paper?

21. A DNA strand consists of a sequence of nitrogen bases. Each is one of cytosine, guanine, thymine, or adenine. How many possible strands of length 8 have exactly 3 cytosine or exactly 3 thymine nitrogen bases?

22. A bag contains 4 red, 3 blue, and 5 green balls. How many ways are there to select 5 balls so that either 2 are red or 2 are blue? The 12 balls are considered distinguishable.

Card hands. For Exercises 23 through 26, use the fact that a standard deck of cards contains 52 cards with 13 each of the 4 suits (clubs, diamonds, hearts, and spades). Moreover, each suit contains the three face cards (Jack, Queen, King). In each case, determine how many hands of 5 cards have the specified property.

23. Exactly 2 clubs or exactly 2 spades.

24. At least 4 face cards.

25. At most one heart.

26. Five face cards or five diamonds.

27. Find $\phi(144)$.

28. Find $\phi(245)$.

29. How many of the integers from 100 to 1000 are relatively prime to 1000?

30. How many of the integers from 25 to 1125 are relatively prime to 1125?

31. There are 27! different permutations of $0, \ldots, 26$. Using the conversions $0 =$ " ", $1 = A, 2 = B, \ldots$, any such permutation specifies a scrambling of the alphabet that can be used to send coded messages. In particular, consider the permutation

$x:$ A B C D E F G H I J K L M N O P Q R S T U V W X Y Z

$y:$ C G K O S W   D H L P T X A E I M Q U Y B F J N R V Z

that sends a letter x to a coded letter y.

(a) Decode "AWWBCAW."

(b) As we saw in Section 3.5, some permutations of $0, \ldots, 26$ take the form of a linear cipher $y = (ax + b) \bmod 27$, for choices of integers a and b. Justify that the permutation above is a linear cipher.

(c)* Determine the number of permutations that are linear ciphers. Hint: What are the restrictions and freedoms on a and b?

32. Given $p = 11$, $q = 19$, and $n = 209$, we saw in Section 3.5 that certain choices of an integer a yield an RSA encryption scheme given by the equation $y = x^a \bmod n$. Such a code could be used to encode messages using 209 different characters.

   (a) Given $n = 209$ and $a = 3$, encode each entry in the sequence $2, 7, 1, 8$.

   (b) When $n = 209$, why is 5 not a valid choice for a?

   (c)* How many possible schemes are there when $n = 209$?

33. How many of the integers from 171 to 1771 are relatively prime to 10?

34. How many of the integers from 543 to 5432 are relatively prime to 63?

35. Prove the General Addition Principle for an arbitrary number of disjoint sets.

36. Formulate and prove a version of Corollary 6.8 for arbitrary numbers of disjoint sets.

Ordered dice. For Exercises 37 through 44, consider the task of rolling a die 5 times and recording the sequence of 5 values. In each case, determine the number of outcomes satisfying the given conditions.

37. Sum is 7.

38. Sum is odd.

39. A sum of at most 7.

40. At least 3 sixes.

41. Four or more common values.

42. At most 1 three.

43.* At most 1 pair.

44.* At least 2 fives and at least 2 sixes.

Scrambled dice. For Exercises 45 through 52, consider the task of rolling 4 dice and recording the 4 numbers showing. The dice are considered identical and have no order to them. In each case, determine the number of outcomes satisfying the given conditions.

45. At least 3 sixes.

46. All even numbers.

47. At least 3 fives or at least 3 sixes.

48. Exactly 3 distinct numbers.

49. Two pairs of values.

50. At most 2 different numbers.

51.* At least 3 different numbers.

52.* At least 2 numbers the same.

6.4 Probability

One application of counting is to compute probabilities. To analyze the relative fairness of games of chance, we must understand probability, especially when gambling is involved. Millions buy state lottery tickets despite the extremely low likelihood of winning. In the insurance industry, actuaries study the probabilities of accidents or deaths to determine insurance premiums.

We start with an example of a game of chance, which motivates both our methods and our terminology.

Example 6.31.   Consider the task of tossing two dice and recording the numbers showing. Assume that one die is red and the other is green. Hence, we may record each result as a pair (r, g), where r and g are the values showing on the red and green die, respectively. The set of possible results of our task is thus

$$S = \{(1,1),\ (1,2),\ (1,3),\ (1,4),\ (1,5),\ (1,6),$$
$$(2,1),\ (2,2),\ (2,3),\ (2,4),\ (2,5),\ (2,6),$$
$$(3,1),\ (3,2),\ (3,3),\ (3,4),\ (3,5),\ (3,6),$$
$$(4,1),\ (4,2),\ (4,3),\ (4,4),\ (4,5),\ (4,6),$$
$$(5,1),\ (5,2),\ (5,3),\ (5,4),\ (5,5),\ (5,6),$$
$$(6,1),\ (6,2),\ (6,3),\ (6,4),\ (6,5),\ (6,6)\}.$$

Of course, $|S| = 6^2 = 36$. What is particularly important is the fact that each of the 36 possible results has the same likelihood. That is, for each pair (r, g), the probability of getting (r, g) is $\frac{1}{36}$. Consider now the sum $r + g$ of the two values showing on the dice, and compute the probability that the sum is 8.

Solution. From the possible results listed in S, we are interested in the subset

$$E = \{(2,6), (3,5), (4,4), (5,3), (6,2)\}$$

consisting of those pairs (r, g) for which $r + g = 8$. The probability of E is given by

$$P(E) = \frac{1}{36} + \frac{1}{36} + \frac{1}{36} + \frac{1}{36} + \frac{1}{36} = \frac{5}{36}.$$

That is, $P(E) = \frac{|E|}{|S|}$.   □

Some general notions can be extracted from Example 6.31. An **experiment** is a specific task of consideration. Each possible result of the experiment is called an **outcome**, and the set S of all possible outcomes is called the **sample space** for the experiment. A subset E of S is said to be an **event**. Given a particular event E, the **probability** of E, denoted $P(E)$, is a value between 0 and 1 that gives the likelihood that an outcome in E will occur if the experiment is performed. The value can be estimated experimentally, but in some cases, as we shall see in Definition 6.4, it can be determined exactly. If E consists of a single outcome $x \in S$, we may more compactly write $P(x)$ for its probability.

In Example 6.31, the experiment is the tossing of two dice. An outcome is a pair (r, g), such that $r, g \in \{1, 2, 3, 4, 5, 6\}$, and the sample space is

$$S = \{(r, g) \ : \ r, g \in \{1, 2, 3, 4, 5, 6\}\}.$$

For the event $E = \{(r, g) \ : \ r + g = 8\}$, we computed the probability of E and got $P(E) = \frac{5}{36}$. That computation depended on the fact that all outcomes from S have the same probability.

Definition 6.3.   Let S be a finite sample space. The outcomes in S are said to be **equally likely** if $\forall \ x, y \in S, P(x) = P(y)$. That is, $\forall \ x \in S, P(x) = \frac{1}{|S|}$.

We consider only finite sample spaces: discrete probability distributions. When we have equally likely outcomes, probability computations reduce to counting.

Definition 6.4.   (Probability When Outcomes Are Equally Likely). If the outcomes in a finite sample space S are all equally likely, then the **probability** of an event E is given by

$$P(E) = \frac{|E|}{|S|}.$$

Many probability questions can be answered using Definition 6.4. Consequently, such probability computations amount to two counting problems; we count the number of elements in E and the number in S. The probability is then given by the quotient $\frac{|E|}{|S|}$.

Example 6.32.   If two fair coins are tossed, then what is the probability of getting exactly one head?

Solution.   To ease our analysis, we regard our two tosses as occurring in sequence, a first coin and then a second. If we use H for heads and T for tails, then the sample space of possible outcomes is given by the set of ordered pairs

$$S = \{TT, TH, HT, HH\}.$$

Since the coins are assumed to be fair, each of the outcomes in S is equally likely. The event of getting exactly one head is given by

$$E = \{TH, HT\}.$$

Hence,

$$P(E) = \frac{|E|}{|S|} = \frac{2}{4} = \frac{1}{2}$$

is the desired probability. □

Why was it helpful to treat the results of the two coin tosses as an *ordered* pair? An alternative analysis of Example 6.32 might consider the outcomes to be the number of heads obtained. Hence, there would be three possible outcomes: 0, 1, or 2. However, these three outcomes are not equally likely, as can be seen by repeating this experiment several times. More precisely, the work in Example 6.32 gives that $P(1) = \frac{1}{2}$ and $P(0) = P(2) = \frac{1}{4}$. Consequently, Definition 6.4 does not apply to this alternative analysis. Hence, the approach taken in Example 6.32 is more convenient.

Example 6.33. (Gender Bias).
There are 8 men and 4 women that work in an office, and the 2 promotions that were recently announced both went to men. If the 2 to be promoted had been randomly selected, then what is the probability they both would have been men?

Solution. Let M denote man and W woman. When workers are selected, there are three possible gender combinations MM, MW, and WW. However, these are not equally likely, even if we further consider our pairs as ordered and separate out MW and WM. The reason is that there are more men than women. Instead, we consider the sample space S of all possible combinations of 2 workers. This gives $|S| = \binom{12}{2}$, and each outcome in S is equally likely.

Our interest is in the event E consisting of all combinations in which both promotions go to men. Since there are 8 men, $|E| = \binom{8}{2}$. Therefore,

$$P(E) = \frac{\binom{8}{2}}{\binom{12}{2}} = \frac{14}{33} \approx 0.424.$$

It was not unlikely for both promotions to go to men. □

In the context of probability, disjoint events are said to be **mutually exclusive**. In general, if events E and F are mutually exclusive, then $P(E \cup F) = P(E) + P(F)$. This fundamental property of probability is the analog of the Addition Principle for counting. Probability also has a Complement Principle that is much like Theorem 6.6 and is similarly useful.

Theorem 6.9. (Probability Complement Principle).
If E is an event in a sample space S, then

$$P(E) = 1 - P(E^c).$$

In the case of equally likely outcomes, the Probability Complement Principle is an easy consequence of Theorem 6.6 and Definition 6.4. A proof in that case is left to the exercises.

Example 6.34. The license plates issued by a certain state consist of 3 letters (A to Z) followed by 4 digits (0–9). If all possible plates are equally likely to be chosen, then what is the probability that a randomly chosen plate will have some letter or digit repeated?

Solution. The sample space S is the set of all possible plates consisting of 3 letters followed by 4 digits. Hence,

$$|S| = 26^3 \cdot 10^4 = 175760000.$$

Let E be the subset of plates in which some letter or digit is repeated. Although it is $P(E)$ that we seek, it is easier to compute $P(E^c)$. Of course, E^c is the subset of plates in which no character is repeated. Since

$$|E^c| = P(26,3) \cdot P(10,4) = 78624000,$$

the desired probability is

$$P(E) = 1 - \frac{78624000}{175760000} = \frac{467}{845} \approx 0.553. \qquad \square$$

Example 6.35. (Day Labor).
At the beginning of the day, 10 day laborers arrive at a construction site seeking work. If each is either retained or sent home with equal likelihood (as might be decided by flipping a fair coin), then what is the probability that at least 3 day laborers will be retained to work?

Solution. At least 3 are retained if exactly 3, exactly 4, . . . , or exactly 10 are retained. It is easier to consider instead the complementary event of retaining at most 2.

The number of ways in which at most 2 day laborers are retained is

$$1 + 10 + \binom{10}{2} = 56.$$

Hence, the desired probability is

$$1 - \frac{56}{2^{10}} = \frac{121}{128} \approx 0.945. \qquad \square$$

Basic Inclusion–Exclusion also has an analog for probability.

Theorem 6.10. (Basic Probability Inclusion–Exclusion). *Given events E and F in a sample space S,*

$$P(E \cup F) = P(E) + P(F) - P(E \cap F).$$

A proof of Theorem 6.10 in the case of equally likely outcomes is left for the exercises.

Example 6.36. (Home Security).
A home security code consists of 4 digits (0–9). If all such codes are equally likely to be chosen, then what is the probability that a randomly chosen code will contain exactly 1 three or exactly 2 sixes?

Solution. Let S be the sample space of all possible 4-digit codes. Let E be the subset of codes containing exactly 1 three, and let F be the subset of codes containing exactly 2 sixes. So $E \cup F$ consists of all codes containing exactly 1 three or exactly 2 sixes. We seek $P(E \cup F)$.
  Since

$$P(E) = \frac{4 \cdot 9^3}{10^4} = \frac{729}{2500},$$

$$P(F) = \frac{\binom{4}{2} \cdot 9^2}{10^4} = \frac{243}{5000}, \quad \text{and}$$

$$P(E \cap F) = \frac{4 \cdot \binom{3}{2} \cdot 8}{10^4} = \frac{6}{625},$$

Theorem 6.10 gives that

$$P(E \cup F) = \frac{729}{2500} + \frac{243}{5000} - \frac{6}{625} = \frac{1653}{5000} = 0.3306. \qquad \square$$

6.4.1 Conditional probability

If we roll two dice, then computations like those used in Example 6.31 give that the probability that the sum is 10 is $\frac{3}{36} = \frac{1}{12}$. However, suppose we are rolling the two dice in sequence, and we know that the value on the first die is at least 5. How does this affect our chances of rolling a sum of 10, when the second die is rolled? Essentially, we are removing from consideration all elements of the 36-element sample space S in Example 6.31 that do not have a 5 or greater in their first coordinate. To estimate our new desired probability by rolling two dice several times, we should simply ignore any outcome from S that achieves a 4 or smaller on the first die. Hence, our new sample space is the subset

$$S' = \{(5,1),\ (5,2),\ (5,3),\ (5,4),\ \mathbf{(5,5)},\ (5,6),$$
$$(6,1),\ (6,2),\ (6,3),\ \mathbf{(6,4)},\ (6,5),\ (6,6)\}$$

of S. Since outcomes in S' must be equally likely, just as they are in S, the new probability is $\frac{2}{12} = \frac{1}{6}$. If we let E be the event in S that the sum is 10 and F be the event in S that the first die has a value of at least 5, then our new probability can be described as the probability of E given that F has occurred. Moreover, its value can be seen to satisfy

$$\frac{1}{6} = \frac{2}{12} = \frac{\frac{2}{36}}{\frac{12}{36}} = \frac{P(E \cap F)}{P(F)}.$$

This enables us to determine our new probability from computations involving the old sample space S and inspires the following definition.

Definition 6.5.   Let E and F be events in a sample space S with $P(F) > 0$. The **conditional probability of E given F**, denoted $P(E \mid F)$, is given by

$$P(E \mid F) = \frac{P(E \cap F)}{P(F)}.$$

Example 6.37. At a company picnic, the children played a soccer game, after which one player's name was randomly drawn to win a prize. The winning team in the soccer game consists of 7 girls and 4 boys, and the losing team consists of 5 girls and 6 boys. Given that the prize winner is a boy, what is the probability that he also comes from the winning soccer team?

Solution. Let S be the set of 22 children that played soccer, let E be the event that the prize winner is from the winning team, and let F be the event that the prize winner is a boy. The probability that we seek is $P(E \mid F)$, the probability that the prize winner comes from the winning team given that he is a boy.

Note that $E \cap F$ is the event that the prize winner is a boy from the winning team. Since

$$P(E \cap F) = \frac{4}{22} = \frac{2}{11} \quad \text{and}$$

$$P(F) = \frac{10}{22} = \frac{5}{11},$$

we conclude that

$$P(E \mid F) = \frac{\frac{2}{11}}{\frac{5}{11}} = \frac{2}{5}. \qquad \square$$

It follows from the formula in Definition 6.5 that

$$P(E \cap F) = P(E \mid F) \cdot P(F). \qquad (6.1)$$

In some circumstances, there is a simpler formula for $P(E \cap F)$.

Definition 6.6.   Two events E and F in a sample space S are said to be **independent** if $P(E \cap F) = P(E) \cdot P(F)$.

If E and F are independent events and $P(E \mid F)$ is defined, then $P(E \mid F) = P(E)$, as we will see in the exercises. In this case, knowing that event F has occurred does not affect the likelihood of event E occurring. This justifies the use of the term *independent*.

Example 6.38.   Consider the experiment of tossing two dice in sequence. Let E be the event that the first die shows a value of at most 3, let F_1 be the event that the values on the two dice are the same, and let F_2 be the event that the sum of the values on the two dice is 5. Note that $P(E) = \frac{18}{36} = \frac{1}{2}$.

(a) Are E and F_1 independent?
   Solution. Since $P(F_1) = \frac{6}{36} = \frac{1}{6}$ and $P(E \cap F_1) = \frac{3}{36} = \frac{1}{12}$, it follows that

$$P(E \cap F_1) = \frac{1}{12} = \frac{1}{2} \cdot \frac{1}{6} = P(E) \cdot P(F_1).$$

So yes, E and F_1 are independent.
Note, in this case, that $P(E \mid F_1) = P(E) = \frac{1}{2}$. $\square$

(b) Are E and F_2 independent?

Solution. Since $P(F_2) = \frac{4}{36} = \frac{1}{9}$ and $P(E \cap F_2) = \frac{3}{36} = \frac{1}{12}$, it follows that

$$P(E \cap F_2) = \frac{1}{12} \neq \frac{1}{2} \cdot \frac{1}{9} = P(E) \cdot P(F_2).$$

So no, E and F_2 are not independent.   □

If a sample space is partitioned into events $F_1, \ldots, F_n$, then we can take advantage of Equation 6.1 and compute the probability of an event E by conditioning on each of the events F_i.

Theorem 6.11.   *Suppose* $E, F_1, \ldots, F_n$ *are events in a sample space* S *with* $P(F_1), \ldots, P(F_n)$ *positive, and* S *is a disjoint union* $S = F_1 \cup \cdots \cup F_n$. *Then*

$$P(E) = \sum_{i=1}^{n} P(E \mid F_i)P(F_i).$$

Proof. Since $E = E \cap S = E \cap (F_1 \cup \cdots \cup F_n) = (E \cap F_1) \cup \cdots \cup (E \cap F_n)$ is a disjoint union,

$$P(E) = \sum_{i=1}^{n} P(E \cap F_i) = \sum_{i=1}^{n} P(E \mid F_i)P(F_i).   □$$

The following corollary to Theorem 6.11 is named after the English minister Thomas Bayes (1702–1761), who discovered it. Its proof is left for the exercises.

Corollary 6.12.   (Bayes' Formula).
Suppose $E, F_1, \ldots, F_n$ are events in a sample space S with $P(E), P(F_1), \ldots, P(F_n)$ positive, and S is a disjoint union $S = F_1 \cup \cdots \cup F_n$. Then for any $1 \leq k \leq n$,

$$P(F_k \mid E) = \frac{P(E \mid F_k)P(F_k)}{\sum_{i=1}^{n} P(E \mid F_i)P(F_i)}.$$

Example 6.39. In a recent election, the majority of female voters preferred a different candidate from the one preferred by the majority of the male voters. Exit polls showed that 75% of female voters chose candidate A, whereas 55% of male voters chose candidate B. Assume that each voter chose either candidate A or candidate B and that an equal number of men and women voted.

(a) Which candidate won the election? Explain.

(b) If a randomly chosen voter is known to have voted for candidate A, then what is the probability that the voter is a female?

Solution. To be simple, let A be the event that a voter chooses candidate A, let B be the event that a voter chooses candidate B, let F be the event that a voter is female, and let M be the event that a voter is male. Based on the exit polls and the assumption that each voter chose A or B, we know that

$$P(A \mid F) = 0.75, \quad P(B \mid F) = 0.25,$$
$$P(A \mid M) = 0.45, \quad P(B \mid M) = 0.55.$$

We are also assuming that

$$P(F) = P(M) = 0.5.$$

Note that F and M form a partition of the sample space of all the voters.

(a) To determine who won, we apply Theorem 6.11. Observe that

$$P(A) = P(A \mid F)P(F) + P(A \mid M)P(M)$$

$$= (0.75)(0.5) + (0.45)(0.5) = 0.6.$$

Since candidate A received 60% of the votes, candidate A won the election.

(b) We seek $P(F \mid A)$, the probability that a voter is female given that she voted for candidate A. For this, Bayes' Formula gives that

$$P(F \mid A) = \frac{P(A \mid F)P(F)}{P(A \mid F)P(F) + P(A \mid M)P(M)}$$

$$= \frac{(0.75)(0.5)}{0.6} = 0.625.$$

That is, there is a 62.5% chance that this voter for candidate A is a female. □

Exercises

1. At the end of each NBA season, only the teams that do not make the playoffs have a chance to get the first overall pick in the draft. Each of these teams is assigned ping pong balls that are placed in a bin, and one ball is randomly selected from this bin to determine which team gets the valuable first pick. The worse a team's record, the more ping pong balls it gets. Table 6.2 reflects the results of the 2014–2015 NBA season. In that lottery, the ping pong ball for the Minnesota Timberwolves was drawn, giving them the first pick. (They chose Karl-Anthony Towns.)

TABLE 6.2
2015 NBA draft lottery

Team	Record	Number of ping pong balls
Minnesota	16–66	250
New York	17–65	199
Philadelphia	18–64	156
L.A. Lakers (CA)	21–61	119
Orlando	25–57	88
Sacramento (CA)	29–53	63
Denver	30–52	43
Detroit	32–50	28
Charlotte	33–49	17
Miami	37–45	11
Indiana	38–44	8
Utah	38–44	7
Phoenix	39–43	6
Oklahoma City	45–39	5

(a) What was the probability of Minnesota winning that lottery?

(b) What was the probability of a team from California winning that lottery?

2. Roulette is a popular casino game in which a ball is dropped onto a spinning wheel. As shown in Figure 6.6, the wheel is split into 38 equal-sized sectors to which numbers and colors are assigned. To 36 of the sectors are assigned the numbers 1 through 36, with half of those

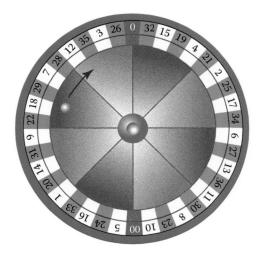

FIGURE 6.6
Roulette wheel.

given the color red (white here) and the other half given the color black (blue here). The remaining spaces are given the color green (gray here), with one assigned the number 0 and the other the number 00 (double zero). Players can bet on the number or color on which the ball will land.

(a) What is the probability that the ball will land on 7?

(b) What is the probability that the ball will land on red?

For Exercises 3 through 6, consider the experiment of rolling two fair standard (6-sided) dice, and determine the probabilities of the given events.

3. A sum of 7.

4. A pair.

5. A sum of at least 9.

6. The two numbers on top differ by at most one.

7. An experiment consists of tossing a coin 3 times and recording the sequence of heads and tails.

(a) Write out the sample space for this experiment.

(b) What is the probability that at least 2 heads occur?

8. An experiment consists of tossing a coin 4 times and recording the sequence of heads and tails.

(a) Write out the sample space for this experiment.

(b) What is the probability that exactly 2 heads occur?

9. The touch pad for a home security code has buttons for the digits $0, 1, 2, 3$. To secure the house, the home owner must set a code that consists of a sequence of 2 digits.

(a) A thief attempting to enter the house plans to consider all possible security codes. Write out the relevant sample space.

(b) To conserve time, the thief decides to only attempt codes for which the second digit is greater than the first. What is the probability that the thief will find the correct code?

10. A computer account has a password consisting of 3 letters. Each letter is one of a, b, or c.

(a) To attempt to break into the account, a hacker should consider all possible passwords. Write out the relevant sample space.

(b) The hacker has neglected to consider passwords that have the same letter occurring exactly twice. What is the probability that the hacker will miss the correct password?

For Exercises 11 through 14, consider the experiment of rolling three fair standard (6-sided) dice, and determine the probabilities of the given events.

11. The values are three consecutive numbers. Note that they need not be rolled in order.

12. All three dice show the same value.

13. All three dice show different values.

14. A pair of one value and an additional distinct value.

Jury selection. For Exercises 15 through 18, we consider a jury trial for which the jury of 12 was selected from a pool containing 20 men and 30 women. Since the case involved a gender discrimination claim, both sides felt that the makeup of the jury was important. As it turned out, the jury consisted of 5 men and 7 women.

15. If the jurors had been randomly selected from the jury pool of 50 people, then what would have been the probability of obtaining the gender balance that occurred?

16. The defense would have preferred a jury of 8 men and 4 women. What would have been the probability of randomly selecting such a jury?

17. (a) If the jurors had been randomly selected from the jury pool, then what would have been the probability that all 12 jurors were the same gender?

(b) There are 50% more women than men in the jury pool. Is it 50% more likely that the jury would consist of all women than that it would consist of all men?

(c) What is the probability that a randomly selected jury would have some of both genders?

18. (a) Since the defendant was male, the defense lawyers would have preferred that there were more men than women on the jury. What would have been the chance of this in a random selection of 12 jurors?

(b) The prosecution lawyers were pleased to have more women than men on the jury. What were the chances of such a gender balance, if the selections were random?

(c) Why do the answers to parts (a) and (b) not add up to 1?

19. The owner of a company needed to hire a president and vice president for her company from the 10 male and 8 female candidates she interviewed. She ended up hiring two women. If two candidates were randomly selected, then what would have been the probability that they both were women?

20. A standard deck contains 26 red cards and 26 black cards. What is the probability that two randomly chosen cards will be the same color?

21. If 8 coins are tossed, then what is the probability that at most 2 heads will be showing?

22. If 4 standard (6-sided) dice are rolled, then what is the probability that at most 1 six will be showing?

23. If 10 coins are tossed, then a common misconception is to believe that a result of 5 heads and 5 tails is very likely.

    (a) What is the true probability that this result will occur?

    (b) Would you bet for or against it?

24. What is the probability that 5 randomly chosen cards from a standard deck (containing 26 red and 26 black) will be all red?

25. What is the probability that a license plate consisting of 6 distinct letters (A to Z) will have its letters in alphabetical order?

26. What is the probability that a license plate consisting of 6 letters (A to Z) will use exactly 5 different letters?

27. If 4 standard (6-sided) dice are thrown, then what is the probability that the sum is even?

28. If 4 cards are randomly chosen from a standard deck (containing 13 each of the four suits: clubs, diamonds, hearts, and spades), then what is the probability that one card of each suit will be obtained?

29.* Prove Theorem 6.9 under the assumption that the outcomes in the sample space are equally likely.

30.* Prove Theorem 6.10 under the assumption that the outcomes in the sample space are equally likely.

31. A baseball player seeking as many at-bats as possible wants to bat in the first 3 positions, since that guarantees an at-bat in the first inning. If you are one of the 9 starting players in a baseball lineup and the batting order is randomly set, then find the probability that you will bat within the first 3 positions.

32. If two integers from 1 to 9 are each randomly selected (repeats allowed), then what is the probability that their sum will be odd?

33. A fun experiment to perform with a class or at a party is called the **Birthday Problem**. Simply determine whether any two people in the group have the same day of the year for their birthday. (Ignore year of birth.) Here, we consider how likely a coincidence this is.

    Ignore leap years and assume that all possible birthdays are equally likely. For $1 \leq n \leq 365$, if n people are randomly chosen, then what is the probability that at least 2 will have the same birthday

(a) When $n = 15$?

(b) When $n = 30$?

(c)* What is the smallest value of n for which it is more likely than not that 2 will have the same birthday?

34. The very popular TV game show *Let's Make a Deal* was hosted by Monty Hall from 1963 to 1986. An analysis of strategies employed on that show has spawned what is known as the **Monty Hall Problem**.

   Game show host Monty Hall has placed a valuable prize behind one of three doors and junk behind each of the other two. First, a contestant selects one of the three doors. Second, since at least one of the unselected doors contains junk, Monty Hall opens one of them to reveal junk. Third, the contestant is given the option to stay with the initially chosen door or to switch to the other unopened door. Determine the probability that the contestant will win the valuable prize

   (a) If the strategy of staying with the original door is employed?

   (b) If the strategy of switching to the other door is employed?

   (c) Which strategy is better for the contestant?

35. If two standard (6-sided) dice are rolled, then what is the probability of obtaining doubles or a sum of 10?

36. A slot machine contains 5 wheels. Equally spaced around each wheel is a star, a moon, a plane, a car, a dog, a cat, a lemon, and an apple. Exactly one of the items can be seen in the display window at a time, and a pull of the arm causes the wheels to spin independently. If the player wins when at least 4 of the wheels show the same item, then what is the probability that a player will win on a single pull?

37. If two integers from 1 to 5 are randomly selected (repeats allowed), then what is the probability that at least one will be even?

38. If two integers from 1 to 10 are randomly selected (repeats allowed), then what is the probability that their product will be odd?

39. If two standard (6-sided) dice are rolled, then what is the probability that their sum will be even or divisible by 3?

40. What is the probability that a randomly chosen 5 digit (0–9) license plate has all of its digits less than 6 or all greater than 3?

41.* Prove that if any number of standard (6-sided) dice are rolled, then the probability that the sum is even is $\frac{1}{2}$.

42.* Is it true in general that if any number of standard (6-sided) dice are rolled, then the probability that the sum is divisible by 3 is $\frac{1}{3}$? Justify your answer.

State lotteries. In Exercises 43 through 48, we explore chances of winning various state lotteries.

43. The New York State Take Five lottery has a drawing at 11:21 each evening. At that time, 5 balls from a set of 39 balls numbered $1, 2, \ldots, 39$ are selected to determine a winner. Determine the probability that a single selection of 5 numbers on a ticket will win this lottery.

44. Each day in Kansas, Nebraska, and North Dakota, $1 will buy a 2 by 2 ticket on which a player selects a pair of distinct red numbers and a pair of distinct white numbers, each chosen from 1 to 26. What is the probability of winning the top prize of $22,000, if it requires that all 4 numbers are selected correctly?

45. In Maine, New Hampshire, and Vermont, drawings for the Tri-State Megabucks Plus lottery occur on Wednesdays and Saturdays. They draw 5 balls from a set of 41 balls numbered $1, 2, \ldots, 41$, and a 6th ball from a set of 6 balls numbered $1, 2, \ldots, 6$. A ticket wins if at least 2 numbers are correct (all 6 numbers earns "The Jackpot"). What is the probability that a ticket will win?

46. On a $1 ticket in the Colorado Cash 5 lottery, a player picks 5 distinct numbers from 1 to 32. That ticket wins $20,000 if all 5 of those numbers are drawn, $200 for 4 numbers, $10 for 3 numbers, and $1 for 2 numbers. What is the probability that a $1 ticket will win more than $1?

47. In the Mega Millions lottery, for $1 a player picks 5 distinct numbers from balls numbered 1 to 75 and a 6th Mega Ball number from 1 to 15. A player wins at least $1 by correctly matching the Mega Ball or at least 3 of the 6 balls.

    (a) What is the probability of winning at least $1?

    (b) What is the probability of winning exactly $1? This happens when a player correctly matches only the Mega Ball.

48. In the Powerball lottery, for $2 a player picks 5 distinct numbers from white balls numbered 1 to 59 and powerball number from red balls numbered 1 to 35. A player wins at least $4 by correctly matching the powerball or at least 3 of the 6 balls.

    (a) What is the probability of winning at least $4?

(b) What is the probability of winning at least $1,000,000? This happens when a player correctly matches at least the 5 white balls.

For Exercises 49 through 54, consider the experiment of rolling 4 fair standard (6-sided) dice. Determine the probability of the given events.

49. A run of 4 consecutive numbers. They need not be rolled in order.

50. 2 fives and 2 sixes.

51. Exactly 3 distinct numbers.

52. At least 3 of a kind.

53.* At least 2 fives or at least 2 sixes.

54. At least a pair.

55. Assume that male and female children are equally likely in a family. If we use B to denote a boy, G to denote a girl, and a pair to denote the genders of the first and second child in a family with two children, then the 4 possible outcomes GG, GB, BG, BB are equally likely. Given that a family with two children has at least one girl, what is the probability that both children are girls?

56. Suppose we toss three fair 6-sided dice in sequence. What is the probability that their sum is 9, given that the first two dice result in a pair?

57. A red box contains 4 black balls and 2 white balls, and a blue box contains 6 black balls and 4 white balls. One of these 16 balls is randomly selected. What is the probability that the ball came from the red box given that the ball is black?

58. A plaintiff is going to choose a lawyer from one of the two law firms in town. Firm 1 employs 5 male and 3 female lawyers, and Firm 2 employs 2 male and 4 female lawyers. Assume that a lawyer is randomly selected. Given that the lawyer is a woman, what is the probability that she is employed by Firm 2?

59. Consider the experiment of selecting a single card from a standard deck. Let E be the event that the suit of the card is hearts, let F_1 be the event that card is a face card, and let F_2 be the event that the card is red.

(a) Are E and F_1 independent? Explain.

(b) Are E and F_2 independent? Explain.

60. Consider the experiment of randomly selecting a number from 1 to 99. Let E be the event that the number is divisible by 3, let F_1 be the event that number is from 1 to 22, and let F_2 be the event that the number is from 23 to 88.

(a) Are E and F_1 independent? Explain.

(b) Are E and F_2 independent? Explain.

61. Suppose E and F are events in a sample space S with $P(E) > 0$ and $P(F) > 0$. Show that E and F are independent if and only if $P(E \mid F) = P(E)$ and $P(F \mid E) = P(F)$.

62. Show that two mutually exclusive events that each have positive probability can never be independent.

63. An insurance company issues car insurance, homeowner insurance, and life insurance. Of the claims the company pays, 50% are car, 40% are homeowner, and 10% are life. The company is primarily concerned with claims over $10,000 that it pays. History has shown that 10% of car claims, 30% of homeowner claims, and 90% of life insurance claims exceed $10,000.

   (a) What percentage of claims paid by this company exceed $10,000?

   (b) Given that this company pays a claim over $10,000, what is the probability that it is a homeowner claim?

64. Juan Rodrigez's batting average against left-handed pitchers is 0.325. That is, when he is facing a left-handed pitcher, the probability that he will get a hit is 0.325. His batting average against right-handed pitchers is 0.285. Suppose that only 15% of the pitchers Juan faces are left-handed.

   (a) What is Juan's overall batting average?

   (b) Given that Juan just got a hit, what is the probability that it occurred off of a right-handed pitcher?

65. A drug company has designed an over-the-counter test for strep throat. The probability of a *false positive* result from this test is 0.05. That is, 5% of the time a patient without strep throat takes the test, it will nonetheless tell the patient that strep throat is present. The probability of a *false negative* result, in which the test reports no strep throat even though it is present, is 0.1. The company has also determined that only 20% of the people choosing to take the strep test actually have strep throat, since patients often believe that they are sicker than they really are. Given that a patient taking the strep test gets a negative result (no strep), what is the probability that this patient actually does not have strep throat?

66. A toy company has a quality control inspector watching the toy cars as they come off the assembly line to separate out defective cars. The probability that this inspector (correctly) pulls out a car when it is defective is 0.9, and the probability that he (mistakenly) pulls out a car when it is nondefective is 0.06. Assume that 4% of the toy cars made

are defective. What percentage of the toy cars pulled out by the inspector are actually nondefective?

67. Market research by TechOrder.com predicts the following satisfaction ratings among all purchases of the Smartphone DS7.

Number of stars	Percentage of all customers
4	50
3	30
2	10
1	5
0	5

However, the likelihood that a customer will enter a ranking on the website depends on the value of the ranking as follows.

Number of stars	Probability that ranking will be entered
4	0.4
3	0.2
2	0.5
1	0.8
0	0.9

(a) Determine the predicted average satisfaction rating among all purchasers of the Smartphone DS7 (regardless of entering it).

(b) Determine the probability that a customer enters a ranking on the website.

(c) For each $0 \leq i \leq 4$, determine the probability of a ranking of i stars given that the ranking is entered on the website.

(d) Determine the average satisfaction rating among those entered on the website.

68. Overall student evaluations rate the clarity of Professor Hartzell as follows.

Number of stars	Percentage of all students
5	70
4	12
3	9
2	3
1	6

The likelihood that a student will submit a rating on RateMyProfessors.com depends on that rating as follows.

Number of stars	Probability that ranking will be entered
5	0.16
4	0.10
3	0.38
2	0.85
1	0.92

(a) Determine Professor Hartzell's average clarity rating among all student evaluations.

(b) Determine the percentage of Professor Hartzell's students that enter ratings on RateMyProfessors.com.

(c) For each $1 \leq i \leq 5$, determine the percentage of submissions on RateMyProfessors.com that rate Professor Hartzell's clarity as i stars.

(d) Determine Professor Hartzell's clarity rating shown on RateMyProfessors.com.

69. Use Definition 6.5 and Theorem 6.11 to prove Bayes' Formula.

70. Prove Bonferroni's inequality: If E and F are events in a sample space S, then $P(E \cap F) \geq P(E) + P(F) - 1$. Hint: $P(E \cup F) \leq 1$.

6.5   Applications of combinations

As can be seen in the examples from the previous sections, combinations are a useful counting tool. Although counting the number of ways to select a subset of a certain size from some fixed set is the most obvious use of this tool, there are many more. In this section, we explore other—and perhaps more subtle—applications of combinations.

6.5.1   Paths in a grid

Example 6.40.   (Delivery Routes).
The grid in Figure 6.7 represents the streets and intersections in a downtown area. A delivery truck must transport goods from the storehouse S to the factory F. To achieve a shortest possible route, the truck will necessarily make a sequence of one block moves, each to the right or down (from the point of view of an observer facing the grid). That is, movements to the left or up are inefficient and avoided. How many routes are there from S to F

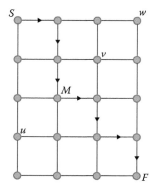

FIGURE 6.7
Routes through a rectangular grid.

(a) with only those restrictions?

(b) with the additional restriction that the route must pass through the displayed point M? Here, M represents an intermediate stop at another factory.

(c) with the additional restriction that the route cannot pass through the displayed point M? Here, M represents a blocked intersection due to a water main break.

Solution. (a) A route from S to F is characterized by a sequence listing each move as either right R or down D. Moreover, any such sequence must have length 7 and consist of exactly 3R's and 4D's. For example, the route marked in Figure 6.7 is characterized by the sequence RDDRDRD. The number of routes from S to F is the number of sequences of length 7 containing 3R's (and by default 4D's). Hence, there are

$$\binom{7}{3} = 35$$

routes from S to F.

(b) Since our routes must pass through M, we can treat each route as a concatenation of two routes. The first goes from S to M, and the second goes from M to F. Paths from S to M have length 3 and contain 1R. Paths from M to F have length 4 and contain 2R's. Therefore, there are

$$\binom{3}{1}\binom{4}{2} = 18$$

routes from S to F that pass through M.

(c) The fastest way to solve this is to use the Complement Principle and the answers from parts (a) and (b). Specifically, there are

$$35 - 18 = 17.$$

routes that do not pass through M.

It is also worthwhile to understand how this problem might be solved directly. Since our routes cannot pass through M, they must pass through one of u, v or w. Since these represent disjoint cases, there are

$$\binom{3}{0}\binom{4}{3} + \binom{3}{2}\binom{4}{1} + \binom{3}{3}\binom{4}{0} = 17$$

routes through u, v or w. Note that the points u, v, w are chosen on an off-diagonal line, since each route can only pass through one such point. They are not chosen on a vertical or horizontal line, since routes may pass through more than one point of such a line. □

The use of binomial coefficients in Example 6.40 can give us a deeper appreciation of Pascal's triangle.

Example 6.41. The locations in Pascal's triangle form a triangular grid, as shown in Figure 6.8. Consider paths that start at the topmost vertex and traverse down this triangular grid. That is, paths consist of a sequence of turns at each vertex to the right R or left L (from the point of view of an observer facing the grid). For each $0 \le k \le n$, let $c_{n,k}$ denote the vertex in the kth entry of the nth row.

$$
\begin{array}{ccccccccc}
& & & & c_{0,0} & & & & \\
& & & c_{1,0} & & c_{1,1} & & & \\
& & c_{2,0} & & c_{2,1} & & c_{2,2} & & \\
& c_{3,0} & & c_{3,1} & & c_{3,2} & & c_{3,3} & \\
c_{4,0} & & c_{4,1} & & c_{4,2} & & c_{4,3} & & c_{4,4} \\
& & & & \vdots & & & &
\end{array}
$$

A path from $c_{0,0}$ to $c_{4,3}$, for example, is characterized by a sequence of length 4 consisting of 3R's (and 1L). Hence, there are $\binom{4}{3}$ such paths. In general, we see that the number of paths from $c_{0,0}$ to $c_{n,k}$ is given by $\binom{n}{k}$. That is, each value in Pascal's triangle counts the number of paths from the topmost location to its location!

If the turns occur randomly with right and left being equally likely, then our analysis here can be used to study the probabilities of a game played on a Plinko board. That study is addressed in the exercises.

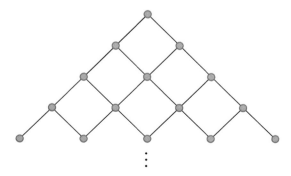

FIGURE 6.8
Pascal's triangular grid.

6.5.2 Poker

We have all of the tools needed to determine the likelihood, and hence the hierarchy, of all of the hands in 5-Card Stud Poker. However, we first need some familiarity with the terminology of Poker and cards in general.

A **standard deck** contains 52 cards. There are 13 of each **suit** (clubs ♣, diamonds ◇, hearts ♡, and spades ♠), and each suit is numbered with the 13 **denominations**

$$2, 3, 4, 5, 6, 7, 8, 9, 10, \text{Jack}, \text{Queen}, \text{King}, \text{Ace}$$

listed here in increasing order of value. Jacks, Queens, and Kings are called **face cards**, since they are decorated with pictures rather than numbers. A **run** of 5 cards is any set of 5 cards whose denominations are either 5 consecutive denominations from the list above or the list Ace, 2, 3, 4, 5.

In 5-Card Stud Poker, each player is dealt 5 cards. Those 5 cards make up the player's **hand**. Of course, there are $\binom{52}{5}$ possible hands. The winner is decided by using the hierarchy of hands listed in Table 6.3 from the highest to the lowest value. The issue of how to break a tie does not concern us here. Note that the definition of a Straight and the definition of a Flush exclude hands that have the higher value of a Straight Flush.

The reason for the ordering in Table 6.3 is justified by the results in Table 6.4. There, the number of ways to obtain each hand and the probability of doing so are given.

TABLE 6.3
Order and description of poker hands

Hand	Description
Straight Flush	A run of 5 cards of the same suit
Four of a Kind	4 cards of one denomination
Full House	3 cards of one denomination and 2 of one other
Flush	5 cards of the same suit, not forming a run
Straight	A run of 5 cards, not all of the same suit
Three of a Kind	3 cards of one denomination and 2 of others
Two Pairs	2 cards each of two denominations and 1 of one other
One Pair	2 cards of one denomination and 3 of others
Nothing	None of the hands listed above

TABLE 6.4
Likelihood of poker hands

Hand	Number possible	Probability (to 8 decimal places)
Straight Flush	40	0.00001539
Four of a Kind	624	0.00024010
Full House	3,744	0.00144058
Flush	5,108	0.00196540
Straight	10,200	0.00392465
Three of a Kind	54,912	0.02112845
Two Pairs	123,552	0.04753902
One Pair	1,098,240	0.42256903
Nothing	1,302,540	0.50117739

Proof of the results in Table 6.4. For each hand, the probability is obtained by dividing the number of ways to achieve that hand by

$$\binom{52}{5} = 2598960.$$

Hence, we focus only on counting the number of ways to achieve each hand. Here, we prove the parts associated with the hands

One Pair, Two Pairs, Straight, and Nothing.

Along the way, we also consider Straight Flushes. The remaining parts are left for the exercises.

(One Pair) There are 13 different denominations that could be involved in the pair. Once the particular denomination has been chosen, there are $\binom{4}{2}$ ways to select which 2 of the 4 cards of that denomination are involved in the pair. The 3 remaining cards cannot match the pair or each other. There are thus $\binom{12}{3}$ ways to select which denominations (outside of the one already used in the pair) are to be used for the 3 remaining cards. Once those denominations have been chosen, the particular suit (of the 4 possible) must be chosen for each of the cards. Therefore, there are

$$13 \cdot \binom{4}{2} \cdot \binom{12}{3} \cdot 4^3 = 1098240$$

ways to get One Pair.

(Two Pairs) We need two denominations to be involved in pairs. There are $\binom{13}{2}$ ways to pick them. Once, those are set, the particular 2 cards of each denomination must be chosen. In both cases, there are $\binom{4}{2}$ ways to do that. The fifth card must not match either of the pairs. Since there are 8 cards in the deck that match the pairs (including the pairs themselves), there are 44 choices for a nonmatching card. Therefore, there are

$$\binom{13}{2} \cdot \binom{4}{2} \cdot \binom{4}{2} \cdot 44 = 123552$$

ways to get Two Pairs.

(Straight) The key is to focus on the highest denomination. Only the 10 denominations $5, 6, \ldots,$ King, Ace can end a Straight. Once that has been set, there are 4 ways to fix the suit of each of the 5 cards. Thus, there are

$$10 \cdot 4^5 = 10240$$

ways to get a run of 5 cards. However, those runs that have the higher value of a Straight Flush must be excluded. Since each such hand is completely determined by its highest card, there are $10 \cdot 4 = 40$ Straight Flushes to exclude. Therefore, there are

$$10,240 - 40 = 10200$$

ways to get a Straight.

(Nothing) In order to count the number of ways to get Nothing, we instead count the number of ways to get something. Then, we can apply the Complement Principle. Hence, this part of the proof should be regarded as being last. Very simply, the counts from the other parts need to be added together, and that sum must be subtracted from $\binom{52}{5}$. Since there are

$$40 + 624 + 3744 + 5108 + 10200 + 54912 + 123552$$
$$+ 1098240 = 1296420$$

ways to get something, there are

$$\binom{52}{5} - 1296420 = 1302540$$

ways to get Nothing. $\qquad\square$

6.5.3   Choices with repetition

We consider here problems involving selections in which an item may be selected multiple times. It turns out that such problems involve combinations. However, their use in these problems is perhaps more subtle than in others we have encountered so far.

Example 6.42.   (Barbecue Orders).
A barbecue is attended by 7 people. Each person has their choice of a hamburger, a piece of barbecued chicken, or a hot dog (but only one in each case) for

their first food item. If the cook will barbecue all 7 items at the same time and does not care who ordered what, then how many different barbecue orders are possible for the cook? We assume that there are ample supplies of each food type.

Solution. This is considered a problem involving choices with repetition, since the cook may receive multiple requests for any of the 3 items

hamburger B, barbecued chicken C, or hot dog D.

A total order would consist of a list of length 7 of B's, C's, and D's, but with no specified number of any particular choice. For example, one order might be CBBDBDB. To help motivate our counting technique, such an order could be recorded in the form of a table.

B	C	D
√√√√	√	√√

Imagine that the cook simply places a √ in the appropriate column of an order sheet as each order is taken. At the end, the order could more efficiently be recorded as a sequence of length 9 consisting of 7 √'s and 2 |'s.

$$\checkmark\checkmark\checkmark\checkmark \mid \checkmark \mid \checkmark\checkmark$$

Here the headers B,C,D are understood. The point is that this is a binary sequence of length 9 containing 7 √'s (and 2 |'s). Note that the number of |'s is one fewer than the number of item choices. We conclude that the number of possible barbecue orders is

$$\binom{9}{7} = 36. \qquad \square$$

The argument used in the solution of Example 6.42 generalizes to give the following result.

Theorem 6.13.   *The number of ways to distribute n identical items into c distinct categories is*

$$\binom{n+c-1}{n}.$$

In Example 6.42, the items are orders and the categories are food types. Of course, the notions of item and category vary from problem to problem.

Example 6.43. (Election Results).

A class of size 30 is choosing among 4 candidates for class president. In how many ways can the vote tallies come out?

Solution. Imagine that the 30 completed ballots are placed in 4 piles according to which candidate is selected. By Theorem 6.13, there are

$$\binom{30+4-1}{30} = \binom{33}{30} = 5456$$

ways for this to result. The fact that there might be a tie and a resulting revote is a separate issue.   □

Exercises

1. A bag contains 4 each of red, blue, green, and yellow balls. How many ways are there to select 8 so that 2 of each color are obtained?

2. How many license plates consisting of 6 digits (0–9) have their digits in strictly increasing order? Compare Exercise 42.

   Delivery routes. Exercises 3 through 10 refer to the traffic grid illustrated in Figure 6.9. Location S is a storehouse from which deliveries are made, and F is a factory to which a large delivery is going. Intersections G and G' contain gas stations to which deliveries are sometimes made and where the delivery truck can get gas if necessary. Since the factory is southeast of the storehouse, we are only interested in efficient delivery routes, which never travel north or west in this grid.

3. How many possible efficient delivery routes are there from the storehouse to the factory? It is not necessary to pass by a gas station.

4. On Monday, an intermediate delivery must be made to gas station G. How many possible efficient delivery

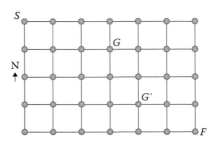

FIGURE 6.9
Routes through a grid.

routes from the storehouse to the factory pass through location G?

5. On Tuesday, an intermediate stop at gas station G' is required. How many possible efficient delivery routes from the storehouse to the factory pass through location G'?

6. On Wednesday, there was a gas leak at gas station G'. Since the wind was blowing to the northeast on that day, G' and all locations north or east of G' had to be avoided. How many possible efficient delivery routes were there from the storehouse to the factory on Wednesday?

7. Sometimes intermediate stops are required at both gas stations G and G'. How many possible efficient delivery routes from the storehouse to the factory accommodate this?

8. On Thursday, the gas leak at gas station G' was stopped, but that intersection itself remained closed for repairs. However, an intermediate delivery needed to be made at location G. Being sure to avoid location G', how many possible efficient delivery routes are there from the storehouse to the factory with an intermediate stop at gas station G?

9. Suppose the delivery truck needs gas. How many possible efficient delivery routes from the storehouse to the factory allow the truck to stop for gas? Hint: It must pass through G or G'.

10. On Friday, there were gas leaks at both gas stations G and G'. The wind blowing to the northeast therefore made it unsafe to travel north of, east of, or through either gas station. How many possible efficient delivery routes were there from the storehouse to the factory on Friday?

11. One of the popular games on the TV game show *The Price is Right* is called **Plinko**. Figure 6.10 shows a small version of the vertical board on which this game is played. A chip is dropped onto the topmost peg and falls, with equal probabilities, either left or right onto a peg in the next row. This process continues until the chip lands in one of the open slots shown on the bottom row.

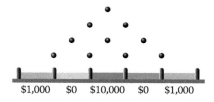

FIGURE 6.10
Plinko board.

(a) Determine the probability of landing in each of those slots.

(b) Below each slot is displayed its monetary reward. What is the probability that a single chip earns $1000?

12. Consider paths in Pascal's Triangular Grid (Figure 6.8) as described in Example 6.41. Given $0 \leq j \leq m$, $0 \leq k \leq n$, $j \leq k$, and $m \leq n$, how many paths are there from $c_{m,j}$ to $c_{n,k}$?

In Exercises 13 through 20, count the number of ways to get the given hand in 5-**Card Stud** Poker.

13. Flush.

14. Four of a Kind.

15. Full House.

16. Three of a Kind.

17. A Straight in red cards. Note that diamonds and hearts are red.

18. A Flush in which the highest card is 10 or better.

19. A hand that is at least as good as a pair of Jacks.

20. A hand that is at least as good as a Jack-high Straight.

In Exercises 21 through 28, consider playing 5-**Card Stud** Poker with *two identical decks* shuffled together. Count the number of ways to get the given hands. Also give the probability to 8 decimal places.

21. Straight Flush.

22. Straight.

23. Flush.

24. Four of a Kind.

25. Two Pairs.

26. Full House.

27. Three of a Kind.

28. One Pair.

29. Ben's mother bought cake and ice cream for his birthday party. While the 10 children were in the dining room, Ben's father was in the kitchen awaiting their dessert orders. If Ben's mother gives each child a choice of cake, ice cream, or both, then how many different dessert orders might she bring to his father? Needless to say, no child skipped dessert.

30. A car dealer is eager to display the newest model car in his lot. Based on the space available, he has decided that he will order 20 cars. However, he is having trouble picking the colors that will sell the best. If the car is available in 5 different colors, then how many different possible purchase orders might the dealer submit?

31. A jury consultant puts people into four categories: white men, white women, minority men, and minority women. She believes that the decisions made by the jury will be

largely influenced by the relative representation of these categories. From this consultant's point of view, how many different juries of 12 people are possible?

32. Mr. Calhoun bought 16 bite-size bars each of Snickers, Mounds, and Butter Fingers to give out on Halloween. He put them in a bowl on his porch and noted that a total of 14 bars had been taken throughout the night. How many possible distributions of leftovers are possible in the bowl?

33. Mrs. Pomfret has decided to make a small fruit basket for her new neighbor. The size of the basket is just right for 8 pieces of fruit, and Mrs. Pomfret has decided that apples, bananas, and oranges are the best types of fruit to consider. Determine the number of different options Mrs. Pomfret now has for her fruit basket. We distinguish types of fruit baskets by the numbers of each type of fruit included.

34. How many different collections of 8 balls are possible from a bag containing 8 each of red, blue, green, yellow, and orange balls? Balls of the same color are indistinguishable.

35. If 4 indistinguishable dice are rolled, then how many different "hands" of 4 values are possible? The order in which the values are rolled is irrelevant.

36.* Determine the number of different *positive* integer solutions e_1, e_2, e_3 there are to the equation $e_1 + e_2 + e_3 = 7$. Hint: Let $f_i = e_i - 1 \geq 0$.

37. Mr. Loomis told Serge that he has 8 coins in his pocket. Serge knows that the only possible coins Mr. Loomis might have are pennies, nickels, dimes, and quarters.

    (a) If coins of the same denomination are considered indistinguishable, then how many possibilities are there for the set of coins in Mr. Loomis' pocket?

    (b) Do distinct possible sets of coins from part (a) necessarily have distinct monetary values? Justify your answer.

38. A class of 100 students is having an election for class president. Greg, Marcia, and Peter are the candidates, and everyone, including the candidates, is required to vote in the election. After the election, the vote tallies will be announced for all three candidates.

    (a) How many different vote tallies are possible?

    (b) Notice that a 3-way tie is not possible. However, if two candidates tie for the most votes, then there will be a runoff between them. How many different possible tallies would lead to a runoff?

39. If 4 balls are selected from a bag containing 3 red, 2 white, and 4 blue, then what is the probability that at least one of each color is chosen?

40. If 4 cards are selected from a standard deck, then what is the probability that there will be 2 each from 2 suits?

41. The game of **Yahtzee** is played by rolling 5 standard (6-sided) dice and keeping track of the resulting "hand" achieved. For example, one hand is 3 twos, a one, and a five. Note that 3 twos, a one, and a six is considered a different hand, even though a player might use either to count for 6 points in the "Twos" box on her or his scorecard. How many different hands are possible?

42. How many license plates consisting of 6 digits (0–9) have their digits in nondecreasing order? Compare Exercise 2.

In Exercises 43 through 51, count the number of ways to get the given hand in **7-Card Stud** Poker. In this case, 7 cards are dealt and 5 are retained to form the best possible hand from the 7 given cards. Be aware that one set of 7 cards might have properties suitable for multiple hands, but the best would be chosen.

43. Four of a Kind.

44.* Full House.

45.* Straight Flush.

46.* Flush.

47.* Three of a Kind.

48.* Two Pairs.

49.* Straight.

50.* Nothing.

51. One Pair.

52. Should the hierarchy of hands for 7-Card Stud Poker agree with that for 5-Card Stud Poker?

In Exercises 53 through 58, we consider the poker game called **Texas Hold'em** that is played by the professionals in the *World Series of Poker*. In it, each player receives two down cards, called their **hole cards**. An additional five cards are dealt face up on the board and are cards common to all players. Each player, using his or her two hole cards plus the five board cards determines his or her hand as in 7-Card Stud. To accommodate betting, the five board cards are dealt by first showing three cards, called the **flop**, then showing one more card, called the **turn**, and finally showing the last card, called the **river**. In each of the given two-player games, determine the probability that player A will win and the probability that player B will win.

53.

Player	Down Cards	Flop	Turn	River
A	$K\spadesuit\ 6\diamond$	$7\heartsuit\ 8\spadesuit\ 5\spadesuit$	$7\diamond$	
B	$Q\spadesuit\ 5\diamond$			

54.

Player	Down Cards	Flop	Turn	River
A	$A\heartsuit\ 8\diamondsuit$	$K\clubsuit\ K\spadesuit\ 5\heartsuit$	$3\diamondsuit$	
B	$10\spadesuit\ 3\clubsuit$			

55.

Player	Down Cards	Flop	Turn	River
A	$K\clubsuit\ 7\diamondsuit$	$3\diamondsuit\ 8\heartsuit\ J\spadesuit$		
B	$2\clubsuit\ 2\diamondsuit$			

56.

Player	Down Cards	Flop	Turn	River
A	$A\diamondsuit\ J\diamondsuit$	$4\spadesuit\ 5\clubsuit\ 9\clubsuit$		
B	$K\spadesuit\ Q\spadesuit$			

For Exercises 57 and 58, use software such as the Poker Odds Calculator tool available at the website http://www.pokerlistings.com to determine the probabilities. An explanation of how and when ties are broken can also be found there.

57.

Player	Down Cards	Flop	Turn	River
A	$A\clubsuit\ K\clubsuit$			
B	$8\diamondsuit\ 8\heartsuit$			

58.

Player	Down Cards	Flop	Turn	River
A	$A\heartsuit\ K\heartsuit$			
B	$A\clubsuit\ A\spadesuit$			

6.6   Correcting for overcounting

A useful counting technique is to start with a rough count that is too large and then make the necessary corrections to obtain the desired count. This is the approach taken in Section 6.2 in our proof that $C(n,k) = \frac{P(n,k)}{k!}$. The quantity $P(n,k)$ is an overcounting of $C(n,k)$, and division by $k!$ is the necessary correction. Overcounting is also at the heart of the Basic Inclusion–Exclusion Principle. In the formula for $|A \cup B|$, the sum $|A| + |B|$ is too large in general; it must be corrected by subtracting $|A \cap B|$. In this section, we consider problems that lend themselves to an overcounting approach.

Example 6.44.   (Seating Arrangements).
How many ways are there to seat 5 girls at a circular table if the particular seat taken by each girl does not matter and what matters to each girl is

(a) who is sitting to her left and who is sitting to her right?

(b) who is sitting next to her (which side does not matter)?

Solution. (a) Temporarily call one seat the head of the table. If we keep track of which girl is seated at the head, then there are 5! ways to seat the girls clockwise around the table. However, 5! is an overcounting of what we want, since we have carried the extra structure of who is seated at the head of the table. Given any such seating, if all of the girls stood up and shifted one position clockwise, then the new seating should be considered the same as the original. Each girl would still have the same neighbor to her left and the same neighbor to her right. Since there are 5 different rotations of any seating, we need to divide the original 5! count by 5. Therefore, there are

$$\frac{5!}{5} = 24$$

different seatings around the table.

(b) When which neighbor is to the left of a girl and which is to the right no longer matters, there are fewer than 24 different seatings. Since only the set of 2 neighbors matters, if we take the mirror reflection of any seating from part (a),

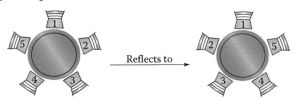

then the reflected seating should now be considered the same as the original. Reflection preserves neighbors (and switches sides). Since each seating from part (a) should be paired with its reflection, we need to divide the answer from part (a) by 2. Thus, there are

$$\frac{24}{2} = \frac{5!}{5 \cdot 2} = 12$$

different seatings in which only neighbors matter.   □

In both parts of Example 6.44 it was necessary to divide out by the size of a set of symmetries. To be specific, assume that a fixed head of the table is assigned seat number 1 and the remaining seats are consecutively numbered clockwise around the table. In part (a), any seating is considered unchanged by the rotations r_0, r_1, r_2, r_3, r_4, where r_i is a clockwise rotation of the seating by $72i$ degrees. That is, r_i represents $\frac{i}{5}$ of a full clockwise 360 degree rotation. The set of rotations $\{r_0, r_1, r_2, r_3, r_4\}$ forms what is called the **cyclic group** of order 5. It is the size of this group that is divided out in part (a). In part (b), we allow the additional symmetries given by reflections (or flips). There, the seatings are also considered unchanged by the flips f_1, f_2, f_3, f_4, f_5, where f_i represents reflection of the seating about the line passing through seat i and the center of the table. The full set $\{r_0, r_1, r_2, r_3, r_4, f_1, f_2, f_3, f_4, f_5\}$ forms what is called the **dihedral group** of order 10. It is by the size of this group that the overcounting of 5! is divided in part (b). Symmetry groups are formally introduced in Section 7.4.

Example 6.45. (Picking Teams).
Eleven players have gathered to play 5-on-5 basketball.

(a) If they have decided that one of the teams will have 6 players and use a substitute, then how many ways are there for them to break up into two teams?

(b) If one player decides to leave and there are only 10 players left, then how many ways are there to break up into two teams of 5?

Solution. (a) Among the 11 players, 5 must be selected to play on the team without a substitute. There are

$$\binom{11}{5} = 462$$

ways to do that.

(b) It is tempting to use the same analysis as that used in part (a) and get $\binom{10}{5}$. However, that is not correct. It would be correct if something distinguished the two teams (like uniforms or a name). Remember that $\binom{10}{5}$ counts the number of ways to select 5 players to be on a team. For example, if we number the players 1 through 10, then $\{2, 4, 7, 9, 10\}$ might be such a choice. Also, $\{1, 3, 5, 6, 8\}$ would be another such choice. However, both of these choices yield the same basketball game. It does not matter which team was "chosen" and which team was "unchosen." We need to equate those two possibilities. Hence, the count $\binom{10}{5}$ must be divided by 2. There are thus

$$\frac{\binom{10}{5}}{2} = 126$$

different possible basketball games for the 10 players. $\square$

FIGURE 6.11
A dodecahedron.

Example 6.46. (Making Dice).
A 12-sided die is to be made by placing the integers 1 through 12 on the faces of a dodecahedron (see Figure 6.11). Such a die is used in the game *Dungeons and Dragons*. How many different such dice are possible? Here, we consider two dice identical if one is a rotation of the other. Reflections are different.

Solution. Since each face must receive a different number, we might start by counting 12! ways to assign the numbers (assuming some fixed ordering or orientation of the faces). However, there is no order to the faces on a die; it may be rolled around into many different orientations. We therefore need to divide out for symmetries of the die. If the die is placed on a table, then any of the 12 faces (say, the one with the number 1 assigned to it) can be rotated to the top position. Further, even after the location of this top face is chosen, there are still 5 ways in which it might be rotated about a line through the centers of the top and bottom faces. That is, adjacent to the top face there are 5 faces from which to designate a front face. Consequently, there are $12 \cdot 5 = 60$ ways to orient any numbering of the faces. Thus, the number of oriented numberings must be divided by 60. It follows that there are

$$\frac{12!}{60} = 7983360$$

different 12-sided dice.   □

In Example 6.46, we divided out by the symmetry group of the oriented dodecahedron. That group is called the alternating group on 5 symbols. A careful description of it is not needed here. Only its order 60 is important for our purposes.

Example 6.47. (Student Population).
A certain school offers biology, chemistry, and physics classes. Its sophomores are required to either take one of these or to take all three of these. If there are 40 sophomores in the biology class, 30 in the chemistry class, 20 in the physics class, and 10 sophomores taking all three classes, then how many sophomores are at this school?

Solution. The sum $40 + 30 + 20$ overcounts the sophomores that took all three classes. Precisely, it counts each of those sophomores three times. Hence, two of those times must be subtracted for each such sophomore. It follows that there are

$$(40 + 30 + 20) - 2(10) = 70$$

sophomores at the school.   □

Example 6.47 foreshadows a counting technique we explore more deeply in Section 7.1.

Exercises

1. The 40 guests at a wedding are going to hold hands in a circle to perform a traditional dance. How many ways are there to do this if what is important to each is

   (a) the guest on the right hand and the guest on the left hand?

   (b) just the guests whose hands are being held?

2. A company makes small plastic rings in 6 colors: red, purple, blue, green, yellow, and orange. A baby rattle is constructed by attaching 6 different-colored small plastic rings around a large white plastic ring. How many such rattles are possible?

3. A set of blocks is to be made by painting wooden cubes in a certain format. Each cube gets two opposite faces painted white. The remaining 4 faces are to be painted red, yellow, blue, and green (one each). How many different resulting blocks are possible?

4. A small company produces 10 kinds of cheese. To promote their cheeses, this company is going to produce several samplers, each filled with 6 sample wedges of cheese.

   To maximize the advertising benefit of the samplers, it is important to the company that each of the 6 sectors receive a different kind of cheese. How many different cheese samplers are possible?

5. Mrs. Polhill is trying to figure out how to seat her dinner guests around the large circular table in the dining

room. Specifically, she needs a seating arrangement for a total of 8 people. How many ways are there to set this arrangement

(a) If only the left and right neighbors of each seat matter?

(b) What if, further, she and Mr. Polhill must sit together?

(c) What if instead they must sit opposite each other?

6. A day care has taken 12 preschoolers to the fair, and the children all want to ride on the merry-go-round together. Making things easier for the day care workers, all of the horses on the merry-go-round are identical and equally spaced. How many ways are there to arrange the children

(a) If the merry-go-round has 12 horses?

(b) If it has 13 horses?

(c) If it has 14 horses?

7. A necklace is to be formed by threading 8 beads on a circle of string. The beads are taken from a bag containing 14 different beads. How many options are there for a necklace if the knot holding the string together is not visible?

8. A craftsman makes bracelets by threading beads on a solid silver circular band. For a special order, a customer has picked out a specific set of 6 beads. Two are identical and the remaining four are all distinct. How many different bracelet options are there for the craftsman that fulfill this customer's wishes?

Making dice. In Exercises 9 through 16, a die is to be made from the specified solid by placing the numbers 1 through n on the faces of the solid, where n is the total number of faces. Determine the number of different dice possible in each case.

9. A cube. Hint: There are 24 different "rotations" of a cube. That is, there are 24 different ways to place the die on the table with one of its 6 sides facing up and then one of the top's 4 sides adjacent sides facing forward.

10. A regular tetrahedron. Hint: There are 12 different "rotations" of a tetrahedron. That is, there are 12 different ways to place the die on the table with a particular side facing forward (and one facing down).

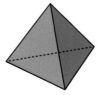

11. An icosahedron. This yields a 20-sided die.

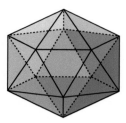

12. A cuboctahedron. This is formed by slicing off the corners of a cube. Specifically, mark the midpoints of each of the square sides of a cube. Then use the three midpoints nearest each corner of the cube to determine an equilateral triangle face produced after the corner is sliced off. There are 14 faces.

13. A triangular prism. It has two opposite faces in the shapes of equilateral triangles with three square faces connecting them.

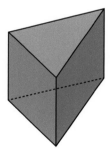

14. A pentagonal prism. It has two opposite faces in the shapes of regular pentagons with five square faces connecting them.

15. A truncated pyramid with a square base. It is pictured viewed from above.

16. A pyramid with a regular pentagon as the base.

    Team sports. In Exercises 17 through 24, we consider examples of splitting people into teams.

17. Six neighborhood children have gathered to play 3-on-3 hockey. In how many ways can they split into teams for a game?

18. Every Saturday, Mr. and Mrs. Nadal and Mr. and Mrs. Federer get together to play doubles tennis. To keep things interesting, they like to vary the pairings. How many weeks does it take to exhaust all of the possible pairings?

19. A basketball camp serves 20 girls. After a day of drills, the girls look forward to an evening of playing competitive games. In how many ways can these girls be split up to form two games of 5-on-5 basketball?

20. An advertising firm needs to be able to handle three clients at a time. Consequently, the boss must hire 12 people to split into 3 teams of 4. Once the boss hires 12 people, in how many different ways might these teams be set?

21. Determine the number of ways that 21 people can be split into 3 teams of 5 and 1 team of 6.

22. Determine the number of ways that 14 people can be split into 4 teams of 2 and 2 teams of 3.

23. A tennis camp for children ages 10–12 has registered 16 children. To get a sense of each child's ability on the first day, the director decides to watch all of the children play singles simultaneously. In how many ways can the 8 singles matches be set for these campers? The particular court on which a match is played does not matter.

24. On Tuesday night, 18 men gathered to play basketball. They decided to split into 2 teams of 5 and 2 teams of 4. In how many different ways might they do this?

25. Fifty women showed up Wednesday morning to play golf in the Bloomsburg Ladies League. The director instructed the women to split into 10 teams of 3 (to play a 3-person scramble tournament) and 10 teams of 2 (to play a best-ball tournament). In how many different ways might these teams be set?

26. On Thursday night, 30 men have gathered and will play 3 separate games of 5-on-5 basketball. Before they can play, teams must be picked and opponents must be decided. In how many ways can these men be split up to form these games?

27.* Kurt and Goldie are planning their wedding reception. They plan to seat their 60 wedding guests at 10 circular tables of 6. If all that matters is the set of neighbors of each guest, then how many ways are there to make the seating arrangements?

28.* As a summer camp arts and crafts project, Sally is going to make necklaces for her four cabinmates. For this she has selected 20 distinct beads. In how many different ways can Sally make four 5-bead necklaces, each with no visible knot?

29. In how many ways can the numbers 1 through 9 be assigned to a 3 × 3 square grid, if we assume that the same grid might be viewed from any of its 4 sides?

30. The ends of a 2 × 5 rectangular grid are pasted together to form a cylinder of height 2. In how many different ways can 10 different colors be assigned to the squares on this cylinder?

31. In how many ways can 24 people be seated on a 12-chair Ferris wheel in which each chair seats 2 people? We do not care which of the two is seated on the right or the left of a chair.

32. In how many ways can 12 skiers be seated on 4 consecutive chairs of a triple-chair lift? That is, each chair seats 3, and we do not care about the order of the 3 skiers in a chair.

33. Ethan bought Snickers, Mars, Mounds, and Crunch bars to give out for Halloween. To each child he gave either one or all four. If Ethan gave out 60 Snickers bars, 40 Mars bars, 50 Mounds bars, and 40 Crunch bars, and 20 children received all four, then how many children got candy from Ethan?

34. Sarah bought Tootsie Pops, Dum Dums, Smarties, and Sweet Tarts to give out for Halloween. To each child she gave either one or two. At the end of the night, Sarah had given out 60 Tootsie Pops, 40 Dum Dums, 50 Smarties, and 40 Sweet Tarts. She had given 10 children both a Tootsie Pop and a Dum Dum, 12 children both a Tootsie Pop and a Smarties, and 8 children both a Tootsie Pop and a Sweet Tarts. If no other combinations were given out, then how many children got candy from Sarah?

35. Several ice cream sundaes were made from choices of strawberry, black raspberry, chocolate, and coffee ice cream. In total, 20 contained strawberry, 10 contained black raspberry, 30 contained chocolate, 20 contained coffee, 6 contained both strawberry and black raspberry, 10 contained both chocolate and coffee, and no other combinations were chosen. How many sundaes were made?

36. Several ice cream sundaes were made from choices of vanilla, strawberry, and chocolate ice cream. If 30 contained vanilla, 20 contained strawberry, 50 contained chocolate, 15 contained all three, and no other combinations were chosen, then how many sundaes were made?

Building mobiles. For Exercises 37 through 40, a mobile is to be constructed in the given configuration (pictured from below). The mobile is hung from a string attached to the central point. The symbol • marks a rotational center, and the symbol ○ marks a location that must be filled with an item from a box containing a sun, a moon, a star, a bird, a plane, and a rocket. Determine the number of different possible mobiles in each case.

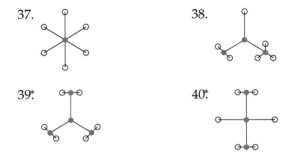

37.

38.

39.*

40.*

6.7 Review problems

1. The New York Mets is a National League baseball team. In the National League, there is no designated hitter and each of the 9 starting players has a position in the batting order. Once the New York Mets manager has determined the 9 starting players, in how many ways can the batting order be set?

2. A certain state makes it easier to identify VIP license plates by requiring them to have a specific form. They are required to consist of 5 letters (A, B or C) followed by 2 digits (0–7), and any other plate is forbidden to have this format. How many VIP's can be accommodated by this scheme?

3. Sam is trying to break into Diane's computer account. All that he knows is that her password consists of two letters (A to Z) followed by 3 distinct digits (0–9). How many such passwords are possible?

4. In a house with a washing machine and a dryer, each taking a half an hour to complete one load of laundry, how many hours are needed to wash and dry 25 loads of laundry?

5. How many of the integers from 1 to 1000 are divisible by 7?

6. How many of the integers from 200 to 4000 are divisible by 3?

7. A license plate is to consist of 7 letters (A to Z). How many possibilities have all of their letters distinct except that the first, fourth, and seventh letters are the same?

8. John and Robert Kennedy's lives were both ended by assassins in the 1960s. How many days after 11/22/1963 (JFK shot) was 6/5/1968 (RFK shot)?

9. Six friends on vacation have asked a stranger to take their picture. In how many ways can they be lined up in a single row (left to right) for this picture?

10. Each of the 50 states has two senators. The intelligence, armed services, and judicial committees are three important Senate committees. Consequently, each should have a different senator as chair.

    (a) How many ways are there to select these chairs?

    (b) What if no two of these chairs can be from the same state?

11. A large company contains a president, 6 vice presidents, 20 managers, and 22 secretaries. Due to financial setbacks, this company needs to downsize to a president, 5 vice presidents, 16 managers, and 17 secretaries. In how many ways can this be accomplished?

12. List all of the combinations of 2 items from the set $\{a, b, c, d, e, f\}$.

13. A couple has quadruplets and wants to send each of them to a different university within the 14 state universities. In how many ways could they assign these 4 children to 4 different state schools?

14. Zahira has 6 quarters and 4 nickels in her purse. Moreover, each coin has a different date on it, making it easily distinguishable from the others. Consequently, in how many different ways can 85¢ be obtained from Zahira's purse?

15. How many different pairs of cards from a standard deck are there that are either both red or both clubs?

16. How many binary numbers of length 8 (leading zeros are allowed) have at least 3 ones?

17. A certain NBA team contains 5 guards, 4 forwards, and 3 centers. The coach needs to name the starting 5 and has decided that it should include exactly 2 guards or exactly 2 centers. Therefore, in how many different ways can the starting 5 be selected?

18. A researcher at a drug company is doing a study of DNA strands that have no cytosine (C) or no guanine (G) nitrogen bases. Recall that each nitrogen base in a DNA strand is one of C, G, A, or T. Determine the number of possible DNA strands of length 8 that are relevant to this researcher's study.

19. How many of the integers from 1 to 507 are divisible by 5 or 7?

20. Find $\phi(10000)$. That is, how many of the integers from 1 to 10,000 are relatively prime to 10,000?

21. How many different hands of 4 cards from a standard deck have at least 3 of the same denomination?

22. **Craps** is a dice game in which two standard 6-sided dice are thrown. The outcome of a game is determined by the sum of the values appearing on the dice. One way in which a player loses on his first turn is if he rolls an 11. What is the probability that the sum is 11?

23. The prosecuting attorney in a high-profile case is upset that the 12-person jury consists only of men. The jury pool contained 50 men and 20 women. The attorney claims that if the jury had been randomly selected, then there would have been only a 1% chance of obtaining such a jury. Is he correct?

24. What is the probability that a sequence of 6 coin flips will result in at least 3 heads?

25. What is the probability that 4 cards selected from a standard deck will be of the same suit?

26. If 5 fair coins are tossed, then what is the probability that there will be more heads than tails?

27. Emma's mother bought a bag of 10 apples, 6 bananas, and 8 oranges for a fruit basket. However, she wants only 9 pieces of fruit in the basket and would like a balance in the types of fruit included. Although Emma insisted on being the one to pick which 9 to include, Emma simply picks them at random from the bag. Therefore, what is the probability that this fruit basket will contain equal numbers of apples, bananas, and oranges?

28. What is the probability of winning the Florida Lotto by correctly matching all 6 numbers chosen from 1 to 53?

29. The Pennsylvania Cash 5 Lottery has players pick 5 numbers from 1 to 43. A winning pick is one in which at least 2 of the numbers are correct. Of course, the reward increases with the number correct. What is the probability of picking a winning ticket?

30. Consider the experiment of randomly drawing two cards from a standard deck. Let E be the event that they form a pair, and let F be the event that they are both face cards.

    (a) Find $P(E \mid F)$.

    (b) Are E and F independent events? Explain.

31. A drug company sells three drugs, labeled A, B, and C, as a cure for a certain disease. It is known that drug A is successful 70% of the time, drug B is successful 80% of the time, and drug C is successful 90% of the time.

However, not all patients can afford drug C, since it is expensive. Instead, within the group of patients with the disease, 20% use drug A, 30% use drug B, and 50% use drug C.

(a) What percentage of the patient group will be cured?

(b) Given that a patient in the group is cured, what is the probability that this patient used drug C?

Navigating city streets. Exercises 32 through 34 refer to the grid of city streets illustrated in Figure 6.12. Mehdi's home is at intersection H, and he works at intersection W. His aunt lives at intersection A. Since Mehdi walks to work every day, he likes to vary his route so that he does not get bored. However, he refuses to make his trip to work any longer than necessary. Consequently, he only travels east or south along the streets.

32. Mehdi works 5 days per week for 48 weeks of the year. Is it possible for Mehdi to walk a different route to work each work day of a calendar year? Justify your answer.

33. On his aunt's birthday, Mehdi decides to stop by her house on the way to work to give her a card. How many possible routes to work will take Mehdi by his aunt's house?

34. One day, there was a huge fire at Mehdi's aunt's house. She escaped with her life and was staying with Mehdi while the firemen worked, but Mehdi still had to go to work. The strong wind blowing the smoke to the southwest that day made it impossible for Mehdi to walk by his aunt's house in the areas south and west of there. Consequently, how many possible routes to work did Mehdi have?

35. In Pascal's triangle, how many downward paths are there from $\binom{0}{0}$ to $\binom{10}{6}$ that pass through $\binom{4}{2}$?

36. In 5-Card Stud Poker, count the number of ways to get a Full House with the denomination for the pair lower than the other denomination.

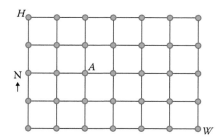

FIGURE 6.12
A grid of city streets.

37.* Using a standard 52-card deck, the number of possibilities for a 5-card hand with all distinct denominations is $\binom{13}{5}4^5$. This is an overcounting of the number of ways to get Nothing in 5-Card Stud Poker. Correct this by excluding hands.

For Exercises 38 through 46, consider **6-Card Stud** Poker in which each player is dealt 6 cards, and 5 cards are chosen to form the best possible 5-card hand from those 6 cards. Count the number of ways of getting each hand.

38. Four of a Kind.   39. Three of a Kind.

40.* Two Pairs.   41.* Full House.

42.* Straight Flush.   43.* Flush.

44.* Straight.   45.* One Pair.

46. Nothing.

47.* Johnny Chan and Daniel Negreanu are playing Texas Hold'em poker. Johnny's down cards are $A\clubsuit$ $9\clubsuit$. The common cards showing on the board are $2\clubsuit$ $9\diamondsuit$ $J\diamondsuit$ $Q\heartsuit$ $K\spadesuit$. Daniel has made a bet that would put Johnny "all in." That is, Johnny will need to bet all of his remaining chips to call the bet. So he needs to think carefully before acting. If all possible sets of down cards for Daniel are considered equally likely, then what is the probability that Daniel will win if Johnny goes "all in"?

48. At the end of each day a fountain is emptied of its coins (pennies, nickels, dimes, quarters, and half dollars), and like coins are grouped together. How many different outcomes are possible if in 1 day each of 100 people toss a coin into the fountain?

49. A computer password consists of a sequence of 5 letters (A to Z). What is the probability that a randomly chosen password will be in alphabetical order (repeats are allowed)?

50. In the showcase showdown on the *Price is Right*, a wheel with 20 sectors containing the values $5, 10, 15, \ldots, 100$ is spun. Each of three contestant tries to obtain a sum of no more than 100 using up to two spins.

(a) How many such wheels are possible?

(b) How many ways are there for a contestant to get 100?

(c) If a contestant gets 65 on her first spin, then what is the probability that adding a second spin will take her over 100?

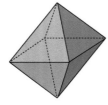

51. An 8-sided die is the shape of an octahedron. In how many ways can the numbers 1 through 8 be assigned to its sides? Note: There are 24 different "rotations" of an octahedron.

52. A high school football team has 22 members. In practice, the coach likes to split them up for an 11-on-11 scrimmage. In how many ways can they be split for this scrimmage?

53. A software company maintains two large and two small work groups to handle the programming projects it gets. There are a total of 24 programmers at this company. In how many ways can they be split up into 2 teams of 8 and 2 teams of 4?

54. The math majors at a certain school take either one or two of the courses linear algebra, numerical methods, and numerical linear algebra, but not all three. In total, 30 took linear algebra, 40 took numerical methods, 25 took numerical linear algebra, and the number of students taking any pair of classes was 10. How many math majors are at this school?

55. Andrew's mother is planning to make a mobile for him in the pictured configuration using the 8 distinct cars that are in a box. The mobile will be hung from a string attached to the central point, and the symbol ● marks each rotational center. If the symbol ○ marks each location that must be filled with one of the cars, then determine the number of different possible mobiles Andrew's mother might make.

56. There are 20 participants in a programming contest. Each is hoping to win one of the first-, second-, or third-place ribbons. How many such outcomes are possible

   (a) With an additional student receiving honorable mention?

   (b) What if not just one but three students get honorable mention?

57. How many possible sequences of 10 coin tosses result in (a) 5 heads? (b) more heads than tails?

58. Find $\phi(729)$. That is, how many integers from 1 to 729 are relatively prime to 729?

59. Determine the number of 8-digit base-3 sequences that have at most 2 zeros. Leading zeros are allowed.

60. A roulette wheel is divided into 38 sectors of equal size, and the numbers

$$0, 00, 1, 2, \ldots, 36$$

are assigned to the sectors. If 0 and 00 must be on opposite sectors, then how many different roulette wheels are possible?

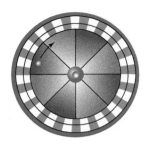

61. How many possible 5-digit (0–9) license plates with distinct digits have their digits in decreasing order?

62. How many of the integers from 1,000 to 10,000 are divisible by 13 or 17?

63. If 4 balls are randomly picked from 8 red, 6 blue, and 5 green, then what is the probability of getting the same number of red balls as blue balls?

64. If 3 fair 6-sided dice are rolled, then what is the probability that the sum is odd?

65. The fans at a game of Texas Hold'em cannot see any of the hole cards held by the players. Otherwise, they might yell out information. Since they can see only the common cards, the first cards the fans see are the 3 cards on the flop. When 3 cards are drawn from a standard 52-card deck, what is the probability that they

    (a) Are all the same color?

    (b) All have the same denomination?

66.* Vanessa Selbst is playing Texas Hold'em. She has two hole cards of the same suit. If she stays in the game, then what is the probability that she will have at least five cards of the same suit, after the five common cards are dealt? This would make a flush or better, which is a great hand. Note that the five common cards could form a "flush" of a suit different from the suit of the hole cards.

7 More Counting

We explore in more depth some of the basic counting techniques from Chapter 6. The Basic Inclusion–Exclusion Principle (Theorem 6.7) is generalized, so that we can analyze the cardinality of the union of any finite number of sets, and not just two. Similarly, binomial coefficients, which we have used in counting combinations (Theorem 6.4) and in the Binomial Theorem (Theorem 4.10), are extended to multinomial versions.

In Section 7.3, we introduce new and powerful counting tools: generating functions, which have a marvelous way of organizing combinatorial information. It is striking to see that elementary algebraic calculations can seemingly magically produce answers to counting problems. A more sophisticated use of algebra leads us to a consideration of Burnside's formula. That result strengthens our abilities from Section 6.6 to correct for overcounting. In the final section, combinatorial proofs are introduced as both a clever proof technique and a way to gain deeper insight into some combinatorial identities.

Applications similar to those in Chapter 6 are encountered throughout this chapter. Additionally, our development of stronger counting tools enables us to go deeper. We count the number of ways to completely mix up the members of a set, as would be required of the members of a group participating in a Secret Santas gift exchange. With the tools in this chapter, we can more easily analyze the splitting of a group of people into several teams. We also consider the number of ways it is possible to build a computer network in a certain configuration, given a variety of available types of computers.

7.1 Inclusion–exclusion

Introduced in Section 6.3, Basic Inclusion–Exclusion enables us to count the number of elements in the union of two sets. It gives the formula

$$|A_1 \cup A_2| = |A_1| + |A_2| - |A_1 \cap A_2|,$$

which is not hard to see from the Venn diagram on the left-hand side of Figure 7.1. However, when we extend

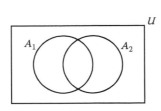

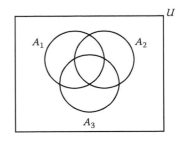

FIGURE 7.1
Venn diagrams for two and three sets.

our consideration to a union of more than two sets, an appropriate general formula for $|A_1 \cup A_2 \cup \cdots \cup A_n|$ is not so obvious. Even for three sets, as we can see in the Venn diagram on the right-hand side of Figure 7.1, the possible intersections among the sets are much more complicated than in the case of two sets.

The appropriate general formula for the cardinality of the union of three sets turns out to be

$$|A_1 \cup A_2 \cup A_3| = |A_1| + |A_2| + |A_3|$$
$$- (|A_1 \cap A_2| + |A_1 \cap A_3| + |A_2 \cap A_3|)$$
$$+ |A_1 \cap A_2 \cap A_3|.$$

As is requested in the exercises, that case can be proven by using the Venn diagram on the right-hand side of Figure 7.1 as a guide. However, such a brute-force method is not reasonable in general. Fortunately, the general formula that handles any finite number of sets has both a reasonable expression and a reasonable proof.

Theorem 7.1. (Generalized Inclusion–Exclusion Principle). *Let n sets $A_1, A_2, \ldots, A_n$ be given. For each $1 \leq i \leq n$, define the sum*

$$S_i = \sum_{1 \leq j_1 < j_2 < \cdots < j_i \leq n} |A_{j_1} \cap A_{j_2} \cap \cdots \cap A_{j_i}|.$$

That is, S_i is the sum of the sizes of the $\binom{n}{i}$ possible intersections of i sets from the collection $\{A_1, A_2, \ldots, A_n\}$. Then

$$|A_1 \cup A_2 \cup \cdots \cup A_n| = \sum_{i=1}^{n} (-1)^{i-1} S_i.$$

The results in Theorem 7.1 in the cases that $n = 2$ or 3 are shown above. To understand the Generalized Inclusion–Exclusion Principle, it may also be helpful to see the formula written out in detail for $n = 4$:

$$|A_1 \cup A_2 \cup_3 \cup A_4| = |A_1| + |A_2| + |A_3| + |A_4|$$
$$- (|A_1 \cap A_2| + |A_1 \cap A_3| + |A_1 \cap A_4|$$
$$+ |A_2 \cap A_3| + |A_2 \cap A_4| + |A_3 \cap A_4|)$$
$$+ (|A_1 \cap A_2 \cap A_3| + |A_1 \cap A_2 \cap A_4|$$
$$+ |A_1 \cap A_3 \cap A_4| + |A_2 \cap A_3 \cap A_4|)$$
$$- |A_1 \cap A_2 \cap A_3 \cap A_4|.$$

In fact, most of the examples we consider do not involve more than four sets.

Proof. Let x be an arbitrary element of $A_1 \cup A_2 \cup \cdots \cup A_n$. We verify the asserted formula by showing that x contributes exactly 1 to $\sum_{i=1}^{n}(-1)^{i-1}S_i$.

Let k be the number of sets $A_1, A_2, \ldots, A_n$ that contain x. By reindexing if necessary, we can assume that x is in the k sets $A_1, A_2, \ldots, A_k$. For each $1 \le i \le n$, the contribution of x to S_i must be computed. Since S_i is the sum of the sizes of intersections of i sets from $A_1, A_2, \ldots, A_n$, that contribution is the number of those intersections that contain x. Since x is only in the k sets $A_1, A_2, \ldots, A_k$, this number is $\binom{k}{i}$, the number of intersections (equivalently, selections) of i sets from $A_1, A_2, \ldots, A_k$. That is, x contributes $\binom{k}{i}$ to S_i. Of course, x makes a nonzero contribution to S_i only if $1 \le i \le k$. Thus, the contribution made by x to $\sum_{i=1}^{n}(-1)^{i-1}S_i$ is

$$\sum_{i=1}^{k}(-1)^{i-1}\binom{k}{i}$$

$$= \binom{k}{1} - \binom{k}{2} + \binom{k}{3} - \cdots + (-1)^{k-1}\binom{k}{k}$$

$$= \left(1 - \binom{k}{0}\right) + \binom{k}{1} - \binom{k}{2} + \binom{k}{3} - \cdots + (-1)^{k-1}\binom{k}{k}$$

$$= 1 - \sum_{i=0}^{k}(-1)^i\binom{k}{i}$$

$$= 1 - (1-1)^k$$

$$= 1.$$

The second to last equality follows from the Binomial Theorem. □

Our first example to which Theorem 7.1 applies involves the union of four sets.

Example 7.1. (Computer Scheduling).
A certain college splits its general education requirements into four categories: Business (B),

Engineering (E), Humanities (H), and Social Science (S). At the beginning of the year, each entering freshman is randomly assigned a sequence of 10 categories (5 for the fall and 5 for the spring). The student's class schedule is then required to satisfy the category sequence so that the first class comes from the first category, the second class comes from the second category, and so on. If each entry in the category sequence has equal probability of being any one of the four categories, then what is the probability that a randomly generated schedule will exclude at least one of the four categories?

Solution. We are considering the set $\mathcal{U}$ of all sequences of length 10 such that each entry in the sequence is one of the categories B, E, H, or S. Since the 4^{10} possible sequences are considered equally likely, to compute the desired probability, we must count the number of sequences in $\mathcal{U}$ that exclude at least one of the categories. So, define the following subsets of $\mathcal{U}$:

A_1 contains those sequences without any B's,

A_2 contains those sequences without any E's,

A_3 contains those sequences without any H's, and

A_4 contains those sequences without any S's.

Consequently, the set whose cardinality we seek is

$$A_1 \cup A_2 \cup A_3 \cup A_4,$$

the set of sequences missing at least one of B, E, H, or S. It is straightforward to compute the sizes

$$|A_j| = 3^{10} \quad \forall\, 1 \le j \le 4,$$

$$|A_{j_1} \cap A_{j_2}| = 2^{10} \quad \forall\, 1 \le j_1 < j_2 \le 4,$$

$$|A_{j_1} \cap A_{j_2} \cap A_{j_3}| = 1 \quad \forall\, 1 \le j_1 < j_2 < j_3 \le 4, \text{ and}$$

$$|A_1 \cap A_2 \cap A_3 \cap A_4| = 0$$

of the various possible intersections. The point is that when a fixed number of categories is excluded, each of the 10 positions can be filled freely with any of the remaining letters. Since, for each fixed value of i, each of the $\binom{10}{i}$ summands in S_i has the same value, we get

$$S_1 = \binom{4}{1} \cdot 3^{10}, \quad S_2 = \binom{4}{2} \cdot 2^{10},$$

$$S_3 = \binom{4}{3} \cdot 1, \quad \text{and} \quad S_4 = 0.$$

By the Generalized Inclusion–Exclusion Principle,

$$|A_1 \cup A_2 \cup A_3 \cup A_4| = \binom{4}{1} \cdot 3^{10} - \binom{4}{2} \cdot 2^{10} + \binom{4}{3} \cdot 1 - 0$$

$$= 4 \cdot 3^{10} - 6 \cdot 2^{10} + 4$$

$$= 230056.$$

Therefore, the probability that a randomly chosen schedule will be missing at least one of the four categories is

$$\frac{230056}{4^{10}} = \frac{28757}{131072} \approx 0.2194. \qquad \square$$

The key to using Inclusion–Exclusion to compute the size of a set A is to express A as a union $A_1 \cup A_2 \cup \cdots \cup A_n$ of simpler sets. What must also be true of the simpler sets is that intersections of the form $A_{j_1} \cap A_{j_2} \cap \cdots \cap A_{j_i}$ have sizes that are reasonable to compute. If A is more naturally expressed as an intersection of the complements $A_1{}^c \cap A_2{}^c \cap \cdots \cap A_n{}^c$ of the simpler sets (as we shall see in Example 7.2), then a similar approach can be taken using the following Corollary to Theorem 7.1. It is a generalization of Corollary 6.8 from Section 6.3, and its proof is left for the exercises.

Corollary 7.2.   Given subsets $A_1, A_2, \ldots, A_n$ of a finite universal set $\mathcal{U}$,

$$|A_1{}^c \cap A_2{}^c \cap \cdots \cap A_n{}^c| = \sum_{i=0}^{n} (-1)^i S_i,$$

where $S_0 = |\mathcal{U}|$.

The **Euler phi-function** ϕ was defined in Section 6.3. Recall that, for each positive integer n, the value $\phi(n)$ is the number of integers from 1 to n that are relatively prime to n. Corollary 7.2 can be applied as a tool for computing values of the Euler phi-function.

Example 7.2.   Compute $\phi(2695)$.

Solution. Let $\mathcal{U} = \{1, 2, \ldots, 2695\}$. Since

$$2695 = 5 \cdot 7^2 \cdot 11,$$

we want the number of integers in $\mathcal{U}$ that are not divisible by any of the primes $5, 7,$ or 11. Define the following subsets of $\mathcal{U}$:

A_5 contains those that are divisible by 5,

A_7 contains those that are divisible by 7, and

A_{11} contains those that are divisible by 11.

Note the wisdom in choosing the subscript of each set to reflect its description. Nothing says we must use $A_1, A_2, \ldots$ every time. Here, we are interested in counting the number of elements in

$$A_5{}^c \cap A_7{}^c \cap A_{11}{}^c,$$

the set of integers in $\mathcal{U}$ that are not divisible by 5, not divisible by 7, and not divisible by 11. Thus, Corollary 7.2 applies. We have

$$|A_5| = \frac{2695}{5} = 539, \quad |A_7| = \frac{2695}{7} = 385,$$

$$|A_{11}| = \frac{2695}{11} = 245,$$

$$|A_5 \cap A_7| = \frac{2695}{5 \cdot 7} = 77, \quad |A_5 \cap A_{11}| = \frac{2695}{5 \cdot 11} = 49,$$

$$|A_7 \cap A_{11}| = \frac{2695}{7 \cdot 11} = 35,$$

and

$$|A_5 \cap A_7 \cap A_{11}| = \frac{2695}{5 \cdot 7 \cdot 11} = 7.$$

Therefore,

$$S_0 = 2695,$$

$$S_1 = 539 + 385 + 245 = 1169,$$

$$S_2 = 77 + 49 + 35 = 161, \text{ and}$$

$$S_3 = 7.$$

By Corollary 7.2,

$$|A_5{}^c \cap A_7{}^c \cap A_{11}{}^c| = 2695 - 1169 + 161 - 7 = 1680.$$

Therefore, 1680 of the integers from 1 to 2695 are relatively prime to 2695. That is,

$$\phi(2695) = 1680. \qquad \square$$

7.1.1 Derangements

Secret Santas is a popular gift exchange method in which each member of some group, such as a group of office workers, is assigned another member of the group to whom a gift must be given. Although it is usually important that the assignments are kept secret, what is most important in our analysis here is that no person can be assigned himself or herself. The assignments might be made by placing everyone's names in a hat and then having each person select a name from the hat. However, if someone selects his or her own name, then this unacceptable outcome must be corrected through some reselection. Consequently, we consider the likelihood that such a problem does not occur.

A **derangement** of a set is a permutation that leaves no element fixed. For example, the permutation 321 of $\{1, 2, 3\}$ leaves the value 2 fixed in its second position and switches values 1 and 3. Hence, 321 is not a derangement. In fact, among the $3! = 6$ permutations of the set $\{1, 2, 3\}$, namely,

$$123, 132, 213, 231, 312, 321,$$

only the two

$$231, 312$$

are derangements. In this case, $\frac{1}{3}$ of the permutations are derangements. Consequently, a group of 3 people selecting names for a Secret Santas gift exchange would do so without a self-selection on the first try only $\frac{1}{3}$ of the time. In general, the percentage of permutations of n items that are derangements has a remarkable formula.

Example 7.3. (Secret Santas).
If each of a set of n people is randomly assigned a member of the set, then what is the probability that no one will be assigned himself or herself?

Solution. We can regard the people as numbered from 1 to n. Let $\mathcal{U}$ be the set of permutations of $\{1, 2, \ldots, n\}$. Of course, $|\mathcal{U}| = n!$ is the size of our sample space. For each $1 \leq j \leq n$, let A_j be the set of those permutations that leave j fixed. That is, j is in the jth position for every permutation in A_j. Observe that $A_1 \cup A_2 \cup \cdots \cup A_n$ is the set of permutations that leave some integer fixed. Therefore,

$$A_1^c \cap A_2^c \cap \cdots \cap A_n^c = (A_1 \cup A_2 \cup \cdots \cup A_n)^c$$

is the set of permutations that leave no integer fixed. It is the size of this set (event) that we need.

For each $1 \leq i \leq n$ and any $1 \leq j_1 < j_2 < \cdots < j_i \leq n$, the intersection $A_{j_1} \cap A_{j_2} \cap \cdots \cap A_{j_i}$ is the set of permutations that leave positions $j_1, j_2, \ldots, j_i$ all fixed. Since only the remaining $n - i$ positions are permuted,

$$|A_{j_1} \cap A_{j_2} \cap \cdots \cap A_{j_i}| = (n - i)!.$$

Therefore, for each $1 \leq i \leq n$,

$$S_i = \binom{n}{i} \cdot (n-i)! = \frac{n!}{i!}.$$

By Corollary 7.2,

$$|A_1^c \cap A_2^c \cap \cdots \cap A_n^c| = \sum_{i=0}^{n} (-1)^i \frac{n!}{i!}.$$

Thus,

$$P(A_1^c \cap A_2^c \cap \cdots \cap A_n^c) = \frac{\sum_{i=0}^{n} (-1)^i \frac{n!}{i!}}{n!} = \sum_{i=0}^{n} \frac{(-1)^i}{i!}$$

is the desired probability.   □

The remarkable fact about the result in Example 7.3 is that as n approaches infinity,

$$\sum_{i=0}^{n} \frac{(-1)^i}{i!} \text{ approaches } \frac{1}{e},$$

the reciprocal of the Euler number $e \approx 2.718$. That is, for large n, the probability that a randomly chosen permutation of $\{1, 2, \ldots, n\}$ fixes no integers is approximately $\frac{1}{e}$. For example, when $n = 12$, both $\sum_{i=0}^{n} \frac{(-1)^i}{i!}$ and $\frac{1}{e}$ equal 0.36787944 to eight decimal places.

Exercises

1. How many base-3 sequences of length 5 are missing one of the digits 0, 1, or 2?

2. In a game at the state fair, a bag contains 7 red, 5 white, and 6 blue balls. A player blindly selects 4 balls and wins if not all colors are obtained. In a single play of this game, what is the probability that at least one color will be missing and hence the player will win the game?

3. In a certain neighborhood on Halloween night, each house is specializing in a single type of candy. There are 6 houses giving out Snickers bars, 10 giving out Mounds bars, and 8 giving out Butterfingers bars. Rodrigo is interested in obtaining each of these types of candy, but his mother has told him that he may go to only 5 houses. How many ways are there to select 5 houses to visit so that each type of candy bar is obtained?

4. A deoxyribonucleic acid (DNA) strand consists of a sequence of nitrogen bases. Each base is one of cytosine (C), guanine (G), thymine (T), or adenine (A). A researcher is studying sequences that contain all of these. How many possible DNA strands of length 6 contain at least one of each type of nitrogen base?

5. A box contains 6 bananas, 7 apples, and 9 oranges. How many possible selections of 6 pieces of fruit to make a fruit basket exclude at least one type of fruit? The 22 pieces of fruit are considered distinguishable.

6. From a committee containing 10 Republicans, 5 Democrats, and 8 Independents, a subcommittee needed to be formed. Although it would have been desirable to have each party represented on the subcommittee, it turned out that the subcommittee contained no Democrats.

   (a) How many possible selections of a subcommittee of size 6 contain at least one member of each party?

   (b) If the subcommittee was selected randomly, then what would have been the probability that it contained no Democrats?

7. The game *Dungeons and Dragons* uses a 4-sided die constructed from a tetrahedron by filing down the vertices and placing the numbers 1 through 4 on these dulled points of the die. Thus, exactly one of 1 through 4 faces up after the die is rolled. If six 4-sided dice are rolled, then what is the probability that

   (a) there will be a number that is not showing on any of the dice?

   (b) the sum is at least 22?

8. What is the probability that at 5-card poker hand contains at least one of each face card denomination (Jack, Queen, King)?

9. What is the probability that at 7-card poker hand (keep all 7) contains at least one of each suit (♣, ◇, ♡, ♠)?

10. What is the probability that at 5-card poker hand is missing some suit?

11. A bag contains 8 quarters, 5 dimes, 4 nickels, and 10 pennies. What is the probability that a random selection of 5 coins from this bag will exclude at least one of these types of coins?

12. A researcher participating in the DNA study discussed in Exercise 4 has found several naturally occurring DNA strands of length 7 that are missing at least one of C, G, A, or T. What is the probability that a randomly generated DNA strand of length 7 would be missing at least one of the possible nitrogen bases?

Exercises 13 through 18 refer to the Euler phi-function ϕ.

13. Compute $\phi(300)$.   14. Compute $\phi(616)$.

15. Compute $\phi(1100)$.   16. Compute $\phi(7000)$.

Exercises 17 through 22 refer to the following formulas for ϕ. From the basic facts

$$\forall k \in \mathbb{Z}^+, \forall \text{ primes } p, \quad \phi(p^k) = p^k(1 - \tfrac{1}{p}) \qquad (7.1)$$

and

$$\forall a, b \in \mathbb{Z}^+, \text{ if } \gcd(a, b) = 1, \text{ then } \phi(ab) = \phi(a)\phi(b) \quad (7.2)$$

it can be proven that

$$\forall n \in \mathbb{Z}^+, \quad \phi(n) = n \prod_{p|n}(1 - \tfrac{1}{p}). \qquad (7.3)$$

A proof of (7.2) is given in Burton (2011), for example.

17. Use the Principle of Inclusion–Exclusion to compute $\phi(2100)$, and confirm that the result agrees with that given by Equation 7.3.

18. Use the Principle of Inclusion–Exclusion to compute $\phi(7700)$, and confirm that the result agrees with that given by Equation 7.3.

19. Prove that (7.3) follows from (7.1) and (7.2).

20. Use the Complement Principle to prove (7.1).

21. Use the Basic Inclusion–Exclusion Principle to prove (7.3) directly, when n has exactly two prime divisors.

22. Use the Generalized Inclusion–Exclusion Principle to prove (7.3) directly, when n has exactly three prime divisors.

23. In 1782, Slovene mathematician Georg (Jurij) Vega (1754–1802) published the prime factorizations of all integers from 1 to 10,500 that are not divisible by 2, 3, or 5. How many such integers are there?

24. Let $S = \{1, 2, \ldots, 120\}$.

    (a) Use the Generalized Inclusion–Exclusion Principle to count the number of integers in S that are not divisible by 2, 3, 5, or 7.

    (b) Now use the Sieve of Eratosthenes to quickly determine the number of primes in S. See Exercise 37 from Section 3.1.

25. How many integers from 1 to 10,000 are divisible by at least one of 3, 13, 23, or 43?

26. How many integers from 1 to 2222 are divisible by at least one of 7, 11, 17, or 31?

27. Compute $\sum_{i=0}^{6} \frac{(-1)^i}{i!}$. To how many decimal places does it agree with $\frac{1}{e}$?

28. List the sample space consisting of all permutations of the set $\{1, 2, 3, 4\}$. List the event consisting of those permutations that do not leave any element fixed. What is the probability of that event?

29. A person applying for jobs accidentally mixed up the 5 applications and corresponding envelopes. Assume that the applications were randomly shuffled before being placed into the 5 envelopes. What is the probability that

    (a) Some application ends up in its correct envelope?

    (b) All applications end up in their correct envelopes?

    (c) At least 3 applications end up in their correct envelopes?

30. Greg, Marcia, Peter, Jan, Bobby, and Cindy are participating in a Secret Santas gift exchange. They have put their names in a hat, and each will randomly select a name from this hat.

    (a) What is the probability that none of them will end up with his or her own name? Hence no reselection is needed.

    (b) What is the probability that exactly three of them will end up with their own names?

    (c) Why would total reselection be warranted in all cases except the case in which none of them end up with his or her own name?

31.* Before playing a card game, it is customary to shuffle the deck. If a standard deck of cards is randomly shuffled, then, to eight decimal places, what is the probability that no card ends up in its original location?

32.* Suppose 20 deranged men throw their hats into a dark closet and then each man randomly selects a hat from the closet. To eight decimal places, what is the probability that some man ends up with his own hat?

33.* Prove Corollary 7.2. Hint: Use De Morgan's Law and the Principle of Complements.

34.* Prove the Generalized Inclusion–Exclusion Principle in the case in which $n = 3$ by mimicking the proof of the Basic Inclusion–Exclusion Principle (Theorem 6.7). Hint: A Venn diagram can motivate the argument.

35. For $n = 5$, write out the sum S_3 defined in the Generalized Inclusion–Exclusion Principle.

36. For $n = 6$, write out the sum S_2 defined in the Generalized Inclusion–Exclusion Principle.

37. Use Corollary 7.2 to obtain a formula for the size of the intersection of three sets A_1, A_2, and A_3 in terms of their complements.

38. Express $|A_1{}^c \cup A_2{}^c \cup A_3{}^c|$ in terms of the sets A_1, A_2, and A_3, without using their complements.

39. How many ways are there to distribute 10 distinct objects into 5 boxes, labeled A through E, so that no box is empty?

40. How many integer solutions to the equation $k_1 + k_2 + k_3 + k_4 = 35$ are there such that $\forall\, i, 0 \le k_i \le 10$? Hint: Let A_i be the number of solutions with $e_i \ge 11$.

41. A child playing with standard fair 6-sided dice is rolling several of them with the hope that, upon a single roll, all possible numbers will be showing.

   (a) What is the probability of success if 6 dice are rolled?

   (b) What is the probability of success if 10 dice are rolled?

   (c)* What is the minimum number of dice that should be rolled so that it is more likely than not that all possible numbers will be showing? Hint: Determine a general formula $p(n)$ for the probability that all possible numbers will be showing when n dice are rolled, and plug in various values for n.

42.* At a birthday party, 4 children are playing a game with their shoes. The children have removed their shoes and thrown them into a pile in a dark closet. Each child will then randomly take two shoes from the closet (not necessarily a matching pair).

   (a) What is the probability that some child gets a matching pair?

   (b) What is the probability that some child gets his or her own pair?

For Exercises 43 through 46, n colors are available, and we are interested in coloring each circle ○ in the given picture. In each case, count the number of ways to do this so that no two circles that are directly joined by a line receive the same color. The resulting expression will be a polynomial in n and is called the **chromatic polynomial** for the figure. Hint: For each i, let A_i be the set of colorings in which both ends of line i receive the same color.

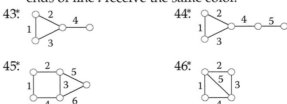

Coloring problems are pursued further in Section 9.5.

7.2 Multinomial coefficients

We have seen in Chapter 6 the utility of binomial coefficients when we need to count the number of ways to select a subset from a set of objects. Specifically, the binomial coefficient $\binom{n}{k}$ counts the number of ways to split a set of n objects into two sets, one with k objects and one with $n - k$ objects. However, it does not handle directly a problem in which the original set of n objects must be split into three or more sets. For that, as we shall see, multinomial coefficients can be employed.

Given nonnegative integers $k_1, k_2, \ldots, k_m$ and $n = k_1 + k_2 + \cdots + k_m$, the **multinomial coefficient** $\binom{n}{k_1, k_2, \ldots, k_m}$ is defined by

$$\binom{n}{k_1, k_2, \ldots, k_m} = \frac{n!}{k_1! k_2! \cdots k_m!}.$$

In fact, binomial coefficients are special instances of multinomial coefficients, since, for $0 \leq k \leq n$,

$$\binom{n}{k} = \frac{n!}{k!(n-k)!} = \binom{n}{k, n-k}.$$

Example 7.4.   (Computing Multinomial Coefficients).

(a) Compute $\binom{10}{3,2,5}$.

Solution.

$$\binom{10}{3, 2, 5} = \frac{10!}{3!2!5!} = 2520. \qquad \square$$

(b) Compute $\binom{10}{4,0,6}$.

Solution.

$$\binom{10}{4, 0, 6} = \frac{10!}{4!0!6!} = 210. \qquad \square$$

In *Mathematica*, the function `Multinomial` can be used to compute multinomial coefficients. For instance, the computation in Example 7.4(a) is entered as `Multinomial[3,2,5]`. Note that the syntax used here is different from that of the function `Binomial[n,k]`, in which the total value n is included. In *Maple*, Example 7.4(a) can be entered as `combinat[multinomial](10,3,2,5)`, and thus n is included there.

The utility of multinomial coefficients in counting comes out of the following generalization of Theorem 6.4.

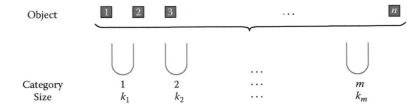

FIGURE 7.2
Splitting n objects into m categories of specified sizes.

Theorem 7.3. *Given nonnegative integers $k_1, k_2, \ldots, k_m$, and*

$$n = k_1 + k_2 + \cdots + k_m,$$

the multinomial coefficient $\binom{n}{k_1, k_2, \ldots, k_m}$ counts the number of ways to split n distinct items into m distinct categories of sizes $k_1, k_2, \ldots, k_m$.

Proof. We want to place n different items into m distinct categories of sizes $k_1, k_2, \ldots, k_m$. The fact that the categories are distinct allows us to regard them as having some fixed ordering in which they are sequentially filled (Figure 7.2). Since there are $\binom{n}{k_1}$ ways to choose k_1 of the n items to fill category 1, then $\binom{n-k_1}{k_2}$ ways to choose k_2 of the remaining $n - k_1$ items to fill category 2, and so forth, in total, the n items can be placed in

$$\binom{n}{k_1}\binom{n - k_1}{k_2}\binom{n - k_1 - k_2}{k_3} \cdots \binom{n - k_1 - k_2 - \cdots - k_{m-1}}{k_m}$$

different ways. The obvious cancellations from the product

$$\frac{n!}{k_1!(n-k_1)!} \frac{(n-k_1)!}{k_2!(n-k_1-k_2)!} \frac{(n-k_1-k_2)!}{k_3!(n-k_1-k_2-k_3)!}$$
$$\cdots \frac{(n-k_1-k_2-\cdots-k_{m-1})!}{k_m!(n-k_1-k_2-\cdots-k_m)!},$$

together with the observation that $n - k_1 - k_2 - \cdots - k_m = 0$ leave the formula for $\binom{n}{k_1, k_2, \ldots, k_m}$. $\qquad\square$

Theorem 7.3 applies to problems in which a group must be divided into a collection of smaller groups.

Example 7.5. (Job Allocation).
The software production company SoftJobs has contracts to write three video games. NBA Dunkfest

requires 4 programmers, Skate Rats requires 3 programmers, and Bug Buster II requires 2 programmers.
In how many ways can the 9 programmers at SoftJobs
be assigned these projects so that no programmer is
working on more than one project?

Solution. The programmers play the role of the items to
be placed, and the projects play the role of the categories
in Theorem 7.3. There are thus

$$\binom{9}{4,3,2} = 1260$$

ways to make the assignments.   □

Example 7.5 could have been computed by using binomial
coefficients and applying the counting technique used in
the proof of Theorem 7.3 to obtain $\binom{9}{4}\binom{5}{3}\binom{2}{2} = 1260$. However,
multinomial coefficients provide a convenient counting tool
for handling this type of problem.

Example 7.6.   (Bridge Party).
To play the card game called bridge, 8 players must split
into 4 teams of 2. The teams then split into two pairs of
teams that compete separately, with one game at each of
two tables. The teams are reflected by chairs of the same
color in Figure 7.3. How many ways are there to have
8 play bridge if

(a) the teams are further assigned the numbers 1
through 4, and it is predetermined that Team 1
plays Team 2 and Team 3 plays Team 4? Note
that here we keep track not only of a player's
teammate but also of the team number.

(b) we are only interested in how the players are
teamed up and how the teams are paired? Here
the teams are not numbered.

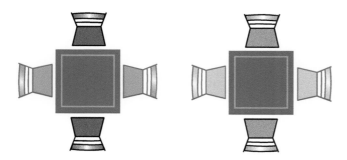

FIGURE 7.3
Two bridge tables.

Solution.

(a) The numbered teams serve as 4 distinct categories. There are thus

$$\binom{8}{2,2,2,2} = 2520$$

ways to set the teams. The games are then predetermined.

(b) Here, the teams reflect *identical* categories. We therefore need to divide out for the order imposed on them in part (a). There are now

$$\frac{\binom{8}{2,2,2,2}}{4!} = 105$$

ways to set the teams. Once the 4 teams are set, there are

$$\frac{\binom{4}{2}}{2} = 3$$

ways to pair them off for competition. Hence, there are

$$105 \cdot 3 = 315$$

ways to set the games. $\square$

The difference between the choices of the teams in parts (a) and (b) of Example 7.6 is significant. It is analogous to the difference between permutations and combinations. In part (a), the order in which the teams of two are selected is important. In part (b), that order is not important and is stripped out by an appropriate division. We have seen such strategies employed in Section 6.2 in the proof that $\binom{n}{k} = \frac{P(n,k)}{k!}$ and in Section 6.6.

There is a generalization of the Binomial Theorem (Theorem 4.10) that involves multinomial coefficients. Before presenting that generalization, we represent the Binomial Theorem using alternate notation that is more conducive to generalizing.

Theorem 7.4. (The Binomial Theorem (Revisited)). *Let $a_1, a_2 \in \mathbb{R}$ and $n \in \mathbb{N}$. Then*

$$(a_1 + a_2)^n = \sum_{\substack{0 \le k_1, k_2 \le n \\ k_1 + k_2 = n}} \binom{n}{k_1, k_2} a_1^{k_1} a_2^{k_2}.$$

Here, the sum is indexed over all ordered pairs of integers k_1, k_2 such that $0 \le k_1, k_2 \le n$ and $k_1 + k_2 = n$.

Theorem 4.10 can be recovered from Theorem 7.4 by setting $k_2 = i$ and $k_1 = n - i$. Since k_1 is determined by i, it can be eliminated from the indexing in the summation in Theorem 7.4 to obtain Theorem 4.10. The appropriate generalization of the Binomial Theorem to multinomial coefficients can now be presented.

Theorem 7.5.   (The Multinomial Theorem).
Let $a_1, a_2, \ldots, a_m \in \mathbb{R}$ and $n \in \mathbb{N}$. Then

$$(a_1 + a_2 + \cdots + a_m)^n$$

$$= \sum_{\substack{0 \le k_1, k_2, \ldots, k_m \le n \\ k_1 + k_2 + \cdots + k_m = n}} \binom{n}{k_1, k_2, \cdots k_m} a_1^{k_1} a_2^{k_2} \cdots a_m^{k_m}.$$

Here, the sum is indexed over all ordered m-tuples of integers $k_1, k_2, \ldots, k_m$ such that $0 \le k_1, k_2, \ldots, k_m \le n$ and $k_1 + k_2 + \cdots + k_m = n$.

Proof. Each unsimplified monomial $a_{i_1} a_{i_2} \cdots a_{i_n}$ in the expanded n-fold product

$$(a_1 + a_2 + \cdots + a_m)(a_1 + a_2 + \cdots + a_m) \cdots (a_1 + a_2 + \cdots + a_m)$$

comes from choosing the value a_{i_1} from the summands in the first factor, a_{i_2} from the summands in the second factor, and so on. The monomial simplifies to $a_1^{k_1} a_2^{k_2} \cdots a_m^{k_m}$ precisely when a_1 is chosen k_1 times, a_2 is chosen k_2 times, and so on. Since there are $\binom{n}{k_1, k_2, \cdots k_m}$ ways to make those choices, the coefficient of $a_1^{k_1} a_2^{k_2} \cdots a_m^{k_m}$ is $\binom{n}{k_1, k_2, \cdots k_m}$. □

Remark 7.1.   It follows from Theorem 6.13 that the number of terms in the expansion of $(a_1 + a_2 + \cdots + a_m)^n$ is $\binom{n+m-1}{n}$. That is, each choice of an m-tuple $0 \le k_1, k_2, \ldots, k_m \le n$ with $k_1 + k_2 + \cdots + k_m = n$ can be recorded as a sequence of n ones and $m - 1$ separators. The number of ones before the first separator is k_1, the number after the first separator and before the second separator is k_2, and so on. For example, the m-tuple $2 + 3 + 0 + 1 = 6$ is recorded as 11 | 111 || 1.

Just as the Binomial Theorem was used in Section 4.6 to expand a power of a sum of two terms, the Multinomial Theorem can be used to expand a power of a sum of two or *more* terms.

Example 7.7. (Expanding Polynomials).

(a) Use the Multinomial Theorem to expand $(x + y + z)^4$.

Solution. Here, the sum in our expansion is indexed over the triples (k_1, k_2, k_3) such that $0 \le k_1, k_2, k_3 \le 4$ and $k_1 + k_2 + k_3 = 4$. That is, the triples

$$(4,0,0), \ (3,1,0), \ (3,0,1), \ (2,2,0), \ (2,1,1),$$
$$(2,0,2), \ (1,3,0), \ (1,2,1), \ (1,1,2), \ (1,0,3),$$
$$(0,4,0), \ (0,3,1), \ (0,2,2), \ (0,1,3), \ (0,0,4)$$

index our sum. In each case, the triple (k_1, k_2, k_3) corresponds to the term $\binom{4}{k_1, k_2, k_3} x^{k_1} y^{k_2} z^{k_3}$. That is, the triple determines both the exponents on x, y, and z and the bottom row of the multinomial coefficient. For example, the triple $(3, 0, 1)$ corresponds to the term $\binom{4}{3,0,1} x^3 y^0 z^1 = 4x^3 z$. Thus, in total, we get

$$(x + y + z)^4$$

$$= \sum_{\substack{0 \le k_1, k_2, k_3 \le 4 \\ k_1 + k_2 + k_3 = 4}} \binom{4}{k_1, k_2, k_3} x^{k_1} y^{k_2} z^{k_3}$$

$$= x^4 + 4x^3 y + 4x^3 z + 6x^2 y^2 + 12x^2 yz$$

$$+ 6x^2 z^2 + 4xy^3 + 12xy^2 z + 12xyz^2 + 4xz^3$$

$$+ y^4 + 4y^3 z + 6y^2 z^2 + 4yz^3 + z^4. \qquad \square$$

(b) Use the Multinomial Theorem to expand $(w + 2x + 3y + 4z)^2$.

Solution. Here, the sum in our expansion is indexed over the ordered 4-tuples (k_1, k_2, k_3, k_4) such that $0 \le k_1, k_2, k_3, k_4 \le 4$ and $k_1 + k_2 + k_3 + k_4 = 2$. That is, the ordered 4-tuples

$$(2,0,0,0), \quad (1,1,0,0), \quad (1,0,1,0),$$

$$(1,0,0,1), \quad (0,2,0,0),$$

$$(0,1,1,0), \quad (0,1,0,1), \quad (0,0,2,0),$$

$$(0,0,1,1), \quad (0,0,0,2)$$

index our sum. Thus, we get

$$(w+2x+3y+4z)^2$$

$$= \sum_{\substack{0 \le k_1, k_2, k_3, k_4 \le 2 \\ k_1+k_2+k_3+k_4 = 2}} \binom{2}{k_1, k_2, k_3, k_4} w^{k_1} (2x)^{k_2} (3y)^{k_3} (4z)^{k_4}$$

$$= w^2 + 4wx + 6wy + 8wz + 4x^2$$

$$+ 12xy + 16xz + 9y^2 + 24yz + 16z^2. \quad \square$$

Example 7.8. (Extracting a Coefficient).
What is the coefficient of $u^2 w^3 x^4 y^2$ in the expansion of $(u + v + 2w + x + 3y + z)^{11}$?

Solution. The expansion includes the summand

$$\binom{11}{2, 0, 3, 4, 2, 0} u^2 v^0 (2w)^3 x^4 (3y)^2 z^0 = 69300 u^2 \cdot 8w^3 x^4 \cdot 9y^2$$

$$= 4989600 u^2 w^3 x^4 y^2.$$

Hence, the desired coefficient is 4,989,600. $\square$

The ability to extract a particular coefficient from a factored polynomial without having to expand the polynomial entirely is needed in Section 7.3, where polynomials are used to solve counting problems.

Exercises

Most of the following Exercises involve some computations. Each can be completed by hand, but some might also be done with the aid of a symbolic calculator, such as *Mathematica* or *Maple*. In those cases, the results should then be compared with hand computations.

1. Compute $\binom{7}{3,2,2}$.

2. Compute $\binom{11}{5,2,4}$.

3. Compute $\binom{15}{2,3,6,4}$.

4. Compute $\binom{12}{3,3,4,2}$.

5. The students know that there is going to be a 12-question multiple-choice quiz on Friday, and each question will have three possible answers: a, b, or c. To tease the students, the teacher further told them that 4 of the questions will have answer a, 3 will have answer b, and 5 will have answer c. How many possible answer keys satisfy these conditions?

6. A small college has 3 physics professors and 9 physics majors. Each of the physics professors needs to be assigned 3 advisees from the physics students. In how many ways can this be done so that each student has an advisor?

7. The housing director at a boarding school needs to assign the 15 resident assistants to the 4 dorms. He needs 5 to go to North Hall, 3 to go to South Hall, 4 to go to East Hall, and 3 to go to West Hall.

   (a) How many ways are there to make the assignments?

   (b) What if Megan, one of the resident assistants, can be assigned to neither South nor West?

   (c) What if Megan must be assigned to North?

8. In the game of bridge, the 52 cards of a standard deck are divided evenly among 4 players. Thus, each player's hand consists of 13 cards. How many different possible hands

   (a) Have all 13 of the denominations represented?

   (b) Have 4 clubs, 3 diamonds, 3 hearts, and 3 spades?

   (c) Have all 13 of the denominations represented and contain 4 clubs, 3 diamonds, 3 hearts, and 3 spades?

Team sports. In Exercises 9 through 14, we consider examples of splitting people into teams.

9. The 12 students in an honors history class will be split into 3 teams of 4 for a quiz bowl competition. How many ways are there to form these teams?

10. Sixteen students have gathered at the basketball court to play 5-on-5 basketball, so they decide to break into two teams of size 5 and one of size 6. (The team with 6 will have a substitute.) How many ways are there to form these teams?

11. The biology teacher needs to split her 22 students into 4 groups for a lab project. How many ways are there to do this so that there are two lab groups of 5 students and two lab groups of 6 students?

12. There are 40 girls at a volleyball camp. At the end of the first week, these girls must be split into teams of 8. How many ways are there to form the 5 teams?

13. A gym class of 16 students is learning how to play tennis. Since there are 4 courts, the teacher decides to have the students split into 4 concurrent doubles matches.

    (a) How many ways are there to set the matches?

    (b) How many ways are there to not only set the matches but also assign courts to the matches?

14. There are 3 basketball courts at the town park, and 30 boys have gathered to play 3 simultaneous games of 5-on-5 basketball.

    (a) How many ways are there to set the games on these courts?

    (b) If we keep track of only teams and opponents, and not which game is on which court, then how many ways are there to set the games?

15. The KiddieCorp day care has taken its 18 children to the fair. In how many different ways can all of the children be put on 3 small merry-go-rounds, each with 6 (single seat) chairs, if

    (a) The children care about which merry-go-round they are on and which child is riding in front of which?

    (b) The children do not care about which merry-go-round they are on, but they do care about which child is riding in front of which?

    (c) The children do not care which merry-go-round they are on, but they do care about who is on the same merry-go-round?

16. The 20 members of a student council need to split into 4 groups of size 5. How many ways are there to do this if

    (a) They are merely forming discussion groups for a common topic?

    (b) They are forming committees, each with a distinct purpose?

    (c) They are forming committees, each with a distinct purpose and a designated chairperson?

17. List all of the integer triples (k_1, k_2, k_3) such that $0 \leq k_1, k_2, k_3$ and $k_1 + k_2 + k_3 = 3$.

18. List all of the integer 4-tuples (k_1, k_2, k_3, k_4) such that $0 \leq k_1, k_2, k_3, k_4 \leq 3$ and $k_1 + k_2 + k_3 + k_4 = 3$.

Expanding polynomials. In Exercises 19 through 26, use the Multinomial Theorem to expand the given expression. The function Expand in *Mathematica*, or expand in *Maple*, might then be used to compare its results with the requested hand computations.

19. $(x - y + z)^2$.

20. $(x + y + z)^3$.

21.* $(x + y + 1)^4$.

22. $(3x + 2y - z)^2$.

23. $(2x + y - z)^2$.

25. $(x + 4y + \frac{1}{2}z)^2$.

24. $(2w + x + y + 2z)^2$.

26. $(w - x + y - z)^2$.

Extracting coefficients. In Exercises 27 through 34, the function Coefficient in *Mathematica*, or coeff in *Maple*, might also be used to compare its results with the requested computations.

27. What is the coefficient of $a^{100}b^{50}c^{40}d^{60}e^{20}f^{30}$ in the expansion of $(a+b+c+d+e+f)^{300}$?

28. What is the coefficient of $a^{20}b^{10}c^{30}d^{10}$ in the expansion of $(a+b+c+d)^{70}$?

29. What is the coefficient of $x^{10}y^{20}z^{30}$ in the expansion of $(x+y+z)^{80}$?

30. What is the coefficient of $x^{30}y^{20}z^{50}$ in the expansion of $(x+2y+3z+4w)^{100}$?

31. What is the coefficient of $x^{25}y^{10}z^{40}w^{25}$ in the expansion of $(x-y+z-w)^{100}$?

32. What is the coefficient of $x^4y^6z^2w^4$ in the expansion of $(w-z+y-x)^{20}$?

33. What is the coefficient of $x^4y^2z^5w^3$ in the expansion of $(x+y-2z+w)^{14}$?

34. What is the coefficient of $x^6y^8z^6$ in the expansion of $(2x-3y+\frac{1}{2}z)^{20}$?

35. How many nonzero terms are there in the expansion of $(x+y-2z+w)^{14}$?

36. How many nonzero terms are there in the expansion of $(2x-3y+\frac{1}{2}z)^{20}$?

37.* Prove Theorem 7.3 by generalizing the proof of Theorem 6.4 given at the end of Section 6.2.

38.* Prove the Multinomial Theorem by induction on m. Use Theorem 7.4 for the base case.

Quick proofs. In Exercises 39 through 42, apply the Multinomial Theorem to give a short proof of the stated identities.

39. Prove: For all integers $n \geq 0$,

$$3^n = \sum_{\substack{0 \leq k_1, k_2, k_3 \leq n \\ k_1 + k_2 + k_3 = n}} \binom{n}{k_1, k_2, k_3}.$$

40. Prove: For all integers $n \geq 0$,

$$0 = \sum_{\substack{0 \leq k_1, k_2, k_3, k_4 \leq n \\ k_1 + k_2 + k_3 + k_4 = n}} (-1)^{k_1+k_2} \binom{n}{k_1, k_2, k_3, k_4}.$$

Hint: $0 = -1 - 1 + 1 + 1$.

41. Prove: For all integers $n \geq 0$,

$$\sum_{\substack{0 \leq k_1, k_2 \leq n \\ k_1 + k_2 \leq n}} \binom{n}{k_1}\binom{n-k_1}{k_2}3^{k_1}2^{k_2} = 6^n.$$

42.* Prove: For all integers $n \geq 0$ and all real numbers x and y,

$$\sum_{\substack{0 \leq k_1, k_2, k_3 \leq n \\ k_1 + k_2 + k_3 = n}} \binom{n}{k_1, k_2, k_3}x^{k_1}y^{n-k_1}$$

$$= \sum_{k=0}^{n} 2^{n-k}\binom{n}{k}x^k y^{n-k}.$$

7.3   Generating functions

We have seen in Chapter 6 that the basic tools for counting are based upon addition and multiplication principles. The fact that rules for addition and multiplication also govern algebraic manipulations of real functions enables functions to store combinatorial information in a powerful and useful way.

For instance, we will see in Example 7.11, that if we have 2 quarters, 4 dimes, and 3 nickels in our pocket, then the number of ways to make 35¢ is the coefficient of x^{35} in the expansion of the polynomial function

$$g(x) = (1 + x^{25} + x^{50})(1 + x^{10} + x^{20} + x^{30} + x^{40})$$
$$\times (1 + x^5 + x^{10} + x^{15}).$$

That is, that there are 4 ways to make 35¢ is reflected by the fact that

$$g(x) = 1 + \cdots + 4x^{35} + \cdots + x^{105}.$$

Consequently, we start this section with a development of some algebraic tools, and then we move on to applying those tools to combinatorial problems.

7.3.1   The algebra

Definition 7.1. The **generating function** for a given sequence $c_0, c_1, c_2, \ldots$ of real numbers is the function

$$g(x) = \sum_{i=0}^{\infty} c_i x^i = c_0 + c_1 x + c_2 x^2 + \cdots,$$

where x is a real variable (and we identify $x^0 = 1$).

If our given sequence is a finite sequence $c_0, c_1, c_2, \ldots, c_n$, then its generating function $g(x) = c_0 + c_1 x + c_2 x^2 + \cdots + c_n x^n$ is indeed a real polynomial function. If the sequence is not finite, then its generating function $g(x) = c_0 + c_1 x + c_2 x^2 + \cdots$ is treated as a **formal power series**. For our purposes, we will not be concerned with the radius of convergence of such power series (as studied in calculus). A simple understanding of the formal operations of addition and multiplication will suffice.

$$\sum_{i=0}^{\infty} a_i x^i + \sum_{i=0}^{\infty} b_i x^i = \sum_{i=0}^{\infty} (a_i + b_i) x^i. \tag{7.4}$$

$$\left(\sum_{i=0}^{\infty} a_i x^i \right) \left(\sum_{i=0}^{\infty} b_i x^i \right) = \sum_{i=0}^{\infty} \left(\sum_{j=0}^{i} a_j b_{i-j} \right) x^i. \tag{7.5}$$

As a special case of (7.5), observe that

$$(a_0 + a_1 x + a_2 x^2)(b_0 + b_1 x + b_2 x^2 + b_3 x^3)$$

$$= a_0 b_0 + (a_0 b_1 + a_1 b_0) x$$

$$+ (a_0 b_2 + a_1 b_1 + a_2 b_0) x^2$$

$$+ (a_0 b_3 + a_1 b_2 + a_2 b_1) x^3$$

$$+ (a_1 b_3 + a_2 b_2) x^4 + a_2 b_3 x^5.$$

In particular, the coefficient $a_0 b_3 + a_1 b_2 + a_2 b_1$ of x^3 on the right-hand side is the sum of all possible products of the form $a_j b_{3-j}$.

Example 7.9.   Use (7.5) to determine the sequence $c_0, c_1, c_2, \ldots$ whose generating function is given by the product

$$(1 + x + x^2 + x^3 + \cdots)(1 + x^2 + x^4 + x^6 + \cdots).$$

Solution. For   each $i \geq 0$,   define $a_i = 1$ and
$b_i = \begin{cases} 1 & \text{if } i \text{ is even,} \\ 0 & \text{if } i \text{ is odd.} \end{cases}$
So, we seek the product $\left(\sum_{i=0}^{\infty} a_i x^i \right) \left(\sum_{i=0}^{\infty} b_i x^i \right)$.
   Let $i \geq 0$. Our aim is to compute the coefficient $c_i = \sum_{j=0}^{i} a_j b_{i-j}$ of x^i in the product above. Very simply, we need to understand how x^i can be obtained from products of the form $x^j x^{i-j}$, where x^j and x^{i-j} come from the first and second factors, respectively, in our product. The key is that the sum of j and $i - j$ is the desired exponent i.

The fact that the powers of x in the second factor are all even warrants the consideration of separate cases based on the parity of i. If i is odd, then the ways to obtain x^i in the product are

$$x^1 x^{i-1}, \ x^3 x^{i-3}, \ x^5 x^{i-5}, \dots, x^{i-2} x^2, \ x^i 1. \tag{7.6}$$

This observation, together with the fact that all of the nonzero coefficients in both factors are 1, implies that $c_i = \frac{i+1}{2}$. That is, there are $\frac{i+1}{2}$ entries in (7.6). Similarly, if i is even, then the ways to obtain x^i in the product are

$$1 x^i, \ x^2 x^{i-2}, \ x^4 x^{i-4}, \dots, x^{i-2} x^2, \ x^i 1.$$

It follows that $c_i = \frac{i+2}{2}$. Consequently, for all $i \geq 0$, a single formula for c_i can be given by $c_i = \lfloor \frac{i+2}{2} \rfloor$. $\square$

The utility of generating functions for counting derives from the fact that algebra, such as that encountered in Example 7.9, can be used to encode combinatorial facts.

Example 7.10. Let $c_0, c_1, c_2, \dots$ be the sequence whose generating function $g(x)$ is given by

$$g(x) = (1 + x^2)(1 + x + x^3)(x^2 + x^3).$$

To find $c_0, c_1, c_2, \dots$, we need to expand $g(x)$ into the standard form consisting of sums of increasing powers of x. However, we also need to understand how the current form provides valuable information about what c_i counts in terms of i. In fact, the encoded counting problem unfolds through a very deliberate expansion of $g(x)$.

To help illuminate a developing pattern, we identify $x^0 = 1$ and write

$$g(x) = (x^0 + x^2)(x^0 + x^1 + x^3)(x^2 + x^3). \tag{7.7}$$

If we expand (7.7) into a sum of monomials and temporarily delay the use of the Laws of Exponents and the combining of like terms, then we obtain

$$\begin{aligned}
g(x) = \ & x^0 x^0 x^2 + x^0 x^1 x^2 + x^0 x^3 x^2 \\
& + x^2 x^0 x^2 + x^2 x^1 x^2 + x^2 x^3 x^2 \\
& + x^0 x^0 x^3 + x^0 x^1 x^3 + x^0 x^3 x^3 \\
& + x^2 x^0 x^3 + x^2 x^1 x^3 + x^2 x^3 x^3.
\end{aligned}$$

Here, each term $x^{e_1} x^{e_2} x^{e_3}$ is a product of a summand x^{e_1} from $(x^0 + x^2)$, a summand x^{e_2} from $(x^0 + x^1 + x^3)$, and a summand x^{e_3} from $(x^2 + x^3)$. That is, the triple e_1, e_2, e_3 satisfies $e_1 \in \{0, 2\}$, $e_2 \in \{0, 1, 3\}$, and $e_3 \in \{2, 3\}$.

By using the Laws of Exponents, we can write

$$g(x) = x^2 + x^3 + x^5 + x^4 + x^5 + x^7$$
$$+ x^3 + x^4 + x^6 + x^5 + x^6 + x^8.$$

In the exponents here, we see the sums $e_1 + e_2 + e_3$ that result from the various choices of allowable triples e_1, e_2, e_3. By finally combining like terms, we get

$$g(x) = x^2 + 2x^3 + 2x^4 + 3x^5 + 2x^6 + x^7 + x^8. \qquad (7.8)$$

That is, $c_0 = 0, c_1 = 0, c_2 = 1, c_3 = 2, c_4 = 2, c_5 = 3, c_6 = 2, c_7 = 1$, and $c_8 = 1$. Moreover, what c_i counts can now be extracted from our work.

First, consider the particular value $c_4 = 2$. Tracing back, we see that the coefficient 2 for x^4 in (7.8) comes from the two terms $x^2 x^0 x^2$ and $x^0 x^1 x^3$ in our initial expansion for $g(x)$. That is, $2 + 0 + 2 = 4$ and $0 + 1 + 3 = 4$ are the only ways to obtain the sum $e_1 + e_2 + e_3 = 4$, when we require that $e_1 \in \{0, 2\}$, $e_2 \in \{0, 1, 3\}$, and $e_3 \in \{2, 3\}$.

In general, for each $i \geq 0$, c_i counts the number of solutions e_1, e_2, e_3 to the equation

$$e_1 + e_2 + e_3 = i$$

under the restrictions that $e_1 \in \{0, 2\}$, $e_2 \in \{0, 1, 3\}$, and $e_3 \in \{2, 3\}$. Table 7.1 displays all such possible solutions. That the coefficients c_i for $g(x)$ count these solutions is reflected by the exponents in the factored form (7.7). The utility of generating functions comes from the fact that the algebraic expansion of $g(x)$, which can be done quickly with the help of any computer algebra system, takes care of what could potentially be a cumbersome process of directly enumerating the

TABLE 7.1
Enumeration for Example 7.10

i	Solutions to $e_1 + e_2 + e_3 = i$
0	None
1	None
2	$0 + 0 + 2$
3	$0 + 1 + 2, 0 + 0 + 3$
4	$2 + 0 + 2, 0 + 1 + 3$
5	$0 + 3 + 2, 2 + 1 + 2, 2 + 0 + 3$
6	$0 + 3 + 3, 2 + 1 + 3$
7	$2 + 3 + 2$
8	$2 + 3 + 3$
9+	None

solutions in Table 7.1. The generating function does not list the solutions, but it counts them.

7.3.2 The combinatorics

With the insight gained from Example 7.10, we can now use a generating function to directly attack a counting problem.

Example 7.11. (Making Change).
For each $i \geq 0$, determine the number of ways c_i to obtain $i \cent$ if we have 2 quarters, 4 dimes, and 3 nickels. Here, coins of like values are considered indistinguishable. That is, different ways to obtain $i \cent$ are distinguished solely by how many of each type of coin they use.

Solution. Table 7.2 reflects the monetary values obtainable from each single type of coin. The numbers in the right-hand column provide the exponents for the factors of our generating function

$$g(x) = (1 + x^{25} + x^{50})(1 + x^{10} + x^{20} + x^{30} + x^{40})$$
$$\times (1 + x^5 + x^{10} + x^{15}).$$

The first, second, and third factors account for the contributions made by the quarters, dimes, and nickels, respectively. In the first factor, for example, 1 reflects the choice of no quarters (since $1 = x^0$), x^{25} reflects the choice of one quarter, and x^{50} reflects the choice of two quarters.

Expanding $g(x)$ yields

$$1 + x^5 + 2x^{10} + 2x^{15} + 2x^{20} + 3x^{25} + 3x^{30}$$
$$+ 4x^{35} + 4x^{40} + 4x^{45} + 4x^{50}$$
$$+ 4x^{55} + 4x^{60} + 4x^{65} + 4x^{70} + 3x^{75}$$
$$+ 3x^{80} + 2x^{85} + 2x^{90} + 2x^{95} + x^{100} + x^{105}.$$

Thus, for example, there are 4 ways to obtain 35¢ and 3 ways to obtain 80¢. It would not be hard to enumerate any one of the cases by hand (in this small example).

TABLE 7.2
Coin Contributions

Coin type	Possible amounts contributed
Quarters	0¢, 25¢, 50¢
Dimes	0¢, 10¢, 20¢, 30¢, 40¢
Nickels	0¢, 5¢, 10¢, 15¢

However, the generating function handles all of the cases at once quite nicely. In general, the number of ways c_i of obtaining $i\cent$ is the coefficient of x^i in the expanded form of $g(x)$.   □

Example 7.12. (Prize Packages).
The first-place winner of a programming contest is allowed to select 6 items from a table containing 3 identical routers, 2 identical printers, and some identical thumb drives.

(a) How many different prize packages are possible if there are 4 thumb drives on the table?

(b) What if there is an unlimited supply of thumb drives available?

(c) What if there is an unlimited supply of thumb drives available, and the winner is required to select at least 2 of them?

Solution. (a) The answer is not $\binom{3+2+4}{6} = 84$, as would be the case if all 9 items were distinct. Here, we seek integer solutions to the equation

$$R + P + T = 6,$$

where R is the number of routers chosen, P is the number of printers chosen, T is the number of thumb drives chosen, $0 \leq R \leq 3$, $0 \leq P \leq 2$, and $0 \leq T \leq 4$. However, it is convenient to simultaneously consider all possible sums $R + P + T$, and not just the sum 6. The relevant generating function is

$$g(x) = (1 + x + x^2 + x^3)(1 + x + x^2)$$
$$\times (1 + x + x^2 + x^3 + x^4). \qquad (7.9)$$

The exponents in the first, second, and third factors reflect the possible numbers of routers, printers, and thumb drives, respectively. Expanding (7.9) gives

$$g(x) = 1 + 3x + 6x^2 + 9x^3 + 11x^4 + 11x^5$$
$$+ 9x^6 + 6x^7 + 3x^8 + x^9.$$

It is now easy to extract the coefficient of x^6 and conclude that there are 9 possible prize packages.

(b) How does our analysis change when there is an unlimited number of thumb drives? In that case, the relevant generating function is

$$f(x) = (1 + x + x^2 + x^3)(1 + x + x^2)(1 + x + x^2 + \cdots).$$

The last factor is an infinite sum. Hence, expansion of $f(x)$ may not easily be handled by a computer algebra system. However, if our interest is only in selecting 6 items, then it suffices to assume that there are only 6 thumb drives available and to use the generating function

$$h(x) = (1 + x + x^2 + x^3)(1 + x + x^2)$$
$$\times (1 + x + x^2 + x^3 + x^4 + x^5 + x^6).$$

In this case, expansion of $h(x)$ yields

$$1 + 3x + 6x^2 + 9x^3 + 11x^4 + 12x^5 + 12x^6$$
$$+ 11x^7 + 9x^8 + 6x^9 + 3x^{10} + x^{11},$$

and we see that there are 12 possible prize packages.

(c) Since we now require that $T \geq 2$, the third factor in the generating function h used in part (b) must be changed so that its minimum exponent is 2. Thus, the relevant generating function here is

$$k(x) = (1 + x + x^2 + x^3)(1 + x + x^2)$$
$$\times (x^2 + x^3 + x^4 + x^5 + x^6)$$
$$= (1 + x + x^2 + x^3)(1 + x + x^2)x^2$$
$$\times (1 + x + x^2 + x^3 + x^4)$$
$$= x^2 g(x).$$

Since the coefficient of x^6 in $k(x)$ must therefore be the same as the coefficient of x^4 in $g(x)$, we conclude that there are 11 possible prize packages. □

The algebraic equation $k(x) = x^2 g(x)$ relating the generating functions in parts (b) and (c) of Example 7.12 reflects a simple combinatorial observation. Selecting a total of 6 items containing anywhere from 2 to 6 thumb drives is equivalent to first selecting 2 thumb drives and then selecting a total of 4 additional items containing anywhere from 0 to 4 thumb drives. The combinatorics is encoded in the algebra!

7.3.3 More algebra

The generating functions used in Example 7.12 were each built from factors of the types $(1 + x + x^2 + \cdots + x^n)$ and $(1 + x + x^2 + \cdots)$. Of course, the second type is the infinite version of the first type. Since these kinds of factors occur frequently in the generating functions we consider, it is important to understand them well.

By Theorem 4.3, we have

$$1 + x + x^2 + \cdots + x^n = \sum_{i=0}^{n} x^i = \frac{1 - x^{n+1}}{1 - x}. \tag{7.10}$$

By a result from calculus (which we accept here without proof), when $|x| < 1$ we have

$$(1 + x + x^2 + \cdots)^n = \frac{1}{(1 - x)^n}. \tag{7.11}$$

With Equations 7.10 and 7.11 in mind, we now give some useful formulas for the coefficients in a few fundamental generating functions.

Theorem 7.6.   *Let $n \in \mathbb{Z}^+$, and let $g(x)$ be the generating function for the sequence $c_0, c_1, c_2, \ldots$. That is, $g(x) = c_0 + c_1 x + c_2 x^2 + \cdots$. Then*

(a) $g(x) = \frac{1 - x^{n+1}}{1 - x}$ *if and only if, for each $0 \le i \le n$, $c_i = 1$, and, for each $i > n$, $c_i = 0$.*

(b) $g(x) = \frac{1}{(1-x)^n}$ *if and only if, for each $i \ge 0$, $c_i = \binom{i+n-1}{i}$.*

(c) $g(x) = (1 + x)^n$ *if and only if, for each $0 \le i \le n$, $c_i = \binom{n}{i}$, and, for each $i > n$, $c_i = 0$.*

Proof.   Part (a) follows from (7.10).

For part (b), observe from (7.11) that $g(x)$ is the n-fold product

$$(1 + x + x^2 + \cdots)(1 + x + x^2 + \cdots) \cdots (1 + x + x^2 + \cdots).$$

This is the generating function for the number of solutions c_i to the equation

$$e_1 + e_2 + \cdots + e_n = i,$$

where each $e_j \ge 0$. That is, c_i counts the number of ways to select i objects from n categories (with repetition allowed). Hence, $c_i = \binom{i+n-1}{i}$.

For part (c), consider expanding the n-fold product

$$(1 + x)(1 + x) \cdots (1 + x)$$

into a sum of 2^n terms (delaying the combining of like terms). Each term results from an n-fold product $a_1 a_2 \cdots a_n$, where each a_j is either 1 or x coming from the jth factor $(1 + x)$. A term x^i is obtained from $n - i$ contributions of a 1 and i contributions of an x. The number of terms x^i, and thus the coefficient of x^i, is $\binom{n}{i}$, the number of ways to choose i factors to contribute an x. A quicker but less enlightening proof follows by applying the Binomial Theorem. $\square$

7.3.4   More combinatorics

Theorem 7.6 can be used to ease the algebraic computations needed to find a particular coefficient in a generating function.

Example 7.13.   (Fruit Baskets).
Since Heather's neighbors just moved in, she thought it would be nice to put together a fruit basket for them in the basket she recently made at summer camp. When Heather's mother returned from the grocery store, she had a bag containing apples, bananas, and oranges. How many different possibilities are there for a fruit basket consisting of 36 pieces of fruit if

(a) there is an ample supply of apples and bananas, but there are only 4 oranges?

(b) there are 16 of each kind of fruit?

(c) there is an ample supply of bananas and oranges, and there are 4 distinguishable apples (one Cortland, one Delicious, one Empire, and one Macintosh)?

In each case, fruit baskets are distinguished solely by how many of each type of fruit they contain.

Solution.   (a) The generating function that accounts for fruit baskets of all possible sizes is

$$g(x) = (1 + x + x^2 + \cdots)(1 + x + x^2 + \cdots)$$
$$\times (1 + x + x^2 + x^3 + x^4).$$

From (7.10) and (7.11), it follows that

$$g(x) = \left(\frac{1}{1-x}\right)^2 \frac{1 - x^5}{1 - x} = \frac{1}{(1-x)^3}(1 - x^5).$$

If we let a_i and b_i denote the coefficients of x^i in $\frac{1}{(1-x)^3}$ and $1 - x^5$, respectively, then the formula for the coefficients in a product given in equation 7.5 and the formula for the coefficients in $\frac{1}{(1-x)^3}$ given by Theorem 7.6(b) with $n = 3$ tell us that the coefficient of x^{36} in $g(x)$ is

$$a_{31}b_5 + a_{36}b_0 = \binom{31 + 3 - 1}{31}(-1) + \binom{36 + 3 - 1}{36}(1)$$

$$= -\binom{33}{31} + \binom{38}{36} = 175.$$

(b) When there are 16 of each type of fruit, the relevant generating function is

$$h(x) = (1 + x + x^2 + \cdots + x^{16})^3$$

$$= \left(\frac{1 - x^{17}}{1 - x}\right)^3$$

$$= \frac{1}{(1 - x)^3}(1 - x^{17})^3$$

$$= \frac{1}{(1 - x)^3}(1 - 3x^{17} + 3x^{34} - x^{51}).$$

Here, the coefficient of x^{36} in $h(x)$ is

$$a_2(3) + a_{19}(-3) + a_{36}(1)$$

$$= \binom{4}{2}(3) + \binom{21}{19}(-3) + \binom{38}{36}(1) = 91.$$

(c) Effectively, there are now 6 types of fruit; each of the 4 apples is its own type. The relevant generating function is

$$f(x) = (1 + x + x^2 + \cdots)^2(1 + x)^4 = \frac{1}{(1 - x)^2}(1 + x)^4.$$

From (7.5) and Theorem 7.6, it follows that the coefficient of x^{36} in $f(x)$ is

$$\binom{33}{32}\binom{4}{4} + \binom{34}{33}\binom{4}{3} + \binom{35}{34}\binom{4}{2}$$

$$+ \binom{36}{35}\binom{4}{1} + \binom{37}{36}\binom{4}{0} = 560. \qquad \square$$

Exercises

In Exercises 1 through 4, find the coefficients in the resulting power series.

1. $(1 + x + x^2 + x^3 + \cdots)^2$.
2. $(1 + x + x^2 + x^3 + \cdots)(1 + x^3 + x^6 + x^9 + \cdots)$.
3. $(1 + x + x^2 + x^3 + \cdots)(1 - x + x^2 - x^3 + \cdots)$.
4. $(1 - x + x^2 - x^3 + \cdots)^2$.

In Exercises 5 through 8, first expand the given function into unsimplified monomials and then group together terms with the same exponent. The point is not merely to obtain the product but, further, to observe the development of each of its coefficients.

5. $(1 + x + x^2)(1 + x^2 + x^4)$.
6. $(1 + x^5 + x^{10})(1 + x + x^2 + x^3 + x^4)$.

7. $(1 + x)(x + x^2)(1 + x^3)$.

8. $(1 + x^{25})(1 + x^{10} + x^{20})(1 + x^5)$.

Making change. In Exercises 9 through 14, for each $i \geq 0$, determine the number of ways c_i of obtaining $i¢$ if you have the given coins. Ways are distinguished solely by how many of each type of coin they contain.

9. 1 dime and 4 nickels.   10. 2 nickels and 5 pennies.

11. 2 dimes, 1 nickel, and 5 pennies.   12. 1 dime, 3 nickels, and 5 pennies.

13. 1 quarter, 2 dimes, and 3 nickels.   14. 4 dimes and 2 nickels.

15. A father promised his son 5 balloons at the fair. When the clown selling balloons approached, she had 3 identical red, 4 identical white, and 2 identical blue balloons. How many different purchases of 5 balloons are possible if purchases are distinguished by how many of each color they contain?

16. How many different collections of 7 pieces of fruit are possible from 4 apples, 3 bananas, and 3 oranges? Collections are distinguished by how many of each type of fruit they contain.

17. Mrs. Coffta bought 6 hamburgers, 6 hot dogs, and 4 pieces of chicken for a barbecue. As Mr. Coffta prepared the grill, Mrs. Coffta took an order of one item from each of their 8 guests, while keeping in mind the available food items. Then she gave Mr. Coffta the list of items to prepare. How many different barbecue orders might Mr. Coffta receive?

18. A father has gone to the store to buy one balloon for each of 6 children attending his daughter's birthday party. For his daughter, in particular, he promised to buy a blue balloon. When he arrived at the store, there were 4 identical red, 4 identical white, and 3 identical blue balloons.

    (a) How many different balloon purchases are possible, given the father's promise?

    (b) If the father simply asks the clerk to grab 6 balloons, and he does so randomly, then what is the probability of obtaining at least one blue balloon?

    (c) In a random selection of 6 balloons, what is the probability that not all colors will be represented?

19. A woman has agreed to bake 8 cakes for a charity picnic. Moreover, she promised to bake at least one of each of the flavors chocolate, vanilla, and marble. When her husband returned from the store he had bought 22 boxes of cake mix, of which 10 were chocolate, 8 were vanilla, and 4 were marble.

(a) In how many possible ways could the woman fulfill her promise?

(b) If 8 of the 22 boxes are randomly selected, then what is the probability that there is at least one of each flavor?

(c) In a random selection of 8 boxes, what is the probability that they are all the same flavor?

20. A diner had 21 pieces of pie ready for the lunch crowd. There were 8 pieces of pumpkin pie, 6 pieces of apple pie, and 7 pieces of blueberry pie. During lunch, 9 customers each ordered a piece of pie. How many different possible collections of pie could remain for the dinner crowd?

21. Let $a \in \mathbb{R}$. For what sequence is $(1 + ax)^n$ a generating function?

22. Let $a \in \mathbb{R}$. For what sequence is $(a + x)^n$ a generating function?

23. For what sequence is $\frac{1}{(1-x)^4}$ a generating function?

24. For what sequence is $\frac{1-x^9}{1-x}$ a generating function?

25. Find the coefficient of x^{100} in $\frac{1}{(1-x)^{50}}$.

26. Find the coefficient of x^{50} in $\frac{1}{(1-x)^{100}}$.

27. Find the coefficient of x^{100} in $\frac{1+x}{(1-x)^{50}}$.

28. Find the coefficient of x^{50} in $\frac{1-x^2}{(1-x)^{100}}$.

29. Find the coefficient of x^{40} in $\frac{(1+x)^2}{(1-x)^{80}}$.

30. Find the coefficient of x^{70} in $\frac{(1+x)^3}{(1-x)^{60}}$.

31. If we ignore suits, then how many different 5-card hands are possible using a standard deck of 52 cards? Note that $3, 5, 7, 8, 8$ and $3, 4, 7, 8, 8$ are considered different, even though they would both have the value of a pair of eights.

32. How many different monetary values can be achieved by selecting 6 stamps from 3 five-cent stamps, 3 twelve-cent stamps, and 3 thirty-cent stamps?

33. How many different collections of 80 balls are possible from 4 red balls, and unlimited numbers of blue, yellow, and green balls?

34. How many different collections of 70 balls are possible from 3 red, 3 blue, and unlimited numbers of yellow and green balls?

35.* For a wedding reception, 60 bottles of wine are needed. The budget can handle any number of bottles of red or white wine, but at most 5 bottles of champagne can be afforded. The alcohol supplier simply needs to know how many bottles of red wine, white wine, and champagne to send to the reception hall.

(a) How many different alcohol orders are possible?

(b) How many possible orders include 5 bottles of champagne?

(c) How many have an equal number of bottles of red and white wine?

36.* In a large neighborhood, more than 120 houses are giving out candy, but from each house only one piece of candy can be obtained. There are 8 houses giving out Snickers bars, 10 giving out Mounds bars, more than 50 giving out Butterfingers, and more than 50 giving out Crunch bars. Malcolm's mother told him that he could collect only 50 pieces of candy. So the only decision Malcolm has to make is how many of each type of candy to acquire.

(a) How many different collections of 50 pieces of candy are possible?

(b) How many contain no Snickers bars?

(c) How many contain at least one of each type of candy bar?

37. There are 8 hamburgers, 8 pieces of chicken, and an unlimited supply of hot dogs. If 40 people each order one item, then how many different barbecue orders are possible for the cook?

38. There are 4 pineapples and unlimited supplies of kiwis and mangos. How many different possibilities are there for a fruit basket consisting of 30 pieces of fruit? Baskets are distinguished by how many of each type of fruit they contain.

39.* A bag contained 30 bronze coins, 40 copper, 50 brass, one silver, one gold, and one platinum. A thief reached in the bag and grabbed 20 coins without looking. How many possibilities are there for the collection of coins stolen from this bag? Collections are distinguished by how many of each type of coin they contain.

40.* At the end of Halloween night, the Jones' bowl of candy contained 12 Snickers, 10 Mounds, 11 Butterfingers, 14 Krackel, one Milkyway, and one Crunch bar. When one last child showed up at the door, they decided to offer her 10 items from the bowl. How many possibilities are there for what she took home?

41.* In the expanded generating function in Example 7.12 the coefficients are symmetric. That is, the first is the same as the last, the second is the same as the second to last, and so on. Explain why this should be expected.

42. Verify the formula $\forall\, n \in \mathbb{Z}^+, \forall\, x \neq 1, \sum_{i=0}^{n} x^i = \frac{1-x^{n+1}}{1-x}$ by multiplying both sides by $1 - x$ and simplifying the left-hand side.

43*. For what sequence is $\frac{1}{(1+x)^n}$ a generating function?

44. For what sequence is $(1 - x)^n$ a generating function?

7.4   Counting orbits

In Section 6.6, we considered counting problems in which symmetries needed to be factored out. Here, we extend the ideas used there and introduce a counting technique based on group theory. Before developing the necessary machinery, we consider a motivating example.

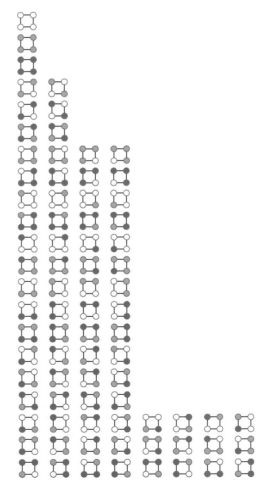

FIGURE 7.4
Fixed dark blue, light blue, and white colorings of a square.

Example 7.14. (Making Necklaces).
If an unlimited supply of dark blue, light blue, and white beads are available, then how many different ways are there to place 4 beads on a loop to form a necklace (which can be freely rotated and flipped to any orientation)?

Solution. Our interest is in coloring each of the vertices of a square dark blue, light blue, or white. To start, we ignore symmetries and list in Figure 7.4 the $3^4 = 81$ colorings of a fixed square (not free to rotate or flip). Since a necklace is unchanged by rotating it or flipping it over, there are not 81 different necklaces. In fact, each pair of colorings from the same row of Figure 7.4 represents different ways of laying the same necklace on a table. Moreover, no two from distinct rows differ by a rotation or flip. For example, the first and second entries in the last row differ by a rotation. The first and fifth entries in the last row differ by a flip (a rotation does not suffice). On the other hand, the first entries in the last and the fourth to last rows do not differ by rotations or flips (even though they have the same numbers of beads of each color). We conclude that the number of different necklaces is 21, the number of rows in Figure 7.4.   □

Unlike results in Section 6.6, the answer 21 in Example 7.14 is not obtained as a quotient of the initial count 81. Instead, a more careful accounting of symmetries is needed. Here, we study symmetries more formally through group actions.

Definition 7.2. A **group** is a set G together with a binary operation $\diamond$ (called composition[*]) such that $\forall\, g, h \in G$, $g \diamond h \in G$ and the conditions

   (i) $\forall\, g, h, k \in G, (g \diamond h) \diamond k = g \diamond (h \diamond k)$
   (ii) $\exists\, e \in G$ such that $\forall\, g \in G, e \diamond g = g \diamond e = g$
   (iii) $\forall\, g \in G, \exists\, g^{-1} \in G$ such that $g^{-1} \diamond g = g \diamond g^{-1} = e$

hold. The composition $g \diamond h$ may be denoted more compactly like multiplication—that is, as gh. Further, exponent notation is used to represent repeated multiplication as usual. Condition (i) is **associativity** of multiplication (composition). The element e required in condition (ii) is called the **identity** element. The element g^{-1} associated to g in condition (iii) is called the **inverse** of g. We identify $g^0 = e$, and for any $n \in \mathbb{Z}^+, g^{-n}$ denotes $(g^{-1})^n$.

---

[*] This need not be the composition of functions from Definition 5.14.

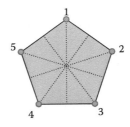 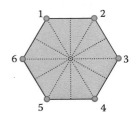

FIGURE 7.5
Labeled n-gons, for $n = 5, 6$.

A deep formal understanding of groups will not be required for our study. Instead, familiarity with a small class of examples is all that is needed. We start with the groups needed for examples like Example 7.14.

Definition 7.3. (Symmetry Groups for Regular n-gons).
Let $n \geq 3$, and let B be a regular n-gon. Let V be the set of vertices of B, which we label 1 to n clockwise around B. Figure 7.5 shows the cases in which $n = 5$ and $n = 6$. For each $i \in \mathbb{Z}$, let r_i denote the clockwise rotation of B about its center by $\frac{360i}{n}$ degrees. If $i < 0$, then the resulting rotation is counterclockwise. Note that $\forall\, i, j \in \mathbb{Z}, r_i = r_j$ if and only if $i \equiv j \pmod{n}$.

We also consider reflections of B about lines through its center. The allowable reflection lines are those that additionally pass through a vertex or a midpoint of a side of B. Those for $n = 5$ and $n = 6$ are pictured as dotted lines in Figure 7.5. Let $f_1, f_2, \ldots, f_n$ denote these n reflections (flips). See Remark 7.2 regarding their order. All compositions are read right to left. For example, the composite $f_1 r_2$ is the symmetry obtained by first applying rotation r_2 to B and then applying reflection f_1 to that result. Note that $\forall\, i, j \in \mathbb{Z},\ r_i r_j = r_{i+j}$.

(a) The set of rotations $\{r_0, r_1, \ldots, r_{n-1}\}$ with the specified composition rule forms a group with identity r_0 and inverses given by $r_i^{-1} = r_{-i}$. This group is called the **cyclic group** of order n and is denoted Z_n.

(b) The set of transformations $\{r_0, r_1, \ldots, r_{n-1}, f_1, f_2, \ldots, f_n\}$ forms a group called the **dihedral group** of order $2n$. It is denoted D_n and contains all of the size- and shape-preserving symmetries of a regular n-gon.

Remark 7.2. For our purposes, the order in which the flips are labeled $f_1, \ldots, f_n$ is unimportant. However, to be consistent, we use the following convention. If n is odd, then for each $1 \leq i \leq n$, we take f_i to be the reflection about the line through vertex i. If n is even, then for each $1 \leq i \leq \frac{n}{2}$, define

f_i as in the odd case, and for each $\frac{n}{2} + 1 \leq i \leq n$, take f_i to be the reflection about the line through the midpoint of the side between the vertices $i - 1$ and i.

Example 7.15.   The dihedral group $D_4 = \{r_0, r_1, r_2, r_3, f_1, f_2, f_3, f_4\}$ consists of all size- and shape-preserving symmetries of a square.

The entire multiplication table for D_4 is shown in Figure 7.6. There, the product $b \diamond a$ is displayed in row b and column a. For example, the fact that $r_3 f_4 = f_1$ can be seen by performing the indicated transformations

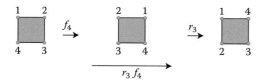

and noting the final locations of the vertex labels. That $f_4 r_3 = f_2$ is shown similarly.

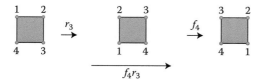

Our interest is in groups that act on sets.

$\diamond$	r_0	r_1	r_2	r_3	f_1	f_2	f_3	f_4
r_0	r_0	r_1	r_2	r_3	f_1	f_2	f_3	f_4
r_1	r_1	r_2	r_3	r_0	f_4	f_3	f_1	f_2
r_2	r_2	r_3	r_0	r_1	f_2	f_1	f_4	f_3
r_3	r_3	r_0	r_1	r_2	f_3	f_4	f_2	f_1
f_1	f_1	f_3	f_2	f_4	r_0	r_2	r_1	r_3
f_2	f_2	f_4	f_1	f_3	r_2	r_0	r_3	r_1
f_3	f_3	f_2	f_4	f_1	r_3	r_1	r_0	r_2
f_4	f_4	f_1	f_3	f_2	r_1	r_3	r_2	r_0

FIGURE 7.6
Multiplication table for D_4.

Definition 7.4. Given a group G with composition $\diamond$ and a set X, we say that G **acts on** X via operation $*$ if $\forall\, g \in G$, $\forall\, x \in X, g * x \in X$ and the conditions

(i) $\forall\, x \in X, e * x = x$

(ii) $\forall\, g, h \in G, \forall\, x \in X, h * (g * x) = (h \diamond g) * x$

hold. The action $g * x$ may also be unambiguously denoted like multiplication—that is, as gx. Condition (ii) tells us that both potential interpretations of hgx yield the same result.

In our study of the colorings of a square in Example 7.14, the relevant symmetry group is D_4. To count the number of different 4-bead necklaces possible, we need to understand the action of D_4 on the set X of colorings shown in Figure 7.4. There, the action of a rotation or a flip from D_4 takes each example of a fixed coloring to another (possibly the same) fixed coloring. For example,

$$r_1 \quad * \quad \text{} \quad = \quad \text{}$$

and

$$f_1 \quad * \quad \text{} \quad = \quad \text{}$$

In general, it is possible to transform between two colorings in Figure 7.4 if and only if they lie on the same row there. This motivates the notion of an orbit of an action.

Definition 7.5. Let a group G act on a set X. For each $x \in X$, the **orbit of** x is the set $\mathrm{Orb}(x) = \{y \,:\, y \in X \text{ and } y = gx \text{ for some } g \in G\}$. An **orbit** is a set $\mathrm{Orb}(x)$ for some $x \in X$.

For each coloring in Figure 7.4, its orbit under the action of D_4 is the subset of colorings in its row. The orbits form equivalence classes.

Theorem 7.7. *If a group G acts on a set X, then the orbits partition X.*

The proof of Theorem 7.7 is presented at the end of this section. As a result of Theorem 7.7, we say that two elements $x, y \in X$ are G-**equivalent** (or just **equivalent** if G is clear in context) if and only if $\mathrm{Orb}(x) = \mathrm{Orb}(y)$. Very simply, counting the number of necklaces in Example 7.14 is the same as counting the number of equivalence classes of the fixed colorings in Figure 7.4. This was done explicitly in Example 7.14.

However, we can now do it using a formula named after the British mathematician William Burnside (1852–1927) that counts the number of orbits in a set with a group action.

Theorem 7.8. *(Burnside's Formula[*]).*
Let a group G act on a set X, and let N be the number of orbits under this action. Then

$$N = \frac{1}{|G|} \sum_{g \in G} |Fix(g)|,$$

where $\forall\, g \in G, Fix(g) = \{x \,:\, x \in X \text{ and } gx = x\}.$

The proof of Theorem 7.8 is sketched at the end of this section. For each $g \in G$, the set $\text{Fix}(g) = \{x \,:\, x \in X \text{ and } gx = x\}$ is called the **fixed points** of g. Problems like Example 7.14 are attacked with Theorem 7.8 by counting sizes of fixed point sets. For our first example of this approach, we revisit Example 7.14, without the benefit of Figure 7.4.

Example 7.16. (Example 7.14 Revisited).
Let X be the set of colorings of the vertices of a square in black, gray, and white. Count the number of orbits of X under the natural action of D_4.

Solution. For each $g \in D_4$, we need to count $|\text{Fix}(g)|$.
  Case 0: $g = r_0$ the identity.
The identity moves nothing. Each coloring is fixed. So $|\text{Fix}(r_0)| = 3^4 = 81$.

  Case 1: $g = r_i$ for $i = 1$ or 3. That is, ↑■↓ or ↓■↑.
The only colorings unchanged by such a rotation are those in which each vertex has the same color (from 3 choices). So $|\text{Fix}(r_i)| = 3$.

  Case 2: $g = r_2$. That is, ⊠.
Such a rotation switches two pairs of opposite vertices. Within each pair, the vertices must have the same color. However, the two pairs are independent. So $|\text{Fix}(r_2)| = 3^2 = 9$.

  Case 3: $g = f_i$ for $i = 1$ or 2. That is, ◪ or ◪.
Such a flip fixes a pair of opposite vertices and switches the other pair. The colorings unchanged by this can have any colors on the fixed vertices, but the two that are switched must have the same color. So $|\text{Fix}(f_i)| = 3^3 = 27$.

---

[*] Burnside's Formula can also be attributed to the work of Augustin-Louis Cauchy and Georg Frobenius.

Case 4: $g = f_i$ for $i = 3$ or 4. That is, $\uparrow\blacksquare\uparrow$ or $\overleftrightarrow{\blacksquare}$.

Such a flip switches two pairs of neighboring vertices. Within each pair, the vertices must have the same color. However, the two pairs are independent. So $|\text{Fix}(f_i)| = 3^2 = 9$.

Let N be the number of orbits of X. Invoking Theorem 7.8 gives

$$N = \frac{1}{|D_4|} \sum_{g \in D_4} |\text{Fix}(g)|$$

$$= \frac{1}{8}(|\text{Fix}(r_0)| + |\text{Fix}(r_1)| + |\text{Fix}(r_3)| + |\text{Fix}(r_2)|$$

$$+ |\text{Fix}(f_1)| + |\text{Fix}(f_2)| + |\text{Fix}(f_3)| + |\text{Fix}(f_4)|)$$

$$= \frac{1}{8}(81 + 3 + 3 + 9 + 27 + 27 + 9 + 9) = \frac{168}{8} = 21.$$

We rediscover the fact that there are 21 different necklaces. $\square$

Having seen the utility of Burnside's Formula, we use it to solve more counting problems.

Example 7.17.   (Building Ferris Wheels).
A Ferris wheel company makes 4 types of chairs. How many different Ferris wheels consisting of 12 chairs could be made by this company? We assume that the chairs face forward in the one direction that the wheel rotates.

Solution. Since there are 4^{12} different Ferris wheels with a fixed position, an explicit listing of them is unreasonable. Instead, we use Theorem 7.8 to count the number N of orbits of these under the action of the cyclic group Z_{12}. Note that D_{12} is not the appropriate group here. In general, the mirror image of a particular Ferris wheel is not the same as the original; a chair that was previously in front of another would now be behind it. Thus, the only relevant symmetries are the rotations in Z_{12}.

For each i, the value $|\text{Fix}(r_i)|$ depends on $\gcd(i, 12)$. In fact,

$$|\text{Fix}(r_i)| = 4^{\gcd(i,12)}.$$

For example, consider the case in which $\gcd(i, 12) = 3$, so $i = 3$ or 9. For r_i not to change the appearance of a Ferris wheel, positions $1, 4, 7, 10$ must hold the same type of chair, as must positions $2, 5, 8, 11$ and positions $3, 6, 9, 12$.

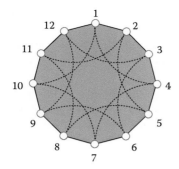

Since there are 4 types of chairs, there are 4^3 Ferris wheels fixed by r_i. The other cases are handled similarly. Thus, there are

$$N = \frac{1}{12}(4(4) + 2(4^2) + 2(4^3)$$
$$+ 2(4^4) + 1(4^6) + 1(4^{12}))$$
$$= \frac{1}{12}(16 + 32 + 128 + 512 + 4096$$
$$+ 16777216) = 1398500.$$

possible Ferris wheels.   □

Example 7.18.   (Making a Die).
A die is to be made from a cube by coloring each face either black or white. How many possibilities are there for such a die?

Solution.   Let G be the symmetry group of a cube. In a symmetry, there are 6 faces that might be moved to the top. Then, there are 4 choices for how that face might be rotated about a line through its center and the center of the cube. We see immediately that $|G| = 6 \cdot 4 = 24$. For each $g \in G$, $|\text{Fix}(g)|$ depends on the "geometry" of g.

Case 0: g moves no faces.
Here, g can only be 1 element, the identity. Since each face may be given either color, $|\text{Fix}(g)| = 2^6$.

Case 1: g fixes a pair of opposite faces and rotates the remaining four by $\pm90°$ around a line through the center of the fixed opposite faces. See Figure 7.7a.

(a)   (b)   (c)   (d)

FIGURE 7.7
Rotations of a die.

Since there are 3 pairs of opposite faces and then 2 possible rotations, g is one of $3 \cdot 2 = 6$ elements. Since each of the fixed opposite faces can receive either color and the other four faces must be all black or all white, $|\text{Fix}(g)| = 2^3$.

Case 2: g fixes a pair of opposite faces and rotates the remaining four by $180°$ around a line through the center of the fixed opposite faces. See Figure 7.7b.

Since there are 3 pairs of opposite faces and then only one possible $180°$ rotation, g is one of 3 elements. Since each of the fixed opposite faces can receive either color and each pair of opposite faces from the other 4 must be both black or both white, $|\text{Fix}(g)| = 2^4$.

Case 3: g reverses the ends of each of a pair of opposite edges. See Figure 7.7c.

Since there are 6 pairs of opposite edges, g is one of 6 elements. Since g reverses 3 pairs of faces, $|\text{Fix}(g)| = 2^3$.

Case 4: g rotates the cube by $\pm 120°$ about a line through a pair of opposite vertices. See Figure 7.7d.

Since there are 4 pairs of opposite vertices and then 2 possible rotations, g is one of $4 \cdot 2 = 8$ elements. Since each vertex on the line of rotation must be the corner of three faces of the same color, $|\text{Fix}(g)| = 2^2$.

The above cases account for all $1 + 6 + 3 + 6 + 8 = 24$ elements of G. By Theorem 7.8, the number N of possible dice is

$$N = \frac{1}{24}(1(2^6) + 6(2^3) + 3(2^4) + 6(2^3) + 8(2^2)) = 10. \qquad \square$$

The counting technique introduced in this section is based on a more general method initiated by the Hungarian mathematician George Pólya (1887–1985). That method is known as Pólya's Method of Counting and additionally employs generating functions. The interested reader can find this developed further in Tucker (2011).

7.4.1 The delayed proofs

We close this section with the promised verifications of Theorems 7.7 and 7.8, regarding a group G acting on a set X.

Proof of Theorem 7.7. We show that the orbits of G partition X.

Certainly, for each $x \in X$, we have $x \in \text{Orb}(x)$. Thus, we need only show that distinct orbits are disjoint. This is accomplished by showing that

$$\forall\, x, y \in X, \text{ if } \text{Orb}(x) \cap \text{Orb}(y) \neq \emptyset, \text{ then } \text{Orb}(x) = \text{Orb}(y).$$
$$(7.12)$$

So suppose we have some $z \in \text{Orb}(x) \cap \text{Orb}(y)$. Hence, we have $g, h \in G$ such that $gx = z = hy$. If $w \in \text{Orb}(x)$, then we have some $k \in G$ such that $w = kx = kg^{-1}hy \in \text{Orb}(y)$. If $w \in \text{Orb}(y)$, then we have some $k \in G$ such that $w = ky = kh^{-1}gx \in \text{Orb}(x)$. Hence, $\text{Orb}(x) = \text{Orb}(y)$. □

Sketch of Proof of Theorem 7.8. We show that the number of orbits is

$$N = \frac{1}{|G|} \sum_{g \in G} |\text{Fix}(g)|.$$

For each $x \in X$, let $S(x) = \{g : g \in G \text{ and } gx = x\}$. The proof follows from a sequence of basic facts.

(i) $\forall\, x \in X, \forall\, y \in \text{Orb}(x), |\{g : gx = y\}| = |S(x)|$.
   Sketch. Fix some $h \in \{g : gx = y\}$. The function

   $$f : \{g : g \in G \text{ and } gx = y\} \longrightarrow S(x)$$

   defined by $f(g) = h^{-1}g$ is a bijection (with inverse $f^{-1}(g) = hg$).

(ii) $\forall\, x \in X, |\text{Orb}(x)| = \frac{|G|}{|S(x)|}$.
   Sketch. Let $k = |S(x)|$. The function $f : G \longrightarrow \text{Orb}(x)$ defined by $f(g) = gx$ is onto. By (i), each $y \in \text{Orb}(x)$ is hit k times. Hence, $|G| = k|\text{Orb}(x)| = |S(x)||\text{Orb}(x)|$.

(iii) $\forall\, x \in X, \forall\, y \in \text{Orb}(x), |S(y)| = |S(x)|$.
   Sketch. This follows from (ii) and (7.12).

Let $x_1, x_2, \ldots, x_N \in X$ be representatives for the N orbits. We have

$$\sum_{g \in G} |\text{Fix}(g)| = \sum_{x \in X} |S(x)| = \sum_{i=1}^{N} |\text{Orb}(x_i)||S(x_i)|$$

$$= \sum_{i=1}^{N} \frac{|G|}{|S(x_i)|}|S(x_i)| = \sum_{i=1}^{N} |G| = N|G|.$$

The first equality holds since, in the leftmost sum, each $x \in X$ is counted $|S(x)|$ times. The remaining equalities follow from conditions (ii) and (iii) above. □

Exercises

1. List all possible colorings of a fixed equilateral triangle for which each vertex is dark blue, light blue, or white.

2. List all possible colorings of a fixed regular pentagon for which each vertex is either black or white.

3. Make a multiplication table for the group Z_4. (Recall that $r_i r_j = r_{i+j}$.)

4. Make a multiplication table for the group Z_6. (Recall that $r_i r_j = r_{i+j}$.)

FIGURE 7.8
Computers networked in a circuit.

5. Consider D_4 and its multiplication table in Figure 7.6.

   (a) Find $f_2 r_2$.

   (b) Find $r_2 f_1 r_3$.

   (c) Is $r_1 f_2 = f_2 r_1$?

6. Make a multiplication table for D_3. (Use the naming conventions from Remark 7.2.)

   (a) Find $f_2 r_2$.

   (b) Find $f_1 r_2 f_2$.

   (c) Is $r_1 f_1 = f_1 r_1$?

Networking computers. In Exercises 7 through 10, we are planning to network together some computers in a circuit configuration, as shown in Figure 7.8. Since different computers with varying speeds and capabilities will be used, and a computer can directly pass information only to an adjacent computer, the behavior of the network will depend not only on which computers are used but also on the order in which they are connected. However, two networks that differ only by a rotation or are mirror images of one another will exhibit exactly the same computational properties and are thus considered the same network type. In each case, determine the number of possible network types for our network of the specified size, given the computers that are available.

7. There are ample supplies of 2 kinds of computers to make a circuit of 4 computers.

8. There are ample supplies of 4 kinds of computers to make a circuit of 4 computers.

9. There are ample supplies of 3 kinds of computers to make a circuit of 3 computers.

10. There are ample supplies of 3 kinds of computers to make a circuit of 5 computers.

11. A company is producing towels based on a simple design. Each towel is formed by sewing together three equal-sized strips of colored cloth to form a rectangle (with three horizontal stripes).

For the stripes, the company is using only the solid colors red, orange, yellow, green, blue, and purple. Also, each towel has no particular orientation distinguishing top from bottom or front from back.

(a) How many different towels are possible?

(b) How many do not use the same color on two adjacent strips?

(c) How many use three different colors for the strips?

12. A cheese company makes wedges of cheddar, muenster, and provolone cheese. Each wedge is packaged in one of the 6 slots of a cheese sampler.

Of course, the appearance of a cheese sampler is considered unchanged by a rotation.

(a) How many different cheese samplers are possible?

(b) What if each sampler is considered the same as its mirror image?

(c) What if mirror images are considered the same and all three kinds of cheese must be present?

Colored blocks and dice. In Exercises 13 through 17, we aim to make colored blocks, some of which might be used as dice.

13. A die is to be made from a cube by coloring each face red, white, or blue. How many possibilities are there for such a die?

14. A triangular prism has two opposite faces in the shapes of equilateral triangles with three square faces connecting them. How many different ways are there to color each of the faces of a triangular prism with one of 3 colors?

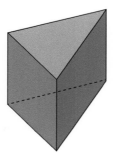

15. How many different ways are there to color each of the faces of a pyramid with square base and equilateral sides either red, blue, yellow, or green?

16. A pentagonal prism has two opposite faces in the shapes of regular pentagons with five square faces connecting them. How many different ways are there to color each of the faces of a pentagonal prism with one of 3 colors?

17.* How many different ways are there to color each of the faces of a tetrahedron red, blue, yellow, or green?

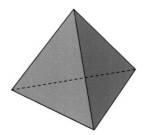

18.* A die is to be made from an octahedron by coloring each face either black or white. How many possibilities are there for such a die?

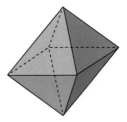

19. A Ferris wheel company makes 3 different types of chairs. How many different Ferris wheels consisting of 8 chairs could be made?

20. A baby rattle is to be made by attaching 6 colored plastic rings (each of which can be red, yellow, blue, or green) to a large plastic ring. How many different rattles are possible?

21. Assume that the end of a Tic-Tac-Toe game results in a 3×3 grid

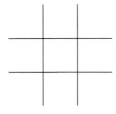

containing five X's and four O's. With the understanding that each complete game might be viewed from any of the four sides of the grid, how many different complete Tic-Tac-Toe grids are possible?

22. The front of a quilt is to be made by attaching, to a plain white back, colored squares in the pictured four-by-four configuration.

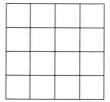

Before sewing begins, a plan is mapped out on the grid. How may different quilts are possible if each square can be red, white, or blue? Note that the quilt is free to be rotated and thus has no specified top edge.

23.* A necklace with an undetectable clasp is to be made by threading 4 black beads and 4 white beads on a loop. How many different necklaces are possible?

24.* Each of 6 black beads and 6 white beads is to be threaded on a loop to form a necklace. If the clasp will be undetectable, then how many possibilities are there for this necklace?

25. A charm bracelet is to be made by attaching 5 small charms on a given band. The jeweler has large supplies of 3 different types of charms.

    (a) How many possibilities are there for this bracelet?

    (b) What if the jeweler additionally attaches a diamond?

    (c) What if, since a diamond has been added, the jeweler feels that at most 2 of the 3 types of charms should be used on the bracelet?

26. A merry-go-round company makes black horses, white horses, tigers, and lions for its merry-go-rounds. This company is planning a series of merry-go-rounds consisting of 8 animals in a cycle, with each animal facing in the direction of rotation.

    (a) How many different merry-go-rounds can there be in this series?

    (b) How many use exactly one lion?

    (c) How many use no lions?

    Making mobiles. For Exercises 27 through 32, a mobile is to be constructed in the given configuration (pictured from below). The mobile is hung from a string attached to the central point. The symbol ● marks a rotational center, and the symbol ○ marks a location that must be filled with a sun, a moon, a star, a bird, or a plane. If there are unlimited supplies of these objects, then determine the number of different mobiles possible.

27.

28.

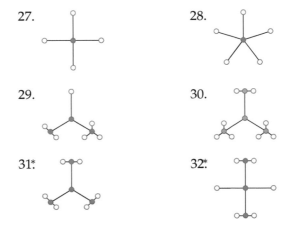

29.

30.

31.*

32.*

33. Let G be a group acting on a set X.
Show: $\forall\, g \in G,\ \text{Fix}(g) = \text{Fix}(g^{-1})$.

34. Let G be a group acting on a set X.
Show: $\forall\, g_1, g_2 \in G,\ \text{Fix}(g_1) \cap \text{Fix}(g_2) \subseteq \text{Fix}(g_1 g_2)$.

Limited computer networks. In Exercises 35 through 38, we are networking computers in a circuit configuration, as in Exercises 7 through 10. However, here there are limited numbers of various types of computers available. In each case, determine the number of possible network types for our network of the specified size, given the computers that are available.

35.* There are 4 each of 3 kinds of computers to make a circuit of 12 computers.

36.* There are 2 each of 4 kinds of computers to make a circuit of 8 computers.

37.* There are 6 each of 2 kinds of computers and 2 more of a third kind with which to make a circuit of 12 computers. There will be two computers left over.

38.* There are 4 each of 2 kinds of computers and 2 more of a third kind with which to make a circuit of 8 computers. There will be two computers left over.

39.* There are 4 black and 4 white squares available to form the sides of a box, which will be used as a decorative piece of furniture. How many different possibilities are there for this box?

40.* There are 3 black and 3 white equilateral triangles available to form the sides of a tetrahedron, which will be filled with sand and used as a paper weight. How many different possibilities are there for this paper weight? See the picture in Exercise 17.

7.5 Combinatorial arguments

There are many beautiful combinatorial identities. For example, we have already encountered Pascal's identity and several identities that are consequences of the Binomial Theorem. Combinatorial identities can be useful tools for solving counting problems or for simplifying their solutions. However, here we turn things around and use counting techniques to prove combinatorial identities. This type of proof is called a **combinatorial proof**.

In this section, we justify identities that have already been proved or could be proved by techniques presented earlier. Our interest is in presenting alternative proofs that can sometimes be very efficient, are often very beautiful, and always provide deeper insight into the identities themselves.

Our first example revisits Pascal's identity.

Example 7.19. Show: For each $1 \leq k \leq n-1$, $\binom{n}{k} = \binom{n-1}{k-1} + \binom{n-1}{k}$.

Proof. The binomial coefficient $\binom{n}{k}$ counts the number of ways to choose a subset of size k from the set $\{1, 2, \ldots, n\}$. Another way to count this involves considering whether or not the element n is chosen. Figure 7.9 illustrates our strategy when $n = 5$ and $k = 3$.

The number of ways to choose a subset of size k that contains the element n is $\binom{n-1}{k-1}$, since the remaining $k - 1$ elements must be chosen from the set $\{1, 2, \ldots, n-1\}$ of $n - 1$ elements. The number of ways to choose a subset of size k that does not contain the element n is $\binom{n-1}{k}$. In this case, all k elements must be chosen from the $n - 1$ element set $\{1, 2, \ldots, n-1\}$.

By the Addition Principle, since each subset of size k either does or does not contain the element n, the number of ways to choose a subset of size k is

$$\binom{n-1}{k-1} + \binom{n-1}{k}. \tag{7.13}$$

$\{1,2,5\}$ $\{1,3,5\}$ $\{1,4,5\}$	$\{1,2,3\}$ $\{1,2,4\}$
$\{2,3,5\}$ $\{2,4,5\}$ $\{3,4,5\}$	$\{1,3,4\}$ $\{2,3,4\}$
$\binom{4}{2}$ subsets contain 5	$\binom{4}{3}$ subsets exclude 5

FIGURE 7.9
$\binom{5}{3}$ subsets of $\{1, 2, 3, 4, 5\}$.

Since $\binom{n}{k}$ and (7.13) both count the same thing, they must be equal.   □

The basic approach of a combinatorial proof is to create a counting problem, to solve that problem in two ways, and to conclude that the two solutions must be equal.

Example 7.20.   Show: $\forall\, n \geq 0, \; \displaystyle\sum_{k=0}^{n} \binom{n}{k} = 2^n.$

Proof. We give two proofs.

(a) Consider the number of ways to form a subset of a set of size n. Since a subset is formed by specifying whether or not each of the n elements is in the subset, there are 2^n subsets. That is, each subset of $\{x_1, x_2, \ldots, x_n\}$ can be characterized by a binary sequence

$$\text{in, out, out, in, in, } \ldots \text{ , out}$$

specifying which elements are *in* the subset and which are *out*. Alternatively, we can separate the subsets according to their size. For each $0 \leq k \leq n$, the number of subsets of size k is $\binom{n}{k}$. Hence, the total number of subsets is $\sum_{k=0}^{n} \binom{n}{k}$. It therefore follows that $\sum_{k=0}^{n} \binom{n}{k} = 2^n$.

(b) Consider the downward paths through Pascal's triangular grid counted in Example 6.41.

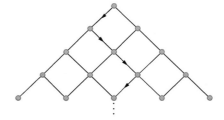

Such a path is characterized by a binary sequence of turns to the right and to the left. We saw in Example 6.41 that the number of paths from $\binom{0}{0}$ to $\binom{n}{k}$ is given by $\binom{n}{k}$. Since the nth row consists of the entries

$$\binom{0}{0}, \ldots, \binom{n}{k}, \ldots, \binom{n}{n},$$

the total number of paths from $\binom{0}{0}$ to the nth row is given by $\sum_{k=0}^{n} \binom{n}{k}$. Alternatively, since each path is characterized by a binary sequence of length n (of right and left turns), there are 2^n such paths. We conclude that $\sum_{k=0}^{n} \binom{n}{k} = 2^n$.   □

The assertion in Example 7.20 was proved by induction in Example 4.26. However, once one gains comfort with combinatorial proofs, the proofs in Example 7.20 have to be seen to be more elegant!

Example 7.21.   Show: $\forall\, n \geq 0,\ \displaystyle\sum_{k=0}^{n} \binom{n}{k} 2^k = 3^n.$

Proof. There are 3^n base-3 sequences of length n. Separate them according to the number of 0's they contain. For each $0 \leq k \leq n$, the number of base-3 sequences of length n with exactly k entries that are not 0's is $\binom{n}{k} 2^k$. This follows since there are $\binom{n}{k}$ ways to pick the k entries that will not be 0's and 2^k ways to fill those entries with 1's or 2's. Hence, the total number of base-3 sequences of length n is $\sum_{k=0}^{n} \binom{n}{k} 2^k$, which must equal 3^n.   □

Note in Example 7.21 that the choice of k entries to not be 0's can also be thought of as a choice of $n - k$ entries to be 0's. Of course, the identity

$$\binom{n}{k} = \binom{n}{n-k} \tag{7.14}$$

gives that the number of such choices is the same. From another point of view, the fact that the number of such choices must be the same gives (7.14).

Example 7.22.   Show: $\forall\, n \geq 0,\ \displaystyle\sum_{k=0}^{n} \binom{n}{k}^2 = \binom{2n}{n}.$

Proof. Our combinatorial argument shows the equivalent identity

$$\sum_{k=0}^{n} \binom{n}{k} \binom{n}{n-k} = \binom{2n}{n}.$$

We consider a counting problem of the type encountered in Example 6.40. Suppose we have an $(n+1)$ by $(n+1)$ square grid of points like the one in Figure 7.10, and we are interested in paths from S to F that move only to the right or down (as we face the grid). By the arguments in Example 6.40, the number of such paths is $\binom{2n}{n}$.

Now separate the paths according to which of the points $d_0, d_1, \ldots, d_n$ in the off diagonal is encountered.

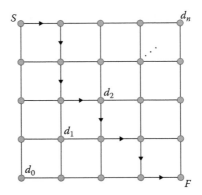

FIGURE 7.10
Routes through a square grid.

For each $0 \leq k \leq n$, the number of paths that pass through d_k is $\binom{n}{k}\binom{n}{n-k}$. Since each path goes through exactly one of the off diagonal points, the total number of paths is

$$\sum_{k=0}^{n} \binom{n}{k}\binom{n}{n-k}. \qquad (7.15)$$

Hence, the sum in (7.15) must equal $\binom{2n}{n}$.   □

College basketball fans can easily remind themselves each March of the following result and proof, in the case that $n = 6$.

Example 7.23. Show: $\forall\, n \geq 1,\ \displaystyle\sum_{k=0}^{n-1} 2^k = 2^n - 1.$

Proof. Consider a binary tournament of the type used for the NCAA basketball tournament. That is, the tournament consists of 2^n teams and has n rounds of games. In each round, the teams are paired off, and only the winner from each pair advances to the next round. Before the tournament starts, the teams are paired off for the first-round matches, and this determines which winners will play each other throughout the tournament. The case in which there are 4 teams and thus 2 rounds of games is bracketed in Figure 7.11.

As the tournament is played, the actual winning teams are listed on the appropriate line of the bracket. We count here the total number of ways to fill in the winning teams.

Round 1 games Round 2 games

Team 1

Winner of 1 vs. 4

Team 4

Champion

Team 2

Winner of 2 vs. 3

Team 3

FIGURE 7.11
Four-team tournament bracket.

For each $1 \leq k \leq n$, the completion of round k yields 2^{n-k} winners. Of course, the round-n winner is declared the champion. Therefore, there are a total of

$$\sum_{k=1}^{n} 2^{n-k} = \sum_{k=0}^{n-1} 2^k$$

victories throughout the tournament. Of course, some teams are winners more than once.

A faster way to count the number of victories is to instead count the number of losses. Since each game has one winner and one loser, these counts are the same. Since the tournament starts with 2^n teams and all but the ultimate champion are losers of exactly one game, there are $2^n - 1$ losses. The number of victories must also be $2^n - 1$ and therefore equals $\sum_{k=0}^{n-1} 2^k$. □

The alternating sum in our final example of this section naturally lends itself to an argument involving the Principle of Inclusion–Exclusion.

Example 7.24. Show: $\forall n \geq 1,\ \sum_{k=1}^{n} (-1)^{k-1} \binom{n}{k} 2^{n-k} = 2^n - 1.$

Proof. Consider a sequence of n coin flips. We count the number of sequences that contain at least one head. Of course, since only the sequence consisting entirely of tails does not contain at least one head, there are $2^n - 1$ sequences containing at least one head.

We could also count this using the Inclusion–Exclusion Principle. For each $1 \leq k \leq n$, let A_k be the set of those sequences that have a head on toss k. Hence, $A_1 \cup A_2 \cup \cdots \cup A_n$ is the set of sequences containing at least one head. For each $1 \leq j_1 < j_2 < \cdots < j_k \leq n$, the intersection $A_{j_1} \cap A_{j_2} \cap \cdots \cap A_{j_k}$ is the set of sequences with heads on each of the k flips $j_1, j_2, \ldots, j_k$. Hence,

$$|A_{j_1} \cap A_{j_2} \cap \cdots \cap A_{j_k}| = 2^{n-k}.$$

Since $S_k = \binom{n}{k} 2^{n-k}$, we get

$$|A_1 \cup A_2 \cup \cdots \cup A_n| = \sum_{k=1}^{n} (-1)^{k-1} \binom{n}{k} 2^{n-k}.$$

Of course, this sum must therefore be equal to $2^n - 1$. $\square$

Exercises

1. Show: $\forall\, n, k \geq 0$, $\sum_{i=0}^{k} \binom{n-i-1}{k-i} = \binom{n}{k}$. Hint: For each $0 \leq i \leq k$, separate out those subsets of size k that contain $1, 2, \ldots, i$ but not $i+1$.

2. Use a subset argument to show: $\forall\, n \geq 0$, $\sum_{k=0}^{n} \binom{n}{k}^2 = \binom{2n}{n}$. Hint: Separate subsets of $\{1, 2, \ldots, 2n\}$ by the size of their intersection with $\{1, 2, \ldots, n\}$.

3. Show: $\forall\, n \geq 2$ and $2 \leq k \leq n-2$, $\binom{n-2}{k-2} + 2 \cdot \binom{n-2}{k-1} + \binom{n-2}{k} = \binom{n}{k}$. Exercise 2 in Section 4.6 requested a computational proof.

4. Show: $\forall\, n \geq m \geq 1$ and $1 \leq k_1, k_2, \ldots, k_m$ with $k_1 + k_2 + \cdots + k_m = n$, that $\binom{n}{k_1, k_2, \ldots, k_m} = \binom{n-1}{k_1 - 1, k_2, \ldots, k_m} + \binom{n-1}{k_1, k_2 - 1, \ldots, k_m} + \cdots + \binom{n-1}{k_1, k_2, \ldots, k_m - 1}$.

5. Show: $\forall\, n \geq 0$, $\sum_{i=0}^{n} \binom{2n}{i} \binom{n}{i} = \binom{3n}{n}$. Hint: Mimic Example 7.22 using a rectangular grid.

6. Show: $\forall\, m, n \geq 0$ and $0 \leq k \leq m, n$, $\sum_{i=0}^{k} \binom{m}{i} \binom{n}{k-i} = \binom{m+n}{k}$. Hint: Consider paths in Pascal's triangle.

7. Show: $\forall\, n \geq 1$, $2 \sum_{k=0}^{n-1} 3^k = 3^n - 1$. Hint: Imagine a game in which 3 teams play and one is declared the winner. (For example, Canadian doubles is a tennis game designed for 3 players.) Consider a tournament for 3^n teams. There will be twice as many losses as victories.

8. Show: $\forall\, n \geq 1,\ 3\sum_{k=0}^{n-1} 4^k = 4^n - 1.$

9. Show: $\forall\, n \geq 1,\ \sum_{i=1}^{n} (-1)^{i-1}\binom{n}{i} 2^{n-i} = 3^n - 2^n.$
   Hint: Use Inclusion–Exclusion to count the number of base-3 sequences of length n that contain at least one 2.

10. Show: $\forall\, n \geq 1,\ \sum_{k=1}^{n} (-1)^{k-1}\binom{n}{k} 3^{n-k} = 4^n - 3^n.$

11. Show: $\forall\, n \geq 1,\ \sum_{k=1}^{n} k\binom{n}{k} = n2^{n-1}.$ Hint: From n players, consider choosing a nonempty team (of any size) with a team captain.

12. Show: $\forall\, n \geq 1,\ \sum_{k=1}^{n} k^2\binom{n}{k} = n(n-1)2^{n-2} + n2^{n-1}.$
    Hint: From n players, consider choosing a team with a team captain and a van driver, noting that the captain might also be the van driver.

13. Show: $\forall\, n \geq 1,\ \frac{\binom{2n}{n}}{2} = \binom{2n-1}{n-1}.$ Hint: If you play n-on-n basketball, then how many ways can you pick your team?

14. Show: $\forall\, n \geq 1,\ \frac{\binom{3n}{n,n,n}}{3!} = \frac{\binom{3n-1}{n-1,n,n}}{2}.$

15. Prove: For all integers $n \geq 0$,
    $$3^n = \sum_{\substack{0 \leq k_1, k_2, k_3 \leq n \\ k_1 + k_2 + k_3 = n}} \binom{n}{k_1, k_2, k_3}.$$
    Hint: Consider base-3 sequences.

16. Prove the Binomial Theorem: For all $n \in \mathbb{Z}^+,\ (a+b)^n = \sum_{i=0}^{n}\binom{n}{i} a^{n-i}b^i.$ Hint: Before like terms are combined, the n-fold product $(a+b)(a+b)\cdots(a+b)$ has 2^n monomials (e.g., for $n = 5$, one would be $baaba = a^3b^2$). For each i, how many equal $a^{n-i}b^i$ (e.g., $abbaa = a^3b^2 = baaba$)?

17*. Consider paths through an $(a+1)$ by $(b+1)$ by $(c+1)$ rectangular grid of points that start at the top back left corner S, end at the bottom front right corner F, and only move right, down, or frontward at each step.

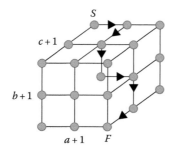

(a) How many such paths are there from S to F?

(b) Use this application to prove: $\forall\ i, j, k \in \mathbb{Z}^+$,
if $n = i + j + k$, then $\binom{n}{i,j,k} = \binom{n-1}{i-1,j,k} + \binom{n-1}{i,j-1,k} + \binom{n-1}{i,j,k-1}$.

18.* Show: $\forall\ n \in \mathbb{Z}^+$,

$$\sum_{\substack{0 \le k_1, k_2, k_3 \le n \\ k_1 + k_2 + k_3 = n}} \binom{n}{k_1, k_2, k_3} \binom{2n}{n-k_1, n-k_2, n-k_3} = \binom{3n}{n,n,n}.$$

Hint: Use the idea in Exercise 17.

19. Show: $\forall\ n \ge 2$, $\sum_{i=1}^{n-1}(i \cdot i!) = n! - 1$. Hint: Count permutations.

20. Show: $\forall\ n \ge 1$, $\frac{\binom{2n}{2,2,\dots,2}}{n!} = (2n-1)(2n-3)\cdots 3 \cdot 1$. Hint: Count the number of ways to form n chess matches from $2n$ players.

21. Show: $\forall\ n \ge 1$, $\sum_{k=1}^{n} \frac{1}{2^k} = 1 - \frac{1}{2^n}$. Hint: If a coin is tossed n times, then the event that a head is obtained can be broken into cases according to the toss in which the first head occurs.

22.* Show: $\forall n \in \mathbb{Z}^+, 2^n \cdot 3^n = 6^n$. Hint: For $s = 2, 3, 6$, let A_s be the set of n-digit base-s numbers, and give a bijection $A_2 \times A_3 \longrightarrow A_6$.

23.* Prove that, given any 6 people, there are either 3 of them who have met each other before or 3 who have never met.

24. Show: $\forall\ k \in \mathbb{Z}^+$, any product of k consecutive integers must be divisible by $k!$. Hint: Let n be the largest integer in the string of k consecutive integers.

For Exercises 25 through 28, for each $n \ge 3$, let T_n be the number of triangulations of a fixed regular n-gon that do not introduce new vertices. For example, $T_5 = 5$, as shown.

25. Find T_4.   26. Find T_6.

27.* Let $n \ge 3$. (a) Show: T_n counts the number of ways to parenthesize a product of $n - 1$ terms. Hint: $a((bc)d)$ corresponds to

(b) Conclude from the assertions in Exercise 45 from Section 4.3 that $T_n = \frac{1}{n-1}\binom{2n-4}{n-2}$.

28.* Show: $\forall\, n \geq 3, (n-1)T_n = 2(2n-5)T_{n-1}$. Hint: Map a triangulation of an n-gon with a nonbase outer edge selected to a triangulation of an $(n-1)$-gon with an edge selected and an end of that edge specified. For example,

Magic tricks. Although Exercises 29 through 32 do not ask for combinatorial proofs per se, they do each ask for a combinatorial explanation of what is being described.

29. Here is a fun task that will keep a young child busy for a while.

> On an 8×8 chess board, place coins in the upper-left and lower-right corners. Assume that a domino can be placed on the board to cover exactly two adjacent squares. Ask the child to cover the remainder of the board with dominoes.

Explain why this task cannot be completed.

30. The following card trick has been performed on television by the magician David Copperfield. Place 9 cards in a 3×3 grid pattern. We shall refer to each position by $c_{i,j}$ for some $1 \leq i,j \leq 3$. A player will be allowed to "walk" over the cards by making horizontal or vertical moves of one step. Ask the player to start on any of the four corners and then to move 4 steps. Remove card $c_{1,2}$. Ask the player to then move 3 steps among the remaining cards. Remove $c_{1,3}$ and $c_{3,1}$. Ask the player to then move 2 steps among the remaining cards. Remove $c_{1,1}$ and $c_{3,3}$. Ask the player to then move 1 step among the remaining cards. Remove all but $c_{2,2}$. This is where the player will be. Moreover, if an entire television audience is playing, then this is where they all will be. Explain why.

31. For a simple card trick, deal 27 cards face up into 9 rows of 3 cards. Keep the cards face up at all times. Be sure to deal each row from left to right and to place the cards in row $i + 1$ on top of the cards in row i so that the denominations and suits of all cards remain visible. Ask a player to secretly pick one of the cards and to only announce the column in which it appears. Slide each of the columns into a pile and stack the piles so that the selected one is on top. Now deal the cards into 9 rows of 3 as before. Again have the player announce the column in which the selected card appears, and stack the piles so that the selected one is on top. Deal the cards into 9 rows of 3 again and have the player announce the column in which the selected card appears. The selected card should be the first card in that column. Explain why this is so.

32. Simple mathematics can be used to fake clairvoyance. On four index cards, write the numbers

$$
\begin{array}{cccccccc}
1 & 3 & 5 & 7 & 9 & 11 & 13 & 15 \\
2 & 3 & 6 & 7 & 10 & 11 & 14 & 15 \\
4 & 5 & 6 & 7 & 12 & 13 & 14 & 15 \\
8 & 9 & 10 & 11 & 12 & 13 & 14 & 15
\end{array}
$$

so that each row appears on exactly one card. Ask a player to secretly pick an integer from 0 to 15 and to announce each of the four cards containing that number. This information is sufficient to quickly determine that number. Explain why.

7.6 Review problems

1. How many possible 8-digit (0–9) license plates exclude at least one of the digits 0, 3, 6, or 9?

2. Compute $\phi(9100)$ using the Principle of Inclusion–Exclusion.

3. How many of the integers from 1 to 4000 are divisible by at least one of 7, 11, or 13?

4. Since a flush is a poker hand containing just one suit, one might call a hand containing all suits an anti-flush. What is the probability that a 5-card hand will contain at least one card from each suit?

5. A cup of Fourth of July holiday candies contains 6 red, 7 white, and 8 blue candies. If 5 are randomly selected, then what is the probability that at most two colors will be represented?

6. Mrs. Hetrick returned from the produce market with a bag containing 8 apples, 5 bananas, and 7 oranges. She then asked her son Luke to use 6 of these pieces of fruit to

make a fruit basket. While Mrs. Hetrick was busy doing other errands, she realized that she had neglected to mention to Luke that he should include in the basket at least one of each type of fruit. How many possible selections of 6 pieces of fruit might Luke make that will contain at least one of each type?

7. While at a club, 4 women left their drinks on the table so that they could dance. When the women returned to their table, each had forgotten which drink was hers, so each of them randomly retrieved one. What is the probability that no woman got her original drink?

8. Compute $\binom{6}{2,1,3}$.

9. A researcher is studying the significance of the DNA strand

$$TGTATACTGT.$$

She wants to know what characteristics are special to this particular ordering and what might be preserved in another sequence of length 10 containing the same numbers of each nitrogen base. How many possible DNA strands of length 10 contain 1 cytosine (C), 2 guanine (G), 5 thymine (T), and 2 adenine (A) nitrogen bases?

10. A witness to a drive-by shooting remembers a partial license plate on the vehicle. She remembered that the plate had exactly 3 fives and 3 sevens. The police are confident that the plate must have had 8 digits, as is standard in their state. To how many possibilities does this information narrow down the license plate?

11. A chemistry teacher is trying to decide how to break his class into groups for a lab the next day. There are 18 students in the class, and the teacher is not sure what group sizes to use. How many ways are there to form

    (a) a group of size 6, one of size 5, one of size 4, and one of size 3?

    (b) three groups of size 6?

    (c) two groups of size 4 and two of size 5?

12. Expand $(x + y + z + w)^2$ using the Multinomial Theorem.

13. Expand $(2x - y + z)^3$ using the Multinomial Theorem.

14. What is the coefficient of $x^{20} y^{50} z^{10}$ in $(3x - 2y + z)^{80}$?

15. What is the coefficient of $x^4 y^5 z^6$ in $(1 + x + y + z)^{20}$?

16. Find the coefficients in the power series
    $(1 - x + x^2 - x^3 + \cdots)(1 + x^2 + x^4 + x^6 + \cdots)$.

17. For each $i \geq 0$, determine the number of ways c_i of obtaining $i\text{¢}$ if there are 2 quarters, 4 dimes, and 2 nickels. Ways

are distinguished by how many of each type of coin they contain.

18. How many possibilities are there for a snack of 5 candies taken from 5 red, 3 white, and 4 blue candies? We distinguish snacks solely by the numbers of candies of each color that they contain.

19. Count all integer solutions a, b, c, d to the equation $a + b + c + d = 12$, such that $0 \leq a \leq 2$, $1 \leq b \leq 4$, $2 \leq c \leq 6$, and $3 \leq d \leq 8$.

20. A grocery store is nearly out of apples and has only 7 left. However, the store has a huge supply of bananas and oranges available. A customer has requested a fruit basket consisting of 8 pieces of fruit. If fruit baskets are distinguished by how many of each type of fruit they contain, then how many possibilities are there for what the store could provide?

21. What is the coefficient of x^{20} in $\frac{1}{(1-x)^{10}}$?

22. What is the coefficient of x^9 in $\frac{x^2 + 3x^4 - x^6}{(1-x)^5}$?

23. How many possibilities are there for a collection of 50 balls taken from 5 red, 60 blue, 50 yellow, and 70 green balls? We distinguish collections solely by the numbers of balls of each color they contain.

24. Make a multiplication table for the group Z_3.

25. In the group D_6 let f be the reflection about the pictured line, and find $f \circ r_2 \circ f$.

26. A programmer wishes to build a computer network consisting of 6 computers connected in a circuit configuration. At work, she has ample supplies of two types of computers available. How many different network types are possible for her?

27. A Ferris wheel company makes 5 different types of chairs. How many different Ferris wheels consisting of 9 chairs could be made? Each chair will face forward.

28.* A company that makes educational toys is planning to sell colored blocks. Each block is to be made from a cube by coloring each face either red, blue, yellow, or green. Moreover, the company wants to sell the blocks in a bag containing exactly one of each possible block, within this coloring scheme. How many blocks will the bag contain?

29. A quilt is to be designed with a back that is solid white and a front that forms a 3-by-3 grid.

If each square on the grid must be colored red, yellow, or blue, then how many different quilts might be designed?

30\. Let G be a group acting on a set X.
Show: $\forall\, g_1, g_2 \in G$, if $\text{Fix}(g_2) = X$ then $\text{Fix}(g_2 g_1) = \text{Fix}(g_1)$.

31\. A mobile is to be constructed in the pictured configuration. The mobile is hung from a string attached to the central point. The symbol • marks a rotational center, and the symbol ○ marks a location that must be filled with a sun, a moon, a star, or a cloud. Determine the number of different possible mobiles.

32\. A child has 3 red, 3 yellow, and 3 blue beads with which to make a necklace, the beads of which will be free to slide around. How many different 9-bead necklaces could this child make?

33. Prove Pascal's Identity $\binom{n+1}{k} = \binom{n}{k-1} + \binom{n}{k}$ by interpreting binomial coefficients as numbers of paths in Pascal's triangular grid.

34. Show: $\forall\, 0 \leq k \leq n$, $\sum_{i=k-1}^{n-1} \binom{i}{k-1} = \binom{n}{k}$. Hint: Separate out the subsets of size k from $\{1, 2, \ldots, n\}$ for which $i + 1$ is the largest item selected.

35\. Show: $\forall\, 0 \leq k \leq n$, $\sum_{i=0}^{k} \binom{2n}{i}\binom{n}{k-i} = \binom{3n}{k}$.

36. Show that the number of ways to balance n pairs of parentheses equals the number of (down and right, only) paths from S to F in an $(n+1)$ by $(n+1)$ square grid that never dip below the diagonal. Hint: ()(()) corresponds to

37. Show: $\forall\, n \geq 1$, $\displaystyle\sum_{\substack{k \text{ even,} \\ 0 \leq k \leq n}} \binom{n}{k} = \sum_{\substack{k \text{ odd,} \\ 0 \leq k \leq n}} \binom{n}{k}$.

Hint: Define a function from the subsets of $\{1, \ldots, n\}$ of even cardinality to those of odd cardinality by mapping each subset A to the subset $A \triangle \{n\}$.

38. What is the coefficient of x^{205} in $(1 + x + x^2 + \cdots + x^{99})^3$?

39. What is the coefficient of x^3 in $(1 + x + x^{99})^{205}$?

40. The game of Bridge starts by dealing 13 cards to each of 4 players labeled North, South, East, and West. How many different initial states are there for a Bridge game?

41.* How many different 8-bead necklaces can be made using 2 red, 2 blue, and 4 white beads?

42. A game at a charity fundraiser involves selecting colored balls from a bag. The bag contains 10 red, 8 white, and 12 blue balls, and 9 are randomly selected. Let N be the number of different outcomes possible, where we distinguish outcomes solely by the numbers of balls of each color that they contain.

    (a) Find N.

    (b) A contestant wins the game by selecting 3 balls of each color. What is the probability of winning?

    (c) Explain why the probability of winning is not $\frac{1}{N}$.

43. Show: $\forall\, k \geq 12$, $\binom{k+3}{k} - 4\binom{k}{k-3} + 6\binom{k-3}{k-6} - 4\binom{k-6}{k-9} + \binom{k-9}{k-12} = 0$ by using the identity $(1 + x + x^2)^4 = (1 - x^3)^4 \frac{1}{(1-x)^4}$. Hint: Consider coefficients.

44.* A quilt is to be made with the back solid white and the front a 4-by-4 grid using 4 red, 4 yellow, 4 blue, and 4 green squares.

    (a) How many ways are there to make such a quilt and hang it on the wall? Note that different orientations matter.

    (b) How many possibilities are there for this quilt? Here it may be freely rotated.

45. How many possibilities are there for a purchase of 8 balloons from a man selling 6 red, 4 white, and 5 blue balloons, if we must purchase at least one balloon of each color? We distinguish purchases solely by the numbers of balloons of each color that they contain.

46. How many solution triples (a, b, c) with $a, b, c \in \mathbb{N}$ are there for the equation $2a + 3b + 4c = 25$?

8 Basic Graph Theory

Graph theory might better be named network theory. It is the study of structures that model such things as computer networks and transportation networks. As a model for a computer network, a graph is built from two kinds of objects. The computers are represented by points, and each cable that provides a direct communication link between a pair of computers is represented by a curve. Adopting terminology from geometry, we call the points vertices and the curves edges. Very simply, a graph consists of a set of vertices, a set of edges, and a map that specifies the pair of vertices linked by each edge.

After providing some motivating examples, the formal definitions, and the basic terminology of graphs, we present several fundamental graphs. Matrices are then studied and seen to be more than merely a storage device for graph adjacencies. Next we discuss when two graphs should be considered equivalent and how to justify when they are not. Graph operations reminiscent of set operations are also studied. In the final section, we consider variations that result from assigning a particular direction to each edge. This leads to a discussion of Markov chains. Because graphs can be used to model many kinds of real-world problems, applications are explored throughout the chapter.

8.1 Motivation and introduction

The term *graph* typically brings to mind the graph of a function $y = f(x)$. However, a different notion of graph is considered here. Before giving a precise definition, we start with some examples that motivate our study.

8.1.1 Motivation

Our introduction to graph theory begins with the Königsberg Bridge Problem; this problem motivated the birth of the subject. In 1736, Königsberg was a city on the Pregel River in Prussia and was renowned for its beautiful bridges. (Königsberg is now Kaliningrad, Russia.) Figure 8.1 depicts the layout of Königsberg, with the bridges labeled 1 through 7 and the land masses labeled A through D.

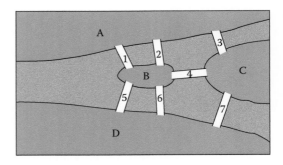

FIGURE 8.1
The seven bridges of Königsberg.

The people of Königsberg considered the following question.

Example 8.1. (Königsberg Bridge Problem).
Is it possible, starting from some place in Königsberg, to go for a walk that passes over each bridge exactly once and returns to the starting place?

In 1736, the Swiss mathematician Leonard Euler (1707–1782) provided a clever solution to the Königsberg Bridge Problem and others like it, by analyzing it in terms of the diagram in Figure 8.2. There, each land mass is represented by a point •, and each bridge is represented by a curve joining the appropriate points. The attached labels correspond to the labels from Figure 8.1. A desired walk through Königsberg could be appropriately traced in Figure 8.2. However, one does not exist. An exhaustive argument using Figure 8.2 to show that no such walk exists is outlined in the exercises. A much more elegant proof is given in Section 9.2, after Euler's Theorem is presented.

Figure 8.2 (without the labels) depicts our first example of a **graph**. Such diagrams arise naturally in many contexts, and often the points represent locations or objects. The next example is motivated by the type of diagram one typically sees in the complementary magazine provided by an airline in the back-of-seat pockets.

Example 8.2. (Airline Service).
Figure 8.3 is a map displaying the direct round-trip flights between major cities offered by a particular airline. The essential information being conveyed by that map is displayed more efficiently in the graph in Figure 8.4. There, points represent cities (or airports) and each curve represents the existence of a two-way direct flight between a pair of cities. To plan a trip involving these cities, one needs to determine which

FIGURE 8.2
Königsberg bridge graph.

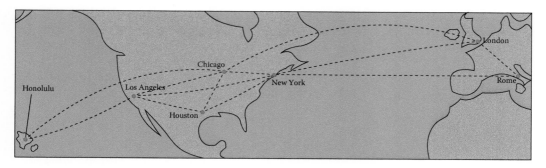

FIGURE 8.3
In-flight magazine ad.

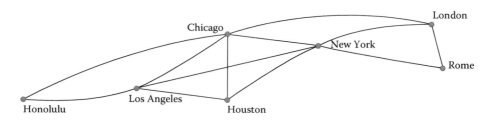

FIGURE 8.4
Direct flights graph.

direct flights to take. Costs and lengths of these flights may also need to be considered. Section 10.3 addresses problems of this sort.

In our next motivating example, the points in our graph do not represent physical locations.

Example 8.3.   (Scheduling Conflicts).
Six classes (Astronomy, Biology, Calculus, Discrete Math, English Composition, and French) need to schedule study group sessions. In the displayed table, an X denotes that the two different classes corresponding to that row and column have a student in common.

	Astr.	Bio.	Calc.	Discr.	Eng.	Fr.
Astr.		X			X	X
Bio.	X		X	X	X	
Calc.		X		X	X	
Discr.		X	X		X	
Eng.	X	X	X	X		X
Fr.	X				X	

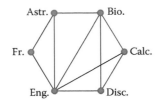

FIGURE 8.5
Study scheduling graph.

Figure 8.5 then depicts the potential scheduling conflicts. There, each class is represented by a point and each curve joins a pair of points corresponding to classes with students in common. Section 9.5 discusses how the graph in Figure 8.5 can be used to find an appropriate schedule.

Example 8.4. (Networking Computers).

Five computers, labeled $1, \ldots, 5$, have been networked together to complete a job in parallel by passing tasks among the computers. In Figure 8.6, a curve between a pair of computers represents the fact that the two computers can directly pass a task between them. The communication links are two-way, so either computer can assign a task to, or receive a task from, the other computer. The loop joining computer 5 to itself denotes that this computer may assign a task to itself. Figure 8.7 displays a graph that reflects the task-passing capabilities of this computer network. Each computer is represented by a point, and each curve represents the ability to pass a task directly between the computers reflected by its endpoints.

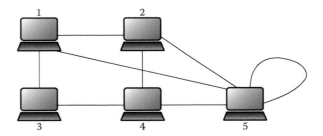

FIGURE 8.6
Task passing in a computer network.

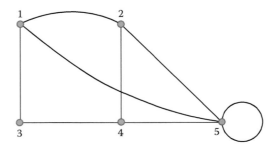

FIGURE 8.7
Computer network graph.

8.1.2   Introduction

Informally, we have seen some examples of graphs in Figures 8.2, 8.4, 8.5, and 8.7. From those examples, we see that the two important features of a graph are the points and the curves joining them. Adopting terminology from geometry, we shall call the points **vertices** and the curves **edges**. For the most part, each edge we have seen joins two distinct vertices. Thus, it is natural to assign such an edge e to the 2-element set $\{u, v\}$ containing this pair of vertices. However, we must take care to allow the possibility that two edges may join the same pair of vertices, as in the Königsberg bridge graph in Figure 8.2. Also, we must handle the case in which an edge joins a vertex to itself, as in the computer network graph in Figure 8.5. In that case, we assign such an edge e to the 1-element set $\{u\}$ containing the vertex joined to itself by e. Letting V denote the set of vertices, we see that each edge is assigned to either a 1-element or 2-element subset of V.

For each natural number k, it will be convenient to denote the set of k-element subsets of a set V by $\mathcal{P}_k(V)$. Note, in particular, that each element of $\mathcal{P}_1(V) \cup \mathcal{P}_2(V)$ may be written in the form $\{u, v\}$ for some $u, v \in V$; if $u = v$, then $\{u, v\} = \{u\} \in \mathcal{P}_1(V)$, and otherwise $\{u, v\} \in \mathcal{P}_2(V)$. Letting E denote the set of edges, we see that each $e \in E$ must therefore be assigned to a set of the form $\{u, v\}$, where $u, v \in V$ are the vertices we understand to be joined by e.

We are now ready to formally define a graph G in terms of a vertex set V, an edge set E, and a function ϵ specifying the vertices joined by each edge. Despite the formalism required for this, it is nonetheless important to keep in mind our informal understanding of the structure of a graph, as developed in Examples 8.1 through 8.4.

Definition 8.1.

(a) A **graph** G consists of a pair of sets V_G and E_G together with a function $\epsilon_G : E_G \longrightarrow \mathcal{P}_2(V_G) \cup \mathcal{P}_1(V_G)$. We write $G = (V_G, E_G)$, and, rather than writing $\epsilon_G(e) = \{u, v\}$, we write $e \mapsto \{u, v\}$. An element of V_G is called a **vertex** of G, and an element of E_G is called an **edge** of G. When a particular graph G is clear in context, the subscripts are dropped from V, E, and ϵ.

(b) If $e \mapsto \{u, v\}$, then we say that the vertices u and v are the **endpoints** of the edge e, that u and v are **adjacent**, and that v is a **neighbor** of u. We say that an edge is **incident** with its endpoints.

(c) An edge e such that $e \mapsto \{v\}$, for some $v \in V$, is called a **loop**; it has a single endpoint. Two or more edges

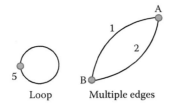

FIGURE 8.8
Features not in a simple graph.

assigned the same set of endpoints are called **multiple edges** (or **parallel edges**).

(d) A **simple graph** is a graph $G = (V, E)$ that has no loops and no multiple edges.

In the computer network graph G in Figure 8.7, the vertex set V is the set of computers $\{1, 2, 3, 4, 5\}$, and the edge set E is the set of direct links enabling the passing of a task from one computer to another. That is, two computers are adjacent in G if and only if a task can be passed directly between them. As shown on the left-hand side of Figure 8.8, G has a loop at vertex 5, reflecting the fact that computer 5 may assign a task to itself. Because of the presence of that loop, G is not a simple graph.

In the Königsberg bridge graph in Figure 8.2, the vertex set V is the set of land masses $\{A, B, C, D\}$, and the edge set E is the set of bridges $\{1, 2, 3, 4, 5, 6, 7\}$. Thus, the endpoints of a bridge are the land masses it joins. As shown on the right-hand side of Figure 8.8, G has multiple edges joining vertices A and B, reflecting the multiple bridges joining land masses A and B. Because of the presence of multiple edges, G is not a simple graph.

The direct flights graph in Figure 8.4 and the study scheduling graph in Figure 8.5 are simple graphs. Both contain neither loops nor multiple edges, as can be seen in their pictures.

Having given the formal definition of a graph and interpreted our earlier examples in terms of it, we should also understand how a graph is presented directly in terms of Definition 8.1. Thus, we consider some examples presented via a vertex set V, an edge set E, and an assignment of edges to their endpoints. Throughout this book, only finite graphs $G = (V, E)$ are considered. That is, V and E are finite.

Remark 8.1.   When a graph $G = (V, E)$ has no multiple edges, we typically regard E as a subset of $\mathcal{P}_1(V_G) \cup \mathcal{P}_2(V_G)$ and understand the assignment of endpoints to be given by the identity $\epsilon(\{u, v\}) = \{u, v\}$. That is, when $E \subseteq \mathcal{P}_1(V_G) \cup \mathcal{P}_2(V_G)$, we shall understand that each edge is the set $\{u, v\}$ of its endpoints. In particular, this is done for simple graphs.

Example 8.5.   In each case below, $G = (V, E)$ is an example of a graph.

(a) Let $V = \{1, 2, 3, 4, 5, 6, 7\}$ and
$E = \{\{1, 2\}, \{2, 3\}, \{1, 4\}, \{2, 5\}, \{4, 5\}, \{4, 6\}, \{5, 6\}, \{5, 7\}\}$.

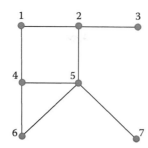

We identify each edge here with a pair of vertices. This is a simple graph, since it contains neither loops nor multiple edges.

(b) Let $V = \{1,2,3,4,5,6,7\}$ and
$E = \{\{1,2\}, \{1,4\}, \{2,5\}, \{4,5\}, \{4,6\}, \{5,6\}, \{3,7\}\}$.

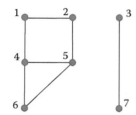

This is a simple graph. Note that this particular graph comes in multiple pieces, a notion we shall formalize in Definition 8.5.

(c) Let $V = \{1,2,3,4,5\}$ and
$E = \{\{1,2\}, \{1,5\}, \{1,3\}, \{2,4\}, \{2,5\}, \{3,4\},$
$\{4,5\}, \{5\}\}$.

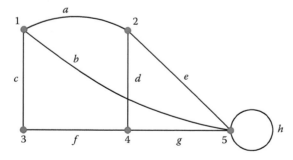

In the picture, we have $a = \{1,2\}$, $b = \{1,5\}$, $c = \{1,3\}$, $d = \{2,4\}$, $e = \{2,5\}$, $f = \{3,4\}$, $g = \{4,5\}$, and $h = \{5\}$. Of course, this is same as the computer network graph in Figure 8.7. Because of the loop edge h, it is not a simple graph.

(d) Let $V = \{1, 2, 3, 4, 5\}$ and $E = \{a, b, c, d, e, f, g, h\}$, where
$a \mapsto \{1, 2\}$, $\quad b \mapsto \{1, 2\}$, $\quad c = \{1, 3\}$, $\quad d = \{2, 4\}$,
$e = \{2, 5\}$, $\quad f = \{3, 4\}$, $\quad g = \{4, 5\}$, $\quad h \mapsto \{5\}$, and
$i \mapsto \{5\}$.

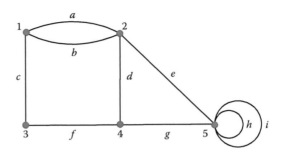

Here, we appreciate the value of the notation $\mapsto$ assigning endpoints to an edge. Both edges a and b need the same endpoints, but we want a and b to be distinct. The same is true of h and i. Because of these multiple edges, the graph is not simple.

We have presented several drawings of graphs so far. What properties must such a drawing have? Notice in the picture of the graph in Example 8.5(c) that two edges may cross at a point that does not represent a vertex. The formal restrictions on a drawing are specified as follows.

Definition 8.2. A **drawing** of a graph $G = (V, E)$ in the plane is a one-to-one assignment of the vertices to points in the plane and, for each edge, the assignment of a curve joining the ends of the edge in such a way that

(i) The only vertex points hit by a curve are the endpoints of the edge it represents,

(ii) Each curve is one-to-one (i.e., does not intersect itself) with the exception that the ends of a loop edge are assigned to a common point, and

(iii) The images of curves associated with two distinct edges intersect in at most finitely many points.

An intersection of two curves outside of their endpoints is called a **crossing**.

Figures 8.2, 8.4, 8.5, and 8.7 and all of the pictures in Example 8.5 are drawings. Only the one in Figure 8.7 (repeated in Example 8.5(c)) contains a crossing. However,

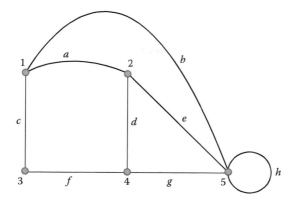

FIGURE 8.9
Alternate drawing for Example 8.5(c).

Figure 8.9 shows a different drawing of that same graph without a crossing. Some graphs can be drawn without crossings but others cannot, as we shall explore in Section 9.4.

8.1.3   Parts of a graph

In some situations, we may be interested only in part of a given graph. In the direct flights graph in Figure 8.4, for example, we may be interested only in traveling within the United States, or we may seek a vacation itinerary that includes only some other subset of cities. We may aim to avoid certain flights, such as the long flight between Honolulu and Chicago. In these cases, we choose to focus on a subgraph of a larger graph.

Definition 8.3.   We say that a graph $H = (W, F)$ is a **subgraph** of a graph $G = (V, E)$ if $W \subseteq V$, $F \subseteq E$, and the endpoints of edges in F all lie in W and are the same as they are in G. Given a subset W of the vertex set V for a graph $G = (V, E)$, the **subgraph induced by** W is the subgraph whose edge set is

$$\{e \ : \ e \in E \text{ and the ends of } e \text{ are in } W\}.$$

The subgraph of the direct flights graph whose vertices are all of the U.S. cities and whose flights are all of the flights between U.S. cities except for the flight between Honolulu and Chicago is shown in Figure 8.10.

Example 8.6.   Let $G = (V, E)$ be the graph from Example 8.5(d). In each of the following pictures, the described subgraph H is shown in bold.

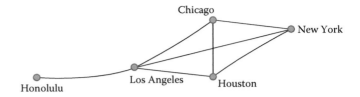

FIGURE 8.10
U.S. vacation subgraph.

(a) Let $W = \{1,2,4,5\}$ and $F = \{b,d,e\}$. So, $H = (W,F)$ is a subgraph of G.

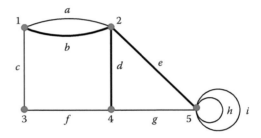

(b) Let H be the subgraph of G induced by $W = \{1,2,4,5\}$.

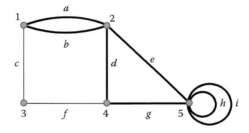

The subgraph in part (b) is induced by its vertex set; it contains every edge that has both ends in W. The subgraph in part (a) is not an induced subgraph, since the edges a, g, h, and i are not included and their endpoints are in W.

Given two vertices in a graph, we might seek the ways to travel between them in the graph. In the direct flights graph in Figure 8.4, we may want to explore our options for traveling from Los Angeles to London, or we may consider possible vacations that start in and return to Los Angeles. In these cases, we seek various kinds of walks in our graph.

Definition 8.4.

(a) A **walk** in a graph $G = (V, E)$ is an alternating list of vertices and edges

$$v_0, e_1, v_1, e_2, v_2, e_3, \ldots, v_{n-1}, e_n, v_n$$

with $n \geq 0$ that **starts** at vertex v_0, **ends** at vertex v_n, and, in which, for each $1 \leq i \leq n$, $e_i \mapsto \{v_{i-1}, v_i\}$. The **length** of a walk is the number of edges it contains (counting multiple occurrences of the same edge), here n. In a graph without multiple edges, a walk can more efficiently be expressed by a list of vertices $v_0, v_1, \ldots, v_n$ such that for each $1 \leq i \leq n$, $\{v_{i-1}, v_i\} \in E$. There, we identify E with the subset $\epsilon(E) \subseteq \mathcal{P}_2(V) \cup \mathcal{P}_1(V)$.

(b) A **circuit** is a walk of positive length that starts and ends at the same vertex.

(c) A **trail** is a walk with no repeated edges. In a graph with multiple edges, distinct multiple edges may be included.

(d) A **path** is a walk with no repeated vertices.

(e) A **cycle** is a circuit in which the only vertex repetition is $v_n = v_0$.

(f) The **distance** between two vertices u and v in G, denoted $\text{dist}_G(u, v)$, is the length of the shortest walk in G between u and v. If there is no walk, then we assign $\text{dist}_G(u, v) = \infty$. When G is clear in context, the subscripts may be dropped.

The relationships among the various kinds of walks in Definition 8.4 are reflected by the Venn diagram in Figure 8.11.

In the direct flights graph, the walk

Los Angeles to New York to London to Rome to New York to Chicago to Los Angeles

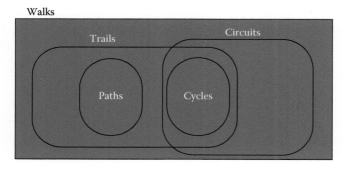

FIGURE 8.11
Kinds of walks.

represents a circuit, since it starts and ends in Los Angeles. Since New York is repeated, this walk is not a cycle. However, since no edge is repeated, it is a trail. Although this walk uses 3 edges to get from Los Angeles to Rome, dist(Los Angeles, Rome) = 2, since there is a path

Los Angeles to New York to Rome

of length 2, and there is no shorter path.

Example 8.7. Given the pictured graph $G = (V, E)$ from Example 8.5(d),

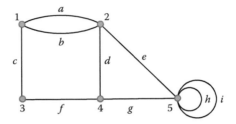

(a) $1, b, 2, e, 5, h, 5, e, 2, d, 4$ is a walk of length 5 that starts at 1 and ends at 4.

(b) $1, b, 2, e, 5, h, 5, g, 4, d, 2, a, 1$ is a circuit of length 6 that starts and ends at 1.

(c) $2, a, 1, b, 2, e, 5, h, 5$ is a trail of length 4 that starts at 2 and ends at 5.

(d) $2, b, 1, c, 3, f, 4$ is a path of length 3 that starts at 2 and ends at 4.

(e) $2, b, 1, c, 3, f, 4, d, 2$ is a cycle of length 4 that starts and ends at 2.

(f) $\text{dist}(2, 4) = 1$, $\text{dist}(3, 5) = 2$, and $\text{dist}(5, 5) = 0$.

Note that the walk in Example 8.7(a) from vertex 1 to vertex 4 is not a path, since it repeats vertex 2, for instance. However, by excising the portion of that walk between those two occurrences of vertex 2, we obtain a shorter walk $1, b, 2, d, 4$ from vertex 1 to vertex 4 that happens to be a path in this case. This surgical technique on walks can be used to prove the following.

Lemma 8.1. Let u and v be vertices in a graph G such that there exists a walk from u to v in G. Then there exists a path from u to v in G.

Proof. Let $v_0, e_1, v_1, e_2, \ldots, v_{n-1}, e_n, v_n$ be a walk in G from $v_0 = u$ to $v_n = v$ with the shortest possible length n. We claim that this walk is our desired path. So suppose to the contrary that it contains a repeated vertex, say, $v_i = v_j$ for some $i < j$. However, $v_0, e_1, \ldots, v_i, e_{j+1}, v_{j+1}, \ldots, e_n, v_n$ forms a shorter walk from u to v. This is a contradiction. Therefore, we have a path from u to v.   □

The proof of the following theorem is left to the exercises. Lemma 8.1 can be used to prove part (c).

Theorem 8.2. (Being Joined by a Path Is an Equivalence Relation). Let u, v, and w be vertices in a graph G.

(a) *There is a path from u to u.*

(b) *If there is a path from u to v, then there is a path from v to u.*

(c) *If there is a path from u to v and a path from v to w, then there is a path from u to w.*

The equivalence relation in Theorem 8.2 partitions the vertex set of a graph into equivalence classes. Each equivalence class of vertices then induces a subgraph comprising a piece of the graph. Some graphs come in one piece and are said to be connected. Others split into more than one piece. In general, the pieces are called components, as the following definition makes precise.

Definition 8.5.

(a) A graph G is **connected** if, for any two vertices, there is a path between them. Otherwise, G is **disconnected**.

(b) A **component** of a graph G is a connected subgraph H that is not properly contained in any other connected subgraph of G; two vertices are in the same component if and only if there is a path between them.

Very simply, a component of a graph is a connected subgraph to which no edges and vertices from the graph can be added while leaving a connected subgraph. In particular, components are always induced subgraphs. However, Example 8.6(b) shows that a connected induced subgraph is not always a component.

Example 8.8.   The graphs in Example 8.5 (a), (c), and (d) are all connected and therefore have a single component. The graph in Example 8.5(b) is disconnected and has two components.

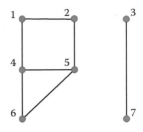

One is induced by $\{1, 2, 4, 5, 6\}$, and the other is induced by $\{3, 7\}$.

Exercises

1. Notre Dame Cathedral is located on the Ile de la Cité in the portion of Paris pictured below. The bridge labeled 9 in this picture is the famous Pont Neuf.

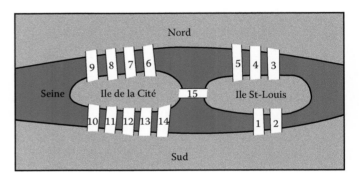

Draw the graph that reflects the bridge crossings in this portion of Paris.

2. Pictured is a portion of Rome through which the Tiber river flows. Vatican City is also near the Tiber river.

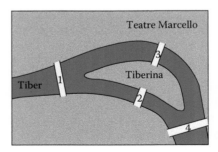

1. Ponte Garibaldi
2. Ponte Crestio
3. Ponte Fabricio
4. Ponte Palatino

Draw the graph that reflects the bridge crossings in this portion of Rome.

3. The classes Algebra, Business, Communications, Differential Equations, Economics, and Finance each need to schedule a study session. In the displayed table, an X

denotes the fact that the two different classes corresponding to that row and column have a student in common.

	Alg.	Bus.	Comm.	Diff. Eq.	Econ.	Fin.
Alg.		X	X		X	X
Bus.	X		X	X		X
Comm.	X	X		X	X	
Diff. Eq.		X	X		X	X
Econ.	X		X	X		X
Fin.	X	X		X	X	

Draw the corresponding study session scheduling graph.

4. At a certain high school, there are several school committees that need to hold meetings: Athletics, Budget, Cleanup,
   Dance, Entertainment, Fund Raising, Class Gift Selection,
   and Homecoming. In the displayed table, an X denotes
   the fact that the two different committees corresponding
   to that row and column have a member in common.

	Ath.	Budg.	Clean.	Dance	Ent.	Fund	Gift	Home.
Ath.		X		X	X	X		
Budg.	X		X			X		X
Clean.		X		X	X		X	
Dance	X		X				X	X
Ent.	X		X			X	X	
Fund	X	X			X			X
Gift			X	X	X			X
Home.		X		X		X	X	

Draw the corresponding committee scheduling graph.

5. Pictured is a map of the subway system in a certain city.
   It contains three separate lines, each specified by a color.
   Transfers can be made at stops sitting on more than one
   line.

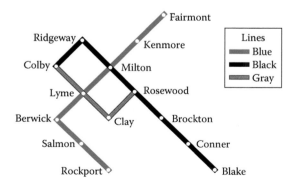

Draw the graph representing the entire subway system.

6. Pictured is a map of the bus system in a certain city. There are two bus routes. However, they do overlap in the busiest part of the city.

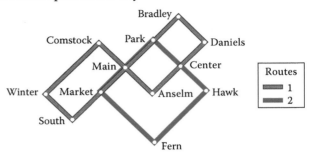

Draw the graph representing the entire bus system.

The following graphs $G = (V, E)$ are referred to throughout the exercises.

(a) $V = \{1, 2, 3, 4, 5\}$ and $E = \{\{1, 3\}, \{1, 5\}, \{2, 3\}, \{2, 4\}, \{3, 4\}, \{3, 5\}\}$.

(b) $V = \{1, 2, 3, 4, 5, 6\}$ and $E = \{\{1, 2\}, \{2, 3\}, \{3, 4\}, \{4, 5\}, \{6\}, \{2, 4\}\}$.

(c) $V = \{1, 2, 3, 4\}$ and $E = \{a, b, c, d, e\}$, where $a \mapsto \{1, 3\}$, $b \mapsto \{2, 3\}$, $c \mapsto \{2, 4\}$, $d \mapsto \{1, 3\}$, and $e \mapsto \{3, 4\}$.

(d) $V = \{1, 2, 3\}$ and $E = \{a, b, c\}$, where $a \mapsto \{1, 2\}$, $b \mapsto \{2, 3\}$, and $c \mapsto \{3, 1\}$.

In Exercises 7 through 10, draw the specified graph and determine whether it is simple. If not, explain why.

7. Graph (a).   8. Graph (b).

9. Graph (c).   10. Graph (d).

The following simple graphs are referred to throughout the remaining exercises. Each edge is taken to be the appropriate 2-element set containing its endpoints.

(e)   (f) 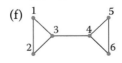   (g) 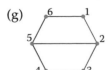   (h)

In Exercises 11 through 14, determine whether a subgraph $H = (W, F)$ is specified. If not, explain why.

11. In graph (e), take $W = \{2, 4, 6\}$ and $F = \{\{2, 4\}, \{4, 6\}\}$.

12. In graph (f), take $W = \{1, 2, 3, 4\}$ and $F = \{\{1, 3\}, \{2, 3\}, \{2, 4\}\}$.

13. In graph (c), take $W = \{1,2,3\}$ and $F = \{a,d,e\}$.

14. In graph (d), take $W = \{1,2,3\}$ and $F = \emptyset$.

In Exercises 15 through 18, determine the edge set for the specified subgraph.

15. In graph (g), the subgraph induced by $W = \{1,2,3,5\}$.

16. In graph (h), the subgraph induced by $W = \{2,3,4,5\}$.

17. In graph (a), the subgraph induced by $W = \{1,3,5\}$.

18. In graph (b), the subgraph induced by $W = \{2,4,6\}$.

19. Consider the study group scheduling problem in Exercise 3. Due to test schedules, during a particular week of the semester, only the classes Algebra, Business, Differential Equations, and Finance need to schedule study sessions. Draw the subgraph induced by those vertices that needs to be considered.

20. Consider the high school committee scheduling problem in Exercise 4. If just the committees on Athletics, Cleanup, Entertainment, Class Gift Selection, and Homecoming need to schedule meetings, then only the subgraph induced by those vertices needs to be considered. Draw that subgraph.

In Exercises 21 through 26, determine if the given picture is the drawing of a graph. If not, then explain why.

21.   22.

23.   24. ●——⬤⬤

25. ◁———▷   26. ◁⬤⬤▷

27. In graph (a), is there a path from 1 to 4?

28. In graph (b), is there a path from 2 to 6?

29. Show that every graph with at least one edge contains walks that are not paths.

30. Show that every graph that is not simple contains a cycle.

31. Does graph (a) contain a cycle?

32. Does graph (b) contain a cycle?

33. In graph (f), find dist$(1,4)$.

34. In graph (g), find dist$(1,5)$.

35. In graph (c), find dist$(1,4)$.

36. In graph (d), find dist$(1,3)$.

37. Consider the city subway system graph from Exercise 5. In that graph, what is the distance between Berwick and Connor? Note that this need not reflect a physical distance.

38. Consider the city bus system graph from Exercise 6. In the corresponding graph, what is the distance between Winter and Hawk? Note that this need not reflect a physical distance.

Exercises 39 through 42 refer to the following pictured simple graph. In this case, walks can be specified by just a list of vertices.

39. Is $5, 1, 2, 3$ a path?   40. Is $2, 5, 1, 2, 3$ a path?

41. Is $4, 1, 2, 5, 1, 4$ a cycle?   42. Is $5, 2, 1, 4, 5$ a cycle?

43. Consider the city subway system graph from Exercise 5.

(a) Express the gray line as a path starting at Colby.

(b) Does the black line similarly form a path?

44. Consider the city bus system graph from Exercise 6.

(a) Express bus route number 2 as a cycle starting with the leg from Main to Market.

(b) Does bus route number 1 similarly form a cycle?

45*. Let u, w be any vertices in a graph. Show: If $\text{dist}(u, w) \geq 1$, then there is a vertex v that is a neighbor of w such that $\text{dist}(u, v) = \text{dist}(u, w) - 1$.

46*. Let u, v, w be arbitrary vertices in a graph. Show:

(a) $\text{dist}(u, u) = 0$.

(b) $\text{dist}(u, v) = \text{dist}(v, u)$.

(c) $\text{dist}(u, w) \leq \text{dist}(u, v) + \text{dist}(v, w)$.

47*. Show: If there are vertices u and v in a graph G such that two distinct paths exist from u to v, then G contains a cycle.

48. Prove Theorem 8.2.

49*. Use Figure 8.2 to show that the Königsberg Bridge Problem has no solution. Hint: You can assume that the walk starts at C. Consider cases while taking advantage of symmetry. Argue that you will get stuck somewhere.

50. In the Königsberg Bridge Problem, what is the greatest number of bridges that can be covered in a circuit that

starts and finishes in the same place without repeating bridges?

51. (a) In Figure 8.4, for what values of n does there exist a path of length n from Honolulu to London?

   (b) What is the distance from Honolulu to London?

52. Which cities in Figure 8.4 have the property that no other city is a distance of more than 2 away?

53. In Example 8.3, which study groups could meet at the same time as the study group for Astronomy?

54. In Example 8.3, is it generally true that if study groups for classes 1 and 2 cannot meet at the same time and study groups for classes 2 and 3 cannot meet at the same time, then study groups for classes 1 and 3 cannot meet at the same time?

8.2   Special graphs

There are several examples of graphs that will be useful to keep in mind as we explore new concepts in subsequent sections. Some provide extreme cases of general results. Some are particularly important when they occur as subgraphs of other graphs. Some are interesting on their own.

Definition 8.6.   Let $n \in \mathbb{Z}^+$, and let $V = \{1, 2, \ldots, n\}$.

   (a) The **path on n vertices** is the graph P_n with vertex set V and edge set $E = \{\{1,2\}, \{2,3\}, \ldots, \{n-1,n\}\}$.

   (b) The **cycle on n vertices** is the graph C_n with vertex set V and edge set $E = \{\{1,2\}, \{2,3\}, \ldots, \{n-1,n\}, \{n,1\}\}$. When $n = 2$, we take C_2 to have two parallel edges.

The graphs C_1 and C_2 are the only cycles that are not simple graphs.

Example 8.9.   Examples of paths and cycles are pictured.

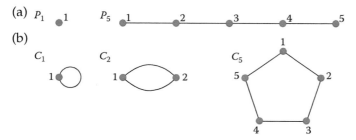

(a) P_1 $\bullet^1$ P_5 $\bullet^1 \underline{\quad\quad} \bullet^2 \underline{\quad\quad} \bullet^3 \underline{\quad\quad} \bullet^4 \underline{\quad\quad} \bullet^5$

(b) C_1 C_2 C_5

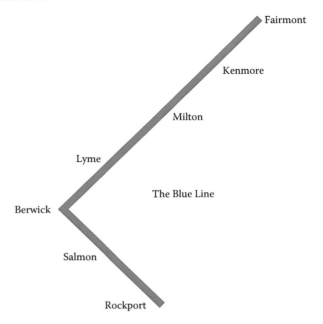

FIGURE 8.12
A line of a subway.

The map for a city subway line shown in Figure 8.12 would be represented by a graph that is a path, namely, P_7. The computer network configuration displayed in Figure 8.13 is reflected by a cycle, namely, C_4. Note the slightly awkward fact that P_n has length $n - 1$. However, C_n has length n if we express it as a trail starting and ending at a particular vertex.

Paths and cycles contain relatively few edges, especially among connected graphs. The most edges a simple graph on n vertices can have are $\binom{n}{2}$.

Theorem 8.3. *Given a simple graph $G = (V, E)$,*

(a) *If $|V| \leq 1$, then $|E| = 0$.*

(b) *If $|V| \geq 2$, then $0 \leq |E| \leq \binom{|V|}{2}$.*

Proof. If $|V| \leq 1$, then $\mathcal{P}_2(V) = \emptyset$. If $|V| \geq 2$, then $|\mathcal{P}_2(V)| = \binom{|V|}{2}$. The result follows from the fact that $E \subseteq \mathcal{P}_2(V)$.   □

At one extreme, we consider simple graphs that contain all possible edges.

FIGURE 8.13
A cyclic computer network.

Definition 8.7. Given an integer $n \geq 1$, the **complete graph on n vertices** is the simple graph K_n with vertex set $V = \{1, 2, \ldots, n\}$ and edge set $E = \mathcal{P}_2(V)$. Any graph in which every pair of distinct vertices is adjacent is said to be **complete**.

Example 8.10. K_5 is pictured.

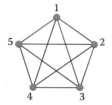

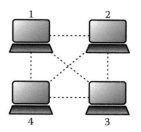

FIGURE 8.14
Wireless computer network.

The graph K_n might represent a computer network in which every pair of computers is directly connected. For example, a wireless network could have this property if all of the computers are close enough together. The wireless computer network configuration displayed in Figure 8.14 is reflected by a complete graph, namely, K_4.

It is also convenient to have notation for graphs without any edges.

Definition 8.8.   Given any $n \in \mathbb{N}$, the **empty graph on n vertices** is the graph Φ_n with vertex set $V = \{1, 2, \ldots, n\}$ and edge set $E = \emptyset$. Any graph in which no pair of vertices is adjacent is said to be **empty**.

In some graphs, the vertex set naturally splits as a disjoint union of two sets.

Definition 8.9.   A graph $G = (V, E)$ is **bipartite** if V can be expressed as a disjoint union $V_1 \cup V_2$ such that each edge of G has one endpoint in V_1 and one in V_2. In this case, the pair (V_1, V_2) is said to form a **bipartition** of G. Note that for $i = 1$ or 2, the subgraph induced by V_i is empty.

Bipartite graphs arise in applications in which the vertices reflect two kinds of objects that need to be "matched." For example, a manager may form a graph in which some vertices represent employees and others represent tasks. An edge then connects an employee to a task if and only if that employee is qualified to perform that task. This graph might be used to determine which employees should be assigned which tasks so that all tasks are accomplished while making the most efficient use of the work force. Such an example for

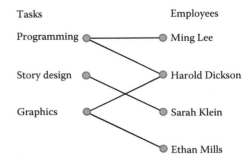

FIGURE 8.15
Possible task assignments.

a small company that designs computer games is shown in Figure 8.15. A study of matching and marriage type problems can be found in Pólya et al. (1983) and Tucker (2011).

Example 8.11.   The graph

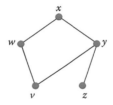

is bipartite. Its bipartition is given by $V_1 = \{v, x, z\}$ and $V_2 = \{w, y\}$. Observe that each edge joins a vertex in V_1 with a vertex in V_2.

An alternate drawing of the graph from Example 8.11 is shown in Figure 8.16. There, the bipartition sets are pictured with one on the left-hand side and the other on the right-hand

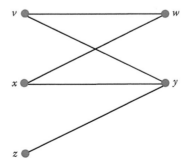

FIGURE 8.16
Standard display of a bipartite graph.

side. The fact that edges only connect vertices from different sides is what makes the graph bipartite.

Certainly, not every graph is bipartite.

Example 8.12. C_3 is not bipartite.

Proof. Here, $V = \{1, 2, 3\}$. Suppose, toward a contradiction, that C_3 is bipartite with bipartition (V_1, V_2). Say, $1 \in V_1$. Since $\{1, 2\}$ and $\{1, 3\}$ are edges, we must have $2, 3 \in V_2$. However, $\{2, 3\}$ is also an edge. Hence, 2 and 3 cannot be in the same set V_2. This contradiction shows that C_3 cannot be bipartite.   □

The result in Example 8.12 is a special case of the following theorem.

Theorem 8.4.   *Let G be any graph. Then G is bipartite if and only if every cycle in G has even length.*

Proof. It suffices to consider only connected simple graphs.

($\rightarrow$) Suppose G is bipartite with bipartition (V_1, V_2), and let

$$v_0, v_1, \ldots, v_{n-1}, v_0$$

be a cycle of length n. Say, $v_0 \in V_2$. By definition of bipartite, the vertices must alternate between V_2 and V_1. Since v_0 follows v_{n-1}, it must be that $v_{n-1} \in V_1$. This implies that $n - 1$ is odd and n is even.

($\leftarrow$) Suppose every cycle in $G = (V, E)$ has even length. Pick a vertex v_0 and define $V_2 = \{v \in V : \text{dist}(v_0, v) \text{ is even}\}$ and $V_1 = V \setminus V_2$. Obviously, $V = V_1 \cup V_2$ is a disjoint union. So it remains to show that there are no edges joining a pair of vertices from the same set V_i.

Suppose to the contrary that u and v are from the same set V_i and that $\{u, v\}$ is an edge. Let P be the shortest path from v_0 to u, and let Q be the shortest path from v_0 to v with the most possible vertices in common with P. Thus, there must be a vertex w such that P and Q are identical from v_0 to w and disjoint after w.

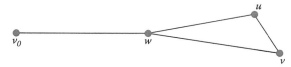

Since the left-hand sides of the two equations

$$\text{dist}(v_0, u) = \text{dist}(v_0, w) + \text{dist}(w, u)$$

$$\text{dist}(v_0, v) = \text{dist}(v_0, w) + \text{dist}(w, v)$$

have the same parity, their sum is even. We therefore see by adding these two equations that $\text{dist}(w, u) + \text{dist}(w, v)$ must be even. Thus, the cycle that follows P from w to u, takes the edge $\{u, v\}$ from u to v, and follows Q backward from v to w has odd length

$$\text{dist}(w, u) + 1 + \text{dist}(w, v).$$

This contradiction establishes our result.   □

At the extreme among simple bipartite graphs are those that contain all possible edges.

Definition 8.10.   Given integers $m, n \geq 1$ and sets

$$V_1 = \{(1, 1), (1, 2), \ldots, (1, m)\} \quad \text{and}$$

$$V_2 = \{(2, 1), (2, 2), \ldots, (2, n)\},$$

the **complete bipartite graph** $K_{m,n}$ has vertex set $V = V_1 \cup V_2$ and edge set $E = \{\{v_1, v_2\} : v_1 \in V_1 \text{ and } v_2 \in V_2\}$.

Example 8.13. $K_{3,4}$ is pictured.

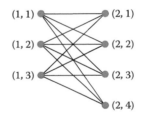

A complete bipartite graph would arise in an example like that shown in Figure 8.15 in which every employee is qualified to perform every task. In that case, some weights may additionally be attached to the edges to reflect preferences or relative abilities of employees to perform certain tasks. Weighted graphs are considered in Section 10.3.

Our final examples are inspired from geometry.

Example 8.14.   (The Platonic Solids).
There are exactly five solids for which each face is a fixed regular polygon and the same number of faces meets at each vertex. These solids are called the **platonic solids**. For each such solid, the graph formed from its vertices and edges is given the same name as the solid. Drawings of these graphs are displayed in Figure 8.17 in perspective to best reflect the connection with the solids. Alternative drawings are encountered later.

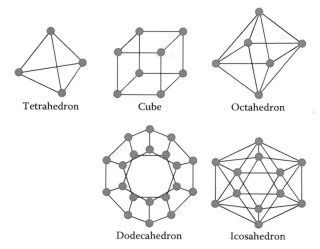

Tetrahedron   Cube   Octahedron

Dodecahedron   Icosahedron

FIGURE 8.17
The platonic solids' graphs.

The cube is the 3-dimensional example from a family of cubes that we now define.

Definition 8.11. The n-**dimensional cube** Q_n is the simple graph whose vertex set is the set of binary sequences of length n and whose edges connect two vertices if and only if they differ in exactly one coordinate.

The drawing of the 3-dimensional cube Q_3 shown in Figure 8.18 is obtained by regarding its vertices as points in $\mathbb{R}^3$. That picture indeed agrees with our drawing of the cube in Figure 8.17. We cannot draw objects of dimensions greater than 3. However, each graph Q_n can be drawn in the plane. Of course, such a drawing need not reflect a corresponding solid nor be esthetically pleasing. A standard drawing of Q_4 is shown in Figure 8.19.

Binary linear codes were introduced in Section 3.2. For instance, in Example 3.18, we considered the binary linear code displayed in Table 8.1. In general, each message, which is expressed as a binary string of some fixed length k, is assigned a code word. The code word is a binary string of some fixed length $n > k$ that contains the message in its first k digits. Each appended digit results from the sum of some of the digits of the message, taken mod 2. What is important here is the observation that the code words are vertices of the graph Q_n, and their spacing in Q_n determines the capability of correcting errors introduced in transmission. Specifically,

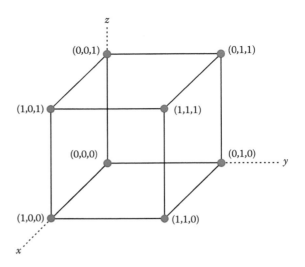

FIGURE 8.18
The graph Q_3.

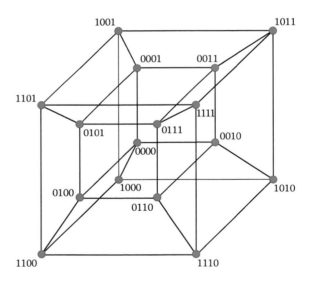

FIGURE 8.19
The graph Q_4.

the **weight** w of the code is the smallest distance between two vertices in Q_n that correspond to code words. A received word of length n also corresponds to a vertex v of Q_n. If v is a distance of no more than $\lfloor \frac{w-1}{2} \rfloor$ from a code word vertex c, then c will be the unique closest vertex to v in Q_n. The method of **nearest neighbor decoding** then uses the vertex c to decode the message. That is, the transmitted message

TABLE 8.1
A binary linear code for
3-digit messages

Message	Code word
000	000000
001	001011
010	010101
011	011110
100	100111
101	101100
110	110010
111	111001

is understood to be the first k digits of c. For example, if the word 010110 is received and the binary linear code in Table 8.1 is being used, then 011110 is seen to be the closest code word vertex in Q_6 to it. Hence, the message 011 is understood to be the intended one.

Exercises

In Exercises 1 through 10, draw the given graph.

1. P_4.   2. C_3.

3. C_6.   4. P_6.

5. K_6.   6. $K_{1,3}$.

7. $K_{3,2}$.   8. K_7.

9. Q_1.   10. Q_2.

11. Consider the subway system graph from Exercise 5 in Section 8.1.

 (a) What graph is the subgraph determined by the black line?

 (b) Does each subway line correspond to a path subgraph?

 (c) Give an example of a path subgraph that does not correspond to a subway line.

12. Consider the bus system graph from Exercise 6 in Section 8.1.

 (a) What graph is the subgraph determined by bus route number 2?

 (b) Does bus route number 1 correspond to a cycle subgraph?

(c) Give an example of a cycle subgraph that does not correspond to a bus route.

13. (a) Is the graph representing the portion of Paris from Exercise 1 in Section 8.1 complete?

(b) For that graph, give an example of a complete subgraph on three vertices.

14. (a) Is the graph representing the portion of Rome from Exercise 2 in Section 8.1 complete?

(b) Is it simple?

15. Is every subgraph of a complete graph complete? What about induced subgraphs?

16. Is every subgraph of an empty graph empty?

17. In the displayed table, an X denotes the fact that the math class is required by the particular major.

	C. S.	Math	Physics	Chemistry
Calculus	X	X	X	X
Diff Eq		X	X	
Discrete Math	X	X		
Linear Algebra	X	X	X	
Group Theory		X		X

Draw the corresponding bipartite graph.

18. In the displayed table, an X denotes the fact that the boy would like to ask the girl to the dance.

	Alice	Beth	Cindy	Dorothy
Albert	X		X	X
Bob		X	X	
Charles			X	
David	X	X	X	X

Draw the corresponding bipartite graph.

19. Is P_n bipartite? Justify your answer.

20. Prove that C_n is bipartite if and only if n is even.

21. Show that subgraphs of bipartite graphs are bipartite.

22. Prove or disprove: Subgraphs of nonbipartite graphs are nonbipartite.

23. Prove or disprove that the pictured graph is bipartite.

24. Prove or disprove that the pictured graph is bipartite.

In Exercises 25 through 34, determine the numbers of vertices and edges in the specified graph.

25. P_n.

26. C_n.

27. K_n.

28. $K_{m,n}$.

29. The cube.

30. The octahedron.

31. The dodecahedron.

32. The icosahedron.

33. Q_4.

34. Q_5.

35. Are 101101 and 101011 adjacent in Q_6?

36. What is the distance between 011011 and 110110 in Q_6?

37. A hospital database encodes the gender and blood type of each of its patients, as in Table 8.2.

   (a) Using the binary linear code in Table 8.3 and nearest neighbor decoding, decode the received word 1111010 to a message.

   (b) Determine the gender and blood type for this patient.

38. To deal with possible errors introduced in communication links, the army is communicating directions to its troops in the field using the symbols in Figure 8.20.

   (a) Using the binary linear code in Table 8.3 and nearest neighbor decoding, decode the received word 1001111 to a message.

   (b) Determine the compass direction being conveyed.

TABLE 8.2
Encoding gender and blood type

Gender, Blood	Message	Gender, Blood	Message
Female, O^-	0000	Male, O^-	1000
Female, O^+	0001	Male, O^+	1001
Female, B^-	0010	Male, B^-	1010
Female, B^+	0011	Male, B^+	1011
Female, A^-	0100	Male, A^-	1100
Female, A^+	0101	Male, A^+	1101
Female, AB^-	0110	Male, AB^-	1110
Female, AB^+	0111	Male, AB^+	1111

TABLE 8.3
A binary linear code for 4-digit messages

Message	Code word	Message	Code word
0000	0000000	1000	1000111
0001	0001011	1001	1001100
0010	0010101	1010	1010010
0011	0011110	1011	1011001
0100	0100110	1100	1100001
0101	0101101	1101	1101010
0110	0110011	1110	1110100
0111	0111000	1111	1111111

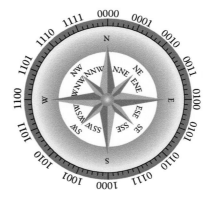

FIGURE 8.20
Direction encoding.

In Exercises 39 through 48, find the maximum possible distance between two vertices of the given graph. This number is called the **diameter** of the graph.

39. P_n.   40. C_n.

41. K_n.   42. $K_{m,n}$.

43. The cube.   44. The octahedron.

45. The dodecahedron. 46. The icosahedron.

47. Q_4.   48. Q_5.

8.3 Matrices

Recall that a matrix is a rectangular array of real numbers. In Section 5.1, we saw that a zero-one matrix can be used to represent a finite relation. In a similar way, matrices can be used to represent graphs. Moreover, matrices are useful for storing the structure of a graph in a computer. However, the value of matrices, as we shall see, goes beyond mere storage of adjacencies.

Definition 8.12.   An **adjacency matrix** for a graph G on n vertices is an n-by-n matrix $A = [a_{i,j}]$ obtained by fixing an ordering of the vertices, say, $v_1, v_2, \ldots, v_n$, and, for each $1 \le i \le n$ and $1 \le j \le n$, taking $a_{i,j}$ to be the number of edges connecting v_i to v_j.

Example 8.15.   If, for the graph

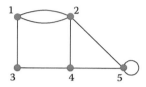

we take the ordering of the vertices to be $1, 2, 3, 4, 5$, then the adjacency matrix obtained is

$$
\begin{array}{c}
\begin{array}{ccccc} 1 & 2 & 3 & 4 & 5 \end{array} \\
\begin{array}{c} 1 \\ 2 \\ 3 \\ 4 \\ 5 \end{array}
\left[
\begin{array}{ccccc}
0 & 2 & 1 & 0 & 0 \\
2 & 0 & 0 & 1 & 1 \\
1 & 0 & 0 & 1 & 0 \\
0 & 1 & 1 & 0 & 1 \\
0 & 1 & 0 & 1 & 1
\end{array}
\right].
\end{array}
\tag{8.1}
$$

For added clarity, labels have been attached to the rows and columns. If, instead, we take the ordering $3, 2, 5, 1, 4$, then a different adjacency matrix is obtained.

$$
\begin{array}{c}
\begin{array}{ccccc} 3 & 2 & 5 & 1 & 4 \end{array} \\
\begin{array}{c} 3 \\ 2 \\ 5 \\ 1 \\ 4 \end{array}
\left[
\begin{array}{ccccc}
0 & 0 & 0 & 1 & 1 \\
0 & 0 & 1 & 2 & 1 \\
0 & 1 & 1 & 0 & 1 \\
1 & 2 & 0 & 0 & 0 \\
1 & 1 & 1 & 0 & 0
\end{array}
\right].
\end{array}
\tag{8.2}
$$

However, there is an important relationship between these two matrices.

The vertex list $3, 2, 5, 1, 4$ is a permutation of the initial list $1, 2, 3, 4, 5$. Moreover, the permutation $3, 2, 5, 1, 4$ can be described by saying that 3 goes to position 1, 2 goes to position 2, 5 goes to position 3, 1 goes to position 4, and 4 goes to position 5. We write

$$\begin{array}{ccccc}
3 & 2 & 5 & 1 & 4 \\
\downarrow & \downarrow & \downarrow & \downarrow & \downarrow \\
1 & 2 & 3 & 4 & 5
\end{array}.$$

In fact, if we first permute the rows of matrix (8.1) according to this permutation, and second permute the columns of the resulting matrix according to this same permutation, then we obtain matrix (8.2).

$$\begin{bmatrix}
0 & 2 & 1 & 0 & 0 \\
2 & 0 & 0 & 1 & 1 \\
1 & 0 & 0 & 1 & 0 \\
0 & 1 & 1 & 0 & 1 \\
0 & 1 & 0 & 1 & 1
\end{bmatrix}
\begin{array}{c}
\text{permute} \\
\text{rows as} \\
3, 2, 5, 1, 4 \\
\longrightarrow
\end{array}
\begin{bmatrix}
1 & 0 & 0 & 1 & 0 \\
2 & 0 & 0 & 1 & 1 \\
0 & 1 & 0 & 1 & 1 \\
0 & 2 & 1 & 0 & 0 \\
0 & 1 & 1 & 0 & 1
\end{bmatrix}$$

$$\begin{array}{c}
\text{permute} \\
\text{columns as} \\
3, 2, 5, 1, 4 \\
\longrightarrow
\end{array}
\begin{bmatrix}
0 & 0 & 0 & 1 & 1 \\
0 & 0 & 1 & 2 & 1 \\
0 & 1 & 1 & 0 & 1 \\
1 & 2 & 0 & 0 & 0 \\
1 & 1 & 1 & 0 & 0
\end{bmatrix}$$

A general observation that we can take from Example 8.15 is that different orderings of the vertices may yield distinct adjacency matrices. However, any two differ only by a fixed permutation applied to both the rows and the columns. Also observe that adjacency matrices for graphs are always symmetric (i.e., $\forall\, i, j$, $a_{i,j} = a_{j,i}$), since $a_{i,j}$, the number of edges connecting v_i to v_j, is the same as $a_{j,i}$, the number of edges connecting v_j to v_i.

With a knowledge of matrix multiplication, we can gain a greater appreciation for adjacency matrices.

Definition 8.13.   (Matrix Multiplication).
Given an m-by-n matrix $A = [a_{i,j}]$ and an n-by-p matrix $B = [b_{i,j}]$, their product AB is defined to be the m-by-p matrix $C = [c_{i,j}]$ such that for each $1 \le i \le m$ and $1 \le j \le p$, $c_{i,j} = \sum_{k=1}^{n} a_{i,k} b_{k,j}$.

Our interest here is in multiplying two square matrices. That is, for some n, both factors will be n-by-n matrices, and, consequently, so will their product.

Example 8.16.

(a) $\begin{bmatrix} 2 & 4 \\ 6 & 8 \end{bmatrix} \begin{bmatrix} 3 & 7 \\ 5 & 9 \end{bmatrix} = \begin{bmatrix} 26 & 50 \\ 58 & 114 \end{bmatrix}$

since

$$2 \cdot 3 + 4 \cdot 5 = 26, \ \ 2 \cdot 7 + 4 \cdot 9 = 50$$
$$6 \cdot 3 + 8 \cdot 5 = 58, \ \ 6 \cdot 7 + 8 \cdot 9 = 114.$$

(b) $\begin{bmatrix} 0 & 0 & 1 & 0 & 0 \\ 0 & 1 & 0 & 0 & 0 \\ 0 & 0 & 0 & 0 & 1 \\ 1 & 0 & 0 & 0 & 0 \\ 0 & 0 & 0 & 1 & 0 \end{bmatrix} \begin{bmatrix} 0 & 2 & 1 & 0 & 0 \\ 2 & 0 & 0 & 1 & 1 \\ 1 & 0 & 0 & 1 & 0 \\ 0 & 1 & 1 & 0 & 1 \\ 0 & 1 & 0 & 1 & 1 \end{bmatrix}$

$= \begin{bmatrix} 1 & 0 & 0 & 1 & 0 \\ 2 & 0 & 0 & 1 & 1 \\ 0 & 1 & 0 & 1 & 1 \\ 0 & 2 & 1 & 0 & 0 \\ 0 & 1 & 1 & 0 & 1 \end{bmatrix}$

For example, the entry in the 4th row and 2nd column of the product is 2, since

$$1 \cdot 2 + 0 \cdot 0 + 0 \cdot 0 + 0 \cdot 1 + 0 \cdot 1 = 2.$$

Mathematical software can be helpful for computing matrix products, especially when the matrices are large. In *Mathematica*, matrices are entered by listing their rows in list notation, and a period is used for multiplication. For instance, Example 8.16(a) is handled as follows.

```
In[1]  :=  A = {{2,4},{6,8}}
In[2]  :=  B = {{3,7},{5,9}}
In[3]  :=  A . B
Out[3] =   {{26,50},{58,114}}
```

The function `MatrixForm` can be used to display a matrix in standard form as an array with rows and columns. In *Maple*, Example 8.16(a) is handled similarly.

```
> A := Matrix([[2,4],[6,8]]);
> B := Matrix([[3,7],[5,9]]);
> A . B ;
```

$$\begin{bmatrix} 26 & 50 \\ 58 & 114 \end{bmatrix}$$

In this case, matrices are automatically displayed as arrays.
The matrix on the left-hand side of the product in Example 8.16(b) is an example from a useful class of matrices.

Definition 8.14. Let $n \in \mathbb{Z}^+$.

(a) The *n*-by-*n* **identity matrix** is the matrix $I_n = [a_{i,j}]$ such that

$$a_{i,j} = \begin{cases} 1 & \text{if } i = j, \\ 0 & \text{otherwise.} \end{cases}$$

(b) Given a permutation $p_1, p_2, \ldots, p_n$ of the integers $1, 2, \ldots, n$, the corresponding **permutation matrix** is the n-by-n matrix $P = [a_{i,j}]$ such that

$$a_{i,j} = \begin{cases} 1 & \text{if } p_i = j, \\ 0 & \text{otherwise.} \end{cases}$$

That is, P is obtained from I_n by permuting its rows according to the permutation $p_1, p_2, \ldots, p_n$.

Example 8.17.

(a) $I_2 = \begin{bmatrix} 1 & 0 \\ 0 & 1 \end{bmatrix}$.

(b) The permutation matrix corresponding to the permutation $3, 2, 5, 1, 4$ is the matrix

$$P = \begin{bmatrix} 0 & 0 & 1 & 0 & 0 \\ 0 & 1 & 0 & 0 & 0 \\ 0 & 0 & 0 & 0 & 1 \\ 1 & 0 & 0 & 0 & 0 \\ 0 & 0 & 0 & 1 & 0 \end{bmatrix}.$$

Letting A be the adjacency matrix (8.1) from Example 8.15, we see in Example 8.16(b) that the result of the product PA is to permute the rows of A according to the permutation $3, 2, 5, 1, 4$. Observe that the transpose P^T of P is the same as the matrix obtained by permuting the columns of I_5 according to the permutation $3, 2, 5, 1, 4$. Moreover, for any 5-by-5 matrix B, the result of the product BP^T is to permute the columns of B according to the permutation $3, 2, 5, 1, 4$. Putting this together with our preceding computation, it follows that the result of the product PAP^T is to permute both the rows and the columns of A according to the permutation $3, 2, 5, 1, 4$. The result is the adjacency matrix (8.2), constructed in terms of the ordering $3, 2, 5, 1, 4$ of the vertices.

The general observation to be taken from Example 8.17(b) is that if A is an adjacency matrix for a graph with respect to one ordering of its vertices and P is the permutation matrix corresponding to a permutation that takes this ordering of the vertices to another ordering, then PAP^T is the adjacency matrix for the graph with respect to this other ordering. Again, the rows and columns of A are simply permuted according to the permutation P.

As another application of matrix multiplication, a remarkable result is obtained when powers of an adjacency matrix are considered.

Example 8.18.   Let A be the adjacency matrix for the graph from Example 8.15, in terms of the ordering $1, 2, 3, 4, 5$ of its vertices.

$$\begin{array}{c} \\ 1 \\ 2 \\ 3 \\ 4 \\ 5 \end{array} \begin{array}{c} 1\ 2\ 3\ 4\ 5 \\ \begin{bmatrix} 0 & 2 & 1 & 0 & 0 \\ 2 & 0 & 0 & 1 & 1 \\ 1 & 0 & 0 & 1 & 0 \\ 0 & 1 & 1 & 0 & 1 \\ 0 & 1 & 0 & 1 & 1 \end{bmatrix} \end{array} = A$$

Observe that

$$A^2 = \begin{bmatrix} 0 & 2 & 1 & 0 & 0 \\ 2 & 0 & 0 & 1 & 1 \\ 1 & 0 & 0 & 1 & 0 \\ 0 & 1 & 1 & 0 & 1 \\ 0 & 1 & 0 & 1 & 1 \end{bmatrix} \begin{bmatrix} 0 & 2 & 1 & 0 & 0 \\ 2 & 0 & 0 & 1 & 1 \\ 1 & 0 & 0 & 1 & 0 \\ 0 & 1 & 1 & 0 & 1 \\ 0 & 1 & 0 & 1 & 1 \end{bmatrix} = \begin{bmatrix} 5 & 0 & 0 & 3 & 2 \\ 0 & 6 & 3 & 1 & 2 \\ 0 & 3 & 2 & 0 & 1 \\ 3 & 1 & 0 & 3 & 2 \\ 2 & 2 & 1 & 2 & 3 \end{bmatrix}.$$

Letting $A^2 = [b_{i,j}]$, it is straightforward to check, for each i, j, that $b_{i,j}$ is the number of walks of length 2 in the graph from i to j. For example, $b_{2,3} = 3$ and

$$2, a, 1, c, 3$$
$$2, b, 1, c, 3$$
$$2, d, 4, f, 3$$

are the walks of length 2 from vertex 2 to vertex 3.

Example 8.18 illustrates a particular instance of a general result.

Theorem 8.5.   *Let A be the adjacency matrix for a graph G obtained from the ordering $v_1, v_2, \ldots, v_n$ of its vertices, and let $m \in \mathbb{N}$. Then the mth power of A, say, $A^m = [b_{i,j}]$, has the property that for each $1 \le i \le n$ and $1 \le j \le n$, the entry $b_{i,j}$ is the number of walks in G of length m from v_i to v_j.*

Proof. We proceed by induction on m. For $m = 0$, we have $A^0 = I$, the n-by-n identity matrix (ones on the diagonal and zeros elsewhere). This reflects the fact that a walk of length 0 is only possible from a vertex to itself. So suppose $m \ge 0$ and A^m has the asserted property. Say, $A = [a_{i,j}]$, $A^m = [b_{i,j}]$, and $A^{m+1} = [c_{i,j}]$.

Suppose $1 \leq i, j \leq n$. For each $1 \leq k \leq n$, the number of paths of length $m + 1$ from v_i to v_j that first visit v_k is $a_{i,k} b_{k,j}$. Hence, $\sum_{k=1}^{n} a_{i,k} b_{k,j}$ is the total number of paths of length $m + 1$ from v_i to v_j. Since $A^{m+1} = A \cdot A^m$, it follows that $c_{i,j}$ is this number. □

Since powers of matrices can be tedious to compute, mathematical software can be very helpful. In *Mathematica*, the mth power of a matrix A is computed with the function `MatrixPower[A,m]`. In *Maple*, we use the function `LinearAlgebra[MatrixPower](A,m)`.

Having seen how an adjacency matrix stores the structure of a graph, we consider another means of storage. This one is more efficient but lacks wonderful consequences like Theorem 8.5.

Definition 8.15. Let $v_1, v_2, \ldots, v_n$ be the vertices of a graph G without multiple edges. For each $1 \leq i \leq n$, an **adjacency list** for vertex v_i is a list of all of the neighbors of v_i. A listing, for each vertex, of its adjacency list forms the **adjacency lists** for G.

Example 8.19. The adjacency lists for the pictured graph are shown.

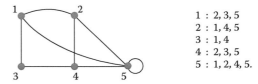

1 : 2, 3, 5
2 : 1, 4, 5
3 : 1, 4
4 : 2, 3, 5
5 : 1, 2, 4, 5.

For added clarity, each row is marked by the vertex whose adjacency list follows.

Remark 8.2. The adjacency lists for a graph can be used to produce a linked list for storing the graph in a computer. For each vertex, the adjacency list then specifies the vertices to which pointers should be directed.

Exercises

1. Find an adjacency matrix for the study group scheduling graph from Exercise 3 in Section 8.1. Note the obvious connection between the table presented there and the adjacency matrix constructed here.

2. Find an adjacency matrix for the high school committee scheduling graph from Exercise 4 in Section 8.1. Note the

obvious connection between the table presented there and the adjacency matrix constructed here.

The following graphs are referred to throughout the remaining exercises.

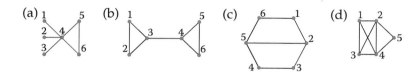

In Exercises 3 through 6, find the adjacency matrix A for the given graph, using the natural ordering $1, 2, 3, \ldots$ of its vertices.

3. Graph (a).   4. Graph (b).

5. Graph (c).   6. Graph (d).

In Exercises 7 through 12, make an adjacency matrix for the given graph using the specified ordering of the vertices.

7. Graph (a) with vertices ordered $2, 5, 1, 4, 6, 3$.

8. Graph (b) with vertices ordered $3, 4, 1, 6, 5, 2$.

9. K_5 with vertices ordered $1, 2, 3, 4, 5$.

10. Q_3 with vertices in binary order.

11. $K_{2,4}$ with vertices ordered $(1, 1), (1, 2),$ $(2, 1), (2, 2), (2, 3), (2, 4)$.

12. C_5 with vertices ordered $1, 2, 3, 4, 5$.

13. Five computers are connected in a network, and the given matrix reflects the direct cable connections between computers in this network.

$$
\begin{array}{c}
 \\
1 \\
2 \\
3 \\
4 \\
5
\end{array}
\begin{array}{c}
1\ 2\ 3\ 4\ 5 \\
\begin{bmatrix}
0 & 1 & 1 & 1 & 0 \\
1 & 0 & 1 & 0 & 1 \\
1 & 1 & 0 & 0 & 0 \\
1 & 0 & 0 & 1 & 1 \\
0 & 1 & 0 & 1 & 0
\end{bmatrix}
\end{array}.
$$

Draw the graph with this adjacency matrix that reflects the network configuration.

14. Five power stations are connected to form a power grid. Redundant connections are used to maintain a backup for important connections that might fail. The given matrix reflects direct connections between the power stations.

$$
\begin{array}{c}
\begin{array}{ccccc} 1 & 2 & 3 & 4 & 5 \end{array} \\
\begin{array}{c} 1 \\ 2 \\ 3 \\ 4 \\ 5 \end{array}
\left[\begin{array}{ccccc}
0 & 2 & 0 & 0 & 0 \\
2 & 0 & 1 & 1 & 1 \\
0 & 1 & 0 & 0 & 1 \\
0 & 1 & 0 & 0 & 1 \\
0 & 1 & 1 & 1 & 0
\end{array} \right].
\end{array}
$$

Draw the graph with the given adjacency matrix that reflects the power grid.

The matrix products and powers in Exercises 15 through 28 should be computed by hand to gain familiarity with those operations. However, mathematical software such as *Mathematica* or *Maple* might also be used to confirm the results obtained by hand. In Exercises 15 through 22, compute the given matrix product.

15. $\begin{bmatrix} 3 & 2 \\ 4 & 1 \end{bmatrix} \begin{bmatrix} 4 & 1 \\ 2 & 5 \end{bmatrix}.$

16. $\begin{bmatrix} 3 & 5 \\ 1 & 2 \end{bmatrix} \begin{bmatrix} 2 & 3 \\ 4 & 1 \end{bmatrix}.$

17. $\begin{bmatrix} 5 & 7 \\ 8 & 6 \end{bmatrix} \begin{bmatrix} 0 & 1 \\ 1 & 0 \end{bmatrix}.$

18. $\begin{bmatrix} 0 & 1 \\ 1 & 0 \end{bmatrix} \begin{bmatrix} 5 & 7 \\ 8 & 6 \end{bmatrix}.$

19. $\begin{bmatrix} 2 & 7 & 0 \\ 0 & 4 & 1 \\ 3 & 0 & 8 \end{bmatrix} \begin{bmatrix} 0 & 1 & 0 \\ 0 & 0 & 1 \\ 1 & 0 & 0 \end{bmatrix}.$

20. $\begin{bmatrix} 0 & 5 & 7 \\ 1 & 0 & 9 \\ 6 & 3 & 0 \end{bmatrix} \begin{bmatrix} 0 & 0 & 1 \\ 0 & 1 & 0 \\ 1 & 0 & 0 \end{bmatrix}.$

21. $\begin{bmatrix} 0 & 0 & 3 \\ 0 & 2 & 0 \\ 5 & 0 & 0 \end{bmatrix} \begin{bmatrix} 6 & 1 & 7 \\ 4 & 2 & 5 \\ 0 & 8 & 3 \end{bmatrix}.$

22. $\begin{bmatrix} 0 & 4 & 0 \\ 0 & 0 & 7 \\ 6 & 0 & 0 \end{bmatrix} \begin{bmatrix} 6 & 1 & 7 \\ 4 & 2 & 5 \\ 0 & 8 & 3 \end{bmatrix}.$

23. Let A be the matrix from Exercise 3.

   (a) Find the permutation matrix P corresponding to the permutation $2, 5, 1, 4, 6, 3$.

   (b) Compute PA.

   (c) Compute PAP^T, and confirm, in this case, the observation made after Example 8.17.

24. Let A be the matrix from Exercise 4.

   (a) Find the permutation matrix P corresponding to the permutation $3, 4, 1, 6, 5, 2$.

   (b) Compute PA.

   (c) Compute PAP^T, and confirm, in this case, the observation made after Example 8.17.

In Exercises 25 through 28, compute the square A^2 of the given adjacency matrix A, and confirm the result in Theorem 8.5 that each entry b_{ij} in A^2 counts the number of walks of length 2 in the corresponding graph from vertex i to j.

25. The matrix A from Exercise 3.

26. The matrix A from Exercise 4.

27. The matrix A from Exercise 5.

28. The matrix A from Exercise 6.

In Exercises 29 through 32, let A denote the adjacency matrix for the given graph G, using the natural ordering $1, 2, 3, \ldots$ of its vertices. Use Theorem 8.5 to find the entry in row 1 and column 1 of A^m, for each given m. Hint: Just count walks.

29. $G = C_6$, for $m = 3, 4$, and 5.

30. $G = C_5$, for $m = 3, 4$, and 5.

31. $G = P_5$, for $m = 4, 5$, and 6.

32. $G = K_7$, for $m = 2, 3$, and 4.

33.* Let G be any graph on n vertices, and let A be an adjacency matrix for G. Show: G is connected if and only if the matrix $I + A + A^2 + \cdots + A^{n-1}$ has no zero entries. Note that matrices are added by adding corresponding entries.

34.* State a result like the one in Exercise 33 that can be used to partition the vertices into the different components of a graph.

In Exercises 35 through 40, give the adjacency lists for the given graph.

35. Graph (a).

36. Graph (b).

37. Graph (c).

38. Graph (d).

39. Q_3.

40. $K_{2,3}$.

41. Seven power stations are connected to form a power grid. The given adjacency lists reflect direct connections between the power stations. Draw the graph whose adjacency lists are given that reflects the power grid.

$$1 : 2, 3, 4, 7$$
$$2 : 1, 6, 7$$
$$3 : 1, 6$$
$$4 : 1, 5, 6$$
$$5 : 4, 6$$
$$6 : 2, 3, 4, 5$$
$$7 : 1, 2.$$

42. Six computers are connected in a network. The given adjacency lists reflect the direct cable connections between computers in this network. Draw the graph whose adjacency lists are given that reflects this network configuration.

$$1 : 2, 4$$

$$2 : 1, 3, 4, 5$$

$$3 : 2, 4, 5, 6$$

$$4 : 1, 2, 3, 5$$

$$5 : 2, 3, 4, 6$$

$$6 : 3, 5.$$

8.4 Isomorphisms

Drawings for two graphs, (a) and (b), are shown in Figure 8.21. Are they pictures of the same graph? The transformation of the drawing of (a) into the drawing of (b) shown in Figure 8.22 suggests that the answer should be yes.

A key observation can be made in the last step in Figure 8.22. There we see that the vertices of (a) match up with the vertices of (b) according to the assignments specified on the left-hand side of Figure 8.23. It is also important that the edges are matched up likewise, as specified on the right-hand side of Figure 8.23. What is critical here is that an edge correspondence like $b \rightarrow m$ fits with the vertex correspondence on the endpoints. That is, the endpoint assignments $b \mapsto \{1, 3\}$ and $m \mapsto \{7, 10\}$ reflect the vertex correspondences $1 \rightarrow 7$ and $3 \rightarrow 10$. In fact, in a case in which just one edge joins two endpoints, the vertex correspondence for the endpoints completely determines the edge correspondence. That is, the decision that $1 \rightarrow 7$ and $3 \rightarrow 10$ forces the choice that $b \rightarrow m$. However, when we have multiple edges, such as e and f, we need to specify that $e \rightarrow k$ and $f \rightarrow j$ or, alternatively, that $e \rightarrow j$ and $f \rightarrow k$.

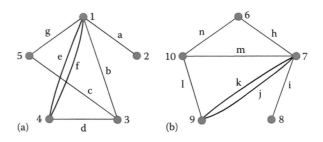

FIGURE 8.21
Two graphs to compare.

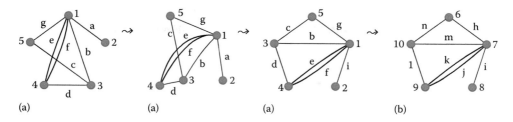

FIGURE 8.22
Transforming one graph into another.

$V_{(a)}$	$V_{(b)}$		$E_{(a)}$	$E_{(b)}$
1	→ 7		a	→ i
2	→ 8		b	→ m
3	→ 10		c	→ n
4	→ 9		d	→ l
5	→ 6		e	→ k
			f	→ j
			g	→ h

FIGURE 8.23
Matching vertices and edges in Figure 8.21.

Having seen the issues involved in showing that a particular pair of graphs is equivalent, we now define the notion of a graph isomorphism in general.

Definition 8.16.   Let $G = (V_G, E_G)$ and $H = (V_H, E_H)$ be graphs.

(a) A **graph isomorphism** from G to H, denoted $f : G \longrightarrow H$, is a pair of bijections $f_V : V_G \longrightarrow V_H$ and $f_E : E_G \longrightarrow E_H$ such that for each $e \in E_G$, the bijection f_V maps the endpoints of e to the endpoints of $f_E(e)$. Since there is no intersection between the domains of f_V and f_E, both are more simply denoted by f. If G and H are simple graphs, then f is completely determined by a function $f_V : V_G \longrightarrow V_H$, since f_E must consequently take $\{u, v\} \in E_G$ to $\{f_V(u), f_V(v)\} \in E_H$.

(b) We say that G is **isomorphic** to H, written $G \cong H$, if there exists a graph isomorphism from G to H.

(c) A **graph automorphism** on G is a graph isomorphism from G to itself. A **nontrivial automorphism** is an automorphism $f : G \longrightarrow G$ that is not the identity map. The set of automorphisms on G is denoted Aut(G).

$$
A_{(a)} = \begin{array}{c|ccccc}
 & 1 & 2 & 3 & 4 & 5 \\
\hline
1 & 0 & 1 & 1 & 2 & 1 \\
2 & 1 & 0 & 0 & 0 & 0 \\
3 & 1 & 0 & 0 & 1 & 1 \\
4 & 2 & 0 & 1 & 0 & 0 \\
5 & 1 & 0 & 1 & 0 & 0
\end{array}
\qquad
A_{(b)} = \begin{array}{c|ccccc}
 & 6 & 7 & 8 & 9 & 10 \\
\hline
6 & 0 & 1 & 0 & 0 & 1 \\
7 & 1 & 0 & 1 & 2 & 1 \\
8 & 0 & 1 & 0 & 0 & 0 \\
9 & 0 & 2 & 0 & 0 & 1 \\
10 & 1 & 1 & 0 & 1 & 0
\end{array}
$$

FIGURE 8.24
Adjacency matrices for Figure 8.21.

We can understand graph isomorphisms in terms of adjacency matrices. For example, using increasing order for the vertices, the adjacency matrices $A_{(a)}$ and $A_{(b)}$ for the graphs (a) and (b), respectively, in Figure 8.21 are shown in Figure 8.24. Although $A_{(a)} \neq A_{(b)}$, if we apply the permutation

$$
\begin{aligned}
6 &\rightarrow 7 \\
7 &\rightarrow 8 \\
8 &\rightarrow 10 \\
9 &\rightarrow 9 \\
10 &\rightarrow 6
\end{aligned}
$$

to the rows and columns of $A_{(b)}$, then we obtain a new adjacency matrix $A'_{(b)}$ for (b) that has exactly the same entries as $A_{(a)}$. Of course, the ordering $7, 8, 10, 9, 6$ is that determined by the vertex map in Figure 8.23. That is, the vertex map tells us how to consistently order the vertices for (a) and (b) so that the corresponding adjacency matrices are the same.

Example 8.20. (Equivalent Network Configurations). The two seemingly different network configurations shown in Figure 8.25 have been proposed to implement a complex parallel program. Show that these networks are in fact the same configuration.

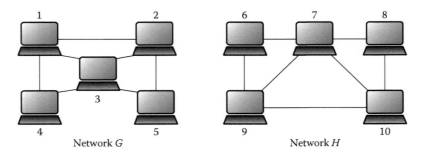

Network G   Network H

FIGURE 8.25
Equivalent computer networks.

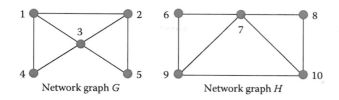

FIGURE 8.26
Isomorphic network graphs.

Solution. The aim is to show that the corresponding graphs displayed in Figure 8.26 are isomorphic. Since G and H are simple graphs, it suffices to specify a vertex bijection $f : V_G \longrightarrow V_H$ that maps endpoints to endpoints and establishes an edge bijection between E_G and E_H. It is straightforward to check that defining $f(1) = 9$, $f(2) = 10$, $f(3) = 7$, $f(4) = 6$, and $f(5) = 8$ gives the desired isomorphism. □

Example 8.21. We consider some graph isomorphisms with domain K_4. In each case, the isomorphism is described in terms of the given labeled pictures. Also, since K_4 is a simple graph, the edge mapping is completely determined by the vertex mapping.

(a) Below are pictured what amounts to two different ways of drawing K_4. An isomorphism $f : K_4 \longrightarrow H$ is given by

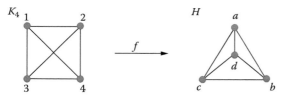

$f(1) = a, f(2) = d, f(3) = c$, and $f(4) = b$. Hence, $K_4 \cong H$. Moreover, if the labels a, b, c, d on H are replaced by $1, 4, 3, 2$, respectively, then the picture of H is seen to be an alternative drawing of K_4.

(b) A nontrivial automorphism of K_4 is given by the graph isomorphism $f : K_4 \longrightarrow K_4$,

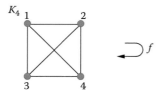

where $f(1) = 2, f(2) = 4, f(3) = 1$, and $f(4) = 3$. That is, f is the permutation

$$\begin{array}{cccc} 1 & 2 & 3 & 4 \\ \downarrow & \downarrow & \downarrow & \downarrow \\ 2 & 4 & 1 & 3 \end{array}$$

of the vertices. Since $f(1) \neq 1$, f is not the identity.

There was a great deal of choice available for the automorphism of K_4 considered in Example 8.21(b). That is, we might also have chosen $f(1) = 4$ and then $f(2) = 2$, $f(3) = 3$, and $f(4) = 1$. In fact, any vertex value for $f(1)$ can be chosen when building an automorphism f of K_4. In this sense, we say that all of the vertices of K_4 look the same.

Definition 8.17. A graph $G = (V, E)$ is said to be **vertex transitive** if, for any $u, v \in V$, there is a graph automorphism f such that $f(u) = v$.

Example 8.22. For any positive integer n, the complete graph K_n is easily seen to be vertex transitive. The point is that, for K_n, graph automorphisms correspond to permutations of $\{1, 2, \ldots, n\}$. For instance, the automorphism in Example 8.21(b) can be thought of as an isomorphism f constructed to send $u = 1$ to $v = 2$.

Example 8.23. Displayed are three different drawings of a graph known as the **Petersen graph**. Verification that these graphs are indeed isomorphic is left for the exercises.

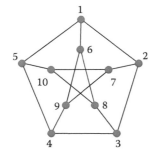 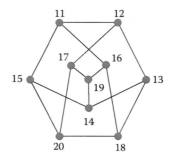 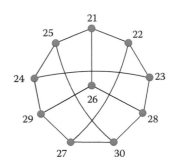

That the Petersen graph is vertex transitive is also left for the exercises.

Theorem 8.6. (Graph Isomorphism Is an Equivalence Relation). For all graphs G, H, and K,

(a) $G \cong G$.

(b) If $G \cong H$, then $H \cong G$.

(c) If $G \cong H$ and $H \cong K$, then $G \cong K$.

Proof. For part (a), the identity maps on V_G and E_G yield the desired isomorphism. Parts (b) and (c) are left for the exercises. $\qquad \square$

Remark 8.3. We have specified several names of graphs so far, such as $P_n, C_n, K_n, K_{m,n}, Q_n$, the tetrahedron, the octahedron, the dodecahedron, the icosahedron, and the Petersen graph. Although specific vertex sets, such as $\{1, \ldots, n\}$ for P_n, C_n, and K_n, have been used in their definitions, we only mean to define these graphs up to isomorphism. That is, any graph that is isomorphic to one of these will also be referred to by the appropriate name. For example, the graph H in Example 8.21(a) can also be called K_4.

The following result establishes that Aut(G) has more structure than that of just a set.

Theorem 8.7. *For any graph G, Aut(G) forms a group under composition.*

Proof. We must establish the three defining properties of a group.

(Associativity) Suppose $f, g, h \in$ Aut(G). Since f_V, g_V, h_V are functions, the associativity of function composition (Theorem 5.7) gives that $(f_V \circ g_V) \circ h_V = f_V \circ (g_V \circ h_V)$. The same holds for f_E, g_E, h_E, and we conclude that $(f \circ g) \circ h = f \circ (g \circ h)$.

(Identity) It is easy to see that the identity function id_G defined by $\text{id}_V(v) = v$ and $\text{id}_E(e) = e$ gives an automorphism of G. Moreover, for any automorphism f, we have $\text{id}_G \circ f = f \circ \text{id}_G = f$.

(Inverses) Let $f \in$ Aut(G). Since f_V and f_E are bijections, it follows from Theorem 5.10 that $g_V = f_V^{-1}$ and $g_E = f_E^{-1}$ are bijections. Letting e be an arbitrary edge, say, $e \mapsto \{u, v\}$, we must confirm that g_V maps the endpoints of e to the endpoints of $g_E(e)$. Since f_E is onto, we have some edge e' such that $f_E(e') = e$. Since f is a graph isomorphism, we must further have $e' \mapsto \{u', v'\}$ with $f_V(u') = u$ and $f_V(v') = v$. From the fact that $g_V = f_V^{-1}$ and $g_E = f_E^{-1}$, it now follows that $g_E(e) = e'$, $g_V(u) = u'$, and $g_V(v) = v'$. That is, g_V maps the endpoints of

e to the endpoints of $g_E(e)$. Hence, g_V and g_E determine an automorphism g of G. Moreover, $g \circ f = f \circ g = \mathrm{id}_G$, and we see that $g = f^{-1}$. □

In general, an automorphism f of a graph $G = (\{v_1, \ldots, v_n\}, E)$ corresponds to a permutation $f(v_1), \ldots, f(v_n)$ of its vertex set such that the adjacency matrix A for G, using the ordering $v_1, \ldots, v_n$, is the same as the adjacency matrix using the ordering $f(v_1), \ldots, f(v_n)$. In terms of matrices, if we let P denote the permutation matrix corresponding to the permutation $f(v_1), \ldots, f(v_n)$, then f is an automorphism of G if and only if $PAP^T = A$. In particular, the set of automorphisms $\mathrm{Aut}(G)$ is a subset of the set of permutations of $v_1, \ldots, v_n$.

Example 8.24. The automorphisms of the graph

are certain permutations of $1, 2, 3, 4$. Specifically, they are

$1, 2, 3, 4$ the identity,
$1, 4, 3, 2$ switch vertices 2 and 4 only,
$3, 2, 1, 4$ switch vertices 1 and 3 only,
$3, 4, 1, 2$ switch vertices 2 and 4 and switch vertices 1 and 3.

In total, there are thus four automorphisms. One is trivial and the other three are nontrivial.

If we remove from the definition of a graph isomorphism the requirement of bijective functions, then we obtain the more general notion of a map between graphs.

Definition 8.18. A **graph map** f from a graph $G = (V_G, E_G)$ to a graph $H = (V_H, E_H)$, denoted $f : G \longrightarrow H$, is a pair of functions $f_V : V_G \longrightarrow V_H$ and $f_E : E_G \longrightarrow E_H$ such that for each edge $e \in E_G$, the function f_V maps the endpoints of e to the endpoints of $f_E(e)$.

Graph maps send vertices to vertices and edges to edges. A graph map $f : G \longrightarrow H$ may be visualized as a folding of G into H.

Example 8.25.   In both instances below, graphs G and H are pictured with labels given, and the values of a graph map $f : G \longrightarrow H$ are specified.

(a) The pictured graph map f must be specified both on the vertices and the edges.

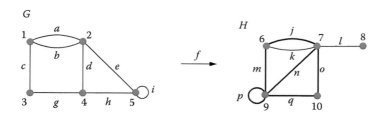

Here, $f(1) = 6, f(2) = 7, f(3) = 9, f(4) = 10,$
$f(5) = 9,$ $f(a) = j, f(b) = j, f(c) = m, f(d) = o,$
$f(e) = n, f(g) = q,$ $f(h) = q,$ and $f(i) = p.$ The image of f is the subgraph of H shown in bold.

(b) The pictured graph map f need only be specified on the vertices.

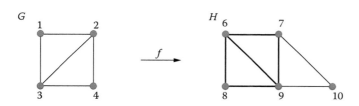

Here, $f(1) = 7, f(2) = 9, f(3) = 6,$ and $f(4) = 8.$ The image of f is shown in bold. Very simply, f rotates G clockwise and places it on a portion of H.

Exercises

Graph isomorphisms are requested throughout the exercises. In the cases that the graphs are simple, it suffices to specify a vertex bijection that behaves appropriately with respect to the edges, as described in Definition 8.16. Otherwise, the edge bijection should also be given.

1. Show that the two pictured graphs are isomorphic.

2. Show that the two pictured graphs are isomorphic.

3. Show that $P_3 \cong K_{1,2}$.

4. Show that $C_4 \cong K_{2,2}$.

5. Show that the two pictured graphs are isomorphic.

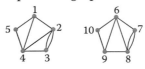

6. Show that the two pictured graphs are isomorphic.

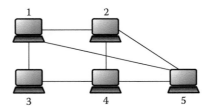

7. The New York office for a certain company contains a network with the configuration shown below on the left. Its Los Angeles office contains the network configuration shown on the right. A programmer from the New York office has traveled to the Los Angeles office and wishes to demonstrate a program that has been tested on the New York network.

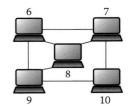

Show that the two network configurations pictured yield equivalent networks, and hence, the program can be made to work in Los Angeles.

8. A company needs to network 5 computers using 7 cable connections. Two competing network designers have come up with the pictured network configurations, and each is asserting the superiority of his design.

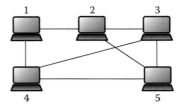
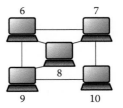

Show that the two network configurations pictured actually yield equivalent networks.

9. Pictured are two design plans for a power grid, in which the symbol ○ denotes a power station. One represents a grid in Nashville, Tennessee, and the other is in Syracuse, New York. In Nashville, they have some software that has been very effective at managing power transfers among their power stations.

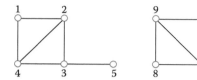

Show that the power grids are equivalent and, hence, Syracuse can employ the same management software as that used in Nashville.

10. Pictured are two design plans for a power grid, in which the symbol ○ denotes a power station. The one on the left is a grid in Philadelphia, Pennsylvania, and the one on the right is in Pittsburgh, Pennsylvania. Philadelphia has paid for a comprehensive analysis of the ability of its grid to handle the failures of some of its power stations in storms.

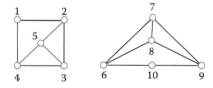

Show that the power grids are equivalent and, hence, Pittsburgh can directly benefit from the grid analysis purchased by Philadelphia.

11.* A certain university is making its summer schedule based upon predetermined student interest in its courses. The Department of Science and Technology is offering classes in Algebra, Biology, C++, Databases, Earth Science, and Functional Analysis. The Department of Languages and Cultures is offering classes in German, History, Indochina, Japanese, Kuwait, and Latin. In the displayed tables, an X denotes the fact that the two different classes corresponding to that row and column have a student in common.

	Alg.	Bio.	C++.	Data.	EaSc.	Func.
Alg.		X	X		X	X
Bio.	X		X	X		X
C++.	X	X		X	X	
Data.		X	X		X	X
EaSc.	X		X	X		X
Func.	X	X		X	X	

	Ger.	His.	Indo.	Jap.	Kuw.	Lat.
Ger.		X	X	X		X
His.	X		X	X	X	
Indo.	X	X			X	X
Jap.	X	X			X	X
Kuw.		X	X	X		X
Lat.	X		X	X	X	

(a) Show that the corresponding class scheduling graphs are isomorphic. Hence, a workable schedule for one department can be converted to one for the other department.

(b) Use this isomorphism to convert the schedule

Time period	Committee meeting
1	Algebra, Databases
2	Biology, Earth Science
3	C++, Functional Analysis

from the Department of Science and Technology to the Department of Languages and Cultures.

12.* The Senate needs to schedule meetings for its committees: Appropriations (A), Budget (B), Commerce (C), Energy (D), Environment (E), Foreign Relations (F), Homeland Security and Governmental Affairs (G), and Health (H). On the other side of the Capitol building, the House of Representatives needs to schedule meetings for its committees: International Relations (I), Judicial (J), Agriculture (K), Armed Services (L), Financial Services (M), Science (N), Standards of Official Conduct (O), and Printing (P). In the displayed tables, an X denotes the fact that the two different committees corresponding to that row and column have a member in common.

	A	B	C	D	E	F	G	H
A		X		X	X	X		
B	X		X			X		X
C		X		X	X		X	
D	X		X				X	X
E	X		X			X	X	
F	X	X			X			X
G			X	X	X			X
H		X		X		X	X	

	I	J	K	L	M	N	O	P
I			X		X		X	X
J			X	X			X	X
K	X	X				X	X	
L		X				X	X	X
M	X		X			X		X
N			X	X	X		X	
O	X	X		X		X		
P	X	X			X	X		

(a) Show that the corresponding committee scheduling graphs are isomorphic. Hence, a workable schedule for the Senate can be converted to one for the House of Representatives.

(b) Use this isomorphism to convert the schedule

Time period	Committee meeting
1	A, C, H
2	B, D, E
3	F, G

from the Senate to the House of Representatives.

Special graphs. Exercises 13 through 18 give alternative presentations of some of the special graphs from Section 8.2, particularly some of the platonic solids.

13. Show that $K_{3,3}$ is isomorphic to the pictured graph.

14. Show that the two pictured graphs are isomorphic.

They are both called the dodecahedron.

15. Show that the two pictured graphs are isomorphic.

They are both called the cube.

16. Show that the two pictured graphs are isomorphic.

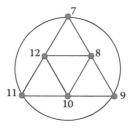

They are both called the octahedron.

17. Show that the three graphs pictured in Example 8.23 are isomorphic.

18. Show that the two pictured graphs are isomorphic.

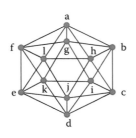
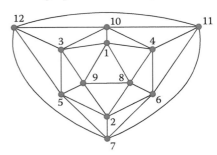

They are both called the icosahedron.

The following graphs are referred to throughout the remaining exercises.

(a) 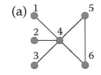 (b) 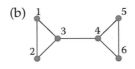 (c) 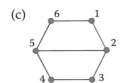 (d)

In Exercises 19 through 26, determine the number of automorphisms of the specified graph. Recall that automorphisms correspond to permutations of the vertices that preserve the structure of the graph.

19. Graph (a).   20. Graph (b).

21. Graph (c).   22. Graph (d).

23. C_n.   24. P_n.

25. $K_{n,n}$.   26. $K_{m,n}$, where $m \neq n$.

27. Show that graph isomorphisms send paths to paths. That is,
    if $v_0, e_1, v_1, \ldots, v_n$ is a path in a graph G and $f : G \longrightarrow H$ is an isomorphism, then $f(v_0), f(e_1), f(v_1), \ldots, f(v_n)$ is a path in H.

28. Show that graph isomorphisms send subgraphs to isomorphic subgraphs. That is, if K is a subgraph of a graph G and $f : G \longrightarrow H$ is an isomorphism, then $f(K)$ is a subgraph of H and $K \cong f(K)$. Conclude that f maps cycles to cycles (of the same length), f maps complete subgraphs to complete subgraphs (of the same size), and f maps induced subgraphs to induced subgraphs.

29. Let $V = \mathcal{P}_2(\{1, 2, 3, 4, 5\})$ and put edges between disjoint sets. Show that the resulting graph is isomorphic to the Petersen graph.

30\. Show that the Petersen graph is vertex transitive. Hint: Use Exercise 29.

31\. Show that for any cycle C of length 5 in the Petersen graph, there is an automorphism of the Petersen graph that sends C to the cycle induced by vertices $1, 2, 3, 4, 5$ in Example 8.23.

32\. Show that for any path P of length 3 in the Petersen graph, there is an automorphism of the Petersen graph that sends P to the path induced by vertices $1, 2, 3$ in Example 8.23.

33. Show that C_n is vertex transitive.

34. Show that $K_{n,n}$ is vertex transitive.

35. Prove Theorem 8.6(b).

36. Prove Theorem 8.6(c).

37\. Show that a graph $G = (V, E)$ is vertex transitive if and only if there is $u \in V$ such that for all $v \in V$, there exists an automorphism f of G such that $f(u) = v$.

38\. Prove or disprove: If a graph G is vertex transitive, then every induced subgraph H of G will also be vertex transitive.

39\. Show: For all graphs G and H, if $G \cong H$, then there are drawings of G and H that have the same image (i.e. occupy the same points) in $\mathbb{R}^2$.

40\. Let $G = (\{v_1, \ldots, v_n\}, E_G)$ and $H = (\{u_1, \ldots, u_n\}, E_H)$ be any graphs without multiple edges.
Show that $G \cong H$ if and only if there is a bijection $f : \{v_1, \ldots, v_n\} \longrightarrow \{u_1, \ldots, u_n\}$ such that the adjacency lists for H using the vertex ordering $f(v_1), \ldots, f(v_n)$ are the same as the adjacency lists obtained by applying f to each entry in the adjacency lists for G.

In Exercises 41 through 50, to specify graph maps between the given simple graphs, it suffices to specify a vertex map that behaves appropriately with respect to the edges, as described in Definition 8.18.

41. Give an example of a graph map from graph (a) to graph (b).

42. Give an example of a graph map from graph (c) to graph (d).

43\. Show that there is no graph map from C_3 to C_4.

44\. Show that there is no graph map from C_5 to K_2.

45. Does there exist a graph map from graph (b) to graph (c)? Explain.

46. Does there exist a graph map from graph (d) to graph (a)? Explain.

47. Show that, $\forall\, m, n \in \mathbb{Z}^+$, there exists a graph map from $K_{m,n}$ to P_2.

48. Show that, $\forall\, n \in \mathbb{Z}^+$, there exists a graph map from C_n to K_3.

49. For which pairs of integers m, n does there exist a graph map from K_m to K_n? Explain.

50. For which pairs of integers m, n does there exist a graph map from Q_m to Q_n? Explain.

8.5 Invariants

In Example 8.20 in Section 8.4, we considered two network configurations on five computers and seven cable connections that, at a casual glance, appeared to be different in structure. However, by establishing an isomorphism between the graphs that model those networks, we saw that they were in fact equivalent in structure. Since surely not all networks on five computers and seven cable connections are equivalent, we need a means of distinguishing two genuinely different network configurations. That is, we want to be able to distinguish two non-isomorphic graphs.

How does one prove that two graphs are not isomorphic? If, for each graph G, we assign a value $i(G)$, in such a way that, for any two isomorphic graphs G and H, we have $i(G) = i(H)$, then we say that i is a **graph invariant**. That is, graph invariants are preserved under isomorphisms. Consequently, if, for two graphs G and H and some invariant i, we have $i(G) \neq i(H)$, then we can conclude that $G \not\cong H$.

The numbers of vertices and edges provide very simple examples of graph invariants. For instance, $|V_G|$ is the invariant $i(G)$ that gives the number of vertices in a graph G.

Theorem 8.8. *Let G and H be graphs. If $G \cong H$, then*

(i) $|V_G| = |V_H|$.

(ii) $|E_G| = |E_H|$.

Proof. Suppose $G \cong H$. Thus, we have a graph isomorphism that consists of bijections $f_V : V_G \longrightarrow V_H$ and $f_E : E_G \longrightarrow E_H$. These bijections show that V_G and V_H have the same cardinality and that E_G and E_H have the same cardinality. $\square$

Certainly, if two graphs do not have the same numbers of vertices and the same numbers of edges, then they cannot be isomorphic. In fact, the invariants $|V|$ and $|E|$ for a graph $G = (V, E)$ are likely the first values one checks when considering a potential isomorphism.

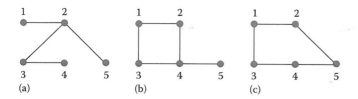

FIGURE 8.27
Non-isomorphic graphs on 5 vertices.

Example 8.26. Graphs (a) and (b) in Figure 8.27 are not isomorphic. Although they have the same number of vertices, they do not have the same number of edges. Similarly, graphs (a) and (c) are seen to be not isomorphic.

Theorem 8.8 is not sufficient to establish that graphs (b) and (c) in Figure 8.27 are not isomorphic, since those graphs do have the same number of vertices and the same number of edges. Consequently, we need a more powerful invariant in this case. In fact, immediately following the definition of a graph isomorphism in Section 8.4, we observed that adjacency matrices provide extremely powerful invariants, as we now formulate precisely.

Theorem 8.9. Let $G = (V_G, E_G)$ and $H = (V_H, E_H)$ be graphs. Then $G \cong H$ if and only if there are orderings of V_G and V_H such that the corresponding adjacency matrices A_G and A_H are equal. In this case, we say that G and H have a *common adjacency matrix*.

The problem with Theorem 8.9 is that it is too powerful. It can work well as a tool to show that two graphs are isomorphic, if we can find a common adjacency matrix for them. However, it is too cumbersome in general to argue that no orderings of V_G and V_H yield a common adjacency matrix. If $|V_G| = |V_H| = n$ and we fix an ordering of V_G, then there are $n!$ possible orderings of V_H to consider. For large values of n, this is an intractable task.

Adjacency matrices for graphs (b) and (c) from Figure 8.27 are displayed in Figure 8.28. What we might notice there that can help us avoid unnecessary work is that row 4 of matrix $A_{(b)}$ contains three 1's. Not only does no row of matrix $A_{(c)}$ has that property, but also, no matter what vertex ordering is used for graph (c), no row of the corresponding adjacency matrix for (c) will ever have three 1's in it. Consequently, we need not consider entire adjacency matrices. Instead, we obtain some

$$A_{(b)} = \begin{array}{c c c c c c} & 1 & 2 & 3 & 4 & 5 \\ 1 & 0 & 1 & 1 & 0 & 0 \\ 2 & 1 & 0 & 0 & 1 & 0 \\ 3 & 1 & 0 & 0 & 1 & 0 \\ 4 & 0 & 1 & 1 & 0 & 1 \\ 5 & 0 & 0 & 0 & 1 & 0 \end{array} \qquad A_{(c)} = \begin{array}{c c c c c c} & 1 & 2 & 3 & 4 & 5 \\ 1 & 0 & 1 & 1 & 0 & 0 \\ 2 & 1 & 0 & 0 & 0 & 1 \\ 3 & 1 & 0 & 0 & 1 & 0 \\ 4 & 0 & 0 & 1 & 0 & 1 \\ 5 & 0 & 1 & 0 & 1 & 0 \end{array}$$

FIGURE 8.28
Some adjacency matrices for Figure 8.27.

less powerful, but still effective, invariants by considering the local structure of a graph at each vertex. Basically, for each vertex v, we keep track of the number of times v serves as an endpoint for an edge.

Definition 8.19.   Let $G = (V, E)$ be a graph.

(a) The **degree of a vertex** v, denoted $\deg(v)$, is the number of non-loop edges incident with v plus twice the number of loops incident with v. That is, each incident loop edge contributes 2 to the degree, and all other incident edges contribute 1.

(b) The **maximum degree** (respectively, **minimum degree**) of G, denoted $\Delta(G)$ (respectively, $\delta(G)$), is the maximum (respectively, minimum) degree among all vertices in G.

(c) A **degree sequence** for G is a sequence

$$\deg(v_1), \deg(v_2), \dots, \deg(v_n)$$

obtained from some ordering $v_1, v_2, \dots, v_n$ of V. A standard convention is to write a degree sequence in non-increasing order.

(d) If G has a constant degree sequence, then G is said to be **regular**. If each vertex has degree r, the G is called r-**regular**.

(e) A vertex of degree 0 is said to be an **isolated vertex**.

(f) A vertex of degree 1 is said to be a **pendant vertex**, or a **leaf**.

Example 8.27.   For the pictured graph G,

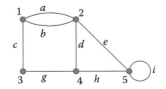

we have $\deg(1) = 3$, since $a, b,$ and c are the edges incident with vertex 1. Similarly, we see that $\deg(2) = 4$, $\deg(3) = 2$, and $\deg(4) = 3$. We have $\deg(5) = 4$, since $e, h,$ and i are incident with vertex 5 and the loop edge i counts twice. A degree sequence for G is thus $4, 4, 3, 3, 2$. In particular, $\Delta(G) = 4$ and $\delta(G) = 2$.

The Petersen graph from Example 8.23 is 3-regular. That is, each of its vertices has degree 3. For each integer $n \geq 1$, C_n is 2-regular and K_n is $(n-1)$-regular. The graph $P_1 = \Phi_1$ consists of a single isolated vertex $\bullet$, and, for each integer $n \geq 2$, P_n contains two pendant vertices.

The utility of the notions in Definition 8.19 for providing graph invariants comes from the fact that, at an individual vertex in a graph, the local structure of the graph reflected by the degree of that vertex must be preserved by an isomorphism.

Lemma 8.10. Let $f : G \longrightarrow H$ be a graph isomorphism and v a vertex of G. Then $\deg(f(v)) = \deg(v)$.

Proof. Let F be the set of edges incident with v, and let F' be the set of edges incident with $f(v)$. If $e \in F$, then $f(e) \in F'$. Conversely, if $e' \in F'$, then $f^{-1}(e') \in F$. It follows that $|F| = |F'|$. If e is a loop edge in F, then $f(e)$ must be a loop edge in F'. Conversely, if e' is a loop edge in F', then $f^{-1}(e')$ must be a loop edge in F. Consequently, the number of loop edges in F must be the same as the number of loop edges in F'. From the definition of the degree of a vertex, it now follows that $\deg(v) = \deg(f(v))$. $\square$

Example 8.28. Graphs (b) and (c) in Figure 8.27 are not isomorphic, since (b) has a vertex of degree 3 and (c) does not. That is, suppose to the contrary that there is an isomorphism f from (b) to (c). In (b), vertex 4 has degree 3. So it follows from Lemma 8.10 that in (c), vertex $f(4)$ must have degree 3. However, (c) has no vertices of degree 3, and a contradiction is obtained.

Theorem 8.11. (Degree Invariants).
Let G and H be graphs. If $G \cong H$, then

   (i) *G and H have a common degree sequence.*

  (ii) $\Delta(G) = \Delta(H)$.

 (iii) $\delta(G) = \delta(H)$.

Proof. Suppose f is an isomorphism from G to H. Let $v_1, v_2, \ldots, v_n$ be a listing of the vertices of G. So $f(v_1), f(v_2), \ldots, f(v_n)$ is a listing of the vertices of H. It follows from Lemma 8.10 that the associated lists of degrees must be the same. Hence, part (i) holds. Parts (ii) and (iii) follow easily from (i). □

Since $\Delta(G)$ and $\delta(G)$ are particular entrees in a degree sequence for G, the most powerful result in Theorem 8.11 is part (i).

Example 8.29. The pictured graph H,

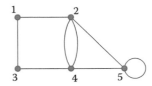

is not isomorphic to the graph G in Example 8.27, since $4, 4, 4, 2, 2$ is a degree sequence for H and it cannot possibly be reordered to give the degree sequence $4, 4, 3, 3, 2$ listed for G. Note, however, that $\Delta(G) = \Delta(H)$ and $\delta(G) = \delta(H)$, despite the lack of an isomorphism between G and H.

The degree of an individual vertex in a graph does not provide information about the global structure of the graph. However, if we add up the degrees of all of the vertices, then a remarkable result occurs.

Theorem 8.12. *For any graph* $G = (V, E)$, *we have* $\sum_{v \in V} \deg(v) = 2|E|$.

Proof. Let $e \in E$. What impact does the presence of e in G have on the sum $\sum_{v \in V} \deg(v)$? If $e \mapsto \{v_1, v_2\}$ is not a loop edge, then e contributes 1 to each term $\deg(v_1)$ and $\deg(v_2)$. If $e \mapsto \{v\}$ is a loop edge, then e contributes 2 to the term $\deg(v)$. Thus, each $e \in E$ contributes 2 to the sum $\sum_{v \in V} \deg(v)$. It follows that $\sum_{v \in V} \deg(v) = \sum_{e \in E} 2 = 2|E|$. □

The first of our corollaries of Theorem 8.12 follows immediately from the definition of an adjacency matrix. The proofs of the others are left for the exercises.

Corollary 8.13. Let $A = [a_{i,j}]$ be the adjacency matrix for a loopless graph $G = (V, E)$ obtained from the ordering $v_1, v_2, \ldots, v_n$ of its vertices. Then, for any $1 \leq k \leq n$, the sum of the entries in the kth row of A and the sum of the entries in the kth column of A both equal the degree of vertex v_k. That is,

$$\sum_{j=1}^{n} a_{k,j} = \sum_{i=1}^{n} a_{i,k} = \deg(v_k).$$

Moreover, the sum of all of the entries in A is twice the number of edges. That is,

$$\sum_{i=1}^{n} \sum_{j=1}^{n} a_{i,j} = 2|E|.$$

Corollary 8.14. In any graph, there must be an even number of odd degree vertices.

By Corollary 8.14, there is no graph with degree sequence $5, 3, 2, 2, 2, 1$, for example, since this sequence has an odd number of odd entries.

Corollary 8.15. If $G = (V, E)$ is r-regular, then $r|V| = 2|E|$.

It follows from Corollary 8.15 that if $G = (V, E)$ is a 4-regular graph, then $|E| = 2|V|$. An example of a graph with these properties is the octahedron graph.

8.5.1 Graph operations

We close this section with some graph operations reminiscent of set operations.

Definition 8.20. The **complement** of a simple graph $G = (V_G, E_G)$ is the graph G^c with $V_{G^c} = V_G$ and $E_{G^c} = E_G{}^c$, where the complement of E_G is taken inside $\mathcal{P}_2(V_G)$.

Example 8.30. Pictured are a simple graph G and its complement G^c.

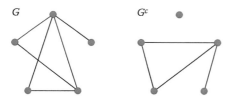

The vertices are the same in G and G^c, and the edges in G^c are precisely those missing from G.

The complement of a graph is invariant under isomorphism.

Theorem 8.16.   *Let G and H be simple graphs. If $G \cong H$, then $G^c \cong H^c$.*

The proof of Theorem 8.16 is left for the exercises. Theorem 8.16 is effective for showing that two graphs G and H are not isomorphic if it is easier to observe that G^c and H^c are not isomorphic. This tool is useful in the exercises.

Definition 8.21.   Let $G = (V_G, E_G)$ and $H = (V_H, E_H)$ be any two graphs.

   (a) To construct the **disjoint union** of G and H, the graph denoted $G + H$, we consider two cases. If V_G and V_H are disjoint, then we simply define $G + H = (V_G \cup V_H, E_G \cup E_H)$. If V_G and V_H are not disjoint, then we construct graphs $G' \cong G$ and $H' \cong H$ such that $V_{G'}$ and $V_{H'}$ are disjoint and define $G + H = (V_{G'} \cup V_{H'}, E_{G'} \cup E_{H'})$.

   (b) If G and H are both subgraphs of the same graph, then

      (i) The **union** of G and H, denoted $G \cup H$, is the graph $(V_G \cup V_H, E_G \cup E_H)$.

      (ii) The **intersection** of G and H, denoted $G \cap H$, is the graph $(V_G \cap V_H, E_G \cap E_H)$.

   (c) The **product** of G and H, denoted $G \times H$, is the graph with vertex set $V_G \times V_H$ and edge set $(E_G \times V_H) \cup (V_G \times E_H)$. An edge's endpoints are determined as follows.
If $e \in E_G$, $v \in V_H$, and $e \mapsto \{x, y\}$, then $(e, v) \mapsto \{(x, v), (y, v)\}$.
If $v \in V_G$, $e \in E_H$, and $e \mapsto \{x, y\}$, then $(v, e) \mapsto \{(v, x), (v, y)\}$.

The graph operations in Definition 8.21 have properties like their analogous set operations, as will be explored in the exercises. Here, we gain some familiarity with how these operations work.

Example 8.31.   The graph $K_4 + C_4$ is pictured.

Note that the standard vertex set $\{1,2,3,4\}$ for C_4 has been changed to $\{5,6,7,8\}$ to make it disjoint from the vertex set $\{1,2,3,4\}$ of K_4.

In terms of pictures, the disjoint union of two graphs is obtained by putting their pictures side by side in disjoint regions.

Example 8.32. Let G and H be the subgraphs of K_5 given by $V_G = \{1,2,4,5\}$, $E_G = \{\{1,2\},\{1,4\},\{1,5\},\{2,4\},\{4,5\}\}$, $V_H = \{1,2,3,4\}$, and $E_H = \{\{1,2\},\{1,4\},\{1,3\},\{2,3\}\}$. Their intersection and union are pictured.

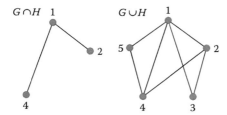

They are subgraphs of K_5.

Example 8.33. The product of the pictured graphs

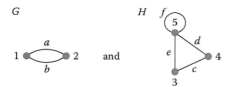

is the displayed graph $G \times H$.

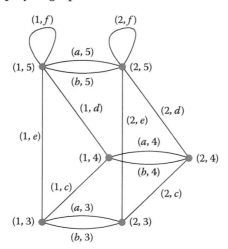

Each vertex of $G \times H$ is a pair (u, v) determined by a vertex u of G and a vertex v of H. There are two kinds of edges. One has the form (u, β), where u is a vertex of G and β is an edge of H. Such an edge joins the vertices (u, v_1) and (u, v_2), where v_1 and v_2 are the vertices of H joined by β. The other has the form (α, v), where α is an edge of G and v is a vertex of H. Such an edge joins the vertices (u_1, v) and (u_2, v), where u_1 and u_2 are the vertices of G joined by α.

Notice that $G \times H$ contains $|V_H| = 3$ disjoint copies of G, as subgraphs. Similarly, it contains $|V_G| = 2$ disjoint copies of H.

Example 8.34. $P_2 \times P_2 \cong Q_2 \cong C_4$. Here, we use $\{0, 1\}$ for the vertex set for P_2 and picture $P_2 \times P_2$.

$$
\begin{array}{ll}
(0,1) \;\bullet\!\!-\!\!-\!\!-\!\!\bullet\; (1,1) \\
(0,0) \;\bullet\!\!-\!\!-\!\!-\!\!\bullet\; (1,0)
\end{array}
$$

Its vertex set is the set of ordered pairs with entries from $\{0, 1\}$. Edges must then join pairs that differ in exactly one coordinate. Thus, the graph Q_2 is obtained.

Since $P_2 \cong Q_1$, Example 8.34 is a special case of the following general result, which allows us to express cubes as products.

Theorem 8.17.   *We have* $Q_0 \cong K_1$, $Q_1 \cong P_2$, *and, for* $n \geq 2$,

$$Q_n \cong Q_{n-1} \times P_2.$$

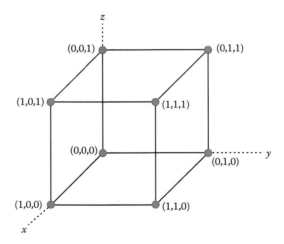

FIGURE 8.29
The graph Q_3.

A proof of Theorem 8.17 is left for the exercises. However, observe, for example, that the graph Q_3 displayed in Figure 8.29 can be seen to contain $|V_{P_2}| = 2$ copies of the graph Q_2 displayed in Example 8.34. One is the top square, each vertex of which has last coordinate 1. The other is the bottom square, each vertex of which has last coordinate 0. Similarly, in Figure 8.19 in Section 8.2, the graph Q_4 is seen as two copies of Q_3 joined together.

Exercises

The following graphs are referred to throughout the exercises.

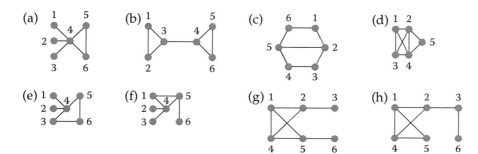

In Exercises 1 through 4, show that the specified graphs are not isomorphic.

1. Graphs (a) and (d).   2. Graphs (a) and (b).

3. Graphs (a) and (c).   4. Graphs (b) and (d).

In Exercises 5 through 8, find a degree sequence for the specified graph.

5. Graph (a).   6. Graph (b).

7. Graph (c).   8. Graph (d).

In Exercises 9 through 12, show that the specified graphs are not isomorphic.

9. Graphs (b) and (g).   10. Graphs (a) and (e).

11. Graphs (e) and (f).   12. Graphs (g) and (h).

13. Observe that the two pictured graphs have common degree sequences.

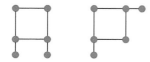

However, show that they are not isomorphic. Hint: Consider the vertices of degree 3.

14. Observe that the two pictured graphs have common degree sequences.

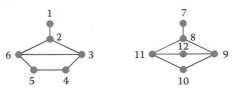

However, show that they are not isomorphic. Hint: Consider the vertices of degree 2.

15. A company needs to network 5 computers using 6 cable connections. Two competing network designers have come up with the pictured network configurations, and each is asserting the superiority of her design.

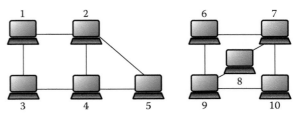

Show that the two network configurations pictured do not yield equivalent networks. Hence, there is a genuine decision to be made.

16. The New York office for a certain company contains a network with the configuration shown below on the left. Its Los Angeles office contains the network configuration shown on the right. A programmer from the New York office has traveled to the Los Angeles office and wishes to demonstrate a program that has been tested on the New York network.

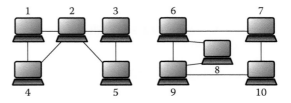

Show that the two network configurations pictured do not yield equivalent networks. Hence, the program may not work in Los Angeles.

17. Pictured are two design plans for a power grid, in which the symbol ○ denotes a power station. The one on the left is a grid in Philadelphia, Pennsylvania, and the one on the right is in Pittsburgh, Pennsylvania. Philadelphia has paid

for a comprehensive analysis of the ability of its grid to handle the failures of some of its power stations in storms.

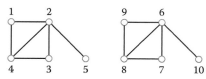

Show that the power grids are not equivalent and, hence, Pittsburgh cannot directly benefit from the grid analysis already purchased by Philadelphia.

18. Pictured are two design plans for a power grid in which the symbol ○ denotes a power station. One represents a grid in Nashville, Tennessee, and the other is in Syracuse, New York. In Nashville, they have some software that has been very effective at managing power transfers among their power stations.

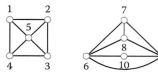

Show that the power grids are not equivalent and, hence, Syracuse may not be able to employ the same management software.

19. A certain university is making its summer schedule based upon predetermined student interest in its courses. The Department of Science and Technology is offering classes in Algebra, Biology, C++, Databases, Earth Science, and Functional Analysis. The Department of Languages and Cultures is offering classes in German, History, Indochina, Japanese, Kuwait, and Latin. In the displayed tables, an X denotes the fact that the two different classes corresponding to that row and column have a student in common.

	Alg.	Bio.	C++.	Data.	EaSc.	Func.
Alg.		X	X		X	X
Bio.	X		X	X		X
C++.	X	X			X	X
Data.		X	X		X	X
EaSc.	X		X	X		X
Func.	X	X			X	X

	Ger.	His.	Indo.	Jap.	Kuw.	Lat.
Ger.		X		X		X
His.	X		X	X	X	
Indo.		X			X	X
Jap.	X	X			X	X
Kuw.		X	X	X		X
Lat.	X		X	X	X	

Show that the corresponding class scheduling graphs are not isomorphic. Hence, a workable schedule for the Department of Science and Technology cannot be converted to one for the Department of Languages and Cultures.

20. The Senate needs to schedule meetings for its committees: Appropriations (A), Budget (B), Commerce (C), Energy (D), Environment (E), Foreign Relations (F), Homeland Security and Governmental Affairs (G), and Health (H). On the other side of the Capitol building, the House of Representatives needs to schedule meetings for its committees: International Relations (I), Judicial (J), Agriculture (K), Armed Services (L), Financial Services (M), Science (N), Standards of Official Conduct (O), and Printing (P). In the displayed tables, an X denotes the fact that the two different committees corresponding to that row and column have a member in common.

	A	B	C	D	E	F	G	H
A		X		X	X	X		
B	X		X			X		X
C		X		X	X		X	
D	X		X				X	X
E	X		X			X	X	
F	X	X			X			X
G			X	X	X			X
H		X		X		X	X	

	I	J	K	L	M	N	O	P
I			X		X		X	X
J			X	X		X	X	X
K	X	X			X	X		
L		X				X		X
M	X		X			X		X
N		X	X	X	X		X	
O	X	X				X		
P	X	X		X	X			

Show that the corresponding committee scheduling graphs are not isomorphic. Hence, a workable schedule for the Senate cannot be converted to one for the House of Representatives.

21. List all non-isomorphic simple graphs on 4 or fewer vertices.

22. List all non-isomorphic simple graphs on 5 vertices.

23. Find all non-isomorphic simple graphs on 6 vertices and 3 edges.

24. Find all non-isomorphic simple graphs on 6 vertices and 4 edges.

In Exercises 25 through 30, list all non-isomorphic simple graphs (if any) with the given degree sequence.

25. $5, 3, 2, 2, 2, 2$.

26. $4, 4, 4, 4$.

27. $4, 3, 3, 2, 1$.

28. $3, 3, 2, 2$.

29.* $4, 2, 2, 2, 1, 1$.

30.* $5, 4, 3, 3, 3, 2$.

31. Prove Corollary 8.14.

32. Prove Corollary 8.15.

33. Determine the numbers of vertices and edges in Q_n.

34. Show: If G has n vertices and m edges, then $\delta(G) \leq \lfloor \frac{2m}{n} \rfloor$. Is that upper bound achievable? Hint: Consider average degree.

In Exercises 35 through 38, find the complement of the specified graph.

35. Graph (a).   36. Graph (b).

37. Graph (c).   38. Graph (d).

39. Show that $C_5{}^c \cong C_5$.

40. Show that $P_4{}^c \cong P_4$.

41. What graph is the complement of the pictured graph?

42. What graph is the complement of the pictured graph?

43. Observe that the two pictured graphs have common degree sequences.

However, show that they are not isomorphic by considering their complements.

44. Observe that the two pictured graphs have common degree sequences.

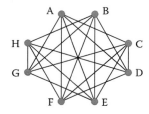
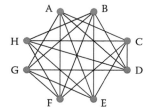

However, show that they are not isomorphic by considering their complements.

45. Prove Theorem 8.16.

46. Show for any simple graph $G = (V, E)$ that $G^c \cup G \cong K_{|V|}$.

47. Show: For any simple graph G, $\text{Aut}(G) = \text{Aut}(G^c)$. Note that here we regard elements of $\text{Aut}(G)$ as functions (permutations) $f : V \longrightarrow V$.

48. Show: Any graph G is vertex transitive if and only if G^c is vertex transitive.

49. Prove or disprove: Any graph that is vertex transitive must be regular.

50. Prove or disprove: Any graph that is regular must be vertex transitive.

In Exercises 51 through 60, draw the specified graph. Regard the graphs (a),(b),(c),(d) defined at the beginning of these exercises as subgraphs of K_6.

51. $(a) + (d)$.

52. $(b) + (c)$.

53. $(a) \cap (c)$.

54. $(a) \cup (c)$.

55. $(b) \cup (c)$.

56. $(b) \cap (c)$.

57. $P_3 \times P_3$.

58. $P_2 \times P_4$.

59. $P_2 \times K_{1,3}$.

60. $C_3 \times P_4$.

61.* **Distributive Laws.** Let M be any graph. Show that, for any subgraphs G, H, K of M,

(a) $G \cap (H \cup K) = (G \cap H) \cup (G \cap K)$.

(b) $G \cup (H \cap K) = (G \cup H) \cap (G \cup K)$.

62.* **De Morgan's Law.** Let M be any graph. Show that, for any subgraphs G, H, K of M,

(a) $(G \cap H)^c = G^c \cup H^c$.

(b) $(G \cup H)^c = G^c \cap H^c$.

63. Prove that the graph pictured in Exercise 41 is isomorphic to $C_3 \times P_2$. In general, the graph $C_n \times P_2$ is called the n-**prism**.

64. Prove Theorem 8.17.

65. Assuming that V_G and V_H are nonempty, show that $G \times H$ has a subgraph isomorphic to G and a subgraph isomorphic to H.

66.* Show G and H are bipartite if and only if $G \times H$ is bipartite.

8.6   Directed graphs and Markov chains

In Section 5.1, the notion of a directed graph, or digraph, for a relation on a set was introduced. For example, Figure 8.30 displays the directed graph for the "is a proper divisor of" relation on $\{1, 2, \ldots, 8\}$. Having studied graphs in this chapter, we can now see that such digraphs are very much like graphs, except that directions are assigned to the edges.

Like graphs, directed graphs arise in many and varied applications. Before giving a formal definition of a directed graph, analogous to Definition 8.1 for ordinary graphs, we consider another example in which the relevant edges have a specified direction.

Example 8.35.   (Rounding the Bases).
Figure 8.31 shows the possible states of a baseball player (who is either at bat or on base) before and after a plate appearance (by either the player himself or a teammate). There, an arrow points from the state prior to the plate appearance to the state afterward. For example, it is possible for a player starting at second (2nd) base to score if batted in. However, it is impossible for a player starting at 2nd base to move to first (1st) base. A diagram such as Figure 8.31 with additional statistics could be used to determine strategy or to keep track of probabilities for a baseball video game.

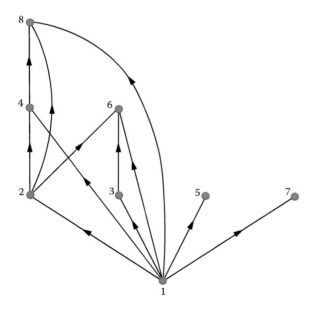

FIGURE 8.30
Digraph for "proper divisor" relation.

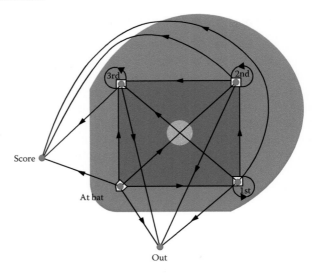

FIGURE 8.31
Plate appearance.

In an ordinary graph, each edge is assigned a set $\{u, v\}$ of endpoints from the vertex set V. In a directed graph, we want this edge assignment to be, more specifically, *from* a vertex u *to* a vertex v. Essentially, we aim to order the endpoints so that u is first and v is second. Although the elements in a set $\{u, v\}$ do not have an ordering, the coordinates in an ordered pair (u, v) certainly do. We use this feature of ordered pairs to give a direction to each edge. Recall that the elements of a product $V^2 = V \times V$ are ordered pairs.

Definition 8.22.

(a) A **directed graph**, or **digraph**, G consists of a **vertex** set V_G, an **edge** set E_G, and a function $\epsilon : E_G \longrightarrow V_G \times V_G$. We write $G = (V_G, E_G)$, and, rather than writing $\epsilon_G(e) = (u, v)$, we write $e \mapsto (u, v)$. As in ordinary graphs, the subscripts may be dropped from our notation.

(b) If $e \mapsto (u, v)$, then the vertex u is called the **initial endpoint** or **tail** of e and the vertex v is called the **terminal endpoint** or **head** of e. We say that e goes from u to v.

(c) An edge e such that $e \mapsto (v, v)$ for some $v \in V$ is called a **loop**, and two or more edges assigned the same initial and terminal endpoints are called **multiple edges**.

(d) A **simple directed graph** is a directed graph $G = (V, E)$ that has no loops and no multiple edges. In this case, we can take E to be a subset of $V^2 \setminus \{(v, v) \ : \ v \in V\}$.

Example 8.36. In the following examples, $G = (V, E)$ is a directed graph.

(a) Let $V = \{1, 2, 3, 4, 5\}$ and $E = \{a, b, c, d, e, f, g\}$, where $a = (1, 2)$, $b = (2, 1)$, $c = (3, 1)$, $d = (4, 2)$, $e = (2, 5)$, $f = (3, 4)$, and $g = (5, 4)$.

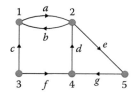

This is indeed a simple directed graph. The edges $(1, 2)$ and $(2, 1)$ are distinct elements of the product V^2.

(b) Let $V = \{1, 2, 3, 4, 5\}$ and $E = \{a, b, c, d, e, f, g, h\}$, where $a \mapsto (1, 2)$, $b \mapsto (1, 2)$, $c = (3, 1)$, $d = (4, 2)$, $e = (2, 5)$, $f = (3, 4)$, $g = (5, 4)$, and $h = (5, 5)$.

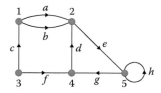

This directed graph is not simple. The edges a and b are multiple edges, since they are assigned to the same element of the product V^2. Also, the edge $h = (5, 5)$ is a loop.

A **subgraph** of a directed graph is defined in the same way that it is for ordinary graphs (in Definition 8.3). However, extra care is taken in defining walks.

Definition 8.23. A **walk** in a directed graph $G = (V, E)$ is an alternating list of vertices and edges

$$v_0, e_1, v_1, e_2, v_2, e_3, \ldots, v_{n-1}, e_n, v_n$$

with $n \geq 0$ that starts at vertex v_0, ends at vertex v_n, and, in which, for each $1 \leq i \leq n$, $e_i \mapsto (v_{i-1}, v_i)$. That is, edges must be traversed from tail to head.

Based on the definition of a walk in a directed graph, the other terms (**length, start, end, circuit, trail, path, cycle,**

distance) and notation from Definition 8.4 are defined in the same way for directed graphs.

Example 8.37. In the directed graph from Example 8.36(a), an example of a walk is given by the list $5, g, 4, d, 2, b, 1, a, 2$. However, that same list is not a walk in the directed graph from Example 8.36(b). There, the portion $2, b, 1$ is not allowed, since b cannot be traversed from head to tail.

Consider the directed graph from Example 8.36(a). Since e is an edge that directly connects 2 *to* 5, we have $\text{dist}(2, 5) = 1$. However, $\text{dist}(5, 2) = 2$, since there is not a single edge from 5 to 2. Instead, $5, g, 4, d, 2$ is the shortest path from 5 to 2.

Example 8.38. (Street Plans).
To reduce traffic and increase parking, city planners have proposed that the neighborhood shown on the left-hand side of Figure 8.32 convert to one-way streets the portions of the streets marked with arrows. The resulting traffic flow is thus reflected by the directed graph shown on the right-hand side of Figure 8.32. In that digraph, intersections are represented by vertices, which we have additionally labeled for easy reference. Each edge represents a possible direction of traffic between two intersections. Hence, any street remaining two-way is reflected by two edges, one in each direction. Note that there are walks from the corner of Spruce and Third (labeled 9) to the corner of Maple and First (labeled 1), such as $9, 8, 5, 1$. However, there is no return walk from 1 to 9. This presents a major flaw in the plan that we shall address further after Definition 8.25.

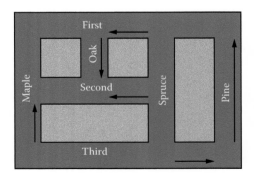
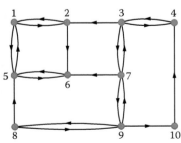

FIGURE 8.32
Planned one-way streets in a neighborhood.

When looking at a directed graph, it is natural to see the ordinary graph underneath; it is obtained by removing the directions from the edges.

Definition 8.24. Given a directed graph $G = (V, E)$, its **underlying graph**, denoted $\underline{G}$, is the ordinary graph with the same vertex set V and with the edge set $\underline{E}$ containing one undirected edge $\underline{e}$ for each directed edge $e \in E$. The ends of $\underline{e}$ are taken to be the head and tail of e.

Example 8.39. Shown are a directed graph G and its underlying graph $\underline{G}$.

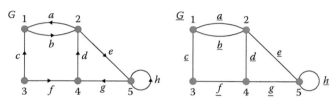

Even if G is a simple directed graph, $\underline{G}$ could have multiple edges, as is the case in Example 8.39.

Since, in a directed graph, it is possible for there to be a path from u to v but not one from v to u, there are two natural ways to define components in a directed graph.

Definition 8.25. Let G be a directed graph.

(a) G is said to be **strongly connected** if, for any two vertices u and v, there is a path from u to v and there is a path from v to u.

(b) G is said to be **weakly connected** if $\underline{G}$ is connected.

(c) A **strong component** of G is a strongly connected subgraph H that is not contained in any other strongly connected subgraph of G.

(d) A **weak component** of G is a subgraph H such that $\underline{H}$ is a component of $\underline{G}$.

Example 8.40. The pictured directed graph G

is weakly connected, since its underlying graph is connected. However, G is not strongly connected; it has 2 strong components. One is the isolated vertex $\{1\}$, and the other is the subgraph induced by $\{2, 3, 4\}$. From each

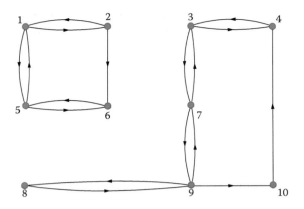

FIGURE 8.33
Isolated areas of the neighborhood.

vertex in $\{2, 3, 4\}$, there exists a path to any other vertex in $\{2, 3, 4\}$, but no path to vertex 1.

The notion of a strong component enables us to better explain the failing of the street plan in Figure 8.32.

Example 8.41. Displayed in Figure 8.33 are the strong components of the directed graph from Figure 8.32. One is the city block induced by the set of vertices $\{1, 2, 5, 6\}$, and the other is induced by $\{3, 4, 7, 8, 9, 10\}$. The point is that, from any of the intersections 1, 2, 5, or 6, it is not possible to get out of that city block and into another portion of the neighborhood. Although it happens to be possible to get to that city block from anywhere else, getting out is impossible. Thus, the proposed plan in Figure 8.32, in which the neighborhood would not be strongly connected, is not a good one.

Directed graphs are compared in much the same way as are ordinary graphs. However, additionally, in directed graphs, the direction of each edge must be preserved by a graph map. This is important in determining isomorphisms.

Example 8.42. The two pictured directed graphs are isomorphic.

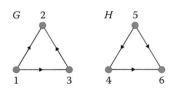

An isomorphism $f : G \longrightarrow H$ is given by $f(1) = 5$, $f(2) = 6$, and $f(3) = 4$. Note, for example, that this causes the edge $(1, 2)$ to be mapped to the edge $(5, 6)$, as opposed to the nonexistent edge $(6, 5)$. That is, $f(1) = 6$, $f(2) = 5$, and $f(3) = 4$ would not provide an isomorphism for the directed graphs here, even though it would do so for their underlying ordinary graphs.

Graph isomorphisms necessarily preserve invariants such as the numbers of vertices and edges. For a directed graph, the **in-degree** of a vertex v, denoted $\text{indeg}(v)$, is the number of edges whose head is v, and the **out-degree**, denoted $\text{outdeg}(v)$, is the number of edges whose tail is v. For a vertex v in a directed graph, we define $\deg(v) = \text{indeg}(v) + \text{outdeg}(v)$. That is, $\deg(v)$ is the degree of v in $\underline{G}$.

Example 8.43.

(a) In the directed graph

we have $\text{indeg}(1) = 0$, $\text{indeg}(2) = 2$, $\text{indeg}(3) = 2$, $\text{indeg}(4) = 1$, $\text{outdeg}(1) = 2$, $\text{outdeg}(2) = 0$, $\text{outdeg}(3) = 0$, and $\text{outdeg}(4) = 3$. For example, vertex 1 has two edges pointing out of it and none pointing in. Also, note that the loop at vertex 4 contributes once to each of the in-degree and out-degree of vertex 4.

(b) The two pictured directed graphs are not isomorphic.

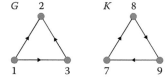

In G, vertex 1 has $\text{indeg}(1) = 0$, whereas each vertex v in K has $\text{indeg}(v) = 1$. Hence, there is no potential vertex with in-degree 0 to be the image of 1 in an isomorphism $f : G \longrightarrow K$.

Example 8.44. (New Street Plans).
If we change the traffic plans in Figure 8.32 by reversing the direction of the portion of Second Street that is one-way, then the resulting directed graph is strongly connected. Consequently, this new plan is not isomorphic to the old plan, even though the underlying graph remains the same. The new plan is certainly better for the residents of the neighborhood, especially those living in the city block induced by $\{1, 2, 5, 6\}$. Otherwise, they would have been stuck there.

As in ordinary graphs, there is a connection between the sum of the degrees of the vertices and the number of edges in a directed graph.

Theorem 8.18. *For any directed graph $G = (V, E)$, we have*

$$\sum_{v \in V} indeg(v) = \sum_{v \in V} outdeg(v) = |E|.$$

The proof of Theorem 8.18 is left for the exercises.

Definition 8.26. The **adjacency matrix** for a directed graph G on the ordered list of vertices $v_1, v_2, \ldots, v_n$ is the n-by-n matrix $A = [a_{i,j}]$ such that $a_{i,j}$ is the number of edges from v_i to v_j.

Example 8.45. If, for the directed graph

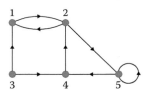

we take the ordering of the vertices to be $1, 2, 3, 4, 5$, then the adjacency matrix obtained is

$$
\begin{array}{c}
\\
1\\
2\\
3\\
4\\
5
\end{array}
\begin{array}{c}
\begin{array}{ccccc}1 & 2 & 3 & 4 & 5\end{array}\\
\left[
\begin{array}{ccccc}
0 & 1 & 0 & 0 & 0\\
1 & 0 & 0 & 0 & 0\\
1 & 0 & 0 & 1 & 0\\
0 & 1 & 0 & 0 & 0\\
0 & 0 & 0 & 1 & 1
\end{array}
\right].
\end{array}
$$

Adjacency matrices for directed graphs need not be symmetric, as they are for ordinary graphs.

8.6.1 Markov chains

A directed graph can be used to model a random process. Vertices are used to represent the states of the process, and each edge (u, v) reflects a possible transition from state u to state v in one iteration of the process. For example, this is the case in Figure 8.31, in which the vertices represent the states of a baseball player and the edges point to possible subsequent states. (We expand on this example in Exercise 63.) In general, each edge (u, v) might be assigned the probability that a direct transition from state u to state v will occur, with $p(u, v) = 0$ if (u, v) is not an edge. Of course, for a fixed state u, since an iteration of the process must result in a transition to some state v, we must have $\sum_{v \in V} p(u, v) = 1$. If, for all states u and v, the probability $p(u, v)$ does not change from iteration to iteration, then the process is called a **Markov chain**. Such processes are named after the Russian mathematician Andrei Markov (1856–1922). In keeping with our consideration of finite graphs only, we shall consider only Markov chains with finitely many states.

Definition 8.27. A **Markov chain graph** is a directed graph $G = (V, E)$ without multiple edges, for which each edge (u, v) is assigned a value $p(u, v)$ in the interval $[0, 1]$. Moreover, for each vertex u, the sum of the values assigned to the edges with tail u must be 1.

Example 8.46. At an extremely competitive private school, each semester every student is assigned one of three states. The student has normal status if everything is fine. The student is assigned probationary status if the previous semester's grades are poor. The student is suspended for a semester if grades are not improved while on probation. We assume that students always return to normal status after a semester of being suspended.

By tracking students for several years, the school has determined the likelihood of the possible status changes. Using v_1 for normal status, v_2 for probationary status, and v_3 for suspended status, the pictured Markov chain graph

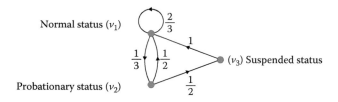

reflects the probabilities of moving from one status to another at this school. For example, from normal status, there is a $\frac{2}{3}$ chance of retaining normal status, a $\frac{1}{3}$ chance of receiving probationary status, and no chance of getting suspended for the subsequent semester.

A matrix similar to an adjacency matrix can be used to keep track of the state-to-state transition probabilities.

Definition 8.28.   Given a Markov chain graph and an ordering of its vertices, $v_1, v_2, \ldots, v_n$, its **transition matrix** is the n-by-n matrix $M = [p(v_i, v_j)]$. That is, the (i, j)th entry of M is the value $p(v_i, v_j)$, which may be denoted more compactly as $p(i, j)$.

Example 8.47.   The transition matrix M for the Markov chain graph in Example 8.46 is displayed.

$$
\begin{array}{c}
 & \begin{array}{ccc} v_1 & v_2 & v_3 \end{array} \\
\begin{array}{c} v_1 \\ v_2 \\ v_3 \end{array} &
\left[\begin{array}{ccc}
\frac{2}{3} & \frac{1}{3} & 0 \\
\frac{1}{2} & 0 & \frac{1}{2} \\
1 & 0 & 0
\end{array} \right]
\end{array}
$$

More than merely storing the transition probabilities for one iteration of a process, the transition matrix can be used to compute analogous transition probabilities for multiple iterations of the process.

Theorem 8.19.   *Let M be the transition matrix for a Markov chain graph G obtained from the ordering $v_1, v_2, \ldots, v_n$ of the vertices, and let $m \in \mathbb{N}$. Then, the mth power of M, say, $M^m = [q_{i,j}]$, has the property that for each $1 \le i \le n$ and $1 \le j \le n$, the entry $q_{i,j}$ is the probability of moving in G from v_i to v_j in a sequence of exactly m steps.*

Example 8.48.   The second and third powers of the transition matrix M for the Markov chain graph in Example 8.46 are shown.

$$
M^2 = \left[\begin{array}{ccc}
\frac{11}{18} & \frac{2}{9} & \frac{1}{6} \\
\frac{5}{6} & \frac{1}{6} & 0 \\
\frac{2}{3} & \frac{1}{3} & 0
\end{array} \right]
\qquad
M^3 = \left[\begin{array}{ccc}
\frac{37}{54} & \frac{11}{54} & \frac{1}{9} \\
\frac{23}{36} & \frac{5}{18} & \frac{1}{12} \\
\frac{11}{18} & \frac{2}{9} & \frac{1}{6}
\end{array} \right]
$$

For example, we see that there is a $\frac{1}{6}$ chance that a student with normal status will be suspended after two semesters. Also, there is a $\frac{23}{36}$ chance that a student on probation will return to normal status after three semesters.

If we consider higher and higher powers of M, then we see that they converge to a fixed matrix with special properties. In this case, a strong trend is already apparent by the 7th power, and it turns out that these powers are approaching the limiting matrix, denoted M^∞, displayed below.

$$M^7 \approx \begin{bmatrix} 0.667 & 0.223 & 0.110 \\ 0.665 & 0.222 & 0.113 \\ 0.668 & 0.221 & 0.112 \end{bmatrix} \quad M^\infty = \begin{bmatrix} \frac{2}{3} & \frac{2}{9} & \frac{1}{9} \\ \frac{2}{3} & \frac{2}{9} & \frac{1}{9} \\ \frac{2}{3} & \frac{2}{9} & \frac{1}{9} \end{bmatrix}$$

Moreover, the rows of M^∞ are all the same, and thus, for $j = 1, 2, 3$, the jth entry in a row of M^∞ gives the long-term probability of being in state v_j, regardless of the initial state. That is, in the long run, roughly $\frac{2}{3}$ of the students have normal status, $\frac{2}{9}$ are on probation, and $\frac{1}{9}$ are suspended.

To better explain the convergence properties for Example 8.46 observed in Example 8.48, we need some terminology. A state v_i in a Markov chain graph is said to have **period** q if $q \in \mathbb{Z}^+$ and the length of every circuit starting at v_i is a multiple of q. We say that v_i is **periodic** if $q > 1$ and **aperiodic** if $q = 1$. Although the shortest circuit starting at v_3 in the Markov chain graph from Example 8.46 has length 3, the presence of the loop at v_1 enables us to form circuits of any length greater than or equal to 3 starting from any vertex. Thus, every state in Example 8.46 is aperiodic.

A **class** in a Markov chain is a set of states that corresponds to the vertex set for a strong component in the corresponding graph. If a Markov chain has just one class, then the Markov chain is said to be **irreducible**. A finite Markov chain is said to be **regular** if it is irreducible and every state is aperiodic. There is a simple test for regularity, which we state without proof.

Lemma 8.20. A finite Markov chain with transition matrix M is regular if and only if there exists some $m \in \mathbb{Z}^+$ such that M^m has all positive entries.

In light of the value M^3 displayed in Example 8.48, it follows from Lemma 8.20 that the Markov chain in Example 8.46 is regular. Of course, it is not hard to see this directly as well, since the graph can be seen to be strongly connected, and we observed above that each state is aperiodic. The following result, which we also state without proof, explains the convergence behavior demonstrated for Example 8.46.

Theorem 8.21.   *If M is a transition matrix for a finite regular Markov chain, then as m increases, the powers M^m of M converge to a matrix, which we denote by M^∞, all of whose rows are the same. Moreover, for each j, the entry in the jth column of every row of M^∞ is the long-term probability of being in state v_j, independent of the initial state.*

To appreciate properties that irregular Markov chains might exhibit, such as periodicity, we consider another example.

Example 8.49.   (Tied in Tennis).
When a tennis game has a score of deuce, it means that the game is tied and the first player to gain a two-point lead will be declared the winner of that game. A player with a one-point lead is said to have the advantage, but the state of deuce may be revisited several times before the game is completed. Since a game has a fixed server and a fixed receiver, the possible states are deuce (D), advantage server (AS), advantage receiver (AR), a win for the server (WS), and a win for the receiver (WR). Assuming that the server has a 60% chance of winning any given point and the receiver has a 40% chance, the relevant Markov chain graph and its transition matrix are displayed in Figure 8.34 and are accompanied by its transition matrix.

By computing the value of M^2, shown in Figure 8.35, we can see what might happen after two more points of the game. For example, from deuce (D), there is a 0.36 chance that the server will win (WS) and a 0.16 chance that the receiver will win (WR). From advantage server (AS), there is a 0.6 chance that the server will win (WS). Of course, that win will have occurred after the first point, and the second iteration will have cycled back onto the state of victory in that case.

The Markov chain is not irreducible; it has three classes. One class consists of the states D, AS, and AR.

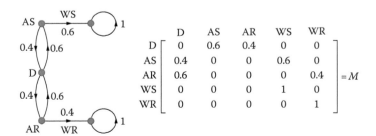

FIGURE 8.34
Tennis end-game analysis.

$$M^2 = \begin{bmatrix} 0.48 & 0 & 0 & 0.36 & 0.16 \\ 0 & 0.24 & 0.16 & 0.6 & 0 \\ 0 & 0.36 & 0.24 & 0 & 0.4 \\ 0 & 0 & 0 & 1 & 0 \\ 0 & 0 & 0 & 0 & 1 \end{bmatrix} \qquad M^{30} \approx \begin{bmatrix} 0 & 0 & 0 & 0.692 & 0.308 \\ 0 & 0 & 0 & 0.877 & 0.123 \\ 0 & 0 & 0 & 0.415 & 0.585 \\ 0 & 0 & 0 & 1 & 0 \\ 0 & 0 & 0 & 0 & 1 \end{bmatrix}$$

FIGURE 8.35
Ongoing tennis.

Each of these states has period 2. Each of WS and WR forms a class by itself and has period 1. Although the Markov chain is not regular and Theorem 8.21 therefore does not apply, we can still observe some meaningful long-term behavior here. In this case, the powers of M do not change much after the 30th power, shown in Figure 8.35. Although the rows are not all the same, as results in the regular case, they still store long-term probabilities. The first row shows that, starting from deuce (D), the chance that the server wins is approximately 0.692 and the chance that the receiver wins is approximately 0.308. The second and third rows give similar probabilities starting from advantage server (AS) and advantage receiver (AR), respectively. The last two rows reflect the fact that once a player has won the game, there are no further transitions to a different state in that game.

In a Markov chain, a state v_i is said to be **absorbing** if $p(i, i) = 1$. That is, in the Markov chain graph, there is a loop edge at v_i with assigned probability 1. In Example 8.49, WS and WR are the absorbing states. Note in the convergence behavior observed there that it is in those columns that the nonzero values congregate; all initial states are absorbed into those absorbing states.

A class is said to be **ergodic** if no edge in the Markov chain graph points from a vertex inside the class to a vertex outside the class. In this case, each of the states in the class is also said to be ergodic, or **recurrent**. In Example 8.49, the states WS and WR are recurrent. A class for which there is an edge in the Markov chain graph pointing from a vertex inside the class to a vertex outside the class is said to be **transient**, as is each of its states. In Example 8.49, the states D, AS, and AR form a class of transient states.

A Markov chain is said to be **absorbing** if each state is either absorbing or transient, so what is special about the Markov chain in Example 8.49 is that it is absorbing. In general, if a Markov chain with transition matrix M is absorbing, then the powers M^m of M converge to a matrix M^∞, which contains long-term transition probabilities. For a proof of this and the other unproved results on Markov chains stated in this section, the reader is referred to Ross (2010).

Exercises

1. When there is a lot of snow on the ground, the residents of the pictured neighborhood have complained that some of the streets are too narrow for two cars to pass. Hence, town planners have proposed that during the winter months, the portions of the streets marked with arrows should be converted to one-way streets.

   Draw a directed graph that reflects the resulting traffic flow.

2. To reduce traffic and increase parking, city planners have proposed that the pictured neighborhood convert to one-way streets the portions of the streets marked with arrows.

   Draw a directed graph that reflects the resulting traffic flow.

3. In the pictured computer network configuration, some of the direct cable connections only allow one-way communication from one computer to another. On each line, an arrow points in a direction of possible communication along that line.

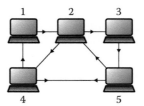

   Draw the directed graph that reflects this network.

4. Pictured is a map of the bus system in a certain city. The direction of travel of each bus is marked on the map. Each bus route runs a circuit.

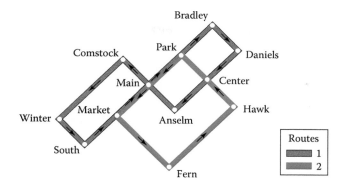

Draw the directed graph representing the entire bus system.

The following directed graphs $G = (V, E)$ are referred to throughout the exercises.

(a) $V = \{1, 2, 3, 4, 5\}$ and $E = \{(1, 3), (1, 5), (2, 3), (4, 2),$
$(3, 4), (3, 5)\}$.

(b) $V = \{1, 2, 3, 4, 5, 6\}$ and
$E = \{(1, 3), (2, 1), (3, 2), (4, 3), (4, 6), (5, 4),$
$(6, 1), (6, 5)\}$.

(c) $V = \{1, 2, 3, 4\}$ and $E = \{(1, 3), (2, 1), (2, 3), (3, 4), (4, 1),$
$(2, 2)\}$.

(d) $V = \{1, 2, 3, 4\}$ and $E = \{(1, 2), (1, 3), (1, 4), (2, 3), (4, 1)\}$.

In Exercises 5 through 8, draw the specified directed graph and determine whether it is simple. If not, explain why.

5. Graph (a).   6. Graph (b).

7. Graph (c).   8. Graph (d).

9. In graph (c), is there a path from 2 to 4?

10. In graph (d), is there a path from 2 to 4?

11. Consider the computer network from Exercise 3. Find two possible paths through which computer 1 could send a message to computer 5.

12. Consider the city bus system from Exercise 4. Find two possible paths one could use to travel from South to Center.

In Exercises 13 through 16, draw the underlying graph of the specified directed graph.

13. Graph (a).   14. Graph (b).

15. Graph (c).   16. Graph (d).

In Exercises 17 through 22, determine the strong components of the specified directed graph.

17. Graph (a).   18. Graph (b).

19. Graph (c).   20. Graph (d).

21. Consider the neighborhood in Exercise 1. From each point in the neighborhood, is it possible to travel to any other? If not, then describe the portions in which travel is restricted.

22. Consider the neighborhood in Exercise 2. From each point in the neighborhood, is it possible to travel to any other? If not, then describe the portions in which travel is restricted.

In Exercises 23 through 26, prove that the two pictured directed graphs are isomorphic.

23.

24.

25.

26.

In Exercises 27 through 32, list the in- and out-degrees for each vertex of the specified directed graph.

27. Graph (a).   28. Graph (b).

29. Graph (c).   30. Graph (d).

31.   32.

In Exercises 33 through 36, prove that the two pictured directed graphs are not isomorphic.

33.

34.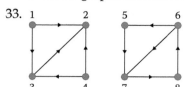

35.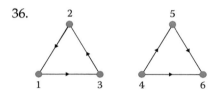

36.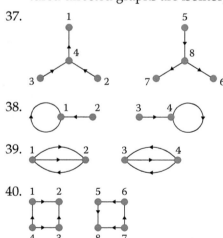

In Exercises 37 through 40, prove or disprove that the pictured directed graphs are isomorphic.

37.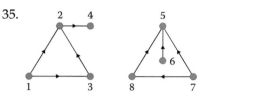

38.

39.

40.

41. List all non-isomorphic simple directed graphs on 3 or fewer vertices.

42. List all non-isomorphic simple directed graphs on 4 vertices and 3 or fewer edges.

43.* List all non-isomorphic directed graphs G such that $\underline{G} \cong K_4$.

44.* List all non-isomorphic directed graphs G such that $\underline{G} \cong K_{2,3}$.

45. Let G be any directed graph. Show: If G is strongly connected, then G is weakly connected.

46. Prove Theorem 8.18.

47. Let G and H be any directed graphs. Show: If $G \cong H$, then $\underline{G} \cong \underline{H}$. Does the converse hold?

48. Show that the strong components and the weak components are the same in any directed graph $G = (V, E)$ satisfying the condition

$$\forall\, u, v \in V, \quad (u, v) \in E \leftrightarrow (v, u) \in E.$$

In Exercises 49 and 50, (a) find the adjacency matrix A for the given graph, and (b) compute A^2 by hand.

49.

50.

51.* If G is a directed graph without loops and A is its adjacency matrix, then express the adjacency matrix for $\underline{G}$ in terms of A.

52.* Under what conditions will the adjacency matrix for a directed graph be symmetric?

53. State a version of Corollary 8.13 for directed graphs.

54. State a version of Corollary 8.15 for directed graphs.

55. Is the pictured directed graph a Markov chain graph? If not, then explain why.

56. Is the pictured directed graph a Markov chain graph? If not, then explain why.

57. For the given Markov chain graph,

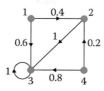

(a) find the transition matrix M and (b) find M^2.

58. For the given Markov chain graph,

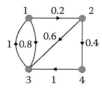

(a) find the transition matrix M and (b) find M^2.

For Exercises 59 through 68, use mathematical software such as *Mathematica* or *Maple* to compute powers of matrices.

59. **Full tennis game**. In a game of tennis, a player must win by at least two points. However, the points are not called $0, 1, 2$, etc.; they are called 0 (or Love), 15, 30, 40, Game. The convention is to state the score as a pair, with the server's score listed first. Once the smaller of the two scores reaches at least 30, any tie score is called deuce, and a player that needs only one more point to win the game is said to have the advantage.

(a) Draw a Markov chain graph whose vertices reflect the possible scores in a tennis game. Note that Figure 8.34 displays a subgraph of this graph. However, our graph will not include any probability labels.

(b) How many classes are there?

(c) How many are transient?

60. **Winning by three**. Players A and B are playing a game in which points are determined by a coin toss and the first player to lead by three points wins. If the coin comes up heads, then player A wins a point. If the coin comes up tails, then player B wins a point. Since only the difference in points matters, a score of $(2, 1)$ is identified with the score $(1, 0)$, for example. In general, (a, b) is identified with $(a - \min\{a, b\}, b - \min\{a, b\})$. Hence, there are only finitely many possible states.

(a) Assume that the coin is fair, and make a Markov chain graph that reflects this game.

(b) What are the absorbing states?

(c) What is the period of the transient states?

(d) What is the likelihood that the player who wins the first toss will win the game? Hint: Consider M^∞.

61. Let M be the transition matrix for a Markov chain graph. Show that for each $1 \le i \le n$, the sum of the entries in the ith row of M is 1. That is, $\sum_{j=1}^{n} p(i,j) = 1$. Must the column sums be 1?

62. Let G be any Markov chain graph on n vertices, and let M be a transition matrix for G. Show that G is strongly connected if and only if every entry of the matrix $I + M + M^2 + \cdots + M^{n-1}$ is positive.

63. **Baseball analysis**. Form a Markov chain graph from the directed graph in Figure 8.31 as follows. Label the vertices $0, 1, 2, 3, 4, 5$, where 0 is At Bat, 1 is First Base, 2 is Second Base, 3 is Third Base, 4 is Score, and 5 is Out. Also, add loops at vertices 4 and 5. Let the following be the transition matrix.

$$
\begin{array}{c}
 \\
0 \\
1 \\
2 \\
3 \\
4 \\
5
\end{array}
\begin{array}{c}
\begin{array}{cccccc}
0 & 1 & 2 & 3 & 4 & 5
\end{array} \\
\left[
\begin{array}{cccccc}
0 & 0.2 & 0.05 & 0.025 & 0.025 & 0.7 \\
0 & 0.55 & 0.125 & 0.125 & 0.075 & 0.125 \\
0 & 0 & 0.65 & 0.05 & 0.2 & 0.1 \\
0 & 0 & 0 & 0.7 & 0.25 & 0.05 \\
0 & 0 & 0 & 0 & 1 & 0 \\
0 & 0 & 0 & 0 & 0 & 1
\end{array}
\right] = M
\end{array}
$$

(a) Compute M^2 and M^3. If a player starts from first base, then what is the probability that he will have scored after 2 more at bats? After 3 more at bats?

(b) Compute a large power of M (say, M^{30}). This reflects a convergence and suggests ultimate likelihood of scoring versus getting out from a certain base. What is the ultimate likelihood of scoring from first base? Does the ultimate likelihood of scoring vary with the starting point?

64. **Population migration**. Each year the inhabitants of four islands, labeled A, B, C, and D, migrate among these islands. The probabilities of the various migrations are shown in the pictured Markov chain graph.

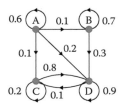

(a) Give the transition matrix M for this Markov chain graph. (b) Compute M^2 and M^3. What is the probability that an inhabitant of island A will end up an inhabitant of island D after 2 years? After 3 years? (c) Compute a large power of M (say, M^{30}). This reflects a convergence and suggests an ultimate balance of the population among the islands. What is the probability that an inhabitant of island A will ultimately end up as an inhabitant of island D? Does the starting island matter in this ultimate likelihood?

65. **Calculus remediation**. At a competitive college, students who do not pass calculus and wish to remain at the college must first take algebra and then take trigonometry (though passing these remedial courses is not required) before they can retake calculus. Many end up withdrawing (WD) from school out of frustration. The transition matrix M for this Markov chain is displayed.

$$
\begin{array}{c}
\quad\; \text{Calc \; Trig \; Alg \; WD \; Pass} \\
\begin{array}{c} \text{Calc} \\ \text{Trig} \\ \text{Alg} \\ \text{WD} \\ \text{Pass} \end{array}
\left[
\begin{array}{ccccc}
0 & 0 & 0.3 & 0.1 & 0.6 \\
0.5 & 0 & 0 & 0.5 & 0 \\
0 & 0.8 & 0 & 0.2 & 0 \\
0 & 0 & 0 & 1 & 0 \\
0 & 0 & 0 & 0 & 1
\end{array}
\right] = M
\end{array}
$$

(a) Draw the corresponding Markov chain graph.

(b) For each state, determine its period.

(c) Compute a large power of M (say, M^{20}). What percentage of students starting in calculus ultimately pass calculus? What percentage of students starting in algebra ultimately withdraw from school?

66. **Doubles tennis rally**. Teams A and B are playing doubles tennis. Team A consists of players A and a, and Team B consists of players B and b. For each player, statistics have been kept on which member of the opponent's team tends to receive his shot. We are assuming that none of the players will miss a shot, and thus an arbitrarily long rally can be maintained.

$$
\begin{array}{c}
\quad\; \text{A \quad a \quad B \quad\; b} \\
\begin{array}{c} \text{A} \\ \text{a} \\ \text{B} \\ \text{b} \end{array}
\left[
\begin{array}{cccc}
0 & 0 & 0.3 & 0.7 \\
0 & 0 & 0.6 & 0.4 \\
0.2 & 0.8 & 0 & 0 \\
0.5 & 0.5 & 0 & 0
\end{array}
\right] = M
\end{array}
$$

(a) Draw the corresponding Markov chain graph.

(b) Compute M^2, and observe that every state has period 2.

(c) This Markov chain does not have a single matrix M^∞ to which the powers of M converge. Instead, because every state has period 2, the even powers of M converge to one matrix M^{even} and the odd powers of M converge to another matrix M^{odd}. Estimate those matrices by computing large powers of M.

67. **Video poker.** Keith is trying to outsmart a video poker machine in a casino. After hours of observation, he has witnessed the transition probabilities listed in the following matrix. That is, given the action of the computer on one turn, the matrix reflects the probable action of the computer on its subsequent turn.

	Bet	Check	Fold	Win
Bet	0.4	0.2	0.1	0.3
Check	0.3	0.2	0.3	0.2
Fold	0	0	1	0
Win	0	0	0	1

$= M$

Of course, the game is over when either the computer folds (equivalently, the computer loses) or the computer wins. (a) Draw the corresponding Markov chain graph. (b) Estimate M^∞. (c) For each possible action of the computer on its first turn, determine the likelihood that the computer goes on to win that game. (d) Is Keith able to find any advantage against the computer as a result of this analysis?

68. **Doubles tennis point.** Here, we consider the same basic setup of a doubles tennis match as that in Exercise 66. However, we now deal with the possibility that a player may fail to complete a shot. Thus, for each state x that represents a player's shot, we add a state x' representing the failing of that player to complete his shot. Of course, once one of these new states is reached, the point is over.

	A	a	B	b	A'	a'	B'	b'
A	0	0	0.2	0.6	0.2	0	0	0
a	0	0	0.4	0.2	0	0.4	0	0
B	0.1	0.6	0	0	0	0	0.3	0
b	0.3	0.4	0	0	0	0	0	0.3
A'	0	0	0	0	1	0	0	0
a'	0	0	0	0	0	1	0	0
B'	0	0	0	0	0	0	1	0
b'	0	0	0	0	0	0	0	1

$= M$

(a) Draw the corresponding Markov chain graph. (b) Estimate M^∞. If player A serves, then what is the probability that player A is the one that loses the point in the end? (c)Which team is likely to win the point? Does it matter which player serves (starts the point)?

8.7   Review problems

1. Pictured is a portion of the city of Zahlenberg through which the Stetig river flows.

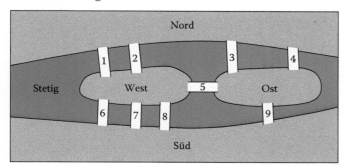

Draw the graph that reflects the bridge crossings in this city.

2. A small airline in upstate New York provides the round-trip service shown in the following table.

	Binghamton	Buffalo	Ithaca	Rochester	Syracuse
Binghamton	No	Yes	No	No	Yes
Buffalo	Yes	No	Yes	No	Yes
Ithaca	No	Yes	No	Yes	No
Rochester	No	No	Yes	No	No
Syracuse	Yes	Yes	No	No	No

A travel agent wants to represent this service in a diagram. Draw a graph that reflects this service.

3. Let $V = \{1,2,3,4\}$, $E = \{\{1,4\}, \{2,4\}, \{3,4\}\}$, and $G = (V,E)$.

   (a) Draw $G = (V,E)$.

   (b) Determine the edge set for the subgraph of G induced by $W = \{1, 2, 3\}$.

   (c) What is the distance from 2 to 3?

   (d) Is G bipartite?

4. Let $V = \{1,2,3,4,5\}$, $E = \{a,b,c,d,e\}$, $a = \{1,3\}$, $b \mapsto \{1,4\}$, $c \mapsto \{1,4\}$, $d = \{2,5\}$, $e = \{3,4\}$, and $G = (V,E)$.

   (a) Draw G.

   (b) Give an example of a cycle in G.

   (c) Is G connected?

   (d) Find the components of G.

5. Determine whether each of the following pictures is the drawing of a graph?

   (a)

   (b)

6. Let G be the pictured graph.

(a) Is G a simple graph?

(b) Does $4,5,2,3$ specify a path?

(c) Does $3,4,5,6,3$ specify a cycle?

7. Draw $K_{3,1}$.

8. Draw P_6.

9. (a) Draw K_4. (b) Find a trail in K_4 that is neither a path nor a cycle.

10. For the purposes of a psychological study, the binary linear code in Table 8.4 is used to encode, for each participant, his or her gender and the hand with which he or she writes. Consequently, the code words can be regarded as vertices of Q_4.

(a) Draw Q_4.

(b) In Q_4, find dist $(0110, 0011)$.

(c) Is there a unique code word that is nearest to the vertex 1101? To what gender and writing hand does it correspond?

(d) Give an example of a vertex in Q_4 for which there is not a unique nearest code word. Such a vertex could not be decoded.

11. Let $G = (V, E)$ be a graph, $W \subseteq V$, and H the subgraph of G induced by W. Show: If G is complete, then H is complete.

12. Can a graph be bipartite if it has (a) a loop? (b) Multiple edges? (c) A cycle?

13. Show that the cube is bipartite.

TABLE 8.4
A binary linear code for a psychological study

Gender	Writing hand	Message	Code word
Male	Right	00	0000
Male	Left	01	0101
Female	Right	10	1011
Female	Left	11	1110

14. Is the dodecahedron bipartite? Justify your answer.

15. A sixth-grade teacher is keeping track of the friendships within her class with a graph. Each student is represented by a vertex, and an edge joins two students if and only if they are friends. Will such a graph always be bipartite? Explain.

16. Let G be the pictured graph.

   (a) Give the adjacency matrix, using the vertex ordering $1, 2, 3, 4$.

   (b) Give the adjacency lists.

17. Compute the product $\begin{bmatrix} 7 & 3 \\ 0 & 6 \end{bmatrix} \begin{bmatrix} 0 & 1 \\ 1 & 0 \end{bmatrix}$.

18. Let A be the adjacency matrix from Exercise 16(a).

   (a) Find the permutation matrix P corresponding to the permutation $2, 4, 3, 1$.

   (b) Compute PA.

   (c) How is the matrix PAP^T related to the graph in Exercise 16?

19. Let A denote the adjacency matrix for the graph P_5, using the ordering $1, 2, 3, 4, 5$ of its vertices. Without computing all of A^4, find the entry in row 1 and column 3 of A^4. Hint: Count walks.

20. Six power stations have been connected to form a power grid. Redundant connections are used to maintain a backup for important connections that might fail. The given matrix reflects direct connections between the power stations.

	NE	E	SE	SW	W	NW
NE	0	1	0	2	0	1
E	1	0	2	0	1	0
SE	0	2	0	1	0	1
SW	2	0	1	0	1	0
W	0	1	0	1	0	2
NW	1	0	1	0	2	0

Draw the graph with the given adjacency matrix that reflects the power grid.

21. Show that the two pictured graphs are isomorphic.

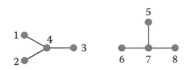

22. Show that the two pictured graphs are isomorphic.

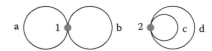

Observe that a vertex map is not sufficient in this case.

23. Pictured are two design plans for a power grid, in which the symbol ○ denotes a power station. One represents a grid at MIT and the other is at Harvard.

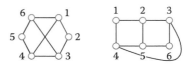

Show that the power grids are equivalent and, hence, that Harvard can employ the same diagnostic software as that used at MIT.

24. Show that Q_n is vertex transitive.

25. Are the graphs in Exercise 23 vertex transitive?

26. Is the pictured graph isomorphic to the Petersen graph?

27. Show that there is a graph map from C_4 to the pictured graph.

28. How many automorphisms does the graph pictured in Exercise 27 have?

29.* Find $|\text{Aut}(Q_3)|$. Compare this result with the number of automorphisms of a die studied in Example 7.18 of Section 7.4.

30. Is there a graph map from the tetrahedron to the octahedron? Explain.

31. Show that if $G = (V, E)$ is an r-regular graph and r is odd, then $|V|$ is even.

32. Show that the two pictured graphs are not isomorphic.

33. Pictured are two design plans for a power grid at Yale, in which the symbol ○ denotes a power station. They seem to have one major design difference around the interior power station on the right. Are they really different?

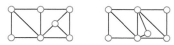

Show that the two graphs that model these grids are in fact not isomorphic.

34. Show that the two pictured graphs are not isomorphic.

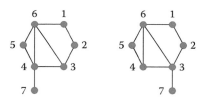

35. Determine whether the two pictured graphs are isomorphic.

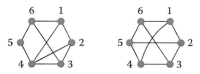

36.* List all non-isomorphic simple graphs with degree sequence $3, 2, 2, 2, 1$.

37.* List all non-isomorphic simple graphs with degree sequence $4, 4, 2, 2, 2, 2$.

38. Show that the complement of the graph in Exercise 23 is isomorphic to the pictured graph.

39. For every simple graph G, show that $(G^c)^c = G$.

40. Draw each of the following graphs. (a) $P_3 + C_3$. (b) $P_3 \times C_3$.

41. In Q_3, let G be the subgraph induced by vertices $000, 001, 010, 011$, and let H be the subgraph induced by vertices $000, 010, 100, 110$.

(a) Find $G \cap H$.   (b) Find $G \cup H$.

42. Let G and H be any graphs. Show: $G \times H \cong H \times G$

43. A small airline in upstate New York now provides the one-way service shown in the following table.

From \To	Binghamton	Buffalo	Ithaca	Rochester	Syracuse
Binghamton	No	Yes	No	No	No
Buffalo	No	No	No	No	Yes
Ithaca	No	Yes	No	No	No
Rochester	No	No	Yes	No	No
Syracuse	Yes	No	No	Yes	No

Draw a directed graph that reflects this service.

44. Let $V = \{1, 2, 3, 4\}$, $E = \{(1,2), (1,4), (3,2), (3,4)\}$, and $G = (V, E)$.

(a) Draw G.

(b) Draw $\underline{G}$.

(c) Determine the weak components of G.

(d) Determine the strong components of G.

(e) List the in- and out-degrees of the vertices.

45. Is the pictured directed graph simple?

46. Pictured are two seemingly different designs for a computer network on six computers using eight cables in which each cable connection allows message passing in only one direction.

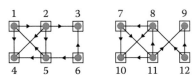

Show that the two directed graphs that model these networks are isomorphic.

47. Show that the two pictured directed graphs are not isomorphic.

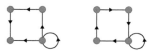

48. Determine whether the two pictured directed graphs are isomorphic.

49. List all non-isomorphic simple directed graphs on 2 vertices.

50. List all directed graphs G for which $\underline{G}$ is the pictured graph.

51. Give the adjacency matrix for the pictured graph.

52. Is the directed graph in Exercise 45 strongly connected? Describe its strong components.

53. Is the pictured directed graph a Markov chain graph?

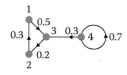

54. Give the transition matrix for the Markov chain whose graph is pictured.

55. **Turtle migration.** Each year the turtles inhabiting the four islands, A, B, C, and D, migrate among these islands. The probabilities of the various migrations are given in the displayed transition matrix.

$$
\begin{array}{c}
\begin{array}{cccc} A & B & C & D \end{array} \\
\begin{array}{c} A \\ B \\ C \\ D \end{array}
\left[
\begin{array}{cccc}
0 & 0.2 & 0 & 0.8 \\
0 & 0.6 & 0 & 0.4 \\
0.3 & 0 & 0.2 & 0.5 \\
0 & 0 & 1 & 0
\end{array}
\right]
\end{array}
$$

(a) Draw the corresponding Markov chain graph.

(b) What is the probability that a turtle on island A will end up on island D after 2 years?

(c) Is this Markov chain irreducible? Regular?

(d) What is the long-term probability that a turtle starting on island A will inhabit island C?

9 Graph Properties

In Chapter 8, we were introduced to some basic graph invariants such as the number of vertices and the maximum degree. Those invariants are numerical values. However, more generally, invariants include global properties such as whether or not a graph is connected.

This chapter focuses more keenly on properties of graphs, starting with connectivity. Beyond just whether or not a graph is connected, the connectivity measures how close the graph is to becoming disconnected should portions be removed. This is of particular interest when a graph models a network whose components may fail. Next, we consider whether or not a graph contains certain special circuits. Euler circuits are those that traverse every edge exactly once. Hamiltonian cycles hit every vertex exactly once. Such circuits and cycles are of interest if we want to plow all of the streets in a neighborhood or make deliveries to each of its addresses.

Our final two sections build up to one of the most famous theorems in graph theory and mathematics in general, the Four-Color Theorem. First, we determine which graphs can be drawn in the plane without edge crossings. Such graphs are said to be planar. Second, we consider the general problem of determining the minimum number of colors needed to color the vertices of a graph so that neighbors have different colors. The fact that planar graphs never need more than four colors is the famous result.

The determination of whether or not a graph is planar is important in circuit design. Finding the minimum number of colors needed to color a graph turns out to have applications in scheduling problems. The fact that planar graphs need at most four colors means that a coloring of the states or countries in a map never requires more than four colors.

9.1 Connectivity

Suppose a graph represents a power grid. The failure of power stations or cables in that grid can be reflected by the deletion of the corresponding vertices or edges in the graph. The new graph obtained can then be used to study the power delivery capabilities and incapabilities of the new grid. Such analysis is important, since storms or other incidents can damage a grid.

We consider the vulnerability of a graph to the removal of vertices and edges. Initially, the focus is on removing vertices. We start with a motivating example.

Example 9.1.   The pictured graph represents a power grid in which it is possible for each of eight power stations to transfer power to all others.

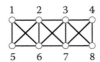

What is the minimum number of power stations whose failure would leave a grid in which a remaining station would be unable to transfer power to some other?

Solution. Two power stations.
      The failure of stations 3 and 7 leaves a grid

in which stations 1 and 4 cannot transfer power to one another. Further, it is easy to check that if just a single station fails in the original grid, then the remaining stations can still transfer power among themselves.   □

The issues arising in Example 9.1 are relevant to graphs in general.

Definition 9.1.   Let $G = (V, E)$ be any graph.

(a) Given subsets $W \subseteq V$ and $F \subseteq E$, the graph resulting from the **removal** of $W \cup F$, denoted[*] $G \setminus (W \cup F)$, is the subgraph of G whose vertex set is $V \setminus W$ and whose edge set is $E \setminus (F \cup \{e \in E : e$ is incident with some $v \in W\})$.

(b) A **disconnecting set** for G is a set D of vertices such that $G \setminus D$ is disconnected.

(c) The **connectivity** of G, denoted $\kappa(G)$, is the minimum number of vertices whose removal results in either a disconnected graph or a single vertex.

---

[*] The notion of a union of graphs, sets, or otherwise is not intended in $W \cup F$. What is important is that each of W and F is removed from V and E, respectively.

(d) A κ-**set** for G is a set of $\kappa(G)$ vertices whose removal results in either a disconnected graph or a single vertex.

Remark 9.1. If a graph $G = (V, E)$ has a pair of distinct non-adjacent vertices, then $\kappa(G)$ is the minimum possible size of a disconnecting set. Only in the case that every pair of distinct vertices is adjacent is $\kappa(G)$ the minimum number of vertices whose removal results in a single vertex. In that case, $\kappa(G) = |V| - 1$. This gives $\kappa(K_n) = n - 1$ and also handles any graph that might be constructed from K_n by adding extra loops or multiple edges.

Example 9.2. Let G be the pictured graph, $W = \{1, 3\}$, and $F = \{g\}$.

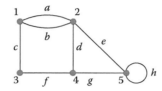

Then the graph $G \setminus (W \cup F)$ is pictured below.

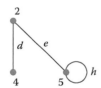

Notice that, in addition to edge g, all edges incident with vertex 1 or vertex 3 have been removed.

If G is not complete, then to establish that $\kappa(G)$ has a particular value k takes two steps. First, a disconnecting set of size k must be specified. This gives $\kappa(G) \le k$. Then it must be argued that no disconnecting set of size smaller than k exists. That gives $\kappa(G) \ge k$, and equality follows. The arguments in Example 9.1 thus showed that the graph presented there has connectivity 2.

We have seen that graphs can be used to model computer networks such as those shown in Figure 9.1. Suppose n is the number of computers for which a network is to be formed. If there are just $n - 1$ cables with which to connect them, then they might be connected to form the path P_n (see Exercise 18). If there are n cables, then C_n is the logical choice (see Exercise 17). The following theorem thus reflects the vulnerability of these types of networks to failure.

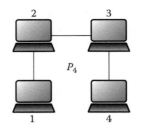 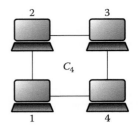

FIGURE 9.1
Network configurations.

Theorem 9.1.   *Let n be an integer.*

(a) *If $n \geq 2$, then $\kappa(P_n) = 1$.*

(b) *If $n \geq 3$, then $\kappa(C_n) = 2$.*

Proof. Our vertices are labeled $1, 2, \ldots, n$ as in Definition 8.6.

(a) Suppose $n \geq 2$. Since P_n is connected (and has more than one vertex), it follows that $\kappa(P_n) \geq 1$. If $n = 2$, then $P_n \setminus \{2\}$ is a single vertex. If $n > 2$, then $P_n \setminus \{2\}$ is disconnected. Thus, $\kappa(P_n) \leq 1$.

(b) Suppose $n \geq 3$. If we remove one vertex, say, vertex i, then the remaining graph forms the path $i + 1, \ldots, n, 1, \ldots, i - 1$, which is obviously connected. Thus, $\kappa(C_n) \geq 2$. Since removing $\{1, 3\}$ results in either a single vertex (if $n = 3$) or a disconnected graph (if $n > 3$), it follows that $\kappa(C_n) \leq 2$.   □

One easy way to disconnect a graph is to remove all of the neighbors of some vertex. It follows that the connectivity of a graph cannot exceed its minimum degree.

Theorem 9.2.   *For any graph G, we have $\kappa(G) \leq \delta(G)$.*

Proof. Let G be any graph. Let v be a vertex with $\deg(v) = \delta(G)$, and let W be the set of neighbors of v distinct from v. So $|W| \leq \delta(G)$. The removal of W from G leaves v without neighbors. Since $G \setminus W$ must then be either a disconnected graph or a single vertex, $\kappa(G) \leq |W| \leq \delta(G)$.   □

Equality is possible in Theorem 9.2, as can be seen in Theorem 9.1.   Namely, $\kappa(P_n) = \delta(P_n) = 1$ and $\kappa(C_n) = \delta(C_n) = 2$. Obviously, each cycle C_n is connected and 2-regular. It turns out that these are the only connected and 2-regular graphs.

Theorem 9.3.   *Let $G = (V, E)$ be any graph. If G is connected and 2-regular, then $G \cong C_n$, where $n = |V|$. So $\kappa(G) = 2$.*

Proof. Suppose G is connected and 2-regular. Let T be a longest possible trail in G. Say $x_0, x_1, \ldots, x_n$ is a listing of the vertices encountered along T.

Suppose toward a contradiction that v is a vertex of G not on T. Since G is connected, there is a path P from x_n to v. Let e be the last edge on P that is also on T. Since v is not on T, there is an edge f in P immediately following e. Thus, e and f share an endpoint x_i for some $0 \le i \le n$.

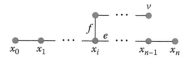

If $i = 0$ or n, then adjoining f to T forms a trail in G longer than T. If $0 < i < n$, then $\deg(x_i) \ge 3$. These contradictions show that f cannot exist and hence neither can v.

We conclude that T contains all of the vertices of G. Moreover, since the argument above shows that there can be no edge f outside of T that is incident with x_0 or x_n, we must have $x_0 = x_n$ and $G \cong C_n$. By Theorem 9.1(b), $\kappa(G) = 2$. $\quad\square$

In light of Theorem 9.3, one might expect that if a graph G is connected and 3-regular, then $\kappa(G) = 3$. However, that is not the case.

Example 9.3.   The pictured graph G

is connected and 3-regular with $\kappa(G) = 2$.

Proof. It is easy to see that the removal of a single vertex does not disconnect the graph. It is also easy to find two vertices whose removal disconnects the graph. This 3-regular graph thus has connectivity 2. $\quad\square$

Thus far, we have computed the connectivities for a few important classes of graphs, namely, complete graphs, paths, and cycles. We now add complete bipartite graphs to our list.

Theorem 9.4.   Let $m, n \in \mathbb{Z}^+$. Then, $\kappa(K_{m,n}) = \min\{m, n\}$.

Proof. We have $V = V_1 \cup V_2$, where

$$V_1 = \{(1,1), (1,2), \ldots, (1,m)\}, \quad V_2 = \{(2,1), (2,2), \ldots, (2,n)\},$$

and $E = \{\{v_1, v_2\} : v_1 \in V_1 \text{ and } v_2 \in V_2\}$. Observe that $K_{m,n} \setminus V_1$ and $K_{m,n} \setminus V_2$ are disconnected.

We claim that any disconnecting set for $K_{m,n}$ must contain either all of V_1 or all of V_2. Suppose not. So we have a disconnecting set W whose removal leaves an edge $\{v_1, v_2\}$ with $v_1 \in V_1$ and $v_2 \in V_2$. Now let x and y be any pair of distinct vertices in $V \setminus W$. If x and y are not both in the same set V_i, then x, y gives a single edge path from x to y. For $i = 1$ or 2, if $x, y \in V_i$, then x, v_{3-i}, y gives a path in $K_{m,n} \setminus W$ from x to y. The fact that $K_{m,n} \setminus W$ is connected is a contradiction.

We conclude that the smaller of the two sets V_1 and V_2 is a κ-set for $K_{m,n}$. That is, $\kappa(K_{m,n}) = \min\{|V_1|, |V_2|\} = \min\{m, n\}$.

$\square$

9.1.1 Edge connectivity

Our focus now shifts to the removal of edges from a graph. If computer cables could fail but the computers themselves could not, then this would be our concern. For example, inexpensive fiber optic cables might be fragile and the primary cause of downtime for a network.

Definition 9.2. Let $G = (V, E)$ be any graph.

(a) A **disconnecting set of edges** for G is a set F of edges such that $G \setminus F$ is disconnected.

(b) The **edge connectivity** of G, denoted $\lambda(G)$, is the minimum number of edges whose removal results in either a disconnected graph or an isolated vertex.

(c) A λ-**set** for G is a set of $\lambda(G)$ edges whose removal results in either a disconnected graph or an isolated vertex.

Remark 9.2.

(a) Connectivity alone always refers to the notion in Definition 9.1(c), which we might also call *vertex* connectivity.

(b) If a graph $G = (V, E)$ has $|V| = 1$, then $\lambda(G) = |E|$.

The following theorem extends Theorem 9.2 and relates vertex connectivity, edge connectivity, and minimum degree.

Theorem 9.5. *For any graph G, we have $\kappa(G) \leq \lambda(G) \leq \delta(G)$.*

Proof. Let G be any graph. First, we show that $\lambda(G) \leq \delta(G)$. Let v be a vertex with $\deg(v) = \delta(G)$. Thus, the set F of edges incident with v has size at most $\delta(G)$. Since $G \setminus F$ is either a disconnected graph or the isolated vertex v, it follows that $\lambda(G) \leq |F| \leq \delta(G)$.

It remains to show that $\kappa(G) \leq \lambda(G)$. It is easy to see that if $\lambda(G) \leq 1$, then $\kappa(G) = \lambda(G)$. (Do Exercise 39.) So it suffices to assume that $\lambda(G) \geq 2$. We further assume that G is a simple graph and handle general graphs that may not be simple at the end.

Let $m = \lambda(G)$, and let F be a λ-set for G. Say $F = \{e_1, e_2, \ldots, e_m\}$ and $e_m = \{u_m, v_m\}$. For each $1 \leq i \leq m-1$, let x_i be an endpoint of e_i distinct from u_m and v_m. Note that $X = \{x_i : 1 \leq i \leq m-1\}$ has at most $m-1$ vertices.

If $G \setminus X$ is disconnected, then $\kappa(G) \leq m-1 < \lambda(G)$. Otherwise, $G \setminus X$ is connected. Note that the edges $e_1, e_2, \ldots, e_{m-1}$ are removed from G when X is removed. Since the additional removal of e_m results in either a disconnected graph or an isolated vertex, so does the additional removal of u_m. Hence, $\kappa(G) \leq |X \cup \{u_m\}| \leq m = \lambda(G)$.

In the general case that G is a graph, let G' be a simple graph formed from G by removing all loops and by retaining a single edge between any pair of vertices connected by one or more edges. It is easy to see that $\kappa(G') = \kappa(G)$ and $\lambda(G') \leq \lambda(G)$. Since we have established that $\kappa(G') \leq \lambda(G')$, it follows that $\kappa(G) = \kappa(G') \leq \lambda(G') \leq \lambda(G)$. $\square$

Theorem 9.5 can be used to help compute one of the parameters κ or λ when the other is known. For example, it follows from Theorems 9.1 and 9.5 that $\lambda(P_n) = 1$ and $\lambda(C_n) = 2$. However, it is not always the case that the three parameters κ, λ, and δ are the same.

Example 9.4. The pictured graph G represents a fiber optic network.

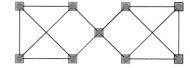

We leave for the exercises the proofs that $\kappa(G) = 1$, $\lambda(G) = 2$, and $\delta(G) = 3$. Here, we interpret what this tells us about our network.

This network is vulnerable to the failure of the central computer. However, if the only likely failures would occur in the fiber optic cables, then this network is not as vulnerable as the vertex connectivity suggests. Its edge connectivity tells us that at least two cables need to fail to disconnect the network. If the failure of two cables at the same time is extremely unlikely, then this design might be reasonable.

Although Example 9.4 shows that κ and λ do not in general coincide, there are conditions under which they do.

Theorem 9.6. *For any graph G, if G is 3-regular, then* $\kappa(G) = \lambda(G)$.

Proof. Let G be a 3-regular graph. By Theorem 9.5, $\kappa(G) \leq \lambda(G)$. Thus, it remains to show that $\lambda(G) \leq \kappa(G)$. Since $\kappa(K_4) = 3 = \lambda(K_4)$ and K_4 is the only 3-regular complete simple graph, the case in which G is complete is covered. Now we will consider the case in which G is not complete.

Let D be a κ-set for G, and let H and K be two distinct components of $G \setminus D$. Let D'' be the set of those vertices in D with neighbors in D, and let $D' = D \setminus D''$. By Exercise 23, each vertex in D has neighbors in both H and K. Since every vertex has degree 3, each vertex in D has either only one neighbor in H or only one neighbor in K. Let D'_K be the set those vertices in D' with only one neighbor in K, and let $D'_H = D' \setminus D'_K$. Note that the vertices in D'_H have only one neighbor in H.

Form a set F of edges as follows:

(i) For each $v \in D'' \cup D'_K$, put into F the unique edge from v to K.

(ii) For each $v \in D'_H$, put into F the unique edge from v to H.

So $|F| = |D|$ and there are no paths from H to K in $G \setminus F$. Since $G \setminus F$ is disconnected, $\lambda(G) \leq |F| = |D| = \kappa(G)$. $\square$

The **connectivity of a directed graph** G, denoted $\kappa(G)$, is the minimum number of vertices whose removal results in a directed graph that either is not strongly connected or is a single vertex. The edge connectivity $\lambda(G)$ is defined analogously.

The relationship between the connectivity of a directed graph G and that of its underlying graph $\underline{G}$ is that $\kappa(G) \leq \kappa(\underline{G})$. Also, $\lambda(G) \leq \lambda(\underline{G})$. The straightforward proofs of these observations are requested in the exercises.

Exercises

Exercises 1 through 4 refer to the pictured graph G.

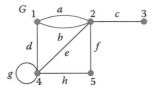

In each case, find $G \setminus (W \cup F)$ for the given sets W and F.

1. $W = \{2, 4\}$ and $F = \emptyset$.

2. $W = \emptyset$ and
   $F = \{b, c, e, h\}$.

3. $W = \{1, 4\}$ and
   $F = \{a, b, c\}$.

4. $W = \{2, 5\}$ and
   $F = \{a, b, g\}$.

Subsequent exercises refer to the following pictured graphs:

(a)   (b) 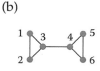 (c)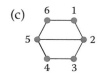

(d)   (e)   (f)

In Exercises 5 through 12, determine the connectivity κ of the specified graph and prove your result.

5. Graph (a).

6. Graph (b).

7. Graph (c).

8. Graph (d).

9. Graph (e).

10. Graph (f).

11. The dodecahedron graph.

12. The Petersen graph.

13. Pictured is a power grid on six stations. In addition to the outer 6-cycle, three cables connecting diametrically opposite stations are included to decrease the power grid's vulnerability to station failures.

(a) What special graph is it that models this power grid?

(b) What is the connectivity of this graph? Justify your answer.

(c) Can we remove a cable from this grid and retain the same connectivity? Explain.

14. Pictured is a computer network on five computers using seven cable connections.

(a) What is the connectivity of the graph that models this network? Prove your result.

(b) Can some cable connections be removed from this network, while retaining the same connectivity? How many?

(c) What special graph is obtained when we remove as many edges as possible, while retaining the original connectivity?

15. In a computer network, if there is a computer that is directly linked to only one other computer, then this leaves a vulnerable point in the network. Let G be any graph. Show: If G is connected and $\delta(G) = 1$, then $\kappa(G) = 1$.

16. The results in Exercises 11, 12, and 13 suggest a tempting conjecture about the connectivity of 3-regular graphs that happens to be false. What connectivity value is shared by these graphs? Give an example of a 3-regular simple graph G with $\kappa(G) = 1$.

17. Given n computers and n cables with which to link pairs of computers, the best network configuration in terms of connectivity is that of the cycle C_n.

(a) Show that C_n has the highest possible connectivity among graphs on n vertices and n edges.

(b) Show that C_n is the only graph (up to isomorphism) on n vertices and n edges that achieves this maximum.

(c) Find all possible isomorphism types of connected networks on 5 computers and 5 cables.

(d) Which of the network configurations in part (c) is the worst? Why?

18. Given n power stations and $n - 1$ cables with which to link pairs of stations, a best power grid configuration in terms of connectivity is that of the path P_n.

(a) Show that P_n has the highest possible connectivity among graphs on n vertices and $n - 1$ edges.

(b) Show that if $n \geq 4$, then P_n is not the only graph (up to isomorphism) on n vertices and n edges that achieves this maximum.

(c) What connected power grid configuration on 5 vertices and 4 edges is the worst? Why?

19. We need to build a power grid on n power stations and have m cables with which to join pairs of stations. If m is relatively small as a function of n, then it is not possible to build a grid with very high connectivity. Show that if a

graph G has n vertices and m edges with $m < \lceil \frac{3n}{2} \rceil$, then $\kappa(G) \leq 2$.

20.* We need to build a computer network on n computers and have m cables with which to join pairs of computers. If m is relatively large as a function of n, then any graph will have a connectivity that is not far from the maximum possible. Show that if a graph G has n vertices and m edges with $m = \binom{n}{2} - 2$, then $\kappa(G) = n - 2$ or $n - 3$.

21. Prove the following corollary of Theorem 9.3. Let G be any graph. Show: If G is 2-regular, then G is isomorphic to a disjoint union of cycles.

22. Characterize 1-regular graphs.

23.* Show: For any graph $G = (V, E)$ and any disconnecting set $D \subset V$, if $\kappa(G) = |D|$, then each $v \in D$ has neighbors in each component of $G \setminus D$.

24. It should be intuitively obvious that removing cables from a computer network can only reduce or maintain its connectivity and that adding cables to a computer network can only increase or maintain its connectivity. Let H and G be any graphs. Show: If H is a subgraph of G with $V_H = V_G$, then $\kappa(H) \leq \kappa(G)$.

25. One might suspect that designing a power grid so that it is connected and each power station is directly linked to several others would make it relatively invulnerable to the failure of any single station. However, there are examples that prove otherwise. Show that, even with the restriction that $\kappa = 1$, it is possible to have δ arbitrarily large. That is, show: For each integer $d \geq 1$, there exists a graph G with $\kappa(G) = 1$ and $\delta(G) = d$. Hint: Glue some complete graphs onto a common vertex.

26. Give an example of a graph G with $\kappa(G) = 2$ and $\delta(G) = 4$.

27.* Show: For any graph G, $\kappa(G \times P_2) \geq 1 + \kappa(G)$.

28.* Show: For all $n \geq 1$, $\kappa(Q_n) = n$. Hint: Induction on n. Use the fact that $Q_n \cong Q_{n-1} \times P_2$.

In Exercises 29 through 36, determine the edge connectivity λ of the specified graph and prove your result.

29. Graph (a).   30. Graph (b).

31. Graph (c).   32. Graph (d).

33. Graph (e).   34. Graph (f).

35. The dodecahedron   36. The Petersen graph.
    graph.

37. Consider the power grid from Exercise 13. Determine the edge connectivity of the graph that models this grid, and prove your result. Is there a unique λ-set?

38. Consider the computer network from Exercise 14. Determine the edge connectivity of the graph that models this network, and prove your result. Is there a unique λ-set?

39. An edge in a graph whose removal disconnects the graph is called a cut-edge or bridge. The presence of such an edge strongly ties the connectivity of the graph with its edge connectivity. Let G be any graph. Show: If $\lambda(G) \leq 1$, then $\kappa(G) = \lambda(G)$.

40. We saw in Theorem 9.6 that 3-regular graphs have edge connectivity equal to their vertex connectivity. Prove or disprove: If G is any 4-regular graph, then $\lambda(G) = \kappa(G)$.

41. Show: $\lambda(K_n) = n - 1$.

42. Prove the assertions in Example 9.4.

43. Give an example of a simple graph G for which $\kappa(G) < \lambda(G) = \delta(G)$.

44. Give an example of a simple graph G for which $\kappa(G) = \lambda(G) < \delta(G)$.

In Exercises 45 through 50, determine (a) the connectivity κ and (b) the edge connectivity λ of the specified directed graph $G = (V, E)$.

45. $V = \{1, 2, 3, 4, 5\}$ and $E = \{(1, 3), (5, 1), (2, 3), (4, 2), (3, 4), (3, 5)\}$.

46. $V = \{1, 2, 3, 4, 5, 6\}$ and $E = \{(1, 3), (2, 1), (3, 2), (4, 3), (4, 6), (5, 4), (6, 1), (6, 5)\}$.

47. $V = \{1, 2, 3, 4\}$ and $E = \{(1, 3), (2, 1), (3, 2), (3, 4), (4, 1), (2, 2)\}$.

48. $V = \{1, 2, 3, 4\}$ and $E = \{(1, 2), (1, 3), (1, 4), (2, 3), (4, 3)\}$.

49. In the pictured neighborhood, the portions of the streets marked with arrows are one-way streets, whereas the others are two-way. Let G be the directed graph that reflects the resulting traffic flow.

50. In the pictured neighborhood, the portions of the streets marked with arrows are one-way streets, whereas the others are two-way. Let G be the directed graph that reflects the resulting traffic flow.

51. Show for any directed graph G that $\kappa(G) \le \kappa(\underline{G})$ and $\lambda(G) \le \lambda(\underline{G})$.

52. For a directed graph G, is there an inequality of the form $\kappa(G) \le \lambda(G) \le \delta(G)$? To what should $\delta(G)$ refer?

9.2 Euler circuits

The highlight of this section is the beautiful theorem used by Euler to solve the Königsberg Bridge Problem (Example 8.1 in Section 8.1). In terms of graph theory, this focuses our study on trails in graphs.

Recall from Definition 8.4 that a trail is a walk with no repeated edges; it contains, in particular, a list of edges. To simplify notation in this section, we specify a trail just by its edges. The edges in this list and their endpoints are said to be **covered** by the trail. Also recall that trails and circuits may repeat vertices. Of course, a circuit must start and end at the same vertex.

Definition 9.3. Let G be a graph or a digraph. An **Euler circuit** (respectively, **Euler trail**) in G is a circuit (respectively, trail) that covers every edge exactly once and covers every vertex. Of course, unless G has isolated vertices, covering every edge will imply that every vertex is covered. If G contains an Euler circuit, then G is said to be **Eulerian**.

We focus our attention here on graphs and consider digraphs in the exercises.

Example 9.5.

(a) In the pictured graph,

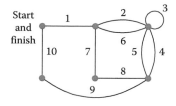

the edges of an Euler circuit have been numbered consecutively.

(b) In the pictured graph,

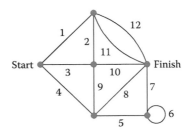

the edges of an Euler trail that is not an Euler circuit have been numbered consecutively.

(c) The graph on the left,

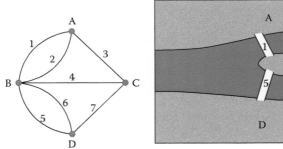 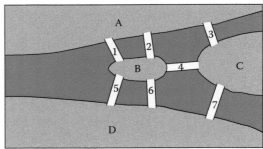

which reflects the seven bridges of Königsberg, shown on the right, has neither an Euler trail nor an Euler circuit. An exhaustive argument was suggested in Exercise 49 of Section 8.1.

Euler's solution to the Königsberg Bridge Problem comes out of the following general result.

Theorem 9.7.   (Euler's Theorem).
Given any graph G,

(a) *G has an Euler circuit if and only if G is connected and every vertex of G has even degree.*

(b) *G has an Euler trail that is not an Euler circuit if and only if G is connected and has exactly two vertices with odd degree.*

The results in Example 9.5 can now be understood in terms of Euler's Theorem. The graph in part (a) is connected, and every vertex has even degree. The graph in part (b) is connected and has exactly two vertices with odd degree, namely, the start and the finish. The graph in part (c) is connected, but more than two vertices have odd degree.

Applying Euler's Theorem to that example gives an elegant solution to the Königsberg Bridge Problem.

Before proving Euler's Theorem, we consider an example that shows the relevance of having vertices of even degree.

Example 9.6. (Motivation for the proof of Euler's Theorem).

Since the pictured graph G is connected and every vertex has even degree,

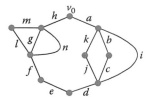

Euler's Theorem guarantees an Euler circuit for G. In fact, the Euler circuit can be found as follows.

From some vertex, say, v_0, we start walking along a trail until we get stuck. For example, the trail

$$a, b, c, d, e, f, g, h$$

is a cycle C that ends back at v_0. It cannot be extended to a longer trail since there are no edges outside of C incident with v_0. Removing the edges of C together with any resulting isolated vertices leaves the following graph H:

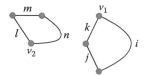

It can be decomposed into the circuits i, j, k and l, m, n starting at v_1 and v_2, respectively. An Euler circuit for G is obtained by inserting the circuits from H into the circuit C from G. First, follow C from v_0 to v_1. Then, follow the circuit i, j, k that returns to v_1. Next, follow C from v_1 to v_2. Then, follow the circuit l, m, n that returns to v_2. Finally, follow C from v_2 back to v_0. The resulting circuit

$$a, i, j, k, b, c, d, e, f, l, m, n, g, h$$

is an Euler circuit for G.

Proof of Euler's Theorem. (a) ($\rightarrow$) Suppose G has an Euler circuit $e_1, e_2, \ldots, e_m$, and let v_0 be the vertex at which it starts and ends. For each $1 \le i \le m - 1$, let v_i be the endpoint common to e_i and e_{i+1}. So the vertex set of G is $\{v_0, v_1, \ldots, v_{m-1}\}$, although this listing may have repetitions. Since any pair of vertices v_i and v_j are connected by a portion of the Euler circuit, G is connected. Since each loop edge contributes 2 to the degree of its endpoint and returns us back to that endpoint, it suffices to assume that $e_1, e_2, \ldots, e_m$ are not loop edges. The edges e_m and e_1 contribute 2 to $\deg(v_0)$, and, for each $1 \le i \le m-1$, the edges e_i and e_{i+1} contribute 2 to $\deg(v_i)$. Since the Euler circuit lists each edge exactly once, the degree of each vertex v_i must be even. Hence, every vertex of G has even degree.

($\leftarrow$) We proceed by strong induction on the number m of edges in G. Obviously, the unique graph with one edge and a single vertex is a loop at that vertex. Thus, it has an Euler circuit. So, suppose $m \ge 1$ and that every connected graph on m or fewer edges, all of whose vertices have even degree, has an Euler circuit. Let G be any connected graph on $m + 1$ edges, all of whose vertices have even degree. Let v_0 be a vertex of G, and form a maximal trail starting at v_0. That is, form a trail $e_1, e_2, \ldots, e_k$ that ends at some vertex v and cannot be extended to a longer trail.

Suppose toward a contradiction that $v \ne v_0$. For each $1 \le i \le k-1$ for which v is the endpoint common to e_i and e_{i+1}, a degree contribution of 2 is used up by the trail. Since the last edge e_k then uses a degree contribution of 1, the total degree contributed to v by the trail is odd. Since v has even degree in G, there must be another edge incident with v that can extend the trail. This contradiction shows that $v = v_0$. That is, our maximal trail is a cycle C that starts and ends at v_0. We can also see that, at each vertex of C, the degree contribution from edges in C must be even.

Remove from G all of the edges of C and any resulting isolated vertices to form a subgraph H. Since G has only vertices of even degree and the same is true of the subgraph C, the vertices of H must all have even degree. Since each component of H has no more than m edges, each has an Euler circuit. Since G is connected, for each component of H, we can select a vertex in common with C. Order the components $H_1, H_2, \ldots, H_n$ so that their selected vertices $v_1, v_2, \ldots, v_n$ are in the order in which they occur in C. For each $1 \le i \le n$, regard the Euler circuit for H_i as starting at v_i. These circuits can then be inserted into C (as in Example 9.6) to form an Euler circuit for G.

The proof of part (b) is left for the exercises. $\square$

How do we find an Euler circuit? Euler's Theorem gives a nice characterization of Eulerian graphs, but its statement does not explain how to find an Euler circuit when we know one exists. For those directions, we need to refer to the proof.

In fact, a systematic procedure for finding an Euler circuit was illustrated in Example 9.6. Very simply, when we know that an Euler circuit exists, we can start from any vertex we wish. From there, we walk around the graph, making sure not to repeat an edge, until we find ourselves stuck back at the starting vertex with no unused edges along which to exit. Call this circuit C_1. If C_1 covers every edge, then we are done. Otherwise, there is a vertex of C_1 incident with an unused edge along which we can start another circuit without repeating edges. As described in Example 9.6, we can insert this new circuit into C_1 to form a circuit C_2 containing all of the edges used so far. If C_2 covers every edge, then we are done. Otherwise, we repeat this process until all edges are covered and an Euler circuit is formed.

The beauty of this process is that no foresight is needed to form the Euler circuit. Getting stuck is not a failure. It is merely the cue to make a renewed start from elsewhere. The new work can be seamlessly folded into the old work. Of course, with practice on small examples, we can minimize the number of times we get stuck, and opportunities for such practice are given in the exercises. However, for large examples warranting the use of a computer program, it is important that a systematic algorithm exists.

Examples like the Königsberg Bridge Problem are certainly not the only type to which Euler circuits apply. We consider now the problem of efficiently removing snow or other debris from the streets of a neighborhood.

Example 9.7.   (Snowplow Routes).
The department of transportation plans the routes for its snowplows well in advance of any snow storms. The pictured neighborhood is a new development that has gone up since last winter, and an efficient snowplow route needs to be designed for it.

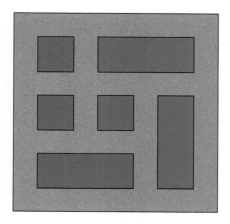

Each street is two-way, and one pass of the plow is required for each side of the street. Also, one pass is sufficient to clear the road surface on one side of a street. To be efficient, once the snowplow enters the neighborhood, it should not pass a second time over a section of road that has already been plowed. Find a maximally efficient route that clears the entire road surface.

Solution. Although the graph pictured below on the left might seem a natural model for our neighborhood, it is the graph on the right with doubled edges that we want.

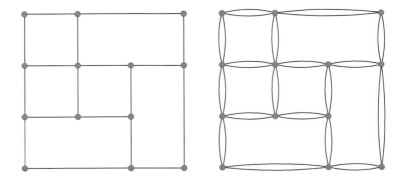

The doubled edges reflect the need to plow the two sides of the road. Consequently, a maximally efficient snowplow route for the neighborhood corresponds to an Euler circuit in the graph on the right. In fact, the doubled edges in this graph cause all of the vertices to have even degree, and thus Euler's Theorem applies. On this graph, we find an Euler circuit and transfer it back to a specified route through the neighborhood.

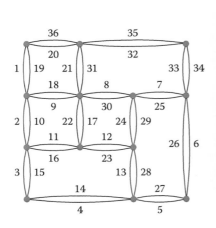

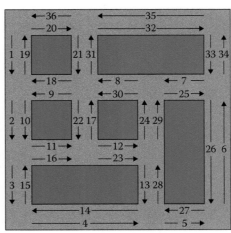

The route that we have found in this case does not always have the plow traveling on the correct side of the street. However, with a minor adjustment to our analysis, we can always find a route that does conform to the traffic laws. Such an adjustment is left for the exercises and can be based on Theorem 9.8.   □

We close this section with a version of Euler's Theorem for directed graphs.

Theorem 9.8.   *Given any directed graph G on n vertices,*

(a) *G has an Euler circuit if and only if G is strongly connected and every vertex v of G has indeg(v) = outdeg(v).*

(b) *G has an Euler trail that is not an Euler circuit if and only if every vertex v of G has indeg(v) = outdeg(v) except that one vertex v_1 has outdeg(v_1) = 1 + indeg(v_1) and another vertex v_n has indeg(v_n) = 1 + outdeg(v_n) and the directed graph obtained from G by adding the edge (v_n, v_1) is strongly connected.*

The proof of Theorem 9.8 is left for the exercises.

Exercises

Subsequent exercises refer to the following pictured graphs:

(a)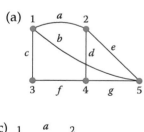

(b)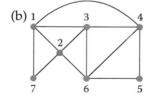

(c)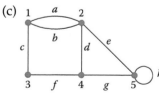

(d)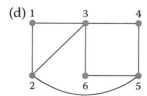

(e)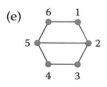

(f)

(g) 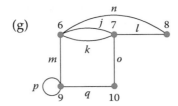 (h)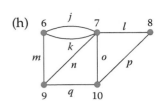

In Exercises 1 through 10, determine whether the indicated graph has an Euler circuit, an Euler trail, or neither. If so, then specify it. If not, then justify why.

1. Graph (a).

2. Graph (b).

3. Graph (c).

4. Graph (d).

5. Graph (e).

6. Graph (f).

7. Graph (g).

8. Graph (h).

9. The octahedron graph.

10. The dodecahedron graph.

11. Notre Dame Cathedral is located on the Ile de la Cité in the portion of Paris pictured below.

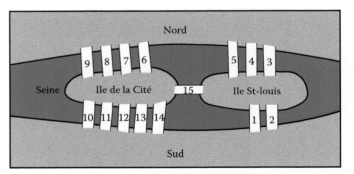

With its many bridges, this section might appeal to tourists wanting to cross over each bridge. Determine whether this portion of Paris has an Euler circuit, an Euler trail, or neither.

12. Pictured is a portion of Rome through which the Tiber river flows.

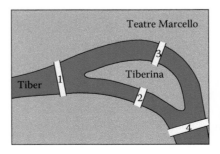

1. Ponte Garibaldi
2. Ponte Crestio
3. Ponte Fabricio
4. Ponte Palatino

For a tourist interested in its bridges, determine whether this portion of Rome has an Euler circuit, an Euler trail, or neither.

13. A street sweeper needs to sweep each side of the two-way streets in the pictured neighborhood.

To be maximally efficient, the route should cover each side of each street exactly once. Find a maximally efficient route through this neighborhood for the street sweeper.

14. To protect against cracking pavement, a sealant needs to be applied to the surfaces of the streets in the pictured neighborhood.

With one pass, a truck can spray the sealant on one side of these two-way streets. Find a route through the neighborhood for the sealant truck so that it covers each side of each street exactly once.

15. Suppose a graph G has an Euler trail and is not Eulerian. Is it possible to add a single edge to G to make the resulting graph Eulerian? Explain.

16. Suppose a graph G has an Euler trail and the two vertices of odd degree are endpoints of an edge e. Must $G \setminus \{e\}$ be Eulerian? Explain.

17. Suppose a graph G has c components and exactly two vertices u and v of odd degree. Can u and v be in different components? What is the minimum number of edges that must be added to G to obtain a graph with an Euler trail?

18. Suppose a graph G has c components and each vertex has even degree. What is the minimum number of edges that must be added to G to obtain an Eulerian graph? Does it matter where the edges are attached?

19. A standard assignment for a janitor is to mop and wax the hallways in some building. Of course, each floor area should be mopped first and waxed second. Also, it is preferred that the janitor should not walk on a floor once it has been waxed. Consequently, the janitor should enter the building, mop and wax all of the floors, and then exit the building, leaving the floors to dry overnight. Prove that, no matter what the layout of the hallways

in the building, the janitor can always find a plan that accomplishes this.

20. Once a building has been constructed, its walls need to be painted. Suppose one man is assigned the job of painting the walls of all of the hallways in a single color. Since the ladder and other materials are cumbersome to move around, the man would like to complete the job in one continuous route around the building, a route that does not require him to carry his materials down a hallway both of whose walls have already been painted. Prove that, no matter what the layout of the hallways in the building, the painter can always find a plan that accomplishes this.

21.* Prove that the equivalent conditions in Theorem 9.7(a) are also equivalent to the condition that the edges of the graph G can be split into a disjoint union of cycles.

22. Prove Theorem 9.7(b).

23. For which integers n is Q_n Eulerian?

24. For which integers n is K_n Eulerian?

25. For which pairs of integers m, n is $K_{m,n}$ Eulerian?

26. Let $G = (V, E)$ be a graph with $|V|$ odd such that both G and G^c are connected. Show: G is Eulerian if and only if G^c is Eulerian.

27. What is the minimum possible number of edge repetitions in a circuit that covers every edge of $K_{2,3}$?

28. What is the minimum possible number of edge repetitions in a circuit that covers every edge of K_4?

29. Consider the layout of the portion of Paris shown in Exercise 11. A tourist desires a walk that crosses every bridge and returns to its starting point.

    (a) What is the smallest possible number of bridges that would need to be crossed more than once in order for the tourist to cross every bridge?

    (b) Suppose that the bridge connecting Ile de la Cité to Ile St-Louis is being repaired and cannot be crossed. What is the smallest number of bridge repetitions needed in that case?

30. Consider the layout of the portion of Rome shown in Exercise 12. A tourist desires a walk that never crosses a bridge more than once and returns to its starting point.

    (a) Suppose one of the bridges crossing the Tiber is out of service. What is the maximum number of bridge crossings that can be accomplished?

    (b) If, instead, one of the bridges to the island is out of service, then what is the maximum number of possible bridge crossings?

Subsequent exercises refer to the following directed graphs:

(i)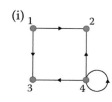

(j)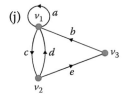

(k)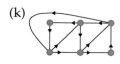

(l)

(m)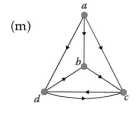

(n)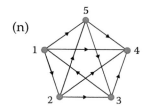

In Exercises 31 through 36, determine whether the indicated directed graph has an Euler circuit, an Euler trail, or neither.

31. Graph (i).

32. Graph (j).

33. Graph (k).

34. Graph (l).

35. Graph (m).

36. Graph (n).

37. Let $G = (V, E)$ be a directed graph, and form a new edge set E' as follows. For each edge $e \in E$, if $e \mapsto (u, v)$, then put an edge $e' \in E'$ with $e' \mapsto (v, u)$. Show that if G is weakly connected, then the directed graph $G' = (V, E \cup E')$ is Eulerian.

38. Suppose G is a directed Eulerian graph. Show that $\underline{G}$ must then be an Eulerian graph.

39.* Prove Theorem 9.8(a).

40. Prove Theorem 9.8(b).

41. The route for the snowplow that we found in Example 9.7 has the problem that it sometimes directs the snowplow to drive on the left-hand side of the road. That route is therefore undesirable, since not only does the plow most naturally throw the snow off to its right side but also the plow could cause major traffic problems by disobeying the traffic laws. Explain how Theorem 9.8 can be used to solve snowplow problems like the one in Example 9.7 in such a way that the traffic laws may be obeyed by the driver.

42. We have seen in our discussions of the Königsberg Bridge Problem that a tourist could not take a tour of Königsberg that travels over each bridge exactly once. Short of this, the tourist may simply desire a route that covers each bridge at least once with the smallest possible number of repetitions. Show that the minimum number of bridge repetitions needed in the Königsberg Bridge Problem is 2.

9.3   Hamiltonian cycles

In Section 9.2, our primary interest was in finding a circuit that covered every edge exactly once. If our graph represents a neighborhood whose streets we wish to plow, then that pursuit is appropriate. However, if we need to deliver mail to the houses in the neighborhood, then our focus shifts. If each address is represented by a vertex, then our interest is in finding a circuit that covers every vertex exactly once.

Example 9.8.   Find a mail route through the pictured neighborhood that hits every mailbox exactly once and returns to its start.

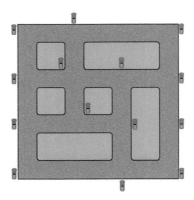

Solution. Make a graph that models the neighborhood in such a way that the vertices correspond to the mailboxes, and the edges reflect streets.

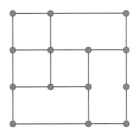

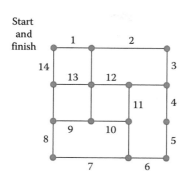

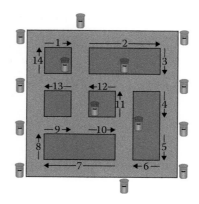

FIGURE 9.2
Designing a mail route.

In the copy of this graph on the left-hand side of Figure 9.2, we have consecutively labeled the edges along a cycle with the desired properties; it visits each vertex exactly once. How one might find such a cycle is addressed after Example 9.9. Here, we simply observe that this cycle corresponds to a desired mail route through the given neighborhood, as labeled on the right-hand side of Figure 9.2; it visits each mailbox exactly once.   □

The special cycle used in Example 9.8 is one of an important type.

Definition 9.4.   Let G be a graph or a digraph. A **Hamiltonian cycle** (respectively, **Hamiltonian path**) in G is a cycle (respectively, path) that covers every vertex. By definition, each vertex must be covered exactly once, with the exception that the starting and ending vertex of a Hamiltonian cycle is covered twice. If G contains a Hamiltonian cycle, then G is said to be **Hamiltonian**.

Hamiltonian cycles are named after the Irish mathematician William Rowan Hamilton (1805–1865), who was one of the first—although not *the* first—to study them.

Example 9.9.   The vertices in the pictured graph have been numbered consecutively along a Hamiltonian cycle.

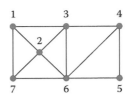

How can we find a Hamiltonian cycle? Since a Hamiltonian cycle ultimately visits every vertex, if there is a Hamiltonian cycle to be found in a graph, then it can be found from any chosen starting point. From there, we might simply start walking through the graph in an attempt to visit every vertex exactly once. If at some point we discover that our walk cannot possibly be completed to a Hamiltonian cycle, then we can retreat on our walk to an earlier position from which we hope an alternative choice of direction will yield better results. If we are systematic in this process and explore every possible direction from every encountered fork in the road, then this exhaustive method always finds a Hamiltonian cycle when one exists. This method is called backtrack searching and will be discussed in Section 10.2, where we will also see that it is extremely inefficient—and thus unreasonable in all but small examples.

Having previously studied Euler circuits, one might expect an analog of Euler's Theorem for characterizing Hamiltonian graphs. However, no such characterization is known. Moreover, there is no known efficient algorithm for finding a Hamiltonian cycle in a graph. Since algorithm efficiency is discussed in Chapter 10, we shall revisit and delve deeper into this issue there. Here, we must understand that the general problem of finding Hamiltonian cycles is quite hard. Hence, we primarily focus on small problems that can be analyzed by hand. Additionally, we pursue special conditions under which the existence of Hamiltonian cycles can be determined. For instance, there is a relationship between Hamiltonicity and connectivity.

Theorem 9.9.   *Let G be any graph. If G is Hamiltonian, then* $\kappa(G) \geq 2$.

Proof. A Hamiltonian cycle is a subgraph isomorphic to a cycle C_n covering all of the vertices of G. The result in Exercise 24 from Section 9.1 then tells us that $\kappa(G) \geq \kappa(C_n)$. By Theorem 9.1, $\kappa(C_n) = 2$, and the result follows.   □

The contrapositive of Theorem 9.9 says that, for any graph G,

if $\kappa(G) \leq 1$, then G is not Hamiltonian.

This can be used to show that a graph does not have a Hamiltonian cycle.

Example 9.10.   The pictured graph G represents a computer network.

The presence of a Hamiltonian cycle in such a network might be of advantage to a programmer employing a parallel algorithm. In that case, the Hamiltonian cycle could provide a route along which each computer could pass a process to the next, much like an assembly line. However, this graph G is not Hamiltonian, since $\kappa(G) = 1$.

The method used in Example 9.10 to show that a graph is not Hamiltonian does not apply very generally. The following theorem provides more useful tools.

Theorem 9.10.   *If C is a Hamiltonian cycle in a graph $G = (V, E)$, then*

  (i) *C covers exactly two edges incident with each vertex, and*

  (ii) *C has no subgraph that is a cycle on fewer than $|V|$ vertices.*

In the case that G is directed, for each vertex v, C must cover exactly one edge whose head is v and one whose tail is v.

Example 9.11.   Show that the pictured graph G does not have a Hamiltonian cycle.

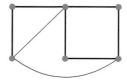

Proof. Suppose G does have a Hamiltonian cycle C. The edges shown in bold would have to be included in C, since each is incident with a vertex of degree 2. However, this bold subgraph cannot be completed to a

Hamiltonian cycle, for two reasons. First, it contains a vertex of (bold) degree 3. Second, it contains a 4-cycle. Of course, either one of these reasons alone suffices to contradict the existence of C.   □

Having seen ways to show that a Hamiltonian cycle does not exist, we ought to consider conditions sufficient to guarantee that one does. There is a bound on the minimum degree $\delta(G)$ of a graph G that implies Hamiltonicity.

Theorem 9.11.   *For any simple graph $G = (V, E)$ with $|V| \geq 3$, if $\delta(G) \geq \frac{|V|}{2}$, then G is Hamiltonian.*

Proof. Suppose not. Let n be the smallest number of vertices on which there is a simple graph H for which $\delta(H) \geq \frac{n}{2}$ but H has no Hamiltonian cycle. Further, among all simple graphs H on n vertices with $\delta(H) \geq \frac{n}{2}$ and no Hamiltonian cycle, pick a graph $G = (V, E)$ with the largest number of edges. Consequently, the addition of any new edge to G would cause G to have a Hamiltonian cycle.

Let v_1 and v_n be two nonadjacent vertices in G, and let $v_1, v_2, \ldots, v_n, v_1$ be an ordering of the vertices in a Hamiltonian cycle in the graph obtained from G by adding an edge $e_n \mapsto \{v_n, v_1\}$. Hence, $v_1, v_2, \ldots, v_n$ is a Hamiltonian path in G.

We claim that there exists an index k such that there exist edges $e_k \mapsto \{v_k, v_n\}$ and $e_{k+1} \mapsto \{v_{k+1}, v_1\}$ in G,

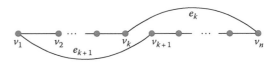

and, consequently, the listing

$$v_1, v_2, \ldots, v_k, v_n, v_{n-1}, \ldots, v_{k+1}, v_1$$

gives a Hamiltonian cycle in G. Let

$$K = \{k \ : \ 1 \leq k \leq n-1 \text{ and}$$
$$\exists\, e \in E \text{ such that } e \mapsto \{v_n, v_k\}\} \quad \text{and}$$
$$K' = \{k \ : \ 1 \leq k \leq n-1 \text{ and}$$
$$\exists\, e \in E \text{ such that } e \mapsto \{v_1, v_{k+1}\}\}.$$

Note   that $|K \cup K'| \leq n - 1$, $|K| \geq \deg(v_n) \geq \frac{n}{2}$,   and $|K'| \geq \deg(v_1) \geq \frac{n}{2}$. Hence, $|K \cap K'| = |K| + |K'| - |K \cup K'| \geq \frac{n}{2} + \frac{n}{2} - (n - 1) = 1$. Since $|K \cap K'| \geq 1$, the sets K and K' must have a value k in common. This k is the desired index that enables the construction of a Hamiltonian cycle in G. Thus, we have a contradiction. □

Example 9.12. It follows from Theorem 9.11 that the pictured graph $G = (V, E)$ is Hamiltonian, since $\delta(G) = 4 = \frac{|V|}{2}$.

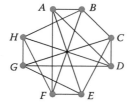

Of course, it is not hard in this case to find a Hamiltonian cycle, such as

$$A, B, C, D, H, G, E, F, A.$$

The result in Theorem 9.11 is best possible, since the graph $G = (V, E)$ in Example 9.11 is not Hamiltonian and satisfies $\delta(G) = \frac{|V|}{2} - 1$.

Although the n-cubes Q_n have relatively low minimum degree and they are hard to visualize for large n, it is rather straightforward to show that they are Hamiltonian.

Example 9.13. For each integer $n \geq 2$, the n-cube Q_n is Hamiltonian.

Proof. Clearly, the 2-cube $Q_2 \cong C_4$ has a Hamiltonian cycle. We proceed by induction on n. So suppose that $n \geq 2$ and Q_n has a Hamiltonian cycle. We identify Q_{n+1} with $Q_n \times P_2$. Say $\{1, 2\}$ is the vertex set for P_2, and let C be a Hamiltonian cycle for Q_n. Say C starts at vertex u and that vertex v is the last on C before returning to u. Let P be the path from u to v along $C \setminus \{\{v, u\}\}$, and let Q be the reverse of P. The case in which $n = 2$ is pictured in Figure 9.3. The cycle

$$P \times \{1\}, \{(v, 1), (v, 2)\}, Q \times \{2\}, \{(u, 2), (u, 1)\}$$

gives the desired Hamiltonian cycle for Q_{n+1}. □

Having focused on the presence of Hamiltonian cycles in ordinary graphs, we now turn our attention to directed graphs. A directed graph can be used to reflect the results of a series of head-to-head competitions. In that case, we shall see that Hamiltonian paths may be of interest.

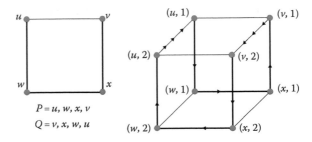

FIGURE 9.3
A Hamiltonian cycle in Q_3 from a Hamiltonian path in Q_2.

Definition 9.5. A **tournament** is a directed graph whose underlying graph is complete.

Example 9.14. The pictured directed graph is a tournament.

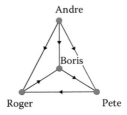

Its underlying graph is K_4. In this case, the tournament reflects the results in Table 9.1 of a tennis competition. For instance, the edge from Andre to Boris represents the victory of Andre over Boris in their match (1a).

In general, a tournament on K_n might reflect the results of a tennis tournament in which every possible pair of players competed in a match. An edge (u, v) would be present if and only if player u defeated player v. At the end of the tennis tournament, a way to rank the players could be provided by

TABLE 9.1
Tennis tournament

Match	Results
1a	Andre defeated Boris
1b	Pete defeated Roger
2a	Andre defeated Pete
2b	Roger defeated Boris
3a	Andre defeated Roger
3b	Boris defeated Pete

a Hamiltonian path. However, this may not always be the best way. In the exercises, we shall see that the results of a round robin tournament do not generally yield a clear champion.

In the tournament in Example 9.14, the path

Andre, Boris, Pete, Roger

is a Hamiltonian path and might be considered a ranking of those tennis players. The fact that such a path exists is true in general for tournaments.

Theorem 9.12.   *Every tournament has a Hamiltonian path.*

Proof. We proceed by induction on the number n of vertices in the tournament. It is easy to see that any tournament on $n = 2$ vertices has a Hamiltonian path. So suppose $n \geq 2$ and that any tournament on n vertices has a Hamiltonian path. Let $K = (V, E)$ be a tournament on $n + 1$ vertices, and let v_{n+1} be a vertex of K. Since $K \setminus \{v_{n+1}\}$ is a tournament on n vertices, it has a Hamiltonian path $v_1, v_2, \ldots, v_n$. If $(v_n, v_{n+1}) \in E$, then $v_1, v_2, \ldots, v_n, v_{n+1}$ is a Hamiltonian path in K. If $(v_{n+1}, v_1) \in E$, then $v_{n+1}, v_1, v_2, \ldots, v_n$ is a Hamiltonian path in K. Otherwise, both (v_{n+1}, v_n) and (v_1, v_{n+1}) are in E, and we can let k be the smallest index less than n such that (v_k, v_{n+1}) and (v_{n+1}, v_{k+1}) are both in E. That is, k is the largest index such that for each $1 \leq i \leq k$, we have $(v_i, v_{n+1}) \in E$. See Figure 9.4. Since $(v_n, v_{n+1}) \notin E$, it follows that $k < n$. A Hamiltonian path for K is then given by $v_1, \ldots, v_k, v_{n+1}, v_{k+1}, \ldots, v_n$. $\square$

Example 9.15.   The pictured tournament must have a Hamiltonian path by Theorem 9.12.

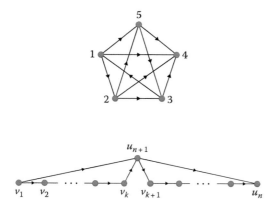

FIGURE 9.4
Finding a Hamiltonian path.

Moreover, the proof of Theorem 9.12 can help us find one. Suppose we have been able to find the Hamiltonian path $1, 2, 3, 4$ on the subtournament induced by those vertices. The proof of Theorem 9.12 then tells us how to insert vertex 5 and obtain the Hamiltonian path $1, 2, 5, 3, 4$.

Exercises

Subsequent exercises refer to the following pictured graphs:

In Exercises 1 through 6, find and shade a Hamiltonian cycle in the specified graph.

1. Graph (a).   2. Graph (b).

3. The tetrahedron graph.   4. The octahedron graph.

5. The dodecahedron   6. The icosahedron graph.
   graph.

7. The mailman needs to deliver mail to each of the mail-boxes in the pictured neighborhood. He wants to enter and exit the neighborhood from the same location. To be efficient, he does not want to pass by a mailbox to which he has already delivered the mail.

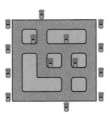

Find a mail route that hits every mailbox exactly once.

8. The UPS truck needs to make deliveries to each of the marked locations in the pictured neighborhood. To be as fast as possible, the truck should not pass by a location to which it has already made a delivery.

Find a route for the UPS truck that starts and finishes in the same location and hits every delivery location exactly once.

In Exercises 9 through 14, show that there is no Hamiltonian cycle in the specified graph.

9. Graph (c).   10. Graph (d).

11. $K_{3,2}$.   12. P_n.

13.   14.

The following pictured graphs are referred to in the subsequent exercises:

(e) 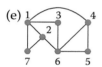   (f) 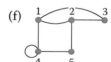   (g) 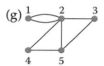   (h)

In Exercises 15 through 18, determine whether the specified graph has a Hamiltonian cycle. Justify your answer.

15. Graph (e).   16. Graph (f).

17. Graph (g).   18. Graph (h).

19. A sales representative is based in Syracuse, NY, and her sales territory includes the same cities in upstate New York covered by a small airline, whose round-trip service is shown in the following table.

	Binghamton	Buffalo	Ithaca	Rochester	Syracuse
Binghamton	No	Yes	No	No	Yes
Buffalo	Yes	No	Yes	No	Yes
Ithaca	No	Yes	No	Yes	No
Rochester	No	No	Yes	No	Yes
Syracuse	Yes	Yes	No	Yes	No

Is it possible for this sales representative to use this airline for a single trip that visits each city in her territory exactly once before immediately returning home? Justify your answer.

20. The ferry service among the eight islands in a popular vacation destination is pictured below.

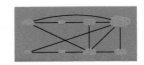

Is it possible for a tourist visiting these islands to take a sequence of ferry rides that visits each island exactly once and immediately returns to its starting point? Justify your answer.

21. Characterize the pairs (m, n) for which $K_{m,n}$ is Hamiltonian.

22. Show that the Petersen graph contains cycles of length n for $n = 5, 6, 8, 9$. See Exercise 31.

In Exercises 23 through 26, regard a Hamiltonian cycle as a subgraph that happens to be a cycle with no particular order specified for its vertices. That is, we distinguish Hamiltonian cycles by their edge sets. In each case, determine the number of different Hamiltonian cycles that the given graph has.

23.

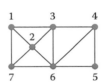

24.

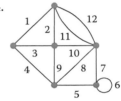

25. K_n.

26. $K_{n,n}$.

27. Let G be any graph. Theorem 9.9 tells us that if G has a Hamiltonian cycle, then $\kappa(G) \geq 2$. Show that there is no integer k such that if $\kappa(G) \geq k$, then G is Hamiltonian.

28. What is the maximum value of k such that the removal of any k edges from K_n yields a Hamiltonian graph?

29. Let $n \geq 3$, and let G be any graph on n vertices. Show that if $\kappa(G) \geq \frac{n}{2}$, then G is Hamiltonian.

30. Show that assuming that G is simple is necessary in Theorem 9.11.

31.* Prove or disprove: The Petersen graph is Hamiltonian. Hint: Use Exercise 32 from Section 8.4.

32. Prove or disprove: The pictured graph is Hamiltonian.

33. Show for all integers $n \geq 2$ that $C_n \times P_2$ is Hamiltonian.

34. Show for all integers $n \geq 2$ that $P_n \times P_2$ is Hamiltonian.

35.* Show for all integers $n \geq 2$ that $C_n \times P_3$ is Hamiltonian.

36.* Show for all integers $n \geq 2$ that $C_3 \times P_n$ is Hamiltonian.

37. Find a Hamiltonian cycle in the pictured directed graph.

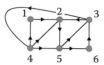

38. Find a Hamiltonian path in the pictured directed graph.

39. Find a Hamiltonian path in the pictured tournament.

40. Find a Hamiltonian path in the pictured tournament.

41. Show that the pictured directed graph is not Hamiltonian.

42. Show that the pictured directed graph is not Hamiltonian.

43. Each of five schools sent its best chess player to the County Speed Chess Tournament. This tournament is a daylong event that follows a round robin format. The schedule and results for the tournament are displayed

TABLE 9.2
Chess tournament

Time	Results
9 a.m.	Ann defeats Dan
	Bob defeats Cari
10 a.m.	Ann defeats Bob
	Cari defeats Ed
11 a.m.	Ann defeats Ed
	Dan defeats Bob
Noon	Lunch
1 p.m.	Ann defeats Cari
	Dan defeats Ed
2 p.m.	Bob defeats Ed
	Cari defeats Dan
3 p.m.	Awards

in Table 9.2. From those results, the organizers need to decide upon a ranking of the competitors from first place to fifth place. The only constraint is that each player must have been defeated by the player ranked immediately above him or her.

(a) Can anyone other than Ann be declared the overall champion?

(b) Suppose that the 11 a.m. game had instead resulted in Ed defeating Ann. Could Ed be declared the champion?

(c) If Ed had defeated Ann, would each competitor have had the possibility of being declared the champion?

44. The Knotty Oaks Tennis Club has set up a 3-day round robin tennis tournament for its female members over the age of sixty. The results of the matches are displayed in Table 9.3. The awards committee now needs to declare a ranking of the players from first place to fourth place

TABLE 9.3
Tennis tournament

Time	Results
Day 1	Beth defeats Alice
	Ceil defeats Dee
Day 2	Alice defeats Ceil
	Dee defeats Beth
Day 3	Alice defeats Dee
	Ceil defeats Beth

in such a way that each woman must have defeated the woman ranked immediately below her.

(a) Find a ranking in which Dee is declared the champion.

(b) Does each player have a chance of being declared the champion?

(c) If Beth is declared the champion, can Dee come in last place?

45.* Show that every tournament has a vertex from which each vertex is a distance of at most 2.

46. Give an example of a tournament with no Hamiltonian cycle.

9.4 Planar graphs

If a graph represents the layout of a circuit board, then edge crossings reflect wire connections that are difficult to accommodate. A desired layout is one in which there are no crossings. In general, we prefer a closed circuit to be implemented without wire crossings.

> **Example 9.16.**   The closed electric circuit on the left-hand side of Figure 9.5 contains a battery ⊣⊢ and some resistors ⌐\/\/⌐. An intersection of wires not marked by a node • is a wire crossing, which we would like to avoid. On the right-hand side of Figure 9.5 is a graph that models this circuit, with edges representing wires. Since that graph can alternatively be drawn with no edge crossings, as shown on the right-hand side of Figure 9.6, the circuit can similarly be constructed without wire crossings, as shown on the left-hand side of Figure 9.6.

Applications like Example 9.16 motivate our study of graphs that can be drawn without crossings.

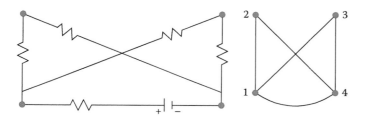

FIGURE 9.5
An electric circuit with a wire crossing.

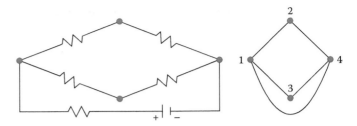

FIGURE 9.6
An electric circuit without wire crossings.

Definition 9.6.

(a) A **planar embedding** of a graph is a drawing of the graph such that the images of distinct edges do not intersect outside of their endpoints. That is, there are no crossings.

(b) A graph is said to be **planar** if it has a planar embedding.

Example 9.17.   A planar embedding of a graph is pictured.

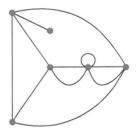

The fact that a graph may have a drawing that is not a planar embedding does not mean that the graph is not planar. Although the graph in Figure 9.5 is pictured with a crossing, the alternative drawing without crossings in Figure 9.6 shows that it is planar. In Example 8.21, we saw two different drawings of K_4,

one with a crossing and one without crossings. The existence of the drawing without crossings implies that K_4 is planar. One might next try to find a planar embedding of K_5, but the result in Example 9.18 cannot be improved.

Example 9.18. K_5 can be drawn with one crossing.

However, simply trying for a while and failing to find a planar embedding does not prove that K_5 is not planar. We shall prove that K_5 has no planar embedding later in this section (Proposition 9.15), after we develop stronger tools.

Example 9.19.   The diagram on the left-hand side of Figure 9.7 reflects the need to connect three homes to each of three utilities. However, for safety reasons, it is undesirable to have lines from these utilities crossing. Since the graph $K_{3,3}$ models this utility problem, the drawing of $K_{3,3}$ on the right-hand side of Figure 9.7 reflects a good layout for these connections. It contains only one crossing. In Proposition 9.17, we will prove that $K_{3,3}$ is not planar and, hence, that the utility connections required here must accommodate at least one crossing.

The machinery we use to prove that K_5 and $K_{3,3}$ are not planar is built from the following tools.

Definition 9.7.   Given a planar embedding of a graph $G = (V_G, E_G)$,

   (a) A **region** is a maximal connected subset of the complement of the image of the embedding.

   (b) We use R_G, or just R, to denote the set of regions.

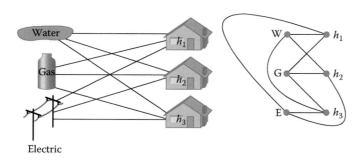

FIGURE 9.7
Connecting three utilities to three homes.

(c) The **dual graph**, denoted $D(G)$, is the graph with vertex set R_G and edge set E_G for which the endpoints of each edge e are taken to be the regions that, in the embedding, share the image of e as part of their boundary.

Although the regions and the dual depend not only on the graph but also on the embedding, a particular embedding will always be fixed in context and is therefore not reflected in the notation. In fact, the isomorphism type of the dual may depend upon the chosen embedding. (See Exercise 15.)

Example 9.20.   The regions of the graph G pictured in Example 9.17 are labeled here A, B, C, D, E, F, O.

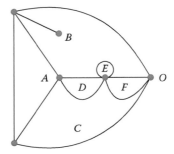

Note that O is the unique unbounded region.

The standard way of drawing the dual of a graph based on a given embedding is shown in the following example.

Example 9.21.   We construct the dual of the graph G from Example 9.17. In Figure 9.8, a vertex represented

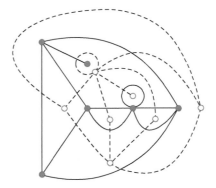

FIGURE 9.8
A graph and tts dual.

by an open point ○ has been placed inside each region of G, and the dotted lines join regions that share an edge as part of their boundaries. Note that each dotted edge crosses exactly one solid edge, and vice versa. The dual graph D(G) can then be drawn by itself in the usual fashion, with solid vertices and edges.

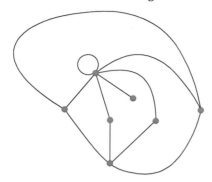

Notice that for any graph G with a planar embedding, D(G) is always planar and $D(D(G)) \cong G$. For example, the dual of the dotted graph in Figure 9.8 can be seen to be the solid graph in that picture.

A graph $G = (V, E)$ with a planar embedding has not only a vertex set V and an edge set E but also a region set R. It turns out that the three numbers $|V|$, $|E|$, and $|R|$ are related in such a way that knowing two of them forces the value of the third.

Theorem 9.13. (Euler's Formula).
Given any planar embedding of a connected graph $G = (V, E)$, we have

$$|V| - |E| + |R| = 2.$$

Proof. We proceed by induction on $n = |V|$. If $n = 1$, then all edges must be loops, and it is straightforward to argue that $|R| = |E| + 1$ in this case. (See Exercise 16.) So assume that $n \geq 1$ and that Euler's formula holds for all graphs on n vertices.

Suppose that $G = (V, E)$ is a graph on $n + 1$ vertices with a planar embedding. Since G is connected, there must be some non-loop edge e, say, $e \mapsto \{u, v\}$. Form a new graph $G' = (V', E')$ from G by removing e and identifying u and v to a single vertex v'.

$$\bullet\!\!-\!\!\overset{u \quad e \quad v}{\underset{\smile}{\bullet\!\!-\!\!\bullet}}\!\!<\quad \rightarrow \quad \bullet\!\!-\!\!\overset{v'}{\underset{\circ}{\bullet}}\!\!<$$

The planar embedding of G can easily be converted (as shown) into a planar embedding for G' with $|R'| = |R|$.

Since $|V'| = |V| - 1 = n$, the induction hypothesis gives that $|V'| - |E'| + |R'| = 2$. Since $|E'| = |E| - 1$, substitution gives that $|V| - |E| + |R| = 2$. □

It follows from Theorem 9.13 that, for a fixed graph, the number of regions in any planar embedding will always be the same. Specifically, that number is $|E| - |V| + 2$ and is a function of only $|V|$ and $|E|$.

The more edges a graph has, the less likely that a planar embedding for it exists. There is, in fact, an upper limit on the number of edges in a planar simple graph, in terms of the number of vertices.

Corollary 9.14.   Given any planar simple graph $G = (V, E)$ with $|V| \geq 3$, we have

$$|E| \leq 3|V| - 6.$$

Proof. It suffices to assume that G is connected, because a disconnected graph has fewer edges than the connected graph obtained by joining its components with additional edges. Since the dual graph $D(G)$ has vertex set R and edge set E, Theorem 8.12 tells us that $\sum_{r \in R} \deg(r) = 2|E|$. Since G is simple, it has no regions bounded by just one or two edges. That is, for each $r \in R$, $\deg(r) \geq 3$. Hence, $3|R| \leq 2|E|$. Theorem 9.13 then gives that

$$2 = |V| - |E| + |R| \leq |V| - |E| + \frac{2}{3}|E| = |V| - \frac{1}{3}|E|.$$

Thus, $6 \leq 3|V| - |E|$, and the result follows. □

Corollary 9.14 can be used to give a simple proof of the following result.

Proposition 9.15. K_5 *is not planar.*

Proof. The values $|V| = 5$ and $|E| = 10$ do not satisfy the inequality in Corollary 9.14. □

The bound on the number of edges in a planar simple graph is further lowered for triangle-free graphs.

Corollary 9.16.   Given any planar simple graph $G = (V, E)$ with $|V| \geq 3$ and no triangles (i.e. no 3-cycles), we have

$$|E| \leq 2|V| - 4.$$

Corollary 9.16 applies, in particular, to planar simple bipartite graphs. Its proof is left for the exercises.

Proposition 9.17. $K_{3,3}$ *is not planar.*

Proof. The values $|V| = 6$ and $|E| = 9$ do not satisfy the inequality in Corollary 9.16. □

Using Corollaries 9.14 and 9.16, the results that K_5 and $K_{3,3}$ are not planar follow with relative ease. The importance of those results is their role in a beautiful characterization of planar graphs. Its presentation requires the introduction of another notion.

Definition 9.8. Let $G = (V, E)$ be a graph.

(a) Given an edge $e \in E$, a new graph $G' = (V', E')$ is said to be obtained by **subdividing** $e = \{u, v\}$ if $V' = V \cup \{w\}$, where w is a new vertex not in V, and $E' = (E \setminus \{e\}) \cup \{e', e''\}$, where $e' \mapsto \{u, w\}$ and $e'' \mapsto \{w, v\}$ are new edges not in E. That is, e is subdivided by the new vertex w.

(b) We say that G' is a **subdivision** of G, or a G-**subdivision**, if G' is obtained from G by a (possibly empty) sequence of edge subdivisions. We also say that two graphs G' and G'' are **homeomorphic** if there is a graph G such that both G' and G'' are G-subdivisions.

Example 9.22. Pictured is a sequence of edge subdivisions.

The following result, which we owe to the Polish mathematician Casimir Kuratowski (1896–1980), characterizes planar graphs in terms of two kinds of forbidden subgraphs.

Theorem 9.18. (Kuratowski's Theorem).
A graph is not planar if and only if it contains a subgraph that is a subdivision of either K_5 or $K_{3,3}$. Equivalently, G is not planar if and only if G contains a subgraph homeomorphic to K_5 or $K_{3,3}$.

The straightforward proof that any graph containing a subdivision of either K_5 or $K_{3,3}$ is not planar is left for the exercises.

The proof of the converse of this is much more involved and is not included in this book. The interested reader can find a proof in Harary (1969), for example. Here, we focus on applications of Kuratowski's Theorem.

Example 9.23. The Petersen graph is not planar. The left-hand picture below is the second drawing of the Petersen graph from Example 8.23.

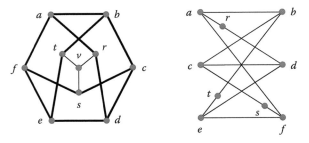

Although the Petersen graph contains no triangles, Corollary 9.16 does not help us here, since indeed $|E| = 15 \leq 16 = 2|V| - 4$. However, Kuratowski's Theorem does the trick since the Petersen graph can be seen to contain a $K_{3,3}$-subdivision (shown in bold above). That is, $G - \{v\}$ is seen in the right-hand picture above to be a $K_{3,3}$-subdivision.

Corollaries 9.14 and 9.16 are not generally useful for proving that graphs are not planar. Those corollaries made their contributions toward showing that K_5 and $K_{3,3}$ are not planar, which is needed in Kuratowski's Theorem. In general, one should seek the help of Kuratowski's Theorem to show that a graph is not planar.

Example 9.24. The graph pictured on the left below is a layout for a power grid for eight stations.

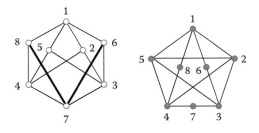

Despite the fact that it may be desirable to lay out this grid without any cable crossings, this graph is not planar. The subgraph obtained by deleting the two bold edges is seen in the right-hand picture above to be a sub-division of K_5. (Check it.)

A significant result on planar graphs known as the Four-Color Theorem is discussed in Section 9.5.

9.4.1 Crossing number

When a planar embedding is not possible, it is natural to seek a drawing of the graph with as few crossings as possible.

Definition 9.9. The **crossing number** of a graph G, denoted $v(G)$, is the minimum possible number of crossings in a drawing of G.

A graph G is planar if and only if $v(G) = 0$. Since K_5 and $K_{3,3}$ are not planar and Examples 9.18 and 9.19 show drawings of them with one crossing, it follows that $v(K_5) = v(K_{3,3}) = 1$. In applications such as network or circuit design, when the number of wire crossings must be minimized, an embedding that achieves the crossing number of the corresponding graph is sought.

Example 9.25. The layout for the power grid in Example 9.24 has crossing number 1. That the crossing number is at least 1 is established in Example 9.24 by showing that the graph modeling this grid is not planar. That the grid can be laid out with just 1 crossing is shown in the following picture:

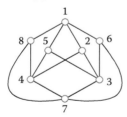

Exercises

1. Show that the pictured computer network configuration can be laid out without cable crossings.

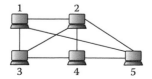

2. Pictured is a design plan for a power grid, in which the symbol o denotes a power station. Show that there is a layout for this power grid without cable crossings.

3. When connecting the components of a personal computer, one typically ends up with a mess of tangled cables stuffed behind the computer. Show that the connections in the pictured computer system can be accomplished without crossings.

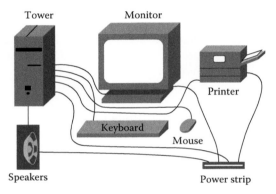

4. A camp is being built to contain three buildings. There are four utilities available, but it is not necessary for each building to be connected to each utility. Show that the required connections shown in the diagram below can be accommodated without cable crossings.

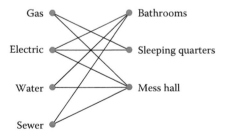

5. Show that $K_{2,3}$ is planar by giving a drawing without crossings.

6. **Two-utility problem**. If there are just 2 utilities and any number n of homes, then each home can be connected to each utility without any cable crossings. Show that for any $n \in \mathbb{Z}^+$, $K_{2,n}$ is planar.

In Exercises 7 through 10, draw the dual of the pictured embedding of a graph.

7.

8.

9.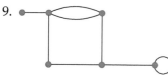

10.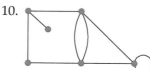

11. Exercises 15 and 16 from Section 8.4 show that the cube and the octahedron are planar graphs. Show that the dual of the cube is the octahedron, and vice versa.

12. Exercises 14 and 18 from Section 8.4 show that the dodecahedron and the icosahedron are planar graphs. Show that the dual of the dodecahedron graph is the icosahedron graph, and vice versa.

13. Given a planar embedding of a graph $G = (V, E)$ with regions R, if $|V| = 10$ and $|E| = 16$, then find $|R|$.

14. Given a planar embedding of a graph $G = (V, E)$ with regions R, if $|V| = 14$ and $|R| = 10$, then find $|E|$.

15* Give an example of a planar graph G and two planar embeddings of G that give rise to non-isomorphic duals.

16* Without using Euler's Formula, prove, for any planar embedding of a graph $G = (V, E)$ with $|V| = 1$, that $|R| = |E| + 1$. Hint: Use induction on $|E|$. Each new edge must split a region in two.

17. What is the appropriate version of Euler's Formula for a planar graph with c components?

18* Show that $\forall\, n, m, r \in \mathbb{N}$ with $n \geq 1$ and $n - m + r = 2$, there exists a connected planar graph G with $|V| = n$, $|E| = m$, and $|R| = r$.

19. Prove the easy direction of Kuratowski's Theorem. That is, show that any graph containing a subdivision of either K_5 or $K_{3,3}$ is not planar. Hint: Subdivisions of planar graphs are planar.

20* Prove Corollary 9.16.

21. Show: For all $n \geq 5$, K_n is not planar.

22. Show: For all $m, n \in \mathbb{Z}^+$, $K_{m,n}$ is planar if and only if $\min\{m, n\} \leq 2$.

23. Determine whether the pictured electric circuit can be constructed without wire crossings. Justify your answer.

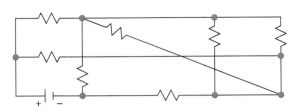

24. Determine whether the pictured electric circuit can be constructed without wire crossings. Justify your answer.

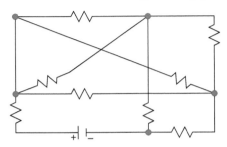

Subsequent exercises refer to the following pictured graphs:

(a) 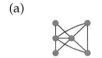   (b)   (c)   (d)

(e)    (f)   (g)   (h)

In Exercises 25 through 32, prove or disprove that the specified graph is planar.

25. Graph (a).   26. Graph (b).

27. Graph (c).   28. Graph (d).

29. Graph (e).   30. Graph (f).

31. Graph (g).   32. Graph (h).

33. Find all nonplanar simple graphs on 5 vertices.

34. Find all nonplanar simple graphs on 6 vertices that contain neither K_5 nor $K_{3,3}$ explicitly (subdivisions are permitted).

35.* Show: Any planar simple graph G has a vertex of degree at most 5. That is, $\delta(G) \leq 5$. Hint: Use Corollary 9.14 in a proof by contradiction.

36* Show: Any planar graph G with no triangles has a vertex of degree at most 3. That is, $\delta(G) \leq 3$.

37. Show that the bounds given in Corollary 9.14 are best possible. That is, for each $n \in \mathbb{Z}^+$, give an example of a planar simple graph $G = (V, E)$ with $|V| = n$ and $|E| = 3n - 6$. Hint: To C_{n-2} add one point inside and one outside.

38. Show that the bounds given in Corollary 9.16 are best possible. That is, for each $n \in \mathbb{Z}^+$, give an example of a triangle-free planar simple graph $G = (V, E)$ with $|V| = n$ and $|E| = 2n - 4$.

39. The pictured graph represents a power grid in which the power stations are connected in a cycle. Beyond the cables used for the outer cycle, other cables have been added to allow the bypassing of power stations. Thus, a cyclic layout can be maintained despite the failure of any single station.

Prove that this graph is planar but that any simple graph obtained by adding an edge to it is not planar.

40. The pictured graph models a computer network. A layout is sought for this network in which no communication cables cross. Can this be accomplished?

That is, prove or disprove that the pictured graph is planar.

In Exercises 41 through 48, find the crossing number of the specified graph, and justify your result.

41. Graph (a).   42. Graph (b).

43. Graph (c).   44. Graph (d).

45. Graph (e).   46. Graph (f).

47. Graph (g).   48. Graph (h).

49. Prove that the crossing number of the Petersen graph is at most 2.

50. Prove that the crossing number of the pictured graph is at most 3.

51. $\nu(K_6) = 3$. Find a drawing of K_6 with 3 crossings.

52. $\nu(K_{4,4}) = 4$. Find a drawing of $K_{4,4}$ with 4 crossings.

53. $\nu(Q_4) = 8$. Find a drawing of Q_4 with 8 crossings.

54. $\forall\, n \geq 2$, $\nu(K_{3,n}) = n - 2$. Find a drawing of $K_{3,n}$ with $n - 2$ crossings.

55. The pictured layout for a planned power grid is undesirable, since there are too many cable crossings.

Find a layout for this grid using as few crossings as possible, and prove that no fewer crossings can be achieved.

56. The graph pictured in Exercise 40 models a computer network that is laid out with two wire crossings. Determine the smallest possible number of wire crossings for that network, and prove your result.

57. Prove or disprove: A subdivision of an Eulerian graph is Eulerian.

58. Prove or disprove: A subdivision of a Hamiltonian graph is Hamiltonian.

9.5 Chromatic number

If the countries on a map are going to be colored, then they ought to be colored in such a way that no two adjacent countries have the same color. As we will see, to each map there corresponds a graph in which the vertices represent the countries and the edges reflect adjacencies. The construction of this graph is essentially the same as that used to produce duals in Section 9.4. Our desire here to color the map that yields this graph leads us to a consideration of graph colorings, with a focus on minimizing the number of colors used.

Definition 9.10. Let G be a graph.

(a) A **coloring** of G is an assignment of colors to the vertices of G in such a way that no two adjacent vertices have the same color.

(b) A **color class** for a coloring is a set of all the vertices of one color. The vertices are partitioned by the color classes.

(c) For any $k \in \mathbb{Z}^+$, a k-**coloring** of G is a coloring that uses k different colors.

(d) We say that G is k-**colorable** if there exists a coloring of G that uses at most k colors.

(e) The **chromatic number** of G, denoted $\chi(G)$, is the minimum possible number of colors in a coloring of G.

Remark 9.3.

(a) It is not possible to color a graph containing loops. However, the existence of multiple edges has no impact on colorings.

(b) We typically use integers as the "colors" in our colorings.

(c) If a graph G has a k-coloring, then $\chi(G) \le k$.

Example 9.26. A graph G is pictured below at the left followed by two different colorings for G.

The coloring shown in the middle is a 5-coloring, while that shown on the right is a 4-coloring. Since G is thus seen to be 4-colorable, we conclude that $\chi(G) \le 4$. In fact, $\chi(G) = 4$, as we shall see after Theorem 9.21. In the coloring on the right, the color classes are the sets $\{s, x, z\}$, $\{t, w\}$, $\{u, y\}$, and $\{v\}$. In particular, $\{s, x, z\}$ is the set of vertices of color 1 in that coloring.

In this section, we will see applications of graph colorings to scheduling problems as well as map colorings. However, to start, it may be natural to think of graph coloring as a purely recreational pursuit. Given a graph, what is the least number of colors we can use in a coloring? Could someone come along and beat our coloring, or can we prove that ours is the best? We start with some general results that both develop our coloring skills and provide tools for proving that certain colorings cannot be improved.

The straightforward proofs of the following two theorems are left for the exercises. The first theorem equates bipartite graphs with those which are 2-colorable. The second implies that the pursuit of an optimal coloring need focus on only one component at a time.

Theorem 9.19.   *Let G be any graph. Then G is bipartite if and only if $\chi(G) \leq 2$.*

Theorem 9.20.   *Given any graphs G and H without loops, the chromatic number of their disjoint union is given by $\chi(G + H) = \max\{\chi(G), \chi(H)\}$.*

When trying to color the vertices of a graph one at a time with as few colors as possible, it is generally wise to reuse colors as much as possible. The most obvious feature of a graph that forces us to resort to unused colors is a collection of mutually adjacent vertices.

Definition 9.11.   Let G be a graph.

(a) A **clique** in G is a subgraph that is complete.

(b) The **clique number** of G, denoted $\omega(G)$, is the maximum number of vertices in a clique of G.

Example 9.27.   The pictured graph G

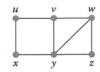

has $\omega(G) = 3$. The vertices v, w, and y form a clique of size 3, and there are no cliques on 4 or more vertices.

The clique number provides a lower bound for the chromatic number.

Theorem 9.21.   *Let G be any graph without loops. Then, $\chi(G) \geq \omega(G)$.*

Proof. In any coloring, the vertices in a clique must receive different colors.   □

The graph G in Example 9.26 has $\omega(G) = 4$; vertices t, u, v, x form a largest clique. It now follows from Theorem 9.21 that $\chi(G) \geq 4$. Since a 4-coloring for G is provided in Example 9.26, this gives $\chi(G) = 4$.

Example 9.28. Show that the graph G from Example 9.27 has $\chi(G) = 3$.

Solution. Theorem 9.21 and Example 9.27 together give that $\chi(G) \geq 3$. The pictured coloring of G

shows that $\chi(G) \leq 3$. Therefore, $\chi(G) = 3$. □

For the graphs in Examples 9.26 and 9.28, we used Theorem 9.21 to prove that our coloring was optimal. However, that technique does not always work. There are graphs G for which $\chi(G) > \omega(G)$.

Example 9.29. The pictured graph G

is called the **Grötzsch graph** and satisfies $\chi(G) = 4 > 2 = \omega(G)$. A proof that $\chi(G) = 4$ is left for the exercises.

Example 9.30. (Accommodating Scheduling Conflicts). In Example 8.3, a graph G

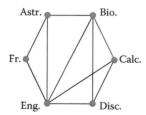

was used to reflect scheduling conflicts among classes in Astronomy, Biology, Calculus, Discrete Math, English Composition, and French, each needing to schedule a study session. That is, in G an edge joins two vertices precisely when the two corresponding class groups cannot meet simultaneously. We can now see how this graph can be used to obtain a schedule that accommodates all of the study sessions in a minimum number of time periods.

Solution. An allowable schedule corresponds to a coloring of the vertices of G with time periods numbered $1, 2, \ldots$. For example, the coloring

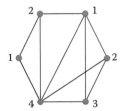

corresponds to the schedule

Time period	Study groups
1	Biology, French
2	Astronomy, Calculus
3	Discrete Math
4	English Composition

that uses four time periods. In fact, a smaller number of time periods is not possible, since G cannot be colored with fewer than four colors. Here, the fact that $\chi(G) = 4$ follows easily from the displayed coloring with four colors and the fact that vertices Bio., Calc., Disc., and Eng. form a clique of size four.   □

For scheduling problems such as Example 9.30, the important result is that for the corresponding scheduling graph G, the minimum number of meeting periods required to accommodate all of the meetings is $\chi(G)$.

Since each pair of vertices in a color class must be nonadjacent, we are interested in subgraphs that are empty.

Definition 9.12. Let $G = (V, E)$ be a graph.

(a) An **independent set** in G is a subset W of V in which no two vertices are adjacent. That is, the subgraph induced by W is empty.

(b) The **independence number** of G, denoted $\alpha(G)$, is the maximum number of vertices in an independent set in G.

A complementary relationship exists between independent sets and cliques.

Theorem 9.22.   *For any simple graph G, we have* $\alpha(G^c) = \omega(G)$ *and* $\omega(G^c) = \alpha(G)$.

The proof of Theorem 9.22 is left for the exercises. The independence number can be used to obtain another lower bound for the chromatic number.

Theorem 9.23.   *For any graph* $G = (V, E)$ *without loops, if* $|V| = n$, *then* $\chi(G) \geq \frac{n}{\alpha(G)}$.

Proof. Take a coloring of G with $\chi(G)$ colors, and partition V into color classes. The size of the largest class is at most $\alpha(G)$. Since there are $\chi(G)$ classes, $\chi(G)\alpha(G) \geq n$. $\square$

Although we have computed several chromatic numbers and, in the process, provided optimal colorings for several graphs, we have not addressed the problem of finding good colorings for graphs in general. In fact, there is no known efficient algorithm that, for every graph G, yields a $\chi(G)$-coloring. However, there are efficient algorithms that give nearly optimal colorings a significant portion of the time.

The algorithm that we consider is an example of a **greedy algorithm**. Such an algorithm is one that attempts to solve a global problem by treating it as a sequence of local problems for which local conditions should be optimized. In the case of our Greedy Coloring Algorithm, we color the vertices of our graph one at a time. At each vertex, we only pay attention to the best possible color choice relative to the neighbors of that vertex. As we shall see, these are not always the best choices in the long run.

Algorithm 9.1.   (Greedy Coloring Algorithm).
Let a graph G on n vertices with no loops be given, together with an ordering $v_1, v_2, \ldots, v_n$ of its vertices. In that order, color the vertices with positive integers so that, for each $1 \leq i \leq n$, vertex v_i is given the smallest possible color not assigned to a neighbor v_j of v_i with $j < i$. Note that v_1 has color 1.

Remark 9.4.   A typical choice of the vertex ordering is in nonincreasing order of degree $\deg(v_1) \geq \deg(v_2) \geq \cdots \geq \deg(v_n)$. However, that is not required.

Example 9.31.   Use the Greedy Coloring Algorithm to color the pictured graph given the ordering $v_1, v_2, v_3, v_4, v_5, v_6, v_7$ of its vertices.

Solution. The resulting coloring is shown in the right-hand picture below.

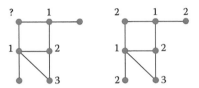

The left-hand picture shows the stage after v_4 has been colored with color 3. (Note that v_4 is adjacent to the colors 1 and 2 used earlier.) At this point, v_5 gets color 2, since it is currently adjacent only to vertices of color 1. The entire coloring that results turns out to be optimal in this case; it is straightforward to confirm that the chromatic number is indeed 3.   □

To demonstrate that the Greedy Coloring Algorithm does not always give an optimal coloring, as it did in Example 9.31, we consider another example.

Example 9.32.   The Greedy Coloring Algorithm applied to the graph pictured at the left with the ordering $v_1, v_2, v_3, v_4, v_5, v_6, v_7, v_8$ of its vertices yields the coloring pictured at the right.

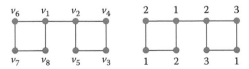

Even though the vertices are colored in nonincreasing order of their degree, which is often a good choice, the resulting coloring is not optimal. The graph is bipartite and thus has chromatic number 2.

In the 1987 movie *Wall Street*, Michael Douglas' character, Gordon Gekko, asserts that "greed, for lack of a better word, is good. Greed is right. Greed works." However, Example 9.32 shows that the Greedy Coloring Algorithm is not perfect. Nonetheless, it can be used to obtain reasonably good upper bounds on the chromatic number of a graph.

Theorem 9.24.   *For any graph G without loops, if $v_1, v_2, \ldots, v_n$ is a listing of its vertices, then $\chi(G) \leq 1 + \max\{\min\{\deg(v_i), i\} : 1 \leq i \leq n\}$.*

Proof. Use the Greedy Coloring Algorithm for the list $v_1, v_2, \ldots, v_n$. For each $1 \leq i \leq n$, the color of vertex v_i is at most $1 + \min\{\deg(v_i), i\}$. The resulting coloring is thus a k-coloring with $k = \max\{1 + \min\{\deg(v_i), i\} : 1 \leq i \leq n\}$. So $\chi(G) \leq k$. □

The general result in Theorem 9.24 is perhaps most easily digested in a weaker but simpler form that is an immediate corollary.

Corollary 9.25.   *For any graph G, we have $\chi(G) \leq \Delta(G) + 1$.*

Proof. For each $1 \leq i \leq n$, we have $\min\{\deg(v_i), i\} \leq \deg(v_i) \leq \Delta(G)$. □

Inspired by Corollary 9.25, we further explore the relationship between the chromatic number and the maximum degree of a graph.

> **Example 9.33.**   There are graphs for which the upper bound in Corollary 9.25 is achieved.
>
> (a) For any $n \in \mathbb{Z}^+$, $\chi(K_n) = n = \Delta(K_n) + 1$.
> (b) For any odd $n \in \mathbb{Z}^+$, $\chi(C_n) = 3 = \Delta(C_n) + 1$.
>
> However, K_n and C_n are the only graphs for which equality holds.

The British mathematician R. Leonard Brooks (1916–1992) improved upon the upper bound in Corollary 9.25.

Theorem 9.26.   (Brooks' Theorem).
Let G be any graph without loops. If G is not complete and not an odd cycle, then $\chi(G) \leq \Delta(G)$.

Proof. Suppose G is a graph that is not complete and not an odd cycle. Let $n = |V_G|$. It suffices to assume that G is connected.

Case 0: G has a vertex v_n with $\deg(v_n) < \Delta(G)$.

This means that G is not regular. Choose an ordering $v_1, v_2, \ldots, v_n$ of the vertices in nonincreasing order of their distance from v_n. By Exercise 45 from Section 8.1, for each $1 \leq i < n$, vertex v_i has a neighbor v_j with $j > i$. Consequently, for each $1 \leq i \leq n$, vertex v_i has at most $\Delta(G) - 1$ neighbors v_j with $j < i$. The Greedy Coloring Algorithm then uses at most $\Delta(G)$ colors.

In light of case 0, it suffices to assume that G is r-regular with $r = \Delta(G)$.

Case 1: $\kappa(G) = 1$.

The proof in this case is left for Exercise 51.

Case 2: $\kappa(G) = 2$.

If $\Delta(G) = 2$, then G is a cycle. So we may assume that $\Delta(G) \geq 3$. For a disconnecting set $\{u, v\}$, let H be some but not all of the components of $G \setminus \{u, v\}$, let H_1 be the subgraph of G induced by $V_H \cup \{u, v\}$, and let H_2 be the subgraph of G induced by $V_G \setminus V_H$. Since both u and v have neighbors outside of $\{u, v\}$ in each of H_1 and H_2, they both have degree at most $\Delta(G) - 1$ in each of H_1 and H_2.

Subcase 2a: In both H_1 and H_2, either u or v has degree at most $\Delta(G) - 2$.

By applying the argument in case 0 to the smaller degree vertex, we can find a coloring of H_1 that not only uses at most $\Delta(G)$ colors but also gives u and v different colors. We can then do the same for H_2, while forcing u and v to have the same colors there as they do in H_1. Hence, these two colorings can be joined together to form a coloring for G using at most $\Delta(G)$ colors.

Subcase 2b: We are not in subcase 2a.

We may assume that in H_1, both u and v have degree $\Delta(G) - 1$. Observe that u and v cannot be adjacent, and v must have a unique neighbor w in H_2. Moreover, since $\kappa(G) \geq 2$, w is not the neighbor of u in H_2. In fact, w has degree 1 in the subgraph of G induced by $H_1 \cup \{w\}$. We can now apply subcase 2a to the subgraphs of G induced by $H_1 \cup \{w\}$ and $H_2 \setminus \{w\}$ determined by the disconnecting set $\{u, w\}$.

Case 3: $\kappa(G) \geq 3$.

It must be that $r \geq 3$. Since G is not complete, there must be a vertex v_n with neighbors v_1 and v_2 such that v_1 and v_2 are not adjacent. Since $\kappa(G) \geq 3$, the graph $G \setminus \{v_1, v_2\}$ is connected and its vertices may be ordered $v_3, v_4, \ldots, v_n$ in nonincreasing order of their distance in $G \setminus \{v_1, v_2\}$ from v_n. For each $1 \leq i < n$, vertex v_i has a neighbor v_j with $j > i$. Apply the Greedy Coloring Algorithm to the ordering $v_1, v_2, v_3, v_4, \ldots, v_n$. The vertices $v_1, v_2, \ldots, v_{n-1}$ require at most r colors. Since v_1 and v_2 both receive color 1, vertex v_n is adjacent to at most $r - 1$ different colors. Thus, at most $r = \Delta(G)$ colors suffice for G.

$\square$

Corollary 9.27.   For any graph G without loops and any integer $r \geq 3$, if G is r-regular and no component of G is complete, then $\chi(G) \leq r$.

The upper bound on $\chi(G)$ given by Brooks' Theorem is not the best possible in general, as the following example illustrates.

Example 9.34. $\chi(K_{2,3}) = 2 < 3 = \Delta(K_{2,3})$.

There is a remarkably simple relationship between the chromatic number of a product and the chromatic numbers of its factors.

Theorem 9.28.   *Given any graphs G and H without loops, the chromatic number of their product is given by* $\chi(G \times H) = \max\{\chi(G), \chi(H)\}$.

Proof. Let $f_G : V_G \longrightarrow \{1, \ldots, \chi(G)\}$ and $f_H : V_H \longrightarrow \{1, \ldots, \chi(H)\}$ be functions that optimally color G and H, respectively. Let $m = \max\{\chi(G), \chi(H)\}$. Define the m-coloring $f : V_{G \times H} \longrightarrow \{0, \ldots, m-1\}$ by $f((u,v)) = (f_G(u) + f_H(v)) \bmod m$. Suppose that (u_1, v_1) and (u_2, v_2) are adjacent in $G \times H$. If $u_1 = u_2$, then v_1 and v_2 are adjacent in H and $f_H(v_1) \neq f_H(v_2)$. If $v_1 = v_2$, then u_1 and u_2 are adjacent in G and $f_G(u_1) \neq f_G(u_2)$. In any case, $f((u_1, v_1)) \neq f((u_2, v_2))$. Since f provides an m-coloring, $\chi(G) \leq m$. Since $G \times H$ contains isomorphic copies of G and H, $\chi(G) \geq m$ too.   □

9.5.1   Coloring maps

We close this section with the application of graph colorings to map colorings mentioned at the beginning of the section. A map of South America is shown in Figure 9.9. Suppose we wish to color its countries with as few colors as possible such that no two countries that share a border receive the same color. We can convert this into a graph coloring problem by taking the dual graph to the map. This is accomplished by placing a vertex in each country and joining a pair of vertices by an edge precisely when their corresponding countries are adjacent. The construction is virtually the same as that of the dual of a planar graph, except that we choose to not place a vertex in the outer region when that does not represent a country. The dual graph for our map of South America is shown on the left side of Figure 9.10. An optimal coloring for this graph yields an optimal coloring for the map.

FIGURE 9.9
South America.

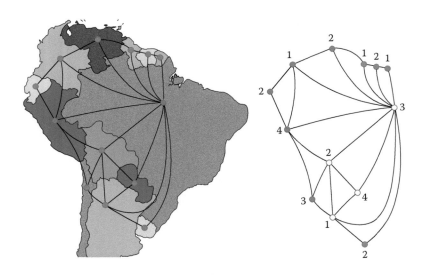

FIGURE 9.10
Coloring of dual graph for South America.

FIGURE 9.11
Coloring of South America.

It is straightforward to check that our dual graph has chromatic number 4 and that an optimal coloring is thus shown on the right side of Figure 9.10. In fact, the clique formed by the vertices corresponding to Argentina, Bolivia, Brazil, and Paraguay (shown with open points) forces the need for four colors. The optimal coloring of the dual graph can now be used to obtain an optimal coloring of the countries in South America. It is shown in Figure 9.11.

The fact that no more than four colors were needed is a consequence of a famous result.

Theorem 9.29.   (Four-Color Theorem).
If G is any planar graph, then $\chi(G) \leq 4$.

Although it was probably believed much earlier by map makers, the Four-Color Theorem was first formally conjectured in 1852 by an Englishman, Francis Guthrie (1831–1899). Twenty-seven years later, an erroneous proof was published by the English mathematician Arthur Kempe (1849–1922). The error was not caught until 1890 when another English mathematician, Percy Heawood (1861–1955), showed that Kempe's arguments could only be made to prove a Five Color Theorem.

The Four-Color Theorem was first correctly proved in 1976 at the University of Illinois by the American mathematician Kenneth Appel (1932–2013) and the German-born American mathematician Wolfgang Haken (1928–). Their proof required hundreds of pages of arguments, over 1200 hours of computer time, and ultimately the consideration of 1478 reducible configurations. It was the first computer-aided proof and was quite controversial at the time, since mathematicians could not check their argument by hand.

Although the Appel and Haken proof is now generally accepted, it is in no small part due to a significantly improved proof of approximately 40 pages published in 1997 by Neil Robertson, Daniel Sanders, Paul Seymour, and Robin Thomas. This better proof required less than 4 hours of computing time and just 633 reducible configurations. However, the importance of the 1976 proof still stands, as many other computer-aided proofs of mathematical results have followed and broken significant ground.

Exercises

1. Do the assignments given by the displayed labels represent a coloring of the pictured graph?

2. Do the assignments given by the displayed labels represent a coloring of the pictured graph?

Subsequent exercises refer to the following graphs:

(a)   (b)   (c)   (d)

In Exercises 3 through 8, find the chromatic number of the specified graph, and prove your result.

3. The graph from Exercise 1.

4. The graph from Exercise 2.

5. Graph (a). 6. Graph (b).

7. Graph (c). 8. Graph (d).

9. Prove Theorem 9.19: G is bipartite if and only if $\chi(G) \leq 2$.

10. Prove Theorem 9.20: $\chi(G + H) = \max\{\chi(G), \chi(H)\}$.

In Exercises 11 through 18, specify the given values.

11. $\omega(C_5)$. 12. $\omega(P_5)$.

13. $\omega(K_6)$. 14. $\omega(K_{3,4})$.

15. $\alpha(C_5)$. 16. $\alpha(P_5)$.

17. $\alpha(K_6)$. 18. $\alpha(K_{3,4})$.

19.* Prove that the Grötzsch graph has chromatic number 4, as asserted in Example 9.29.

20.* Show that the pictured graph is not 3-colorable.

21. Show that the removal of any vertex from the Grötzsch graph leaves a graph with chromatic number 3.

22. Show that $\chi(C_n) = \begin{cases} 2 & \text{if } n \text{ is even,} \\ 3 & \text{if } n \text{ is odd.} \end{cases}$

23. Six classes from the Department of Languages and Cultures (German, History, Indochina, Japanese, Kuwait, and Latin) need to schedule study groups. In the displayed table, an X denotes the fact that the two different classes corresponding to that row and column have a student in common.

	Ger.	His.	Indo.	Jap.	Kuw.	Lat.
Ger.		X	X		X	X
His.	X		X	X		X
Indo.	X	X		X	X	
Jap.		X	X		X	X
Kuw.	X		X	X		X
Lat.	X	X		X	X	

Assuming that each group needs to meet for 1 hour, find a schedule that accomplishes these study sessions in the least possible number of hours. Prove that your schedule is best possible.

24. Eight Senate committees—Appropriations (A), Budget (B), Commerce (C), Energy (D), Environment (E), Foreign Relations (F), Homeland Security and Governmental Affairs (G), and Health (H)—need to schedule meetings. In the displayed table, an X denotes the fact that the two different committees corresponding to that row and column have a member in common.

	A	B	C	D	E	F	G	H
A			X		X		X	X
B			X	X			X	X
C	X	X			X	X		
D		X				X	X	X
E	X		X			X		X
F			X	X	X		X	
G	X	X				X		
H	X	X		X	X			

Assuming that each committee needs to meet for 1 hour, find a schedule that accomplishes these meetings in the least possible number of hours. Prove that your schedule is best possible.

25. Prove that C_5 is "uniquely" 3-colorable. That is, for any two colorings, there is an automorphism of C_5 that sends the color classes in one coloring to the color classes in the other.

26. Prove that for every even integer n, C_n is "uniquely" 2-colorable. See Exercise 25.

27. Show that the pictured graph is not uniquely 3-colorable.

28. Show that $K_{2,1} + K_{2,1}$ is not uniquely 2-colorable.

In Exercises 29 through 34, determine, with proof, the chromatic number of the specified graph.

29. The Petersen Graph.   30. The octahedron.

31.* The icosahedron.   32. The dodecahedron.

33.*    34.*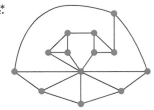

35. Prove Theorem 9.22.

36. Prove that $\chi(G)$ is the minimum value of n for which there exists a graph map from G to K_n.

37. We need to store chemicals in some cabinets, while being careful not to put in the same cabinet two chemicals that might cause a dangerous chemical reaction. If acids are mixed with bleach or sulfides, then a toxic gas can be produced. If acids come in contact with hydrogen peroxide, then an explosion can result. Mixing bleach with ammonia can also produce a toxic gas. What is the smallest number of cabinets needed to store these chemicals, based upon the described dangers? Justify your answer using graph colorings.

38. The camp councillor at math camp needs to break her girls up into groups to work on some proofs. However, some of the girls are not getting along with each other and need to be kept in separate groups. Betty is not getting along with Mary, Amy, Sally, and Cindy. Wendy does not like Cindy, Sally, and Kelly. Sally cannot stand Amy and Kelly. And Mary and Amy have been fighting since camp started. What is the smallest number of groups into which these girls can be split? Justify your answer using graph colorings.

Subsequent exercises refer to the following graphs:

(e)

(f)

(g)
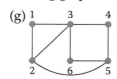

(h)
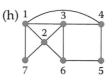

In Exercises 39 through 42, apply the Greedy Coloring Algorithm to the specified graph, using the ordering $1, 2, \ldots$ of the vertices.

39. Graph (e).   40. Graph (f).

41. Graph (g).   42. Graph (h).

43. Show that G and H are bipartite if and only if $G \times H$ is bipartite. Hint: Use bipartite and product coloring results.

44. Does $\chi(G) \leq 4$ imply that G is planar? Hint: Consider 2-colorable graphs.

45. The **wheel** W_n is the graph obtained from C_{n-1} by adding a new vertex n and joining n to each vertex $1, 2, \ldots, n-1$ of C_{n-1}. Note that $W_4 = K_4$.

    (a) Draw W_5.

    (b) Prove that the chromatic number of W_n is 4 if n is even and is 3 if n is odd.

46. The n-**prism** is the graph $C_n \times P_2$. Note that the 4-prism is the cube Q_3.

    (a) Draw the 3-prism.

    (b) Prove that the chromatic number of the n-prism is 2 if n is even and is 3 if n is odd.

47. Suppose a graph G has a unique vertex of maximum degree, and let d_2 be the second highest degree in G. Show that $\chi(G) \le d_2 + 1$.

48. Suppose a graph G has adjacent vertices v_1 and v_2 of degrees d_1 and d_2, respectively, such that $d_1 \ge d_2 > \deg(v)$, for all vertices $v \notin \{v_1, v_2\}$. Show that $\chi(G) \le d_2$.

49.* Suppose that, from a given graph G, a graph G' is obtained by combining two nonadjacent vertices of G into one.

    (a) Show: $\chi(G) \le \chi(G')$.

    (b) Give an example for which $\chi(G) < \chi(G')$.

50.* Suppose that, from given a graph G, a graph G' is obtained by combining two nonadjacent vertices of G into one.

    (a) Show: $\chi(G') \le \chi(G) + 1$.

    (b) Give an example for which $\chi(G') < \chi(G) + 1$.

51. Prove Brooks' Theorem in the case that G is a regular graph with $\kappa(G) = 1$. Hint: For each component H of $G \setminus \{v\}$, optimally color the subgraph induced by $H \cup \{v\}$.

52. Show that there always is some ordering of the vertices for which the Greedy Coloring Algorithm yields an optimal coloring. Hint: Let $c = \chi(G)$. Pick a coloring with c colors and color classes $A_1, \ldots, A_c$ such that $|A_1| \le \cdots \le |A_c|$ and A_1 is smallest possible, A_2 is smallest possible given A_1, A_3 is smallest possible given A_2, and so on. Argue that for each i, each vertex of A_i is adjacent to every color j with $j > i$. Otherwise, A_i could be made smaller. Order the vertices so that $A_c, \ldots, A_1$.

53. Pictured are the voting districts in a certain county.

    The county commissioners want to include a color picture of this map in a pamphlet, so that colors clearly distinguish voting districts. Since each additional color adds expense to the production of this pamphlet, they want to minimize the number of colors needed. What is the small-est number of colors that can be used?

54. Pictured are the police precincts in a portion of a major city.

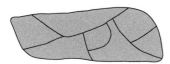

The mayor wants a color version of this map, with colors distinguishing precinct boundaries, using as few colors as possible. How many colors are required?

55. Find a map of the 48 contiguous states in the United States.

   (a) Find a coloring of the map using as few colors as possible so that no two adjacent states have the same color.

   (b) How many colors are needed?

   (c) Prove that no fewer colors would suffice.

56. Find a map of the 54 countries in Africa.

   (a) Find a coloring of the map using as few colors as possible so that no two adjacent countries have the same color.

   (b) How many colors are needed?

   (c) Prove that no fewer colors would suffice.

9.6   Review problems

1. Let G be the pictured graph.

   (a) Prove $\kappa(G) = 1$.   (b) Prove $\lambda(G) = 2$.

   For Exercises 2 through 5, determine the given values. Justify your answers.

2. $\kappa(K_{5,7})$.   3. $\lambda(K_5)$.

4. $\kappa(P_5)$.   5. $\lambda(C_5)$.

6. Pictured is a power grid on eight stations. In addition to the outer 8-cycle, five extra cables connecting stations are included to decrease the power grid's vulnerability to station failures.

Determine the smallest possible number of power stations whose failure would leave a remaining grid that is not connected. Prove your result.

7. Prove or disprove that there exists a graph G with $\kappa(G) = 3$ and $\lambda(G) = 2$.

8. Prove or disprove that there exists a graph G with $\kappa(G) = 2$ and $\lambda(G) = 4$.

9. Let G be a simple graph on n vertices and $\binom{n}{2} - 1$ edges. Show that $\kappa(G) = n - 2$.

10. Two cyclic computer networks have been joined together with the aim of forming a stronger network. The designers are concerned with the possible failures of both computers and cable connections.

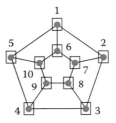

Prove that at least three computers or at least three cables must fail to disconnect this network. That is, for the graph G modeling this network, prove that $\kappa(G) = \lambda(G) = 3$.

11. Determine $\lambda(K_{m,n})$ and prove your result.

12. Specify an Euler circuit in the pictured graph.

13. Prove that no Euler trail exists in the pictured graph.

14. A tourist visiting the city of Zahlenberg is interested in crossing over each of its bridges.

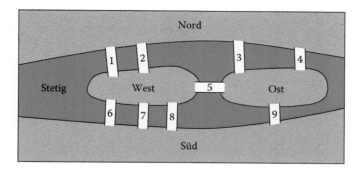

Determine whether this portion of Zahlenberg has an Euler circuit, an Euler trail, or neither. Justify your answer.

For Exercises 15 through 21, determine whether the specified graph has an Euler circuit, an Euler trail, or neither. If so, then specify one. If not, then justify why.

15. $K_{2,3}$.   16. Q_3.

17. K_5.   18. P_6.

19.   20.

21.

22. It is easy to find a Hamiltonian cycle C_n in K_n. For some values of n, the remaining edges form an Euler circuit. For which integers n does $C_n{}^c$ have an Euler circuit? Justify your answer.

23. Since several potholes developed over the winter months, the streets in the pictured neighborhood need to be repaved.

So that traffic does not need to be completely stopped in this neighborhood, the workers pave one side of each of these two-way streets at a time. Find a route that takes the workers over each side of each street exactly once and leaves the neighborhood from the same point it entered.

24. The pictured neighborhood is a new development that has gone up since last winter, and consequently an efficient snowplow route needs to be designed for it.

Each street is two-way, and one pass of the plow clears one side of the street. To be efficient, once the snowplow enters the neighborhood, it should not pass a second time over a section of road that has already been plowed. Find a maximally efficient route that clears the entire road surface and keeps the plow on the right-hand side of the street at all times.

25. Find a Hamiltonian cycle in $K_{4,4}$.

26. A Hamiltonian cycle in Q_3 determines a subgraph isomorphic to C_8. How many different subgraphs of Q_3 are isomorphic to C_8?

27. Show that the pictured graph is not Hamiltonian.

28. Determine whether or not the pictured graph is Hamiltonian.

29. Let G be any graph.
Show: If G is Hamiltonian, then $\delta(G) \geq 2$.

30. FedEx needs to make deliveries to the marked locations in the pictured neighborhood.

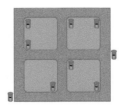

To be maximally efficient, FedEx does not want its truck to pass by a delivery site to which a delivery has already been made, and it wants the truck to exit the neighborhood from the same location at which the truck entered. Does such a maximally efficient delivery route exist? Justify your answer.

31. Determine whether the pictured directed graph is Hamiltonian.

32. The pictured graph is a tournament.

It reflects the results of a round robin Scrabble tournament.

(a) Who won the match between Jack and Zed?

(b) Find a Hamiltonian path through the tournament.

(c) Have the results of this tournament determined a clear ranking of its competitors? Justify your answer.

33. Show that if a directed graph is Hamiltonian, then it is strongly connected.

34. Given a planar embedding of a graph $G = (V, E)$ with regions R, if $|E| = 23$ and $|R| = 10$, then find $|V|$.

35. Show that the Grötzsch graph is not planar.

36* Prove that any planar graph $G = (V, E)$ with all of its cycles of length at least 5 has $|E| \le \frac{5}{3}(|V| - 2)$.

37. Use the result in Exercise 36 to show that the Petersen graph is not planar.

38. Without using Kuratowski's Theorem, prove that all graphs on 4 or fewer vertices are planar.

39. Draw the dual of the planar graph from Exercise 13.

40. Since the tetrahedron graph is isomorphic to K_4, we have seen that it is planar. Show that the dual of the tetrahedron is itself.

41. Does every nonplanar graph contain K_5 or $K_{3,3}$ as a subgraph? Explain.

42. Pictured is a computer network that needs to be laid out without any cable crossings.

Show that this is indeed possible by presenting such a layout.

43. Pictured is the layout for a circuit board that needs to be constructed.

Although a layout without any wire crossings is preferred, show that no such layout exists.

44. The pictured power grid needs to be laid out, but the presence of the displayed cable crossings is undesirable.

Can this grid be laid out without any cable crossings? Justify your answer.

45. What is the minimum possible number of wire crossings required in a layout of the circuit board from Exercise 43? Justify your answer.

46. What is the minimum possible number of cable crossings required in a layout of the power grid from Exercise 44. Justify your answer.

47.* Determine, with proof, the chromatic number of the pictured graph.

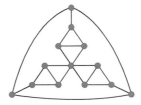

48. $v(K_{3,4}) = 2$. Give a drawing of $K_{3,4}$ with 2 crossings.

49. Find each of $\omega(G)$, $\alpha(G)$, and $\chi(G)$, where G is the graph that models the computer network in Exercise 42.

50. Find each of $\omega(G)$, $\alpha(G)$, and $\chi(G)$, where G is the circuit board graph in Exercise 43.

51. The yearbook needs group pictures of each of these clubs: Archery, Chess, Math, National Honor Society, and Student Council. There are two photographers who can work at the same time, but some of the clubs have overlapping membership, as shown in the following table.

	Arch.	Chess	Math	NHS	StCo.
Arch.					X
Chess			X	X	
Math		X		X	X
NHS		X	X		X
StCo.	X		X	X	

With these restrictions, determine the minimum number of sessions required to complete all of these club pictures. Justify your answer.

52. Use the Greedy Coloring Algorithm to color the vertices of the pictured graph,

using the ordering $1, \ldots, 7$ of the vertices. Is that coloring optimal? Justify your answer.

53. If G is a simple graph on n vertices, then how are $\Delta(G)$ and $\delta(G^c)$ related?

54. Displayed are the school districts in a certain county.

We would like to color these districts so that adjacent districts receive distinct colors. Find a coloring that uses the smallest possible number of colors.

55. Find a map of the political regions (provinces and territories) of Canada.

(a) Find a coloring of the map using as few colors as possible so that no two adjacent provinces have the same color.

(b) How many colors are needed?

(c) Prove that no fewer colors would suffice.

Trees and Algorithms

An important special class of graphs is made up of those that are connected and contain no cycles. Such graphs are called trees. After exploring basic properties and applications of trees, we consider algorithms for finding breadth-first search trees and depth-first search trees. Not only do these algorithms provide systematic means of traversing trees, but they also yield useful spanning trees for general graphs. The Depth-First Search Algorithm in particular is the basis for a general searching technique called backtrack searching.

In the case that weights are attached to the edges of a graph, we study Kruskal's Algorithm and Prim's Algorithm for finding a spanning tree with the minimum possible total weight. An additional algorithm that we owe to Dijkstra enables us to find, from a fixed vertex, the shortest path to any other. In the case that the graph represents a potential computer network, in which each edge is weighted with the cost of the link it reflects, a minimum spanning tree represents a cheapest possible connected network. If, instead, each edge is weighted with the time delay across the link it reflects, then a shortest path tree from a specified vertex represents a network with the fastest response time from the corresponding specified computer.

The last two sections initiate a study of the analysis of algorithms in general. For that, we focus primarily on algorithms that search for a value in a given array and those that sort an array. Our study of the complexities and efficiencies of algorithms leads us to a study of function growth and, in particular, big-O notation. Finally, through our consideration of decision trees for algorithms, we see how trees provide a valuable tool for analyzing algorithms.

10.1 Trees

In Section 5.1, we considered the "is the father of" relation, as reflected by a family tree. In Section 6.1, we used tree diagrams to analyze counting problems. In light of our studies in Chapters 8 and 9, we see that these diagrams are examples of graphs. In fact, they are examples of a specialized yet extremely useful kind of graph. A **tree** is a graph that is connected and contains no cycles. More generally, a **forest** is a graph that contains no cycles; each component of a forest is a tree. This section explores the graph-theoretical properties of trees and forests as well as some applications.

A relatively simple class of trees is provided by paths. That is, for each $n \geq 1$, the path P_n is a tree. A great variety of other trees (and forests) exists as well.

Example 10.1.

(a) The graph pictured in the left-hand box of Figure 10.1 is a tree. We can see that this graph is connected and contains no cycles.

(b) The graph pictured in the right-hand box of Figure 10.1 is a forest that consists of two trees. That graph contains no cycles and has two components.

When working with forests, a pendant vertex (a vertex of degree 1) is usually called a **leaf**. The vertices of degree at least two are said to be **internal vertices**. In Example 10.1(a) the vertices $3, 6, 7$ are leaves, whereas in (b) the leaves are $2, 3, 5, 6, 7$.

Example 10.2.   (Chemical Structures).
Graphs in general can be used to model molecules; we represent each atom by a vertex and each chemical bond by an edge. Figure 10.2 displays the chemical structure of ethane, C_2H_6, whose corresponding graph

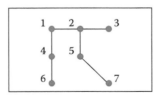 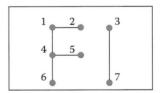

FIGURE 10.1
A tree and a forest.

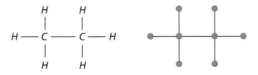

FIGURE 10.2
Ethane C_2H_6 and its corresponding tree.

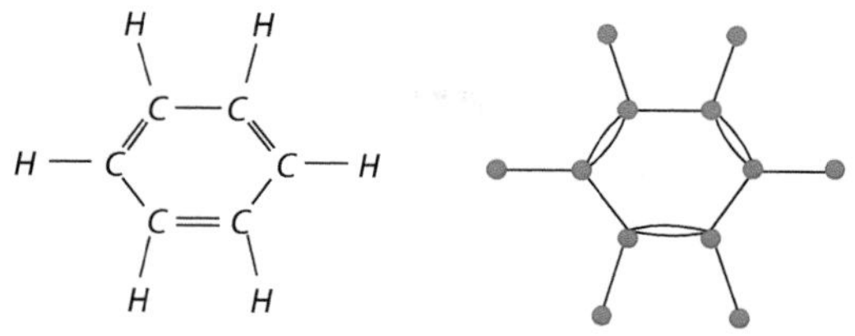

FIGURE 10.3
Benzene C_6H_6 and its corresponding graph.

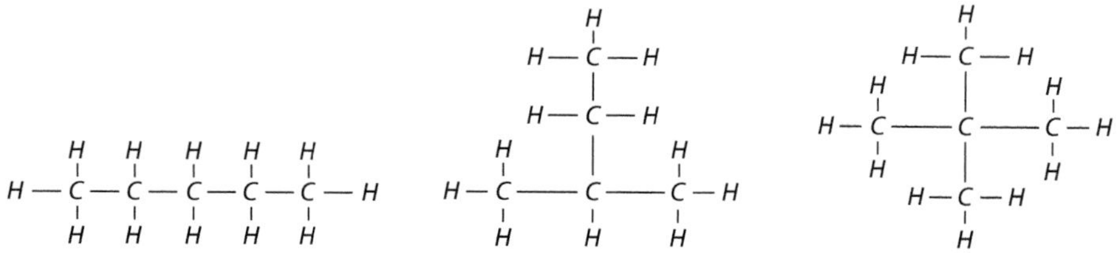

FIGURE 10.4
The three pentane C_5H_{12} isomers.

is a tree. Figure 10.3 shows benzene C_6H_6, whose graph is not a tree. That graph contains a cycle on the vertices corresponding to the carbon atoms, as well as some doubled edges representing double bonds. Ethane and benzene are examples of **hydrocarbons**, which are molecules built exclusively from atoms of hydrogen, H, and carbon, C. An important subset of these is the **saturated hydrocarbons**, whose chemical formulas have the form C_nH_{2n+2} for some positive integer n. We highlight them here, since their structures always correspond to graphs that are trees.

Although ethane is the only hydrocarbon with formula C_2H_6, in general, the chemical formula alone does not determine the chemical structure. For example, there are three different pentanes, C_5H_{12}, as shown in Figure 10.4. In general, the different structures associated with a fixed formula are called its **isomers**, and each has its own chemical and physical properties. The structure of an isomer of a saturated hydrocarbon C_nH_{2n+2} is determined by the subtree induced by the n carbon atoms.[*] That subtree must be a tree of maximum

_____

[*] Here we ignore possible differences among optical isomers. These occur among saturated hydrocarbons when a fixed subtree determined by the carbon atoms can be built in distinct ways that differ by non-superimposable mirror images of portions of the tree. Distinct optical isomers have bond angles that are oriented differently but have mostly identical physical properties.

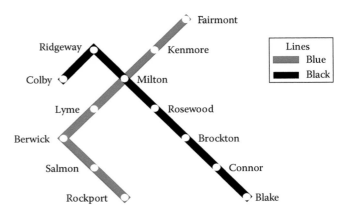

FIGURE 10.5
A city subway system.

degree at most 4. Then the entire tree is the unique tree in which the carbon atoms form the set of internal vertices, each with degree 4, and the hydrogen atoms form the set of leaves. For example, when $n = 4$, the isomers butane and isobutane of C_4H_{10} correspond to the carbon trees P_4 and $K_{1,3}$, respectively. In general, the number of possible isomers of C_nH_{2n+2} is the number of non-isomorphic trees on n vertices with maximum degree at most 4. The problem of determining this number was studied by the English mathematician Arthur Cayley (1821–1895). In 1875, he determined this number for n vertices in the cases that $n \leq 13$. In the exercises, we consider some of those values. However, no convenient general formula is known.

We have seen that a graph can be used to represent a transportation network, such as the subway system displayed in Figure 10.5. In that example, the resulting graph is a tree. By inspection, we can also see that, within this particular subway system, there is a unique path between any pair of its stops. This is a general property held by trees.

Theorem 10.1.   *Let G be a graph. If G is a tree, then there is a unique path between any pair of vertices in G.*

Proof. Let G be a tree, and let u and v be two vertices of G. Suppose toward a contradiction that there are distinct paths P and Q from u to v. Since P and Q must be different at some point after u, let y be the first vertex in P that is not in Q. So the vertex x in P immediately preceding y must also be in Q. See Figure 10.6. Since P and Q become different after vertex x

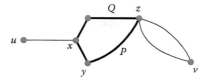

FIGURE 10.6
Supposed distinct paths in a tree.

but both end up at vertex v, let z be the first vertex in P that is after x (and y) and is common to P and Q. Note that the portion of P strictly between x and z has nothing in common with Q. Therefore, the walk that follows P from x to z and then follows Q backward from z to x forms a cycle in G (bold in Figure 10.6). This contradicts the fact that G is a tree.   □

A proof that the converse of Theorem 10.1 also holds is left for the exercises. Here, we establish a lower bound on the number of leaves in a tree. It can be seen to hold in the examples considered so far.

Theorem 10.2.   *Every tree on two or more vertices has at least two leaves.*

Proof. Let T be a tree, and let P be a longest possible path in T. The start and end vertices of P must be distinct and must be leaves. Otherwise, P could be made longer.   □

The result in Theorem 10.2 is as strong as possible, since, for $n \geq 2, P_n$ has exactly two leaves. Additionally, observe that the number of edges in the tree P_n is one fewer than the number of vertices. That result is true in general for trees.

Theorem 10.3.   *Let $G = (V, E)$ be a graph. If G is a tree, then*
$$|E| = |V| - 1.$$

Proof. We prove by induction that for every $n \geq 1$, every tree on n vertices has $n - 1$ edges. Since every tree on one vertex must have no edges, the asserted result holds when $n = 1$. So suppose $n \geq 1$ and every tree on n vertices has $n - 1$ edges. Let $G = (V, E)$ be a tree with $|V| = n + 1$. Since $n + 1 \geq 2$, it follows from Theorem 10.2 that G has a leaf v. Let e be the one edge incident with v, and observe that $G' = (V \setminus \{v\}, E \setminus \{e\})$ is a tree on n vertices. By the induction hypothesis,

$$|V| - 1 = |V \setminus \{v\}| = |E \setminus \{e\}| - 1 = |E| - 1 - 1.$$

Therefore, $|V| = |E| - 1$.   □

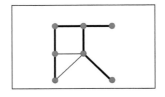 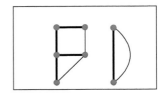

FIGURE 10.7
In bold, a spanning tree and a spanning forest.

A proof that the converse of Theorem 10.3 holds for connected graphs is left for the exercises.

Spanning trees. Although trees are special kinds of graphs, they are nonetheless useful for studying graphs in general. In a sense, a tree can serve as a skeleton for a larger graph. A **spanning tree** for a graph $G = (V, E)$ is a subgraph $H = (V, F)$ that is a tree on the same vertex set V. It may be specified uniquely by its edge set F. More generally, a subgraph of a graph G is called a **spanning forest** if its intersection with each component of G is a spanning tree for that component.

A spanning tree for the graph pictured in the left-hand box of Figure 10.7 is shown there in bold. In the right-hand box of Figure 10.7, a spanning forest for the graph there is shown in bold. Since that graph is not connected it cannot have a spanning tree. However, every graph that is connected has at least one spanning tree.

Theorem 10.4.   *Let $G = (V, E)$ be a graph. If G is connected, then G has a spanning tree.*

Proof. We proceed by strong induction on the number n of vertices in the graph. Any graph with only one vertex requires no edges for its spanning tree. So suppose $n \geq 1$ and any connected graph with n or fewer vertices has a spanning tree. Let G be an arbitrary connected graph with $n + 1$ vertices, and select a vertex v of G. Since every component of $G \setminus \{v\}$ has n or fewer vertices, each has a spanning tree. Let F_0 be the union of their edges. Since each component of $G \setminus \{v\}$ must contain a neighbor of v, we can add to F_0 an edge from v to that neighbor. The edges in the resulting set F must then form a spanning tree for G. (Think about why there are no cycles.)   □

A proof that the converse of Theorem 10.4 holds and a proof of the following corollary are left for the exercises.

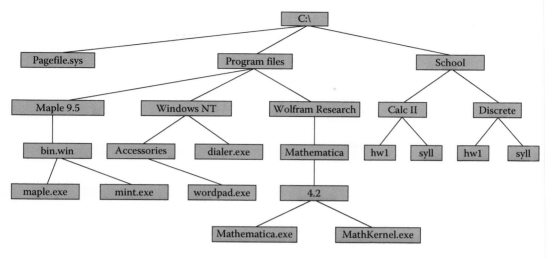

FIGURE 10.8
Files and folders in a directory system.

Corollary 10.5.   Every graph has a spanning forest.

A connected graph may have more than one spanning tree. In the next two sections, we consider algorithms that yield particular spanning trees.

Rooted trees. A directory or file system in a computer takes the form of a tree with additional structure. For example, the topmost directory is known as the root directory. When there are directories within directories, the directory immediately above a particular directory is known as its parent directory. An example of a portion of such a file system is displayed in Figure 10.8. In Definitions 10.1 and 10.2, we shall see that the terminology used to describe computer directory systems corresponds to the terminology of rooted trees.

Definition 10.1.   A **rooted tree** is a pair (T, v) where T is a tree and v is a vertex of T. The distinguished vertex v is said to be the **root** of T.

The directory tree in Figure 10.8 has the structure of a rooted tree with root C:\. Note in this application of rooted trees that files are always leaves and internal vertices are always folders. However, a folder, if empty, can also be a leaf.

The graph in Figure 10.9 with vertex v_1 designated as the root is an example of a rooted tree. Drawings of rooted trees

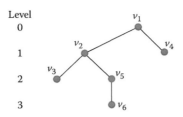

FIGURE 10.9
A rooted tree.

typically show the prominence of the root by having the other vertices trail away from it. We shall usually draw the root as the topmost vertex. With such conventions in mind, there is a variety of terminology associated with rooted trees.

Definition 10.2.   Let T be a rooted tree with root v.

(a) The **level** of a vertex u in T is its distance from v, namely, $\text{dist}(v, u)$.

(b) A **child** of a vertex u in T is a neighbor of u at a level greater than that of u; its level is always one greater.

(c) The **parent** of a vertex u in T is the unique neighbor of u with level less than that of u; its level is always one less. The root v is the only vertex in T without a parent.

(d) The **height** of T is the maximum level among its vertices.

(e) If each vertex of T has at most m children, then we say that T is an m-**ary tree**. If each internal vertex has exactly m children, then T is said to be a **full m-ary tree**. A 2-ary tree is also called a **binary tree**.

(f) If T has height h and all of its leaves are at levels h or $h-1$, then T is said to be **balanced**.

In the directory system displayed in Figure 10.8, the directory `Windows NT` has parent `Program Files` and children `Accessories` and `dialer.exe`. In this case, one child is a folder and an internal vertex, whereas the other is a file and a leaf.

Consider the rooted tree (T, v_1) displayed in Figure 10.9. Vertex v_2 is at level 1, since $\text{dist}(v_1, v_2) = 1$, and v_6 is at level 3, since $\text{dist}(v_1, v_6) = 3$. Vertex v_2 has children v_3 and v_5, and the parent of v_2 is v_1. The height of T is 3, and T is a binary tree that is neither full nor balanced. Figure 10.10 shows a full and balanced rooted 3-ary tree (T', v_1) of height 2.

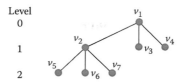

Level
0
1
2

FIGURE 10.10
A full and balanced rooted 3-ary tree.

If an *m*-ary tree is full, then a certain portion of the vertices must be internal and the remaining portion must, of course, be leaves.

Theorem 10.6.   *Let T be a full m-ary rooted tree with n vertices, l leaves, and i internal vertices. Then,*

(i) $n = i + l$, *and*

(ii) $n = mi + 1$.

It follows from Theorem 10.6 that knowing two of the three values n, l, and i for a full *m*-ary tree determines the third. The proof of Theorem 10.6 and the proofs of some useful consequences of it are left for the exercises. The next theorem will play a significant role in our analyses of algorithms in Sections 10.4 and 10.5.

Theorem 10.7.   *Let T be an m-ary rooted tree of height h with l leaves. Then*

(i) $l \leq m^h$, *and*

(ii) $h \geq \lceil \log_m l \rceil$.

Proof. Our proof is by induction on h. If $h = 0$, then $l = 1$ and we have equalities in our desired results, for any choice of m. So suppose that $h \geq 1$, that T is an *m*-ary tree of height h with l leaves, and that our desired results hold for all trees with smaller height. Let v_0 be the root of T, and let $v_1, \ldots, v_d$ be its children. Note that $d \leq m$.

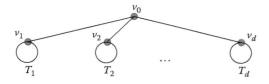

If we remove v_0, then there remain d rooted trees. For each $1 \leq j \leq d$, let T_j be the tree with root v_j, height h_j, and l_j leaves. So $h_j < h$, and the inductive hypothesis applies to T_j. Hence,

$$l = l_1 + \cdots + l_d \leq m^{h_1} + \cdots + m^{h_d} \leq d \cdot m^{h-1} \leq m^h.$$

Taking $\log_m$ gives $\log_m l \leq h$. Since $h \in \mathbb{Z}$, it follows that $\lceil \log_m l \rceil \leq h$. □

In practice, rooted trees often have additional structure given by an ordering of the children for each internal vertex.

Definition 10.3.

(a) An **ordered rooted tree** is a rooted tree in which each internal vertex has its set of children ordered.

(b) An **ordered binary tree** is a binary tree in which each child of an internal vertex is designated as either the **left child** or the **right child**, but not both. Moreover, at each internal vertex, the corresponding **left subtree** (respectively, **right subtree**) is the subtree rooted at its left child (respectively, right child).

The rooted tree displayed in Figure 10.10 may be more strongly regarded as an ordered rooted tree, if we take the ordering on the set of children for each internal vertex given by the natural ordering on the subscripts of the vertices, for example. Notice that this ordering agrees with the left-to-right placement of the vertices in the picture in Figure 10.10. In fact, the convention for displaying ordered rooted trees is to place the children in order from left to right. In ordered binary trees, we simply display the left child to the left of its parent and the right child to the right.

There is more than one way in which we might provide the binary tree in Figure 10.9 with the structure of an ordered binary tree. The two shown in Figure 10.11 are distinguished by the choice of regarding vertex v_6 as either the left or the right child of its parent v_5.

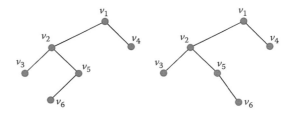

FIGURE 10.11
Two different ordered binary trees.

An ordered binary tree can be used to define a convenient data structure in which to store items indexed by identifying keys and in which to look up those items. A **binary search tree** is an ordered binary tree such that the key at each vertex is greater than all of the keys in its left subtree and less than all of the keys in its right subtree. We build a binary search tree for a given list of items using the following recursive construction. The initial item from the list is designated as the root vertex. Each subsequent item in turn is added to the binary search tree built so far by first focusing our attention at the root. If the new item's key is the same as that of the root, then this is a duplicate of an item already in the tree and need not be added. If its key is less than that of the root, then we move our attention to the left child of the root. If its key is greater than that of the root, then we move our attention to the right child of the root. After this comparison, if we are attempting to consider a (child) vertex that does not yet exist in our tree, then we add our new item by creating that new position in the tree. Otherwise, we have focused on a higher-level vertex of the tree, and we now treat that vertex as we did the root, repeating our comparisons until the new item is either found or added. This process is illustrated in the next example.

Example 10.3.   Suppose we wish to keep track of which words occur in a journal article. Under the usual dictionary ordering for words, each word can serve as its own key. However, the words in an article are certainly not expected to occur in alphabetical order. For example, say the first sentence is

Discrete mathematics is my favorite course.

Its six words are entered into the corresponding binary search tree in the steps shown in Figure 10.12. After step (3), the word *my* is compared first to *discrete* and then to *mathematics*. Since *my* is greater than *mathematics*, and

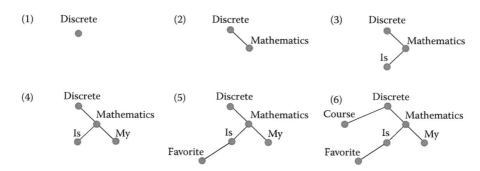

FIGURE 10.12
Building a binary search tree.

mathematics has no right child in step (3), *my* is added as the right child of *mathematics* in step (4).

If we wish to look up a word in this tree (such as *cat*), then essentially the same procedure is employed. The word *cat* is compared first to *discrete* and then to *course*. Since *cat* is less than *course*, and *course* has no left child, we conclude that *cat* is not in the tree.

Ordered rooted trees can be used to encode two-dimensional black-and-white images. The **quadtree** for a given image is an ordered full 4-ary tree constructed by placing a square bounding box around the image and recursively subdividing boxes within the image into four quadrants. Each box that is neither entirely black nor entirely white in color is itself subdivided into four quadrants, and this process is repeated until all boxes are monochromatic. Each box that gets subdivided is designated the parent of its resulting four quadrants, and an ordering for the quadrants is given by the numbering convention shown in Figure 10.13. The procedure is illustrated in Example 10.4.

FIGURE 10.13
Quadrant order in quadtrees.

Example 10.4. The image displayed on the left-hand side of Figure 10.14 might simply represent a spiral-like design, or it might represent the letter "J" in some crude font. The necessary quadrant subdivisions are shown on the right-hand side of Figure 10.14. The quadtree for this image is displayed in Figure 10.15, where *B* represents a black square and *W* represents a white square. The size

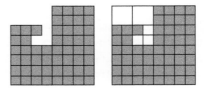

FIGURE 10.14
An image and its corresponding subdivision.

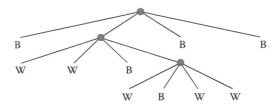

FIGURE 10.15
Quadtree for the letter "J".

and location of each square is determined by the level and position order of the corresponding leaf in the tree.

Note that higher and higher levels in the quadtree represent greater and greater levels of detail in the corresponding image. Consequently, one could obtain a courser version of a given image by pruning its quadtree above some level. That is, one could remove all of the vertices above some fixed level and then, for each vertex that changes from an internal vertex to a leaf, substitute either a B or a W according to some rule. This rule might be determined by the majority of the children, or it could simply replace all such vertices by a W, for instance.

FIGURE 10.16
A courser version of the "J" from Figure 10.14.

For example, suppose we wish to print a smaller version of the "J" from Figure 10.14 and decide to decrease the level of detail. We could remove all of the vertices above level 2 in the quadtree from Figure 10.15. At the resulting new leaf in level 2, we could substitute a W, based on the majority of the former children of that vertex. The resulting courser version of our "J" is shown in Figure 10.16.

Exercises

1. Is the pictured graph a tree?

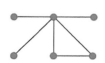

2. Is the pictured graph a tree?

3. For which pairs (m, n) is the graph $K_{m,n}$ a tree?

4. For $n \in \mathbb{Z}^+$, is the empty graph Φ_n a forest?

5. The hydrocarbon ethylene, C_2H_4, is used to accelerate the ripening of bananas and other fruit. Its chemical structure is pictured.

$$\begin{array}{ccc} H & & H \\ \diagdown & & \diagup \\ & C = C & \\ \diagup & & \diagdown \\ H & & H \end{array}$$

Is its corresponding graph a tree?

6. Ammonia, NH_3, is a cleaning agent and is also used to produce fertilizer. Its chemical structure is pictured.

$$\begin{array}{cc} & H \\ & \diagup \\ H - N & \\ & \diagdown \\ & H \end{array}$$

Is its corresponding graph a tree?

7. Could a graph with a loop edge be a tree?

8. Could a graph with multiple edges be a tree?

9. Let T be a tree. Is $T \times \Phi_n$ a forest?

10. Under what conditions on graphs G and H will $G \times H$ be a forest?

11. Hexanes are used in the preparation of adhesives and inks. They also serve as solvents. Find the number of possible isomers of the hexanes, C_6H_{14}.

12. Heptanes are used in dry cleaners and pesticides. Find the number of possible isomers of the heptanes, C_7H_{16}.

13. Pictured is the bus system for a certain city. It consists of two bus routes, each of which is a path.

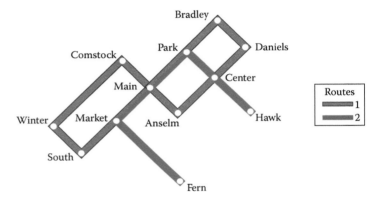

Consider the graph representing the coverage by this bus system.

(a) Is there a unique path from Fern to Market?

(b) Is there a unique path from Hawk to Park?

(c) The graph is not a tree, so in general, paths in it are not unique. However, is there a unique shortest path between any pair of stops in this city? For this, we assume that all edges in our graph have the same length. Explain your answer.

14. Pictured are the roads in a particular neighborhood.

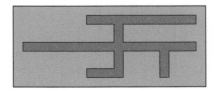

Consider the graph in which intersections and dead ends are represented by vertices, and streets joining them are represented by edges.

   (a) Observe that the graph is a tree. For any pair of intersections that sit in this neighborhood, must there be a unique path between them?

   (b) What features of the tree represent the dead ends?

   (c) What features of the tree represent the intersections?

15.* Prove the converse of Theorem 10.1.

16.* Prove the converse of Theorem 10.3 for connected graphs.

17.* Let G be an arbitrary graph without loops. Show: G has only one spanning tree if and only G is a tree.

18. Find all trees with degree sequence $4, 4, 2, 2, 1, 1$.

19. Let $G = (V, E)$ be an arbitrary graph. Show: If G is connected and $|E| = |V| - 1$, then G is a tree. Hint: Consider a spanning tree.

20.* Use Theorem 10.3 to prove Theorem 10.2. Hint: Consider the degree sequence.

21.* Show that a tree has exactly two leaves if and only if it is a path.

22. What is the degree sequence of any tree with exactly 3 leaves?

23. How many edges are there in a forest on n vertices with c components?

24. Show that a forest with c components and no isolated vertices must have at least $2c$ leaves.

25. What trees have $n - 1$ leaves, where n is the number of vertices?

26. What trees have $n - 2$ leaves, where n is the number of vertices?

27.* Prove Theorem 10.3 by using Euler's Formula.

28.* Prove that any graph G with $\delta(G) \geq 2$ must contain a cycle. Hint: Use Theorem 10.3.

29. In a tree on n vertices, how many different subgraphs are there that are isomorphic to a path?

30. Prove that trees are bipartite.

31. Find a spanning tree for C_n.

32. Find a spanning tree for K_n.

33. Find a spanning tree for Q_n.

34. Find a spanning tree for $K_{m,n}$.

35. Prove the converse of Theorem 10.4.

36. Prove Corollary 10.5.

FIGURE 10.17
A tree for Exercises 37 through 40.

37. Take 1 to be the root of the tree in Figure 10.17.
    (a) List the children of 2.
    (b) Which vertex is the parent of 5?
    (c) What is the level of 3?
    (d) What is the height of this rooted tree?
    (e) Is this a binary tree?
    (f) Is it balanced?

38. Take 7 to be the root of the tree in Figure 10.17.
    (a) List the children of 2.
    (b) Which vertex is the parent of 5?
    (c) What is the level of 3?
    (d) What is the height of this rooted tree?
    (e) Is this a binary tree?
    (f) Is it balanced?

39. Take 2 to be the root of the tree in Figure 10.17.
    (a) List the children of 1.
    (b) Which vertex is the parent of 3?
    (c) What is the level of 4?
    (d) What is the height of this rooted tree?
    (e) Is this a binary tree?
    (f) Is it full?

40. Take 6 to be the root of the tree in Figure 10.17.
    (a) List the children of 2.
    (b) Which vertex is the parent of 1?
    (c) What is the level of 3?
    (d) What is the height of this rooted tree?
    (e) Is this a binary tree?
    (f) Is it full?

41. Billy Creasey has stored some music files on a CD according to the displayed file system. It makes it easier for him to find his songs.

   ◊ CD Drive (E:)
      △ Country
         □ Black.mpg
         □ Reba.mpg
      △ Hip Hop
         □ Fifty.mpg
         □ JLo.mpg
      △ Rock
         △ Classic Rock
            □ Lola.mpg
            □ Satisf.mpg
            □ USSR.mpg
         △ Modern Rock
            □ GrnDay.mpg

   Consider the tree that corresponds to this structure.

   (a) Is Lola.mpg a child of Rock?

   (b) The "is the folder containing" relation in this example correspond to either the "is the parent of" relation or the "is a child of" relation on trees. Which one?

(c) The music (.mpg) files in this example correspond to what tree feature?

(d) What is the height of the tree?

(e) Is the tree balanced?

42. Displayed is the family tree for the set of living male members of a certain family. This same example was considered in Example 5.5 of Section 5.1.

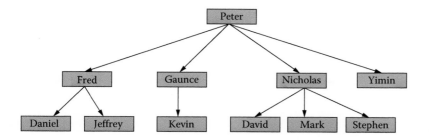

(a) In trees, as in families, the child of a child is called a grandchild. How many grandchildren does Peter have?

(b) The men with children in this example correspond to what tree feature?

(c) Is the tree 3-ary? 4-ary?

(d) Is the tree balanced?

43. If a tree is balanced with a particular vertex taken as the root, then must it be balanced no matter which vertex is taken as the root?

44. If a tree is a full m-ary tree with a particular vertex taken as the root, then must it be a full m-ary tree no matter which vertex is taken as the root?

45. Prove or disprove: Every tree has a vertex that can be chosen as the root to make the tree balanced.

46. Prove or disprove: Every tree has a vertex that can be chosen as the root to make the tree full.

For Exercises 47 through 50, let T be a given full m-ary tree. Let n denote the number of vertices, l the number of leaves, and i the number of internal vertices.

47.* Prove Theorem 10.6.

48. Express i and l in terms of n and m.

49. Express n and i in terms of l and m.

50. Express n and l in terms of i and m.

51. Beth Robbins made a family tree consisting of the women in her family, starting with her grandmother. After doing

so, she noticed that each mother has exactly three daughters and precisely ten of the women are mothers. How many women are in Beth's family?

52. Luther Foote is trying to make a family tree consisting of the men in his family, starting with his father. If precisely eight of the men do not have sons but every father has exactly two sons, then how many men are in the Foote family?

53. Let T be a full m-ary tree of height h in which all l leaves are at level h. Show that $l = m^h$.

54*. Let T be a full and balanced m-ary tree of height h with l leaves. Show that $h = \lceil \log_m l \rceil$.

In Exercises 55 through 58, construct the binary search tree for the words in the given sentence. Use the usual dictionary order to compare words.

55. All students take calculus.

56. Long live the king.

57. Eat, drink, and be merry.

58. Three is the magic number.

In Exercises 59 through 62, construct the quadtree for the given image. Note that grid lines have been drawn over the image for reference and are not themselves part of the image.

59.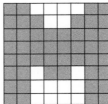

60.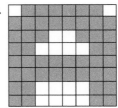

61.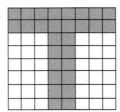

62.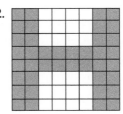

In Exercises 63 through 66, draw the image that corresponds to the given quadtree. In each case, the image represents a letter.

63.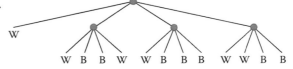

64.

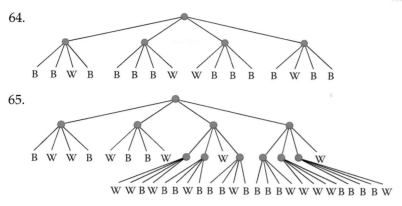

B B W B   B B B W   W B B B   B W B B

65.

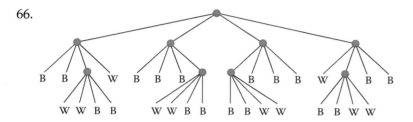

B W W B   W B   B W   W   W

W W B W B B W B B B W B B B B W W W W B B B B W

66.

B B   W B B B   B B B W   B B

W W B B   W W B B   B B W W   B B W W

10.2   Search trees

Recall that an algorithm is a finite list of simple instructions leading to a desired result. In this section, we consider two algorithms that yield spanning trees by traversing a given connected graph from a specified vertex. The first is called **Breadth-First Search** and reaches out as widely as possible to nearby neighbors before moving on to less immediate neighbors. The second is called **Depth-First Search** and reaches into the graph as long as possible until it is forced to back up and resume reaching out in another direction. Although these two algorithms generally produce different spanning trees, they are more significantly distinguished by the orders in which vertices are visited. Consequently, each algorithm outputs not only a set F of edges for its spanning tree, but also a list L imposing an ordering on the vertices. To aid in the management within each algorithm, we keep track of vertices that have been visited so far in a list M and regard those vertices as "marked." Lists of vertices are stored as one-dimensional arrays $[v_1, v_2, \ldots, v_n]$, with $[\,]$ denoting the empty list.

Applications of our algorithms are also explored. In particular, we consider ways in which a computer can interpret and evaluate expressions. This further leads us to a consideration of some additional tree traversal algorithms. Our study of the Depth-First Search Algorithm also leads us to a general searching strategy called backtrack searching.

Algorithm 10.1 Breadth-First Search

Let $G = (V, E)$ be a connected graph on n vertices and v a vertex of G.

Algorithm.
   Let $M = [\ ]$, $L = [v]$, and $F = \emptyset$.
   While $|F| < n - 1$,
      \begin
      Let t be the first vertex in L and not in M.
      Add t to the end of M.
      While t has neighbors that are not in L,
         \begin
         Let e be an edge such that one end is t
               and the other is a vertex u outside of L.
         Let $F = F \cup \{e\}$.
         Add u to the end of L.
         \end.
      \end.
   Return $T = (V, F)$.

In particular, we see how this technique applies to a search for a Hamiltonian cycle in a graph.

10.2.1   Breadth-first search trees

The tree T constructed by Algorithm 10.1[*] is called the **breadth-first search tree** for G starting from a given vertex v. It is a spanning tree and may be regarded as a rooted tree with root v. In Breadth-First Search, the neighbors of v are visited first, then any unvisited neighbors of those neighbors are visited, and so on. Each vertex is entered into the list L when it is first visited. A vertex is marked just prior to seeking its unvisited neighbors.

Remark 10.1.   Breadth-First Search and Depth-First Search construct spanning trees by selecting edges. Since, at some stages in these algorithms, there may be more than one reasonable edge to select, an initial ordering of the edges would need to be specified to ensure unique choices. For a simple graph, such an ordering can be specified by giving an initial ordering of the vertices. In that case, the resulting lexicographic, or "alphabetical," ordering of the edges is used, as described in Section 5.2. That is, we compare two edges by comparing the two lower-valued endpoints and using the higher-valued endpoints to break a tie. General sorting algorithms are discussed in Section 10.5.

---

[*] For an explanation of the pseudocode used to present algorithms throughout this book, see Appendix B.

In our first example, we perform Breadth-First Search on a tree. Thus, we can focus our attention not on the resulting tree, which is obviously the entire graph in this case, but rather on the imposed ordering L on the vertices.

Example 10.5. Perform Breadth-First Search on the pictured graph starting from vertex v_1, given the ordering $v_1, v_2, v_3, v_4, v_5, v_6$ of the vertices.

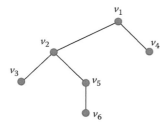

Keep a running table of values for the list of marked vertices M, the imposed (ordered) list of vertices L, and the set of edges F.

Solution. Starting from the initial values, we make a new row in our table each time either M changes or both F and L change. To aid in our trace, we also record our current position in the graph at vertex t.

M	L	F	t
[ ]	$[v_1]$	$\emptyset$	
$[v_1]$	$[v_1]$	$\emptyset$	v_1
$[v_1]$	$[v_1,v_2]$	$\{\{v_1,v_2\}\}$	v_1
$[v_1]$	$[v_1,v_2,v_4]$	$\{\{v_1,v_2\}, \{v_1,v_4\}\}$	v_1
$[v_1,v_2]$	$[v_1,v_2,v_4]$	$\{\{v_1,v_2\}, \{v_1,v_4\}\}$	v_2
$[v_1,v_2]$	$[v_1,v_2,v_4,v_3]$	$\{\{v_1,v_2\}, \{v_1,v_4\}, \{v_2,v_3\}\}$	v_2
$[v_1,v_2]$	$[v_1,v_2,v_4,v_3,v_5]$	$\{\{v_1,v_2\}, \{v_1,v_4\}, \{v_2,v_3\}, \{v_2,v_5\}\}$	v_2
$[v_1,v_2,v_4]$	$[v_1,v_2,v_4,v_3,v_5]$	$\{\{v_1,v_2\}, \{v_1,v_4\}, \{v_2,v_3\}, \{v_2,v_5\}\}$	v_4
$[v_1,v_2,v_4,v_3]$	$[v_1,v_2,v_4,v_3,v_5]$	$\{\{v_1,v_2\}, \{v_1,v_4\}, \{v_2,v_3\}, \{v_2,v_5\}\}$	v_3
$[v_1,v_2,v_4,v_3,v_5]$	$[v_1,v_2,v_4,v_3,v_5]$	$\{\{v_1,v_2\}, \{v_1,v_4\}, \{v_2,v_3\}, \{v_2,v_5\}\}$	v_5
$[v_1,v_2,v_4,v_3,v_5]$	$[v_1,v_2,v_4,v_3,v_5,v_6]$	$\{\{v_1,v_2\}, \{v_1,v_4\}, \{v_2,v_3\}, \{v_2,v_5\}, \{v_5,v_6\}\}$	

For example, the first time that the outermost while loop is entered (so $M = [\ ]$, $L = [v_1]$, and $F = \emptyset$), we set $t = v_1$ and $M = [v_1]$. The inner while loop then adds v_2 and v_4 to L and adds $\{v_1, v_2\}$ and $\{v_1, v_4\}$ to F. In the end, we see that the final imposed ordering on the vertices

$$L = [v_1, v_2, v_4, v_3, v_5, v_6]$$

need not agree with the original given ordering. □

In our second example, we apply Breadth-First Search to a graph that is not a tree. Thus, the particular spanning tree produced should also be a focus of our attention.

Example 10.6. Use Breadth-First Search to find the breadth-first search tree for the pictured graph starting from the vertex v_1, given the initial ordering $v_1, v_2, v_3, v_4, v_5, v_6, v_7$ of its vertices.

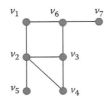

Shade the edge set F for the resulting breadth-first search tree, and report the new ordering L imposed on the vertices.

Solution. The final value of the vertex list is

$$L = [v_1, v_2, v_6, v_3, v_4, v_5, v_7],$$

and the breadth-first search tree is shown in bold in the right-hand picture.

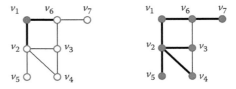

The left-hand picture shows the point in the algorithm when the outermost while loop is entered with $|F| = 2$. There, F is shown in bold, only the vertices in M are darkened, and $L = [v_1, v_2, v_6]$. At that point, we get $t = v_2$, add v_2 to M, add v_3, v_4, v_5 to L, and add the edges joining those vertices with v_2 to F.   □

The list L provided by Breadth-First Search lists the vertices in nondecreasing order of their distance from v. The reverse of such a listing was utilized in Section 9.5 in our proof of Brooks' Theorem.

10.2.2 Depth-first search trees

The tree T constructed by Algorithm 10.2 is called the **depth-first search tree** for G starting from a given vertex v. It is a spanning tree and may be regarded as a rooted tree with

Algorithm 10.2 Depth-First Search

Let $G = (V, E)$ be a connected graph on n vertices and v a vertex of G.

Algorithm.
>   Let $M = [v]$, $L = [\ ]$, and $F = \emptyset$.
>   Let $t = v$.
>   While $t \neq$ null,
>>   \begin
>>   While t has neighbors that are not in M,
>>>   \begin
>>>   Let e be an edge such that one end is t
>>>   and the other is a vertex u outside of M.
>>>   Let $F = F \cup \{e\}$.
>>>   Add u to the end of M.
>>>   Let $t = u$.
>>>   \end.
>>   Add t to the end of L.
>>   Replace t by the parent of t.
>>   \end.
>   Return $T = (V, F)$.

---

root v. In Depth-First Search, first a neighbor of v is visited, then an unvisited neighbor of that neighbor is visited, and so on. When a vertex has no unvisited neighbors, we return to the previously visited vertex and continue our process from there. Each vertex is marked when it is first visited. A vertex is entered into the list L after all of its neighbors have been marked. This list L is thus said to be in **postorder**. Alternatively, a list in **preorder** is obtained if we list vertices when they are first visited.

Since, at each stage of the Depth-First Search Algorithm, F determines a v-rooted tree on the vertices of M, each vertex in M, except for v, has a well-defined parent. For v, we assign its parent to be the value **null**, assumed not to be in the vertex set.

Note the difference between the innermost while loops for Breadth-First Search and Depth-First Search. This while loop in Depth-First Search ends with the assignment "Let $t = u$." It is this feature that sends Depth-First Search deep into a graph, as opposed to holding it near its starting point as long as possible.

In our first example using Depth-First Search, we apply it to the same tree that we first considered using Breath-First Search. Thus, we can focus our attention on how the imposed ordering L on the vertices develops under Depth-First Search in contrast to how it developed under Breath-First Search.

Example 10.7. Perform Depth-First Search on the pictured graph starting from v_1, given the initial ordering $v_1, v_2, v_3, v_4, v_5, v_6$ of its vertices.

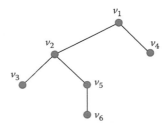

Keep a running table of values for the list of marked vertices M, the imposed (ordered) list of vertices L, and the set of edges F.

Solution. Starting from the initial values, we make a new row in our table each time either M and F both change or L changes. To aid in our trace, we also record our current position in the graph at vertex t.

M	L	F	t
$[v_1]$	$[\,]$	$\emptyset$	v_1
$[v_1, v_2]$	$[\,]$	$\{\{v_1, v_2\}\}$	v_2
$[v_1, v_2, v_3]$	$[\,]$	$\{\{v_1, v_2\}, \{v_2, v_3\}\}$	v_3
$[v_1, v_2, v_3]$	$[v_3]$	$\{\{v_1, v_2\}, \{v_2, v_3\}\}$	v_2
$[v_1, v_2, v_3, v_5]$	$[v_3]$	$\{\{v_1, v_2\}, \{v_2, v_3\}, \{v_2, v_5\}\}$	v_5
$[v_1, v_2, v_3, v_5, v_6]$	$[v_3]$	$\{\{v_1, v_2\}, \{v_2, v_3\}, \{v_2, v_5\}, \{v_5, v_6\}\}$	v_6
$[v_1, v_2, v_3, v_5, v_6]$	$[v_3, v_6]$	$\{\{v_1, v_2\}, \{v_2, v_3\}, \{v_2, v_5\}, \{v_5, v_6\}\}$	v_5
$[v_1, v_2, v_3, v_5, v_6]$	$[v_3, v_6, v_5]$	$\{\{v_1, v_2\}, \{v_2, v_3\}, \{v_2, v_5\}, \{v_5, v_6\}\}$	v_2
$[v_1, v_2, v_3, v_5, v_6]$	$[v_3, v_6, v_5, v_2]$	$\{\{v_1, v_2\}, \{v_2, v_3\}, \{v_2, v_5\}, \{v_5, v_6\}\}$	v_1
$[v_1, v_2, v_3, v_5, v_6, v_4]$	$[v_3, v_6, v_5, v_2]$	$\{\{v_1, v_2\}, \{v_2, v_3\}, \{v_2, v_5\}, \{v_5, v_6\}, \{v_1, v_4\}\}$	v_4
$[v_1, v_2, v_3, v_5, v_6, v_4]$	$[v_3, v_6, v_5, v_2, v_4]$	$\{\{v_1, v_2\}, \{v_2, v_3\}, \{v_2, v_5\}, \{v_5, v_6\}, \{v_1, v_4\}\}$	v_1
$[v_1, v_2, v_3, v_5, v_6, v_4]$	$[v_3, v_6, v_5, v_2, v_4, v_1]$	$\{\{v_1, v_2\}, \{v_2, v_3\}, \{v_2, v_5\}, \{v_5, v_6\}, \{v_1, v_4\}\}$	Null

For example, the first time that the innermost while loop is entered (so $M = [v_1]$, $L = [\,]$, $F = \emptyset$, and $t = v_1$), we add v_2 to M, add $\{v_1, v_2\}$ to F, and set $t = v_2$. We then effectively start a Depth-First Search from v_2 and, upon completing that, reset t back to v_1. In the end, we see that the final imposed ordering on the vertices,

$$L = [v_3, v_6, v_5, v_2, v_4, v_1]$$

is quite different from that obtained by Breadth-First Search in Example 10.5.   □

In our second example, we apply Depth-First Search to the graph on which Breadth-First Search was earlier applied in Example 10.6. Here, we see that a different spanning tree results.

Example 10.8. Use Depth-First Search to find the depth-first search tree for the pictured graph starting from the vertex v_1, given the initial ordering $v_1, v_2, v_3, v_4, v_5, v_6, v_7$ of its vertices.

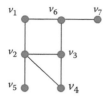

Shade the edge set F for the resulting depth-first search tree, and report the new ordering L imposed on the vertices.

Solution. The final value of the vertex list is

$$L = [v_4, v_7, v_6, v_3, v_5, v_2, v_1],$$

and the depth-first search tree is shown in bold in the right-hand picture.

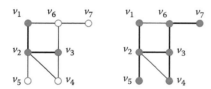

The left-hand picture shows the first point in the algorithm when the outermost while loop is entered with $t = v_3$. There, F is shown in bold, only the vertices in M are darkened, and L is empty. At that point, we add v_4 to M and let $t = v_4$. Then we add v_4 to L, and let $t = v_3$ again. □

10.2.3 Applications

Managing algebraic operations. The list L provided by Depth-First Search provides a useful way to encode algebraic expressions. An algebraic expression involving the operations addition $+$, subtraction $-$, multiplication $*$, and division $\div$ can be represented by an ordered binary tree. For example, the expression

$$(a + b * (c - 1)) \div 2 \tag{10.1}$$

is represented by the leftmost tree in Figure 10.18. There, the leaves are the variables and numbers, and the internal vertices

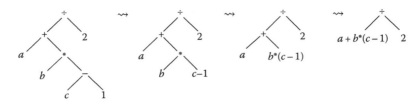

FIGURE 10.18
Tree representing $(a + b * (c - 1)) \div 2$.

are the binary operations. Each operation acts on the results at its left child and its right child, in the order pictured. The steps in the evaluation process are also shown in Figure 10.18.

Expression (10.1) lists all of the vertices in the corresponding tree, together with the parentheses needed to uniquely characterize the resulting value. Alternatively, **postfix** notation can be used to represent the same expression with no parentheses and only the items in the vertices. It is obtained by listing the vertices in the order L given by Depth-First Search (using a left-to-right ordering of the children as pictured). For example, the leftmost tree in Figure 10.18 results in the postfix expression

$$a \ b \ c \ 1 \ - \ * \ + \ 2 \ \div. \tag{10.2}$$

Its value is computed by reading from left to right. When an operation is encountered, it is performed on the previous two values. The three positions involved are then replaced by a single position containing the result of that operation. This process corresponds exactly to the tree pruning displayed in Figure 10.18. Expression (10.2) gets evaluated as shown in Figure 10.19.

There is another way to describe postfix notation that leads to two related notations. To accomplish this, we define three traversal methods for an ordered binary tree T, each of which produces a list (and hence an ordering) of the vertices in T. All three behave the same on a tree with at most one vertex and are otherwise defined recursively.

$$
\begin{array}{ccccccccc}
a & b & c & 1 & - & * & + & 2 & \div \\
 & & & & \downarrow & & & & \\
 & a & b & c-1 & * & + & 2 & \div \\
 & & & & \downarrow & & & \\
 & & a & b*(c-1) & + & 2 & \div \\
 & & & & \downarrow & & \\
 & & & a+b*(c-1) & 2 & \div \\
 & & & & & \downarrow \\
 & & & & (a+b*(c-1))\div 2
\end{array}
$$

FIGURE 10.19
Evaluating postfix notation.

Definition 10.4. (Ordered Binary Tree Traversals).
Let T be an ordered binary tree. If T has no vertices, then nothing is listed. If T consists only of the root v, then just v is listed. Otherwise, let T_L denote the left subtree of the root v, and let T_R denote the right subtree, noting that one of T_L or T_R could have no vertices.

(a) A **postorder traversal** of T is accomplished by first performing a postorder traversal on T_L, second performing a postorder traversal on T_R, and third listing the root v for T.

(b) A **preorder traversal** of T is accomplished by first listing the root v for T, second performing a preorder traversal on T_L, and third performing a preorder traversal on T_R.

(c) An **inorder traversal** of T is accomplished by first performing an inorder traversal on T_L, second listing the root v for T, and third performing an inorder traversal on T_R.

In Definition 10.4, note that the list obtained from a postorder traversal is the same as that obtained from a depth-first search. Consequently, given an algebraic expression, the postfix notation for it is the list obtained from a postorder traversal of the ordered binary tree representing the algebraic expression. Similarly, the **prefix** notation is the list obtained from a preorder traversal, and the **infix** notation is the list obtained from an inorder traversal. The infix notation turns out to be the original algebraic expression with the parentheses removed and is hence ambiguous in general. Prefix notation, on the other hand, gives an unambiguous expression that can be evaluated and is explored in the exercises.

An application of inorder traversal relates to binary search trees. Recall that a binary search tree stores a set of items, based on an order relation on the set. It turns out that an inorder traversal of a binary search tree lists the items in order, based on the order relation.

Example 10.9. An inorder traversal of the binary search tree that resulted in Example 10.3,

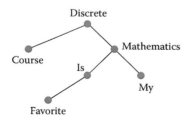

yields the list

course, discrete, favorite, is, mathematics, my.

Note that since the dictionary ordering was used to generate the binary search tree, the resulting list is in dictionary order.

Backtrack searches. The approach of a Depth-First Search can be used to search for a Hamiltonian cycle in a graph G. The idea is to start at a vertex a in G and to walk through the graph until a Hamiltonian cycle is found. If at some point in the walk we get "stuck" and cannot possibly complete a Hamiltonian cycle from the path we have followed so far, then we back up one step in our path and check whether we can alternatively complete a Hamiltonian cycle from that point. Consequently, this approach is also called **backtrack searching**. Precisely, we are performing a Depth-First Search, not in G, but in a related graph H, which we now describe.

If we fix a vertex a in a given graph G, then there is a graph H in which each vertex is either a path or a cycle in G starting at a. Two vertices P and Q in H are taken to be adjacent if and only if one is the other followed by an additional vertex. The utility of H is that G is Hamiltonian if and only if H contains a vertex that is a Hamiltonian cycle. Moreover, when G is Hamiltonian, there must be a path in H from the trivial path a to a Hamiltonian cycle. Although H is, in general, a very large graph, the advantage of H is that traversing it requires only knowledge of its local structure, and that structure is determined by G. Moreover, not only is it unnecessary to store the entire vertex and edge sets for H, but also these sets are unknown prior to the start of the search. Otherwise, we would already know whether G is Hamiltonian by knowing whether or not one of the vertices in H is a Hamiltonian cycle.

Example 10.10.   Backtrack Searching for a Hamiltonian Cycle.

(a) The pictured graph G

yields the pictured graph H, in which our searching occurs.

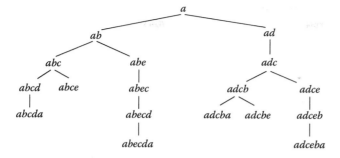

In this case, a Depth-First Search of H from a would lead us to find the vertex *abecda*, which is a Hamiltonian cycle in G.

(b) The pictured graph G

yields the pictured graph H, in which our searching occurs.

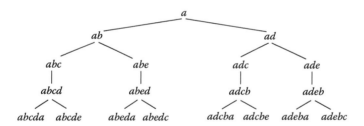

In this case, a Depth-First Search of H from a would complete without having found a vertex corresponding to a Hamiltonian cycle. We could then conclude that G is not Hamiltonian.

The method used in Example 10.10 to determine whether a given graph is Hamiltonian is very inefficient in general. However, there is no known efficient general method for solving this problem. Moreover, it is widely believed that no efficient general method exists.

The problem of simply determining whether or not a graph is Hamiltonian is related to a famous problem known as the **traveling salesman problem**. In that problem, a salesman wishes to visit each location in his territory exactly once before returning home. Moreover, there is some cost (in time or money) associated with each move from one location to another, and the salesman wants the route with the cheapest total cost. The traveling salesman problem is an example of an NP-complete problem, as will be discussed at the end of Section 10.4. In the exercises here, we shall consider the problem of finding a route for a salesman, but we do not consider associated costs.

Exercises

The following graphs are referred to throughout the exercises.

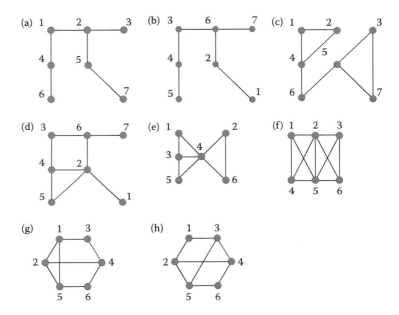

In Exercises 1 through 8, perform Breadth-First Search on the given graph starting at vertex 1. Use the ordering $1, 2, 3, \ldots$ of the vertices. In each case, specify the resulting list L ordering the vertices. If the graph is not a tree, then shade the resulting tree.

1. Graph (a).   2. Graph (b).

3. Graph (c).   4. Graph (d).

5. Graph (e).   6. Graph (f).

7. Graph (g).   8. Graph (h).

In Exercises 9 through 16, perform Depth-First Search on the given graph starting at vertex 1. Use the ordering

$1, 2, 3, \ldots$ of the vertices. In each case, specify the resulting list L ordering the vertices. If the graph is not a tree, then shade the resulting tree.

9. Graph (b).   10. Graph (a).

11. Graph (d).   12. Graph (c).

13. Graph (f).   14. Graph (e).

15. Graph (h).   16. Graph (g).

17. Does there exist a graph and an ordering of its vertices for which the breadth-first search tree and the depth-first search tree are distinct? Justify your answer.

18. Does there exist a graph on more than one vertex and an ordering of its vertices for which the breadth-first search tree and the depth-first search tree are the same? Justify your answer.

19. Given a fixed root, does the input ordering of the vertices affect the tree T resulting from Breadth-First Search? See Remark 10.1. Justify your answer.

20. Given a fixed root, does the input ordering of the vertices affect the tree T resulting from Depth-First Search? See Remark 10.1. Justify your answer.

21.* Given the ordered list L resulting from Depth-First Search applied to a tree T, can T be recovered from L alone? Explain.

22.* Given the ordered list L resulting from Breadth-First Search applied to a tree T, can T be recovered from L alone? Explain.

23. In graph (g), what shortest path from 1 to 4 is provided by Breadth-First Search? Is that the only shortest path from 1 to 4?

24. In graph (h), what shortest path from 1 to 6 is provided by Breadth-First Search? Is that the only shortest path from 1 to 6?

25.* True or false? Given a fixed root, the breadth-first search tree has the minimum height among all spanning trees for a graph. Explain.

26.* True or false? Given a fixed root, the depth-first search tree has the maximum height among all spanning trees for a graph. Explain.

27.* Here is a potential algorithm for producing a spanning tree for a given connected graph $G = (V, E)$.

```
Let F = E.
While the subgraph (V, F) contains a cycle,
    \begin
    Let e be an edge in a cycle.
    Remove e from F.
    \end.
Return T = (V, F).
```

Will T always be a spanning tree? Justify your answer.

28.* Here is a potential algorithm for producing a spanning tree for a given connected graph $G = (V, E)$.

```
Let F = Ø.
Let v be a vertex of G.
Let S = {v}.
While there is a vertex u in V \ S,
    \begin
    Find a path P from v to u.
    Add the edges of P into F.
    Add the vertices of P into S.
    \end.
Return T = (V, F).
```

Will T always be a spanning tree? Justify your answer.

29. Write an algorithm using Breadth-First Search that determines whether or not a given graph is connected.

30. Write an algorithm using Depth-First Search that counts the number of components in a given graph.

31.* Given the ordered list L resulting from Depth-First Search applied to a full m-ary tree T, can T be recovered from L together with the knowledge that T is a full m-ary tree? Explain. (Compare Exercise 21.)

32.* Given the ordered list L resulting from Breadth-First Search applied to a full m-ary tree T, can T be recovered from L together with the knowledge that T is a full m-ary tree? Explain. (Compare Exercise 22.)

33. If an investment is earning **simple interest**, then interest is only being earned on the initial investment (the principle) and not on subsequent interest. That is, given a periodic interest rate i and an initial investment of P dollars, the future value F of the investment after n periods is given by $F = P(1 + ni)$. Construct the tree that represents the following expression for the value of F.

$$P * (1 + (n * i))$$

34. If there are n people in a room and each person makes sure to shake hands with every other person exactly once, then a total of $\frac{n(n-1)}{2}$ handshakes occur. That is, there are $\binom{n}{2} = \frac{n(n-1)}{2}$ possible pairs of people in the room. Construct the

tree that represents the following expression for the number of handshakes.

$$(n * (n - 1)) \div 2$$

35. The calculus teacher gives a quiz every week, but at the end of the semester, she drops each student's lowest quiz grade before computing the average. That is, for each student, she computes the sum S of the quiz grades and subtracts the lowest quiz score L. If there were a total of n quizzes, then the quiz average is $\frac{S-L}{n-1}$. Write the ordinary expression

$$(S - L) \div (n - 1)$$

(a) in postfix notation. (b) in prefix notation. (c) in infix notation.

36. If resistors of resistances R_1 and R_2 are connected in parallel in a circuit, then the resistance R that results from this combination is given by $R = \frac{R_1 R_2}{R_1 + R_2}$. Write the ordinary expression

$$(R_1 * R_2) \div (R_1 + R_2)$$

(a) in postfix notation. (b) in prefix notation. (c) in infix notation.

37. Evaluate the following postfix expression:
6 5 3 − ÷ 1 + .

38. Evaluate the following postfix expression:
2 3 * 1 4 + − .

39. Evaluate the following postfix expression:
8 2 ÷ 3 − 6 + .

40. Evaluate the following postfix expression:
7 1 4 + 5 − * .

In Exercises 41 and 44, note that prefix expressions are evaluated from right to left. When an operation is encountered, it is performed on the two values to its right. The three positions involved are then replaced by a single position containing the result of that operation.

41. Evaluate the following prefix expression:
− 5 ÷ + 8 6 7 .

42. Evaluate the following prefix expression:
* − 5 + 2 1 3 .

43. Evaluate the following prefix expression:
− 3 + 1 * 4 2 .

44. Evaluate the following prefix expression:
÷ * 4 4 − 5 3 .

45. The standings in the American League East Division were used as the ordering when its member teams were put into the following binary search tree.

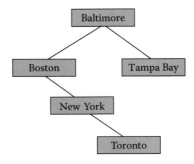

Perform an inorder traversal of the binary search tree to determine the standings (order of the teams) in the American League East Division.

46. The itinerary for a European trip was used as the ordering when its highlight cities were put into the following binary search tree.

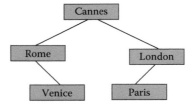

Perform an inorder traversal of the binary search tree to determine the itinerary (order of the cities) for the European trip.

Subsequent exercises refer to the following graphs. Each graph G reflects the sales territories (displayed below it) for a particular salesman. As discussed in Section 9.5, the edges in this dual graph G reflect the adjacencies of the territories, which are represented by vertices. Since the salesman can travel only from one territory to an adjacent one, these graphs can be used in the analysis required for Exercises 47 through 54.

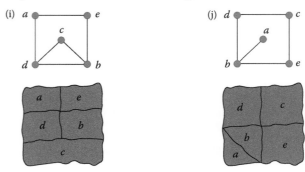

(k)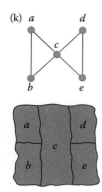

(l)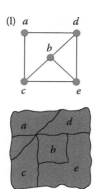

In Exercises 47 through 50, for the salesman with the specified sales territory graph G, search for a travel route that starts in territory a and visits each territory exactly once before immediately returning to a. Do this by mimicking Example 10.10 to search for a Hamiltonian cycle in G. That is, draw the graph H that would be searched with Depth-First Search to determine if G is Hamiltonian. If G is Hamiltonian, then specify the Hamiltonian cycle that would be found in this way.

47. Graph (i).   48. Graph (j).

49. Graph (k).   50. Graph (l).

In Exercises 51 through 54, we are interested in coloring the sales territory map reflected by the specified graph G. Since we would like to color the map with the fewest possible colors so that no two adjacent territories receive the same color, we want to know if just 2 colors would suffice. Do a backtrack search to determine if G is 2-colorable. Specifically, color the vertices in alphabetical order, and color each vertex with the smallest possible color in $\{1, 2\}$ that has yet to be tried on that vertex given the coloring built so far. If G is 2-colorable, then specify the coloring that would be found in this way. If not, then list the attempted incomplete colorings.

51. Graph (i).   52. Graph (j).

53. Graph (k).   54. Graph (l).

55.* Given integers $1 \leq k \leq n$, explain how Depth-First Search could be used to generate all of the permutations of k elements from $\{1, 2, \ldots, n\}$.

56.* Given integers $1 \leq k \leq n$, explain how Depth-First Search could be used to generate all of the combinations of k elements from $\{1, 2, \ldots, n\}$.

10.3 Weighted trees

In the previous sections, we considered spanning trees for connected graphs and saw that there are, in general, many choices for a spanning tree. If, however, there is additional structure dictating that some edges in a graph are more "costly" or "weighty" than others, then some choices may now be considered better than others.

A **weighted graph** is a graph $G = (V, E)$ for which each edge has been assigned a positive real number called the **weight** *of the edge*. Formally, this assignment is given by a function $\omega : E \longrightarrow \mathbb{R}^+$. In the examples we consider, the weights are displayed by attaching them as labels on the edges in a drawing of the graph. Figure 10.20 displays an example of a weighted graph. In all of the examples we consider, the weights are positive integers.

10.3.1 Minimum spanning trees

Figure 10.21 shows the layout of a large company that needs to set up a computer network among all of its buildings. Suppose the vertices in the graph in Figure 10.20 represent the buildings from Figure 10.21, each edge represents a potential direct link between two buildings, and the weight assigned to that edge reflects the cost in thousands of dollars of constructing that link. To enable all of the buildings to communicate with each other, the links in the network must correspond

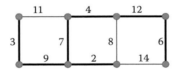

FIGURE 10.20
A weighted graph (for a company setting up a network).

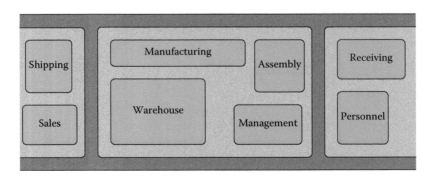

FIGURE 10.21
Company building map.

(at least) to the edges in a spanning tree for the graph. The cheapest possible network is thus obtained by choosing a spanning tree of least possible total weight (cost).

Definition 10.5.   Let G be a weighted graph.

(a) The **weight** *of a subgraph* is the sum of the weights of the edges in that subgraph.

(b) A **minimum spanning tree** for G is a spanning tree with the minimum weight among all spanning trees.

Example 10.11.   For the weighted graph in Figure 10.20, the edges shown in bold with weights $2, 3, 4, 6, 7, 9, 12$ form a minimum spanning tree. It has weight 43, and no other spanning tree for that graph has smaller weight. Consequently, the company whose layout is displayed in Figure 10.21 will need to spend at least \$43,000 to set up its computer network.

The preceding discussion has shown the utility of a minimum spanning tree. Consequently, it is desirable to have a method for finding one. We now present two algorithms for finding a minimum spanning tree. After both are presented, the proof that they indeed always find a minimum spanning tree will be given.

Our first algorithm is presented in Algorithm 10.3 and is named after the American mathematician Joseph Bernard Kruskal (1928–2010). In **Kruskal's Algorithm**, the step of adding an edge of minimum possible weight that does not cause the formation of a cycle is repeated until a spanning

---

Algorithm 10.3 Kruskal's Algorithm

Let $G = (V, E)$ be a weighted connected graph on $n > 1$ vertices.

Algorithm.
    Let $F = \{e\}$, where e is a non-loop edge of minimum possible weight.
    While $|F| < n - 1$,
        \begin
        Let e be an edge of minimum possible weight
            among all edges in $E \setminus F$ for which $F \cup \{e\}$
            contains no cycle.
        Let $F = F \cup \{e\}$.
        \end.
    Return $T = (V, F)$.

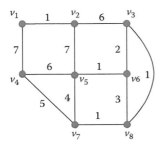

FIGURE 10.22
Possible power grid connections with costs.

tree is built. Note that, at intermediate stages, the edges chosen so far form a forest that may not be a tree.

At any stage of an implementation of Kruskal's Algorithm, if there is a tie among edges that may be chosen next, then any one such choice will suffice. However, if the edge set additionally comes with some ordering, then that ordering can be used to force a unique choice at each stage. For example, it can be helpful to presort the edges into some order of nondecreasing weight. Hence, the algorithms for sorting items in nondecreasing order that will be discussed in Section 10.5 could be employed here. To be consistent in all of the tree algorithms considered in this section, we shall break ties between edges of the same weight by using the lexicographic ordering induced by a given initial ordering of the vertices, as discussed in Remark 10.1.

Example 10.12. The graph in Figure 10.22 represents an apartment complex for which electric power must be supplied. Each vertex represents an apartment building, each edge represents a potential direct power cable between two buildings, and the weight assigned to an edge reflects the cost of establishing that link. Consequently, the cheapest power grid that connects all of the apartment buildings is reflected by a minimum spanning tree.

Use Kruskal's Algorithm to find a minimum spanning tree for the graph in Figure 10.22. List the set of edges F in the spanning tree in the order in which they are obtained by the algorithm, and shade the spanning tree.

Solution. The minimum spanning tree has edge set

$$F = \{\{v_1, v_2\}, \{v_3, v_8\}, \{v_5, v_6\}, \{v_7, v_8\}, \{v_3, v_6\},$$

$$\{v_4, v_7\}, \{v_2, v_3\}\}$$

and is shown in bold in the following right-hand picture.

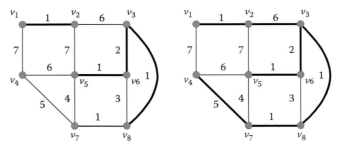

The left-hand picture shows the stage after all of the edges of weight 1 or 2 have been chosen. At that point, the edge of weight 3 and the edge of weight 4 cannot

---

Algorithm 10.4 Prim's Algorithm

Let $G = (V, E)$ be a weighted connected graph on $n > 1$ vertices.

Algorithm.

Let $F = \{e\}$, where e is a non-loop edge of minimum possible weight.

While $|F| < n - 1$,

\begin{verbatim}\begin{verbatim}
\begin{verbatim}
\end{verbatim}

Let e be an edge of minimum possible weight among all edges in $E \setminus F$ that connect an endpoint of an edge from F to a vertex that is not an endpoint of an edge from F.

Let $F = F \cup \{e\}$.

\end.\end.

Return $T = (V, F)$.

---

be chosen, since each, together with what was chosen previously, would complete a cycle. Consequently, the edge of weight 5 is chosen next. After that, note that although there are two edges of weight 6, only one of them is a possible addition to our tree.

The weight of our spanning tree is 17 and reflects the cheapest possible cost of a power grid for the apartment complex. ☐

The reader can now verify that the bold portion of the graph in Figure 10.20 is the minimum spanning tree produced by Kruskal's Algorithm.

Our second spanning tree algorithm is presented in Algorithm 10.4; it is named after the American mathematician Robert Clay Prim (1921–). In **Prim's Algorithm**, at each step, we add an edge of minimum possible weight. However, unlike in Kruskal's Algorithm, here the stronger requirement that the set of chosen edges always forms a tree further restricts our possible choices. The process continues until this growing tree forms a spanning tree.

Example 10.13. The graph in Figure 10.23 represents the same apartment complex considered in Example 10.12, but now the weighted edges reflect potential gas line connections and their costs. So here a minimum spanning tree reflects the cheapest possible system of gas lines joining the apartment buildings.

Use Prim's Algorithm to find a minimum spanning tree for the graph in Figure 10.23. List the set of edges F in the spanning tree in the order in which they are

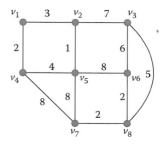

FIGURE 10.23
Possible gas line connections with costs.

obtained by the algorithm, and shade the minimum spanning tree.

Solution. The spanning tree has edge set

$$F = \{\{v_2, v_5\}, \{v_1, v_2\}, \{v_1, v_4\}, \{v_2, v_3\}, \{v_3, v_8\},$$

$$\{v_6, v_8\}, \{v_7, v_8\}\}$$

and is shown in bold in the right-hand picture below.

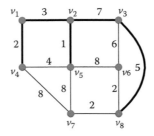 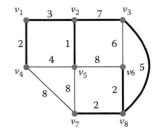

The left-hand picture shows the stage after edges of weights $1, 3, 2, 7, 5$ are chosen (in that order). It is at this point, after some higher-weight edges have been chosen, that the remaining two edges of weight 2 are added.

The weight of our spanning tree is 22 and reflects the cheapest possible cost for installing gas lines for the apartment complex. □

The reader can verify that the minimum spanning tree in Figure 10.20 is also that which is produced by Prim's Algorithm. The proof that the results from Kruskal's Algorithm and Prim's Algorithm generally agree is left for the exercises. Here, we now prove the correctness of both algorithms.

Theorem 10.8. *Let G be any weighted connected graph. Each of Kruskal's Algorithm and Prim's Algorithm yields a minimum spanning tree for G.*

Since the proofs for both algorithms are essentially the same, they are presented in parallel. However, it may be best to focus on just one algorithm for a first reading and on the other for a second.

Proof of Theorem 10.8. Let T be the spanning tree for G produced by either Kruskal's Algorithm or Prim's Algorithm. Select a minimum spanning tree T' for G satisfying certain conditions. In the case of Kruskal's Algorithm, choose T'

with the maximum possible number of edges in common with T. In the case of Prim's Algorithm, choose T' containing the longest possible initial string $S : e_1, e_2, \ldots, e_k$ of edges chosen by the algorithm. Suppose toward a contradiction that $T' \neq T$.

Let e be the first edge chosen for T that is not in T'. In the case of Prim's Algorithm, note that $e = e_{k+1}$. In any case, $T' \cup \{e\}$ contains a cycle C. Select an edge e' from C satisfying certain conditions. In the case of Kruskal's Algorithm, choose e' not in T. In the case of Prim's Algorithm, choose e' not in S but with exactly one endpoint in S. In any case, e' is in T', and $T' \setminus \{e'\} \cup \{e\}$ is a spanning tree for G. By comparing the weights of e and e', we obtain our desired contradictions.

If $\omega(e') < \omega(e)$, then e' would have been chosen for T instead of e. If $\omega(e') > \omega(e)$, then $T' \setminus \{e'\} \cup \{e\}$ has smaller weight than T'. Thus, it suffices to assume that $\omega(e') = \omega(e)$. In the case of Kruskal's Algorithm, $T' \setminus \{e'\} \cup \{e\}$ has more edges in common with T than does T'. In the case of Prim's Algorithm, $T' \setminus \{e'\} \cup \{e\}$ has a longer initial string of edges chosen by the algorithm.   □

10.3.2 Shortest path trees

The cities in Table 10.1 are the office locations of a large corporation based in New York. The costs therein are the airfares for the direct flights available between pairs of these cities. In Figure 10.24, we have represented the information from Table 10.1 in a weighted graph. Its vertices represent the office locations. Each edge and its weight represent the existence and cost, respectively, of a direct flight between a pair of cities. A corporate leader in New York might like to know, for each of the corporation's office locations, the cheapest way to get there. In this case, it is not a minimum spanning tree for our weighted graph that is desired, but something similar.

TABLE 10.1
Airfares between major cities

	Chicago	Honolulu	Houston	Los Angeles	London	New York	Rome
Chicago	—	$400	$200	$200	$400	$100	—
Honolulu	$400	—	—	$300	—	—	—
Houston	$200	—	—	$100	—	$400	—
Los Angeles	$200	$300	$100	—	—	$400	—
London	$400	—	—	—	—	$300	$100
New York	$100	—	$400	$400	$300	—	$500
Rome	—	—	—	—	$100	$500	—

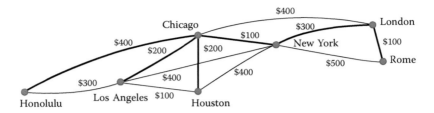

FIGURE 10.24
Travel costs among cities.

Definition 10.6. Let $G = (V, E)$ be a weighted graph.

(a) The **weighted distance** between two vertices u and v in G, denoted $\text{Dist}_G(u, v)$, is the minimum weight of a path in G from u to v. If there is no path in G from u to v, then we assign $\text{Dist}_G(u, v) = \infty$. When G is clear in context, the subscripts may be dropped. Note that capital D is used here in the notation for the weighted distance where lower-case d was used for ordinary distance in Definition 8.4(f).

(b) Suppose that a vertex v is specified in G. A **shortest path tree** for G from v is a spanning tree T such that, for each vertex w in G, the path in T from v to w has the minimum weight among all paths in G from v to w.

A shortest path tree need not be a minimum spanning tree. An example of a graph in which they are different is requested in the exercises.

Example 10.14. For the weighted graph in Figure 10.24, the edges shown in bold form a shortest path tree from New York. Consequently, for each office location, the cheapest way to fly from New York to that location follows the path shown in bold. For example, it is cheaper to fly from New York to Chicago to Los Angeles than it is to fly direct from New York to Los Angeles.

Having seen the utility of a shortest path tree, we now consider an algorithm for finding one that was discovered by the Dutch computer scientist Edsger W. Dijkstra (1930–2002). **Dijkstra's Algorithm** is presented in Algorithm 10.5, and the proof of its correctness is given at the end of this section.

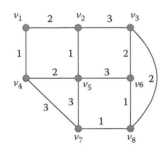

FIGURE 10.25
Possible water line connections with pressure losses.

Example 10.15. The graph in Figure 10.25 again represents the apartment complex considered in Example 10.12, but now the edges reflect potential water lines,

Algorithm 10.5 Dijkstra's Algorithm

Let G be a weighted connected graph on $n > 1$ vertices and v a vertex of G.

Algorithm.

Let $F = \{e\}$, where e is a non-loop edge incident with v
of minimum possible weight.
While $|F| < n - 1$,
\begin
Let e be an edge such that
one endpoint t is an endpoint of an edge from F, the other
endpoint is not, and $\text{Dist}_T(v, t) + \omega(e)$ is as small as possible.
Let $F = F \cup \{e\}$.
\end.
Return $T = (V, F)$.

and the weights reflect the relative difficulty of pumping water between buildings. A large pumping station has been installed beneath the apartment building represented by vertex v_1. Hence, water must be pumped from there to the other buildings. Consequently, the system that accomplishes this with the lowest loss of water pressure for each building is reflected by a shortest path tree.

For the graph in Figure 10.25, use Dijkstra's Algorithm to find a shortest path tree starting from vertex v_1. List the set of edges F and the vertices W in the shortest path tree in the order in which they are added in the algorithm, and shade the shortest path tree.

Solution. The shortest path tree has edge set

$$F = \{\{v_1, v_4\}, \{v_1, v_2\}, \{v_2, v_5\}, \{v_4, v_7\}, \{v_7, v_8\},$$

$$\{v_2, v_3\}, \{v_6, v_8\}\},$$

and is shown in bold in the right-hand picture below. The vertices are encountered in the order $v_1, v_4, v_2, v_5, v_7, v_8, v_3, v_6$.

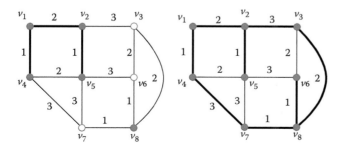

The left-hand picture shows the stage after the first three edges have been chosen. The three vertices pictured as open circles ○ are then the candidates for a vertex to be added along one of the edges of weight 3. Since vertex v_7, via edge $\{v_4, v_7\}$ together with the bold path from v_4 to v_1, is the closest to v_1, it is added next with edge $\{v_4, v_7\}$.

By using the shortest path tree to determine which water lines to establish, we can ensure that each apartment building will have the highest possible water pressure it can receive from the pumping station provided.   □

The reader can confirm that the bold portion of the graph in Figure 10.24 is the shortest path tree from New York produced by Dijkstra's Algorithm. We now establish that Dijkstra's Algorithm is always successful.

Theorem 10.9.   *Let G be a weighted connected graph and v a vertex of G. Dijkstra's Algorithm yields a shortest path tree for G from v.*

Proof. Let W be the set of ends of edges in F as T is growing. We prove inductively that, at each stage of the algorithm, the paths in $T = (W, F)$ from v to the vertices in W are of minimum weight in G. Obviously, this holds for $W = \{v\}$. So consider the case in which the algorithm adds a new edge $e \mapsto \{t, u\}$ to T, where $t \in W$ and $u \in V \setminus W$.

Suppose toward a contradiction that the distance in G from v to u is smaller than $\mathrm{Dist}_T(v, t) + \omega(e)$. Let P be a minimum-weight path in G from v to u with the greatest possible number of edges in common with T. Consequently, there must be an edge e' that is in P but not in T such that

(i) $e' \mapsto \{t', u'\}$,

(ii) P and T agree from v to t' but not afterward, and

(iii) $u' \notin W$.

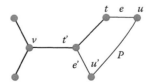

Since the algorithm chooses e, we have $\mathrm{Dist}_T(v, t) + \omega(e) \leq \mathrm{Dist}_T(v, t') + \omega(e')$. That is, $\mathrm{Dist}_T(v, t') + \mathrm{Dist}_T(t', t) + \omega(e) \leq \mathrm{Dist}_T(v, t') + \omega(e')$. So

$$\mathrm{Dist}_T(t', t) + \omega(e) - \omega(e') \leq 0. \tag{10.3}$$

Since P has smaller weight than $\text{Dist}_T(v, t) + \omega(e)$, we have

$$\text{Dist}_T(v, t') + \omega(e') + \text{Dist}_P(u', u)$$
$$< \text{Dist}_T(v, t') + \text{Dist}_T(t', t) + \omega(e).$$

Hence,

$$\text{Dist}_P(u', u) < \text{Dist}_T(t', t) + \omega(e) - \omega(e'). \qquad (10.4)$$

Since $\text{Dist}_P(u', u) \geq 0$, inequalities (10.3) and (10.4) contradict each other. □

Exercises

The following graphs are referred to throughout the exercises.

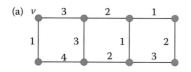

(a)

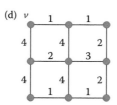

(b)

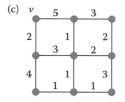

(c)

(d)

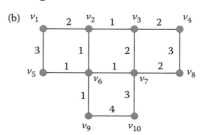

(e)

(f)

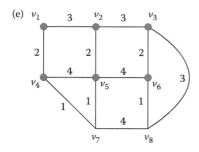

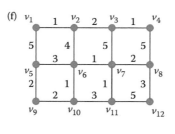

(g)

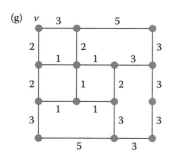

(h)

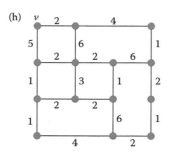

In Exercises 1 through 8, we need to establish a network of gas lines in a condominium development. The vertices in the specified graph represent the condominiums in the development; each edge represents a possible gas line connection, and its weight gives the cost of making that connection in tens of thousands of dollars. Consequently, a cheapest possible network of gas lines corresponds to a minimum spanning tree.

(a) Use Kruskal's Algorithm to find and shade a minimum spanning tree for the specified graph.

(b) What is the cheapest possible cost for a network of gas lines for the condominium development?

1. Graph (a).   2. Graph (b).

3. Graph (c).   4. Graph (d).

5. Graph (e).   6. Graph (f).

7. Graph (g).   8. Graph (h).

In Exercises 9 through 16, we need to establish a power grid in an apartment complex. The vertices in the specified graph represent the apartments in the complex; each edge represents a possible power line connection, and its weight gives the cost of establishing that line in tens of thousands of dollars. Consequently, a cheapest possible power grid corresponds to a minimum spanning tree.

(a) Use Prim's Algorithm to find and shade a minimum spanning tree for the specified graph. Also, list the weights of the edges in the order that they are added in the algorithm.

(b) What is the cheapest possible cost for a power grid for the apartment complex?

9. Graph (b).   10. Graph (a).

11. Graph (d).   12. Graph (c).

13. Graph (f).   14. Graph (e).

15. Graph (h).   16. Graph (g).

17. Does a minimum spanning tree always contain an edge of smallest weight? Justify your answer.

18. Let G be a weighted graph with all weights distinct. Suppose C is a cycle in G. Is it possible that the largest-weight edge of C is in a minimum spanning tree for G? Justify your answer.

19.* Suppose G is a weighted graph with all weights distinct. Prove or disprove that G must have a unique minimum spanning tree.

20. Let G be a weighted graph, and form a weighted graph G' by either adding some fixed constant to all of the weights of G or multiplying all of the weights of G by some fixed positive constant. Show that, as a subgraph, a tree is a minimum spanning tree for G if and only if it is for G'.

21. A college wishes to set up a closed circuit television network on its campus. The following table gives the costs determined by a contractor to connect pairs of buildings on the campus.

	Bartle	Evans	Guild	Miles	Pine	Taft	Waller
Bartle	—	$5000	$3000	—	—	—	—
Evans	$5000	—	$5000	$2000	—	—	—
Guild	$3000	$5000	—	—	$2000	$9000	—
Miles	—	$2000	—	—	$4000	—	$3500
Pine	—	—	$2000	$4000	—	$6000	$4500
Taft	—	—	$9000	—	$6000	—	—
Waller	—	—	—	$3500	$4500	—	—

Find the cheapest cost of establishing a TV network that services all of the buildings.

22. A new summer camp needs to provide electricity to each of its buildings. The following table gives the costs of connecting pairs of buildings in this summer camp.

	Assembly	Cafeteria	Councilors	Dorm 1	Dorm 2	Dorm 3
Assembly	—	$4000	$5000	—	—	$5000
Cafeteria	$4000	—	$3000	$3000	—	—
Councilors	$5000	$3000	—	—	$4000	—
Dorm 1	—	$3000	—	—	$2000	$3000
Dorm 2	—	—	$4000	$2000	—	$2000
Dorm 3	$5000	—	—	$3000	$2000	—

Find the cheapest cost of providing power to all of the buildings in the camp.

23. Prove: If the edges of a weighted connected graph are ordered and, when there is a tie, the lower-numbered edge is chosen, then both Kruskal's Algorithm and Prim's Algorithm give the same tree.

24.* Prove: Given any minimum spanning tree T for a connected weighted graph $G = (V, E)$, there is some ordering for E such that Prim's Algorithm will produce T.

In Exercises 25 through 32, we need to establish a network of water lines in a new industrial park. The vertices in the specified graph represent the business sites in the park; each edge represents a possible water line connection, and its weight gives the loss of water pressure (in pounds per square inch) when water is pumped between the sites at its endpoints. The specified vertex represents the site on which a water tower has been constructed to serve the park. Consequently, a shortest path tree from the specified water tower site reflects a network that provides the minimum possible water pressure loss for each business site.

(a) Use Dijkstra's Algorithm to find and shade a shortest path tree for the specified graph and vertex.

(b) What is the largest water pressure loss that must be experienced by a business site in the industrial park?

25. Graph (a) from v.

26. Graph (b) from v_1.

27. Graph (c) from v.

28. Graph (d) from v.

29. Graph (e) from v_1.

30. Graph (f) from v_1.

31. Graph (g) from v.

32. Graph (h) from v.

33.* Prove or disprove: If the edges in any graph are all given weight 1, then Dijkstra's Algorithm always yields a depth-first search tree.

34.* Prove or disprove: If the edges in any graph are all given weight 1, then Dijkstra's Algorithm always yields a breadth-first search tree.

35. Give an example of a weighted graph G and a selected vertex v such that a shortest path tree starting at v is not a minimum spanning tree.

36.* Can Dijkstra's Algorithm be altered to find a longest distance tree? Explain.

37. In a graph with no loops, does Dijkstra's Algorithm always select an edge of smallest weight? Justify your answer.

38. Does Dijkstra's Algorithm ever select an edge of largest weight? Justify your answer.

39. Let G be a weighted graph, and form a weighted graph G' by multiplying all of the weights of G by some fixed positive constant. Let v be any vertex. Prove or disprove that, as a subgraph, a tree is a shortest path tree for G if and only if it is for G'.

40. Let G be a weighted graph, and form a weighted graph G' by adding some fixed constant to all of the weights of G by some fixed positive constant. Let v be any vertex. Prove or disprove that, as a subgraph, a tree is a shortest path tree for G if and only if it is for G'.

41. A fire station serving a remote town needs to minimize its response time to each home in the town. Since many of the homes are on dirt roads of low quality, the fire chief wants to plan in advance the routes that will be used to respond to certain key locations in the town. The chief has driven the fire truck on each road, measured the number of minutes it takes to travel between the key locations, and recorded those times in the following table.

	Station	Adams	Johnson	Kennedy	Lincoln	Nixon	Polk
Station	—	—	8	5	7	6	—
Adams	—	—	6	7	—	—	—
Johnson	8	6	—	2	—	—	—
Kennedy	5	7	2	—	10	—	—
Lincoln	7	—	—	10	—	6	—
Nixon	6	—	—	—	6	—	12
Polk	—	—	—	—	—	12	—

Find the minimum possible response time from the fire station to each of the key locations in the town.

42. A campus pizza shop claims the fastest delivery time to each dorm in the college. To minimize delivery times, it plans in advance the shortest route to each dorm. The following table lists the travel times in minutes between campus locations.

	Shop	Columbia	Elwell	Lycoming	Smith	Wells
Shop	—	—	—	3	4	—
Columbia	—	—	3	—	—	4
Elwell	—	3	—	5	—	1
Lycoming	3	—	5	—	3	—
Smith	4	—	—	3	—	2
Wells	—	4	1	—	2	—

Find the minimum possible delivery time from the pizza shop to each of the dorms on campus.

43. Give an example of a graph G and two vertices v_1 and v_2 such that two corresponding shortest path trees are distinct, as subgraphs.

44. Given a graph G and a vertex v, is a shortest path tree always unique? What if the weights are distinct? Justify your answers.

45.* Here is a potential algorithm for producing a minimum spanning tree for a given connected graph $G = (V, E)$.

Let $F = E$.
While the subgraph (V, F) contains a cycle,
    \begin
    Let e be an edge of maximum weight in a cycle.
    Remove e from F.
    \end.
Return $T = (V, F)$.

Will T always be a minimum spanning tree? Justify your answer.

46.* Here is a potential algorithm for producing a shortest path tree for a given connected graph $G = (V, E)$.

Let $F = \emptyset$.
Let v be a vertex of G.
Let $S = \{v\}$.
While there is a vertex u in $V \setminus S$,
    \begin
    Find a shortest path P from v to u.
    Add the edges of P into F.
    Add the vertices of P into S.
    \end.
Return $T = (V, F)$.

Will T always be a shortest path tree? Justify your answer.

47. An army commander in one unit needs to send messages to the units in the field by messenger, since electronic communications are insecure. If he knows the travel times between each pair of units and wants to determine the fastest possible routes by which to send messages, then what kind of tree is relevant, a minimum spanning tree or a shortest path tree? Explain.

48. A farmer has a power source in his barn and needs to supply power to several locations on his farm. He has determined the amount of cable needed to join each pair of locations and wants to use the least possible amount of cable to provide all locations with power. What kind of tree is relevant, a minimum spanning tree or a shortest path tree? Explain.

10.4 Analysis of algorithms (Part 1)

In the previous two sections, various tree-finding algorithms were considered. To fully appreciate the value of a particular algorithm, one must understand its efficiency, especially in comparison with other algorithms that perform the same task. Since a detailed study of the efficiency of tree-finding algorithms is more involved than we shall get in this book, the reader interested in such pursuits is referred to Cormen et al. (2009) or Sedgewick and Wayne (2011).

Algorithm 10.6 Sequential Search

Let x be a real number whose index is sought in an array A of length n. The index of x is returned in a variable called **location**. If x is not in A, then the algorithm returns location $= 0$.

Algorithm.
> Let location $= 0$.
> Let $i = 1$.
> While location $= 0$ and $i \leq n$,
> > `\begin`
> > If $A[i] = x$, then
> > > Let location $= i$.
> >
> > Otherwise,
> > > Let $i = i + 1$.
> >
> > `\end`.
>
> Return location.

Instead, our goal is a basic introduction to analyzing the efficiency of algorithms. Consequently, we pursue some fundamental examples that provide a good setting for this. In this section, we initiate our study by focusing on algorithms that search for a value in an array. In the next section, we consider algorithms that sort an array.

10.4.1   Search algorithms

A common programming need is to search for an item in a collection of data, such as a database. Perhaps a suspect with a certain DNA signature is sought, or a store wants to know whether a particular item is still in stock. To simplify our discussion, we assume that the data being searched have been stored as numbers in an array.

The first of the two algorithms we consider takes a natural approach. It runs through the array in its given order and checks each entry as it goes. It is called **Sequential Search** and is presented in Algorithm 10.6.

Example 10.16. Given $x = 4$, $A = \boxed{\begin{array}{c|c|c} 6 & 1 & 4 \end{array}}$ $\boxed{\begin{array}{c|c} 7 & 3 \end{array}}$, and $n = 5$, Sequential Search runs through the values listed in the following table

i	Location
1	0
2	0
3	3

and returns location $= 3$. That is, $4 = A[3]$.

Algorithm 10.7 Binary Search

Let x be a real number whose index is sought in an *ordered* array A of length n. The index of x is returned in a variable called **location**. If x is not in A, then the algorithm returns location $= 0$. The variables **low** and **high** store the indices of the first and last entries, respectively, of the portion of A we are considering.

Algorithm.
    Let location $= 0$.
    Let low $= 1$.
    Let high $= n$.
    While low $<$ high,
        \begin
        Let mid $= \lfloor \frac{\text{low}+\text{high}}{2} \rfloor$.
        If $A[\text{mid}] < x$, then
            Let low $=$ mid $+ 1$.
        Otherwise,
            Let high $=$ mid.
        \end.
    If $A[\text{low}] = x$, then
        Let location $=$ low.
    Return location.

Our second search algorithm takes a more sophisticated approach requiring that the entries in the array be given in nondecreasing order. (Sorting algorithms are examined in Section 10.5.) Very simply, we repeat the process of cutting a portion of the array into two portions and continuing our search only on the portion that could possibly contain the desired value. This algorithm is called **Binary Search** and is presented in Algorithm 10.7.

Example 10.17. Given $x = 3$, $A =$

0	2	2

3	4	6	6	7	8	8

, and $n = 10$, Binary Search runs through the following stages and returns location $= 4$. That is, $3 = A[4]$. In our display we have used l, m, and h for the values of the variables low, mid, and high, respectively. In each row, we also display only the portion of the array, from index low to index high, that is under current consideration.

10.4.2   Complexity of algorithms

Given an ordered array of size n, both Sequential Search and Binary Search could be used to determine the location of a particular value. Although Binary Search may seem more complicated to implement, it is the more efficient of the two. To make this assertion precise, we need to specify our measure of efficiency.

In this book, we consider only the **time complexity** of an algorithm. That is, we are interested in the length of time it takes for an algorithm to run. One might also consider the notion of **space complexity** and be concerned with the amount of memory used by a program implementing the algorithm. However, we are concerned with time only.

Since run-time varies from machine to machine, we in fact focus on the number of operations required in a running of the algorithm. Consequently, within a particular context, exactly what operations are counted must be specified. Among the potential operations to be considered are arithmetic operations, such as addition ($+$) and multiplication ($\cdot$); comparisons, such as $<$, $>$, and $=$; and more.

Definition 10.7. The **worst-case complexity** of an algorithm is a function $f(n)$ of the size n of the input to the algorithm. For each n, the value of $f(n)$ is the maximum number of operations performed in a run of the algorithm on an input of size n.

For search algorithms, we take the size of the input to be the length n of the array being searched. Also, the only operations we consider are comparisons of an array entry with some other value (possibly another array entry). This is reasonable since it can be shown that the time a search algorithm takes to run is indeed proportional to the number of comparisons.

Another type of complexity that can be considered for an algorithm is **average-case complexity**. It is defined, for each n, to be the average number of operations performed, considering all possible inputs of size n. This measure can give a better sense of the expected run-time for a randomly chosen input. However, it is generally complicated to compute, and we therefore focus our attention on worst-case analysis. Consequently, we get a sense of the longest possible run-time.

Example 10.18.   The worst-case complexity of Sequential Search is n.

Proof. The comparison $A[i] = x$ is performed once each time through the while loop. If the desired value x is in location k, then the number of comparisons performed is

$$\underbrace{1 + 1 + \cdot + 1}_{k \text{ times}} = k \le n.$$

The worst case is, of course, when $k = n$. All n comparisons are also done when x is not in A.   □

Example 10.19.   The worst-case complexity of Binary Search is $1 + \lceil \log_2 n \rceil$.

Proof. For each $n \ge 1$, let c_n be the maximum number of comparisons needed in a Binary Search of an array of length n (counting the last comparison $A[low] = x$). So $c_1 = 1$. In the while loop, in the worst case, we move from considering a portion of A of length $k = high - low + 1$ to a portion of length $\lceil \frac{k}{2} \rceil$. So $\forall \, n \ge 2$, $c_n = 1 + c_{\lceil \frac{n}{2} \rceil}$. From this, we shall prove by strong induction that $\forall \, n \ge 1$, $c_n = 1 + \lceil \log_2 n \rceil$.

Since the base case is easy, we move immediately to the inductive step. Suppose $k \ge 1$ and that $c_i = 1 + \lceil \log_2 i \rceil$, for all $1 \le i \le k$. Observe that

$$c_{k+1} = 1 + c_{\lceil \frac{k+1}{2} \rceil}$$

$$= 1 + 1 + \lceil \log_2 \lceil \frac{k+1}{2} \rceil \rceil$$

$$= \begin{cases} 2 + \lceil \log_2 \left(\frac{k+2}{2} \right) \rceil = 1 + \lceil \log_2(k+2) \rceil & \text{if } k \text{ is even,} \\ 2 + \lceil \log_2 \left(\frac{k+1}{2} \right) \rceil = 1 + \lceil \log_2(k+1) \rceil & \text{if } k \text{ is odd,} \end{cases}$$

$$= 1 + \lceil \log_2(k+1) \rceil.$$

The last equality follows from Exercise 13.   □

Let us compare the relative efficiencies of Sequential Search and Binary Search by comparing their worst-case complexities, n and $1 + \lceil \log_2 n \rceil$, respectively. We can see in a table of values

n	$1 + \lceil \log_2 n \rceil$
2	2
10	5
100	8
10^6	21

or in the graph in Figure 10.26 (in which we dropped the ceiling function to simplify the picture) that eventually n gets large more quickly than does $1 + \lceil \log_2 n \rceil$. This leads us to conclude that Binary Search is more efficient than Sequential Search. However, we first need to clarify the notion of one function getting large more quickly than another.

10.4.3   Growth of functions

The graph in Figure 10.26 suggests that the function $y = x$ grows faster than the function $y = 1 + \log_2(x)$. This statement is made precise by introducing **big-O notation**.

Definition 10.8.   Given a real function $g(x)$, **big-O of $g(x)$**, denoted $O(g(x))$, is the set of real functions $f(x)$ such that there exist positive constants C and d for which

$$\forall\, x > d, \ \ |f(x)| \leq C|g(x)|.$$

That the inequality $|f(x)| \leq C|g(x)|$ holds for all x greater than some fixed constant d can be described by saying that it holds **for large** x, or **eventually**.

Figure 10.26 displays the fact that $1 + \log_2(x) \in O(x)$ (use $C = 1$ and $d = 2$). In general, the assertion $f(x) \in O(g(x))$ is the precise meaning of the statement that $f(x)$ grows no more quickly than does $g(x)$ or that $g(x)$ grows at least as quickly as does $f(x)$. Equivalently, we say that the growth rate of $f(x)$ is no faster than the growth rate of $g(x)$ or that the growth rate of $g(x)$ is no slower than the growth rate of $f(x)$.

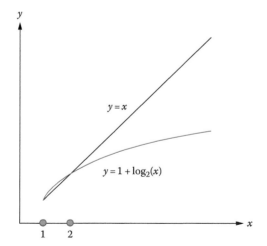

FIGURE 10.26
Comparing the growth of x and $1 + \log_2(x)$.

Since $O(g(x))$ is a set, the subset relation $\subseteq$ provides another convenient way of comparing the growth rates of two functions. This is reflected in the following lemma, whose proof is left for the exercises.

Lemma 10.10.   Let f and g be real functions. Then

$$f(x) \in O(g(x)) \quad \text{if and only if} \quad O(f(x)) \subseteq O(g(x)).$$

As should be expected, a higher power of x grows at least as fast as a lower power of x.

Lemma 10.11.   Let $r_1, r_2 \in \mathbb{Q}$ with $r_1 \leq r_2$. Then

$$\forall\, x > 1, \ x^{r_1} \leq x^{r_2}.$$

Proof. Write $r_1 = \frac{m_1}{b}$ and $r_2 = \frac{m_2}{b}$ for some $m_1, m_2, b \in \mathbb{Z}$ with $b > 0$. For any fixed $x \geq 1$, it is straightforward to prove inductively that $\forall\, n \in \mathbb{N}, x^n \geq 1$. Consequently, $x^{m_2 - m_1} \geq 1$, and we can multiply both sides by x^{m_1} to obtain $x^{m_1} \leq x^{m_2}$. Taking the bth root gives $x^{r_1} = x^{\frac{m_1}{b}} \leq x^{\frac{m_2}{b}} = x^{r_2}$. $\square$

If $r_1, r_2 \in \mathbb{Q}$ and $r_1 \leq r_2$, then it follows from Lemma 10.11 and Definition 10.8 (using $C = 1$ and $d = 1$) that $x^{r_1} \in O(x^{r_2})$. That result, which we may also express by writing $O(x^{r_1}) \subseteq O(x^{r_2})$, is reflected in the graphs in Figure 10.27. Further, Lemma 10.11 can be used to prove a general result about polynomial growth.

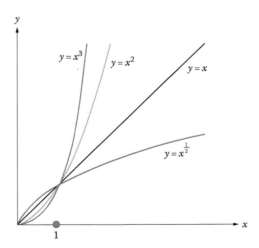

FIGURE 10.27
Comparing the growth of powers of x.

Theorem 10.12. *Let $m \in \mathbb{N}$ and let $f(x) = c_m x^m + c_{m-1} x^{m-1} + \cdots + c_0$ be a polynomial of degree at most m. Then $f(x) \in O(x^m)$.*

Proof. For $x > 1$,

$$|f(x)| \leq |c_m| x^m + |c_{m-1}| x^{m-1} + \cdots + |c_0|$$
$$\leq |c_m| x^m + |c_{m-1}| x^m + \cdots + |c_0| x^m$$
$$= (|c_m| + |c_{m-1}| + \cdots + |c_0|) x^m.$$

The first inequality follows from the Triangle Inequality (Exercise 39 in Section 2.5), and the second follows from Lemma 10.11. Let $d = 1$ and $C = |c_m| + |c_{m-1}| + \cdots + |c_0|$, and observe that $\forall\, x > d$, $|f(x)| \leq C|x^m|$. $\qquad\square$

Just as big-O notation provides a way to compare the relative growth rates of functions, **big-Θ notation** provides a way to say that two functions have the same growth rate.

Definition 10.9. Given a real function $g(x)$, **big-Θ of** $g(x)$, denoted $\Theta(g(x))$, is the set of real functions $f(x)$ such that $f(x) \in O(g(x))$ and $g(x) \in O(f(x))$. Equivalently, $f(x) \in \Theta(g(x))$ if and only if there exist positive constants C_1, C_2, and d for which

$$\forall\, x > d, \quad C_1|g(x)| \leq |f(x)| \leq C_2|g(x)|.$$

In this case, we say that $f(x)$ has the same **order** (or **order of growth**) as $g(x)$.

A proof of the following lemma is left for the exercises.

Lemma 10.13. Let f and g be real functions. Then

$$f(x) \in \Theta(g(x)) \quad \text{if and only if} \quad \Theta(f(x)) = \Theta(g(x))$$
$$\text{if and only if} \quad O(f(x)) = O(g(x)).$$

Example 10.20. $3x^2 + 5x + 2 \in \Theta(x^2)$.
This is justified by establishing two facts.

(i) $3x^2 + 5x + 2 \in O(x^2)$.
This follows from Theorem 10.12, but it can also be seen directly, since
$\forall\, x > 1$, $|3x^2 + 5x + 2| \leq 10|x^2|$.

(ii) $x^2 \in O(3x^2 + 5x + 2)$.
This holds since $\forall\, x > 0$, $|x^2| \leq |3x^2 + 5x + 2|$.

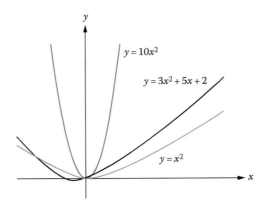

FIGURE 10.28
Comparing the growth of x^2 and $3x^2 + 5x + 2$.

The result in Example 10.20 is reflected in Figure 10.28. That x^2 and $3x^2 + 5x + 2$ have the same order of growth can also be expressed by writing $O(x^2) = O(3x^2 + 5x + 2)$. In general, polynomials of the same degree have the same order of growth. In fact, it is the allowance of the constant C in the inequality $|f(x)| \leq C|g(x)|$ characterizing $f(x) \in O(g(x))$ and the absence of requiring $C = 1$ that allows us to equate the growth rates of polynomials of the same degree. Indeed, it is the degree that most significantly distinguishes the growth rates in Figure 10.27.

We are interested in instances when one function $f(x)$ has a strictly slower rate of growth than another function $g(x)$. For this, we want precisely that $f(x) \in O(g(x))$ and $g(x) \notin O(f(x))$. It should not be surprising that this comparison, which is similar to strict inequality $<$ on rates of function growth, is reflected by the proper subset relation $\subset$. A proof of the following lemma is left for the exercises.

Lemma 10.14. Let f and g be real functions. Then

$$f(x) \in O(g(x)) \text{ and } g(x) \notin O(f(x))$$

$$\text{if and only if} \quad O(f(x)) \subset O(g(x)).$$

The following lemma, together with Lemma 10.11, allows us to conclude that strictly bigger powers of x have strictly faster growth.

Lemma 10.15. Let $r_1, r_2 \in \mathbb{Q}$ with $r_1 < r_2$. Then $x^{r_2} \notin O(x^{r_1})$.

Proof. Suppose to the contrary that there are positive constants C and d such that $\forall\, x > d$, $|x^{r_2}| \leq C|x^{r_1}|$. Since this would remain true for any bigger value of C, we may assume that $C > \frac{d^{r_2-r_1}}{2}$. Note then that $(2C)^{\frac{1}{r_2-r_1}} > d$. For $x > d$, we have $\frac{x^{r_2}}{x^{r_1}} \le C$ and hence $x^{r_2-r_1} \le C$. Plugging in $x = (2C)^{\frac{1}{r_2-r_1}}$ then gives $2C \le C$, a contradiction. $\square$

10.4.4 Order of algorithms

Since our interest is in comparing worst-case complexities of algorithms, we restrict our attention to functions of $\mathbb{Z}^+$ (or some subset thereof).

Definition 10.10. Let $g(n)$ be a function. An algorithm is said to be $O(g(n))$ (respectively, $\Theta(g(n))$) if its worst-case complexity $f(n)$ is in $O(g(n))$ (respectively, $\Theta(g(n))$).

Example 10.18 tells us that Sequential Search is $O(n)$ (moreover, $\Theta(n)$), and Example 10.19 tells us that Binary Search is $O(1 + \lceil \log_2 n \rceil)$ (moreover, $\Theta(1 + \lceil \log_2 n \rceil)$). In fact, since it is established in the exercises that $O(1 + \lceil \log_2 n \rceil) = O(\log_2 n)$, we can say more compactly that Binary Search is $O(\log_2 n)$ (moreover, $\Theta(\log_2 n)$).

In general, there is an important difference between asserting that an algorithm is $O(f(n))$ and asserting that it is $\Theta(f(n))$. An algorithm is $O(f(n))$ if, for large n, no input of size n requires a number of operations more than some fixed multiple of $f(n)$. To further assert that the algorithm is $\Theta(f(n))$, there must be, for large n, some input of size n for which the number of operations is at least some fixed multiple of $f(n)$. For example, it is true that Sequential Search is $O(n^2)$, but it is not true that it is $\Theta(n^2)$.

The following theorem compares the relative growth rates of worst-case complexities of many standard algorithms.

Theorem 10.16. Let $b \in \mathbb{R}$ with $b > 1$ and $r \in \mathbb{Q}$ with $r > 2$. $O(1) \subset O(\log_b n) \subset O(n) \subset O(n \log_b n) \subset O(n^2) \subset O(n^r) \subset O(b^n) \subset O(n!)$.

Portions of the proof of Theorem 10.16 are left for the exercises. Here, we prove a special case of Theorem 10.16.

Example 10.21. Show: $O(n^2) \subset O(3^n)$.

Proof. First, we must show that $n^2 \in O(3^n)$. We do this by proving by induction that $\forall n \ge 1$, $n^2 \le 3^n$. Since $1^2 \le 3^1$ and $2^2 \le 3^2$, suppose that $k \ge 2$ and $k^2 \le 3^k$. It follows that

$$(k+1)^2 = k^2 + 2k + 1 \le k^2 + k^2 + k^2 = 3k^2 \le 3 \cdot 3^k = 3^{k+1}.$$

Hence, $n^2 \in O(3^n)$.

Second, we must show that $3^n \notin O(n^2)$. So suppose to the contrary that there are constants C and d such that $\forall\, n > d$, $3^n \le Cn^2$. Let N be an integer larger than both C and d. So, in particular, $3^N \le CN^2$. However, by Exercise 6 from Section 4.3 and our choice of N, it follows that $3^N \ge N^3 > CN^2$. This is a contradiction. Hence, $3^n \notin O(n^2)$.   □

An algorithm that is $\Theta(\log_b n)$ for some $b > 1$ is said to be a **logarithmic**. Those that are $\Theta(n^m)$ for some $m \in \mathbb{N}$ are said to be **polynomial**, and those that are $\Theta(b^n)$ for some $b > 1$ are called **exponential**. Problems that can be solved by polynomial algorithms are said to be in class **P**. Problems for which a proposed solution can be checked for correctness with a polynomial algorithm are said to be in class **NP**, the class of non-deterministic polynomial algorithms. Although it is known that $P \subseteq NP$, it is a famous outstanding problem in computer science to determine whether or not $P = NP$. Moreover, there are classes of problems called **NP-complete** problems that are not known to be in P, but if any one problem from such a class can be shown to be in P, then all problems in that class will be known to be in P. The traveling salesman problem, as described at the end of Section 10.2, is perhaps the most well-known example of an NP-complete problem. The reader is referred to Cormen et al. (2009) or Sedgewick and Wayne (2011) for further discussion of this.

To get a sense of the relative differences in run-times associated with the various complexities from Theorem 10.16, suppose a computer has a speed of 4 gigahertz (4 GHz). This means in theory that such a machine can perform approximately 4 billion instructions per second. If we assume that each instruction corresponds to the operations being counted in the complexity of an algorithm, then Table 10.2 reflects the relative run-times associated with the various complexities from Theorem 10.16 in the cases of inputs of size $n = 100$ and $n = 100{,}000$. In particular, we see the great advantage of a polynomial algorithm over an exponential one.

Exercises

1. How many comparisons are done by Sequential Search in seeking the value $x = 1$ in the array

$A =$

| 5 | 2 | 1 | 6 | 4 | ?

2. How many comparisons are done by Sequential Search in seeking the value $x = 2$ in the array

$A =$

| 9 | 3 | 6 | 1 | 4 | 7 | ?

TABLE 10.2
Run-times for some complexities on a 4 GHz machine

Number of instructions	Run-time for $n = 100$	Run-time for $n = 100{,}000$
$\log_2 n$	0.0000000017 seconds	0.0000000042 seconds
n	0.000000025 seconds	0.000025 seconds
$n \log_2 n$	0.00000017 seconds	0.00042 seconds
n^2	0.0000025 seconds	2.5 seconds
n^3	0.00025 seconds	25,0000 seconds $\approx$ 2.9 days
2^n	3.2×10^{20} seconds	$2.5 \times 10^{30{,}093}$ seconds
	$\approx 1.0 \times 10^{13}$ years	$\approx 7.9 \times 10^{30{,}085}$ years
$n!$	2.3×10^{148} seconds	$7.1 \times 10^{456{,}563}$ seconds
	$\approx 7.4 \times 10^{140}$ years	$\approx 2.2 \times 10^{456{,}556}$ years

In Exercises 3 through 6, show the main states of a Binary Search for the specified value x in the given array A.

3. $x = 5$ and $A = \boxed{\begin{array}{|c|c|c|c|c|c|} 1 & 3 & 4 & 5 & 7 & 8 \end{array}}$.

4. $x = 4$ and $A = \boxed{\begin{array}{|c|c|c|c|c|} 0 & 1 & 4 & 6 & 7 \end{array}}$ $\boxed{\begin{array}{|c|c|} 8 & 9 \end{array}}$.

5. $x = 3$ and $A = \boxed{\begin{array}{|c|c|c|c|c|} 0 & 1 & 2 & 4 & 5 \end{array}}$ $\boxed{\begin{array}{|c|c|} 7 & 8 \end{array}}$.

6. $x = 6$ and $A = \boxed{\begin{array}{|c|c|c|c|c|c|} 1 & 2 & 4 & 5 & 8 & 9 \end{array}}$.

7. If an array A contains the value x in more than one location, then which location is returned by Sequential Search?

8. If an ordered array A contains the value x in more than one location, then which location is returned by Binary Search?

9. Suppose a computer has a speed of 2 GHz. That is, it can perform 2 billion instructions per second.

   (a) Assume that comparisons are the only time-consuming instructions in Binary Search. What is the longest possible time it would take this computer to use Binary Search to find a value in an array of one million entries?

   (b) What is the longest possible time it would take this computer to run an algorithm of worst-case complexity n^2 on an input of size $n = 10^6$?

10. Suppose a computer has a speed of 3 GHz. That is, it can perform 3 billion instructions per second.

   (a) Assume that comparisons are the only time-consuming instructions in Sequential Search. What is the longest possible time it would take this computer to use Sequential Search to find a value in an array of one million entries?

(b) What is the longest possible time it would take this computer to run an algorithm of worst-case complexity $n^{\frac{3}{2}} \lfloor \log_2 n \rfloor$ on an input of size $n = 10^6$?

11. What is the worst-case complexity for Sequential Search restricted to ordered arrays?

12. Suppose we consider only ordered arrays and use the following search algorithm. First compare the desired value with the middle value in the array. Then perform Sequential Search on the left or right half of the array as appropriate. What is the worst-case complexity for this algorithm?

13.* Show that if k is even, then $\lceil \log_2(k+2) \rceil = \lceil \log_2(k+1) \rceil$.

14. Show: $\forall \, x \in \mathbb{R}^+, 1 + \log_2 x \leq 1 + \lceil \log_2 x \rceil < 2 + \log_2 x$.

15. Write an (efficient as possible) algorithm, called **Maximum**, that finds the location of a largest element of a given array A of length n. Only one location is returned, even if the maximum value occurs in multiple locations. Note that the elements of the array need not be sorted in advance.

16. Write an (efficient as possible) algorithm, called **Count**, that counts the number of occurrences of a given value x in a given array A of length n. Note that the elements of the array need not be sorted in advance.

17. Determine the worst-case complexity for Maximum. See Exercise 15.

18. Determine the worst-case complexity for Count. See Exercise 16.

In Exercises 19 through 21, let $g(x)$ be an arbitrary real function.

19. Show that $g(x) \in O(g(x))$. 20. Show that $0 \in O(g(x))$.

21. Let $c \neq 0$. Show that $O(cg(x)) = O(g(x))$.

22. Prove Lemma 10.10.

In Exercises 23 through 26, decide whether the given statement is true or false.

23. $x^{\frac{3}{2}} \in O\left(\frac{3}{2}x\right)$.

24. $x^2 + 5x + 7 \in O(x^2)$.

25. $O(x^3 + x^2) \subseteq O(x^4)$.

26. $O\left(\frac{1}{10}x^3 - x\right) \subseteq O(x^2)$.

27. Suppose $f_1(x), f_2(x) \in O(g(x))$.
Show: $f_1(x) + f_2(x) \in O(g(x))$.

28. Suppose $f_1(x) \in O(g_1(x))$ and $f_2(x) \in O(g_2(x))$.
Show: $f_1(x)f_2(x) \in O(g_1(x)g_2(x))$.

29. Show that if $f(x) \in O(g(x))$, then $f(x) + g(x) \in O(g(x))$.

30. Show that if $O(f(x)) = O(1)$, then $O(f(x)g(x)) = O(g(x))$.

31. Let $c \in \mathbb{R}$, and suppose $1 \in O(g(x))$.
Show: $c + g(x) \in O(g(x))$.

32. Suppose $1 \in O(g(x))$. Show: $O(\lceil g(x) \rceil) = O(g(x))$.

33. Suppose P dollars have been invested in an account earning a monthly interest rate i. If the account earns **simple
interest**, then the amount in the account after n months
is $P(1 + in)$. If the account earns **compound interest**, then
the amount in the account after n months is $P(1 + i)^n$. In
general, the investor is better off with compound interest,
since $(1 + i)^n$ has a higher growth rate than $(1 + in)$. However, suppose a bank is offering a compound interest rate
of 2% ($i = 0.02$) or a simple interest rate of 4% ($i = 0.04$),
so the interest rates are not equal.

(a) Which option is better for a 5-year investment?

(b) Which option is better for a 6-year investment?

(c) Find the smallest number of months for which the
compound interest option would be the better choice.

34. A new hire is negotiating her salary with her employer,
who has provided two salary options. Under option 1,
her salary for year n will be $1000(20\sqrt{n + 2})$. Under option
2, her salary for year n will be $1000(16n - 1)$. Notice that
option 2 has a higher growth rate than option 1.

(a) Which option has a better second-year salary?

(b) Which option has a better fourth-year salary?

(c) If this new hire plans to stay at this job exactly 4 years
before moving on, then which option is better for her?

35.* Show: $O(1 + \lceil \log_2 n \rceil) = O(\log_2 n)$.

36.* Show: $O(n + n \log_2 n) = O(n \log_2 n)$.

37.* Show: $O\left(\frac{n}{2} \log_2 \frac{n}{2}\right) = O(n \log_2 n)$.

38.* Show: $O(100 + 3 \cdot 2^n) = O(2^n)$.

39. Prove the assertion in Definition 10.9. That is, show
that $f(x) \in O(g(x))$ and $g(x) \in O(f(x))$ if and only if
there exist positive constants C_1, C_2, and d for which
$\forall x > d$, $C_1|g(x)| \le |f(x)| \le C_2|g(x)|$.

40. Prove that $f(x) \in \Theta(g(x))$ if and only if $g(x) \in \Theta(f(x))$.

41. Prove Lemma 10.13.

42. Prove Lemma 10.14.

In Exercises 43 through 46, decide whether the given statement is true or false.

43. $n \log_2 n \in O(n)$.

44. $n^4 \in O(2^n)$.

45. $O(\log_2 n) = O(\log_2(n^2))$.

46. $O(n^2) \subseteq O((\log_2 n)^2)$.

47. A programmer has written two programs for sorting a list of numbers. One uses $64\lfloor \log_2 n \rfloor + 108n + 18$ operations and the other uses $2n^2 + 4n + 8$ operations. She is trying to determine which program is faster for certain applications and has noticed that both use the same number of operations to sort a list of $n = 55$ items.

   (a) Which is faster at sorting a list of fewer than 55 items?

   (b) Which is faster at sorting a list of more than 55 items?

   (c) Which program has a smaller order of growth?

48. A programmer has written two programs for finding the median value in an unsorted list of numbers (of length $n \geq 21$). One uses $n^3 - 20n^2 - 6n + 1$ operations and the other uses $80n^2 + 64n - 3$ operations. He is trying to determine which program is faster for certain applications.

   (a) Which is faster for a list of at most 100 items?

   (b) What is the smallest length of a list for which the second program is faster?

   (c) Which program has a smaller order of growth?

49. Show that $\forall\, b > 1$, $O(\log_b n) = O(\log_2 n)$.
   Hint: $\log_b n = \frac{\log_2 n}{\log_2 b}$.

50. Is it true that $\forall\, b > 1$, $O(b^n) = O(2^n)$?

51. Show: $O(1) \subset O(\log_2 n)$.

52. Show: $O(\log_2 n) \subset O(n)$.
   Hint: $\log_2 n \leq n$ if and only if $n \leq 2^n$, so use induction.

53. Show: $O(n) \subset O(n \log_2 n)$.
   Hint: Use work from Exercise 51.

54. Show: $O(n \log_2 n) \subset O(n^2)$.
   Hint: Use work from Exercise 52.

55.* Show: $O(n^2) \subset O(2^n)$. Hint: Use induction.

56.* Show: $O(2^n) \subset O(n!)$. Hint: Use induction.

10.5   Analysis of algorithms (Part 2)

In Section 10.1, we studied trees and their applications. In Sections 10.2 and 10.3, we considered various algorithms that find spanning trees in connected graphs. Having shifted our interest to the analysis of algorithms in Section 10.5, we now come full circle and use trees to study algorithms.

10.5.1   Decision trees

One tool used to analyze an algorithm is a **decision tree**. Such a tree reflects the paths an algorithm can follow based on the various decision tests performed throughout a running of the algorithm. Here, we focus on binary decision trees in which the decision tests are comparisons. The vertices of the tree record the states of the algorithm as it is running. Vertices at which comparisons are made are connected to children distinguished by the result of the comparison (true or false).

A representation of the decision tree for Sequential Search is shown in Figure 10.29. For each $1 \leq i \leq n$, there is an internal vertex reflecting the test whether $A[i] = x$. The leaves reflect the potential return values of the location of x. The height of that tree is n.

The decision tree for Binary Search (restricted to arrays of length 3) is represented in Figure 10.30. There, the vertices similarly reflect comparisons made and locations returned by the algorithm. In this case, the vertices display only the portion of the array under consideration at each stage. The height of this tree is $3 = 1 + \lceil \log_2 3 \rceil$.

Note in Figures 10.29 and 10.30 that the height of the tree is equal to the worst-case complexity of the algorithm. This is true in general, since the worst case of the algorithm corresponds to the longest path from the root.

While the internal vertices in a decision tree represent the intermediate states of the algorithm at which decisions are made, it is the leaves that represent the potential final outcomes. A search for a value x in an array of length n has $n + 1$ possible outcomes. If x is found, then some index from

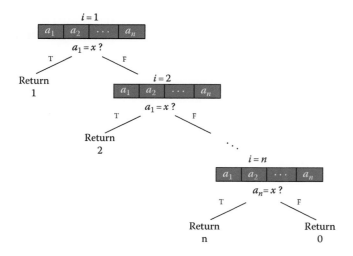

FIGURE 10.29
Decision tree for Sequential Search.

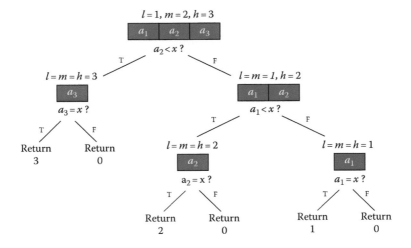

FIGURE 10.30
Decision tree for Binary Search (for $n = 3$).

1 to n is returned. If x is not found, then 0 is returned. Thus, a decision tree for any search algorithm must have at least $n + 1$ leaves. By Theorem 10.7, its height must be at least $\lceil \log_2(n + 1) \rceil$. That is, the smallest possible complexity of a search algorithm is $\Theta(\lceil \log_2(n + 1) \rceil) = \Theta(\log_2 n)$. Therefore, Binary Search has the least possible complexity. In this sense, no other search algorithm is more efficient.

10.5.2 Sorting algorithms

We continue our study of the analysis of algorithms by moving to another class of algorithms. Here, we are interested in putting the elements of a given array into nondecreasing order. This will certainly be of interest to us if we frequently need to search for elements in the array. For example, we typically alphabetize our files so that we can subsequently find files more easily. Similarly, the records in a database might be sorted according to some identification number. This would provide a convenient way to access records more quickly. Although the keys to be sorted might be text and not numbers, it suffices to consider algorithms that sort numbers, since any type of key could alternatively be assigned a numerical value.

Our first sorting algorithm does so by sequentially running through the entries of the array. Each entry is inserted into its correct position relative to the already sorted prior portion of the array. It is called **Insertion Sort** and is presented in Algorithm 10.8.

Algorithm 10.8 Insertion Sort

Let A be an array of length n that needs to be sorted (into non-decreasing order). Variable i is the index of the array entry to be inserted into its correct position relative to the previously sorted portion $A[1, \ldots, i - 1]$, and j runs through that portion in search of this correct position.

Algorithm.

> For $i = 2$ to n,
> > \begin
> > Let $j = 1$.
> > While $j < i$ and $A[j] < A[i]$,
> > > Let $j = j + 1$.
> >
> > If $j < i$, then
> > > \begin
> > > Let temp $= A[i]$.
> > > For $k = i$ down to $j + 1$,
> > > > Let $A[k] = A[k - 1]$.
> > >
> > > Let $A[j] = $ temp.
> > > \end.
> >
> > \end.
>
> Return A.

Example 10.22. Given $A = \boxed{3} \boxed{1} \boxed{5} \boxed{6} \boxed{4} \boxed{2}$, as Insertion Sort is performed, the main states of the array (based on the outermost for loop) are as follows.

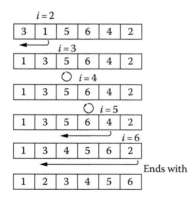

For each i, the arrows point from position i to the position to which that entry is moved.

As with search algorithms, the operations of interest in sorting algorithms are comparisons involving array entries. Also, the size of the input is considered to be the length of the array. It is these factors that are used to determine the worst-case complexity of a sorting algorithm.

Example 10.23. The worst-case complexity of Insertion Sort is $\frac{n(n-1)}{2}$.

Proof. Keep in mind that the worst performance occurs when A starts out sorted. For each value of i, there are at most $i - 1$ comparisons $A[j] < A[i]$ made in the while loop. Note that a comparison $A[i] < A[i]$ is never made, because the test $i < i$ fails before that point would be reached. The number of comparisons performed is at most

$$\sum_{i=2}^{n} \left(\sum_{j=1}^{i-1} 1 \right) = \sum_{i=2}^{n} (i-1) = \frac{n(n-1)}{2}.$$

This number is achieved when the input starts in non-decreasing order. ☐

A representation of the decision tree for Insertion Sort (on arrays of length $n = 3$) is shown in Figure 10.31. The height of that tree is $3 = \frac{3(2)}{2}$.

Our second sorting algorithm does so by splitting the given array in half. It sorts each half and then merges the two sorted halves back into one sorted array. This recursive algorithm is called **Merge Sort** and is presented in Algorithm 10.9. The algorithm is based primarily on an algorithm called **Merge**, which is presented in Algorithm 10.10. That algorithm combines two sorted arrays into one sorted array.

Example 10.24. Given $A = \boxed{\begin{array}{|c|c|c|c|} 3 & 1 & 5 & 6 \end{array}}$ $\boxed{\begin{array}{|c|c|} 4 & 2 \end{array}}$, the implementation of Merge Sort is displayed in Figure 10.32. The tree above the dotted line shows how the array gets repeatedly split until only pieces of length 1 remain. The tree below the dotted line

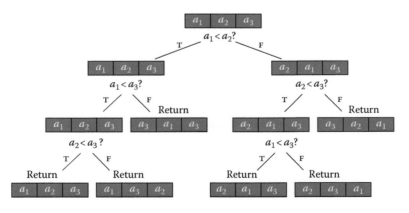

FIGURE 10.31
Decision tree for Insertion Sort (for $n = 3$).

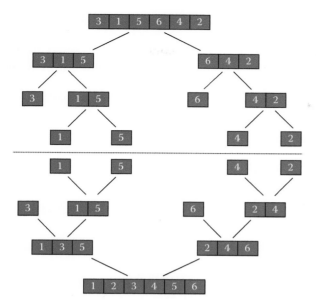

FIGURE 10.32
Merge Sort (for $n = 6$).

Algorithm 10.9 Merge Sort

Let A be an array of length n that needs to be sorted (into nondecreasing order). Arrays A_1 and A_2 are used to store the first and second halves of A, respectively, after they have been sorted by this same algorithm.

Algorithm.
    If $n \leq 1$, then
        Return A.
    Otherwise,
        \begin
        Let $A_1 = \text{Merge\_Sort}(A[1, \ldots, \lfloor \frac{n}{2} \rfloor], \lfloor \frac{n}{2} \rfloor)$.
        Let $A_2 = \text{Merge\_Sort}(A[\lfloor \frac{n}{2} \rfloor + 1, \ldots, n], \lceil \frac{n}{2} \rceil)$.
        Return $\text{Merge}(A_1, \lfloor \frac{n}{2} \rfloor, A_2, \lceil \frac{n}{2} \rceil)$.
        \end

---

then shows how the pieces get merged back together to finally put the array in order. Note that these are not decision trees.

Example 10.25.   Merge Sort is $O(n \log_2 n)$.

Proof. Let c_n be the maximum number of comparisons needed in a Merge Sort of an array of length n.

Algorithm 10.10 Merge

Given a sorted array A_1 of length n_1 and a sorted array A_2 of length n_2, this algorithm produces a sorted array A of length $n_1 + n_2$.

Algorithm.
Let $i_1 = i_2 = i = 1$.
While $i_1 \leq n_1$ and $i_2 \leq n_2$,
\begin
If $A_1[i_1] < A_2[i_2]$, then
\begin
Let $A[i] = A_1[i_1]$.
Let $i_1 = i_1 + 1$.
\end
Otherwise,
\begin
Let $A[i] = A_2[i_2]$.
Let $i_2 = i_2 + 1$.
\end
Let $i = i + 1$.
\end.
If $i_1 > n_1$, then
While $i_2 \leq n_2$,
\begin
Let $A[i] = A_2[i_2]$.
Let $i = i + 1$ and $i_2 = i_2 + 1$.
\end
If $i_2 > n_2$, then
While $i_1 \leq n_1$,
\begin
Let $A[i] = A_1[i_1]$.
Let $i = i + 1$ and $i_1 = i_1 + 1$.
\end
Return array A of length $n_1 + n_2$.

So $c_0 = c_1 = 0$ and $c_2 = 1$. A Merge Sort of an array of length n is accomplished by

1. merge sorting an array of length $\lfloor \frac{n}{2} \rfloor$,

2. merge sorting an array of length $\lceil \frac{n}{2} \rceil$, and

3. performing the at most $n - 1$ comparisons in a merge.

So $\forall n \geq 2, c_n = c_{\lfloor \frac{n}{2} \rfloor} + c_{\lceil \frac{n}{2} \rceil} + (n - 1)$. It can now be shown by strong induction that $\forall n \geq 1, c_n \leq 2n \log_2 n$. This is left for Exercise 37. $\square$

Theorem 10.17.   *The worst-case complexity of every sorting algorithm is at least $\frac{n}{2} \log_2 \frac{n}{2}$.*

Proof. Since there are $n!$ possible permutations of an array of length n, the decision tree for any sorting algorithm must have at least $n!$ leaves. Theorem 10.7 then tells us that the height of this tree must be at least $\lceil \log_2 n! \rceil$. Consequently, the worst-case complexity is at least

$$\lceil \log_2 n! \rceil \geq \log_2 n! = \sum_{i=1}^{n} \log_2 i$$

$$\geq \sum_{i=\lceil \frac{n}{2} \rceil}^{n} \log_2 i = \log_2 \lceil \tfrac{n}{2} \rceil + \cdots + \log_2 n$$

$$\geq (n - \lceil \tfrac{n}{2} \rceil + 1) \log_2 \lceil \tfrac{n}{2} \rceil = (\lfloor \tfrac{n}{2} \rfloor + 1) \log_2 \lceil \tfrac{n}{2} \rceil$$

$$\geq \lceil \tfrac{n}{2} \rceil \log_2 \lceil \tfrac{n}{2} \rceil$$

$$\geq \tfrac{n}{2} \log_2 \tfrac{n}{2}. \qquad \square$$

Since Exercise 37 from Section 10.4 tells us that $O(\frac{n}{2} \log_2 \frac{n}{2}) = O(n \log_2 n)$, it follows that Merge Sort is $\Theta(n \log_2 n)$. Moreover, no sorting algorithm can be more efficient than Merge Sort, in the sense of big-Θ. However, recall that we have considered only time complexity and not space complexity. The recursive nature of Merge Sort requires more memory to be used each time the procedure calls itself. Insertion Sort, on the other hand, is an in-place sort. That is, the elements of the array A to be sorted are simply shuffled around A itself. There is no need to generate copies of subarrays of A, as is required in the recursive Merge Sort.

In the exercises, we introduce three additional sorting algorithms: **Bubble Sort**, **Selection Sort**, and **Quick Sort**. Including those, the time complexities of the sorting algorithms considered in this section are listed in Table 10.3, where the missing values in that table are left for the exercises. The unproved average-case complexities listed in Table 10.3

TABLE 10.3
Time complexities for sorting algorithms

Sorting algorithm	Insertion	Bubble	Selection	Merge	Quick
Worst case	$\Theta(n^2)$			$\Theta(n \log_2 n)$	
Average case	$\Theta(n^2)$	$\Theta(n^2)$	$\Theta(n^2)$	$\Theta(n \log_2 n)$	$\Theta(n \log_2 n)$

are included to give a sense of the expected run-times on a random array in need of sorting. Also, the average-case complexity of Quick Sort helps to justify its name.

Exercises

1. Make the decision tree for Binary Search for arrays of length 4.

2. Make the decision tree for Binary Search for arrays of length 5.

3. Make a decision tree for Maximum (from Exercise 15 in Section 10.4) for $n = 4$.

4. Make a decision tree for Count (from Exercise 16 in Section 10.4) for $n = 3$.

In Exercises 5 through 8, for the given matrix A, show the main states of Insertion Sort.

5. $A = \boxed{\begin{array}{c|c|c|c} 7 & 5 & 3 & 1 \end{array}}$.

6. $A = \boxed{\begin{array}{c|c|c|c|c|c|c|c} 3 & 8 & 4 & 1 & 2 & 6 & 5 & 7 \end{array}}$.

7. $A = \boxed{\begin{array}{c|c|c|c|c|c} 4 & 2 & 6 & 1 & 5 & 3 \end{array}}$.

8. $A = \boxed{\begin{array}{c|c|c|c|c} 3 & 8 & 6 & 1 & 4 \end{array}}$.

9. Make the decision tree for Insertion Sort for arrays of length 2.

10. Make the portion of the decision tree for Insertion Sort for arrays of length 4, in the case that $A[4]$ is the smallest value.

In Exercises 11 through 14, for the given matrix A, make a tree diagram of Merge Sort that shows first how the matrix is split into pieces and then how the pieces are merged back together in order.

11. $A = \boxed{\begin{array}{c|c|c|c} 7 & 5 & 3 & 1 \end{array}}$.

12. $A = \boxed{\begin{array}{c|c|c|c|c|c|c|c} 3 & 8 & 4 & 1 & 2 & 6 & 5 & 7 \end{array}}$.

13. $A = \boxed{\begin{array}{c|c|c|c|c|c} 4 & 2 & 6 & 1 & 5 & 3 \end{array}}$.

14. $A = \boxed{\begin{array}{c|c|c|c|c} 3 & 8 & 6 & 1 & 4 \end{array}}$.

The sorting algorithm **Bubble Sort** is presented in Algorithm 10.11. In Exercises 15 through 18, for the given matrix A, show the main states of Bubble Sort.

15. $A = \boxed{\begin{array}{c|c|c|c|c|c} 4 & 2 & 6 & 1 & 5 & 3 \end{array}}$.

16. $A = \boxed{\begin{array}{c|c|c|c|c} 5 & 8 & 1 & 3 & 6 \end{array}}$.

17. $A = \boxed{\begin{array}{c|c|c|c|c} 7 & 5 & 4 & 2 & 1 \end{array}}$.

18. $A = \boxed{\begin{array}{c|c|c|c|c|c} 8 & 6 & 5 & 3 & 2 & 1 \end{array}}$.

The sorting algorithm **Selection Sort** is presented in Algorithm 10.12. In Exercises 19 through 22, for the given matrix A, show the main states of Selection Sort.

Algorithm 10.11 Bubble Sort

Variable i is the index of the entry to be determined based on the assumption that the portion $A[i + 1, \ldots, n]$ is already set, while j runs through the lower indices to filter up the correct value for the ith spot.

Algorithm.

> For $i = n$ down to 2,
>> For $j = 1$ to $i - 1$,
>>> If $A[j] > A[j + 1]$, then
>>>> \begin
>>>> Let temp $= A[j + 1]$.
>>>> Let $A[j + 1] = A[j]$.
>>>> Let $A[j] =$ temp.
>>>> \end.
>
> Return A.

Algorithm 10.12 Selection Sort

Variable i is the index of the entry to be determined based on the assumption that the portion $A[1, \ldots, i - 1]$ is already sorted, while j runs through the upper indices to find the correct value for the ith spot. Variable min stores the index of the current best candidate for the ith spot.

Algorithm.

> For $i = 1$ to $n - 1$,
>> \begin
>> Let min $= i$.
>> For $j = i + 1$ to n,
>>> If $A[j] < A[\min]$, then
>>>> Let min $= j$.
>> If min $\neq i$, then
>>> \begin
>>> Let temp $= A[i]$.
>>> Let $A[i] = A[\min]$.
>>> Let $A[\min] =$ temp.
>>> \end.
>> \end.
>
> Return A.

19. $A = \boxed{\begin{array}{|c|c|c|c|c|} 3 & 8 & 6 & 1 & 4 \end{array}}$.

20. $A = \boxed{\begin{array}{|c|c|c|c|c|c|} 2 & 8 & 3 & 1 & 5 & 4 \end{array}}$.

21. $A = \boxed{\begin{array}{|c|c|c|c|c|c|} 6 & 5 & 4 & 3 & 2 & 1 \end{array}}$.

22. $A = \boxed{\begin{array}{|c|c|c|c|c|} 7 & 5 & 4 & 2 & 1 \end{array}}$.

23. Determine the worst-case complexity for Bubble Sort.

24. Determine the worst-case complexity for Selection Sort.

25. Make a decision tree for Bubble Sort for $n = 3$.

26. Make a decision tree for Bubble Sort for $n = 2$.

27. Make a decision tree for Selection Sort for $n = 2$.

28. Make a decision tree for Selection Sort for $n = 3$.

29. Is Bubble Sort a maximally efficient sorting algorithm in the sense of big-Θ?

30. Is Selection Sort a maximally efficient sorting algorithm in the sense of big-Θ?

31. There is a sorting algorithm that is sometimes called Bridge Sort, because it works the way a card player may sort the cards in her hand when playing bridge. First, she moves the highest card to the rightmost position in her hand. Then, she moves the second highest card to the second rightmost position in her hand, and so on. This technique coincides with one of the sorting algorithms considered in this section. Which one is it?

32. Suppose a worker needs to find the medical record for patient Smith. So he goes to the drawer labeled "S" and pulls it open. First, he fingers the file about halfway back in the drawer and finds that it is the file for patient Shears. So he moves to about three-quarters back in the drawer and fingers the file for patient Stevens. Then, he moves forward, and so on. This approach is similar to one of the searching algorithms considered in Section 10.4. Which one is it?

For Exercises 33 through 36, if there are duplicated entries in an array A, determine whether the specified sorting algorithm will always leave them in their original order. If not, explain.

33. Insertion Sort.

34. Merge Sort.

35. Bubble Sort.

36. Selection Sort.

37.* Let c_n be the maximum number of comparisons needed in a Merge Sort of an array of length n. Complete the proof of Example 10.25 by showing that $\forall\, n \geq 1$, $c_n \leq 2n \log_2 n$. Hint: Prove by strong induction that $\forall\, n \geq 1, 2^{c_n} \leq 2^{2n \log_2 n} = n^{2n}$.

38.* Observe that, for any fixed k, finding the top k entries in an array of length n is $O(n)$, yet sorting is $O(n \log_2 n)$. Explain how this can be.

39. Show that Maximum (from Exercise 15 in Section 10.4) is $\Theta(n)$.

40. Show that Count (from Exercise 16 in Section 10.4) is $\Theta(n)$.

The sorting algorithm **Quick Sort** is presented in Algorithm 10.13. It is based primarily on the algorithm **Split** presented in Algorithm 10.14.

In Exercises 41 through 44, for the given matrix A, show the main states of Quick Sort. In particular, each time Split

---

Algorithm 10.13 Quick Sort

Let A be an array of length n that needs to be sorted (into nondecreasing order). The variable **mid** will be a position chosen by the function Split that cuts the array into two pieces and partially sorts it so that
$A[1], \ldots, A[\text{mid}-1] < A[\text{mid}] \leq A[\text{mid}+1], \ldots, A[n]$.

Algorithm.

    If $n \leq 1$, then

        Return A.

    Otherwise,

        \begin

        Let $(A, \text{mid}) = \text{Split}(A, n)$.

        Let $A[1, \ldots, \text{mid}-1] =$

        Quick_Sort$(A[1, \ldots, \text{mid}-1], \text{mid}-1)$.

        Let $A[1, \ldots, \text{mid}+1] =$

        Quick_Sort$(A[\text{mid}+1, \ldots, n], n-\text{mid})$.

        Return A.

        \end

Algorithm 10.14 Split

Let A be an array of length n that needs to be reordered so that $A[1]$ is moved to the position, index **mid**, into which it will end up when the array is entirely sorted. When this function completes with A reordered, we want no larger entries than $A[\text{mid}]$ prior to position mid and no smaller entries than $A[\text{mid}]$ after position mid.

Algorithm.

    Let mid $= 1$.

    For $i = 2$ to n,

        If $A[i] < A[1]$, then

            \begin

            Let mid $=$ mid $+ 1$.

            Let temp $= A[\text{mid}]$. Let $A[\text{mid}] = A[i]$.

            Let $A[i] =$ temp.

            \end

    Let temp $= A[\text{mid}]$. Let $A[\text{mid}] = A[1]$.

        Let $A[1] =$ temp.

    Return (A, mid).

is called, report the returned value of mid. Also, follow the calls of Quick Sort down to those on subarrays of length one.

41. $A =$ | 3 | 5 | 6 | 1 | 4 | .

42. $A =$ | 2 | 1 | 4 | 3 | 6 | 5 | .

43. $A =$ | 1 | 2 | 3 | 4 | .

44. $A =$ | 2 | 7 | 1 | 3 | 5 | .

45. Determine the worst-case complexity for Quick Sort. Note that this occurs when the given array A starts in nondecreasing order. That is, each call of Split returns the value mid $= 1$. Exercise 43 is an example of this worst case. Warning: The worst-case complexity is not always of the same order as the average-case complexity.

46.* Determine the number of comparisons done in Quick Sort if each call of Split returns the value mid $= 2$. Note that Exercise 42 is an example of this case.

10.6 Review problems

The following graphs are referred to throughout the exercises.

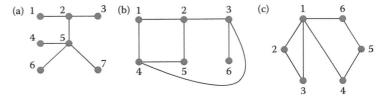

1. Let $G = (V, E)$ be a graph. Show: If $|E| < |V| - 1$, then G is not connected. Hint: Make use of spanning trees or forests, and prove the contrapositive.

2. List all possible degree sequences for a tree on n vertices with 4 leaves.

3. Gasoline that is high in octane, C_8H_{18}, can handle compression well and thus has a smoother burn with less engine knocking. How many different isomers of the hydrocarbon C_8H_{18} are possible?

4. How many vertices are there in a graph with 4 components whose spanning forest has 26 edges?

5. We have seen that, for a fixed n, there may be many different isomers of the saturated hydrocarbon C_nH_{2n+2}. However, any two isomers will have the same number of chemical bonds. How many chemical bonds are there in a saturated hydrocarbon C_nH_{2n+2}?

6. Show that, in a tree, any two vertices u and v are joined by a path in which all of the vertices, except possibly u and v, are internal vertices.

7. Take 2 to be the root of the tree in graph (a).

    (a) List the children of 5.   (e) Is this a binary tree?

    (b) Which vertex is the
        parent of 3?   (f) Is it balanced?

    (c) What is the level of 5? (g) Is it full?

    (d) What is the height of
        this tree?

8. Can a spanning forest for a graph $G = (V, E)$ be characterized as a subgraph on V that is a forest? Explain.

9. Let T be a full m-ary tree with n vertices, l leaves, and i internal vertices. Express m and i in terms of n and l.

10. The file directory system on Sarah's A:\ drive contains 9 folders, not counting A:\ itself. If each folder, including A:\ itself, has exactly 4 children, and no folder is empty, then how many files are on this disk?

11.* Prove or disprove: The diameter of any rooted tree is at most twice its height.

12. Using the usual dictionary order to compare words, construct the binary search tree for the words in the sentence

       Time flies when you are having fun.

13. Construct the quadtree for the given image. Note that grid lines have been drawn over the image for reference and are not themselves part of the image.

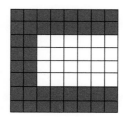

14. Draw the image that corresponds to the given quadtree. It represents a letter.

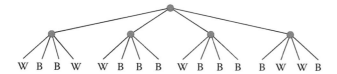

In Exercises 15 through 20, use the ordering of the vertices on the specified graph given by the natural ordering of the labels, and perform the specified algorithm starting at vertex 1. In each case, give the resulting ordering L on the vertices. If the graph itself is not a tree, then shade the resulting spanning tree.

15. Breadth-First Search on graph (a).

16. Breadth-First Search on graph (b).

17. Breadth-First Search on graph (c).

18. Depth-First Search on graph (a).

19. Depth-First Search on graph (b).

20. Depth-First Search on graph (c).

21. The ideal gas law for n moles of gas at pressure P, volume V, and temperature T is $PV = nrT$, where r is the universal gas constant obtained by multiplying Boltzmann's constant by Avogadro's number. Solving for pressure we obtain the expression $P = \frac{nrT}{V}$. Construct the tree that represents the expression

$$((n * r) * T) \div V$$

for the pressure P.

22. A trapezoid with height h and base lengths b_1 and b_2 has its area given by the formula $A = \frac{b_1 + b_2}{2} \cdot h$. Write the expression for the area A of this trapezoid

$$((b_1 + b_2) \div 2) * h$$

(a) In postfix notation.

(b) In prefix notation.

(c) In infix notation.

23. Compute the value of the postfix expression.
    4 5 1 + 2 ÷ *

24. Compute the value of the prefix expression.
    + 1 * − 8 5 2

25. Five major cities of the United States were put into the following binary search tree, using longitude for comparisons.

Perform an inorder traversal of the binary search tree to determine the west-to-east ordering for these cities.

26. If a Hamiltonian cycle is sought in the graph

using a backtrack search that starts from the trivial path v, then what are the next two paths considered after the path vwx is encountered?

Subsequent exercises refer to the following graphs.

(d)

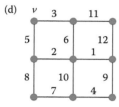

(e)

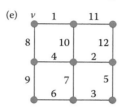

(f)

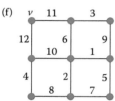

(g)

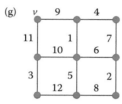

In Exercises 27 through 32, use the ordering of the vertices on the specified graph given by the natural ordering of the labels, and find the tree resulting from the specified algorithm.

27. Kruskal's Algorithm for graph (d).

28. The vertices in graph (e) represent buildings in an apartment complex under construction. Each edge represents a potential power line joining two buildings, and its weight represents the cost of establishing that line in tens of thousands of dollars. Consequently, the cheapest possible way to establish a power grid joining all of the buildings corresponds to a minimum spanning tree.

    (a) Use Kruskal's Algorithm to find a minimum spanning tree for graph (e).

    (b) What is the cheapest possible cost of establishing a power grid for this apartment complex?

29. Prim's Algorithm for graph (f). Also, list the order in which the edges are added.

30. The vertices in graph (g) represent the condominiums in a new development. Each edge represents a potential sewer line joining two condominiums, and its weight represents the cost of establishing that line in tens of thousands of

dollars. Consequently, the cheapest possible way to establish a sewer system joining all of the condominiums corresponds to a minimum spanning tree.

(a) Use Prim's Algorithm to find a minimum spanning tree for graph (g). Also, list the order in which the edges are added.

(b) What is the cheapest possible cost of establishing a sewer system for this condominium development?

31. Dijkstra's Algorithm for graph (d) starting from vertex v.

32. The vertices in graph (e) represent critical police patrol locations under the jurisdiction of the police station at vertex v. Each edge represents a road joining two locations, and its weight gives the travel time in minutes between those locations. Consequently, the fastest response time from the police station to each location is reflected by a path on a shortest path tree for graph (e) from v.

(a) Use Dijkstra's Algorithm to find a shortest path tree for graph (e) starting from vertex v.

(b) What is the longest required response time from the police station to a critical location under its jurisdiction?

33. Suppose we have a weighted graph in which all of the weights are distinct. Determine whether each of the following statements is true or false.

(a) The spanning trees produced by Kruskal's Algorithm and Prim's Algorithm are always the same.

(b) The tree produced by Dijkstra's Algorithm is independent of the edge ordering.

34. The cities in the following table are the business locations for a company, based in Syracuse, New York, that serves the upstate New York area.

	Binghamton	Buffalo	Ithaca	Rochester	Syracuse
Binghamton	—	$200	—	—	$200
Buffalo	$200	—	$150	—	$450
Ithaca	—	$150	—	$300	—
Rochester	—	—	$300	—	$300
Syracuse	$200	$450	—	$300	—

The costs therein are the airfares for the direct flights available between pairs of these cities.

(a) What is the greatest necessary cost to travel from Syracuse to one of the other company locations?

(b) What kind of spanning tree is relevant to this problem?

35. A college wishes to set up a computer network joining the residence halls on its campus. The following table gives the costs in thousands of dollars determined by a contractor to connect pairs of residence halls.

	Arsenio	Fawn	Monty	Philip	Rich
Arsenio	—	11	6	7	15
Fawn	11	—	8	9	12
Monty	6	8	—	5	13
Philip	7	9	5	—	13
Rich	15	12	13	13	—

(a) Find the cheapest possible cost of establishing a computer network that serves all of the residence halls.

(b) What kind of spanning tree is relevant to this problem?

36. Given $x = 5$ and $A = \boxed{1\ 2\ 4\ 6\ 8}$, show the main states of Binary Search.

37. Suppose that a personal computer has a speed of 1 GHz. That is, this computer can perform one billion instructions per second. If we assume that comparisons are the only time-consuming instructions in Sequential Search, then what is the longest possible time it would take this computer to run Sequential Search on an array of length 5×10^{10}?

Subsequent exercises refer to the following algorithm for finding a location for the smallest value in a given array A of length n.

Minimum.
> Let location $= 1$.
> Let $i = 2$.
> While $i \leq n$,
> > \begin
> > If $A[i] < A[\text{location}]$, then
> > > Let location $= i$.
> > Otherwise,
> > > Let $i = i + 1$.
> > \end.
> Return location.

38. Determine the worst-case complexity of Minimum.
For Exercises 39 through 43, decide whether each of the following statements is true or false.

39. $3x \in O(x-1)$.

40. $x^2 + 1 \in O(25x)$.

41. $x^3 - x^2 + 7 \in \Theta(x^4)$.

42. $x^2 \in \Theta(4x^2 + 5x)$.

43. $O(n^2) \subseteq O(n \log_2 n)$.

44. A newly hired employee must decide which of two retirement plans to accept. Both options simply pay out a lump sum upon retirement based on the number n of years in employment. The first plan pays $30n^3 + 500n$ dollars, and the second pays $600n^2 + 10{,}000$ dollars.

   (a) Which plan has a greater growth rate?

   (b) Which plan would be better, based on 15 years of employment?

   (c) What is the least number of years of employment for which the first plan is better?

45. Let $f(x)$ be a bounded real function. Show that $f(x) \in O(1)$.

46. Prove that $O(\log_2 n) \subset O(n^{\frac{3}{2}})$. Hint: We know the relative growth rates of $\log_2 n$ versus n, and we know how powers of n compare.

47. Is it true that Minimum is $O(n^2)$? Explain.

48. Is it true that Minimum is $\Theta(n)$? Explain.

49. Make a decision tree for Minimum for $n = 3$.

50. Given $A = \boxed{6\,|\,3\,|\,8\,|\,2\,|\,5}$, show the main states of Insertion Sort.

51. Given $A = \boxed{4\,|\,3\,|\,2\,|\,1}$, make a tree diagram of Merge Sort that shows first how the matrix is split into pieces and then how the pieces are merged back together in order.

52. Given $A = \boxed{4\,|\,2\,|\,1\,|\,3}$, show the main states of Bubble Sort.

53. Given $A = \boxed{2\,|\,4\,|\,3\,|\,1}$, show the main states of Selection Sort.

54. What ordering of an array yields the most comparisons in Insertion Sort?

55. If the array starts out ordered, will Bubble Sort use fewer comparisons than otherwise?

56. Given $A = \boxed{3\,|\,1\,|\,6\,|\,4}$, show the main states of Quick Sort.

57. Is it true that Quick Sort is $O(n \log_2 n)$? Explain.

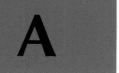

A

Assumed Properties of $\mathbb{Z}$ and $\mathbb{R}$

For easy reference, we list here some basic properties of real numbers and integers that may be used freely (without proof) throughout the book. The properties listed here are sufficient to characterize the ordered field of real numbers. However, they are not intended to be completely independent of one another. Also, properties that are straightforward consequences of these, but might nonetheless be considered "basic" by some, may not be listed.

1. Let S be any of the sets $\mathbb{Z}$, $\mathbb{Z}^+$, $\mathbb{N}$, $\mathbb{R}$, or $\mathbb{R}^+$. Then S is closed under addition and multiplication. That is,

$$\forall\, x, y \in S, \quad x + y \in S \quad \text{and} \quad x \cdot y \in S.$$

Note that the product $x \cdot y$ may also be denoted xy.

2. Addition and multiplication are associative and commutative. That is,

$$\forall\, x, y, z \in \mathbb{R} \quad \text{we have}$$

(a) $(x + y) + z = x + (y + z)$
and we can unambiguously write $x + y + z$ for both expressions.

(b) $(x \cdot y) \cdot z = x \cdot (y \cdot z)$
and we can unambiguously write $x \cdot y \cdot z$ for both expressions.

(c) $x + y = y + x$.

(d) $x \cdot y = y \cdot x$.

3. The additive identity is $0 \in \mathbb{R}$, and the multiplicative identity is $1 \in \mathbb{R}$. Further, inverses exist and behave appropriately. That is,

$$\forall\, x \in \mathbb{R}$$

(a) $0 + x = x$.

(b) There exists $-x \in \mathbb{R}$ such that $x + (-x) = 0$.
Further, $-(-x) = x$.

(c) $1 \cdot x = x$.

(d) If $x \neq 0$, then $\exists\ \frac{1}{x} \in \mathbb{R}$ such that $x \cdot \left(\frac{1}{x}\right) = 1$.
Further, $\frac{1}{\frac{1}{x}} = x$.

Note that, for any $y \in \mathbb{R}$,
the difference $y - x$ is defined as $y + (-x)$, and,
for $x \neq 0$, the quotient $\frac{y}{x}$ is defined as $y \cdot \left(\frac{1}{x}\right)$.

4. The Distributive Law holds. That is,

$$\forall\, x, y, z \in \mathbb{R} \quad x \cdot (y + z) = x \cdot y + x \cdot z.$$

5. For $x \in \mathbb{R}$ and $n \in \mathbb{Z}^+$, we use exponent notation

$$x^n = \underbrace{x \cdot x \cdot \cdots \cdot x}_{n \text{ copies}},$$

$$x^{-n} = \frac{1}{x^n} \quad \text{if } x \neq 0,$$

$$x^0 = 1 \quad \text{if } x \neq 0,$$

and the **Laws of Exponents** hold. That is,

$$\forall\, x \in \mathbb{R} \text{ and } m, n \in \mathbb{Z}, \quad x^m \cdot x^n = x^{m+n} \quad \text{and}$$

$$(x^m)^n = x^{mn}.$$

6. The **Zero Multiplication Property** holds for $\mathbb{R}$. That is,

$$\forall\, x, y \in \mathbb{R}, \text{if } x \cdot y = 0, \text{ then either } x = 0 \text{ or } y = 0.$$

7. The **Trichotomy Law**. Every pair of real numbers is uniquely comparable. That is,

$$\forall\, x, y \in \mathbb{R}, \text{ exactly one of the following holds:}$$

$$x < y \quad \text{or} \quad x = y \quad \text{or} \quad x > y.$$

Of course, $x \leq y$ means that either $x < y$ or $x = y$.
Note that $x > y$ may also be written as $y < x$.

8. We have transitivity of inequalities. That is,

$$\forall\, x, y, z \in \mathbb{R}, \text{ if } x < y \text{ and } y < z, \text{ then } x < z.$$

Of course, the same holds for $\leq$. In mixed cases, the strict inequality $<$ overrides.

9. Inequalities are preserved under addition. That is,

$$\forall\, x, y, z, w \in \mathbb{R}, \text{ if } x < y \text{ and}$$

$$z < w, \text{ then } x + z < y + w.$$

Of course, the same holds for $\leq$. In mixed cases, the strict inequality $<$ overrides.

10. Multiplication of an inequality by a positive real number preserves it, whereas multiplication by a negative real number reverses it. That is,

$$\forall\, x, y, z \in \mathbb{R} \text{ such that } x < y,$$

(a) if $z > 0$, then $zx < zy$.

(b) if $z < 0$, then $zx > zy$.

11. Squares are nonnegative. That is,

$$\forall\, x \in \mathbb{R}, \text{ if } x \neq 0 \text{ then } x^2 > 0.$$

Of course, $0^2 = 0$.

12. Note that $\mathbb{Z}^+$ and $\mathbb{N}$ are subsets of $\mathbb{Z}$. Further, $\mathbb{Z}$ and $\mathbb{R}^+$ are subsets of $\mathbb{R}$. Thus, most of the properties of $\mathbb{R}$ described above restrict down to these subsets, with the following exceptions. All of the properties quantified above for $x, y, z, w \in \mathbb{R}$ hold for $x, y, z, w \in \mathbb{Z}$ except 3(d) (since $\frac{1}{x} \notin \mathbb{Z}$ unless $x = \pm 1$). Also, 3(b) does not hold for $\mathbb{Z}^+, \mathbb{N}$, and $\mathbb{R}^+$.

13. The **Least Upper Bound Property**. Given any nonempty subset $A \subseteq \mathbb{R}$ for which $U = \{u \in \mathbb{R} : \forall\, a \in A, a \leq u\}$ is nonempty, there exists a unique real number $l \in U$ such that $\forall\, u \in U, \; l \leq u$. We denote this value l by $\mathrm{lub}(A)$.

14. The **Archimedean Principle**.

$$\forall\, x \in \mathbb{R}, \exists\, n \in \mathbb{Z} \text{ such that } x < n.$$

15. The **Density of the Rationals**.

$$\forall\, x, y \in \mathbb{R}, \text{if } x < y, \text{ then } \exists\, q \in \mathbb{Q} \text{ such that } x < q < y.$$

16. For each $b \in \mathbb{R}$ and $n \in \mathbb{Z}^+$, if either

(i) $b > 0$ or

(ii) $b < 0$ and n is odd,

then there exists a unique $y \in \mathbb{R}$ such that $y^n = b$ and y has the same sign as b. We denote this value y by $\sqrt[n]{b}$ or $b^{\frac{1}{n}}$.

17. There is a unique value b^a associated to any pair a, b of real numbers with $b > 0$ in such a way that

(a) $\forall\, m \in \mathbb{Z}, n \in \mathbb{Z}^+, \; b^{\frac{m}{n}} = \sqrt[n]{b^m} = (\sqrt[n]{b})^m$,

(b) $1^a = 1$,

(c) if $\qquad\qquad b > 1, \qquad\qquad$ then

$\qquad b^a = \mathrm{lub}(\{b^{\frac{m}{n}} \; : \; m \in \mathbb{Z}, n \in \mathbb{Z}^+, \frac{m}{n} \leq a\})$,

(d) if $\qquad\qquad b < 1, \qquad\qquad$ then

$\qquad b^a = \mathrm{lub}(\{b^{\frac{m}{n}} \; : \; m \in \mathbb{Z}, n \in \mathbb{Z}^+, \frac{m}{n} \geq a\})$,

(d) $\forall\, a_1, a_2 \in \mathbb{R}, \; b^{a_1} b^{a_2} = b^{a_1 + a_2}$, and

(f) $\forall\, a_1, a_2 \in \mathbb{R}, \; (b^{a_1})^{a_2} = b^{a_1 a_2}$.

B Pseudocode

Pseudocode is an algorithm presentation format suggestive of the code for many programming languages but lacking the formal syntax rules of any one language. This appendix is designed to serve as a quick introduction to and easy reference for the pseudocode used to present algorithms throughout this book.

To describe our pseudocode, we make use of some key words. Each algorithm uses **variables**, whose values can be changed by the algorithm. However, at any point in the implementation of an algorithm, each variable will have a fixed value. Similarly, an **expression**, which may depend upon constants, variables, or the results of an algorithm, will also have a fixed value. Expressions that take on the value True or False are called **conditions**.

Each algorithm in this book is preceded by a description of its assumed input data. This includes the input variables as well as properties of those variables. The start of the algorithm itself is signified by the word **Algorithm**.

The most basic element of an algorithm is an **instruction**. More generally, an **instruction block** is a list of instructions preceded by \begin and followed by \end, as shown in Figure B.1. However, an instruction block can consist of just one instruction.

Each algorithm in this book ends with an instruction of the form

$$\text{Return } \textit{variable.}$$

The value of *variable* (or *variables*) at that ending point of the algorithm is what we consider to be the result of the algorithm.

The most basic type of instruction that we use is an **assignment instruction**, as shown in Figure B.2. It is with such an instruction that we assign a value to a variable.

For example, the instruction

$$\text{Let } x = 2.$$

assigns the value 2 to the variable x. If the variable on the left-hand side of an assignment instruction also appears

---

\begin
instruction 1.
$\vdots$
instruction n.
\end.

FIGURE B.1
Instruction block.

Let *variable = expression.*

FIGURE B.2
Assignment instruction.

on the right-hand side, then the value of the variable prior to the instruction is used to evaluate the expression on the right-hand side. The resulting value is then assigned to the variable on the left-hand side. For example, the instruction block

$$\begin{aligned}&\texttt{\textbackslash begin}\\&\text{Let } x = 2.\\&\text{Let } x = 3x.\\&\texttt{\textbackslash end}.\end{aligned}$$

If *condition*, then
   instruction block 1.
Otherwise,
   instruction block 2.

FIGURE B.3
Conditional instruction.

results in the value 6 being assigned to the variable x.

A **conditional instruction** has the form shown in Figure B.3. If this instruction is entered with the value of *condition* being True, then *instruction block* 1 is performed (and *instruction block* 2 is skipped). If the value of *condition* is False, then *instruction block* 2 is performed (and *instruction block* 1 is skipped). If the "Otherwise, *instruction block* 2" portion of a conditional instruction is omitted, then nothing is performed when the value of *condition* is False. For example, if the instruction

If $x < 4$, then
   Let $x = x + 1$.
Otherwise,
   Let $x = x - 1$.

is entered with the value $x = 2$, then it results in the value 3 being assigned to x. If it is entered with the value $x = 6$, then it results in the value 5 being assigned to x. In contrast, if the instruction

If $x < 4$, then
   Let $x = x + 1$.

For *variable* = *start value* to
   end value,
   instruction block.

FIGURE B.4
For loop.

is entered with the value $x = 6$, then no action results (and the value $x = 6$ is maintained).

A **loop instruction** is an instruction that enables us to iterate a step multiple times in an algorithm. Moreover, the number of iterations can depend on the values of variables set in the algorithm. We consider two types of loop instructions.

A **for Loop** has the form shown in Figure B.4, where *start value* and *end value* are integers with *start value* no bigger than *end value*. When it is entered, *variable* is assigned the value *start value* and the following two steps are performed: (1) *instruction block* and (2) increase of the value of *variable* by 1. As long as the value of *variable* is no greater than *end value*, these two steps are repeated. For example, suppose we have $x = 1$ and $n = 3$ when we enter the instruction

For $i = 1$ to n,
   Let $x = 2x$.

Since $n = 3$, the step "Let $x = 2x$" will be performed three times. The first will result in $x = 2$, the second in $x = 4$, and the third in $x = 8$. Thus, the final result is that the value 8 is assigned to x. Note that, when *start value* is no smaller than *end value*, there is an analog of the for loop in Figure B.4 that uses "down to" in place of "to" and decreases the value of *variable* by 1 upon each iteration.

A **while loop** has the form shown in Figure B.5. When it is entered, if *condition* is True, then *instruction block* is performed. (Note that *instruction block* may cause the value of *condition* to change.) As long as *condition* remains True, *instruction block* is repeated. The while loop ends when *condition* becomes False. For example, suppose $x = 2.78$ when we enter the instruction

> While $x > 1$,
>> Let $x = x - 1$.

Since $2.78 > 1$, we let $x = 2.78 - 1 = 1.78$. Since $1.78 > 1$, we next let $x = 1.78 - 1 = 0.78$. Since $0.78 \not> 1$, the final result is that $x = 0.78$.

While *condition*,
    instruction block.

FIGURE B.5
While loop.

C Answers to Selected Exercises

Here we include the answers to the odd-numbered exercises from each section and all of the exercises from the review sections. However, each answer listed may be more brief than that required by its exercise. Students are expected to flesh out the answers given to obtain the answers requested.

Introduction

1. 10.
3. 23.
5. 46.
7. 75.
9. 171.

11. 111011.
13. 1010100.
15. 1110101.
17. 100110000.
19. 10000000000.

21. Using T for Tails and H for Heads, we see the 16 possibilities.

```
T T T T
T T T H
T T H T
T T H H
T H T T
T H T H
T H H T
T H H H
H T T T
H T T H
H T H T
H T H H
H H T T
H H T H
H H H T
H H H H
```

23. 115.
25. 1679.
27. 16,712.
29. 3529.

31. 23,166.
33. 50.
35. 73.
37. 165.

39. 214.

41. $3b$.

43. 75.

45. *acdc*.

47. (a) 1463. (b) 333.

49. (a) 54,613. (b) 598b.

51. 100111.

53. 101011001100.

55. 2^m.

57. A one followed by $2n$ zeros.

59. It is divisible by 4 if and only if it ends in 00.

61. $8^m - 1$.

63. The number is divisible by 7 if and only if the sum of its digits is divisible by 7.

65. First rewrite the number in binary, and then group the digits into blocks of size 4 to convert to hexadecimal.

Chapter 1

Section 1.1

1. A true statement.

3. Not a statement.

5. Not a statement.

7.
p	q	$p \vee \neg q$
F	F	T
F	T	F
T	F	T
T	T	T

9.
p	q	r	$\neg p \to (q \wedge r)$
F	F	F	F
F	F	T	F
F	T	F	F
F	T	T	T
T	F	F	T
T	F	T	T
T	T	F	T
T	T	T	T

11.
p	q	r	$(p \to q) \vee r$
F	F	F	T
F	F	T	T
F	T	F	T
F	T	T	T
T	F	F	F
T	F	T	T
T	T	F	T
T	T	T	T

13. They differ in the two rows in which q is true.

15. They differ in the two rows in which p is true and q is false.

17.

p	q	$p \vee q$	$p \rightarrow p \vee q$
F	F	F	T
F	T	T	T
T	F	T	T
T	T	T	T

19. (a)

p	$\neg p$	$\neg\neg p$
F	T	F
T	F	T

(b)

$\underline{t}$	$\neg\underline{t}$	$\underline{f}$	$\neg\underline{f}$
T	F	F	T

21.

p	$\underline{t}$	$\underline{t} \rightarrow p$	$p \rightarrow \underline{t}$
F	T	F	T
T	T	T	T

23.

p	q	$p \vee q$	$\neg(\neg p \wedge \neg q)$
F	F	F	F
F	T	T	T
T	F	T	T
T	T	T	T

25.

p	q	$p \rightarrow p \vee q$	$\underline{t}$
F	F	T	T
F	T	T	T
T	F	T	T
T	T	T	T

27.

p	q	r	$(p \wedge q) \wedge r$	$p \wedge (q \wedge r)$	$(p \vee q) \vee r$	$p \vee (q \vee r)$
F	F	F	F	F	F	F
F	F	T	F	F	T	T
F	T	F	F	F	T	T
F	T	T	F	F	T	T
T	F	F	F	F	T	T
T	F	T	F	F	T	T
T	T	F	F	F	T	T
T	T	T	T	T	T	T

29.

p	q	r	$(p \oplus q) \oplus r$	$p \oplus (q \oplus r)$
F	F	F	F	F
F	F	T	T	T
F	T	T	T	T
F	T	F	F	F
T	F	F	T	T
T	F	T	F	F
T	T	T	F	F
T	T	F	T	T

31.

p	q	r	$p \wedge (q \vee r)$	$(p \wedge q) \vee (p \wedge r)$	$p \vee (q \wedge r)$	$(p \vee q) \wedge (p \vee r)$
F	F	F	F	F	F	F
F	F	T	F	F	F	F
F	T	F	F	F	F	F
F	T	T	F	F	T	T
T	F	F	F	F	T	T
T	F	T	T	T	T	T
T	T	F	T	T	T	T
T	T	T	T	T	T	T

33.

p	q	$\neg (p \wedge q)$	$\neg p \vee \neg q$	$\neg (p \vee q)$	$\neg p \wedge \neg q$
F	F	T	T	T	T
F	T	T	T	F	F
T	F	T	T	F	F
T	T	F	F	F	F

35.

p	q	$\neg (p \oplus q)$	$p \leftrightarrow q$
F	F	T	T
F	T	F	F
T	F	F	F
T	T	T	T

37. They differ when p is false, q is true, and r is false.

39. They differ when p is true, q is false, and r is true.

41.

p	q	$p \oplus q$	$\neg p \oplus \neg q$
F	F	F	F
F	T	T	T
T	F	T	T
T	T	F	F

43. (a) $\neg q \rightarrow p$. (b) $q \rightarrow \neg p$. (c) $\neg p \rightarrow q$. (d) $p \wedge q$.

45. (a) $r \rightarrow p \wedge \neg q$. (b) $\neg r \rightarrow \neg p \vee q$. (c) $\neg p \vee q \rightarrow \neg r$. (d) $p \wedge \neg q \wedge \neg r$.

47. (a) If Ted has a failing grade, then Ted's average is less than 60. (b) If Ted has a passing grade, then Ted's average

is at least 60. (c) If Ted's average is at least 60, then Ted has a passing grade. (d) Ted's average is less than 60, and Ted has a passing grade.

49. (a) If George is going to a movie or going dancing, then George feels well. (b) If George is not going to a movie and not going dancing, then George does not feel well. (c) If George does not feel well, then George is not going to a movie and not going dancing. (d) George feels well, and George is not going to a movie and not going dancing.

51. $\neg p \wedge q$

53. $p \vee (\neg q \wedge r)$.

55. Helen's average is less than 90, or Helen is not getting an A.

57. (a)

p	q	r	$(p \wedge q) \to r$	$\neg p \vee (q \to r)$
F	F	F	T	T
F	F	T	T	T
F	T	F	T	T
F	T	T	T	T
T	F	F	T	T
T	F	T	T	T
T	T	F	F	F
T	T	T	T	T

(b) $p \wedge q \to r \equiv \neg(p \wedge q) \vee r \equiv (\neg p \vee \neg q) \vee r \equiv \neg p \vee (\neg q \vee r) \equiv \neg p \vee (q \to r)$.

59. $p \wedge (q \vee r \vee s) \equiv p \wedge (q \vee (r \vee s)) \equiv (p \wedge q) \vee (p \wedge (r \vee s)) \equiv (p \wedge q) \vee ((p \wedge r) \vee (p \wedge s)) \equiv (p \wedge q) \vee (p \wedge r) \vee (p \wedge s)$.

61. $(p \wedge q \wedge \neg r) \vee (p \wedge \neg q \wedge r) \equiv (p \wedge (q \wedge \neg r)) \vee (p \wedge (\neg q \wedge r)) \equiv p \wedge ((q \wedge \neg r) \vee (\neg q \wedge r)) \equiv p \wedge (q \oplus r)$.

63. $p \wedge (\neg(q \wedge r)) \equiv p \wedge (\neg q \vee \neg r) \equiv (p \wedge \neg q) \vee (p \wedge \neg r)$.

65. Since $p \wedge q \wedge r \to p \wedge q$ is a tautology, the result follows from the Absorption Rule.

67. (i) $p \vee q \equiv \neg(\neg p \wedge \neg q)$ and $p \to q \equiv \neg(p \wedge \neg q)$.
(ii) $p \wedge q \equiv \neg(\neg p \vee \neg q)$ and $p \to q \equiv \neg p \vee q$.
(iii) $p \wedge q \equiv \neg(p \to \neg q)$ and $p \vee q \equiv \neg p \to q$.

69. (a) $(P \wedge Q) \vee \neg Q = S$.

(b)

P	Q	S
0	0	1
0	1	0
1	0	1
1	1	1

(c) Yes.

71. (a) $P \vee (\neg Q \wedge R) = S$.

(b)

P	Q	R	S
0	0	0	0
0	0	1	1
0	1	0	0
0	1	1	0
1	0	0	1
1	0	1	1
1	1	0	1
1	1	1	1

(c) No.

73.

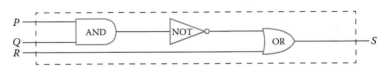

75.

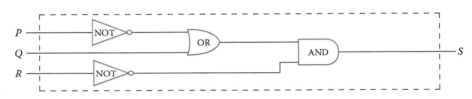

77. (a) $(P \vee Q) \wedge \neg(P \wedge Q) = S$. (b) $(P \wedge \neg Q) \vee (\neg P \wedge Q) = S$. The definition uses fewer gates.

Section 1.2

1. True.

3. True.

5. False.

7. True.

9. $\{2, 4, 6\}$

11. $\{\{1\}, \{4\}\}$

13. $\{x : x \in \mathbb{R}$ and $x^3 - 4x^2 + 5x - 6 = 0\}$, which happens to equal $\{3\}$.

15. $\{n : n \in \mathbb{Z}$ and $n < -10\}$.

17. $(0, \infty)$.

19. $[0, 0]$.

21. $(1, \infty)$.

23. $(-1, 1)$.

25. True.

27. False.

29. True.

31. True.

33. $\subset, \subseteq$.

35. $\subseteq, =$.

37. $\in, \subset, \subseteq$.

39. Finite. $|A| = 5$.

41. Infinite.

43. Finite. $|E| = 9$.

45. Finite. $|G| = 2$.

47. If yes, then he should not. If no, then he should. Hence, either way, there is a contradiction.

49. False.

51. 2.

53. true.

55. 1.

Section 1.3

1. $\forall\, x \in \mathbb{R}, x^2 + 1 > 0$.

3. $\exists\, n \in \mathbb{Z}$ such that $\frac{1}{n} \in \mathbb{Z}$.

5. $\exists\, n \in \mathbb{N}$ such that $\forall\, x \in \mathbb{R}, x^n \geq 0$.

7. $\exists\, x \in \mathbb{R}$ such that $\forall\, y \in \mathbb{R}$, if $2 \leq y \leq 3$, then $1 \leq xy < 2$. Also, $\exists\, x \in \mathbb{R}$ such that $\forall\, y \in [2,3], 1 \leq xy < 2$.

9. $\exists\, x, y \in \mathbb{R}$ such that $x + y \in \mathbb{Z}$ and $xy \notin \mathbb{Z}$.

11. $\forall\, x, y \in \mathbb{R}$, if $x < y$, then $e^x < e^y$.

13. $\exists\, x \in [-2, 2]$ such that $x^3 \notin [0, 8]$. Negation is true.

15. $\exists\, x \in \mathbb{R}^+$ such that $x^2 > 4$ and $x \leq 2$. Original is true.

17. $\forall\, n \in \mathbb{Z}, \exists\, m \in \mathbb{Z}$ such that $nm \geq 1$. Original is true.

19. $\exists\, m, n \in \mathbb{Z}$ such that $m + n \notin \mathbb{Z}$. Original is true.

21. $\forall\, n \in \mathbb{Z}, \frac{1}{n} \notin \mathbb{Z}$.

23. $\exists\, x \in \mathbb{R}$ such that $x^2 + 1 \leq 0$.

25. $\forall\, n \in \mathbb{N}, \exists\, x \in \mathbb{R}$ such that $x^n < 0$.

27. $\forall\, x \in \mathbb{R}, \exists\, y \in \mathbb{R}$ such that $2 \leq y \leq 3$ and $xy \notin [1, 2)$.

29. $\forall\, x, y \in \mathbb{R}, x + y \notin \mathbb{Z}$ or $xy \in \mathbb{Z}$.

31. There is a student at Harvard University whose age is at most 17.

33. There exists a truly great accomplishment that is immediately possible.

35. There is such a thing as bad publicity.

37. (a) A real function f is not constant iff $\forall\, c \in \mathbb{R}$, $\exists\, x \in \mathbb{R}$ such that $f(x) \neq c$. (b) $\exists\, x, y \in \mathbb{R}$ such that $f(x) \neq f(y)$.

39. $\exists\, x, y \in \mathbb{R}$ such that $x < y$ and $f(x) \geq f(y)$.

41. $\exists\, x, y \in \mathbb{R}$ such that $x \leq y$ and $f(x) > f(y)$.

43. $\forall\, M \in \mathbb{R}, \exists\, x \in \mathbb{R}$ such that $f(x) > M$.

45. True. An if-then statement is true whenever its hypothesis is false.

47. Say $\mathcal{U} = \{a, b\}$. The statement is equivalent to "$p(a) \wedge p(b)$."

49. Say $\mathcal{U} = \{a, b\}$. The logical equivalences are equivalent to $\neg[p(a) \wedge p(b)] \equiv \neg p(a) \vee \neg p(b)$ and $\neg[p(a) \vee p(b)] \equiv \neg p(a) \wedge \neg p(b)$.

Section 1.4

1. $A^c = \{4\}$, $B^c = \{1,2\}$, $A \cap B = \{3\}$, $A \cup B = \{1,2,3,4\}$, $A \backslash B = \{1,2\}$, $B \backslash A = \{4\}$, and $A \triangle B = \{1,2,4\}$.

3. $A^c = (-\infty, -1] \cup [1, \infty)$, $B^c = (-\infty, 0) \cup (1, \infty)$, $A \cap B = [0,1)$, $A \cup B = (-1,1]$, $A \backslash B = (-1,0)$, $B \backslash A = \{1\}$, and $A \triangle B = (-1,0) \cup \{1\}$.

5. $A^c = \mathbb{Z}^-$, $B^c = \mathbb{Z}^- \cup \{0\}$, $A \cap B = \mathbb{Z}^+$, $A \cup B = \mathbb{N}$, $A \backslash B = \{0\}$, $B \backslash A = \emptyset$, and $A \triangle B = \{0\}$.

7. $A^c = [3, \infty)$, $B^c = (0,2)$, $A \cap B = [2,3)$, $A \cup B = (0, \infty)$, $A \backslash B = (0,2)$, $B \backslash A = [3, \infty)$, and $A \triangle B = (0,2) \cup [3, \infty)$.

9. Yes.

11. No.

13. $\{(1,2),(1,4),(3,2),(3,4)\}$.

15. $\{(3,5),(5,5),(7,5),(9,5)\}$.

17.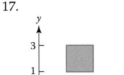

19.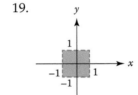

21.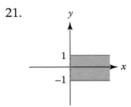

23.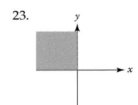

25. $\{(1,2,1),(1,2,2),(1,4,1),(1,4,2),(3,2,1),(3,2,2), (3,4,1),(3,4,2)\}$.

27. $\{(a,a,a),(a,a,b),(a,b,a),(a,b,b),(b,a,a),(b,a,b),(b,b,a), (b,b,b)\}$.

29. $\{\emptyset, \{a\}, \{b\}, \{c\}, \{a,b\}, \{a,c\}, \{b,c\}, \{a,b,c\}\}$.

31. $\{\emptyset, \{1\}, \{2\}, \{3\}, \{4\}, \{1,2\}, \{1,3\}, \{1,4\}, \{2,3\}, \{2,4\}, \{3,4\}, \{1,2,3\}, \{1,2,4\}, \{1,3,4\}, \{2,3,4\}, \{1,2,3,4\}\}$.

33. 1024.

35. $\emptyset, \{\pi\}, (-2,7]$, and $\mathbb{Z}$.

37. False.

39. False.

41. $(A \triangle B) \triangle C = A \triangle (B \triangle C)$.

43. $A \cap (B \cup C) = (A \cap B) \cup (A \cap C)$ and $A \cup (B \cap C) = (A \cup B) \cap (A \cup C)$.

45.

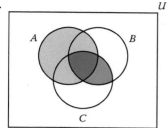
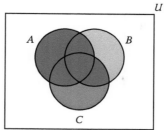

$A \cup (B \cap C)$ is the shaded portion of the left diagram. $(A \cup B) \cap (A \cup C)$ is the dark shaded portion of the right diagram. Both portions are the same.

47.

$A \cup B$ is the shaded portion of the left diagram. B is shaded in the right diagram. Both portions are the same.

49. $A \cap (A^c \cup B \cup C) = (A \cap A^c) \cup (A \cap B) \cup (A \cap C) = \emptyset \cup (A \cap B) \cup (A \cap C) = (A \cap B) \cup (A \cap C)$.

51. $(A \cap B \cap C^c) \cup (A \cap B^c \cap C) = (A \cap (B \cap C^c)) \cup (A \cap (B^c \cap C)) = A \cap ((B \cap C^c) \cup (B^c \cap C)) = A \cap ((B \setminus C) \cup (C \setminus B)) = A \cap (B \bigtriangleup C)$.

53. $A \cap ((B \cap C)^c) = A \cap (B^c \cup C^c) = (A \cap B^c) \cup (A \cap C^c)$.

55. Since $A \cap B \cap C \subseteq A \cap B$, the Absorption Rule yields the desired result.

57. Yes, as sets, but not as lists.

59. `disjoint[x_, y_] := (Intersection[x, y] == {}).`

61. `symmDiff[x_, y_] := Union[Complement[x, y], Complement[y, x]].`

63. In both cases, `intersect` is performed before `union`.

65. They test whether A is a subset of B. The first should be more efficient, since the power set can be much larger than the given set and thus expensive to compute.

67. `compU := x -> U minus x;.`

Section 1.5

1.

p	q	$p \rightarrow q$	$\neg q$	$\neg p$
F	F	T	T	T
F	T	T	F	
T	F	F	T	
T	T	T	F	

3.

p	q	r	$p \to r$	$q \to r$	$p \vee r$	r
F	F	F	T	T	F	
F	F	T	T	T	F	
F	T	F	T	F	T	
F	T	T	T	T	T	T
T	F	F	F	T	T	
T	F	T	T	T	T	T
T	T	F	F	F	T	
T	T	T	T	T	T	T

5.

p	q	$p \wedge q$	p
F	F	F	
F	T	F	
T	F	F	
T	T	T	T

7.

p	q	$p \wedge q$
F	F	
F	T	
T	F	
T	T	T

9. Invalid. Consider when p is false and q is true.

11. Valid.

13. Invalid. Consider when p is true and q is true.

15. Valid.

17. (a)

p	q	r	$p \to r$	$q \to r$	$p \vee q \to r$
F	F	F	T	T	T
F	F	T	T	T	T
F	T	F	T	F	
F	T	T	T	T	T
T	F	F	F	T	
T	F	T	T	T	T
T	T	F	F	F	
T	T	T	T	T	T

(b)

	Statement form	Justification
1.	$p \to r$	Given
2.	$q \to r$	Given
3.	$p \vee q$	Given
4.	$p \vee q \to r$	(1), (2), Part (a)
5.	$\therefore$ r	(3), (4), Direct implication

(c)

	Statement form	Justification
1.	$p \to s$	Given
2.	$q \to s$	Given
3.	$r \to s$	Given
4.	$p \vee q \vee r$	Given
5.	$p \vee q \to s$	(1), (2), Part (a)
6.	$p \vee q \vee r \to s$	(3), (5), Part (a)
7.	$\therefore$ s	(4), (6), Direct implication

19. Invalid, since its argument form $p \vee q$; $\therefore p$ is invalid.

21. Invalid, since its argument form $p \vee \neg p$; $\neg p$; $\therefore q$ is invalid.

23.

Statement form	Justification
1. $p \rightarrow q$	Given
2. $q \rightarrow r$	Given
3. p	Given
4. $p \rightarrow r$	(1), (2), Transitivity of $\rightarrow$
5. $\therefore r$	(3), (4), Direct implication

25.

Statement form	Justification
1. $p \rightarrow r$	Given
2. $p \wedge q$	Given
3. p	(2), In particular
4. $\therefore r$	(1), (3), Direct implication

27.

Statement form	Justification
1. $p \wedge (q \vee r)$	Given
2. $(p \wedge q) \rightarrow s$	Given
3. $(p \wedge r) \rightarrow s$	Given
4. $(p \wedge q) \vee (p \wedge r)$	(1), Distributivity
5. $\therefore s$	(2), (3), (4), Two separate cases

29.

Statement form	Justification
1. $p \rightarrow q$	Given
2. p	Given
3. $\neg p \vee q$	(1), Substitution of equivalent
4. $\neg\neg p$	(2), Double negative
5. $\therefore q$	(3), (4), Eliminating a possibility

31.

Statement form	Justification
1. $\forall x \in \mathcal{U}, p(x) \rightarrow q(x)$	Given
2. $a \in \mathcal{U}$	Given
3. $\neg q(a)$	Given
4. $p(a) \rightarrow q(a)$	(1), (2), Principle of Specification
5. $\therefore \neg p(a)$	(3), (4), Contrapositive implication

33.

Statement form	Justification
1. $\forall x \in \mathcal{U}, p(x) \rightarrow q(x)$	Given
2. $\forall x \in \mathcal{U}, \neg q(x)$	Given
3. Let $a \in \mathcal{U}$ be arbitrary.	Assumption
4. $p(a) \rightarrow q(a)$	(1), (3), Principle of Specification
5. $\neg q(a)$	(2), (3), Principle of Specification
6. $\neg p(a)$	(4), (5), Contrapositive implication
7. $\therefore \forall x \in \mathcal{U}, \neg p(x)$	(3), (6), Principle of Generalization

35.

Statement form	Justification
1. $\forall x \in \mathcal{U}, p(x)$	Given
2. $\forall x \in \mathcal{U}, q(x)$	Given
3. $a \in \mathcal{U}$	Given
4. $p(a)$	(1), (3), Principle of Specification
5. $q(a)$	(2), (3), Principle of Specification
6. $\therefore\ p(a) \wedge q(a)$	(4), (5), Obtaining and

37.

Statement form	Justification
1. $\forall x \in \mathcal{U}, p(x)$	Given
2. $\forall x \in \mathcal{U}, q(x)$	Given
3. Let $a \in \mathcal{U}$ be arbitrary.	Assumption
4. $p(a) \wedge q(a)$	(1), (2), (3), Exercise 35
5. $\therefore\ \forall x \in \mathcal{U}, p(x) \wedge q(x)$	(3), (4), Principle of Generalization

39.

Statement form	Justification
1. $\forall x \in \mathcal{U}, p(x) \vee q(x)$	Given
2. $a \in \mathcal{U}$	Given
3. $q(a) \rightarrow r(a)$	Given
4. $p(a) \vee q(a)$	(1), (2), Principle of Specification
5. $p(a) \rightarrow p(a) \vee r(a)$	Tautology
6. $r(a) \rightarrow p(a) \vee r(a)$	Tautology
7. $q(a) \rightarrow p(a) \vee r(a)$	(3), (6), Transitivity of $\rightarrow$
8. $\therefore\ p(a) \vee r(a)$	(4), (5), (7), Separate cases

41. Let $\mathcal{U} = \mathbb{Z}$, $p(n) = $ "$n^2 < 0$," and $q(n) = $ "$n^2 \geq 0$."

43. Let $\mathcal{U} = \mathbb{R}$, $p(x) = $ "$x \geq 0$," $q(x) = $ "$x \leq 0$," and $a = 0$.

45. Invalid. The form of the argument

$$\forall x \in \mathcal{U}, p(x) \rightarrow q(x)$$

$$r$$

$$q(2)$$

$$\therefore\ p(2)$$

is invalid.

47. Valid. The form of the argument

$$\forall x \in \mathcal{U}, p(x) \rightarrow q(x)$$

$$a \in \mathcal{U}$$

$$p(a) \vee r(a)$$

$$\therefore\ q(a) \vee r(a)$$

is valid. Note that $\mathcal{U} = \mathbb{Z}$, $a = -2$, $p(n) = $ "$n < 0$," $q(n) = $ "$-n > 0$," and $r(n) = $ "$n = 0$."

49. If $\forall x, y \in \mathcal{U}, p(x, y)$ holds and $a, b \in \mathcal{U}$, then $p(a, b)$ holds.

Review

1. It is a true statement.

2.

p	q	$p \to \neg q$
F	F	T
F	T	T
T	F	T
T	T	F

3.

p	q	$p \vee q$	$(\neg p \wedge q) \vee p$
F	F	F	F
F	T	T	T
T	F	T	T
T	T	T	T

4.

p	q	$\neg (p \wedge \neg q)$	$\neg p \vee q$
F	F	T	T
F	T	T	T
T	F	F	F
T	T	T	T

5.

p	q	$(p \to q) \vee (q \to p)$
F	F	T
F	T	T
T	F	T
T	T	T

6. Yes.

7. Yes.

8. (a) $p \vee \neg q \to p$. (b) $\neg p \wedge q \to \neg p$. (c) $\neg p \to \neg p \wedge q$. (d) $\underline{f}$.

9. If the program compiles, then the program does not contain a syntax error.

10. $p \wedge (\neg q \vee r)$.

11. Steve is not doing his homework and Steve is going to the basketball game.

12. $\neg p \wedge (q \vee \neg r) \equiv (\neg p \wedge q) \vee (\neg p \wedge \neg r) \equiv (\neg p \wedge q)$
    $\vee \neg (p \vee r)$.

13. $(p \wedge q \wedge \neg r) \vee (\neg p \wedge q \wedge \neg r) \equiv (p \wedge (q \wedge \neg r))$
    $\vee (\neg p \wedge (q \wedge \neg r)) \equiv (p \vee \neg p) \wedge (q \wedge \neg r) \equiv \underline{t} \wedge (q \wedge \neg r) \equiv$
    $q \wedge \neg r$.

14. $((P \wedge Q) \vee \neg Q) \wedge R = S$.

P	Q	R	S
0	0	0	0
0	0	1	1
0	1	0	0
0	1	1	0
1	0	0	0
1	0	1	1
1	1	0	0
1	1	1	1

15.

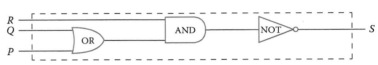

16.

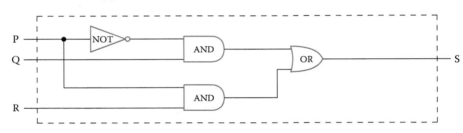

17. True.

18. True.

19. False.

20. False.

21. True.

22. False.

23. True.

24. False.

25. False.

26. True.

27. False.

28. $\{4, 6, 8, 10, 12\}$.

29. $\{x \ : \ x \in \mathbb{R} \text{ and } x^5 + x^4 + x^3 + x^2 + x + 1 = 0\}$. It happens to be $\{-1\}$.

30. $\{x \ : \ x \in \mathbb{R} \text{ and } -3 < x \leq -1\}$.

31. 2.

32. Read Example [1.22.]

33. $\forall \, n \in \mathbb{Z}, \ 2^n \in \mathbb{Z}$.

34. $\exists \, n \in \mathbb{Z}$ such that $2^n > 1000$.

35. $\forall \, x, y \in \mathbb{R}$, if $y \neq 0$, then $\frac{x}{y} \in \mathbb{R}$.

36. $\forall \, x \in \mathbb{R}$, if $x \in (1, 4]$, then $\frac{1}{x} \in [\frac{1}{4}, 1)$.

37. $\forall \, m, n \in \mathbb{Z}, m + n \in \mathbb{Z}$.

38. $\forall \, x \in \mathbb{R}, \exists \, y \in \mathbb{R}$ such that $y^3 = x$.

39. $\exists\, x \in \mathbb{N}$ such that $x^2 \notin \mathbb{N}$ or $\frac{1}{2x} \in \mathbb{N}$.

40. $\forall\, x \in \mathbb{R}, x^2 - x + 1 \neq 0$.

41. $\exists\, x \in \mathbb{R}$ such that $x^3 < 0$ and $x \geq 0$.

42. $\exists\, x, y \in \mathbb{R}$ such that $(x + y)^2 \neq x^2 + 2xy + y^2$.

43. $\forall\, n \in \mathbb{Z}, n < 0$ and $n^2 - 1 \leq 0$.

44. $\exists\, x \in \mathbb{R}$ such that $\forall\, n \in \mathbb{Z}, x^n \leq 0$.

45. Truth is always popular, or it is sometimes wrong.

46. (a) $\{2\}$. (b) $\{1, 2, 3, 5\}$. (c) $\{5\}$. (d) $\{1, 3, 5\}$. (e) $\{(1, 2), (1, 5),$
    $(2, 2), (2, 5), (3, 2), (3, 5)\}$. (f) $\{\emptyset, \{2\}, \{5\}, \{2, 5\}\}$.

47. $(-\infty, -1]$.

48. $\{0\}$.

49. $\{1, 2, 3, 4, 6, 8\}$.

50. $[4, 5]$.

51. $\{a, b, c, d, e\}$.

52. No.

53. $\{(x, p), (x, q), (y, p), (y, q), (z, p), (z, q)\}$.

54.

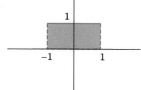

55. $\{(1, 1)\}$.

56. $\{\emptyset, \{x\}, \{y\}, \{z\}, \{x, y\}, \{x, z\}, \{y, z\}, \{x, y, z\}\}$.

57. $(A^c \cap B^c)^c = (A^c)^c \cup (B^c)^c = A \cup B$.

58. $A^c \cap (B \cup C^c) = (A^c \cap B) \cup (A^c \cap C^c) = (A^c \cap B) \cup$
    $(A \cup C)^c$.

59. $(A \cap B^c) \cup (A \cap B) = A \cap (B^c \cup B) = A \cap \mathcal{U} = A$.

60. $(A^c \cap B \cap C^c) \cup (A^c \cap (B^c \cup C)) = (A^c \cap (B \cap C^c)) \cup$
    $(A^c \cap (B^c \cup C)) = A^c \cap ((B \cap C^c) \cup (B^c \cup C)) =$
    $A^c \cap ((B \cap C^c) \cup (B \cap C^c)^c) = A^c \cap \mathcal{U} = A^c$.

61.

p	q	r	$p \wedge q$	$p \rightarrow r$	$q \wedge r$
F	F	F	F	T	
F	F	T	F	T	
F	T	F	F	T	
F	T	T	F	T	
T	F	F	F	F	
T	F	T	F	T	
T	T	F	T	F	
T	T	T	T	T	T

62.

p	q	r	$p \to q$	$q \to r$	$r \to p$	$p \leftrightarrow r$
F	F	F	T	T	T	T
F	F	T	T	T	F	
F	T	F	T	F	T	
F	T	T	T	T	F	
T	F	F	F	T	T	
T	F	T	F	T	T	
T	T	F	T	F	T	
T	T	T	T	T	T	T

63. Valid.

64. Valid.

65. Invalid. Consider when p is false, q is true, and r is false.

66. The argument is invalid, since its form $p \vee q$; q; $\therefore \neg p$ is invalid.

67.

	Statement form	Justification
1.	$p \to (q \vee r)$	Given
2.	$\neg q \wedge \neg r$	Given
3.	$\neg(q \vee r)$	(2), De Morgan's Law
4.	$\therefore \ \neg p$	(1), (3), Contrapositive implication

68.

	Statement form	Justification
1.	$\neg r$	Given
2.	$p \to q$	Given
3.	$q \to r$	Given
4.	$p \to r$	(2), (3), Transitivity of $\to$
5.	$\therefore \ \neg p$	(1), (4), Contrapositive implication

69.

	Statement form	Justification
1.	$\forall\, x \in \mathcal{U}, p(x) \wedge q(x)$	Given
2.	Let $a \in \mathcal{U}$ be arbitrary	Assumption
3.	$p(a) \wedge q(a)$	(1), (2), Principle of Specification
4.	$p(a)$	(3), In Particular
5.	$\therefore \ \forall\, x \in \mathcal{U}, p(x)$	(2), (4), Principle of Generalization

70.

	Statement form	Justification
1.	$\forall\, x \in \mathcal{U}, p(x) \vee q(x)$	Given
2.	$a \in \mathcal{U}$	Given
3.	$\neg q(a)$	Given
4.	$p(a) \vee q(a)$	(1), (2), Principle of Specification
5.	$\therefore \ p(a)$	(3), (4), eliminating a possibility

71.

	Statement form	Justification
1.	$\forall\, x \in \mathcal{U}, p(x) \vee \neg q(x)$	Given
2.	$\forall\, x \in \mathcal{U}, q(x)$	Given
3.	Let $a \in \mathcal{U}$ be arbitrary	Assumption
4.	$p(a) \vee \neg q(a)$	(1), (3), Principle of Specification
5.	$q(a)$	(2), (3), Principle of Specification
6.	$\neg\neg q(a)$	(5), Double negative
7.	$p(a)$	(4), (6), Eliminating a possibility
8.	$\therefore\ \forall\, x \in \mathcal{U}, p(x)$	(3), (7), Principle of Generalization

72. It has the form

$$\forall\, x \in \mathcal{U}, p(x) \vee q(x)$$

$$\forall\, x \in \mathcal{U}, q(x) \vee r(x)$$

$$\therefore\ \forall\, x \in \mathcal{U}, p(x) \vee r(x)$$

which can be seen to be invalid.

Chapter 2

Section 2.1

1. Use $y = 3x - 5$.

3. $A = \{2, 4\}$.

5. $A = B = \{1\}$.

7. $n = -3$.

9. $m = -3, n = 2$.

11. $A = B = \mathbb{Z}$.

13. $x = -5$.

15. *Proof.* The polynomial $x^2 - 1$ factors as $(x + 1)(x - 1)$. From the zero multiplication property, the solutions to the equation $(x + 1)(x - 1) = 0$ occur when $x + 1 = 0$ or $x - 1 = 0$. That is, $x = -1$ and $x = 1$ are the two distinct real roots of $x^2 - 1$. $\square$

17. $\frac{2 \pm \sqrt{-16}}{2} \notin \mathbb{R}$. Note that $x^2 - 2x + 5 > 0$, for all $x \in \mathbb{R}$.

19. (a) \$12,366.19. (b) Let $P = 4{,}900$, and note that $4{,}900(1.075)^{10} = 10{,}099.05 > 10{,}000$.

21. $A = \emptyset$.

23. $(-2, -1) \cup (1, 2)$ is not an interval.

25. Counterexample: $(\frac{1}{2})^2 \leq \frac{1}{2}$.

27. Counterexample: $3^2 > 2^3$.

29. $x = -11$.

31. $0 < 2\pi$ but $\sin(0) \not< \sin(2\pi)$.

33. False, since 0 is nonnegative but not positive.

35. False. $A = \{1\}, B = \{2, 3\}$.

37. Let $A = \{1, 2\}, B = \{1, 3\}, C = \{1\}$.

39. *Proof.* We have $\emptyset \cup \{3\} = \{3\} = \emptyset \bigtriangleup \{3\}, \{1\} \cup \{3\} = \{1, 3\} = \{1\} \bigtriangleup \{3\}$, and $\{1, 2\} \cup \{3\} = \{1, 2, 3\} = \{1, 2\} \bigtriangleup \{3\}$. □

41. Observe that $9(2) = 18$ and $1 + 8 = 9$; $9(3) = 27$ and $2 + 7 = 9$; $9(4) = 36$ and $3 + 6 = 9$; $9(5) = 45$ and $4 + 5 = 9$; $9(6) = 54$ and $5 + 4 = 9$; $9(7) = 63$ and $6 + 3 = 9$; $9(8) = 72$ and $7 + 2 = 9$; $9(9) = 81$ and $8 + 1 = 9$.

43.

n	$11n$	Hundreds − tens + ones
19	209	$2 - 0 + 9 = 11 = 11(1)$
20	220	$2 - 2 + 0 = 0 = 11(0)$
21	231	$2 - 3 + 1 = 0 = 11(0)$
22	242	$2 - 4 + 2 = 0 = 11(0)$
23	253	$2 - 5 + 3 = 0 = 11(0)$
24	264	$2 - 6 + 4 = 0 = 11(0)$
25	275	$2 - 7 + 5 = 0 = 11(0)$
26	286	$2 - 8 + 6 = 0 = 11(0)$
27	297	$2 - 9 + 7 = 0 = 11(0)$
28	308	$3 - 0 + 8 = 11 = 11(1)$
29	319	$3 - 1 + 9 = 11 = 11(1)$

45. 3, 7, 31, and 127 are prime.

47. $6 \mapsto 3 \mapsto 10 \mapsto 5 \mapsto 16 \mapsto 8 \mapsto 4 \mapsto 2 \mapsto 1$.

Section 2.2

1. *Proof.* Let $x \in \mathbb{R}^+$. So $x > 0$. Multiplying by -1 gives $-x < 0$. So $-x \in \mathbb{R}^-$. □

3. *Proof.* Suppose $x \in \mathbb{R}$ and $x \in (2, 4)$. That is, $2 < x < 4$. Multiplication by 2 gives $4 < 2x < 8$. Thus, $2x \in (4, 8)$. □

5. Counterexample: When $x = \frac{1}{4}$, we have $\sqrt{\frac{1}{4}} \not< \frac{1}{4}$.

7. Counterexample: When $x = -3$, we have $x < 2$ and $x^2 \geq 4$.

9. *Proof.* Suppose $x \in \mathbb{R}$ and $x < -2$. Since $x < 0$, multiplication by x gives $x^2 > -2x$. Since $-2 < 0$, multiplying $x < -2$ by -2 gives $-2x > (-2)^2$. Transitivity of $>$ gives $x^2 > (-2)^2$. That is, $x^2 > 4$. □

11. *Proof.* Suppose $0 < x < y$. Since $x > 0$, we have $x^2 = x \cdot x < x \cdot y$. Since $y > 0$, we have $x \cdot y < y \cdot y = y^2$. Hence, $x^2 < x \cdot y < y^2$. □

13. *Proof.* Suppose $R > 2$. So $2I < RI = 10$. Division by 2 gives $I < 5$. $\square$

15. *Proof.* Suppose f is a constant real function. So we have $c \in \mathbb{R}$ such that $\forall x \in \mathbb{R}$, $f(x) = c$. Observe that $\forall x \in \mathbb{R}$, $(2f)(x) = 2f(x) = 2c$. So $2f$ is constant. $\square$

17. *Proof.* Suppose f is a periodic real function. So we have $p \in \mathbb{R}^+$ such that $\forall x \in \mathbb{R}$, $f(x + p) = f(x)$. Observe that $\forall x \in \mathbb{R}$, $f^2(x + p) = [f(x + p)]^2 = [f(x)]^2 = f^2(x)$. So f^2 is periodic. $\square$

19. *Proof.* Suppose f and g are nondecreasing real functions. Suppose $x \leq y$ are real numbers. Observe that $(f+g)(x) = f(x) + g(x) \leq f(y) + g(y) = (f+g)(y)$. So $f+g$ is nondecreasing. $\square$

21. *Sketch.* If $\quad \forall x \in \mathbb{R}$, $f(x) = c \quad$ and $\quad g(x) = d, \quad$ then $\forall x \in \mathbb{R}$, $(fg)(x) = f(x)g(x) = cd$. $\square$

23. *Proof.* Suppose f is periodic. So we have $p \in \mathbb{R}^+$ such that $\forall x \in \mathbb{R}$, $f(x + p) = f(x)$. Observe that $\forall x \in \mathbb{R}$, $(f + c)(x + p) = f(x + p) + c = f(x) + c = (f + c)(x)$. So $f + c$ is periodic. $\square$

25. *Proof.* Suppose f is increasing. Suppose $x < y$ are real numbers. So $f(x) < f(y)$. Since $c > 0$, multiplication by c gives $cf(x) < cf(y)$. That is, $(cf)(x) < (cf)(y)$. So cf is increasing. $\square$

27. *Proof.* Let A be a square. So A is a rectangle. Hence, A is a parallelogram. $\square$

29. *Proof.* Let A be a rectangle and a rhombus. Hence, A is a rectangle with all sides congruent. That is, A is a square. $\square$

31. *Proof.* Suppose $a \in \mathbb{Z} \cap \mathbb{R}^+$. So $a \in \mathbb{Z}$ and $a > 0$. Thus, $a \in \mathbb{N}$. $\square$

33. *Proof.* Suppose $x \in \mathbb{R}^+$. So $x > 0$. That is, $x \not< 0$. In particular, $x \not< 0$. Hence, $x \notin \mathbb{R}^-$. That is, $x \in (\mathbb{R}^-)^c$. $\square$

35. *Proof.* Suppose $\quad x \in A$. Hence, $\quad x \in A \quad$ or $\quad x \in B$. So $x \in A \cup B$. $\square$

37. *Proof.* Suppose $A \subseteq A \cap B$ and suppose $x \in A$. It follows that $x \in A \cap B$. That is, $x \in A$ and $x \in B$. In particular, $x \in B$. Therefore, $A \subseteq B$. $\square$

39. *Proof.* Suppose $A \subseteq B$. Suppose $x \in A \cap C$. So $x \in A$ and $x \in C$. Since $x \in A$ and $A \subseteq B$, we get $x \in B$. So $x \in B$ and $x \in C$. Thus, $x \in B \cap C$. $\square$

41. $x \in (A \cap B) \cap C \quad$ iff $\quad (x \in A \wedge x \in B) \wedge x \in C \quad$ iff $x \in A \wedge (x \in B \wedge x \in C)$ iff $x \in A \cap (B \cap C)$.

43. $x \in A \cap B$ iff $x \in A \wedge x \in B$ iff $x \in B \wedge x \in A$ iff $x \in B \cap A$.

45. Counterexample: Let $A = \{1\}$ and $B = \{2\}$. Since $A \times B = \{(1, 2)\}$ and $B \times A = \{(2, 1)\}$, we see that $A \times B \neq B \times A$.

47. $x \in A \cup (B \cap C)$ iff $x \in A \vee (x \in B \wedge x \in C)$ iff
$(x \in A \vee x \in B) \wedge (x \in A \vee x \in C)$ iff $x \in (A \cup B) \cap (A \cup C)$.

49. $x \in (A \cap B)^c$ iff $\neg(x \in A \cap B)$ iff $\neg(x \in A \wedge x \in B)$ iff
$\neg(x \in A) \vee \neg(x \in B)$ iff $x \in A^c \vee x \in B^c$ iff $x \in A^c \cup B^c$.

Section 2.3

1. *Proof.* ($\rightarrow$) Suppose $x \in \mathbb{R}^-$. So $x < 0$. Multiplication by -1 gives $-x > 0$. That is, $-x \in \mathbb{R}^+$. ($\leftarrow$) Suppose $-x \in \mathbb{R}^+$. So $-x > 0$. Multiplication by -1 gives $x = (-1)(-x) < 0$. That is, $x \in \mathbb{R}^-$. $\square$

3. *Sketch.* $x = 2x \leftrightarrow 0 = 2x - x = x$. $\square$

5. *Proof.* ($\rightarrow$) Suppose $x^3 > 0$. Note $x \neq 0$ (since $0^3 = 0$). So $x^2 > 0$ and $\frac{1}{x^2} > 0$. Multiplying both sides of $x^3 > 0$ by $\frac{1}{x^2}$ gives $x > 0$. ($\leftarrow$) Suppose $x > 0$. Since $x^2 > 0$, multiplication by x^2 gives $x^3 > 0$. $\square$

7. *Sketch.* $4 - x < 2$ iff $-x < -2$ iff $x > 2$. $\square$

9. *Sketch.* $x^4 - 16 = (x^2 + 4)(x^2 - 4)$ and $x^2 + 4 \neq 0$. $\square$

11. *Proof.* ($\rightarrow$) Done in Exercise 15 from Section 2.2. ($\leftarrow$) Suppose $2f$ is constant. So we have $c \in \mathbb{R}$ such that $\forall x \in \mathbb{R}$, $2f(x) = c$. Thus, $\forall x \in \mathbb{R}$, $f(x) = \frac{c}{2}$. So f is constant. $\square$

13. *Proof.* ($\rightarrow$) Suppose f is bounded above. So we have $M \in \mathbb{R}$ such that $\forall x \in \mathbb{R}$, $f(x) \leq M$. Observe that $\forall x \in \mathbb{R}$, $(f + 1)(x) = f(x) + 1 \leq M + 1$. So $f + 1$ is bounded above. ($\leftarrow$) Suppose $f + 1$ is bounded above. So we have $M \in \mathbb{R}$ such that $\forall x \in \mathbb{R}$, $(f + 1)(x) \leq M$. So, $\forall x \in \mathbb{R}$, $f(x) + 1 \leq M$. So $\forall x \in \mathbb{R}$, $f(x) \leq M - 1$. Thus, f is bounded above. $\square$

15. *Sketch.* ($\rightarrow$) If $\forall x \in \mathbb{R}$, $L \leq f(x) \leq M$, then $\forall x \in \mathbb{R}$, $f^2(x) \leq \max\{L^2, M^2\}$. ($\leftarrow$) If $\forall x \in \mathbb{R}$, $f^2(x) \leq M$, then $\forall x \in \mathbb{R}$, $-\sqrt{M} \leq f(x) \leq \sqrt{M}$. $\square$

17. *Sketch.* If f is periodic with period p, then $(2f)(x + p) = 2 \cdot f(x + p) = 2 \cdot f(x) = (2f)(x)$. If $2f$ is periodic with period p, then $f(x + p) = \frac{1}{2} \cdot (2f)(x + p) = \frac{1}{2} \cdot (2f)(x) = f(x)$. $\square$

19. *Sketch.* If 20 points are scored, then the number of touchdowns is at most 2. Considering the possibilities of 0, 1, or 2 touchdowns, only 2 is possible. $\square$

21. $x > 0$ and $-2 \leq x \leq 2$ if and only if $0 < x \leq 2$.

23. *Proof.* ($\subseteq$) Suppose $x \in \mathbb{N} \setminus (-1, 1)$. So $x \in \mathbb{Z}$, $x \geq 0$, and $x \notin (-1, 1)$. Hence, $x \geq 1$, and $x \in \mathbb{Z}^+$. ($\supseteq$) Suppose $x \in \mathbb{Z}^+$. So $x \in \mathbb{Z}$ and $x \geq 1$. Hence, $x \in \mathbb{N} \setminus (-1, 1)$. $\square$

25. $1 \leq x < 3$ and $2 \leq x < 4$ if and only if $2 \leq x < 3$.

27. Suppose $A \subseteq B$. Show that $A \cap B \subseteq A$ and $A \subseteq A \cap B$.

29. *Proof.* ($\subseteq$) Suppose $x \in (A\backslash C) \cap B$. So $x \in A\backslash C$ and $x \in B$. Hence, $x \in A$, $x \notin C$, and $x \in B$. Since $x \in B$ and $x \notin C$, we have $x \in B\backslash C$. Thus, $x \in A$ and $x \in B\backslash C$. That is, $x \in A \cap B\backslash C$. ($\supseteq$) Essentially, reverse the previous argument. $\square$

31. Suppose $A \cap B = A \cap C$. Show that $A \cap B \cap C \subseteq A \cap B$ and $A \cap B \subseteq A \cap B \cap C$.

33. *Proof.* ($\subseteq$) Suppose $x \in (A\backslash B)\backslash C$. So $x \in A\backslash B$ and $x \notin C$. So $x \in A$ and $x \in B^c \cap C^c$. Since $x \notin (B^c \cap C^c)^c$, De Morgan's Law tells us that $x \notin B \cup C$. Since $x \in A$ and $x \notin B \cup C$, we have $x \in A\backslash(B \cup C)$. ($\supseteq$) Essentially, reverse the previous argument. $\square$

35. *Proof.* Suppose $A \subseteq B$ and $(x, y) \in A^2$. Since $x, y \in A$ and $A \subseteq B$, it follows that $x, y \in B$. Hence, $(x, y) \in B^2$. $\square$

37. *Proof.* ($\subseteq$) Suppose $(x, y) \in (\mathcal{U} \times B)\backslash(A \times B)$. So $(x, y) \in \mathcal{U} \times B$ and $(x, y) \notin A \times B$. We have $(x \in \mathcal{U}$ and) $y \in B$. Since $(x, y) \notin A \times B$, it must be that $x \notin A$. That is, $x \in A^c$. Hence $(x, y) \in A^c \times B$. ($\supseteq$) Suppose $(x, y) \in A^c \times B$. So $x \in A^c$ and $y \in B$. Since $x \notin A$, we get $(x, y) \notin A \times B$. Since $(x, y) \in \mathcal{U} \times B$, we have $(x, y) \in (\mathcal{U} \times B)\backslash(A \times B)$. Thus, $(\mathcal{U} \times B)\backslash(A \times B) = A^c \times B$. $\square$

39. $(x, y) \in A \times (B \cap C)$ iff $x \in A \wedge y \in B \wedge y \in C$ iff $x \in A, y \in B \wedge x \in A, y \in C$ iff $(x, y) \in (A \times B) \cap (A \times C)$.

41. *Proof.* ($\rightarrow$) Suppose $A \times C = B \times C$. ($\subseteq$) Suppose $x \in A$. Since $(x, c) \in A \times C$, it follows that $(x, c) \in B \times C$. So $x \in B$. ($\supseteq$) Similar. ($\leftarrow$) Suppose $A = B$. ($\subseteq$) Suppose $(x, y) \in A \times C$. So $x \in A$ and $y \in C$. Since $A = B$, we have $x \in B$. So $(x, y) \in B \times C$. ($\supseteq$) Similar. $\square$

43. *Proof.* Suppose $S \in \mathcal{P}(A^c)\backslash\{\emptyset\}$. So $S \subseteq A^c$ and $S \neq \emptyset$. Since $S \neq \emptyset$, we have some $x \in S$. Since $S \subseteq A^c$, we have $x \in A^c$. That is, $x \in S$ and $x \notin A$. So $S \not\subseteq A$. Therefore, $S \notin \mathcal{P}(A)$. That is, $S \in \mathcal{P}(A)^c$. $\square$

45. *Sketch.* Suppose $S \subseteq A \cap B$. So $S \subseteq A$. Similarly, $S \subseteq B$. Thus, $S \in \mathcal{P}(A) \cap \mathcal{P}(B)$. $\square$

47. Let r be the average speed over the entire trip, let r_1 be the average speed over the first lap, and let r_2 be the average speed over the second lap. Observe that $\frac{2}{r} = \frac{1}{r_1} + \frac{1}{r_2}$. (a) $r = 48$ mph. (b) $\frac{2r_1 r_2}{r_1 + r_2} = r = 60$ if and only if $r_1 = \frac{30 r_2}{r_2 - 30} > 30$.

Section 2.4

1. *Proof.* Suppose not. Let s be the smallest element of $(1, 2)$. Observe that $\frac{s+1}{2}$ is a smaller element of $(1, 2)$. (Think about it.) This is a contradiction. $\square$

3. *Sketch.* If L were the largest element, then $L + 1$ would be larger. $\square$

5. *Proof.* Let A be a set, and suppose $A \cap \emptyset \neq \emptyset$. So we have an element $x \in A \cap \emptyset$. Hence, $x \in A$ and $x \in \emptyset$. However, $x \in \emptyset$ is impossible. This is a contradiction. So it must be that $A \cap \emptyset = \emptyset$. $\square$

7. *Sketch.* Suppose $(0,1]$ has finite cardinality n. The list $1, \frac{1}{2}, \ldots, \frac{1}{n}, \frac{1}{n+1}$ of numbers in $(0,1]$ is then too long. $\square$

9. *Sketch.* Suppose $\{(x,y) : x, y \in \mathbb{R} \text{ and } y = \sqrt{x}\}$ has finite cardinality n. The list of elements $(1,1), (4,2), \ldots,$ $((n+1)^2, n+1)$ is then too long. $\square$

11. *Proof.* Suppose $(1,0) \neq \emptyset$. So there is a real number x such that $1 < x < 0$. In particular, $1 < 0$. This is a contradiction. $\square$

13. *Proof.* Suppose $\mathbb{R}^+ \cap \mathbb{R}^- \neq \emptyset$. So there is a real number x such that $x \in \mathbb{R}^+$ and $x \in \mathbb{R}^-$. However, it is impossible to have both $x > 0$ and $x < 0$. $\square$

15. (a) *Sketch.* Suppose to the contrary that Tracy wins the election. So every other candidate must have also received fewer than $\frac{1}{n}$ of the votes. However, the total of the fractions of the votes for the n candidates would then be less than $n \cdot \frac{1}{n} = 1$, which is impossible. $\square$
(b) *Sketch.* Suppose to the contrary that Tracy comes in last. So every other candidate must have also received more than $\frac{1}{n}$ of the votes. However, the total of the fractions of the votes for the n candidates would then be more than $n \cdot \frac{1}{n} = 1$, which is impossible. $\square$

17. *Proof.* Suppose $b > a$ are real numbers. Since $\frac{a+b}{2} \in (a,b)$, we see that $(a,b) \neq \emptyset$. $\square$

19. *Proof.* Suppose $A \neq \emptyset$. So we have an element $x \in A$. Thus, $(x,x) \in A^2$. Hence, $A^2 \neq \emptyset$. $\square$

21. *Proof.* Suppose $A \subseteq B$. Suppose $(x,y) \in A^2$. Since $x \in A$ and $y \in A$ and $A \subseteq B$, we get $x \in B$ and $y \in B$. That is, $(x,y) \in B^2$. $\square$

23. *Proof.* Suppose $A \times B \neq \emptyset$. So we have some element $(x,y) \in A \times B$. In particular, $x \in A$. So $A \neq \emptyset$. $\square$

25. If $A = B$, then $A \subseteq B$.

27. If $A = B$, then $\mathcal{P}(A) = \mathcal{P}(B)$.

29. If $|A| = m$ and $|B| = n$, then $|A \times B| = mn$.

31. *Proof.* Suppose to the contrary that $x \in A$, $x \notin A \cap B$, and $x \in B$. Thus, $x \in A \cap B$ and $x \notin A \cap B$, a contradiction. $\square$

33. *Proof.* Suppose to the contrary that there are two distinct lines l and m that intersect in two or more points. That is, we have distinct points P and Q in their intersection. Since both l and m contain P and Q, we must have $l = m$, by the uniqueness assertion in Euclid's First Postulate. $\square$

35. (a) If we have $x, y \in \mathbb{R}$ with $x < y$ and $0 < f(x) \le f(y)$, then $f^2(x) \le f^2(y)$. (b) Yes.

37. If we have $M \in \mathbb{R}$ such that $\forall x \in \mathbb{R}$, $f(x) \le M$, then $\forall x \in \mathbb{R}$, $(f + 100)(x) \le M + 100$.

39. *Proof.* Suppose f is increasing. (Goal: f is not periodic.) Suppose $p \in \mathbb{R}^+$. Since $f(0 + p) > f(0)$, it cannot be that f is periodic. $\square$

41. The contrapositive, "If f is constant, then f^2 is constant," is easy to prove.

43. We prove the contrapositive. If $x \ne 0$, then $x^2 > 0$. In particular, $x^2 \ne 0$.

45. *Proof.* Suppose not. So, we have some $x > 0$ with $\frac{1}{x} < 0$. Hence, $1 = x \cdot \frac{1}{x} < 0$. This is a contradiction. $\square$

47. *Proof.* Suppose have some $0 < x < y$ with $0 < \frac{1}{x} \le \frac{1}{y}$. Hence, $1 = \frac{1}{x} \cdot x < \frac{1}{x} \cdot y$ and $\frac{1}{x} \cdot y \le \frac{1}{y} \cdot y = 1$. So $1 < 1$, a contradiction. $\square$

49. *Proof.* Suppose not. So, we have some $x \in (-\infty, -1) \cap (1, \infty)$. That is, $x < -1$ and $x > 1$. Hence, $1 < x < -1$, which is a contradiction. $\square$

Section 2.5

1. *Proof.* Suppose $A \subseteq B$ and $C \subseteq D$. (Goal: $A \cup C \subseteq B \cup D$.) Suppose $x \in A \cup C$. That is, $x \in A$ or $x \in C$. *Case 1:* $x \in A$. Since $A \subseteq B$, we get $x \in B$. So $x \in B$ or $x \in D$. Hence, $x \in B \cup D$. *Case 2:* $x \in C$. Since $C \subseteq D$, we get $x \in D$. So $x \in B$ or $x \in D$. Hence, $x \in B \cup D$. In either case, $x \in B \cup D$. Thus, $A \cup C \subseteq B \cup D$. $\square$

3. *Sketch.* Suppose $A \subseteq B$. ($\subseteq$) Suppose $x \in A \cup B$. If $x \in A$, then $x \in B$. If $x \in B$, then $x \in B$. ($\supseteq$) Suppose $x \in B$. Hence, $x \in A \cup B$. $\square$

5. *Proof.* ($\subseteq$) $A \cup A^c \subseteq \mathcal{U}$ since everything is in $\mathcal{U}$. ($\supseteq$) Suppose $x \in \mathcal{U}$. *Case 1:* $x \in A$. We have $x \in A$ or $x \in A^c$. So $x \in A \cup A^c$. *Case 2:* $x \notin A$. So $x \in A^c$. We have $x \in A$ or $x \in A^c$. So $x \in A \cup A^c$. In either case, $x \in A \cup A^c$. Hence, $\mathcal{U} \subseteq A \cup A^c$. $\square$

7. *Sketch.* $A \cup \mathcal{U} \subseteq \mathcal{U}$ can be seen through cases. $\mathcal{U} \subseteq A \cup \mathcal{U}$ is straightforward. $\square$

9. *Sketch.* ($\rightarrow$) Suppose $A \cup B \subseteq C$. So $A \subseteq A \cup B \subseteq C$ and $B \subseteq A \cup B \subseteq C$. ($\leftarrow$) Suppose $A \subseteq C$, $B \subseteq C$ and $x \in A \cup B$. If $x \in A$, then $x \in C$. If $x \in B$, then $x \in C$. $\square$

11. *Proof.* Suppose $x \in A \bigtriangleup B$. We use the characterization of $\bigtriangleup$ displayed after Definition 1.18. That is, $x \in (A \backslash B) \cup (B \backslash A)$. So, $x \in A \backslash B$ or $x \in B \backslash A$. If $x \in A \backslash B$, then $x \in A$. If $x \in B \backslash A$, then $x \in B$. In either case, $x \in A \cup B$. $\square$

13. *Sketch.* Suppose $x \in (A \cup B) \backslash C$. If $x \in A$, then $x \in A$. If $x \in B$, then $x \in B \backslash C$. $\square$

15. *Sketch.* $(\subseteq)$ Suppose $x \in A \cup (B \backslash C)$. *Case 1:* $x \in A$. We have $x \in A \cup B$ and $x \notin C \backslash A$. *Case 2:* $x \in B \backslash C$. We have $x \in A \cup B$ and $x \notin C \backslash A$. $(\supseteq)$ Suppose $x \in (A \cup B) \backslash (C \backslash A)$. Thus, $x \in A$ or $x \in B$, and $x \notin C$ or $x \in A$. *Case 1:* $x \in A$. We have $x \in A \cup (B \backslash C)$. *Case 2:* $x \notin A$. We have $x \in A \cup (B \backslash C)$. $\square$

17. $(x, y) \in (A \times B)^c$ iff $x \in A^c \vee y \in B^c$ iff $(x, y) \in (A^c \times \mathcal{U}) \cup (\mathcal{U} \times B^c)$.

19. $x \in A \cap (B \triangle C)$ iff $x \in A \wedge (x \in B \oplus x \in C)$ iff $x \in A \cap B \oplus x \in A \cap C$ iff $x \in (A \cap B) \triangle (A \cap C)$.

21. *Sketch.* Suppose $S \in \mathcal{P}(A) \cup \mathcal{P}(B)$. That is, $S \subseteq A$ or $S \subseteq B$. In either case, $S \subseteq A \cup B$. $\square$

23. *Sketch.* If bridge 2 is taken next, then bridge 3 must follow, with bridge 4 after that, leaving the tourist on the wrong side of the Tiber with no way to return. It bridge 4 is taken next, then bridge 3 must follow, with bridge 2 after that, leaving the tourist on the wrong side of the Tiber with no way to return. $\square$

25. *Sketch.* $-x \geq 0$ iff $x \leq 0$. Also, $-x < 0$ iff $x > 0$. Note that $-(-x) = x$. $\square$

27. *Sketch.* $1 - 2x \geq 0$ iff $x \leq \frac{1}{2}$. Also, $1 - 2x < 0$ iff $x > \frac{1}{2}$. In the second case, $-(1 - 2x) = 2x - 1$. $\square$

29. *Sketch.* $x^2 + 2ax + a^2 = 0$ iff $x = \frac{-2a \pm \sqrt{4a^2 - 4a^2}}{2} = -a$. Also, $x + 3 = 0$ iff $x = -3$. Either $-a = -3$ or not. $\square$

31. *Sketch.* If $x \geq 0$, then $|x|^2 = x^2$. If $x < 0$, then $|x|^2 = (-x)^2 = x^2$. $\square$

33. *Sketch.* If $x, y \geq 0$, then $|xy| = xy = |x||y|$. If $x \leq 0$ and $y \geq 0$, then $|xy| = -xy = (-x)y = |x||y|$. If $x \geq 0$ and $y \leq 0$, then $|xy| = -xy = x(-y) = |x||y|$. If $x, y \leq 0$, then $|xy| = xy = (-x)(-y) = |x||y|$. $\square$

35. *Sketch.* $(\rightarrow)$ Prove the contrapositive. $(\leftarrow)$ If $x < -1$ or $x > 1$, then $x^2 > 1$, whence $x^4 > 1$. $\square$

37. *Sketch.* Suppose $x^2 - y^2 = 0$. So $x + y = 0$ or $x - y = 0$. Therefore, $x = \pm y$. $\square$

39. *Sketch.* If $xy \geq 0$, then $|x + y| = |x| + |y|$. If $xy < 0$, then $|x + y| \leq \max\{|x|, |y|\} \leq |x| + |y|$. $\square$

41. *Sketch.* Let $A_n = \frac{180(n-2)}{n}$. If $n \geq 7$, then $128 < A_n < 180$ and no multiple of A_n can equal 360. If $n = 5$, then no multiple of $A_5 = 108$ can equal 360. Equilateral triangles ($n = 3$), squares ($n = 4$), and regular hexagons ($n = 6$) certainly do tile the floor as shown. $\square$

Review

1. *Sketch.* $x^2 + y^2 = 25$ fits each point. $\square$

2. *Sketch.* $x^4 - 2x^2 - 8 = (x^2 + 2)(x^2 - 4)$ and $x^2 + 2 \neq 0$. $\square$

3. *Sketch.* Let $m = 20, n = 5$. $\square$

4. *Sketch.* Let $x = 2$. $\square$

5. *Sketch.* Let $A = \emptyset, B = C = \mathbb{Z}$. $\square$

6. *Sketch.* Let $A = \{1, 2\}$. $\square$

7. An 85 on the third test yields an average of 75.

8. False. Consider $f(x) = \begin{cases} 1 & \text{if } x \geq 0, \\ -1 & \text{if } x < 0. \end{cases}$

9. *Sketch.* Let $A = B = \mathbb{Z}, C = \emptyset$. $\square$

10. *Sketch.* Let $A = B = \{1\}, C = \emptyset$. $\square$

11. *Proof.* Observe that $(-1)^4 = 1 = (-1)^2$, $0^4 = 0 = 0^2$, and $1^4 = 1 = 1^2$. $\square$

12. *Proof.* If $A = \{1\}$, then $A^3 = \{(1,1,1)\}$ and $|A^3| = 1$. If $A = \{2\}$, then $A^3 = \{(2,2,2)\}$ and $|A^3| = 1$. $\square$

13. *Proof.* Since $(-1, 0) \in \mathbb{Z} \times \mathbb{N}$ and $(-1, 0) \notin \mathbb{N} \times \mathbb{Z}$, it follows that $\mathbb{Z} \times \mathbb{N} \neq \mathbb{N} \times \mathbb{Z}$. $\square$

14. *Proof.* Let $n \in \mathbb{Z}^+$. So $n \geq 1$. Hence, $n \cdot n \geq n \cdot 1$. That is, $n^2 \geq n$. $\square$

15. If $2 \leq x \leq 4$, then $4 = 2^2 \leq x^2 \leq 4^2 = 16$.

16. If $\forall\, x \in \mathbb{R}$, $f(x) = c$, then $\forall\, x \in \mathbb{R}$, $(f + 1)(x) = c + 1$.

17. *Proof.* Suppose f is periodic and g is constant. So we have $p \in \mathbb{R}^+$ and $c \in \mathbb{R}$ such that $\forall\, x \in \mathbb{R}$, $f(x + p) = f(x)$ and $g(x) = c$. Observe that $\forall\, x \in \mathbb{R}$, $(f + g)(x + p) = f(x + p) + g(x + p) = f(x) + c = f(x) + g(x) = (f + g)(x)$. So $f + g$ is periodic. $\square$

18. *Sketch.* Suppose $\forall\, x \in \mathbb{R}$, $f(x) \leq M$ and $g(x) \geq L$. So $\forall\, x \in \mathbb{R}$, $(f - g)(x) = f(x) - g(x) = f(x) + (-g(x)) \leq M + (-L) = M - L$. $\square$

19. *Proof.* Let t_1, t_2, t_3 represent the test scores, in order. Suppose $t_1 \leq 40$. Since $t_2 \leq 100$ and $t_3 \leq 100$, we have an average of at most $\frac{40 + 100 + 100}{3} = 80$. $\square$

20. *Proof.* Suppose $A \subseteq C$. Suppose $x \in A \cap B$. So $x \in A$ and $x \in B$. Since $x \in A$ and $A \subseteq C$, we get $x \in C$. Thus, $A \cap B \subseteq C$. $\square$

21. *Proof.* Suppose $x \in A \backslash B$. So $x \in A$ and $x \notin B$. In particular, $x \in A$. $\square$

22. *Sketch.* $(A \backslash B)^c = (A \cap B^c)^c = A^c \cup (B^c)^c = A^c \cup B$. $\square$

23. *Proof.* Suppose $A \subset B$. Hence, we have $x \in B$ with $x \notin A$. That is, $x \in B \backslash A$. So $B \backslash A \neq \emptyset$. $\square$

24. *Sketch.* $1 \le x \le 2$ iff $2 \le 2x \le 4$. $\square$

25. *Sketch.* $3x - 2 \in (1, 4)$ iff $x \in (1, 2)$ iff $5 - 2x \in (1, 3)$. $\square$

26. *Sketch.* $x^2 - y^2 = 0$ iff $x + y = 0$ or $x - y = 0$. Since $x, y \in \mathbb{R}^+$, it is not possible that $x + y = 0$. $\square$

27. *Sketch.* The Trichotomy Law in Appendix A tells us that $\forall\, x, y \in \mathbb{R},\ x = y \oplus x > y \oplus y > x$. From this, it follows that $\forall\, x, y \in \mathbb{R},\ x \ne y \leftrightarrow x > y$ or $y > x$. Negating both sides of this equivalence gives the desired result. $\square$

28. Here, it is more convenient to use the characterization of constant functions given in Exercise 37(b) from Section 1.3 (and proven in Exercise 12 from Section 2.3).

29. *Proof.* $(\rightarrow)$ Done in Exercise 17 from Section 2.2. $(\leftarrow)$ Suppose f^2 is periodic. So we have $p \in \mathbb{R}^+$ such that $\forall\, x \in \mathbb{R},\ f^2(x + p) = f^2(x)$. Since f is nonnegative, $\forall\, x \in \mathbb{R},\ f(x + p) \ge 0$ and $f(x) \ge 0$. From Exercise 26, it follows that $\forall\, x \in \mathbb{R},\ f(x + p) = f(x)$. So f is periodic. $\square$

30. *Proof.* $(\rightarrow)$ Suppose $A^2 = B^2$. $(\subseteq)$ Suppose $x \in A$. So $(x, x) \in A^2 = B^2$. Hence, $x \in B$. $(\supseteq)$ Similar. So $A = B$. $(\leftarrow)$ Suppose $A = B$. $(\subseteq)$ Suppose $(x, y) \in A^2$. So $x \in A = B$ and $y \in A = B$. Hence, $(x, y) \in B^2$. $(\supseteq)$ Similar. So $A^2 = B^2$. $\square$

31. *Proof.* $(\rightarrow)$ Suppose $A \backslash B \subseteq C$. Suppose $x \in A$. *Case 1:* $x \in B$. We have $x \in B \cup C$. *Case 2:* $x \notin B$. So $x \in A \backslash B$. Hence, $x \in C$. We have $x \in B \cup C$. In both cases, $x \in B \cup C$. $(\leftarrow)$ Suppose $A \subseteq B \cup C$. Suppose $x \in A \backslash B$. So $x \in A$ and $x \notin B$. Since $x \in A$, we have $x \in B \cup C$. Since $x \notin B$, it must be that $x \in C$. Hence, $A \backslash B \subseteq C$. $\square$

32. *Proof.* Let $t_1, t_2, \dots, t_n$ be the test scores. $(\rightarrow)$ Suppose some test score t_k is less than 100. Then the average $\frac{t_1 + t_2 + \cdots + t_n}{n}$ is at most $\frac{(n-1)100 + t_k}{n} < \frac{(n-1)100 + 100}{n} = 100$. $(\leftarrow)$ Suppose $t_1 = t_2 = \cdots = t_n = 100$. The average is then $\frac{n(100)}{n} = 100$. $\square$

33. *Sketch.* $(\subseteq)$ Suppose $x \in (A \cap B) \backslash C$. Since $x \in A$ and $x \notin C$, we have $x \in A \backslash C$. Since $x \in B$ and $x \notin C$, we have $x \in B \backslash C$. $(\supseteq)$ Suppose $x \in (A \backslash C) \cap (B \backslash C)$. Since $x \in A$ and $x \in B$, we have $x \in A \cap B$. Also note that $x \notin C$. $\square$

34. *Sketch.* $(\subseteq)$ Suppose $(x, y) \in A \times (B \cup C)$. *Case 1:* $y \in B$. We get $(x, y) \in A \times B$. *Case 2:* $y \in C$. We get $(x, y) \in A \times C$. $(\supseteq)$ Suppose $(x, y) \in (A \times B) \cup (A \times C)$. *Case 1:* $(x, y) \in A \times B$. Since $y \in B$, $(x, y) \in A \times (B \cup C)$. *Case 2:* $(x, y) \in A \times C$. Since $y \in C$, $(x, y) \in A \times (B \cup C)$. $\square$

35. *Sketch.* Suppose $S \in \mathcal{P}(A) \cup \mathcal{P}(B)$. *Case 1:* $S \in \mathcal{P}(A)$. Since $A \subseteq A \cup B$, we get $S \subseteq A \cup B$. *Case 2:* $S \in \mathcal{P}(B)$. Since $B \subseteq A \cup B$, we get $S \subseteq A \cup B$. $\square$

36. *Sketch.* Suppose not. Let L be the largest element. Observe that $\frac{L}{2}$ is a larger element of $\mathbb{R}^-$. (Think about it.) This is a contradiction. $\square$

37. *Proof.* Suppose not. Let s be the smallest element of $(-1, 1)$. However, $\frac{-1+s}{2}$ is a smaller element of $(-1, 1)$. This is a contradiction. $\square$

38. *Proof.* Suppose $p \in \mathbb{R}^+$. Observe that $f(0 + p) = p \neq 0 = f(0)$. So f cannot be periodic. $\square$

39. *Sketch.* Suppose $M \in \mathbb{R}$. Let $L = \max\{M, 2\}$. Observe that $f(L) = L^2 > L \geq M$. So f cannot be bounded above. $\square$

40. *Proof.* Suppose $x < 0$. Repeated multiplication by x gives $x^2 > 0$, $x^3 < 0$, $x^4 > 0$, and finally $x^5 < 0$. $\square$

41. If $(x, y) \in A \times \emptyset$, then $y \in \emptyset$.

42. If $x \in A \cap B$, then $x \in A$.

43. If $x \in A$ and $y \in B$, then $(x, y) \in A \times B$.

44. *Proof.* Suppose $B = C$. ($\subseteq$) Suppose $(x, y) \in A \times B$. So $x \in A$ and $y \in B = C$. Hence, $(x, y) \in A \times C$. ($\supseteq$) Similar. Therefore, $A \times B = A \times C$. $\square$

45. *Sketch.* If $|A| = n$, then $|\mathcal{P}(A)| = 2^n$. $\square$

46. The contrapositive is proven in Example 2.11.

47. If $t_1, t_2, t_3, t_4 < 60$, then $\frac{t_1 + t_2 + t_3 + t_4}{4} < \frac{4(60)}{4} = 60$.

48. Mimic the proof of De Morgan's Law for two sets from Exercise 49 in Section 2.2.

49. *Sketch.* $(A \backslash C) \cup (B \backslash C) = (A \cap C^c) \cup (B \cap C^c) = (A \cup B) \cap C^c = (A \cup B) \backslash C$. $\square$

50. *Proof.* Suppose $x \in (A \triangle B) \cap (A \triangle C)$. So $x \in A \triangle B$ and $x \in A \triangle C$. Case 1: $x \in A$. It must be that $x \notin B$ (and $x \notin C$). In particular, $x \notin B \cap C$. Case 2: $x \notin A$. It must be that $x \in B$ and $x \in C$. Hence, $x \in B \cap C$. In both cases, $x \in A \triangle (B \cap C)$. $\square$

51. *Sketch.* If $A \times B \neq \emptyset$, choose $(x, y) \in A \times B$. Then $x \in A$ and $y \in B$, so $A \neq \emptyset$ and $B \neq \emptyset$. $\square$

52. *Proof.* ($\rightarrow$) Suppose $xy > 0$. Observe that $x \neq 0$ and $y \neq 0$. Case 1: $x > 0$. We see that $y = \frac{xy}{x} > 0$. Case 2: $x < 0$. We see that $y = \frac{xy}{x} < 0$. Thus, either $x, y > 0$ or $x, y < 0$. ($\leftarrow$) Suppose $x, y > 0$ or $x, y < 0$. Case 1: $x, y > 0$. We get $xy > 0$. Case 2: $x, y < 0$. We get $xy = (-x)(-y) > 0$. In both cases, $xy > 0$. $\square$

53. $\forall\, x \in \mathbb{R}, x^2 \geq 0$.

54. *Sketch.* $x^2 - 1 < 0$ iff $x^2 < 1$ iff $-1 < x < 1$. Also $-(x^2 - 1) = 1 - x^2$. $\square$

55. *Sketch.* Since $|x|^2 = x^2$ and $|x| \geq 0$, it follows that $|x| = \sqrt{x^2}$. $\square$

56. $x^2 - 6x + 8 = (x - 4)(x - 2)$. Now apply Exercise 52.

57. *Sketch.* $x = \frac{1}{x}$ iff $x^2 = 1$ iff $x^2 - 1 = 0$ iff $(x + 1)(x - 1) = 0$ iff $x + 1 = 0$ or $x - 1 = 0$ iff $x = -1$ or $x = 1$ iff $x = \pm 1$. $\square$

Chapter 3

Section 3.1

1. *Proof.* Let m be even and n be odd. So $m = 2j$ and $n = 2k + 1$ for some $j, k \in \mathbb{Z}$. Observe that $mn = (2j)(2k + 1) = 2(j(2k + 1))$. Since $j(2k + 1) \in \mathbb{Z}$, we see that mn is even. □

3. *Proof.* Suppose n is odd. So $n = 2k + 1$ for some $k \in \mathbb{Z}$. Observe that $n^2 = (2k + 1)^2 = 4k^2 + 4k + 1 = 2(2k^2 + 2k) + 1$. Since $2k^2 + 2k \in \mathbb{Z}$, we see that n^2 is odd. □

5. *Proof.* Suppose n is odd. So $n = 2k + 1$ for some $k \in \mathbb{Z}$. Thus, $\frac{n+1}{2} = k + 1 \in \mathbb{Z}$. □

7. *Sketch.* If $n = 2k$, then $(-1)^n = ((-1)^2)^k = 1^k = 1$. □

9. On. Off $= -1$ and On $= 1$.

11. *Sketch.* $a \cdot 0 = 0$. □

13. *Proof.* Suppose $a \mid 1$. So $1 = ak$ for some $k \in \mathbb{Z}$. Since $a, k \in \mathbb{Z}$, this is only possible if $a = k = \pm 1$. (Under any other conditions, $|ak| > 1$.) □

15. (a) *Sketch.* $a^2 - 1 = (a + 1)(a - 1)$. □ (b) R breaks into two $(a - 1) \times 1$ rectangles and an $(a - 1) \times (a - 1)$ square.

17. *Proof.* Suppose $a \mid b$ and $b \mid a$. So $b = aj$ and $a = bk$ for some $j, k \in \mathbb{Z}$. So $a = bk = a(jk)$. So $1 = jk$. Thus, $k \mid 1$. By Exercise 13, it follows that $k = \pm 1$. Therefore, $a = bk = \pm b$. □

19. *Sketch.* If $n = 2k$, then $n^2 = 4k^2$. □

21. No. Yes.

23. *Proof.* Suppose $a \mid b$ and $a \mid c$. So $b = aj$ and $c = ak$ for some $j, k \in \mathbb{Z}$. Note that $b - c = aj - ak = a(j - k)$. Since $j - k \in \mathbb{Z}$, we see that $a \mid (b - c)$. □

25. Yes.

27. $2, 3, 5, 7, 11, 13, 17, 19, 23, 29, 31, 37, 41, 43, 47, 53, 59, 61, 67, 71$.

29. *Proof.* Let p be a prime with $3 \mid p$. So $p = 3k$ for some $k \in \mathbb{Z}$. In fact, $k > 0$. Since p is prime and $3 \neq 1$, it must be that $k = 1$. Thus, $p = 3$. □

31. *Proof.* Let $p \in \mathbb{Z}$ with $p > 1$. ($\rightarrow$) Suppose p is prime. Suppose $r > 1$ and $s > 1$. Since the only positive divisors of p are 1 and p, we cannot have $rs = p$, so $rs \neq p$. ($\leftarrow$) Suppose $\forall\, r, s \in \mathbb{Z}$, if $r > 1$ and $s > 1$, then $rs \neq p$. Suppose t is a positive divisor of p. So $p = tu$, for some $u \in \mathbb{Z}$. Moreover, $u > 0$. Since $tu = p$, we must have $t \leq 1$ or $u \leq 1$. This forces $t = 1$ or $u = 1$. If $u = 1$, then $t = p$. So $t = 1$ or $t = p$. □

33. Negate the characterization given in Exercise 31.

35. *Sketch.* Suppose $n = rs$, where $r, s > 1$. Suppose to the contrary that both $r, s > \sqrt{n}$. Then $n = rs > n$, a contradiction. $\square$

37.

	2	3	4	5	6	7	8	9	10	11
12	13	14	15	16	17	18	19	20	21	22
23	24	25	26	27	28	29	30	31	32	33
34	35	36	37	38	39	40	41	42	43	44
45	46	47	48	49	50	51	52	53	54	55
56	57	58	59	60	61	62	63	64	65	66
67	68	69	70	71	72	73	74	75	76	77
78	79	80	81	82	83	84	85	86	87	88
89	90	91	92	93	94	95	96	97	98	99
100	101	102	103	104	105	106	107	108	109	110
111	112	113	114	115	116	117	118	119	120	121

The first 30 primes are $2, 3, 5, 7, 11, 13, 17, 19, 23, 29, 31, 37,$ $41, 43, 47, 53, 59, 61, 67, 71, 73, 79, 83, 89, 97, 101, 103, 107,$ $109, 113$.

39. 14.

41. 18.

43. 15.

45. (a) 2. (b) 4. (c) $\gcd(n, k)$.

47. *Proof.* Suppose $d_1, d_2 \in \mathbb{Z}$ both satisfy conditions (i), (ii), and (iii). By conditions (i) and (ii) for d_2 and condition (iii) for d_1 with $c = d_2$, we see that $d_2 \leq d_1$. A similar argument with d_1 and d_2 switched gives $d_1 \leq d_2$. Hence, $d_2 = d_1$. $\square$

49. *Proof.* Let $d = \gcd(m, n)$. So $d \mid m$ and $d \mid n$. Also, $d \mid (-m)$ and $d \mid n$. Suppose $c \mid (-m)$ and $c \mid n$. So $c \mid m$ and $c \mid n$. Thus, $c \mid d$. Hence, $d = \gcd(-m, n)$. $\square$

51. *Sketch.* $\gcd(m, -n) = \gcd(-n, m) = \gcd(n, m) = \gcd(m, n)$. $\square$

53. *Proof.* Let p and q be distinct primes. Suppose $d \in \mathbb{Z}^+$ with $d \mid p$ and $d \mid q$. Since $d \mid p$, we have $d = 1$ or $d = p$. If $d = p$, then $p \mid q$, giving $p = 1$ or $p = q$. Hence, $d \neq p$. Therefore, $d = 1$. Thus, $\gcd(p, q) = 1$. $\square$

55. Write $1 = 2(n) + 1(1 - 2n)$, and mimic the argument in the proof of Lemma 3.3.

57. 168.

59. 540.

61. (a) 16. (b) 30. (c) $\text{lcm}(a, b)$.

63. *Sketch.* Let $l = \text{lcm}(n, m)$. So $l > 0, n \mid l$, and $m \mid l$. If $k \in \mathbb{Z}^+$, $m \mid k$, and $n \mid k$, then $l \leq k$. $\square$

Section 3.2

1. $3 = 11 - 4(2)$.

3. $4 = 12(2) + 20(-1)$.

5. $2^{17} - 1$ is prime, $2^{19} - 1$ is prime, $2^{23} - 1 = 47 \cdot 178{,}481$, and $2^{29} - 1 = 233 \cdot 1103 \cdot 2089$.

7. *Sketch.* If not, then there are fewer than $10^{(10^8)}$ primes. □

9. (a) *Sketch.* $b^n - 1 = (b-1)(b^{n-1} + b^{n-2} + \cdots + b + 1)$. Since $b \geq 3$, both factors are larger than 1. □
   (b) *Proof.* Suppose $n \in \mathbb{Z}^+$ is not prime. So $n = rs$ for some integers $r, s \geq 2$. Since $2^r \geq 3$, it follows from part (a) that $2^n - 1 = (2^r)^s - 1$ is not prime. □

11. (a) 10 remainder 7. (b) 14 remainder 6.

13. (a) $45 = 7(6) + 3$. (b) $-37 = 4(-10) + 3$.

15. 7 and 3.

17. (a) 5 and 2. (b) -6 and 11.

19. 165 div $18 = 9$ full rows. 165 mod $18 = 3$ extra seats.

21. (a) 100111. (b) 127.

23. n and 0.

25. Because $\mathbb{Z}$ does not have a smallest element.

27. *Proof.* Let $a \in \mathbb{Z}$, and let S be a subset of $\mathbb{Z}$ such that $\forall\, x \in S$, $x \geq a$. Let $T = \{t \,:\, t = s - a$ for some $s \in S\}$. So $T \subseteq \mathbb{N}$. By the Well-Ordering Principle, T has a smallest element, say τ. Let $\sigma = \tau + a$. Observe that σ is the smallest element of S. (Think about it.) □

29.

n	$n^3 - n + 2$	$(n^3 - n + 2)$ mod 6
$6k$	$216k^3 - 6k + 2$	2
$6k+1$	$216k^3 + 108k^2 + 12k + 2$	2
$6k+2$	$216k^3 + 216k^2 + 66k + 8$	2
$6k+3$	$216k^3 + 324k^2 + 156k + 26$	2
$6k+4$	$216k^3 + 432k^2 + 282k + 62$	2
$6k+5$	$216k^3 + 540k^2 + 444k + 122$	2

31. *Proof.* Suppose $n \in \mathbb{Z}$ and $3 \nmid n$. So $n = 3q + r$ for some $q \in \mathbb{Z}$ and $r = 1$ or 2. *Case 1:* $r = 1$. Since $n^2 = (3q+1)^2 = 3(3q^2 + 2q) + 1$, we see that n^2 mod $3 = 1$. *Case 2:* $r = 2$. Since $n^2 = (3q+2)^2 = 3(3q^2 + 4q + 1) + 1$, we see that n^2 mod $3 = 1$. □

33. *Sketch.* If $n = 5k + 1$, then $n^4 - 1 = 5(125k^4 + 100k^3 + 30k^2 + 4k)$. If $n = 5k + 2$, then $n^4 - 1 = 5(125k^4 + 200k^3 + 120k^2 + 32k + 3)$. If $n = 5k + 3$, then $n^4 - 1 = 5(125k^4 + 300k^3 + 270k^2 + 108k + 16)$. If $n = 5k + 4$, then $n^4 - 1 = 5(125k^4 + 400k^3 + 480k^2 + 256k + 51)$. □

35. (a) 4. (b) -5. (c) 9. (d) -8.

37. (a) -5. (b) 3. (c) 5. (d) 11.

39. (a) $\lceil \frac{500}{32} \rceil = 16$. (b) $\lceil \frac{c}{m} \rceil$.

41. *Sketch.* Note that (i) $\lfloor x \rfloor + n \in \mathbb{Z}$, (ii) $\lfloor x \rfloor + n \le x + n$, (iii) $x + n < 1 + \lfloor x \rfloor + n$. $\square$

43. *Proof.* Suppose $n \in \mathbb{Z}$. Case 1: $n = 2k$ for some $k \in \mathbb{Z}$. Observe that $k \in \mathbb{Z}$, $\frac{n}{2} \le k$, and $k - 1 < \frac{n}{2}$. Hence, $\lceil \frac{n}{2} \rceil = k = \frac{n}{2}$. Case 2: $n = 2k + 1$ for some $k \in \mathbb{Z}$. Observe that $k + 1 \in \mathbb{Z}$, $\frac{n}{2} \le k + 1$, and $(k + 1) - 1 < \frac{n}{2}$. Hence, $\lceil \frac{n}{2} \rceil = k + 1 = \frac{n+1}{2}$. $\square$

45. *Sketch.* Write $n = 3k + r$, where $r = 0, 1$, or 2. Then $k = \frac{n}{3} - \frac{r}{3} \le \frac{n}{3}$ and $\frac{n}{3} = k + \frac{r}{3} < k + 1$, so $\lfloor \frac{n}{3} \rfloor = k = \frac{n-r}{3}$. $\square$

47. *Sketch.* We have $\lfloor x \rfloor + \lfloor y \rfloor \in \mathbb{Z}$ and $\lfloor x \rfloor + \lfloor y \rfloor \le x + y$. $\square$

49. Counterexample: Let $x = \frac{1}{2}$.

51. *Proof.* Let $x \in \mathbb{R}$. ($\rightarrow$) Suppose $x \notin \mathbb{Z}$. Then $\lfloor x \rfloor \ne x$, and it must be that $\lfloor x \rfloor < x$. Since $x \le \lceil x \rceil$, we get $\lfloor x \rfloor \ne \lceil x \rceil$. ($\leftarrow$) Suppose $x \in \mathbb{Z}$. It follows that $\lfloor x \rfloor = x = \lceil x \rceil$. $\square$

53. *Sketch.* Certainly $\lfloor x \rfloor \in \mathbb{Z}$ and $\lfloor x \rfloor \le \lfloor x \rfloor < \lfloor x \rfloor + 1$. $\square$

55. If n is odd, then $\lfloor \frac{n+1}{2} \rfloor = \frac{n+1}{2} = \lceil \frac{n}{2} \rceil$. If n is even, then $\lfloor \frac{n+1}{2} \rfloor = \frac{n}{2} = \lceil \frac{n}{2} \rceil$.

57. $\text{round}(x) = \lfloor x + \frac{1}{2} \rfloor$.

59. The statement is equivalent to the fact that 1 is the smallest positive integer.

61. *Proof.* Suppose $n_1, n_2 \in \mathbb{Z}$ with $n_1 \le x < n_1 + 1$ and $n_2 \le x < n_2 + 1$. Without loss of generality, say $n_1 \ge n_2$. Adding $n_2 \le x < n_2 + 1$ to $-n_1 - 1 < -x \le -1_2$, we get $n_2 - n_1 - 1 < 0 < n_2 - n_1 + 1$. Adding $n_1 - n_2$ to this inequality gives $-1 < n_1 - n_2 < 1$. So $0 \le n_1 - n_2 < 1$. From Exercise 59, it follows that $n_1 - n_2 = 0$. That is, $n_1 = n_2$. $\square$

63. *Proof.* Let $S = \{s : n = 2^r s$ where $r \in \mathbb{N}$ and $s \in \mathbb{Z}^+\}$. Since $n = 2^0 n$, it follows that $n \in S$ and thus S is nonempty. By the Well-Ordering Principle, S has a smallest element. Call it b. Since $b \in S$, there is some $a \in \mathbb{N}$ such that $n = 2^a b$. If b were even (so $\frac{b}{2} \in \mathbb{Z}$), then $n = 2^{a+1} \frac{b}{2}$, whence $\frac{b}{2}$ would be a smaller element of S than b. Therefore, b must be odd. $\square$

65. Let m and n be integers that are not both zero. We must show that there exists an integer d such that (i) $d > 0$, (ii) $d \mid m$ and $d \mid n$, and (iii) $\forall\, c \in \mathbb{Z}^+$, if $c \mid m$ and $c \mid n$, then $c \le d$. Let $d = \max\{a : a > 0, a \mid m$, and $a \mid n\}$.

67. 4.

69. No.

71. 3.

73. No. That is the ISBN for *How the Grinch Stole Christmas*.

75. *Sketch.* Let a and b be the consecutive digits. Note that $(3a + b) - (3b + a) = 2(a - b)$ is divisible by 10 iff $a - b$ is divisible by 5. $\square$

77. (a) 0110101.

(b)

Message	Code word
0000	0000000
0001	0001001
0010	0010011
0011	0011010
0100	0100110
0101	0101111
0110	0110101
0111	0111100
1000	1000100
1001	1001101
1010	1010111
1011	1011110
1100	1100010
1101	1101011
1110	1110001
1111	1111000

(c) 2. (d) Female, A^+. (e) 1010011 is one digit away from both 0010011 and 1010111.

79. "LQ KZMAMHUIAP."

81. "SELL IMCLONE."

Section 3.3

1. $x = 2, y = -5$.

3. $x = 3, y = -4$.

5. 12.

7. 22.

9. 8.

11. 1.

13. $2 = 14(-1) + 8(2)$.

15. $5 = 50(-2) + 35(3)$.

17. $3 = 81(3) + 60(-4)$.

19. $x = -5$ and $y = 23$.

21. No. For $m = 2, n = 3$, we can use $x = 2, y = -1$ or $x = -1, y = 1$.

23. *Proof.* Let x_0, y_0 be any fixed pair that gives $\gcd(m, n) = mx_0 + ny_0$. Observe that, $\forall k \in \mathbb{Z}$, $\gcd(m, n) = mx_0 + ny_0 = m(x_0 + kn) + n(y_0 - km)$. Therefore, $x = x_0 + kn$ and $y = y_0 - km$ gives a general solution to $\gcd(m, n) = mx + ny$. $\square$

25. No. $6 \mid (2 \cdot 3)$ but $6 \nmid 2$ and $6 \nmid 3$.

27. *Sketch.* Let $m_1 = m_2 = \cdots = m_n = m$. $\square$

29. *Sketch.* Let p be prime. It follows from Corollary 3.19 that $p \mid a \leftrightarrow p \mid a^m$, and $p \mid b \leftrightarrow p \mid b^n$. $\square$

31. *Proof.* Let $d = \gcd(m, n)$. So $d \mid m$ and $d \mid n$. Note that $d \mid m$ and $d \mid (n - m)$. If $c \mid m$ and $c \mid (n - m)$, then $c \mid m$ and $c \mid n$, whence $c \leq d$. So $\gcd(m, n - m) = d$. □

33. Argue that $\min\{mu_1 + nv_1 : mu_1 + nv_1 > 0\} = \min\{nu_2 + mv_2 : nu_2 + mv_2 > 0\}$ by using $(u_2, v_2) = (v_1, u_1)$.

35. From the given characterization, we get $\gcd(k, 0) = \min\{ku + 0v : ku + 0v > 0\} = \min\{ku : ku > 0\} = k \cdot 1 = k$.

37. *Proof.* Suppose $c \in \mathbb{Z}$, $c \mid m$, and $c \mid n$. So $m = ca$ and $n = cb$ for some $a, b \in \mathbb{Z}$. By Theorem 3.13, there are $x, y \in \mathbb{Z}$ such that $\gcd(m, n) = mx + ny$. Since $\gcd(m, n) = cax + cby = c(ax + by)$, we see that $c \mid \gcd(m, n)$. □

39. $(5n + 3)(7) + (7n + 4)(-5) = 1$.

41. (a) $ad - bc = 3(2) - 5(1) = 1$. (b) If the $ad - bc = 1$, then Corollary 3.14 tells us that a and b are relatively prime and that c and d are relatively prime. (c) No.

Section 3.4

1. $5\frac{1}{2} = \frac{11}{2}$.

3. $-13\frac{2}{5} = \frac{-67}{5}$.

5. $5.821 = \frac{5821}{1000}$.

7. $3.\overline{14} = \frac{311}{99}$.

9. $-4.3\overline{21} = \frac{-713}{165}$.

11. $12.75\overline{8} = \frac{11483}{900}$.

13. *Sketch.* If $r = \frac{a}{b}$, then $nr = \frac{na}{b}$. □

15. (a) *Sketch.* If $s = \frac{a}{b}$, then $-s = \frac{-a}{b}$. □
    (b) *Sketch.* If $r = \frac{a}{b}$ and $s = \frac{c}{d}$, then $r - s = \frac{ad - bc}{bd}$. □

17. *Proof.* Suppose $n \geq 0$. We can write $r = \frac{a}{b}$ where $a, b \in \mathbb{Z}$ with $b \neq 0$. Observe that $r^n = \frac{a^n}{b^n}$ and $a^n, b^n \in \mathbb{Z}$ with $b^n \neq 0$. So $r^n \in \mathbb{Q}$. □

19. $\frac{5}{3}$.

21. $\frac{-57}{8}$.

23. $\frac{157}{50}$.

25. $\frac{-189}{500}$.

27. 0.48.

29. $0.\overline{428571}$.

31. $0.5\overline{3}$.

33. Yes.

35. No.

37. When $p \nmid n$.

39. Yes. A fraction $\frac{a}{b}$ in lowest terms has a finite binary decimal expansion iff b is a power of 2.

41. *Proof.* Since $k > 0$, $2k + 1 > 0$. Since $(3k + 1)(-2) + (2k + 1)(3) = 1$, we have $\gcd(3k + 1, 2k + 1) = 1$. □

43. $\sqrt{2}$. It is not in $\mathbb{Q}$.

45. *Sketch.* Write $\sqrt{3} = \frac{a}{b}$ in lowest terms. So $b^2 3 = a^2$. Since $3 \mid a^2$, $3 \mid a$. Write $a = 3c$. So $b^2 3 = a^2 = 9c^2$. So $b^2 = 3c^2$. Since $3 \mid b^2$, $3 \mid b$. So $\gcd(a, b) \geq 3$, a contradiction. $\square$

47. *Sketch.* Write $\sqrt{13} = \frac{a}{b}$ in lowest terms. So $b^2 13 = a^2$. Since $13 \mid a^2$, $13 \mid a$. Write $a = 13c$. So $b^2 13 = a^2 = 13^2 c^2$. So $b^2 = 13c^2$. Since $13 \mid b^2$, $13 \mid b$. So $\gcd(a, b) \geq 13$, a contradiction. $\square$

49. *Sketch.* Write $\sqrt[3]{2} = \frac{a}{b}$ in lowest terms. So $b^3 2 = a^3$. Since $2 \mid a^3$, $2 \mid a$. Write $a = 2c$. So $b^3 2 = a^3 = 2^3 c^3$. So $b^3 = 2^2 c^3$. Since $2 \mid b^3$, $2 \mid b$. So $\gcd(a, b) \geq 2$, a contradiction. $\square$

51. *Sketch.* Write $\sqrt[3]{7} = \frac{a}{b}$ in lowest terms. So $b^3 7 = a^3$. Since $7 \mid a^3$, $7 \mid a$. Write $a = 7c$. So $b^3 7 = a^3 = 7^3 c^3$. So $b^3 = 7^2 c^3$. Since $7 \mid b^3$, $7 \mid b$. So $\gcd(a, b) \geq 7$, a contradiction. $\square$

53. *Sketch.* Write $\log_2 3 = \frac{a}{b}$, with $a, b > 0$. So $2^{\frac{a}{b}} = 3$. So $2^a = 3^b$. Hence, $a = b = 0$, a contradiction. $\square$

55. *Sketch.* Write $\log_3 7 = \frac{a}{b}$, with $a, b > 0$. So $3^{\frac{a}{b}} = 7$. So $3^a = 7^b$. Hence, $a = b = 0$, a contradiction. $\square$

57. No, $\frac{\sqrt{2} + \sqrt{6}}{\sqrt{2 + \sqrt{3}}} = 2 \in \mathbb{Z} \subseteq \mathbb{Q}$.

59. No.

61. *Proof.* Suppose $r = \frac{1 + \sqrt{5}}{2}$ is rational. So $\sqrt{5} = 2r - 1$. However, $2r - 1$ is rational, and $\sqrt{5}$ is irrational. This is a contradiction. $\square$

63. *Proof.* Suppose not. So $r = \frac{15 + 7\sqrt{5}}{4}$ is rational. So $\sqrt{5} = \frac{4r - 15}{7}$. However, $\frac{4r - 15}{7}$ is rational, and $\sqrt{5}$ is irrational. This is a contradiction. $\square$

65. *Sketch.* Suppose $r = \frac{7 - \sqrt{2}}{3 + \sqrt{2}}$ is rational. So $\sqrt{2} = \frac{7 - 3r}{r + 1} \in \mathbb{Q}$, a contradiction. $\square$

67. $-\frac{1}{3}$ and $\frac{3}{2}$.

69. None.

71. *Sketch.* Observe that $\sqrt{10}$ is a root of $f(x) = x^2 - 10$. However, by the Rational Roots Theorem, $f(x)$ has no rational roots. $\square$

73. *Sketch.* Observe that $\sqrt{6} + \sqrt{2}$ is a root of $f(x) = x^4 - 16x^2 + 16$. However, by the Rational Roots Theorem, $f(x)$ has no rational roots. $\square$

75. *Sketch.* Observe that $\sqrt{3 - 2\sqrt{2}}$ is a root of $f(x) = x^4 - 6x^2 + 1$. However, by the Rational Roots Theorem, $f(x)$ has no rational roots. $\square$

77. *Sketch.* Observe that $\frac{\sqrt[4]{3}}{\sqrt{2}}$ is a root of $f(x) = 4x^4 - 3$. However, by the Rational Roots Theorem, $f(x)$ has no rational roots. $\square$

79. *Proof.* Suppose $r = \frac{\pi+1}{2}$ is rational. So $\pi = 2r - 1$. However, $2r - 1$ is rational, and π is irrational. This is a contradiction. $\square$

81. *Sketch.* Suppose $\sqrt{x} = \frac{a}{b}$ is rational. Then $x = \frac{a^2}{b^2}$ is rational. $\square$

83. *Proof.* Suppose $r \in \mathbb{Q}$. So $r = \frac{a}{b}$ for some $a, b \in \mathbb{Z}$ with $b \neq 0$. Observe that r is a root of $f(x) = bx - a$ and is hence algebraic. $\square$

85. *Sketch.* Let $g(x) = r_n x^n + r_{n-1} x^{n-1} + \cdots + r_1 x + r_0$ be a polynomial with rational coefficients. For each $0 \leq i \leq n$, write $r_i = \frac{a_i}{b_i}$, where $a_i, b_i \in \mathbb{Z}$ with $b_i \neq 0$. Define $f(x) = g(x) \prod_{i=1}^{n} b_i$. So $f(x)$ is a polynomial with integer coefficients, and f and g have exactly the same roots. Since the roots of f are algebraic, so are the roots of g. $\square$

87. No.

89. *Proof.* Suppose not. So $2e$ is algebraic. Thus, $2e$ is a root of some polynomial $f(x) = c_n x^n + c_{n-1} x^{n-1} + \cdots + c_1 x + c_0$, where $n \in \mathbb{Z}^+, c_n, c_{n-1}, \ldots, c_1, c_0 \in \mathbb{Z}$. That is, $0 = f(2e) = c_n 2^n e^n + c_{n-1} 2^{n-1} e^{n-1} + \cdots + c_1 2e + c_0$. Define $g(x) = c_n 2^n x^n + c_{n-1} 2^{n-1} x^{n-1} + \cdots + c_1 2x + c_0$. Since $c_n 2^n, c_{n-1} 2^{n-1}, \ldots, c_1 2, c_0 \in \mathbb{Z}$, and $g(e) = 0$, we see that e is algebraic. This is a contradiction. $\square$

Section 3.5

1. True.

3. False.

5. Thursday.

7. 9 p.m.

9. (a) $n \mid (a - a)$. (b) If $n \mid (a - b)$, then n divides $(b - a) = -(a - b)$. (c) If $a - b = nj$ and $b - c = nk$, then $a - c = (a - b) + (b - c) = n(j + k)$.

11. If $n \mid (a_1 - a_2)$, then $n \mid (-a_1 + a_2)$.

13. If $a - b = nk$, then $a^2 - b^2 = (a + b)(a - b) = (a + b)nk$.

15. *Sketch.* ($\rightarrow$) If $n_1 \bmod d = n_2 \bmod d$, then $n_1 \equiv [n_2 \bmod d] \equiv n_2 \pmod{d}$. ($\leftarrow$) If $n_1 \equiv n_2 \pmod{d}$, then $[n_1 \bmod d] \equiv n_2 \pmod{d}$. Note that $0 \leq n_1 \bmod d < d$. $\square$

17. $8763 + 536 \equiv 13 + 11 \equiv 24 \pmod{25}$.

19. 4.

21. 16.

23. 1.

25. 1.

27. Note that $0^3 - 0 - 1 \equiv -1 \equiv 2 \pmod{3}$, $1^3 - 1 - 1 \equiv -1 \equiv 2 \pmod{3}$, and $2^3 - 2 - 1 \equiv 5 \equiv 2 \pmod{3}$.

29. *Sketch.* If $n \equiv 1, 2$, or $4 \pmod{7}$, then $n^3 \equiv 1 \pmod{7}$. If $n \equiv 3, 5$, or $6 \pmod{7}$, then $n^3 \equiv -1 \pmod{7}$. $\square$

31. (a) If n is odd, then $n \equiv 1, 3, 5$, or $7 \pmod{8}$. (b) Multiply part (a) by n.

33. If $n - r = 3k$, then $2^n - 2^r = 2^r(2^{n-r} - 1) = 2^r(8^k - 1)$ and $8^k \equiv 1^k \equiv 1 \pmod{7}$.

35. If $m \mid (a - b)$ and $n \mid (a + b)$, then mn divides $(a + b)(a - b) = (a^2 - b^2)$.

37. 23.

39. 7.

41. "PZSNQRURHGJUX."

43. "BORAT."

45. *Sketch.* Suppose to the contrary that $x \equiv y \pmod{n}$, and say $x > y$. Then $n \mid (x - y)$. However, $0 \le x - y < n$. This is a contradiction. $\square$

47. Apply Lemma 3.31, Lemma 3.29, and Exercise 45.

49. 6.

51. 79.

53. (a) $y = 2$. (b) $y = 14$. (c) Discover. (d) MasterCard.

55. 16.

57. *Sketch.* When $p \mid a$, $a^p \equiv 0 \equiv a \pmod{p}$. When $p \nmid a$, multiply both sides of $a^{p-1} \equiv 1 \pmod{p}$ by a. $\square$

59. $2^{253} \equiv 162 \not\equiv 2 \pmod{253}$.

61. *Sketch.* Observe that $x \in \{1, \ldots, p - 1\}$ and $x^2 \equiv 1 \pmod{p}$ iff $x = 1$ or $p - 1$. Hence, the suggested pairing off of values in the product $(p - 1)!$ gives $(p - 1)! = (p - 1) \cdot 1^{\frac{p-3}{2}} \cdot 1 \equiv -1 \pmod{p}$. $\square$

63. $10d_1 + 9d_2 + 8d_3 + 7d_4 + 6d_5 + 5d_6 + 4d_7 + 3d_8 + 2d_9 + d_{10} = -(d_1 + 2d_2 + 3d_3 + 4d_4 + 5d_5 + 6d_6 + 7d_7 + 8d_8 + 9d_9 + 10d_{10}) + 11(d_1 + d_2 + d_3 + d_4 + d_5 + d_6 + d_7 + d_8 + d_9 + d_{10})$.

65. $x = 3$.

67. $x = 2, y = 5$.

69. $[2]_3$.

71. $[3]_4$.

73. $[1]_{10}$.

75. $\{k : k \equiv a \pmod{n}\} = \{k : k \equiv b \pmod{n}\}$ iff $a \equiv b \pmod{n}$.

77. *Proof.* ($\subseteq$) Suppose $k \in [a]_n + [b]_n$. So $k = s + t$ for some $s \in [a]_n$ and $t \in [b]_n$. Thus, $s \equiv a \pmod{n}$ and

$t \equiv b \pmod{n}$.   Since $s + t \equiv a + b \pmod{n}$,   it   follows   that $k \in [a + b]_n$.   ($\supseteq$)   Suppose $k \in [a + b]_n$.   So $k \equiv a + b \pmod{n}$.   That   is, $n \mid (k - a - b)$.   Note   that $k = a + (k - a)$ and $k - a \equiv b \pmod{n}$.   Therefore, $k \in [a]_n + [b]_n$. $\square$

79. Apply Exercises 77 and 78.

81. *Sketch.* Any $k \equiv a \pmod{n}$ can be written as $k = 0 + k$. $\square$

83. Note   that $n = 10(10^{k-1}a_k + 10^{k-2}a_{k-1} + \cdots + a_1) + a_0$ and $10(10^{k-1}a_k + 10^{k-2}a_{k-1} + \cdots + a_1) \equiv 0 \pmod{5}$.   So $n \equiv 0 + a_0 \equiv a_0 \pmod{5}$.

85. Note   that $10^k a_k + 10^{k-1}a_{k-1} + \cdots + 10^0 a_0 \equiv 1^k a_k + 1^{k-1}a_{k-1} + \cdots + 1^0 a_0 \pmod{9}$. Also, n is divisible by 3 iff $n \equiv 0, 3,$ or $6 \pmod{9}$.

Review

1. If $m = 2j$ and $n = 2k$, then $mn = 2(2jk)$.

2. If $n = 2k$, then $n^2 = 4k^2$.

3. No.

4. *Sketch.* $(a + b)^3 - b^3 = a(a^2 + 3ab + 3b^2)$. $\square$

5. $b = ak$ iff $b = (-a)(-k)$.

6. No. No.

7. Yes.

8. If $n = aj$ and $n + 2 = ak$, then $2 = (n + 2) - n = a(k - j)$.

9. 91.

10. *Proof.* Let $n \in \mathbb{Z}$ with $n \neq 0$. *Case 1:* $n > 0$. So $n > 0$, $n \mid n$, and $n \mid -n$. Also, if $c \mid n$ and $c \mid -n$, then, in particular, $c \mid n$, whence $c \leq n$. So $\gcd(n, -n) = n = |n|$. *Case 2:* $n < 0$. So $-n > 0$, $-n \mid n$, and $-n \mid -n$. Also, if $c \mid n$ and $c \mid -n$, then, in particular, $c \mid -n$, whence $c \leq -n$. So $\gcd(n, -n) = -n = |n|$. $\square$

11. (a) $a = 2, b = 3, m = 2, n = 2$. (b) *Proof.* Let $d = \gcd(a, b)$. Since $d \mid a$ and $d \mid b$, it follows that $d \mid a^m$ and $d \mid b^n$. Therefore, $d \leq \gcd(a^m, b^n)$. $\square$
    (c) $\gcd(a, b) > 1$, and at least one of $m > 1$ or $n > 1$.

12. 840.

13. *Sketch.* Let $i = \max\{j, k\}$. So $m^i > 0$, $m^j \mid m^i$, and $m^k \mid m^i$. If $c \in \mathbb{Z}^+$, $m^j \mid c$, and $m^k \mid c$, then $m^i \mid c$, whence $m^i \leq c$. $\square$

14. 5.

15. False.

16. 12 remainder 5.

17. (a) 6 and 1. (b) -6 and 3.

18. 2 remain, and each have 17.

19. If $n = 3q + r$, then $n^3 - n = 3(9q^3 + 3q^2 r + qr^2 - q) + r^3 - r$. Consider $r = 0, 1,$ or 2.

20. *Proof.* Suppose n is odd. So $n = 2k + 1$ for some $k \in \mathbb{Z}$. So $n^2 = 4k^2 + 4k + 1 = 4(k^2 + k) + 1$. Since $n^2 \bmod 4 = 1 \neq 0$, we see that $4 \nmid n^2$. $\square$

21. (a) 6. (b) -7.

22. (a) 6. (b) -5.

23. If $n = 4k$, then $k = \frac{n}{4} \leq \frac{n+2}{4} < \frac{n}{4} + 1 = k + 1$.

24. *Sketch.* Certainly $\lceil x \rceil \in \mathbb{Z}$ and $\lceil x \rceil - 1 < \lceil x \rceil \leq \lceil x \rceil$. $\square$

25. 3. No.

26. "MDPSZIDXSQ."

27. $\frac{n}{\gcd(b,n)}$.

28. $x = 3$ and $y = -2$.

29. *Sketch.* Let $x = -11, y = 24$. $\square$

30. 22.

31. 5.

32. $x = -1, y = 2$.

33. $x = 5, y = -16$.

34. Apply Corollaries 3.17 and 3.19.

35. *Sketch.* $6 \mid 52n$ iff $3 \mid 26n$ iff $3 \mid n$, by Euclid's Lemma. $\square$

36. No.

37. (a) Since $5 \mid (25x + 10y)$ for all integers x, y, only multiples of 5 can be achieved. (b) 5¢ and 15¢. (c) 1¢, 3¢, and 5¢.

38. $6\frac{3}{4} = \frac{27}{4}$.

39. $1.\overline{414} = \frac{157}{111} > \sqrt{2}$.

40. $1.6\overline{25} = \frac{1609}{990}$.

41. If $r = \frac{a}{b}$, then $\frac{3r}{4} = \frac{3a}{4b}$.

42. *Sketch.* Write $r = \frac{a}{b}$. Get $r^2 = \frac{a^2}{b^2}$. $\square$

43. If $\gcd(a, b) = 1$, then $\gcd(a^2, b^2) = 1$.

44. $0.\overline{45}$.

45. *Sketch.* Write $\sqrt{7} = \frac{a}{b}$ in lowest terms. So $b\sqrt{7} = a$. So $b^2 7 = a^2$. So $7 \mid a^2$. By Corollary 3.19, $7 \mid a$. Write $a = 7c$. So $b^2 7 = a^2 = 49c^2$. So $b^2 = 7c^2$. So $7 \mid b^2$. By Corollary 3.19, $7 \mid b$. So $\gcd(a, b) \geq 7$. This is a contradiction. $\square$

46. *Proof.* Suppose $r = \frac{5 + \sqrt{7}}{3}$ is rational. So $\sqrt{7} = 3r - 5$. However, $3r - 5$ is rational, and $\sqrt{7}$ is irrational. This is a contradiction. Therefore, $r = \frac{5 + \sqrt{7}}{3}$ must be irrational. $\square$

47. *Sketch.* Write $\log_3 11 = \frac{a}{b}$ with $a, b \in \mathbb{Z}^+$. So $3^{\frac{a}{b}} = 11$. So $3^a = 11^b$. By the Fundamental Theorem of Arithmetic, this is impossible. $\square$

48. *Sketch.* (a) Suppose $r = \frac{e^2-4}{3} \in \mathbb{Q}$. However, we get $e^2 = 3r + 4 \in \mathbb{Q}$. (b) Write $\ln 2 = \frac{a}{b}$ for $a, b \in \mathbb{Z}^+$. So $e^{\frac{a}{b}} = 2$. However, $e^a = 2^b \in \mathbb{Z}$. $\square$

49. *Sketch.* Observe that $\sqrt{3 + \sqrt{2}}$ is a root of $f(x) = x^4 - 6x^2 + 7$. By the Rational Roots Theorem, f has no rational roots. $\square$

50. *Sketch.* Observe that $\frac{\sqrt[3]{2}}{\sqrt{5}}$ is a root of $f(x) = 125x^6 - 4$. By the Rational Roots Theorem, f has no rational roots. $\square$

51. *Sketch.* Observe that $\frac{1}{2}\sqrt{2 + \sqrt{2 + \sqrt{2}}}$ is a root of $f(x) = 256x^8 - 512x^6 + 320x^4 - 64x^2 + 2$. By the Rational Roots Theorem, f has no rational roots. $\square$

52. No, it equals 4.

53. Yes, they are the same as the roots of $15x^2 - 8x + 12$.

54. Wednesday.

55. If $a - b = nk$, then $ac - bc = n(kc)$.

56. (a) 7. (b) 11.

57. If n is odd, then $n \equiv 1$ or $3 \pmod 4$. In both cases, $n^2 \equiv 1 \pmod 4$.

58. If $n \not\equiv 0 \pmod 3$, then $n \equiv 1$ or $2 \pmod 3$. In both cases, $n^2 \equiv 1 \pmod 3$.

59. "RSA."

60. -9.

61. 172.

62. (a) 32. (b) "The package has been received." is the message that was sent.

63. 4.

64. $[2]_5$.

65. $[1]_3$.

66. $[4]_7$.

67. *Proof.* Since $a \equiv a \pmod n$, we have $a \in [a]_n$. $\square$

Chapter 4

Section 4.1

1. 3,628,800.

3. 21.

5. 126.

7. $\frac{n!}{k!(n-k)!} = \frac{n!}{(n-k)!(n-(n-k))!}$.

9. False.

11. False.

13. $4, 2, 0, -2$.

15. $6, 12, 40, 180$.

17. $7, 9, 11, 13$.

19. 47.

21. $\forall\, n \geq 1$, $t_n = 2n$.

23. $\forall\, n \geq 0$, $s_n = 3 \cdot 2^n$.

25. $\forall\, n \geq 0$, $s_n = (-1)^n(2n + 1)$.

27. $\forall\, n \geq 1$, $s_n = \frac{1}{n}$.

29. (a) \$6180 after 1 period. \$6365.40 after 2 periods. (b) $1000(1 + i)^2$. No. (c) $\forall\, n \geq 0$, $s_n = P(1 + i)^n$. A geometric sequence.
Note that $s_0 = P$, $s_1 = P + Pi = P(1 + i)$, $s_2 = s_1(1 + i) = P(1 + i)(1 + i) = P(1 + i)^2$, etc. This is a geometric sequence with multiplying factor $r = i$ and $s_0 = P$.

31. $\forall\, n \geq 0$, $t_n = 10^{n-2}$.

33. $\forall\, n \geq 0$, $t_n = 7 + 2n$.

35. $\forall\, n \geq 0$, $t_n = (-1)^n \frac{n}{n+2}$.

37. $4, 10, 28, 82$.

39. $5, 3, 1, -1$.

41. $-1, -4, -19, -94$. Use $s_1 = -\frac{1}{4}$.

43. $t_1 = 2$, and $\forall\, n \geq 2$, $t_n = 2 + t_{n-1}$.

45. $s_0 = 3$, and $\forall\, n \geq 1$, $s_n = 2s_{n-1}$.

47. $s_0 = 1$, and $\forall\, n \geq 1$, $s_n = s_{n-1} + (-1)^n 4n$.

49. $s_1 = 1$, and $\forall\, n \geq 2$, $s_n = s_{n-1}/(1 + s_{n-1})$.

51. (a) $s_2 = 2050$, $s_3 = 3152.50$. (b) 1348.63. (c) $s_0 = 0$, and $\forall\, n \geq 1$, $s_n = (1 + i)s_{n-1} + D$.

53. $s_{k+1} = 3s_k - 2$.

55. $s_{k+1} = s_k - 2$.

57. $s_2 = 8$, $s_3 = -29$, $s_{k+1} = 5s_{k-1} - 3s_k$.

59. $\forall\, n \geq 2$, $s_n = s_{n-1} - 2$.

61. $\forall\, n \geq 2$, $s_n = 5s_{n-2} - 3s_{n-1}$.

63. (a) In *Mathematica*, use

```
In[1]:= AppRt2[n_] := 1 + 1/(1 + AppRt2[n - 1])
In[2]:= AppRt2[0] := 1
```

(b) 1.41421. (c) The 12th.

Section 4.2

1. $\frac{65}{24} \approx 2.708$.

3. 5525.

5. $\sum_{i=1}^{10} i^3 = 3025$.

7. $\sum_{i=0}^{10} 2^i = 2047$.

9. $\sum_{i=1}^{9} (-2)^i = -342$.

11. $\sum_{i=2}^{n} 3i^2 = \frac{n(n+1)(2n+1)}{2} - 3$.

13. $\sum_{i=1}^{n} 4^i = \frac{4}{3}(4^n - 1)$.

15. $\sum_{i=2}^{n} (-3)^i = \frac{9-(-3)^{n+1}}{4}$.

17. (a) $s_2 = D(1 + i) + D,\quad s_3 = [D(1 + i) + D](1 + i) + D = D(1 + i)^2 + D(1 + i) + D.\quad s_4 = [D(1 + i)^2 + D(1 + i) + D](1 + i) + D = D(1 + i)^3 + D(1 + i)^2 + D(1 + i) + D$.
(b) 406.04. (c) $s_n = \sum_{j=0}^{n-1} D(1 + i)^j = D\frac{(1+i)^n - 1}{i}$. (d) 788.49.
(e) Deposit $D = \frac{iF}{(1+i)^N - 1}$.

19. $n^4 + 2n^3 - 2n^2 - 4n = n(n + 2)(n^2 - 2)$.

21. $\frac{2n^3 - 3n^2 + n}{6} = \frac{n(n-1)(2n-1)}{6}$.

23. $\frac{3 - (\frac{1}{3})^n}{2}$.

25. $2^{101} - 2^{10}$.

27. $\frac{5}{3}(4^{65} - 16)$.

29. $2(1 - \frac{1}{3^{100}})$.

31. $\sum_{i=1}^{n} (4i - 3) = 4\sum_{i=1}^{n} i - 3\sum_{i=1}^{n} 1 = 4\frac{n(n+1)}{2} - 3n = n(2n - 1)$.

33. $\sum_{i=1}^{n} (3i^2 - i) = 3\sum_{i=1}^{n} i^2 - \sum_{i=1}^{n} i = 3\frac{n(n+1)(2n+1)}{6} - \frac{n(n+1)}{2} = n^2(n + 1)$.

35. $\sum_{i=1}^{2n} i = \frac{2n(2n+1)}{2} = n(2n + 1)$.

37. $\sum_{i=1}^{n} (2i + 1) = 2\sum_{i=1}^{n} i + \sum_{i=1}^{n} 1 = 2\frac{n(n+1)}{2} + n = n(n + 2)$.

39. $\prod_{i=1}^{n} k$.

41. $2^{\frac{n(n+1)}{2}}$.

43. $\sum_{i=1}^{n} (y_i \prod_{j=1, j\neq i}^{n} \frac{x-x_j}{x_i-x_j})$.

45. Since $2S = \underbrace{(n + 1) + (n + 1) + \cdots + (n + 1)}_{n \text{ times}} = n(n + 1)$, it follows
that $S = \frac{n(n+1)}{2}$.

47. $\frac{n^2(n+1)^2(2n^2+2n-1)}{12}$.

49. (a) s_6

(b) s_5. (c) $a_n = \sum_{i=0}^{n-1} (\frac{1}{4})^i$. (d) $a_n = \frac{4}{3}(1 - \frac{1}{4^n})$.

Section 4.3

1. *Proof. Base case*: ($n = 3$). Note that $3^2 + 1 \geq 3(3)$. *Inductive step*: Suppose $k \geq 3$ and $k^2 + 1 \geq 3k$. (Goal: $(k+1)^2 + 1 \geq 3(k+1)$.) Observe that $(k+1)^2 + 1 = k^2 + 2k + 1 + 1 = (k^2 + 1) + (2k + 1) \geq 3k + (2k + 1) \geq 3k + 3 = 3(k+1)$. $\square$

3. *Sketch.* Check the case when $n = 3$. *Inductive step*: If $k \geq 3$ and $k^2 \geq 2k + 1$, then $(k+1)^2 = k^2 + (2k + 1) \geq 2k + 1 + (2k + 1) = 2k + (2k + 2) \geq 2k + 3 = 2(k+1) + 1$. $\square$

5. *Sketch.* Check the case when $n = 4$. If $k \geq 4$ and $2^k \geq k^2$, then $2^{k+1} = 2 \cdot 2^k \geq 2 \cdot k^2 \geq k^2 + (2k + 1) = (k+1)^2$. The last inequality follows from Exercise 3. $\square$

7. *Sketch.* Check the case when $n = 4$. If $k \geq 4$ and $k! \geq k^2$, then $(k+1)! = (k+1) \cdot k! \geq (k+1) \cdot k^2 \geq (k+1)^2$. Note that $k^2 \geq k + 1$ for $k \geq 4$. $\square$

9. *Sketch.* Check the case when $n = 4$. If $k \geq 4$ and $k! > 2^k$, then $(k+1)! = (k+1) \cdot k! > (k+1) \cdot 2^k \geq 2 \cdot 2^k = 2^{k+1}$. $\square$

11. *Proof. Base case*: ($n = 0$). Note that $3 \mid (4^0 - 1)$. *Inductive step*: Suppose $k \geq 0$ and $3 \mid (4^k - 1)$. So $4^k - 1 = 3c$ for some $c \in \mathbb{Z}$. (Goal: $3 \mid (4^{k+1} - 1)$.) Observe that $4^{k+1} - 1 = 4 \cdot 4^k - 1 = (3 + 1)4^k - 1 = 3 \cdot 4^k + (4^k - 1) = 3 \cdot 4^k + 3c = 3(4^k + c)$. Thus, $3 \mid (4^{k+1} - 1)$. $\square$

13. *Sketch.* Check the case when $n = 0$. Suppose $k \geq 0$ and $4 \mid (6^k - 2^k)$. For some $c \in \mathbb{Z}$, $6^{k+1} - 2^{k+1} = 6 \cdot 6^k - 2 \cdot 2^k = 4 \cdot 6^k + 2(6^k - 2^k) = 4 \cdot 6^k + 2(4c) = 4(6^k + 2c)$. $\square$

15. *Sketch.* Check the case when $n = 0$. Suppose $k \geq 0$ and $k^3 - k = 6c$ for some $c \in \mathbb{Z}$. So $(k+1)^3 - (k+1) = (k^3 - k) + 3k^2 + 3k = 6(c + \frac{k^2+k}{2})$. Since k^2 and k have the same parity, $\frac{k^2+k}{2} \in \mathbb{Z}$. $\square$

17. *Proof. Base case*: ($n = 1$). Note that $3^1 + 1 = 4$. *Inductive step*: Suppose $k \geq 1$ and $s_k = 3^k + 1$. (Goal: $s_{k+1} = 3^{k+1} + 1$.) So $s_{k+1} = 3s_k - 2 = 3(3^k + 1) - 2 = 3^{k+1} + 1$. $\square$

19. *Sketch.* Check the case when $n = 2$. If $k \geq 2$ and $s_k = 9 - 2k$, then $s_{k+1} = s_k - 2 = 9 - 2k - 2 = 9 - 2(k+1)$. $\square$

21. (a) $s_0 = 0$, $s_1 = D$, $s_2 = (i + 2)D$. (b) $(1 + i)(D\frac{(1+i)^{n-1}-1}{i}) + D = D\frac{(1+i)^n - 1}{i}$. (c) Check the case when $n = 0$. If $k \geq 0$ and $s_k = D\frac{(1+i)^k - 1}{i}$, then $s_{k+1} = (1 + i)s_k + D = (1 + i)[D\frac{(1+i)^k - 1}{i}] + D = D\frac{(1+i)^{k+1} - 1}{i}$. $\square$ (d) \$5237.69.

23. (a) $A \cap ((B_1 \cup B_2) \cup B_3) = (A \cap (B_1 \cup B_2)) \cup (A \cap B_3) = (A \cap B_1) \cup (A \cap B_2) \cup (A \cap B_3)$. (b) *Sketch.* Note the case when $n = 1$. Suppose $k \geq 1$ and, for all sets $B_1, B_2, \ldots, B_k$, $A \cap (B_1 \cup B_2 \cup \cdots \cup B_k) = (A \cap B_1) \cup (A \cap B_2) \cup \cdots \cup (A \cap B_k)$. Let $B_1, B_2, \ldots, B_{k+1}$ be any sets. Observe that $A \cap ((B_1 \cup B_2 \cup \cdots \cup B_k) \cup B_{k+1}) = (A \cap (B_1 \cup B_2 \cup \cdots \cup B_k)) \cup (A \cap B_{k+1}) = (A \cap B_1) \cup (A \cap B_2) \cup \cdots \cup (A \cap B_{k+1})$. $\square$ (c) Reverse $\cap$ and $\cup$ in part (b).

25. (a) $\neg((p_1 \vee p_2) \vee p_3) \equiv \neg(p_1 \vee p_2) \wedge \neg p_3 \equiv \neg p_1 \wedge \neg p_2 \wedge \neg p_3$ (b) *Sketch.* Note the case when $n = 1$. Suppose $k \geq 1$ and, for all statement forms $p_1, p_2, \ldots, p_k$, $\neg(p_1 \vee p_2 \vee \cdots \vee p_k) \equiv \neg p_1 \wedge \neg p_2 \wedge \cdots \wedge \neg p_k$. Let p_1, $p_2, \ldots, p_{k+1}$ be any statement forms. Observe that $\neg((p_1 \vee p_2 \vee \cdots \vee p_k) \vee p_{k+1}) \equiv \neg(p_1 \vee p_2 \vee \cdots \vee p_k) \wedge$ $\neg p_{k+1} \equiv \neg p_1 \wedge \neg p_2 \wedge \cdots \wedge \neg p_{k+1}$. $\square$ (c) Reverse $\wedge$ and $\vee$ in part (b).

27. (a) *Sketch.* Note the case when $m = 1$. Suppose $k \geq 1$ and, $\forall a_1, \ldots, a_k, b_1, \ldots, b_k \in \mathbb{Z}$, if $a_1 \equiv b_1 \pmod{n}$, $a_2 \equiv b_2 \pmod{n}$, $\ldots$, $a_k \equiv b_k \pmod{n}$, then $\sum_{i=1}^{k} a_i \equiv \sum_{i=1}^{k} b_i \pmod{n}$. Let $a_1, a_2, \ldots, a_{k+1}, b_1, b_2, \ldots, b_{k+1}$ be any integers. Suppose $a_1 \equiv b_1 \pmod{n}$, $a_2 \equiv b_2 \pmod{n}$, $\ldots$, $a_{k+1} \equiv b_{k+1} \pmod{n}$. Hence, $\sum_{i=1}^{k+1} a_i \equiv (\sum_{i=1}^{k} a_i) + a_{k+1} \equiv (\sum_{i=1}^{k} b_i) + b_{k+1} \equiv \sum_{i=1}^{k+1} b_i \pmod{n}$. $\square$ (b) A similar proof replacing sums with products works here.

29. *Sketch.* Note $\max(\{s_1\}) = s_1$. Suppose $k \geq 1$, and any set S with $|S| = k$ has a maximal element. Suppose $S = \{s_1, s_2, \ldots, s_{k+1}\}$ has $k + 1$ elements. The set $\{s_1, s_2, \ldots, s_k\}$ has a maximal element, say s_j. So
$$\max(S) = \begin{cases} s_j & \text{if } s_j > s_{k+1} \\ s_{k+1} & \text{otherwise.} \end{cases} \quad \square$$

31. *Proof.* Base case: ($n = 1$). Obvious. *Inductive step*: Suppose $k \geq 1$ and $\begin{bmatrix} 1 & 1 \\ 0 & 1 \end{bmatrix}^k = \begin{bmatrix} 1 & k \\ 0 & 1 \end{bmatrix}$. Observe that $\begin{bmatrix} 1 & 1 \\ 0 & 1 \end{bmatrix}^{k+1} =$
$\begin{bmatrix} 1 & 1 \\ 0 & 1 \end{bmatrix}\begin{bmatrix} 1 & 1 \\ 0 & 1 \end{bmatrix}^k = \begin{bmatrix} 1 & 1 \\ 0 & 1 \end{bmatrix}\begin{bmatrix} 1 & k \\ 0 & 1 \end{bmatrix} = \begin{bmatrix} 1 & k+1 \\ 0 & 1 \end{bmatrix}.$ $\square$

33. *Sketch.* In the inductive step, observe that $\begin{bmatrix} 1 & 1 \\ 0 & 2 \end{bmatrix}^{k+1} =$
$\begin{bmatrix} 1 & 1 \\ 0 & 2 \end{bmatrix}\begin{bmatrix} 1 & 1 \\ 0 & 2 \end{bmatrix}^k = \begin{bmatrix} 1 & 1 \\ 0 & 2 \end{bmatrix}\begin{bmatrix} 1 & 2^k - 1 \\ 0 & 2^k \end{bmatrix} = \begin{bmatrix} 1 & 2^{k+1} - 1 \\ 0 & 2^{k+1} \end{bmatrix}.$ $\square$

35. *Sketch.* In the inductive step, observe that
$\begin{bmatrix} \cos\theta & \sin\theta \\ -\sin\theta & \cos\theta \end{bmatrix}^{k+1} = \begin{bmatrix} \cos\theta & \sin\theta \\ -\sin\theta & \cos\theta \end{bmatrix}\begin{bmatrix} \cos\theta & \sin\theta \\ -\sin\theta & \cos\theta \end{bmatrix}^k =$
$\begin{bmatrix} \cos\theta & \sin\theta \\ -\sin\theta & \cos\theta \end{bmatrix}\begin{bmatrix} \cos k\theta & \sin k\theta \\ -\sin k\theta & \cos k\theta \end{bmatrix} =$
$\begin{bmatrix} \cos(k+1)\theta & \sin(k+1)\theta \\ -\sin(k+1)\theta & \cos(k+1)\theta \end{bmatrix}.$ $\square$

37. (a) $\sin(2\theta) = \sin\theta\cos\theta + \cos\theta\sin\theta = 2\sin\theta\cos\theta$.   (b) $\sin 4\theta = 2\sin 2\theta\cos 2\theta = 4\sin 2\theta\cos 2\theta$.   (c)   *Sketch.* If $k \geq 1$ and $\sin 2^k\theta = 2^k \sin\theta \prod_{i=0}^{k-1} \cos 2^i\theta$, then $\sin 2^{k+1}\theta = \sin(2(2^k\theta)) = 2\sin 2^k\theta\cos 2^k\theta = 2^{k+1}\sin\theta\prod_{i=0}^{k}\cos 2^i\theta$. $\square$

39. $\forall n \geq 0, n \geq 1$.

41. *Proof.* Suppose $\forall n \geq a$, $P(n)$. Suppose $k \geq a$ and $P(k)$ holds. Since $k + 1 \geq a$, we have $P(k + 1)$ too. $\square$

43. *Sketch.* Check the case when $n = 0$. If $k \geq 0$ and $8^k \equiv 1 \pmod 7$, then $8^{k+1} \equiv 8 \cdot 8^k \equiv 1 \cdot 1 \equiv 1 \pmod 7$. $\square$

45. (a) $1, 1, 2, 5, 14$. (b) *Sketch.* Check the case when $n = 0$. If $k \geq 0$ and $C_k = \frac{1}{k+1}\binom{2k}{k}$, then $C_{k+1} = \frac{2(2(k+1)-1)}{k+2} C_k = \frac{2(2k+1)}{k+2} \cdot \frac{1}{k+1}\binom{2k}{k} = \frac{1}{k+2}\binom{2(k+1)}{k+1}$. $\square$ (c) $n = 0$: a; $n = 1$: ab; $n = 2$: $(ab)c$, $a(bc)$; $n = 3$: $(ab)(cd)$, $((ab)c)d$, $(a(bc))d$, $a((bc)d)$, $a(b(cd))$; $n = 4$: $(abc)(de)$ in two ways, $(ab)(cde)$ in two ways, $a(bcde)$ in five ways, $(abcd)e$ in five ways.

Section 4.4

1. *Proof. Base case:* ($n = 1$). Note that $\sum_{i=1}^{1} 0 = 0$. *Inductive step:* Suppose $k \geq 1$ and $\sum_{i=1}^{k} 0 = 0$. (Goal: $\sum_{i=1}^{k+1} 0 = 0$.) Observe that $\sum_{i=1}^{k+1} 0 = (\sum_{i=1}^{k} 0) + 0 = 0 + 0 = 0$. $\square$

3. (a) *Sketch.* Check the case when $n = 1$. If $k \geq 1$ and $\sum_{i=1}^{k} i = \frac{k(k+1)}{2}$, then $\sum_{i=1}^{k+1} i = \sum_{i=1}^{k} i + (k+1) = \frac{k(k+1)}{2} + (k+1) = \frac{(k+1)(k+2)}{2}$. $\square$ (b) $n(n+1)$. (c) $\frac{n(n+2)}{4}$. (d) $\frac{n^2-1}{4}$.

5. *Proof. Base case:* ($n = 1$). Note that $\sum_{i=1}^{1}(3i^2 - i) = 2 = 1^2(1+1)$. *Inductive step:* Suppose $k \geq 1$ and $\sum_{i=1}^{k}(3i^2 - i) = k^2(k+1)$. (Goal: $\sum_{i=1}^{k+1}(3i^2 - i) = (k+1)^2(k+2)$.) So $\sum_{i=1}^{k+1}(3i^2 - i) = \sum_{i=1}^{k}(3i^2 - i) + (3(k+1)^2 - (k+1)) = k^2(k+1) + 3(k+1)^2 - (k+1) = (k+1)[k^2 + 3(k+1) - 1] = (k+1)^2(k+2)$. That is, $\sum_{i=1}^{k+1}(3i^2 - i) = (k+1)^2(k+2)$. $\square$

7. *Proof. Base case:* ($n = 1$). Note $\sum_{i=1}^{1}(2i)^3 = 8 = 2(1)(4)$. *Inductive step:* Suppose $k \geq 1$ and $\sum_{i=1}^{k}(2i)^3 = 2k^2(k+1)^2$. (Goal: $\sum_{i=1}^{k+1}(2i)^3 = 2(k+1)^2(k+2)^2$.) Observe that $\sum_{i=1}^{k+1}(2i)^3 = \sum_{i=1}^{k}(2i)^3 + (2(k+1))^3 = 2k^2(k+1)^2 + (2(k+1))^3 = 2(k+1)^2(k+2)^2$. $\square$

9. *Sketch.* Check the case when $n = 1$. If $k \geq 1$ and $\sum_{i=1}^{k}(4i - 3) = k(2k - 1)$, then $\sum_{i=1}^{k+1}(4i - 3) = k(2k - 1) + (4k + 1) = (k + 1)(2(k + 1) - 1)$. $\square$

11. Prove $\sum_{i=1}^{n}(2i + 1) = n(n + 2)$.

13. Prove $\sum_{i=0}^{n} 2^i = 2^{n+1} - 1$.

15. *Sketch.* Check the case when $n = 2$. If $k \geq 0$ and $\sum_{i=2}^{k} i2^i = (k - 1)2^{k+1}$, then $\sum_{i=2}^{k+1} i2^i = (k - 1)2^{k+1} + (k + 1)2^{k+1} = k2^{k+2}$. $\square$

17. *Sketch.* Check the case when $n = 1$. If $k \geq 1$ and $\sum_{i=1}^{k} i^2 2^i = (k^2 - 2k + 3)2^{k+1} - 6$, then $\sum_{i=1}^{k+1} i^2 2^i = [(k^2 - 2k + 3)2^{k+1} - 6] + (k + 1)^2 2^{k+1} = ((k + 1)^2 - 2(k + 1) + 3)2^{k+2} - 6$. $\square$

19. *Sketch.* Check the case when $n = 1$. If $k \geq 1$ and $\sum_{i=1}^{k}(i \cdot i!) = (k + 1)! - 1$, then $\sum_{i=1}^{k+1}(i \cdot i!) = [(k + 1)! - 1] + (k + 1) \cdot (k + 1)! = (k + 2)! - 1$. $\square$

21. *Proof. Base case:* ($n = 1$). Note that $\sum_{i=1}^{2} i = 3 = 1 (2 \cdot 1 + 1)$. *Inductive step:* Suppose $k \geq 1$ and

$\sum_{i=1}^{2k} i = k(2k+1)$. Observe that $\sum_{i=1}^{2(k+1)} i = \sum_{i=1}^{2k} i +$ $(2k+1)+(2k+2) = k(2k+1)+(2k+1)+(2k+2) =$ $(k+1)(2(k+1)+1)$. $\square$

23. *Sketch.* Check the case when $n=1$. If $k \geq 1$ and $\sum_{i=1}^{k} \frac{1}{i(i+1)} = \frac{k}{k+1}$, then $\sum_{i=1}^{k+1} \frac{1}{i(i+1)} = \frac{k}{k+1} + \frac{1}{(k+1)(k+2)} =$ $\frac{k+1}{k+2}$. $\square$

25. *Sketch.* In the inductive step, we have $\sum_{i=1}^{k+1} \frac{1}{2^i} =$ $(1 - \frac{1}{2^k}) + \frac{1}{2^{k+1}} = 1 - \frac{1}{2^{k+1}}$. $\square$

27. *Proof. Base case*: $(n=1)$. Note that $\prod_{i=1}^{1} \frac{i}{i+2} = \frac{1}{3} =$ $\frac{2}{(1+1)(1+2)}$. *Inductive step*: Suppose $k \geq 1$ and $\prod_{i=1}^{k} \frac{i}{i+2} =$ $\frac{2}{(k+1)(k+2)}$.   (Goal: $\prod_{i=1}^{k+1} \frac{i}{i+2} = \frac{2}{(k+2)(k+3)}$.)   Observe that $\prod_{i=1}^{k+1} \frac{i}{i+2} = (\prod_{i=1}^{k} \frac{i}{i+2})(\frac{k+1}{k+3}) = (\frac{2}{(k+1)(k+2)})(\frac{k+1}{k+3}) =$ $\frac{2}{(k+2)(k+3)}$. $\square$

29. *Sketch.* Check the case when $n=1$. If $k \geq 1$ and $\prod_{i=1}^{k} r^{2i} =$ $r^{k(k+1)}$, then $\prod_{i=1}^{k+1} r^{2i} = (r^{k(k+1)})(r^{2(k+1)}) = r^{(k+1)(k+2)}$. $\square$

31. *Sketch.* In the inductive step, we have $x^{2^{k+1}} - y^{2^{k+1}} =$ $(x^{2^k} - y^{2^k})(x^{2^k} + y^{2^k}) = (x-y) \prod_{i=0}^{k}(x^{2^i} + y^{2^i})$. $\square$

33. *Proof. Base case*: $(n=0)$. We have $s_1 \geq 2s_0 \geq s_0$. *Inductive step*: Suppose $k \geq 0$ and $s_{k+1} \geq \sum_{i=0}^{k} s_i$. Observe that $s_{k+2} \geq 2s_{k+1} = s_{k+1} + s_{k+1} \geq s_{k+1} + \sum_{i=0}^{k} s_i = \sum_{i=0}^{k+1} s_i$. $\square$

35. *Proof. Base case*: $(n=1)$. Note that $\sum_{i=1}^{1} \frac{1}{i^2} = 1 = \frac{3}{2} - \frac{1}{2}$. *Inductive step*: Suppose $k \geq 1$ and $\sum_{i=1}^{k} \frac{1}{i^2} \geq \frac{3}{2} - \frac{1}{k+1}$. (Goal: $\sum_{i=1}^{k+1} \frac{1}{i^2} \geq \frac{3}{2} - \frac{1}{k+2}$.) Observe that $\sum_{i=1}^{k+1} \frac{1}{i^2} =$ $\sum_{i=1}^{k} \frac{1}{i^2} + \frac{1}{(k+1)^2} \geq \frac{3}{2} - \frac{1}{k+1} + \frac{1}{(k+1)^2} \geq \frac{3}{2} - \frac{1}{k+2}$. $\square$

37. *Proof. Base case*: $(n=1)$. Note that $\sum_{i=1}^{1} \frac{1}{i^2} = 1 = 2-1$. *Inductive step*: Suppose $k \geq 1$ and $\sum_{i=1}^{k} \frac{1}{i^2} \leq 2 - \frac{1}{k}$. (Goal: $\sum_{i=1}^{k+1} \frac{1}{i^2} \leq 2 - \frac{1}{k+1}$.) Observe that $\sum_{i=1}^{k+1} \frac{1}{i^2} =$ $\sum_{i=1}^{k} \frac{1}{i^2} + \frac{1}{(k+1)^2} \leq 2 - \frac{1}{k} + \frac{1}{(k+1)^2} \leq 2 - \frac{1}{k+1}$. $\square$

39. *Sketch.* We focus on $+$ since $-$ is handled simi-larly.   Suppose $\sum_{i=a}^{b}(s_i + t_i) = \sum_{i=a}^{b} s_i + \sum_{i=a}^{b} t_i$. Then, $\sum_{i=a}^{b+1}(s_i + t_i) = \sum_{i=a}^{b}(s_i + t_i) + s_{b+1} + t_{b+1} =$ $\sum_{i=a}^{b} s_i + \sum_{i=a}^{b} t_i + s_{b+1} + t_{b+1} = \sum_{i=a}^{b+1} s_i + \sum_{i=a}^{b+1} t_i$. $\square$

Section 4.5

1. *Proof. Base cases*: $(n=0,1)$. Note that $0 = 2^0 - 1$ and $1 = 2^1 - 1$. *Inductive step*: Suppose $k \geq 1$ and that, for each $0 \leq i \leq k$, $s_i = 2^i - 1$. (Goal: $s_{k+1} = 2^{k+1} - 1$.) We have $s_{k+1} = 3s_k - 2s_{k-1} = 3(2^k - 1) - 2(2^{k-1} - 1) =$ $3 \cdot 2^k - 3 - 2^k + 2 = 2 \cdot 2^k - 1 = 2^{k+1} - 1$. $\square$

3. *Proof. Base cases*: $(n=0,1)$. Note that $2 = 1+1$ and $11 =$ $4+7$. *Inductive step*: Suppose $k \geq 1$ and that, for each $0 \leq$ $i \leq k$, $s_i = 4^i + 7^i$. (Goal: $s_{k+1} = 4^{k+1} + 7^{k+1}$.) Observe that

$$s_{k+1} = 11s_k - 28s_{k-1} = 11(4^k + 7^k) - 28(4^{k-1} + 7^{k-1}) = 11(4^k) + 11(7^k) - 7(4^k) - 4(7^k) = 4(4^k) + 7(7^k) = 4^{k+1} + 7^{k+1}. \ \square$$

5. *Proof. Base cases*: $(n = 0, 1)$. Note that $1 = 2^0$ and $2 = 2^1$. *Inductive step*: Suppose $k \geq 1$ and that, for each $0 \leq i \leq k$, $s_i = 2^i$. (Goal: $s_{k+1} = 2^{k+1}$.) Observe that $s_{k+1} = 4s_{k-1} = 4 \cdot 2^{k-1} = 2^{k+1}$. $\square$

7. *Sketch.* Check the cases when $n = 0, 1, 2$. If $k \geq 2$ and, for each $0 \leq i \leq k$, $s_i = 5^i - 3^i - 2^i$, then $s_{k+1} = 10s_k - 31s_{k-1} + 30s_{k-2} = 10(5^k - 3^k - 2^k) - 31(5^{k-1} - 3^{k-1} - 2^{k-1}) + 30(5^{k-2} - 3^{k-2} - 2^{k-2}) = 5^{k+1} - 3^{k+1} - 2^{k+1}. \ \square$

9. *Sketch.* Check the cases when $n = 0, 1$. Suppose $k \geq 1$ and that, for each $0 \leq i \leq k$, s_i is odd. We have $c \in \mathbb{Z}$ such that $s_{k+1} = 3s_{k-1} - 2s_k = 3(2c + 1) - 2s_k = 2(3c - s_k + 1) + 1$. $\square$

11. (a) $6, 30, 114$. (b)*Sketch.* Check the cases when $n = 0, 1$. If $k \geq 1$ and, for each $0 \leq i \leq k$, $s_i = 2 \cdot 3^i - 3 \cdot 2^i$, then $s_{k+1} = -6s_{k-1} + 5s_k = -6(2 \cdot 3^{k-1} - 3 \cdot 2^{k-1}) + 5(2 \cdot 3^k - 3 \cdot 2^k) = 6 \cdot 3^k - 6 \cdot 2^k = 2 \cdot 3^{k+1} - 3 \cdot 2^{k+1}. \ \square$
    (c) $s_{n+1} - s_n = 4 \cdot 3^n - 3 \cdot 2^n \geq 3(3^n - 2^n) \geq 0$.

13. (a) $3, 7, 17$. (b) *Sketch.* Check the cases when $n = 0, 1$. If $k \geq 1$ and, for each $0 \leq i \leq k$, $s_i = \frac{1}{2}((1 + \sqrt{2})^i + (1 - \sqrt{2})^i)$, then $s_{k+1} = 2s_k + s_{k-1} = 2(\frac{1}{2})[(1 + \sqrt{2})^k + (1 - \sqrt{2})^k] + \frac{1}{2}((1 + \sqrt{2})^{k-1} + (1 - \sqrt{2})^{k-1}) = \frac{1}{2}((1 + \sqrt{2})^{k+1} + (1 - \sqrt{2})^{k+1})$. $\square$ (c) $7, 0, 2, 2, 2, 0, 5$.

15. *Sketch.* Check the cases when $n = 0, 1$. Suppose $k \geq 1$ and, for each $0 \leq m \leq k$, $\sum_{i=0}^{m}(-1)^i = \begin{cases} 1 & \text{if } m \text{ is even} \\ 0 & \text{if } m \text{ is odd} \end{cases}$.
    Thus, $\sum_{i=0}^{k+1}(-1)^i = \sum_{i=0}^{k-1}(-1)^i + (-1)^k + (-1)^{k+1} = \sum_{i=0}^{k-1}(-1)^i$. Since $k + 1$ and $k - 1$ have the same parity, the result follows. $\square$

17. (a) *Sketch.* Note that $4 and $5 are achievable. Suppose $k \geq 5$ and that, for each $4 \leq i \leq k$, it is possible to attain i with $2 bills and $5 bills. Since, for some $a, b \in \mathbb{N}$, $(k - 1) = a \times \$2 + b \times \$5$, we have $(k + 1) = (a + 1) \times \$2 + b \times \$5$. $\square$ (b) Increase.

19. *Sketch.* Check the cases when $n = 25, 26, 27, 28$. Suppose $k \geq 28$ and that, for each $25 \leq i \leq k$, it is possible to attain i inches from 4-inch bricks and 9-inch bricks and a sheet of plywood 1 inch thick. If $(k - 3)$ inches $= a \times 4$ inches $+ b \times 9$ inches $+ 1$, then $(k + 1)$ inches $= (a + 1) \times 4$ inches $+ b \times 9$ inches $+ 1$. $\square$

21. (a) *Sketch. Base cases*: $4(5\cent) = 20\cent = 2(10\cent)$ and $5(5\cent) = 25\cent = 1(25\cent)$. *Inductive step*: Use $(k + 1)(5\cent) = (k - 1)(5\cent) + 10\cent$. $\square$ (b) $5, 6, 7, 8, 9, 15, 16, 17, 18, 19\cent$.

23. $2^2 \cdot 3^2 \cdot 5 \cdot 7$.

25. $3 \cdot 7 \cdot 13^2$.

27. $2^{10} \cdot 3^5 \cdot 5^2 \cdot 7 \cdot 11$.

29. (a) *Sketch.* Note the case when $n = 2$. If $k \geq 2$ and each integer i with $2 \leq i \leq k$ has a squared prime divisor, then consider the case when $k + 1$ is prime and the case when $k + 1$ is composite. $\square$ (b) *Sketch.* Write $n = p_1^{e_1} \cdot p_2^{e_2} \cdot \ \cdots \ \cdot p_m^{e_m}$. So, $n^2 = (p_1^2)^{e_1} \cdot (p_2^2)^{e_2} \cdot \ \cdots \ \cdot (p_m^2)^{e_m}$. Take $p = p_1$. $\square$

31. Assume $p_1 < p_2 < \cdots < p_m$. Let $d = p_1^{\min\{e_1,f_1\}} p_2^{\min\{e_2,f_2\}} \cdots p_m^{\min\{e_m,f_m\}}$, $a = p_1^{e_1} p_2^{e_2} \cdots p_m^{e_m}$, and $b = p_1^{f_1} p_2^{f_2} \cdots p_m^{f_m}$. So $d > 0$, $d \mid a$, and $d \mid b$. Suppose $c \in \mathbb{Z}^+$, $c \mid a$, and $c \mid b$. Let $c = q_1^{g_1} q_2^{g_2} \cdots q_n^{g_n}$ be the standard factorization. Since $c \mid a$, we have $q_1 \mid a$, whence $q_1 = p_1$. Moreover, $q_1^{g_1} \mid p_1^{e_1}$ and thus $g_1 \leq e_1$. Similarly, $g_1 \leq f_1$. Hence, $g_1 \leq \min\{e_1, f_1\}$. Repeating this argument, we see that $c \mid d$, whence $c \leq d$.

33. Apply the result in Exercise 31.

35. (a) Prove existence and uniqueness separately. For existence, break the inductive step into cases based on the parity of n. In each case, the binary representation $b_m b_{m-1} \cdots b_1 b_0$ for n div 2 can be used to get the binary representation $b_m b_{m-1} \cdots b_1 b_0 r$ for n, where $r = 0$ if n is even, and $r = 1$ if n is odd. (b) Generalize part (a).

37. *Sketch.* In the inductive step, use the fact that

$$\sum_{i=1}^{n} i^{k+1} = \frac{(n+1)((n+1)^{k+1} - 1) - \sum_{j=1}^{k}\left[\binom{k+2}{j}\sum_{i=1}^{n} i^j\right]}{k+2}$$

and n is a factor of $(n+1)^{k+1} - 1$. $\square$

39. *Sketch. Inductive step:* $\sum_{i=0}^{k+1} F_i = (F_{k+2} - 1) + F_{k+1} = F_{k+3} - 1$. $\square$

41. *Sketch. Inductive step:* $\begin{bmatrix} 1 & 1 \\ 1 & 0 \end{bmatrix}^{k+1} = \begin{bmatrix} F_k & F_{k-1} \\ F_{k-1} & F_{k-2} \end{bmatrix}\begin{bmatrix} 1 & 1 \\ 1 & 0 \end{bmatrix} = \begin{bmatrix} F_{k+1} & F_k \\ F_k & F_{k-1} \end{bmatrix}$. $\square$

43. *Sketch.* Certainly $\gcd(F_0, F_1) = 1$. If c divides F_n and F_{n+1}, then c divides $F_{n-1} = F_{n+1} - F_n$, which would be impossible by the inductive hypothesis. $\square$

45. *Proof.* Assume conditions (i) and (ii) in the hypotheses of the theorem. Suppose it is not true that $P(n)$ holds $\forall n \geq a$. Let S be the set of those integers $n \geq a$ for which $P(n)$ does not hold. By our assumptions, S is nonempty. Hence, by the Generalized Well-Ordering Principle, S has a smallest element, say s. Since $P(a), P(a + 1), \ldots, P(b)$ all hold, it must be that $s > b$. Therefore, $s - 1 \geq b$. Since $a, \ldots, s - 1 \notin S$, it follows that $P(a), \ldots, P(s - 1)$ all hold. However, for $k = s - 1$, by condition (ii), $P(k + 1)$ must also hold. That is, $P(s)$ holds. This contradicts the fact that $s \in S$. $\square$

47. *Sketch.* Attempting to write $23 = 5t + 2c$ for $t = 0, \ldots, 4$ shows that it is impossible. Now observe that $24, \ldots, 28$ are achievable. For any number of points $k \geq 29$, we can always score $(k - 5)$ points first and then score another try. $\square$

Section 4.6

1. $\dfrac{(n-1)!}{(k-1)!(n-k)!} \dfrac{n!}{(k+1)!(n-1-k)!} \dfrac{(n+1)!}{k!(n+1-k)!} = \dfrac{(n-1)!}{k!(n-1-k)!} \dfrac{n!}{(k-1)!(n+1-k)!} \dfrac{(n+1)!}{(k+1)!(n-k)!}$.

3. $x^5 + 5x^4y + 10x^3y^2 + 10x^2y^3 + 5xy^4 + y^5$.

5. $729x^6 + 1458x^5y + 1215x^4y^2 + 540x^3y^3 + 135x^2y^4 + 18xy^5 + y^6$.

7. $32x^5 - 80x^4y + 80x^3y^2 - 40x^2y^3 + 10xy^4 - y^5$.

9. $x^n - nx^{n-1} + \binom{n}{2}x^{n-2} - \binom{n}{3}x^{n-3} + \cdots + (-1)^n$.

11. (a) $x^n + \frac{1}{2}nx^{n-1} + \frac{1}{4}\binom{n}{2}x^{n-2} + \frac{1}{8}\binom{n}{3}x^{n-3} + \cdots + \frac{1}{2^n}$.
    (b) $\frac{1}{2^{n-5}}\binom{n}{5}$.

13. $x^8 + 4x^6y^2 + 6x^4y^4 + 4x^2y^6 + y^8$.

15. $243x^{10} + 405x^8y^3 + 270x^6y^6 + 90x^4y^9 + 15x^2y^{12} + y^{15}$.

17. (a) $x^{2n} + nx^{2n-2} + \binom{n}{2}x^{2n-4} + \binom{n}{3}x^{2n-6} + \cdots + nx^2 + 1$.
    (b) $\binom{n}{4}$.

19. $\binom{n}{2}n^{n-2} - \binom{n}{3}n^{n-3} + \cdots + (-1)^{n-1}n^2 + (-1)^n$.

21. (a) 2.7048, 2.7169, 2.7181. (b) $\sum_{i=0}^{n} \frac{1}{i!} \frac{(n-1)(n-2)\cdots(n-i+1)}{n^{i-1}}$.

23. $\binom{100}{40}3^{60}2^{40}$.

25. $\binom{400}{10}2^{10}$.

27. 0.

29. (a) $\frac{15}{128}$. (b) $\frac{105}{512}$. (c) 0.

31. *Proof.* $9^n = (1+8)^n = \sum_{i=0}^{n} \binom{n}{i}1^{n-i}8^i = \sum_{i=0}^{n} \binom{n}{i}8^i$. $\square$

33. Consider $(10 + 2)^n$.

35. Consider $(5 + 3)^n = (2^3)^n$.

37. Consider $(3 - 1)^n$.

39. (a) Expand $(2 + 2^2)^n$ and simplify. (b) $\sum_{i=0}^{n} \binom{n}{i}2^i = 3^n$.

41. Consider $(1 - \frac{1}{3})^n$.

43. *Proof.* Suppose a and b are relatively prime. By Corollary 3.14, we get $ax + by = 1$, for some $x, y \in \mathbb{Z}$. So, $b^n y^n = (1 - ax)^n = 1 - nax + \binom{n}{2}a^2x^2 - \binom{n}{3}a^3x^3 + \cdots + (-1)^n a^n x^n = 1 + x(-na + \binom{n}{2}a^2x - \binom{n}{3}a^3x^2 + \cdots + (-1)^n a^n x^{n-1})$. With $c = -na + \binom{n}{2}a^2x - \binom{n}{3}a^3x^2 + \cdots + (-1)^n a^n x^{n-1}$, we have $b^n y^n = 1 - xc$. That is, $cx + b^n y^n = 1$. By Corollary 3.14, a and b^n are relatively prime. $\square$

Review

1. $8, 64, 320, 1280, 4480$.

2. $3, 5, 21, 437, 190965$.

3. $\forall\, n \geq 1,\ s_n = \frac{n}{2^n}$.

4. $s_0 = -6$, and $\forall\, n \geq 1,\ s_n = s_{n-1} + 12$.

5. $\forall\, n \geq 0,\ s_n = -2(-3)^n$.

6. (a) \$505.01 and \$563.58. (b) $s_t = P(1 + \frac{r}{m})^{mt}$.

7. $\forall\, n \geq 0,\ s_n = 8 \cdot 2^n \binom{n+3}{3}$.

8. $s_k^2 - 4$.

9. $\sum_{i=3}^{n} 3 \cdot 2^{i-1}$.

10. $125{,}250$.

11. $\frac{4^{11}-1}{3}$.

12. (a) 210. (b) 2870.

13. $1{,}010{,}200$.

14. $\frac{n(2n^2 - 9n + 13)}{6}$.

15. $\frac{1 - (-2)^{n+1}}{3}$.

16. 945.

17. (a) \$63,612.07 and \$33,795.91. (b) $M = \$83{,}395.81$.

18. (a) $\frac{2}{3^5} \sum_{i=0}^{198} (\frac{1}{3})^i$. (b) $\frac{1 - (\frac{1}{3})^{199}}{3^4}$.

19. *Sketch.* Check the case when $n = 9$. If $k \geq 9$ and $k! > 4^k$, then $(k+1)! = (k+1)k! > (k+1)4^k \geq 4^{k+1}$. $\square$

20. *Sketch.* Check the case when $n = 6$. If $k \geq 6$ and $k^2 > 4(k+2)$, then $(k+1)^2 = k^2 + 2k + 1 > 4(k+2) + 2k + 1 > 4(k+3)$. $\square$

21. *Sketch.* Check the case when $n = 0$. If $k \geq 0$ and $3^k \geq k^2 + 1$, then $3^{k+1} = 3 \cdot 3^k \geq 3(k^2 + 1) \geq (k+1)^2 + 1$. $\square$

22. *Sketch.* Check the case when $n = 0$. Suppose $k \geq 0$ and $k^3 - 4k + 6 = 3c$ for some $c \in \mathbb{Z}$. Hence, $(k+1)^3 - 4(k+1) + 6 = k^3 + 3k^2 + 3k + 1 - 4k - 4 + 6 = (k^3 - 4k + 6) + (3k^2 + 3k - 3) = 3(c + k^2 + 3k + 1)$. $\square$

23. *Sketch.* Check the case when $n = 0$. Suppose $k \geq 0$ and $7^k - 1 = 6c$ for some $c \in \mathbb{Z}$. Hence, $7^{k+1} - 1 = (6+1) \cdot 7^k - 1 = 6(7^k + c)$. $\square$

24. *Sketch.* Check the case when $n = 0$. Suppose $k \geq 0$ and $5^k - 2^k = 3c$ for some $c \in \mathbb{Z}$. Hence, $5^{k+1} - 2^{k+1} = (3+2) \cdot 5^k - 2 \cdot 2^k = 3(5^k + 2c)$. $\square$

25. (a) $300, 601.50, 904.51$. (b) *Sketch.* If $s_k = 60000(1.005^k - 1)$, then $s_{k+1} = 1.005 s_k + 300 = (1.005)60000(1.005^k - 1) + 300 = 60000(1.005^{k+1} - 1)$. $\square$

26. Use the fact that $a(x_1 + \cdots + x_k + x_{k+1}) = a((x_1 + \cdots + x_k) + x_{k+1}) = a(x_1 + \cdots + x_k) + ax_{k+1}$.

27. *Sketch.* Note that $\min(\{s_1\}) = s_1$. Suppose $k \geq 1$ and, any set S with $|S| = k$ has a minimal element. If $s_j = \min(\{s_1, s_2, \ldots, s_k\})$, then $\min(\{s_1, s_2, \ldots, s_k, s_{k+1}\}) = \begin{cases} s_j & \text{if } s_j < s_{k+1} \\ s_{k+1} & \text{otherwise} \end{cases}$. $\square$

28. *Sketch.* Check the case when $n = 1$. If $k \geq 1$ and $\sum_{i=1}^{k}(\frac{1}{i} - \frac{1}{i+1}) = 1 - \frac{1}{k+1}$, then $\sum_{i=1}^{k+1}(\frac{1}{i} - \frac{1}{i+1}) = (1 - \frac{1}{k+1}) + (\frac{1}{k+1} - \frac{1}{k+2}) = 1 - \frac{1}{k+2}$. $\square$

29. *Sketch.* Check the case when $n = 1$. If $k \geq 1$ and $\sum_{i=1}^{k}(3i + 1) = \frac{k}{2}(3k + 5)$, then $\sum_{i=1}^{k+1}(3i + 1) = \frac{k}{2}(3k + 5) + (3k + 4) = \frac{k+1}{2}(3(k + 1) + 5)$. $\square$

30. *Sketch.* Check the case when $n = 1$. If $k \geq 1$ and $\sum_{i=1}^{k} 3^i = \frac{3}{2}(3^k - 1)$, then $\sum_{i=1}^{k+1} 3^i = \frac{3}{2}(3^k - 1) + 3^{k+1} = \frac{3}{2}(3^{k+1} - 1)$. $\square$

31. *Sketch.* Check the case when $n = 0$. If $k \geq 0$ and $\sum_{j=0}^{k}(i + 1)2^i = k2^{k+1} + 1$, then $\sum_{i=0}^{k+1}(i+1)2^i = k2^{k+1} + 1 + (k+2)2^{k+1} = (k+1)2^{k+2} + 1$. $\square$

32. *Sketch.* Check the case when $n = 1$. If $k \geq 1$ and $\sum_{i=1}^{k}(3i^2 + 5i) = k(k + 1)(k + 3)$, then $\sum_{i=1}^{k+1}(3i^2 + 5i) = k(k + 1)(k + 3) + (3(k + 1)^2 + 5(k + 1)) = (k + 1)((k + 1) + 1)((k + 1) + 3)$. $\square$

33. *Sketch.* Check the case when $n = 1$. If $k \geq 1$ and $\sum_{i=1}^{k} i4^i = \frac{4}{9}[4^k(3k - 1) + 1]$, then $\sum_{i=1}^{k+1} i4^i = \frac{4}{9}[4^k(3k - 1) + 1] + (k + 1)4^{k+1} = \frac{4}{9}[4^{k+1}(3k + 2) + 1]$. $\square$

34. Use the fact that $b^{\sum_{i=1}^{n+1} a_i} = b^{(\sum_{i=1}^{n} a_i) + a_{n+1}} = b^{(\sum_{i=1}^{n} a_i)}b^{a_{n+1}}$.

35. *Sketch.* Check the cases when $n = 0, 1$. Suppose $k \geq 1$ and that, for each $0 \leq i \leq k$, $3 \mid s_i$. We have $c, d \in \mathbb{Z}$ such that $s_{k+1} = 2s_{k-1} + s_k = 2(3c) + (3d) = 3(2c + d)$. $\square$

36. *Sketch.* Check the cases when $n = 0, 1$. Suppose $k \geq 1$ and that, for each $0 \leq i \leq k$, $s_i = 4 \cdot 5^i + 3 \cdot 4^i$. So $s_{k+1} = -20s_{k-1} + 9s_k = -20(4 \cdot 5^{k-1} + 3 \cdot 4^{k-1}) + 9(4 \cdot 5^k + 3 \cdot 4^k) = 4 \cdot 5^{k+1} + 3 \cdot 4^{k+1}$. $\square$

37. *Sketch.* Check the cases when $n = 0, 1$. Suppose $k \geq 1$ and that, for $0 \leq i \leq k$, $s_i = 2^{i+1} + 3 \cdot 2^{2i}$. So $s_{k+1} = 6s_k - 8s_{k-1} = 6(2^{k+1} + 3 \cdot 2^{2k}) - 8(2^k + 3 \cdot 2^{2(k-1)}) = 2^{(k+1)+1} + 3 \cdot 2^{2(k+1)}$. $\square$

38. (a) Four 3¢ stamps and one 8¢ stamp. (b) *Sketch.* $14 = 2(3) + 1(8)$, $15 = 5(3)$, and $16 = 2(8)$. Further, $k + 1 = (k - 2) + 1(3)$. Therefore, if, for some $k \geq 16$, we can obtain $(k - 2)¢$, then an additional 3¢ stamp will yield $(k + 1)¢$. $\square$

39. $7 \cdot 11 \cdot 13$.

40. $2^3 \cdot 3^4 \cdot 11^2$.

41. $2^2 \cdot 5^2 \cdot 7^2 \cdot 11 \cdot 23 \cdot 43 \cdot 47$.

42. Use the fact that $\sum_{i=1}^{n+1} L_i = (\sum_{i=1}^{n} L_i) + L_{n+1} = (L_{n+2} - 3) + L_{n+1} = (L_{n+1} + L_{n+2}) - 3 = L_{n+3} - 3$.

43. *Sketch.* Check the cases when $n = 0, 1$. Suppose $k \geq 1$ and that, for each $0 \leq i \leq k$, $s_i = 3^i(3^i - 1)$. So $s_{k+1} = 3(4s_k - 9s_{k-1}) = 3(4[3^k(3^k - 1)] - 9[3^{k-1}(3^{k-1} - 1)]) = 3^{k+1}(3^{k+1} - 1)$. $\square$

44. $x^4 + 4x^3y + 6x^2y^2 + 4xy^3 + y^4$.

45. $6561x^8 - 69984x^7y + 326592x^6y^2 - 870912x^5y^3 + 1451520x^4y^4 - 1548288x^3y^5 + 1032192x^2y^6 - 393216xy^7 + 65536y^8$.

46. $x^{10} - 5x^8y^2 + 10x^6y^4 - 10x^4y^6 + 5x^2y^8 - y^{10}$.

47. $\binom{100}{10}2^{90}$.

48. $-\binom{100}{25}3^{25}$.

49. 0.

50. Consider $(1 + 5)^n$.

51. Consider $(1 + 2^2)^n$.

52. Consider $(3 + (-4))^n$.

Chapter 5

Section 5.1

1. (a) False. (b) True. (c) False.

3. (a) True. (b) False. (c) True.

5. True.

7. False.

9. (a) No. (b) Brazil, Colombia, Guyana.

11. The "is a son of" relation.

13. $\supseteq$.

15. $\perp$.

17. R itself.

19. (a) GameCo. (b) No. (c) Yes.

21.

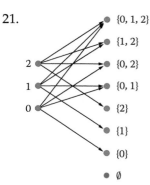

23.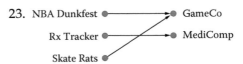

25.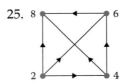

27.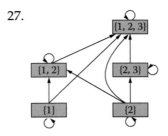

29.
$$\begin{array}{c c} & \begin{array}{c c c c} 0 & 3 & 6 & 9 \end{array} \\ \begin{array}{c} 0 \\ 1 \\ 2 \end{array} & \left[\begin{array}{c c c c} 1 & 0 & 0 & 0 \\ 1 & 1 & 1 & 1 \\ 1 & 0 & 1 & 0 \end{array}\right] \end{array}$$

31.
$$\begin{array}{c c} & \begin{array}{c c c c} \varnothing & \{1\} & \{2\} & \{1,2\} \end{array} \\ \begin{array}{c} \varnothing \\ \{1\} \\ \{2\} \\ \{1,2\} \end{array} & \left[\begin{array}{c c c c} 1 & 1 & 1 & 1 \\ 0 & 1 & 0 & 1 \\ 0 & 0 & 1 & 1 \\ 0 & 0 & 0 & 1 \end{array}\right] \end{array}$$

33.
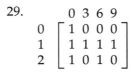
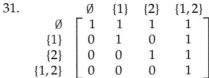

35.
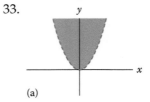
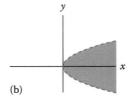
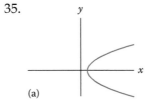
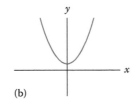

37 through 53. In Exercise 41, $-1 \not{R} -1$, since $\sqrt{-1}$ does not exist.

Exercise	Reflexive	Symmetric	Antisymmetric	Transitive
37	F	T	F	F
39	T	T	F	F
41	F	T	T	T
43	F	F	T	T
45	F	F	T	F
47	T	T	F	T
49	T	F	F	F
51	T	F	F	T
53	T	F	T	T

55. Symmetric.

57. *Sketch.* If R is symmetric, then $x \, R \, y \leftrightarrow y \, R \, x \leftrightarrow x \, R^{-1} \, y$. If $R^{-1} = R$, then $x \, R \, y \leftrightarrow x \, R^{-1} \, y \leftrightarrow y \, R \, x$. $\square$

59. Note that $R \cap R^{-1} \subseteq \Delta$ iff $\forall \, x, y \in X, (x, y) \in R \cap R^{-1} \to (x, y) \in \Delta$ iff $\forall \, x, y \in X, x \, R \, y \wedge y \, R \, x \to x = y$.

Section 5.2

1. The arguments used in Example 5.18 from Section 5.1 show that the "divides" relation on $X = \mathbb{Z}^+$ is reflexive, antisymmetric, and transitive.

3. *Proof.* Let x, y, z be arbitrary elements of X. *Reflexive*: Since $x \, R \, x$, we have $x \, R^{-1} \, x$. *Antisymmetric*: Suppose $x \, R^{-1} \, y$ and $y \, R^{-1} \, x$. That is, $y \, R \, x$ and $x \, R \, y$. Hence, $x = y$. *Transitive*: Suppose $x \, R^{-1} \, y$ and $y \, R^{-1} \, z$. That is, $y \, R \, x$ and $z \, R \, y$. Thus, $z \, R \, x$, and we have $x \, R^{-1} \, z$. $\square$

5. Just the one in Exercise 53.

7. No. It is not antisymmetric.

9. No. It is not reflexive.

11. Yes.

13.

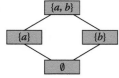

15.

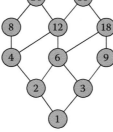

17. (a) No. (b) O^-, O^+, A^-, A^+.

(c)   (d)

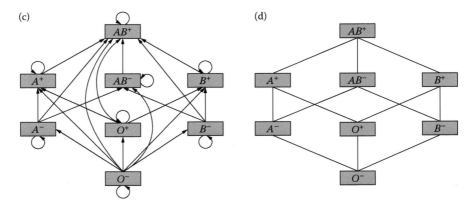

19. One of the lines in the triangle would follow from the other two by transitivity.

21. b12vitamin ◁ b6vitamin.

23. $(2, -3, 1, 0) ◁ (2, -1, 5, 3)$.

25. $a_{n-1}a_{n-2} \cdots a_0 < b_{n-1}b_{n-2} \cdots b_0$ iff $a_{n-1}2^{n-1} + a_{n-2}2^{n-2} + \cdots + a_0 < b_{n-1}2^{n-1} + b_{n-2}2^{n-2} + \cdots + b_0$ iff $a_{n-1} = b_{n-1}, \ldots, a_k = b_k$ and $a_{k-1} < b_{k-1}$.

27. *Sketch.* Reflexive: $\lfloor x \rfloor = \lfloor x \rfloor$. Symmetric: $\lfloor x \rfloor = \lfloor y \rfloor \rightarrow \lfloor y \rfloor = \lfloor x \rfloor$. Transitive: $\lfloor x \rfloor = \lfloor y \rfloor$, $\lfloor y \rfloor = \lfloor z \rfloor \rightarrow \lfloor x \rfloor = \lfloor z \rfloor$. $\square$

29. It all follows from the fact that $=$ is an equivalence relation.

31. *Sketch.* Reflexive: With $c = 1$, we have $cx_1 = x_1$ and $cy_1 = y_1$. Symmetric: $cx_1 = x_2$, $cy_1 = y_2 \rightarrow \frac{1}{c}x_2 = x_1$, $\frac{1}{c}y_2 = y_1$. Transitive: $cx_1 = x_2, cy_1 = y_2, dx_2 = x_3, dy_2 = y_3 \rightarrow cdx_1 = x_3, cdy_1 = y_3$. $\square$

33. For example, it is not reflexive.

35. Just the one in Exercise 47.

37. Since R is symmetric, $x \in [y]$ iff $x \, R \, y$ iff $y \, R \, x$ iff $y \in [x]$.

39. $(m_1 - n_1, 0)$ if $m_1 \geq n_1$, and $(0, n_1 - m_1)$ if $m_1 < n_1$.

41. $\left(\dfrac{x_1}{\sqrt{x_1^2 + y_1^2}}, \dfrac{y_1}{\sqrt{x_1^2 + y_1^2}} \right)$.

43. Yes.

45. No.

47. No.

49. Each integer is either odd or even, and not both.

51. Each $(a, b) \in \mathbb{Z} \times \mathbb{Z}^*$ is in $A_{\frac{a}{b}}$ and no other A_r.

53. $\forall n \in \mathbb{Z}$, let $A_n = [n, n+1)$.

55. $\forall b \in \mathbb{Z}$, let $A_b = \{(m, n) \; : \; m, n \in \mathbb{N} \text{ and } m - n = b\}$.

57. $\forall m \in \mathbb{R}$, let $A_m = \{(x, y) \; : \; y = mx\}$. Additionally, let $A_\infty = \{(x, y) \; : \; x = 0\}$.

59. (a) {apple}, {eat, ear}, {peace}, {car, call}. (b) {apple, peace}, {call}, {eat, car, ear}. (c) {apple, eat, peace, ear}, {car, call}.

61. $m \mathrel{R} n \leftrightarrow m - n$ is even.

63. $(a_1, b_1) \mathrel{R} (a_2, b_2) \leftrightarrow \frac{a_1}{b_1} = \frac{a_2}{b_2}$.

65. (a) "has the same suffix as" or "has the same file type as." (b) "has the same base (or file) name as."

67. *Sketch.* ($\rightarrow$) Suppose $\mathcal{A}$ is the partition of X corresponding to R. Let R' be the equivalence relation on X corresponding to $\mathcal{A}$. Argue that $x \mathrel{R'} y$ iff $\exists\, A \in \mathcal{A}$ such that $x, y \in A$ iff $x \mathrel{R} y$. Hence, $R = R'$. ($\leftarrow$) Suppose R is the equivalence relation on X corresponding to $\mathcal{A}$. Let $\mathcal{A}' = \{[x]_R : x \in X\}$ be the partition of X corresponding to R. Argue that if $A \in \mathcal{A}$, then $A = [y]_R$ for some $y \in A$. Conclude that $\mathcal{A} \subseteq \mathcal{A}'$, and, since $\mathcal{A}$ and $\mathcal{A}'$ are both partitions, $\mathcal{A} = \mathcal{A}'$. □

69. *Sketch.* ($\subseteq$) Suppose $x \in \left(\bigcup_{A \in \mathcal{A}_1} A\right) \cup \left(\bigcup_{A \in \mathcal{A}_2} A\right)$. In the case that $x \in A_1$ for some $A_1 \in \mathcal{A}_1$, we have $x \in \bigcup_{A \in \mathcal{A}_1 \cup \mathcal{A}_2} A$. The symmetric case is handled similarly. ($\supseteq$) Suppose $x \in \bigcup_{A \in \mathcal{A}_1 \cup \mathcal{A}_2} A$. If $x \in A_0$ for some $A_0 \in \mathcal{A}_1$, then $x \in \bigcup_{A \in \mathcal{A}_1} A$. The symmetric case is handled similarly. □

Section 5.3

1. No.

3. It is not.

5. It is not.

7. (a) It is not. (b) It is.

9. It is.

11. (a) It is not. (b) It is not. (c) It is.

13. (a) It is not. (b) It is.

15. Yes.

17. Domain = $\{-3, -2, \ldots, 3\}$ and range = $\{0, 1, 4, 9\}$.

19. Domain = $\{0, 1, \ldots, 4\}$ and range = $\{1, 2, 4, 8, 16\}$.

21. Domain = $\mathbb{R}$ and range = $[-1, \infty)$.

23. Domain = $[1, \infty)$ and range = $[0, \infty)$.

25. Domain = $\mathbb{R} \backslash \{-1\}$ and range = $\mathbb{R} \backslash \{0\}$.

27. $f(0) = f(4) = 4$, $f(1) = f(3) = 1$, $f(2) = 0$, $f(5) = 9$, $f(6) = 16$.

29. *Sketch.* $0 \leq x \leq 2$ iff $0 \leq 3x \leq 6$ iff $-2 \leq 3x - 2 \leq 4$. □

31. *Sketch.* If $0 \leq x \leq 2$, then $0 \leq x^2 \leq 4$. If $0 \leq y \leq 4$, then $0 \leq \sqrt{y} \leq 2$ and $f(\sqrt{y}) = y$. □

33. $(g \circ f)(n) = (n!)^2$.

35. $(g \circ f)(x) = -2 - 9x$.

37. $(g \circ f)(x) = \frac{1}{|x|}$.

39. (a) Yes. (b) 4.

41. The "is the grandfather of" relation.

43. (a) *Sketch.* ($\rightarrow$) Suppose R is transitive. If there is some y such that $(x, y) \in R$ and $(y, z) \in R$, then $(x, z) \in R$. Hence, $R \circ R \subseteq R$. ($\leftarrow$) Suppose $R \circ R \subseteq R$. If $(x, y) \in R$ and $(y, z) \in R$, then $(x, z) \in (R \circ R) \subseteq R$. Hence, R is transitive. $\square$ (b) *Sketch.* Suppose R is reflexive and transitive. If $(x, y) \in R$, then $(x, y) \in (R \circ R)$ (since $(x, x) \in R$ and $(x, y) \in R$). By part (a), the converse holds. $\square$

45. (a) No. (b)

Artist	Music company
MandM	Aristotle Records
Fifty Percent	Bald Boy Records
MandM	Bald Boy Records
M.C. Escher	Aristotle Records

(c) Aristotle Records and Bald Boy Records.

47. (a)

Programmer	Client
Martha Lang	GameCo
Megan Johnson	MediComp
Charles Murphy	GameCo

(b) Only GameCo.

49. (a)

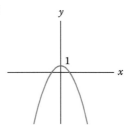

(b) Yes. Domain = $\mathbb{R}$. (c) Range = $(-\infty, 1]$.

51. (a)

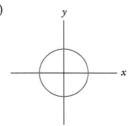

(b) No. (c) None.

53. (a)

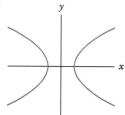

(b) No. (c) None.

55. The relation is a function iff each row has at most one 1.

57. (a) $\forall\, x \in \mathbb{R},$ $((f + g) \circ h)(x) = (f + g)(h(x)) = f(h(x)) + g(h(x)) = (f \circ h)(x) + (g \circ h)(x) = (f \circ h + g \circ h)(x).$
(b) Define $f(x) = x^2$ and $g(x) = h(x) = 1.$

59. (a) $\forall\, x \in \mathbb{R},$ $(c(f \circ g))(x) = c((f \circ g)(x)) = c(f(g(x))) = (cf)(g(x)) = ((cf) \circ g)(x).$ (b)   Let $c = 2,$ $f(x) = x,$ and $g(x) = 1.$

61. Yes.

Section 5.4

1. If $x_1^3 + 8 = x_2^3 + 8,$ then $x_1 = x_2.$

3. *Proof.* Suppose $n_1, n_2 \in \mathbb{Z}^-$ and $1 - n_1^2 = 1 - n_2^2.$ So $n_1^2 = n_2^2.$ Since $n_1, n_2 \in \mathbb{Z}^-,$ we have $n_1 = n_2.$ $\square$

5. $f(0) = f(1),$ but $0 \neq 1.$

7. *Proof.* Suppose $y \in (1, \infty).$ Let $x = \sqrt{y - 1}.$ Observe that $f(x) = (\sqrt{y - 1})^2 + 1 = y.$ $\square$

9. *Proof.* Observe that $f(x) = \frac{1}{2}$ is impossible, since $x^2 = -\frac{1}{2}$ has no solution in $\mathbb{R}.$ $\square$

11. $\forall\, y \in \mathbb{R}^+, f\left(\frac{-1+\sqrt{1+4y}}{2}\right) = y.$

13. (a) $\forall\, k \in \mathbb{Z}, f(k, 1) = k.$ (b) $f(2, 1) = f(1, 2).$

15. *Sketch.* Suppose $x_1 \neq x_2.$ If $x_1 < x_2,$ then $f(x_1) < f(x_2).$ In any case, $f(x_1) \neq f(x_2).$ $\square$

17. (a) If $[n'] = [n]$ then $6 \mid (n' - n),$ so $6 \mid (2n' - 2n).$ Hence, $[2n'] = [2n].$ (b) $f([0]) = f([3]).$ (c) $\gcd(2, 6) \neq 1.$

19. (a) 535. (b) 0000000003216 and 0000000009461.

21. (a) $\forall\, x \in X, p(x, y_0) = x.$ (b) If $(x_1, y_0) = (x_2, y_0),$ then $x_1 = x_2.$ (c) $\forall\, x \in X, (p \circ i)(x) = p(x, y_0) = x.$
(d) $\forall\, x \in X, y \in Y, (i \circ p)(x, y) = (x, y_0).$

23. Suppose $i(a_1) = i(a_2).$ Then $a_1 = a_2.$

25. If $B \subseteq [0, 1],$ then $f(B) = B.$

27. f and g are onto. f and g are not one-to-one.

29. One-to-one.

31. $\forall\, x \in X, (f \circ \mathrm{id}_X)(x) = f(\mathrm{id}_X(x)) = f(x)$ and $(\mathrm{id}_X \circ f)(x) = \mathrm{id}_X(f(x)) = f(x)$.

33. Exercise 1 established that f is one-to-one. Since $f((y-8)^{\frac{1}{3}}) = y$, we see that f is also onto.

35. *Sketch.* $-2 \mapsto 0,\ -1 \mapsto 2,\ 0 \mapsto 4,\ 1 \mapsto 6,\ 2 \mapsto 8.$ □

37. *Sketch.* $\forall\, j \in \mathbb{Z}$, if $-n \le j \le n$, then $1 \le j + 1 + n \le 2n + 1$ and $f(j + 1 + n) = j$. So f is onto. If $k_1 - 1 - n = k_2 - 1 - n$, then $k_1 = k_2$. So f is one-to-one. □

39. (a) If $[k'] = [k]$ then $[mk'] = [mk]$ and $[m + k'] = [m + k]$.
    (b) *Sketch.* ($\rightarrow$) Suppose f is a bijection. Since f is onto, there is some $k \in \mathbb{Z}$ such that $[mk] = f([k]) = [1]$. Since $1 \equiv mk \pmod{n}$, there is $j \in \mathbb{Z}$ such that $km + jn = 1$. So $\gcd(m, k) = 1$. ($\leftarrow$) Suppose $\gcd(m, k) = 1$. So there are $j, k \in \mathbb{Z}$ such that $mk + nj = 1$. That is, $f([k]) = [mk] = [1]$. Hence, for each $y \in \mathbb{Z}, f([ky]) = [y]$. Thus, f is onto. Now suppose $f([k_1]) = f([k_2])$. So $mk_1 \equiv mk_2 \pmod{n}$. Hence, $k_1 \equiv k_2 \pmod{n}$. That is, $[k_1] \equiv [k_2]$. So f is one-to-one. □
    (c) All $m \in \mathbb{Z}$.

41. If every column contains at most one 1, then f is one-to-one.

43. (b) If we have $y \in Y$ such that $g(y) = z$, and we have $x \in X$ such that $f(x) = y$, then $(g \circ f)(x) = z$. (c) This follows from parts (a) and (b).

45. *Sketch.* ($\rightarrow$) Suppose f is symmetric and $x \in X$. Since $(x, f(x)) \in f$, $(f(x), x) \in f$. So $(f \circ f)(x) = f(f(x)) = x$. ($\leftarrow$) Suppose $f \circ f = \mathrm{id}_X$ and $(x, y) \in f$. Since $x = (f \circ f)(x) = f(f(x)) = f(y), (y, x) \in f$. □

47. (a) *Sketch.* Suppose $f(x_1) = f(x_2)$. Thus, $g(f(x_1)) = g(f(x_2))$. Since $g \circ f$ is one-to-one, $x_1 = x_2$. So f is one-to-one. □ (b) $X = Y = Z = (0, \infty)$, $f(x) = 1 + \sqrt{x}$, $g(x) = (x-1)^2$.

49. (a) Apply Exercise 47(a). (b) Apply Exercise 48(a). (c) $X = Y = Z = [0, \infty), f(x) = 1 + \sqrt{x}, g(x) = (x-1)^2$.

51. (a) *Sketch.* Suppose f and g are increasing. If $x < y$, then $f(x) < f(y)$, whence $g(f(x)) < g(f(y))$. □ (b) $\forall\, x \in \mathbb{R}, f(x) = g(x) = -x$.

53. $\forall\, x \in \mathbb{R}, g(f(x)) = g(2x + 5) = \frac{2x + 5 - 5}{2} = x$ and $f(g(x)) = f(\frac{x-5}{2}) = 2(\frac{x-5}{2}) + 5 = x$.

55. $\forall\, x \in \mathbb{R}, g(f(x)) = g(4 - 2x) = 2 - \frac{1}{2}(4 - 2x) = 2 - 2 + x = x$ and $f(g(x)) = f(2 - \frac{1}{2}x) = 4 - 2(2 - \frac{1}{2}x) = 4 - 4 + x = x$.

57. $g(f(2)) = g(1) = 2,\ g(f(3)) = g(3) = 3,\ g(f(4)) = g(6) = 4,$ and $g(f(5)) = g(10) = 5$. One similarly shows that $g \circ f = \mathrm{id}$.

59. $\forall\, r \in \mathbb{Q}^+, (f \circ f)(r) = f(\frac{1}{r}) = \frac{1}{\frac{1}{r}} = r$.

61. $f^{-1}(x) = \sqrt[3]{\frac{x-1}{4}}$.

63.

Phone number	Name
555-3148	Blair, Tina
555-3992	Walsh, Carol
555-4500	Tillman, Paul
555-6301	Jennings, Robert

The function need not be one-to-one and need not be onto.

65. By Theorem 5.10(a), f^{-1} is a function. Since f^{-1} and f are inverses of one another, Theorem 5.10(b) tells us that f^{-1} is a bijection.

67. (a) *Sketch.* Suppose f_1 and f_2 are one-to-one. Suppose that $(f_1 \times f_2)((x'_1, x'_2)) = (f_1 \times f_2)((x_1, x_2))$. Since $f_1(x'_1) = f_1(x_1)$ and $f_2(x'_2) = f_2(x_2)$, $x'_1 = x_1$ and $x'_2 = x_2$. □
(b) *Sketch.* Suppose f_1 and f_2 are onto. Suppose $(y_1, y_2) \in Y_1 \times Y_2$. We have $x_1 \in X_1$ and $x_2 \in X_2$ such that $(f_1 \times f_2)((x_1, x_2)) = (f_1(x_1), f_2(x_2)) = (y_1, y_2)$. □ (c) This follows from parts (a) and (b).

69. (a) Motivation.

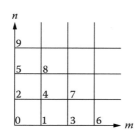

(b) *Sketch.* $\forall\, m \in \mathbb{N}, g(m+1) = m$. So g is onto. If $n_1 - 1 = n_2 - 1$, then $n_1 = n_2$. So g is one-to-one. □ (c) This follows from part (b) and Exercise 67(c). (d) This follows from part (c), part (a), Exercise 65, and Theorem 5.9(c).

71. (a) -2. (b) 4. (c) $\frac{1}{4}$. (d) -1.

73. $\log_2 3$.

75. (a) Since $b^0 = 1$. (b) Since $b^{\log_b y + \log_b z} = b^{\log_b y} b^{\log_b z} = yz$. (c) Since $b^{a \log_b y} = (b^{\log_b y})^a = y^a$.

77. If $a, b \in \mathbb{Z}^+$ and $2^a = 2^b$, then the Fundamental Theorem of Arithmetic tells us that $a = b$.

Section 5.5

1. (a) $\{1, 2, 6\}$. (b) No.

3. $\{-6, -5, 10\}$.

5. $f([-2,2]) = [-3,5]$ since, $\{t \ : \ t \in \mathbb{R} \text{ and } 2s + 1 = t \text{ for some } -2 \leq s \leq 2\} = \{t \ : \ t \in \mathbb{R} \text{ and } -2 \leq \frac{t-1}{2} \leq 2\} = \{t \ : \ t \in \mathbb{R} \text{ and } -3 \leq t \leq 5\}$.

7. $f([1,3]) = [2,10]$. $2 \leq t \leq 10$ iff $1 \leq \sqrt{t-1} \leq 3$ and $f(\sqrt{t-1}) = t$.

9. *Sketch.* ($\subseteq$) Suppose $z \in g(\text{range}(f))$. So we have some $x \in X$ such that $z = g(f(x))$. Hence, $z \in \text{range}(g \circ f)$. ($\supseteq$) Suppose $z \in \text{range}(g \circ f)$. So we have $x \in X$ such that $g(f(x)) = z$. Since $f(x) \in \text{range}(f)$, we see that $z \in g(\text{range}(f))$. $\square$

11. (a) $\{0,1,2,4\}$. (b) $\{3\}$. (c) No. $f^{-1}(\{120\}) = \{5\}$.

13. $\{-2,-1,1,2\}$.

15. $f^{-1}(\{-1\}) = O$, the set of odd integers.

17. The set of relatively prime pairs of positive integers.

19. $f^{-1}([1,4]) = [-2,-1] \cup [1,2]$.   *Sketch.* If $x \in [-2,-1] \cup [1,2]$, then $x^2 \in [1,4]$. If $x \in (-\infty, -2) \cup (-1,1) \cup (2,\infty)$, then $x^2 \in [0,1) \cup (4,\infty)$. $\square$

21. *Sketch.* ($\subseteq$) Suppose $(x,y) \in f(\mathbb{R})$. So $y = x$. Hence, $g(x,y) = 0$ and $(x,y) \in g^{-1}(\{0\})$. ($\supseteq$) Suppose $(x,y) \in g^{-1}(\{0\})$. So $x - y = 0$. Since $y = x$, it follows that $f(x) = (x,y)$ and $(x,y) \in f(\mathbb{R})$. $\square$

23. (a) Eagles and Huskies. Image. (b) KSU, Northwestern, UNH, and Villanova. Inverse Image.

25. (a) The set E of even integers. (b) $\{3,5,7\}$.

27. (a) $\{0\}$, $\{1\}$, and $\{0,1\}$.
$R(\{0,1\}) = \{A \ : \ A \subset \mathbb{Z}, A \cap \{0,1\} \neq \emptyset\}$. (b) $\{0,1,2,3,4,6\}$.

29. (a) No. (b) A wrench and pliers. Inverse image. (c) Susan Brower and Abe Roth. Image.

31. (a) *Sketch.* Suppose $S_1 \subseteq S_2$ and $y \in f(S_1)$. So $y = f(x)$ for some $x \in S_1 \subseteq S_2$. Thus, $y \in f(S_2)$. $\square$ (b) *Sketch.* Suppose $T_1 \subseteq T_2$ and $x \in f^{-1}(T_1)$. So $f(x) \in T_1 \subseteq T_2$. Thus, $x \in f^{-1}(T_2)$. $\square$

33. (a) and (b) follow from the fact that, $\forall \ x \in X, f(x) \in Y$.
(c) $X = Y = \mathbb{R}$ and $f(x) = x^2$.

35. (a) *Sketch.* ($\subseteq$) Suppose $x \in f^{-1}(T_1 \cup T_2)$. So $f(x) \in T_1 \cup T_2$. For $i = 1,2$, if $f(x) \in T_i$, then $x \in f^{-1}(T_i)$. Hence, $x \in f^{-1}(T_1) \cup f^{-1}(T_2)$.   ($\supseteq$)   By   Exercise 31(b),   for $i = 1,2$, $f^{-1}(T_i) \subseteq f^{-1}(T_1 \cup T_2)$.   Hence, $f^{-1}(T_1) \cup f^{-1}(T_2) \subseteq f^{-1}(T_1 \cup T_2)$. $\square$ (b)   *Sketch.* By Exercise 31(b), for $i = 1,2$, $f^{-1}(T_1 \cap T_2) \subseteq f^{-1}(T_i)$. So $f^{-1}(T_1 \cap T_2) \subseteq f^{-1}(T_1) \cap f^{-1}(T_2)$. Now suppose $x \in f^{-1}(T_1) \cap f^{-1}(T_2)$. So for $i = 1,2$, $f(x) \in T_i$. Since $f(x) \in T_1 \cap T_2, x \in f^{-1}(T_1 \cap T_2)$. $\square$

37. (a) *Sketch.* Suppose $x \in S \cap f^{-1}(T)$. Since $x \in S$ and $f(x) \in T$, $f(x) \in f(S) \cap T$. $\square$ (b) *Sketch.* Suppose $x \in S \cap f^{-1}(T)$. By part (a), $f(x) \in f(S) \cap T$. So $x \in f^{-1}(f(S) \cap T)$. $\square$

39. (a) *Sketch.* ($\rightarrow$) Suppose f is one-to-one and $F(A_1) = F(A_2)$. So $f(A_1) = f(A_2)$. Suppose $x \in A_1$. So $f(x') = f(x) \in f(A_1) = f(A_2)$ for some $x' \in A_2$. Thus, $x = x' \in A_2$. By symmetry, $A_1 = A_2$. ($\leftarrow$) Suppose F is one-to-one and $f(x_1) = f(x_2)$. Since $F(\{x_1\}) = f(\{x_1\}) = f(\{x_2\}) = F(\{x_2\})$, $\{x_1\} = \{x_2\}$, whence $x_1 = x_2$. $\square$ (b)   *Sketch.* ($\rightarrow$) Suppose f is onto and $B \in \mathcal{P}(Y)$. Since f is onto, $F(f^{-1}(B)) = f(f^{-1}(B)) = B$. ($\leftarrow$) Suppose F is onto and $y \in Y$. We have some $A \subseteq X$ such that $f(A) = F(A) = \{y\}$. Pick $x \in A$. So, for some $x \in A, f(x) = y$. $\square$

41. $\{0\}$.

43. $[1, \infty)$.

45. $[0, 1]$.

47. $\bigcup_{x \in [3,4)} A_x$.

49. $\bigcap_{\alpha \text{ is a vowel}} A_\alpha$.

51. Generalize the argument from Exercise 36(a).

53. Generalize the argument from Exercise 35(b).

55. (a) *Sketch.* Suppose $x \in \bigcup_{i \in \mathcal{J}} A_i$. So $x \in A_{i_0}$ for some $i_0 \in \mathcal{J} \subseteq \mathcal{I}$. Thus, $x \in \bigcup_{i \in \mathcal{I}} A_i$. $\square$ (b)   *Sketch.* Suppose $x \in \bigcap_{i \in \mathcal{I}} A_i$. Since $\mathcal{J} \subseteq \mathcal{I}$, in particular, $\forall \, i \in \mathcal{J}, x \in A_i$. Thus, $x \in \bigcap_{i \in \mathcal{J}} A_i$. $\square$

57. (a) *Proof.* ($\subseteq$) Suppose $x \in B \cup \bigcap_{i \in \mathcal{I}} A_i$. If $x \in B$, then, $\forall \, i \in \mathcal{I}, x \in B \cup A_i$. So $x \in \bigcap_{i \in \mathcal{I}} (B \cup A_i)$. If $x \in \bigcap_{i \in \mathcal{I}} A_i$, then $\forall \, i \in \mathcal{I}, \ x \in A_i \subseteq B \cup A_i$. So $x \in \bigcap_{i \in \mathcal{I}} (B \cup A_i)$. ($\supseteq$) Suppose $x \in \bigcap_{i \in \mathcal{I}} (B \cup A_i)$. So, $\forall \, i \in \mathcal{I}, x \in B \cup A_i$. If $x \notin B$, then it must be that, $\forall \, i \in \mathcal{I}, x \in A_i$. So $x \in \bigcap_{i \in \mathcal{I}} A_i$. In any case, $x \in B \cup \bigcap_{i \in \mathcal{I}} A_i$. $\square$ (b) *Proof.* ($\subseteq$) Suppose $x \in B \cap \bigcup_{i \in \mathcal{I}} A_i$. So $x \in B$ and $x \in A_{i_0}$ for some $i_0 \in \mathcal{I}$. Thus, $x \in B \cap A_{i_0} \subseteq \bigcup_{i \in \mathcal{I}} (B \cap A_i)$. ($\supseteq$) Suppose $x \in \bigcup_{i \in \mathcal{I}} (B \cap A_i)$. So $x \in B \cap A_{i_0}$ for some $i_0 \in \mathcal{I}$. Hence, $x \in B$ and $x \in A_{i_0} \subseteq \bigcup_{i \in \mathcal{I}} A_i$. Thus, $x \in B \cap \bigcup_{i \in \mathcal{I}} A_i$. $\square$

59. Use Exercise 57(b) and the fact that $S \backslash T = S \cap T^c$.

61. Yes.

Section 5.6

1. 76.

3. 3.

5. 3.

7. 4.

9. The function $f : \{0, 1, \ldots, n\} \longrightarrow \{1, 2, \ldots, n + 1\}$ given by $f(k) = k + 1$ is a bijection.

11. The function $f : \{n^2, n^2 + 1, \ldots, (n + 1)^2\} \longrightarrow \{1, 2, \ldots, 2n + 2\}$ given by $f(k) = k - n^2 + 1$ is a bijection.

13. The function $f : \mathbb{N} \longrightarrow \mathbb{Z}^-$ given by $f(n) = -n - 1$ is a bijection.

15. The function $f : \mathbb{Z}^+ \longrightarrow \{k^2 \ : \ k \in \mathbb{Z}^+\}$ given by $f(m) = m^2$ is a bijection.

17. $f(x) = 2(x - 3) + 1$ is a bijection from $[3, 8]$ to $[1, 11]$ with inverse $g(x) = \frac{1}{2}(x - 1) + 3$.

19. For (a), (b), and (c), the function given by $f(x) = \frac{x-a}{b-a}$ is a bijection.

21. $f(x) = \frac{b-x}{b-a}$ gives a bijection.

23. For (a) and (b), $f(x) = \frac{x-a}{b-x}$ gives a bijection.

25. $f(x) = \begin{cases} \frac{x-a}{2(b-a)} & \text{if } x \in [a, b) \\ \frac{1}{2} + \frac{x-c}{2(d-c)} & \text{if } x \in [c, d) \end{cases}$, gives a bijection.

27. *Proof.* Suppose A has the same cardinality as B. So we have a bijection $f : A \longrightarrow B$. Since $f^{-1} : B \longrightarrow A$ is also a bijection, B has the same cardinality as A. $\square$

29. *Sketch.* Suppose to the contrary that A has cardinality m. So $m \in \mathbb{N}$, and we have a bijection $f : A \longrightarrow \{1, 2, \ldots, m\}$. In particular, f is one-to-one. If $n > m$, then f cannot exist by the Pigeon Hole Principle. If $n < m$, a contradiction occurs with f^{-1}. $\square$

31. Apply the contrapositive of Corollary 5.13.

33. There are 2^{16} possible integers, and $70000 > 2^{16}$.

35. Use Exercise 67(c) from Section 5.4.

37. Assume $m \geq 1$ and $A = \{1, \ldots, m\}$. By induction on n, we can prove that, for any $n \geq 1$, $|A \times \{1, \ldots, n\}| = mn$.

39. Exercise 69(c) from Section 5.4 gives a bijection $\mathbb{Z}^+ \times \mathbb{Z}^+ \longrightarrow \mathbb{Z}^+$.

41. *Sketch.* Let C_1 and C_2 be countable sets. So we have bijections $f : C_1 \longrightarrow \mathbb{Z}^+$ and $g : C_2 \longrightarrow \mathbb{Z}^+$. By Exercise 67(c) from Section 5.4, we know that $f \times g : C_1 \times C_2 \longrightarrow \mathbb{Z}^+ \times \mathbb{Z}^+$ is a bijection. Exercise 39 gives a bijection $h : \mathbb{Z}^+ \times \mathbb{Z}^+ \longrightarrow \mathbb{Z}^+$. The composite $h \circ (f \times g)$ is the desired bijection. $\square$

43. $h(1) = z$ and $\forall \, k \geq 2, h(k) = g(k - 1)$.

45. (a) Show: $\forall \, n \in \mathbb{N}$, for any set B with cardinality n and any subset $A \subseteq B$, that A is finite. *Sketch.* Note the case when $n = 0$, so $B = \emptyset$. Let $k \geq 0$ and suppose, for any set B of cardinality k and any subset $A \subseteq B$, that A is finite. Let B be a set with cardinality $k + 1$, and suppose $A \subseteq B$. If $A = B$, then A has cardinality $k + 1$ and is finite as well. If $A \subset B$, then define $B' = B \backslash \{b\}$, where $b \in B \backslash A$. Since B' has cardinality k and $A \subseteq B'$, A is finite. $\square$ (b) This is the contrapositive of part (a).

47. (a) *Proof.* Suppose $A \subseteq \mathbb{Z}^+$, and define the function $f : \mathbb{Z}^+ \longrightarrow A$ by $f(n) = \min(A \setminus \{f(1), f(2), \ldots, f(n-1)\})$. In fact, f is an increasing function. Hence, f is one-to-one. Now suppose $a \in A \subseteq \mathbb{Z}^+$. Let m be the cardinality of $\{k : k \in A$ and $k \le a\}$. In fact, $f(m) = a$. So f is onto. Therefore, f is a bijection. $\square$ (b) *Proof.* Suppose $g : A \longrightarrow \mathbb{Z}^+$ is one-to-one. Let $A' = g(A)$, and define $g' : A \longrightarrow A'$ by $g'(a) = g(a)$. Since g' is a bijection, A' is an infinite subset of $\mathbb{Z}^+$. By part (a), A' is countably infinite. Hence, A is countably infinite. $\square$

49. *Proof.* Suppose B is countable. If A is finite, then A is countable. So it suffices to assume that A is infinite. We have a bijection $f : B \longrightarrow \mathbb{Z}^+$. Let $i : A \longrightarrow B$ be the inclusion. So $g = f \circ i$ is a one-to-one map $A \longrightarrow \mathbb{Z}^+$. By Exercise 47, A is countably infinite, hence countable. $\square$

51. *Proof.* The bijection $b : \mathbb{Z}^+ \longrightarrow \mathbb{Z}$ given by $b(n) = \begin{cases} \frac{n}{2} & \text{if } n \text{ is even} \\ \frac{1-n}{2} & \text{if } n \text{ is odd} \end{cases}$, together with the results in Exercise 69 from Section 5.4, enables us to construct a bijection $g : \mathbb{Z}^+ \longrightarrow \mathbb{Z} \times \mathbb{Z}^+$. Since $h : \mathbb{Z} \times \mathbb{Z}^+ \longrightarrow \mathbb{Q}$ defined by $h((m, n)) = \frac{m}{n}$ is onto, the composite $h \circ g$ is an onto map $\mathbb{Z}^+ \longrightarrow \mathbb{Q}$. By Exercise 48, $\mathbb{Q}$ is countably infinite. $\square$

53. *Proof.* Since B is countable, we have a bijection $g : \mathbb{Z}^+ \longrightarrow B$. For each $b \in B$, since A_b is countable, we have a bijection $f_b : \mathbb{Z}^+ \longrightarrow A_b$. We claim that $h : \mathbb{Z}^+ \times \mathbb{Z}^+ \longrightarrow \bigcup_{b \in B} A_b$ defined by $h((m, n)) = f_{g(m)}(n)$ is onto. Suppose $a \in \bigcup_{b \in B} A_b$. So $a \in A_{b_0}$ for some $b_0 \in B$. Since g is onto, we have $m \in \mathbb{Z}^+$ such that $g(m) = b_0$. Since f_{b_0} is onto, we have $n \in \mathbb{Z}^+$ such that $f_{b_0}(n) = a$. Thus, $h((m, n)) = f_{b_0}(n) = a$. So h is onto. Since Exercise 69 from Section 5.4 guarantees a bijection $w : \mathbb{Z}^+ \longrightarrow \mathbb{Z}^+ \times \mathbb{Z}^+$, we have an onto map $h \circ w : \mathbb{Z}^+ \longrightarrow \bigcup_{b \in B} A_b$. By Exercise 48, $\bigcup_{b \in B} A_b$ is countable. $\square$

Review

1. (a) Yes. (b) No. (c) No.

2.

	0	1	2
0	0	0	0
1	1	0	0
2	0	1	0
3	0	0	0
4	0	0	1

3. $x R^{-1} y$ iff $y = 2^x$.

4. The transpose of the matrix in Exercise 2.

5. (a) No. (b) Yes. (c) Computer science and mathematics.

6.

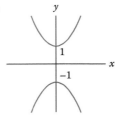

7. (a) NNSE. (b) No.   (c)

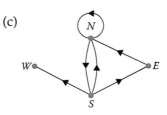

8. (a) No. (b) Yes. (c) No.

9.

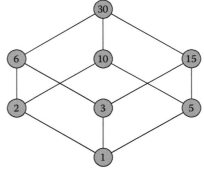

10. $x\,R^{-1}\,y$ iff $x^2 - y^2 = 1$.

11. Reflect the graph from Exercise 9 about the line $y = x$.

12. Not reflexive. Symmetric. Not antisymmetric. Not transitive.

13. Not reflexive. Not symmetric. Antisymmetric. Transitive.

14. That R is reflexive, antisymmetric, and transitive follows from the fact that both R_1 and R_2 are.

15. $=$.

16.

17. $(6, -1, -3, 5, 2) \preceq (6, -1, 2, -4, 7)$.

18. (a) $(40, 16, 4) \prec (40, 18, 2)$ in the lexicographic ordering, but $(40, 16, 4)$ is a better record than $(40, 18, 2)$. (b) Use triples (W, T, L) instead.

19. *Sketch. Reflexive:* $x^2 = x^2$. *Symmetric:* If $x^2 = y^2$, then $y^2 = x^2$. *Transitive:* If $x^2 = y^2$ and $y^2 = z^2$, then $x^2 = z^2$. $\square$

20. (a) $[x] = \{x, -x\}$. (b) $|x|$.

21. $\forall\, m, n \in \mathbb{Z}$, if $m \neq n$, then $(m - 1, m] \cap (n - 1, n] = \emptyset$. Also, $\bigcup_{n \in \mathbb{Z}} (n - 1, n] = \mathbb{R}$.

22. (a) A vertical line through $(x, 0)$. (b) Each point (x, y) lies on a unique vertical line $\{x\} \times \mathbb{R}$.

23. No.

24. $\forall\, x \in [0, \infty)$, let $A_x = \{-x, x\}$.

25. $x\, R\, y$ iff $\lceil x \rceil = \lceil y \rceil$.

26. $(x_1, y_1)\, R\, (x_2, y_2)$ iff $x_1 = x_2$.

27. $f(1)$ is not defined uniquely.

28. No.

29. Domain $= \{2, 3, 4, 5\}$ and range $= \{1, 3, 6, 10\}$.

30. $\forall\, x \in \mathbb{R}, -1 \le x \le 2$ iff $0 \le 2 - x \le 3$.

31. Domain $= \mathbb{R} \backslash \{2\}$ and range $= \mathbb{R} \backslash \{1\}$.

32. $g \circ f : \mathbb{Z}^+ \longrightarrow \mathbb{R}$ is given by $(g \circ f)(n) = n - 2$.

33. (a) *Of Mice and Cats* and *Raisins of Wrath*.

(b)

Publisher	Customer
Book Farm	Raul Cortez
Authority Pubs	Mary Wright
Word Factory	Mary Wright
Book Farm	David Franklin

(c) Raul Cortez and David Franklin.

34. Observe that $x\, R\, y$ and $y\, S\, z$ iff $z\, S^{-1}\, y$ and $y\, R^{-1}\, x$. Hence, $x\, (S \circ R)\, z$ iff $z\, (R^{-1} \circ S^{-1})\, x$.

35. If $3n_1 - 2 = 3n_2 - 2$, then $n_1 = n_2$.

36. $f(0) = f(1)$ but $0 \neq 1$.

37. $f(-1) = 0, f(2) = 3$, and $f(-3) = f(3) = 8$.

38. $f(x) = -11$ is impossible.

39. (a) 84. (b) $036 - 77 - 5484$ and $036 - 77 - 5709$.

40. *Sketch.* (One-to-one) If $\frac{r_1}{n} = \frac{r_2}{n}$, then $r_1 = r_2$. (Onto) If $s \in \mathbb{Q}$, then $f(ns) = s$. $\square$

41. *Sketch.* (One-to-one) If $f(n_1) = f(n_2)$, then n_1 and n_2 must have the same parity, whence must be equal. (Onto) $\forall\, (m, i) \in \mathbb{Z} \times \{0, 1\}$, $f(2m + i) = (m, i)$. $\square$

42. Since f is one-to-one, so is f'. Since $Y' = \text{range}(f) = \text{range}(f'), f'$ is onto.

43. (a) *Proof.* Suppose $f : [0,2] \longrightarrow [0,1]$. Let $x \in [0,2]$. Since $f(x) \in [0,1]$, it follows that $g(f(x)) = f(x)$. Hence, $g \circ f = f$. $\square$
    (b) Define $f(x) = \frac{x}{2}$.

44. Confirm that $\forall\, x \in \mathbb{R}, f(g(x)) = x$ and $g(f(x)) = x$.

45. *Proof.* Suppose $f : X \longrightarrow Y$. Let $g_1 : Y \longrightarrow X$ and $g_2 : Y \longrightarrow X$ be inverses of f. That is, $g_1 \circ f = \text{id}_X$, $f \circ g_1 = \text{id}_Y$, $g_2 \circ f = \text{id}_X$, and $f \circ g_2 = \text{id}_Y$. It follows that $g_2 = g_2 \circ \text{id}_Y = g_2 \circ (f \circ g_1) = (g_2 \circ f) \circ g_1 = \text{id}_X \circ g_1 = g_1$. $\square$

46. 3.

47. 1.

48. (a) $\{-1,0,4\}$. (b) $\{-1,0,1,2\}$.

49. (a) $[0,4]$. (b) $[-\sqrt{3}, \sqrt{3}]$.

50.

S	$f(S)$	T	$f^{-1}(T)$
$\{1\}$	$\{-1\}$	$\{1\}$	$\{0\}$
$[0,1]$	$[-1,1]$	$[1,4)$	$[-1,0]$
$(-1,0)$	$(1,3)$	$(-4,-2)$	$\emptyset$

51. (a) Megan Johnson, Martha Lang, and Abe Roth.
    (b) Inverse image.

52. (a) True. (b) False.

53. (a) False. Let $f(x) = 0$, $S_1 = \{-1\}$, and $S_2 = \{1\}$. (b) True. $f^{-1}(T_1 \bigtriangleup T_2) = f^{-1}(T_1 \cap T_2^c) \cup f^{-1}(T_2 \cap T_1^c) = (f^{-1}(T_1) \cap (f^{-1}(T_2))^c) \cup (f^{-1}(T_1) \cap (f^{-1}(T_2))^c) = f^{-1}(T_1) \bigtriangleup f^{-1}(T_2)$.

54. $(0,5)$.

55. $\emptyset$.

56. $S = \bigcup_{r \in [80,111]} A_r$. Or, $S = A_{80} \cup A_{90} \cup A_{100} \cup A_{110} \cup A_{115}$.

57. *Sketch.* Let $j \in \mathcal{I}$. ($\subseteq$) If $x \in \bigcup_{i \in \mathcal{I}} (B \cup A_i)$, then we have some $i_0 \in \mathcal{I}$ such that $x \in B \cup A_{i_0} \subseteq B \cup \bigcup_{i \in \mathcal{I}} A_i$. ($\supseteq$) If $x \in B \cup \bigcup_{i \in \mathcal{I}} A_i$, then consider the cases when $x \notin B$ and when $x \in B$. $\square$

58. 3.

59. $f : \{-100, -99, \ldots, 200\} \longrightarrow \{1,2,\ldots,301\}$ defined by $f(n) = n + 101$ is a bijection.

60. 3.

61. $f : [-1,0) \longrightarrow (1,7]$ defined by $f(x) = 1 - 6x$ is a bijection.

62. $f : \{2k\ :\ k \in \mathbb{Z}, 0 \le k \le n\} \longrightarrow \{1,2,\ldots,n+1\}$ defined by $f(m) = \frac{m}{2} + 1$ is a bijection.

63. $f : \mathbb{Z} \longrightarrow T$ defined by $f(n) = 10n$ is a bijection.

64. The function $f : A \times B \longrightarrow B \times A$ defined by $(a, b) \mapsto (b, a)$ is a bijection.

65. False. Let $A = B = C = (0, 1)$ and $D = (0, 2)$.

66. *Proof.* Suppose to the contrary that, for some $m \in \mathbb{N}$, there is a bijection $f : [1, 2] \longrightarrow \{1, 2, \ldots, m\}$. Let $i : \{1 + \frac{1}{n} : 1 \leq n \leq m + 1\} \longrightarrow [1, 2]$ be the inclusion of a set of cardinality $m + 1$. The composite $f \circ i$ is a one-to-one map that contradicts the Pigeon Hole Principle. $\square$

67. $f : \{3k : k \in \mathbb{Z}^+\} \longrightarrow \mathbb{Z}^+$ defined by $f(n) = \frac{n}{3}$ is a bijection.

68. *Proof.* Suppose to the contrary that $\mathbb{R}^2$ is countable. Then the subset $\mathbb{R} \times \{0\}$ is countable. Since $\mathbb{R}$ and $\mathbb{R} \times \{0\}$ have the same cardinality, $\mathbb{R}$ is countable. This is a contradiction. $\square$

Chapter 6

Section 6.1

1. 48.

3. 372.

5. 10. The number of choices for the second letter depends on the first letter.

7. (a) 18. (b) No.

9. (a) 17,576,000.
   (b) 17,558,424.
   (c) 17,557,425.

11. 1680 days.

13. 81,250.

15. 436,700,160.

17. 10,000.

19. 156,000,000.

21. 4096.

23. 1,099,511,627,776.

25. 60.

27. 312.

29. 766.

31. 11.

33. 250.

35. 1131.

37. 333.

39. 380.

41. 86.

43. 21,828.

45. 23,957.

47. 31,591.

Section 6.2

1. 24.

3. 720.

5. $P(200,10)$.

7. 116,280.

9. 336.

11. 4845.

13. 312,500.

15. 191,362,500.

17. 239,400.

19. 840.

21. 230,230.

23. (a) 792. (b) 180.

25. 16,134,720.

27. 1,058,400.

29. (a) $P(15, 5)^4 P(15, 4)$.
(b) i. $5! P(15, 5)^3 P(15, 4)$.
ii. $4! P(15, 5)^4$.

31. 720.

33. 1250.

35. 300.

37. 3.

39. 15.

41. 60.

43. 21090–9000.

45. (a) 6 bars. (b) 3 bars will be long.

47. (a) $360. (b) $116,280.
(c) When the horses favored to win do well, the superfecta payoff is less than $116,280.

Section 6.3

1. 69,300.

3. 2430.

5. 176.

7. 22.

9. 834,222,844.

11. 16.

13. 334.

15. 214.

17. 1081.

19. (a) 1,679,616.
(b) 1,679,616.
(c) 3,358,976.

21. 24,976.

23. 1,267,500.

25. 1,645,020.

27. 48.

29. 360.

31. (a) "MEET ME."
(b) $y = 4x + 3$, and 4 is relatively prime to 27.
(c) 486.

33. 641.

35. *Theorem*: If $A_1, A_2, \ldots, A_n$ are disjoint sets, then $|A_1 \cup A_2 \cup \cdots \cup A_n| = |A_1| + |A_2| + \cdots + |A_n|$. *Sketch.* (By induction) Check the cases when $n = 1$ and $n = 2$. Suppose $k \geq 2$ and $A_1, \ldots, A_{k+1}$ are disjoint sets. Observe that $|A_1 \cup A_2 \cup \cdots \cup A_k \cup A_{k+1}| = |A_1 \cup A_2 \cup \cdots \cup A_k| + |A_{k+1}| = |A_1| + |A_2| + \cdots + |A_k| + |A_{k+1}|$. $\square$

37. 15.

39. 21.

41. 156.

43. 4320.

45. 6.

47. 12.

49. 15.

51. 75.

Section 6.4

1. (a) 0.25. (b) 0.182.

3. $\frac{1}{6}$.

5. $\frac{5}{18}$.

7. (a) $\{TTT, TTH, THT, THH, HTT, HTH, HHT, HHH\}$. (b) $\frac{1}{2}$.

9. (a) $\{00, 01, 02, 03, 10, 11, 12, 13, 20, 21, 22, 23, 30, 31, 32, 33\}$.

  (b) $\frac{3}{8}$.

11. $\frac{1}{9}$.

19. $\frac{28}{153}$.

13. $\frac{5}{9}$.

21. $\frac{37}{256}$.

15. $\frac{24279264}{93384347} \approx 0.2600$.

23. (a) $\frac{63}{256} \approx 0.2461$.

17. (a) $\frac{1332603}{1867686940} \approx 0.0007$.

  (b) Bet against it.

  (b) No.

25. $\frac{1}{720}$.

  (c) $\frac{1866354337}{1867686940} \approx 0.9993$.

27. $\frac{1}{2}$.

29. Probability Complement Principle: If E is an event in a sample space S, then $P(E) = 1 - P(E^c)$. Assumption: The outcomes in S are equally likely. *Proof.* $P(E) = \frac{|E|}{|S|} = \frac{|S| - |E^c|}{|S|} = 1 - \frac{|E^c|}{|S|} = 1 - P(E^c)$. $\square$

31. $\frac{1}{3}$.

33. If we let $p(n)$ be the probability that at least two of n people have the same birthday, then $p(n) = 1 - \frac{P(365, n)}{365^n}$. (a) $p(15) \approx 0.253$. (b) $p(30) \approx 0.706$. (c) $n = 23$.

35. $\frac{2}{9}$.

37. $\frac{16}{25}$.

39. $\frac{2}{3}$.

41. *Sketch.* Pretend one die is red and the others are green. Since the total sum is the red plus the sum of the green, an inductive argument works. $\square$

43. $\frac{1}{575757}$.

51. $\frac{5}{9}$.

45. $\frac{253937}{1498796} \approx 0.1694$.

53. $\frac{7}{27}$.

47. (a) $\frac{2,933734}{43148475} \approx 0.068$.

55. $\frac{1}{3}$.

  (b) $\frac{2017169}{43148475} \approx 0.0467$.

57. $\frac{2}{5}$.

49. $\frac{1}{18}$.

59. (a) Yes. (b) No.

61. *Sketch.* ($\rightarrow$) Suppose E and F are independent. So $P(E|F) = \frac{P(E \cap F)}{P(F)} = \frac{P(E)P(F)}{P(F)} = P(E)$. Similarly, $P(F|E) = P(F)$. ($\leftarrow$) Suppose $P(E|F) = P(E)$. So $P(E \cap F) = P(E \mid F)P(F) = P(E)P(F)$. $\square$

63. (a) 26%. (b) $\frac{6}{13} \approx 0.4615$.

65. $\frac{38}{39} \approx 0.974$.

67. (a) 3.15 stars. (b) 0.395. (c) $p(4) = \frac{40}{79}$, $p(3) = \frac{12}{79}$, $p(2) = \frac{10}{79}$, $p(1) = \frac{8}{79}$, and $p(0) = \frac{9}{79}$. (d) $\frac{224}{79} \approx 2.84$ stars.

69. *Sketch.* $P(F_k \mid E) = \frac{P(F_k \cap E)}{P(E)} = \frac{P(E|F_k)P(F_k)}{P(E)}$. Now use Theorem 6.11 to rewrite $P(E)$. $\square$

Section 6.5

1. 1296.

3. 210.

5. 105.

7. 36.

9. 149.

11. (a) $\frac{1}{16}, \frac{1}{4}, \frac{3}{8}, \frac{1}{4}, \frac{1}{16}$.
    (b) $p = \frac{1}{8}$.

13. 5108.

15. 3744.

17. 300.

19. 536,100.

21. $N = 1280$ and
    $p \approx 0.00001392$.

23. $N = 261,840$ and
    $p \approx 0.00284725$.

25. $N = 5,374,512$ and
    $p \approx 0.05844242$.

27. $N = 3,075,072$ and
    $p \approx 0.03343832$.

29. 66.

31. 455.

33. 45.

35. 126.

37. (a) 165. (b) No.

39. $\frac{4}{7}$.

41. 252.

43. $N = 224,848$ and
    $p \approx 0.00168067$.

45. $N = 41,584$ and
    $p \approx 0.00031083$.

47. $N = 6,461,620$ and
    $p \approx 0.04829870$.

49. $N = 6,449,688$ and
    $p \approx 0.04820951$.

51. $N = 58,524,368$ and
    $p \approx 0.43745233$.

53. $P(A \text{ wins}) \approx 0.3864$ and $P(B \text{ wins}) \approx 0.6136$.

55. $P(A \text{ wins}) \approx 0.2828$ and $P(B \text{ wins}) \approx 0.7172$.

57. $P(A \text{ wins}) \approx 0.4754$, $P(B \text{ wins}) \approx 0.5211$,   and
    $P(\text{tie}) \approx 0.0035$.

Section 6.6

1. (a) 39!. (b) $\frac{39!}{2}$.

3. 3.

5. (a) 5040. (b) 1440.
   (c) $6! = 720$.

7. 7,567,560.

9. 30.

11. $\frac{19!}{3}$.

13. 20.

15. 180.

17. 10.

19. 1,466,593,128.

21. 6,844,101,264.

23. 2,027,025.

25. $\dfrac{\binom{50}{3}\binom{47}{3}\cdots\binom{23}{3}\binom{20}{2}\binom{18}{2}\cdots\binom{2}{2}}{(10!)^2}$.

27. $\dfrac{\binom{60}{6}\binom{54}{6}\cdots\binom{6}{6}\left(\frac{5!}{2}\right)^{10}}{10!}$.

29. 90,720.

31. 12,623,055,048,283,680,000.

33. 130.

35. 64.

37. 120.

39. 30.

Review

1. 362,880.

2. 15,552.

3. 486,720.

4. 13 hours.

5. 142.

6. 1267.

7. 7,893,600.

8. 1657.

9. 720.

10. (a) 970,200. (b) 940,800.

11. 765,529,380.

12. $\{a,b\}, \{a,c\}, \{a,d\}, \{a,e\},$
$\{a,f\}, \{b,c\}, \{b,d\}, \{b,e\},$
$\{b,f\}, \{c,d\}, \{c,e\}, \{c,f\},$
$\{d,e\}, \{d,f\}, \{e,f\}.$

13. 24,024.

14. 120.

15. 403.

16. 219.

17. 482.

18. 12,866.

19. 159.

20. 4000.

21. 2509.

22. $\frac{1}{18}$.

23. Approximately yes,
but not exactly 0.01.

24. $\frac{21}{32}$.

25. $\frac{44}{4165}$.

26. $\frac{1}{2}$.

27. $\frac{8,400}{81719} \approx 0.10$.

28. $\frac{1}{22957480}$.

29. $\frac{4,361}{45838} \approx 0.0951$.

30. (a) $\frac{3}{11}$. (b) No.

31. (a) 83%. (b) $\frac{45}{83} \approx 0.5422$.

32. No.

33. 90.

34. 95.

35. 90.

36. 1872.

37. 1,302,540.

38. 14,664.

39. 732,160.

40. 2,532,816.

41. 165,984.

42. 1844.

43. 205,792.

44. 365,772.

45. 9,730,740.

46. 6,608,748.

47. $\frac{236}{495} \approx 0.4768$.

48. 4,598,126.

49. $\frac{5481}{456976} \approx 0.0120$.

50. (a) 19!. (b) 20. (c) $\frac{13}{20}$.

51. 1680.

52. 352,716.

53. 165,646,455,975.

54. 65.

55. 630.

56. (a) 116,280.
(b) 4,651,200.

57. (a) 252. (b) 386.

58. 486.

59. 3072.

60. 36!.

61. 252.

62. 1182.

63. $\frac{75}{323}$.

64. $\frac{1}{2}$.

65. (a) $\frac{4}{17}$. (b) $\frac{1}{425}$.

66. $\frac{69729}{1,059,380} \approx 0.0658$.

Chapter 7

Section 7.1

1. 93.

3. 27,880.

5. 14,637.

7. (a) $\frac{317}{512} \approx 0.62$.

   (b) $\frac{7}{1024} \approx 0.0068$.

9. $\frac{63,713}{111860} \approx 0.5696$.

11. $\frac{271}{351} \approx 0.7721$.

13. 80.

15. 400.

17. 480.

19. $\phi\left(p_1^{k_1} \cdots p_m^{k_m}\right) = \phi\left(p_1^{k_1}\right) \phi\left(p_2^{k_2}\right) \cdots \phi\left(p_m^{k_m}\right) = p_1^{k_1}\left(1 - \frac{1}{p_1^{k_1}}\right) p_2^{k_2}\left(1 - \frac{1}{p_2^{k_2}}\right) \cdots p_m^{k_m}\left(1 - \frac{1}{p_1^{k_1}}\right)$ can be reorganized to give the desired result.

21. *Sketch.* Let p and q be the only prime divisors of n. So $\phi(n) = n - \frac{n}{p} - \frac{n}{q} + \frac{n}{pq} = n\left(1 - \frac{1}{p} - \frac{1}{q} + \frac{1}{pq}\right) = n\left(1 - \frac{1}{p}\right)\left(1 - \frac{1}{q}\right)$. $\square$

23. 2800.

25. 4250.

27. $\frac{53}{144} = 0.3680\overline{5}$ agrees with $\frac{1}{e}$ to 2 decimal places.

29. (a) $\frac{19}{30} \approx 0.633$. (b) $\frac{1}{120}$. (c) $\frac{11}{120}$.

31. 0.36787944.

33. $|A_1^c \cap A_2^c \cap \cdots \cap A_n^c| = |(A_1 \cup A_2 \cup \cdots \cup A_n)^c| = |\mathcal{U}| - |A_1 \cup A_2 \cup \cdots \cup A_n| = |\mathcal{U}| - \sum_{i=1}^{n}(-1)^{i-1}S_i = |S_0| + \sum_{i=1}^{n}(-1)^i S_i = \sum_{i=0}^{n}(-1)^i S_i$.

35. $|A_1 \cap A_2 \cap A_3| + |A_1 \cap A_2 \cap A_4| + |A_1 \cap A_2 \cap A_5| + |A_1 \cap A_3 \cap A_4| + |A_1 \cap A_3 \cap A_5| + |A_1 \cap A_4 \cap A_5| + |A_2 \cap A_3 \cap A_4| + |A_2 \cap A_3 \cap A_5| + |A_2 \cap A_4 \cap A_5| + |A_3 \cap A_4 \cap A_5|$.

37. $|A_1 \cap A_2 \cap A_3| = |\mathcal{U}| - (|A_1^c| + |A_2^c| + |A_3^c|) + (|A_1^c \cap A_2^c| + |A_1^c \cap A_3^c| + |A_2^c \cap A_3^c|) - |A_1^c \cap A_2^c \cap A_3^c|$.

39. 5,103,000.

41. (a) $\frac{5}{324} \approx 0.0153$. (b) $\frac{38045}{139968} \approx 0.2718$. (c) 13.

43. $n^4 - 4n^3 + 5n^2 - 2n = n(n-1)^2(n-2)$.

45. $n^5 - 6n^4 + 14n^3 - 15n^2 + 6n = n(n-1)(n-2)(n^2 - 3n + 3)$.

Section 7.2

1. 210.

3. 6,306,300.

5. 27,720.

7. (a) 1,26,12,600.
   (b) 7,567,560.
   (c) 4,204,200.

9. 5775.

11. 37,642,556,952.

13. (a) 212,837,625. (b) 5,108,103,000.

15. (a) 29,640,619,008,000. (b) 4,940,103,168,000. (c) 2,858,856.

17. $(0,0,3), (0,1,2), (1,0,2), (0,2,1), (1,1,1), (2,0,1), (0,3,0),$
$(1,2,0), (2,1,0), (3,0,0).$

19. $x^2 - 2xy + 2xz + y^2 - 2yz + z^2.$

21. $x^4 + 4x^3y + 4x^3 + 6x^2y^2 + 12x^2y + 6x^2 + 4xy^3 + 12xy^2 +$
$12xy + 4x + y^4 + 4y^3 + 6y^2 + 4y + 1.$

23. $4x^2 + 4xy - 4xz + y^2 - 2yz + z^2.$

25. $4w^2 + 4wx + 4wy + 8wz + x^2 + 2xy + 4xz + y^2 + 4yz +$
$4z^2.$

27. $\binom{300}{100,50,40,60,20,30}.$

29. 0.

31. $-\binom{100}{25,10,40,25}.$

33. $-80,720,640.$

35. 680.

37. There are $n!$ ways to order the n items. We shall understand that the first k_1 go into category 1, the next k_2 go into
category 2, and so forth. Since, within each category, order
is not important, we must divide $n!$ by $k_1!k_2!\cdots k_m!$, the
number of different orderings leaving items within their
categories. We get $\frac{n!}{k_1!k_2!\cdots k_m!}$, which is $\binom{n}{k_1,k_2,\ldots,k_m}.$

39. Consider $3^n = (1+1+1)^n.$

41. Consider $6^n = (3+2+1)^n.$

Section 7.3

1. $1 + 2x + 3x^2 + 4x^3 + \cdots = \sum_{i=0}^{\infty}(i+1)x^i.$

3. $1 + x^2 + x^4 + x^6 + \cdots = \sum_{i=0}^{\infty} x^{2i}.$

5. $1 \cdot 1 + 1 \cdot x^2 + 1 \cdot x^4 + x \cdot 1 + x \cdot x^2 + x \cdot x^4 + x^2 \cdot 1 +$
$x^2 \cdot x^2 + x^2 \cdot x^4 = \quad 1 + x^2 + x^4 + x + x^3 + x^5 + x^2 + x^4 +$
$x^6 = 1 + 2x + 2x^2 + x^3 + x^4 + x^5 + x^6.$

7. $(1 \cdot x + 1 \cdot x^2 + x \cdot x + x \cdot x^2)(1 + x^3) = (x + x^2 +$
$x^2 + x^3)(1 + x^3) = x \cdot 1 + x \cdot x^3 + x^2 \cdot 1 + x^2 \cdot x^3 + x^2 \cdot 1 +$
$x^2 \cdot x^3 + x^3 \cdot 1 + x^3 \cdot x^3 = x + x^4 + x^2 + x^5 +$
$x^2 + x^5 + x^3 + x^6 = x + 2x^2 + x^3 + x^4 + 2x^5 + x^6.$

9. $c_0 = 1, c_5 = 1, c_{10} = 2, c_{15} = 2, c_{20} = 2, c_{25} = 1, c_{30} = 1,$
and, otherwise, $c_i = 0.$

11.

i	0	1	2	3	4	5	6	7	8	9	10	11	12	13	14	15
c_i	1	1	1	1	1	2	1	1	1	1	2	1	1	1	1	2

i	16	17	18	19	20	21	22	23	24	25	26	27	28	29	30
c_i	1	1	1	1	2	1	1	1	1	2	1	1	1	1	1

13. i	0	5	10	15	20	25	30	35	40	45	50	55	60
c_i	1	1	2	2	2	3	2	3	2	2	2	1	1

15. 11.

17. 29.

19. (a) 18.

(b) $\frac{27256}{31977} \approx 0.8524$.

(c) $\frac{23}{159885} \approx 0.00014$.

21. $c_i = \binom{n}{i}a^i$, for $i = 0, 1, \ldots, n$.

23. $\forall\, i \geq 0, c_i = \binom{i+3}{i}$.

25. $\binom{149}{100}$.

27. $\binom{149}{100} + \binom{148}{99}$.

29. $\binom{119}{40} + 2\binom{118}{39} + \binom{117}{38}$.

31. 6175.

33. 15,805.

35. (a) 351. (b) 56. (c) 3.

37. 81.

39. 1602.

41. When there are finitely many items, say n, in general, the number of ways of selecting $n - i$ is the same as the number of ways of de-selecting i.

43. $\forall\, i \geq 0, c_i = (-1)^i \binom{i+n-1}{i}$.

Section 7.4

1.

3. $\circ$	r_0	r_1	r_2	r_3
r_0	r_0	r_1	r_2	r_3
r_1	r_1	r_2	r_3	r_0
r_2	r_2	r_3	r_0	r_1
r_3	r_3	r_0	r_1	r_2

5. (a) f_1. (b) f_3. (c) No.

7. 6.

9. 10.

11. (a) 126. (b) 90. (c) 60.

13. 57.

15. 280.

17. 36.

19. 834.

21. 34.

23. 8.

25. (a) 39. (b) 135. (c) 57.

27. 165.

29. 3375.

31. 1135.

33. *Sketch.* $gx = x$ iff
$g^{-1}gx = g^{-1}x$ iff $x =$
$g^{-1}x$ iff $g^{-1}x = x.$ $\square$

35. 1493.

37. 2420.

39. 6.

Section 7.5

1. Each of the $\binom{n}{k}$ subsets of $\{1, 2, \ldots, n\}$ of size k can be uniquely represented by a binary sequence of length n with k ones, as in Example 7.20(a). For each i, group together those that start with i ones followed by a zero. The size of this group is $\binom{n-i-1}{k-i}$.

3. The subsets of $\{1, 2, \ldots, n\}$ of size k can be split into (i) those containing both 1 and 2, (ii) those containing exactly one of 1 or 2, and (iii) those containing neither 1 nor 2.

5. Of the $\binom{3n}{n}$ paths from $(0, 2n)$ to $(n, 0)$ in the $(2n + 1)$ by $(n + 1)$ rectangular grid of points $([0, n] \times [0, 2n]) \cap (\mathbb{Z} \times \mathbb{Z})$, for each $0 \leq i \leq n$, the number that pass through (i, i) is $\binom{2n}{i}\binom{n}{n-i} = \binom{2n}{i}\binom{n}{i}$.

7. A Canadian doubles tournament starts with 3^n players and ends with $3^n - 1$ losers and 1 champion. For each $1 \leq k \leq n$, round k has 3^{n-k+1} competitors, of which $\frac{2}{3}$ are eliminated as losers.

9. Of the set of base-3 sequences of length n, $3^n - 2^n$ contain at least one 2. For each $1 \leq j \leq n$, let A_j be those that have a 2 in position j. For each $1 \leq j_1 < j_2 < \cdots < j_i \leq n$, $|A_{j_1} \cap A_{j_2} \cap \cdots \cap A_{j_i}| = 2^{n-i}$. By the Principle of Inclusion–Exclusion, $3^n - 2^n = \sum_{i=1}^{n}(-1)^{i-1}\binom{n}{i}2^{n-i}$.

11. For each k, there are $\binom{n}{k}$ choices for a team of size k and then k choices for its captain. All together, there are n choices for a team captain and then 2^{n-1} choices for the remaining team members.

13. The number of ways to split $2n$ people into 2 teams of size n is $\frac{\binom{2n}{n}}{2}$. The number of ways for you to pick your $n - 1$ teammates is $\binom{2n-1}{n-1}$.

15. The 3^n base-3 sequences of length n can be partitioned according to triples (k_0, k_1, k_2), where k_0 is the number of

zeros, k_1 is the number of ones, and k_2 is the number of twos in the sequence.

17. (a) $\binom{a+b+c}{a,b,c}$. (b) Consider an $(i+1)$ by $(j+1)$ by $(k+1)$ rectangular grid of points. In a path from S to F, the final position before reaching F must be exactly one of the 3 pictured points x, y, or z. The number of paths to x is $\binom{n-1}{i-1,j,k}$, the number to y is $\binom{n-1}{i,j-1,k}$, and the number to z is $\binom{n-1}{i,j,k-1}$.

19. The identity permutation of $\{1,2,\ldots,n\}$ moves none of its elements. Of the $n!-1$ nonidentity permutations of $\{1,2,\ldots,n\}$, for each $1 \le i \le n-1$, the number for which $i+1$ is the largest position moved is $|P_i| = i \cdot i!$.

21. If a fair coin is tossed n times, then the probability that at least one head will occur is $1 - \frac{1}{2^n}$. For each $1 \le k \le n$, the probability that the first head occurs on toss k is $\left(\frac{1}{2}\right)^{k-1}\left(\frac{1}{2}\right) = \frac{1}{2^k}$.

23. Assume that you are one of the 6 people. Of the other 5 people, there must be either 3 whom you have met before or 3 whom you have not met before. Without loss of generality, assume that there are 3 whom you have met before. If two of them have met each other before, then you and those two are a set of 3 who have met each other before. Otherwise, those 3 are a set who have never met each other before.

25. 2.

27. (a) List the $n-1$ terms on the nonbase sides of the n-gon. Parentheses go around any pair that sits on a common triangle. Collapse each such triangle to its interior side, write the resulting product from the exterior sides on that interior side, and regard it as a single term. Now repeat this process on the resulting smaller polygon. This establishes a one-to-one correspondence between triangulations of the n-gon and parenthesizations of a product of $n-1$ terms. (b) $T_n = C_{n-2} = \frac{1}{n-2+1}\binom{2(n-2)}{n-2} = \frac{1}{n-1}\binom{2n-4}{n-2}$.

29. The coins are on squares of the same color. That leaves 30 squares of that color and 32 of the other color. Since each domino covers one square of each color, they cannot be used to fill the rest of the board.

31. The first redeal leaves the selected card within the first 3 rows. The second redeal leaves the card in the first row. Thus, the column specifies the card.

Review

1. 92,903,176.

2. 2880.

3. 1123.

4. $\frac{2197}{8330} \approx 0.2637$.

5. $\frac{887}{2907} \approx 0.3051$.

6. 31,150.

7. $\frac{3}{8}$.

8. 60.

9. 7560.

10. 35,840.

11. (a) 514,594,080.
    (b) 2,858,856.
    (c) 192,972,780.

12. $x^2 + 2xy + 2xz + 2xw + y^2 + 2yz + 2yw + z^2 + 2zw + w^2$.

13. $8x^3 - 12x^2y + 12x^2z + 6xy^2 - 12xyz + 6xz^2 - y^3 + 3y^2z - 3yz^2 + z^3$.

14. $3^{20}2^{50}\binom{80}{20,50,10}$.

15. 9,777,287,520.

16. $1 - x + 2x^2 - 2x^3 + 3x^4 - 3x^5 + \cdots$ has $c_i = (-1)^i \lceil \frac{i+1}{2} \rceil$.

17.

i	0	5	10	15	20	25	30	35	40	45	50
c_i	1	1	2	1	2	2	3	3	3	3	3

i	55	60	65	70	75	80	85	90	95	100
c_i	3	3	3	3	2	2	1	2	1	1

18. 17.

19. 49.

20. 44.

21. 10,015,005.

22. 673.

23. 7211.

24.

$\circ$	r_0	r_1	r_2
r_0	r_0	r_1	r_2
r_1	r_1	r_2	r_0
r_2	r_2	r_0	r_1

25. r_4.

26. 13.

27. 217,045.

28. 240.

29. 4995.

30. *Sketch.* Suppose $g_1, g_2 \in G$ and $\mathrm{Fix}(g_2) = X$. ($\subseteq$) If $x \in \mathrm{Fix}(g_2 g_1)$, then $x = g_2 g_1 x = g_1 x$, whence $x \in \mathrm{Fix}(g_1)$. ($\supseteq$) Similar. $\square$

31. 4624.

32. 94.

33. Paths from $c_{0,0}$ to $c_{n+1,k}$ must pass through $c_{n,k-1}$ or $c_{n,k}$.

34. There are $\binom{n}{k}$ subsets of size k from $\{1, 2, \ldots, n\}$. For each $k - 1 \leq i \leq n - 1$, there are $\binom{i}{k-1}$ for which $i + 1$ is the largest element.

35. For $i = 0, 1, \ldots, k$, the number of paths from $S = c_{0,0}$ to $F = c_{3n,k}$ through $c_{2n,i}$ is $\binom{2n}{i}\binom{n}{k-i}$. Each path from S to F goes through exactly one of these.

36. Here, $R = $ "(", $D = $ ")", and our sequences never have more D's than R's at any point.

37. The same formula $A \mapsto A \bigtriangleup \{n\}$ defines both a function and its inverse. This bijection therefore establishes the equality.

38. 4371.

39. 1,414,910.

40. $\binom{52}{13,13,13,13}$.

41. 66.

42. (a) 54. (b) $\frac{896}{8671} \approx 0.1033$. (c) The outcomes are not equally likely.

43. Let c_k be the coefficient of x^k in $(1 + x + x^2)^4 = (1 - x^3)^4 \frac{1}{(1-x)^4}$, and suppose $k \geq 12$. From the left-hand side, it is obvious that $c_k = 0$. Since the right-hand side equals $(1 - 4x^3 + 6x^6 - 4x^9 + x^{12})\frac{1}{(1-x)^4}$, we see also that $c_k = \binom{k+3}{k} - 4\binom{k}{k-3} + 6\binom{k-3}{k-6} - 4\binom{k-6}{k-9} + \binom{k-9}{k-12}$.

44. (a) 871,782,912,000. (b) 15,766,392.

45. 17.

46. 10.

Chapter 8

Section 8.1

1.

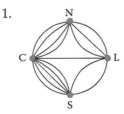

3.

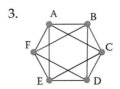

5.

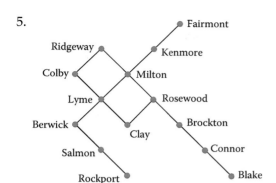

7. The graph is simple.

9. The graph is not simple, because it has multiple edges.

11. Yes.

13. No. Edge e needs vertex 4.

15. $E = \{\{1,2\}, \{2,3\}, \{2,5\}\}$.

17. $E = \{\{1,3\}, \{3,5\}, \{1,5\}\}$.

19.

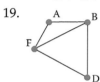

21. No. An endpoint is missing.

23. Yes.

25. No. Two edges intersect in infinitely many points.

27. Yes.

29. Let $e \mapsto \{u, v\}$ be an edge. The walk u, e, v, e, u is not a path.

31. Yes.

33. 2.

35. 2.

37. 5.

39. Yes.

41. No.

43. (a) Colby, Lyme, Clay, Rosewood. (b) Yes.

45. *Sketch.* Suppose $d = \text{dist}(u, w) \geq 1$. Let P be a path of length d from u to w. Let v be the last vertex on P before w. So, $\text{dist}(u, v) \leq d - 1$. If there were a path of length $l < d - 1$ from u to v, then there would be a path of length $l + 1 < d$ from u to w. Thus, $\text{dist}(u, v) = d - 1$. $\square$

47. *Sketch.* Let P and Q be distinct paths from u to v. We can find a portion of P followed by a portion of Q that forms a cycle. $\square$

49. *Sketch.* By symmetry, we must start along edge 3 or edge 4. *Case 1*: If we take edge 3, then without loss of generality, we take edge 1. We must then, without loss of generality, take edge 5. We then finish with $7, 4, 2$ or $7, 4, 6$ or $6, 2$ or $6, 4, 7$ and get stuck without hitting every edge. *Case 2*: If we take edge 4, then without loss of generality, we take edge 1. *Subcase 2a*: Take edge 3 next, then 7, and, without loss of generality, take 5. After taking 2 or 6 we get stuck. *Subcase 2b*: Take edge 2 next, and then, without loss of generality, 5. If we take 6, then we get stuck. Otherwise, we take edge 7, then take 3, and get stuck. □

51. (a) $n = 2, 3, 4, 5,$ or 6. (b) 2.

53. Calculus and discrete math.

Section 8.2

1.

3.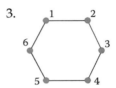

5.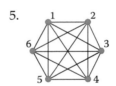

7.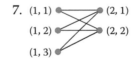

9.

11. (a) P_7. (b) Yes. (c) Colby, Lyme, Milton.

13. (a) No. (b) Just exclude Sud.

15. (a) No. (b) Yes.

17.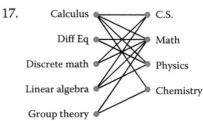

19. Yes. Let V_1 contain the odd-numbered vertices, and V_2 the even.

21. *Sketch.* Let V_1, V_2 bipartition G, and let W be the vertex set of a subgraph H of G. Then, $W \cap V_1, W \cap V_2$ bipartition H.
$\square$

23. It is not bipartite, because it contains a 3-cycle.

25. $|V| = n$ and $|E| = n - 1$.

27. $|V| = n$ and $|E| = \binom{n}{2}$.

29. $|V| = 8$ and $|E| = 12$.

31. $|V| = 20$ and $|E| = 30$.

33. $|V| = 16$ and $|E| = 32$.

35. No. They differ by more than one digit.

37. (a) 1101. (b) Male, A^+.

39. $n - 1$.

41. 1, for $n \geq 2$.

43. 3.

45. 5.

47. 4.

Section 8.3

1.

$$
\begin{array}{c c}
 & \begin{array}{c c c c c c} A & B & C & D & E & F \end{array} \\
\begin{array}{c} A \\ B \\ C \\ D \\ E \\ F \end{array} &
\left[\begin{array}{c c c c c c}
0 & 1 & 1 & 0 & 1 & 1 \\
1 & 0 & 1 & 1 & 0 & 1 \\
1 & 1 & 0 & 1 & 1 & 0 \\
0 & 1 & 1 & 0 & 1 & 1 \\
1 & 0 & 1 & 1 & 0 & 1 \\
1 & 1 & 0 & 1 & 1 & 0
\end{array} \right].
\end{array}
$$

3.

$$
\begin{array}{c c}
 & \begin{array}{c c c c c c} 1 & 2 & 3 & 4 & 5 & 6 \end{array} \\
\begin{array}{c} 1 \\ 2 \\ 3 \\ 4 \\ 5 \\ 6 \end{array} &
\left[\begin{array}{c c c c c c}
0 & 0 & 0 & 1 & 0 & 0 \\
0 & 0 & 0 & 1 & 0 & 0 \\
0 & 0 & 0 & 1 & 0 & 0 \\
1 & 1 & 1 & 0 & 1 & 1 \\
0 & 0 & 0 & 1 & 0 & 1 \\
0 & 0 & 0 & 1 & 1 & 0
\end{array} \right] = A.
\end{array}
$$

5.

$$
\begin{array}{c c}
 & \begin{array}{c c c c c c} 1 & 2 & 3 & 4 & 5 & 6 \end{array} \\
\begin{array}{c} 1 \\ 2 \\ 3 \\ 4 \\ 5 \\ 6 \end{array} &
\left[\begin{array}{c c c c c c}
0 & 1 & 0 & 0 & 0 & 1 \\
1 & 0 & 1 & 0 & 1 & 0 \\
0 & 1 & 0 & 1 & 0 & 0 \\
0 & 0 & 1 & 0 & 1 & 0 \\
0 & 1 & 0 & 1 & 0 & 1 \\
1 & 0 & 0 & 0 & 1 & 0
\end{array} \right] = A.
\end{array}
$$

7.

$$
\begin{array}{c}
\;\; 2\;5\;1\;4\;6\;3 \\
\begin{array}{c} 2 \\ 5 \\ 1 \\ 4 \\ 6 \\ 3 \end{array}
\left[
\begin{array}{cccccc}
0 & 0 & 0 & 1 & 0 & 0 \\
0 & 0 & 0 & 1 & 1 & 0 \\
0 & 0 & 0 & 1 & 0 & 0 \\
1 & 1 & 1 & 0 & 1 & 1 \\
0 & 1 & 0 & 1 & 0 & 0 \\
0 & 0 & 0 & 1 & 0 & 0
\end{array}
\right] = B.
\end{array}
$$

9.

$$
\begin{array}{c}
\;\; 1\;2\;3\;4\;5 \\
\begin{array}{c} 1 \\ 2 \\ 3 \\ 4 \\ 5 \end{array}
\left[
\begin{array}{ccccc}
0 & 1 & 1 & 1 & 1 \\
1 & 0 & 1 & 1 & 1 \\
1 & 1 & 0 & 1 & 1 \\
1 & 1 & 1 & 0 & 1 \\
1 & 1 & 1 & 1 & 0
\end{array}
\right].
\end{array}
$$

11.

$$
\begin{array}{c}
\;\; (1,1)\;(1,2)\;(2,1)\;(2,2)\;(2,3)\;(2,4) \\
\begin{array}{c} (1,1) \\ (1,2) \\ (2,1) \\ (2,2) \\ (2,3) \\ (2,4) \end{array}
\left[
\begin{array}{cccccc}
0 & 0 & 1 & 1 & 1 & 1 \\
0 & 0 & 1 & 1 & 1 & 1 \\
1 & 1 & 0 & 0 & 0 & 0 \\
1 & 1 & 0 & 0 & 0 & 0 \\
1 & 1 & 0 & 0 & 0 & 0 \\
1 & 1 & 0 & 0 & 0 & 0
\end{array}
\right].
\end{array}
$$

13.

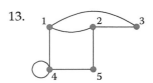

15. $\begin{bmatrix} 16 & 13 \\ 18 & 9 \end{bmatrix}.$

17. $\begin{bmatrix} 7 & 5 \\ 6 & 8 \end{bmatrix}.$

19. $\begin{bmatrix} 0 & 2 & 7 \\ 1 & 0 & 4 \\ 8 & 3 & 0 \end{bmatrix}.$

21. $\begin{bmatrix} 0 & 24 & 9 \\ 8 & 4 & 10 \\ 30 & 5 & 35 \end{bmatrix}.$

23.

(a) $P = \begin{bmatrix}
0 & 1 & 0 & 0 & 0 & 0 \\
0 & 0 & 0 & 0 & 1 & 0 \\
1 & 0 & 0 & 0 & 0 & 0 \\
0 & 0 & 0 & 1 & 0 & 0 \\
0 & 0 & 0 & 0 & 0 & 1 \\
0 & 0 & 1 & 0 & 0 & 0
\end{bmatrix}.$

(b) $PA = \begin{bmatrix} 0 & 0 & 0 & 1 & 0 & 0 \\ 0 & 0 & 0 & 1 & 0 & 1 \\ 0 & 0 & 0 & 1 & 0 & 0 \\ 1 & 1 & 1 & 0 & 1 & 1 \\ 0 & 0 & 0 & 1 & 1 & 0 \\ 0 & 0 & 0 & 1 & 0 & 0 \end{bmatrix}.$

(c) $PAP^T = \begin{bmatrix} 0 & 0 & 0 & 1 & 0 & 0 \\ 0 & 0 & 0 & 1 & 1 & 0 \\ 0 & 0 & 0 & 1 & 0 & 0 \\ 1 & 1 & 1 & 0 & 1 & 1 \\ 0 & 1 & 0 & 1 & 0 & 0 \\ 0 & 0 & 0 & 1 & 0 & 0 \end{bmatrix}.$

25.

$$A^2 = \begin{bmatrix} 1 & 1 & 1 & 0 & 1 & 1 \\ 1 & 1 & 1 & 0 & 1 & 1 \\ 1 & 1 & 1 & 0 & 1 & 1 \\ 0 & 0 & 0 & 5 & 1 & 1 \\ 1 & 1 & 1 & 1 & 2 & 1 \\ 1 & 1 & 1 & 1 & 1 & 2 \end{bmatrix}.$$

27.

$$A^2 = \begin{bmatrix} 2 & 0 & 1 & 0 & 2 & 0 \\ 0 & 3 & 0 & 2 & 0 & 2 \\ 1 & 0 & 2 & 0 & 2 & 0 \\ 0 & 2 & 0 & 2 & 0 & 1 \\ 2 & 0 & 2 & 0 & 3 & 0 \\ 0 & 2 & 0 & 1 & 0 & 2 \end{bmatrix}.$$

29. 0, 6, and 0, respectively.

31. 2, 0, and 5, respectively.

33. *Sketch.* Let i, j be vertices. For each $0 \leq k \leq n - 1$, there is a walk of length k from i to j iff A^k has a positive value in entry i, j. In a graph on n vertices, if there is a path between two vertices, then the distance between them cannot be greater than $n - 1$. $\square$

35. $1:4;$ $2:4;$ $3:4;$ $4:1,2,3,5,6;$ $5:4,6;$ $6:4,5.$

37. $1:2,6;$ $2:1,3,5;$ $3:2,4;$ $4:3,5;$ $5:2,4,6;$ $6:1,5.$

39. $000:001,010,100;$ $001:000,011,101;$ $010:000,011,110;$
    $011:001,010,111;$ $100:000,101,110;$ $101:001,100,111;$
    $110:010,100,111;$ $111:011,101,110.$

41.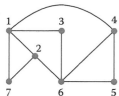

Section 8.4

1. Define $f(1) = 3, f(2) = 4$.

3. Define $f(1) = (2,1), f(2) = (1,1), f(3) = (2,2)$.

5. Define $f(1) = 7, f(2) = 8, f(3) = 9, f(4) = 6, f(5) = 10$. Also match up the parallel edges connecting 1 and 4 with those connecting 6 and 7.

7. Define $f(1) = 9, f(2) = 6, f(3) = 10, f(4) = 7, f(5) = 8$.

9. Define $f(1) = 8, f(2) = 9, f(3) = 6, f(4) = 7, f(5) = 10$.

11. (a) We *can* define $f(A) = G, f(B) = J, f(C) = H, f(D) = K,$ $f(E) = I,$ and $f(F) = L$.

(b)

Time period	Study group meeting
1	German, Kuwait
2	Indochina, Japanese
3	History, Latin

13. Define $f(1,1) = 1, f(1,2) = 3, f(1,3) = 5,$ $f(2,1) = 2,$ $f(2,2) = 4, f(2,3) = 6$.

15. Define $f(0) = 15, f(1) = 12, f(2) = 14, f(3) = 13, f(4) = 11,$ $f(5) = 8, f(6) = 10, f(7) = 9$.

17. From the first graph to the second graph, define $\forall\, i,\ f(i) = i + 10$. The same formula works from the second to the third.

19. 12.

21. 4.

23. $2n$.

25. $2(n!)^2$.

27. *Sketch.* First observe that $f(v_0), f(e_1), f(v_1), \ldots, f(v_n)$ is a walk in H. Second, note that because f_V is a bijection, if there are no vertex repetitions in the list $v_0, v_1, \ldots, v_n,$ then there cannot be any in the list $f(v_0), f(v_1), \ldots, f(v_n)$. □

29.

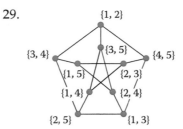

31. *Sketch.* It suffices to show that the cycles $6,8,10,7,9,6$ and $1,2,7,10,5,1$ and $6,8,3,4,9,6$ all work. Rotations handle the rest. □

33. *Sketch.* $\text{Aut}(C_n) = D_n \supseteq Z_n$. Given vertices i, j, rotation r_{j-i} moves i to j. $\square$

35. *Sketch.* Suppose $G \cong H$ by isomorphism $f : G \longrightarrow H$. Define $f^{-1} : H \longrightarrow G$ by taking $f_V^{-1} : V_H \longrightarrow V_G$ to be the inverse of $f_V : V_G \longrightarrow V_H$ and $f_E^{-1} : E_H \longrightarrow E_G$ to be the inverse of $f_E : E_G \longrightarrow E_H$. Then f^{-1} is a graph isomorphism. $\square$

37. *Sketch.* ($\rightarrow$) Obvious. ($\leftarrow$) Suppose $v_1, v_2 \in V$. So there exist automorphisms f_1 and f_2 such that $f_1(u) = v_1$ and $f_2(u) = v_2$. Let $f = f_2 \circ f_1^{-1}$. Observe that f is an automorphism and $f(v_1) = v_2$. $\square$

39. *Sketch.* ($\rightarrow$) Suppose $f : G \longrightarrow H$ is an isomorphism, and let $D : H \longrightarrow \mathbb{R}^2$ be any drawing of H. Then $D \circ f : G \longrightarrow \mathbb{R}^2$ is a drawing of G with the same image. $\square$

41. Define $f(1) = f(2) = f(3) = f(5) = 5, f(4) = 4, f(6) = 6$.

43. *Sketch.* Suppose there is one. Without loss of generality, say $f(1) = 1$ and $f(2) = 2$. If $f(3) = 3$, then $\{1, 3\}$ needs to be an edge of C_4. If $f(3) = 4$, then $\{2, 4\}$ needs to be an edge of C_4. These contradictions show that f cannot exist. $\square$

45. No. Triangle $1, 2, 3$ has no place to go.

47. $\forall i$, define $f(1, i) = 1$ and $f(2, i) = 2$.

49. $m \leq n$.

Section 8.5

1. (a) has 6 vertices, whereas (d) has only 5.

3. (a) has 6 edges, whereas (c) has 7.

5. $5, 2, 2, 1, 1, 1$.

7. $3, 3, 2, 2, 2, 2$.

9. $\delta((b)) = 2$ and $\delta((g)) = 1$.

11. They do not have a common degree sequence.

13. The subgraph induced by the degree-3 vertices is P_2 on the left graph and Φ_2 on the right graph. See Exercise 28 in Section 8.4.

15. The computers of degree 3 are adjacent in the left configurations, but not in the right.

17. The grid on the left has a power station of degree 4, and that on the right does not.

19. All vertices in the first graph have degree 4, while vertex "German" in the second graph has degree 3.

21. $\cdot$, $\vdots$, $]$, $\therefore$, $\dot{\underline{\,}}$, $\underline{}$, $\wedge$, $\triangle$, $\vdots\,\vdots$, $\underline{\underline{}}$, $\sqsubset$, $\supset$, $\boxbslash$, $\boxslash$, $\sqcup$, $\boxtimes$, $\square$, $\boxslash$, $\boxtimes$.

23.

25.

27. None.

29.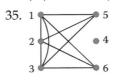

31. *Sketch.* If there is an odd number of odd-degree vertices, then $\sum_{v \in V} \deg(v)$ is odd. However, $\sum_{v \in V} \deg(v) = 2|E|$ is even. $\square$

33. $|V| = 2^n$ and $|E| = n2^{n-1}$.

35.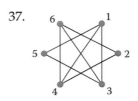

37.

39. Define $f : C_5^c \longrightarrow C_5$ by $f(1) = 1, f(2) = 3, f(3) = 5,$ $f(4) = 2, f(5) = 4.$

41. C_6.

43. The complements C_6 and $C_3 + C_3$, respectively, are not isomorphic. One is connected, and the other is not.

45. *Sketch.* Suppose $G \cong H$ via isomorphism $f : V_G \longrightarrow V_H$. The induced map $f : E_G \longrightarrow E_H$ is a bijection mapping edges to edges iff the induced map $f : E_G^c \longrightarrow E_H^c$ is a bijection mapping nonedges to nonedges. $\square$

47. The idea is in the proof for Exercise 45.

49. True. Apply Lemma 8.10.

51.

53.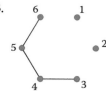

55.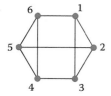

57.

59.

61. For (a), the Distributive Laws for sets give $V_G \cap (V_H \cup V_K) = (V_G \cap V_H) \cup (V_G \cap V_K)$ and $E_G \cap (E_H \cup E_K) = (E_G \cap E_H) \cup (E_G \cap E_K)$. The result now follows from the definitions of graph unions and intersections. Part (b) follows similarly.

63. Define $f(1) = (1,1), f(2) = (1,2), f(3) = (1,3),$ $f(4) = (2,1), f(5) = (2,2), f(6) = (2,3)$.

65. Fix $v \in V_G$ and $w \in V_H$. Then $\{v\} \times H \cong H$ and $G \times \{w\} \cong G$.

Section 8.6

1.

3.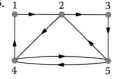

5. A simple directed graph.

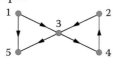

7. Not a simple directed graph.

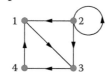

9. Yes.

11. $1, 2, 3, 5$ and $1, 2, 4, 5$.

13.
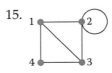

15.
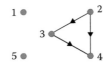

17. Three strong components.

19. Two strong components.

21. Yes.

23. Define $f(1) = 6$, $f(2) = 5, f(3) = 4$.

25. Define $f(1) = 7$, $f(2) = 5, f(3) = 8$, $f(4) = 6$.

27.

Vertex	Indeg	Outdeg
1	0	2
2	1	1
3	2	2
4	1	1
5	2	0

29.

Vertex	Indeg	Outdeg
1	2	1
2	1	3
3	2	1
4	1	1

31.

Vertex	Indeg	Outdeg
1	0	2
2	2	1
3	1	1
4	1	0

33. In the left graph, vertex 2 has in-degree 3, and no vertex in the right graph has that.

35. In the right graph, vertex 5 has in-degree 3, and no vertex in the left graph has that.

37. The left graph has one vertex (namely, 1) with in-degree 1, and the right graph has more (namely, $6, 7, 8$). Hence, they are not isomorphic.

39. Define $f(1) = 4, f(2) = 3$.

41.

43.

45. *Sketch.* For each path P in G, we have the underlying path $\underline{P}$ in $\underline{G}$. $\square$

47. *Sketch.* Suppose $G \cong H$ via isomorphism $f : G \longrightarrow H$. Define $\underline{f} : \underline{G} \longrightarrow \underline{H}$ by $\underline{f}(v) = f(v)$ and $\underline{f}(\underline{e}) = \underline{f(e)}$ for all vertices v and edges e in G. Observe that $\underline{f}$ is an isomorphism. $\square$

49.
$$
\begin{array}{c}
\quad\;\, 1\;2\;3 \\
\begin{array}{c} 1 \\ 2 \\ 3 \end{array}
\left[\begin{array}{ccc} 0 & 0 & 1 \\ 1 & 0 & 0 \\ 0 & 1 & 0 \end{array}\right] = A,
\end{array}
\qquad
\begin{array}{c}
\quad\;\, 1\;2\;3 \\
\begin{array}{c} 1 \\ 2 \\ 3 \end{array}
\left[\begin{array}{ccc} 0 & 1 & 0 \\ 0 & 0 & 1 \\ 1 & 0 & 0 \end{array}\right] = A^2.
\end{array}
$$

51. $A + A^T$.

53. $\sum_{j=1}^{n} a_{k,j} = \text{outdeg}(v_k)$, $\qquad \sum_{i=1}^{n} a_{i,k} = \text{indeg}(v_k)$, $\qquad$ and $\sum_{i=1}^{n} \sum_{j=1}^{n} a_{i,j} = |E|$.

55. No. $p(1,3) + p(1,2) = .5 \neq 1$.

57.
$$
\begin{array}{c}
\quad\;\; 1 \quad 2 \quad 3 \quad 4 \\
\begin{array}{c} 1 \\ 2 \\ 3 \\ 4 \end{array}
\left[\begin{array}{cccc} 0 & 0.4 & 0.6 & 0 \\ 0 & 0 & 1 & 0 \\ 0 & 0 & 1 & 0 \\ 0 & 0.2 & 0.8 & 0 \end{array}\right] = M,
\end{array}
\qquad
\begin{array}{c}
\quad\;\; 1\;2\;3\;4 \\
\begin{array}{c} 1 \\ 2 \\ 3 \\ 4 \end{array}
\left[\begin{array}{cccc} 0 & 0 & 1 & 0 \\ 0 & 0 & 1 & 0 \\ 0 & 0 & 1 & 0 \\ 0 & 0 & 1 & 0 \end{array}\right] = M^2.
\end{array}
$$

59. (a)

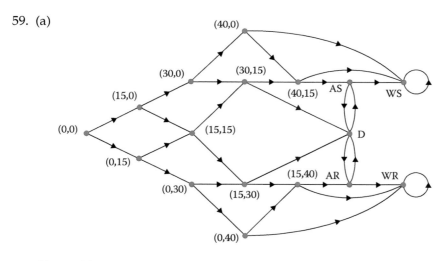

(b) 15. (c) 13.

61. For a fixed i, the sum of the weights of the edges of the form (i, j) must be 1, by the definition of a Markov chain graph. The column sums need not be 1. See Exercise 57, for example.

63. (a) 0.173 after two at bats, and 0.266 after three at bats. (b) 0.590. Start matters, since the values in column 4 of M^∞ vary.

65. (a)

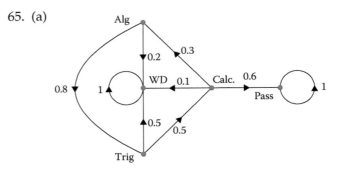

(b) Calculus, trigonometry, and algebra have period 3. Withdraw and pass have period 1. (c) 68.2% go from calculus to pass. 72.7% go from algebra to withdraw.

67. (a)

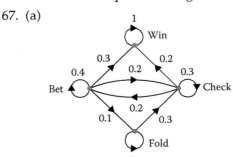

(b)

	Bet	Check	Fold	Win
Bet	0	0	0.333	0.667
Check	0	0	0.5	0.5
Fold	0	0	1	0
Win	0	0	0	1

$\approx M^\infty$

(c) If the computer bets on the first turn, then the computer probably wins. If the computer checks on the first turn, then there is an even chance of winning and losing. (d) No.

Review

1.

Nord

West Ost

Süd

2.

Buff   Rock   Syr

Ith   Bing

3. (a) 1   2

4   3

(b) ∅. (c) 2. (d) Yes.

4. (a)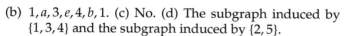

(b) $1, a, 3, e, 4, b, 1$. (c) No. (d) The subgraph induced by $\{1, 3, 4\}$ and the subgraph induced by $\{2, 5\}$.

5. (a) Yes. (b) Yes.

6. (a) Yes. (b) Yes. (c) Yes.

7.

(1, 1) •————→• (2, 1)

(1, 2) •

(1, 3) •

8. 1   2   3   4   5   6
    •———•———•———•———•———•

9. (a)   1   (b)

    3   2 $1, 2, 3, 1, 4$.

10. (a) See Figure 8.19. (b) 2. (c) Yes, 0101, a left-handed male. (d) 0001 is equidistant from 0000 and 0101.

11. *Sketch.* Every pair of vertices in W must be connected by an edge of G that also must be an edge of H. □

12. (a) No. (b) Yes. (c) Yes.

13. Let $V_1 = \{111, 100, 001, 010\}$ and $V_2 = \{000, 011, 110, 101\}$.

14. No. It contains 5-cycles.

15. No. It could contain 3-cycles.

16. (a)

$$
\begin{array}{c}
\quad\quad 1\ 2\ 3\ 4 \\
\begin{array}{c} 1 \\ 2 \\ 3 \\ 4 \end{array}
\left[
\begin{array}{cccc}
0 & 1 & 1 & 0 \\
1 & 0 & 0 & 1 \\
1 & 0 & 0 & 1 \\
0 & 1 & 1 & 1
\end{array}
\right].
\end{array}
$$

(b) $1 : 2, 3; 2 : 1, 4; 3 : 1, 4; 4 : 2, 3, 4$.

17. $\begin{bmatrix} 3 & 7 \\ 6 & 0 \end{bmatrix}$.

18. (a) $\begin{bmatrix} 0 & 1 & 0 & 0 \\ 0 & 0 & 0 & 1 \\ 0 & 0 & 1 & 0 \\ 1 & 0 & 0 & 0 \end{bmatrix} = P.$ (b) $\begin{bmatrix} 1 & 0 & 0 & 1 \\ 0 & 1 & 1 & 1 \\ 1 & 0 & 0 & 1 \\ 0 & 1 & 1 & 0 \end{bmatrix} = PA.$

(c) It is the adjacency matrix relative to the vertex ordering $2, 4, 3, 1$.

19. 3.

20.

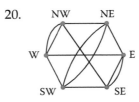

21. Define $f(1) = 5, f(2) = 6, f(3) = 8, f(4) = 7$.

22. Define $f(1) = 2, f(a) = c, f(b) = d$.

23. Define $f(1) = 3, f(2) = 6, f(3) = 5, \quad f(4) = 4, f(5) = 1,$ $f(6) = 2$.

24. *Sketch.* For each vertex $b_1 b_2 \cdots b_n$ in Q_n, we can find an automorphism f of Q_n such that $f(00 \cdots 0) = b_1 b_2 \cdots b_n$. $\square$

25. No.

26. No.

27. Define $f(1) = f(4) = a, f(2) = b, f(3) = c$.

28. 2.

29. 48.

30. No. There is no K_4 in the octahedron.

31. It follows from Corollary 8.14.

32. The graph on the left has maximum degree 4, while that on the right has maximum degree 5.

33. The degree sequences are different.

34. Let U be the set of vertices in the left graph of degree 2 or 3, and similarly define W for the right graph. The subgraphs induced by U and W are not isomorphic, so the graphs cannot be isomorphic.

35. No.

36.

37.

38. Pictured is the complement of the left graph.

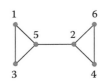

39. *Sketch.* $V_{(G^c)^c} = V_{G^c} = V_G \quad$ and $\quad E_{(G^c)^c} = (E_{G^c})^c = (E_G{}^c)^c = E_G$. $\square$

40. (a) (b)

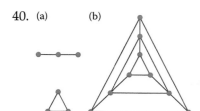

41. (a) The subgraph induced by {000, 010}. (b) The subgraph induced by {000, 001, 010, 011, 100, 110}.

42. *Sketch.* For all $u \in V_G$, $v \in V_H$, $d \in E_G$, and $e \in E_H$, define $f(u, v) = (v, u)$, $f(d, v) = (v, d)$, and $f(u, e) = (e, u)$. This gives an isomorphism $f : G \times H \longrightarrow H \times G$. $\square$

43.

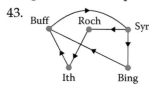

44.

(a) (b) (c) (d)

All of
G is
one.

(e) Vert.	In.	Out.
1	0	2
2	2	0
3	0	2
4	2	0

45. Yes.

46. Define $f(1) = 7$, $f(2) = 11$, $f(3) = 9$, $f(4) = 10$, $f(5) = 8$, $f(6) = 12$.

47. The left graph has a vertex with out-degree 3, and the right graph does not.

48. They are not. The left graph has a vertex with out-degree 0, and the right graph does not.

49.

50.

51.

$$
\begin{array}{c}
1\,2\,3\,4\,5\,6 \\
\begin{array}{c} 1 \\ 2 \\ 3 \\ 4 \\ 5 \\ 6 \end{array}
\left[
\begin{array}{cccccc}
0 & 1 & 0 & 1 & 0 & 0 \\
0 & 0 & 1 & 1 & 1 & 0 \\
0 & 0 & 0 & 0 & 0 & 0 \\
0 & 0 & 0 & 0 & 0 & 0 \\
1 & 0 & 0 & 1 & 0 & 0 \\
0 & 0 & 1 & 0 & 1 & 0
\end{array}
\right].
\end{array}
$$

52. No, its strong components are two isolated vertices and the subgraph induced by the other three.

53. No.

54.

$$\begin{array}{c@{\quad}cccc} & 1 & 2 & 3 & 4 \\ \begin{array}{c} 1 \\ 2 \\ 3 \\ 4 \end{array} & \left[\begin{array}{cccc} 0 & 1 & 0 & 0 \\ 0 & 0 & 0 & 1 \\ 0 & 0.2 & 0.8 & 0 \\ 0.3 & 0 & 0.7 & 0 \end{array}\right] \end{array}.$$

55. (a)

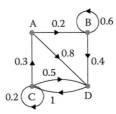

(b) 0.08 is the relevant value in M^2. (c) Irreducible, since M^3 has all nonzero entries. Regular, since all states are aperiodic. (d) $\frac{4}{9} \approx 0.4444$.

Chapter 9

Section 9.1

1. Isolated vertices 1, 3, and 5.

3.

5. $\kappa = 1$. It is connected, and {4} disconnects it.

7. $\kappa = 2$. The removal of one vertex leaves a connected graph, and {1, 5} disconnects it.

9. $\kappa = 1$. It is connected, and {4} disconnects it.

11. $\kappa = 3$. The graph is vertex transitive. The removal of one vertex leaves a graph that no one vertex will disconnect. Also, $\delta(G) = 3$.

13. (a) $K_{3,3}$. (b) $\kappa = 3$, by Theorem 9.4. (c) No.

15. *Proof.* Suppose G is connected and $\delta(G) = 1$. So $1 \le \kappa(G) \le \delta(G) = 1$. Hence, $\kappa(G) = 1$. $\square$

17. (a) *Proof.* Suppose G has n vertices and n edges. Then, $n\delta(G) \le \sum_{v \in V} \deg(v) = 2|E| = 2n$. Thus, $\kappa(G) \le \delta(G) \le 2$. Since $\kappa(C_n) = 2$, the cycle C_n has the highest possible connectivity. $\square$ (b) See Theorem 9.3.
(c) C_5 and

(d) Probably last, since the greatest number of components can be left by removing a single vertex.

19. *Sketch.* Here $n\delta(G) \le \sum_{v\in V} \deg(v) = 2|E| < 2\lceil\frac{3n}{2}\rceil$. It follows that $\delta(G) < 3$. Hence $\kappa(G) \le \delta(G) \le 2$. $\square$

21. By Theorem 9.3, each component must be a cycle. Of course, a graph is the disjoint union of its components.

23. *Sketch.* Let D be a κ-set for G, and let $v \in D$. Suppose v is not adjacent to some component of $G\backslash D$. Then, $D' = D\backslash\{v\}$ disconnects G. Thus, $\kappa(G) \le |D'| < |D|$, a contradiction. $\square$

25. Take two copies of K_{d+1} and let u and v be one vertex from each. Make a new graph G by pasting together u and v to make a new single vertex w that will have degree $2d$. Note that $\delta(G) = d$, and $\{w\}$ is a κ-set.

27. Note that $G \times P_2$ contains subgraphs $G \times \{1\}$ and $G \times \{2\}$ that are isomorphic copies of G. Let D be a κ-set for $G \times P_2$. At least one of the copies of G must be disconnected in $(G \times P_2)\backslash D$. Moreover, D must contain at least one vertex in each copy of G. So at least one of the vertices of D was not needed to disconnect the copy of G that got disconnected. Hence, $\kappa(G) \le \kappa(G \times P_2) - 1$.

29. $\lambda = 1$. The graph is connected, and edge $\{1,4\}$ disconnects it.

31. $\lambda = 2$. It follows from Exercise 7 that $2 = \kappa \le \lambda$. Also, the edges $\{\{6,1\},\{1,2\}\}$ disconnect the graph.

33. $\lambda = 2$. The removal of a single edge will not disconnect the graph, yet $\{\{4,5\},\{5,6\}\}$ will.

35. $\lambda = 3$. Apply Theorem 9.6 and Exercise 13(b).

37. $\lambda = 3$. Apply Theorem 9.6 and Exercise 11. There are multiple λ-sets.

39. For $\lambda(G) \le 1$, consider cases based on whether G is connected or not.

41. $n - 1 = \kappa(K_n) \le \lambda(K_n) \le \delta(K_n) = n - 1$.

43.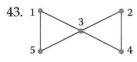

45. (a) $\kappa = 1$. (b) $\lambda = 1$.

47. (a) $\kappa = 1$. (b) $\lambda = 1$.

49. $\kappa = \lambda = 1$.

51. If S disconnects $\underline{G}$, then S disconnects G.

Section 9.2

1. Neither. There are four vertices of odd degree.

3. An Euler trail. $1, c, 3, f, 4, d, 2, a, 1, b, 2, e, 5, h, 5, g, 4$.

5. An Euler trail. $5, 6, 1, 2, 3, 4, 5, 2$.

7. An Euler circuit. $6, n, 8, l, 7, j, 6, k, 7, o, 10, q, 9, p, 9, m, 6$.

9. An Euler circuit. The graph is 4-regular.

11. A (romantic) Euler trail.

13.

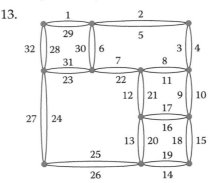

15. Yes. Connect the two odd-degree vertices with an edge.

17. No. c.

19. The layout of the hallways determines a graph in which vertices represent intersections (cross ways) or corners and edges represent hallways. Since each hallway must be both mopped and waxed, each edge is doubled. Thus, every vertex has even degree and Euler's Theorem applies.

21. Vertex repetitions in an Euler circuit determine where to begin and end the cycles. Each cycle will use two edges at each of its vertices.

23. Even n.

25. Both m and n even.

27. 2.

29. (a) 2. (b) 2.

31. Neither.

33. An Euler circuit.

35. Neither.

37. Note that G' is strongly connected and each vertex now has equal in- and out-degrees.

39. Mimic the proof of Theorem 9.7(a).

41. Use a directed graph.

Section 9.3

1.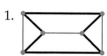

3.

5.

7.

9. The edges incident with degree-2 vertices must be included. However, premature 3-cycles are then formed.

11. Since the edges incident with degree-2 vertices must be included, all edges get included. However, a Hamiltonian cycle is not formed.

13. After all of the edges incident with degree-2 vertices get included, the bottom-middle vertex is then incident with 3 edges of the cycle, which is impossible.

15. It is not Hamiltonian.

17. It is not Hamiltonian.

19. Yes.

21. $K_{m,n}$ is Hamiltonian iff $m = n$.

23. 3.

25. $\forall\, n \geq 3,\ \frac{(n-1)!}{2}$.

27. Consider $K_{k+1,k}$.

29. Apply Theorems 9.2 and 9.11.

31. See Example 8.23, and take advantage of the symmetries. We may assume that $1, 2, 3$ is part of a Hamiltonian cycle. If the cycle further contains $1, 2, 3, 4$, then it suffices to assume that it contains $1, 2, 3, 4, 5$, and we see that this cannot be extended to a Hamiltonian cycle. If the cycle instead contains $1, 2, 3, 8$, then it suffices to assume that it contains $6, 1, 2, 3, 8$, and we see that this cannot be extended to a Hamiltonian cycle.

33. $(1,1), (1,2), \ldots, (1,n), (2,n), (2,n-1), \ldots, (2,1), (1,1)$ is a Hamiltonian cycle.

35. $(1,1), (1,2), \ldots, (1,n), (2,n), (2,n-1), \ldots, (2,2), (3,2),$ $(3,3), \ldots, (3,n), (3,1), (2,1), (1,1)$ is a Hamiltonian cycle.

37. $1, 2, 5, 6, 3, 4, 1$.

39. $1, 4, 3, 2$.

41. The lower-right vertex has out-degree zero.

43. (a) No. (b) Yes. (c) Yes.

45. *Sketch.* Let v be a vertex with the maximum possible out-degree. Let A_{out} be the set of heads of edges with tail v. Let A_{in} be the set of tails of edges with head v. Let u be any vertex in A_{in}. If there is no edge with tail in A_{out} and head u, then u has higher out-degree than v, a contradiction. Hence, u is distance 2 from v. □

Section 9.4

1.

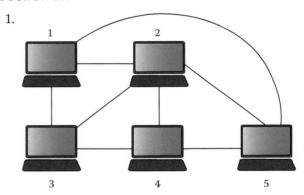

3.

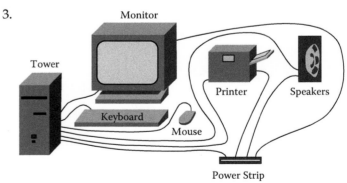

5.

7.

9.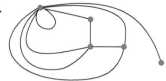

11.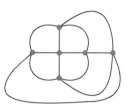

13. $|R| = 8$.

15.

17. $|V| - |E| + |R| = c + 1$.

19. *Proof.* Suppose G contains a subdivision H of K_5 or $K_{3,3}$. Suppose to the contrary that G is planar. Then H is planar, and hence K_5 or $K_{3,3}$ is planar. This is a contradiction. $\square$

21. $\forall\, n \geq 5$, K_n contains K_5 as a subgraph.

23. It is not planar, since it contains a $K_{3,3}$ subdivision.

25. It is planar.

27. This is $K_{3,3}$, which is not planar.

29. It is planar.

31. It is not planar.

33. K_5 is the only one.

35. *Proof.* Let G be a planar graph. Suppose to the contrary that $\delta(G) \geq 6$. Then $3|V| = \frac{1}{2}(6|V|) \leq \frac{1}{2}\sum_{v \in V} \deg(v) = |E| \leq 3|V| - 6$, a contradiction. $\square$

37. Let $V = \{0, 1, \ldots, n-1\}$, $E_0 = \{\{0,1\}, \{0,2\}, \ldots, \{0, n-2\}\}$, $E_1 = \{\{1,2\},\ \{2,3\}, \ldots, \{n-3, n-2\},\ \{n-2, 1\}\}$, $E_2 = \{\{n-1, 1\}, \{n-1, 2\}, \ldots,\ \{n-1, n-2\}\}$, $E = E_0 \cup E_1 \cup E_2$, and $G = (V, E)$. Draw the cycle induced by $\{1, 2, \ldots, n-2\}$ in the unit circle, put 0 at the origin, and put $n+1$ outside of the unit circle. So G can be seen to be planar. Note that $|V| = n$ and $|E| = 3(n-2) = 3n - 6$.

39. A planar embedding is pictured.

A single edge added must be a diameter across the picture. The resulting graph then contains $K_{3,3}$ as a subgraph. So it is not planar.

41. $v = 0$ by Exercise 25.

43. $v = 1$ by Exercise 27 and Example 9.19.

45. $v = 0$ by Exercise 29.

47. $v = 1$. The graph is not planar and can be drawn with one crossing.

49. This drawing has two crossings.

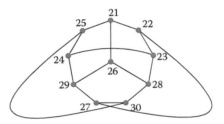

51.

53.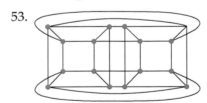

55. One crossing is possible as shown.

Now apply the result in Exercise 39.

57. True. The properties of being connected and having only even degrees are preserved in subdivisions.

Section 9.5

1. No.

3. $\chi = 2$.

5. Since $\omega = 3$ and a 3-coloring is pictured, $\chi = 3$.

7. $\chi = 2$. This graph is $K_{3,3}$.

9. Let (V_1, V_2) be a bipartition. Use color 1 on the vertices in V_1 and color 2 on V_2.

11. 2.

13. 6

15. 2

17. 1

19. It is easy to produce a 4-coloring of the Grötzsch graph. So we establish that $\chi = 4$ by showing that at least 4 colors are needed. *Sketch.* The outer 5-cycle is "uniquely" 3-colorable. This then forces all three colors to be used on the 5 neighbors of the center vertex. Now, the center vertex requires a fourth color. □

21. By symmetry, it suffices to consider three cases.

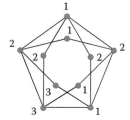 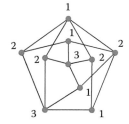 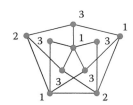

None is bipartite, and each has a 3-coloring.

23.

Time period	Committee meeting
1	German, Japanese
2	History, Kuwait
3	Indochina, Latin

25. Since C_5 is not bipartite, 3 colors are needed. One color class must be of size 1. The other color classes are then forced.

27. The sizes of the color classes do not match up in the pictured colorings.

29. $\chi = 3$. The 5-cycles need 3 colors, and a 3-coloring is easily found.

31. $\chi = 4$. A 4-coloring is easily found. A 3-coloring is seen to be impossible, by trying to construct one.

33. $\chi = 4$. Try starting a 3-coloring from the middle vertex, and observe that it fails.

35. For any subset W of V, the set W is independent in G^c iff W induces a clique in G.

37. The graph with vertices {acid, bleach, sulfides, ammonia, hydrogen peroxide}, whose edges reflect potential dangerous chemical reactions, has chromatic number 2. Putting acids and ammonia in cabinet 1 and bleach, sulfides and hydrogen peroxide in cabinet 2 is safe.

39.

41.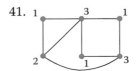

43. *Proof.* ($\rightarrow$) Suppose G and H are bipartite. So $\chi(G), \chi(H) \leq 2$, by Theorem 9.19. Thus, $\chi(G \times H) \leq 2$, by Theorem 9.28. Hence, $G \times H$ is bipartite, by Theorem 9.19. ($\leftarrow$) Suppose $G \times H$ is bipartite. So $\chi(G), \chi(H) \leq \max\{\chi(G), \chi(H)\} = \chi(G \times H) \leq 2$. Hence, G and H are bipartite. □

45. (a)

   (b) Use the result from Exercise 22, together with the observation that vertex n must receive a color different from those of $1, 2, \ldots, n - 1$.

47. Use the Greedy Coloring Algorithm and color the highest-degree vertex first. At most $d_2 + 1$ colors will be used.

49. (a) Say v_1 and v_2 are combined to form v. A $\chi(G')$-coloring of G' gives a $\chi(G')$-coloring of G with v_1 and v_2 the same color. (b) Let G be C_6 and identify two opposite vertices (such as 1 and 4).

51. *Sketch.* Let v be a vertex such that $G \backslash \{v\}$ is disconnected, and let $H_1, \ldots, H_c$ be the components of $G \backslash \{v\}$. Argue that, for each $1 \leq i \leq c$, the subgraph induced by $H_i \cup \{v\}$ can be colored with at most $\Delta(G)$ colors. Further, all of these colorings can be arranged to give v the same color. $\square$

53. 3.

55. (a) By the Four-Color Theorem, the map can be colored using only four colors. (b) 4. (c) The states PA, MD, VA, KY, and OH form a 5-cycle that requires 3 colors by Exercise 22. Since WV is adjacent to each of these, a fourth color is required.

Review

1. (a) G is connected, and the removal of the central vertex disconnects it. (b) No single edge disconnects G, but two do.

2. $5 = \min\{5, 7\}$.

3. 4, since $4 = \kappa \leq \lambda \leq \delta = 4$.

4. $\kappa = 1$ for all paths on 2 or more vertices.

5. 2, since $2 = \kappa \leq \lambda \leq \delta = 2$.

6. $\kappa(G) = 2$, since the top-middle and bottom-middle vertices form a disconnecting set, and no single vertex does.

7. False. $\kappa(G) \leq \lambda(G)$.

8. True.

9. *Sketch.* The complement of the pair of vertices not joined by an edge forms a κ-set. $\square$

10. Since $\kappa(G) \leq \lambda(G) \leq \delta(G) = 3$, it suffices to observe that no two vertices disconnect G.

11. $\lambda = \min\{m, n\}$, since $\min\{m, n\} = \kappa \leq \lambda \leq \delta = \min\{m, n\}$.

12.

13. There are four vertices of odd degree.

14. An Euler circuit exists, since each vertex has even degree and the graph is connected.

15. There is no Euler circuit, since two vertices have odd degree. There is an Euler trail $(1, 1)$, $(2, 1)$, $(1, 2)$, $(2, 2)$, $(1, 1)$, $(2, 3)$, $(1, 2)$.

16. Neither. There are eight vertices of degree 3.

17. $1, 2, 3, 4, 1, 3, 5, 2, 4, 1$ is an Euler circuit.

18. There is no Euler circuit, since two vertices have odd degree. There is an Euler trail $1, 2, 3, 4, 5, 6$.

19. An Euler circuit is shown.

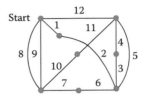

Since there are no vertices of odd degree, there is no Euler trail.

20. An Euler circuit is shown.

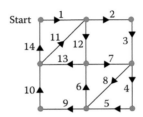

Since each vertex has outdeg = indeg, there is no Euler trail.

21. An Euler trail is shown.

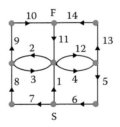

Since vertex S has outdeg = indeg + 1, there is no Euler circuit.

22. Odd $n \geq 5$.

23.

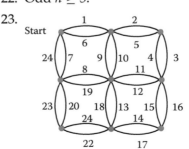

24.

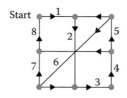

25. $(1, 1), (2, 1), (1, 2), (2, 2), (1, 3), (2, 3), (1, 4), (2, 4), (1, 1)$.

26. 6

27. The edges incident with the degree-2 vertices form a cycle prematurely.

28. No. Edges incident with the degree-2 vertices form a cycle prematurely.

29. *Proof.* Suppose G is Hamiltonian with Hamiltonian cycle C. Since C is a subgraph of G, we have $2 = \delta(C) \leq \delta(G)$. □

30. No. The corresponding graph does not have a Hamiltonian cycle.

31. A Hamiltonian cycle is shown.

32. (a) Zed. (b) Zed, Xia, Quo, Jack. (c) Yes, the Hamiltonian path is unique.

33. From any one vertex, you can follow the Hamiltonian cycle to any other.

34. $|V| = 15$.

35. The subgraph obtained by deleting the center vertex is a subdivision of K_5.

36. *Sketch.* Since $5|R| \leq 2|E|$, we have $2 = |V| - |E| + |R| \leq |V| - |E| + \frac{2}{5}|E|$. □

37. The values $|V| = 10$ and $|E| = 15$ do not satisfy the inequality in Exercise 36.

38. K_4 is planar, and all others are subgraphs of K_4.

39.

40. The face labels on the left correspond to the vertex labels on the right.

41. No. Subdivide an edge of K_5. The result contains neither K_5 nor $K_{3,3}$.

42. A planar embedding is pictured.

43. The following graph is isomorphic to $K_{3,3}$,

and the circuit board contains a subgraph that is a subdivision of it.

44. It is not planar. It is a subdivision of $K_{3,3}$.

45. $\nu = 1$.

46. $\nu = 1$.

47. $\chi = 4$. Try starting a 3-coloring from the middle vertex, and observe that it fails.

48.

49. $\omega = 3, \alpha = 3, \chi = 3$.

50. $\omega = 3, \alpha = 3, \chi = 3$.

51. 3 sessions are needed since Chess, Math, and NHS form a 3-clique. It is possible by the schedule: (1) Archery, NHS, (2) Chess, Student Council, (3) Math.

52. The Greedy Coloring Algorithm gives the 4-coloring on the left, whereas the 3-coloring on the right is optimal.

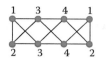 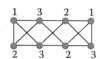

53. $\Delta(G) + \delta(G^c) = n - 1$.

54.

55. (a) It is easy to do in three colors. (b) Three. (c) The regions Alberta, Northwest Territories, and Saskatchewan form a 3-clique that requires three colors.

Chapter 10

Section 10.1

1. No.

3. $\min\{m, n\} = 1$.

5. No.

7. No.

9. Yes.

11. 5.

13. (a) Yes. (b) No. (c) No. There are distinct shortest paths from Center to Main.

15. The existence of paths in G makes G connected. Uniqueness of paths implies that no cycle can exist in G.

17. *Proof.* ($\rightarrow$) Suppose G has a unique spanning tree T. So $V_T = V_G$. Suppose that outside of T there is some edge $e \mapsto \{u, v\}$. Hence, in T there is a path P from u to v. If we let d be the first edge on P, then $(T\backslash\{d\}) \cup \{e\}$ is a spanning tree, different from T. So it must be that $E_T = E_G$, and hence $T = G$. ($\leftarrow$) If G is a tree, then only G itself can be its own spanning tree. $\square$

19. *Sketch.* Suppose G is connected and $|E| = |V| - 1$. Let T be a spanning tree for G. Since $|E_T| = |V_T| - 1 = |V| - 1 = |E|$, it follows that $E_T = E$ and $G = T$ is a tree. $\square$

21. *Proof.* ($\rightarrow$) Suppose T is a tree with exactly 2 leaves, and let P be a longest path in T. Note that the two ends of P must be leaves. Suppose that there are vertices outside of P, and let v be one of greatest possible distance from P. Then v must be a third leaf. Since there can be no vertices outside of P, it follows that $T = P$. ($\leftarrow$) Obvious. $\square$

23. $n - c$.

25. Just $K_{1,n}$.

27. *Proof.* Argue that trees are planar. So $|V| - |E| + 1 = 2$. □

29. $\binom{n}{2} + n$.

31. $E_T = \{\{1,2\}, \{2,3\}, \ldots, \{n-1, n\}\}$.

33. Remove an edge from the Hamiltonian cycle guaranteed in Example 9.13.

35. *Proof.* Suppose G has a spanning tree T. Let u and v be vertices in G. The path from u to v in T is also a path in G. So G is connected. □

37. (a) 3, 5. (b) 2. (c) 2. (d) 3. (e) Yes. (f) Yes.

39. (a) 4. (b) 2. (c) 2. (d) 3. (e) No. (f) No.

41. (a) No, grandchild. (b) Parent. (c) Leaves. (d) 3. (e) Yes.

43. No. The tree in Figure 10.9 is balanced with v_2 as the root, but not with v_1 as the root.

45. False.

47. *Sketch.* (i) Each vertex is either internal or a leaf. (ii) There are $n - 1$ vertices other than the root. Direct the edges from parent to child. Each nonroot vertex is the head of a unique edge with internal tail. Each internal vertex is the tail of m edges. So $mi = n - 1$. □

49. $n = \frac{ml-1}{m-1}$ and $i = \frac{l-1}{m-1}$.

51. 31.

53. *Sketch.* Alter the proof of Theorem 10.7, changing inequalities to equalities. □

55.

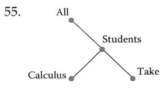

57.

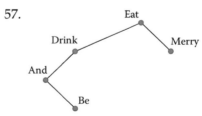

59.

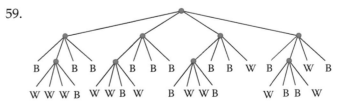

61.

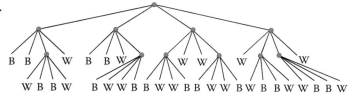

B   B   W   B   B   W   W   W   W   W

W B   B W   B W W B   B W W B   B W W B W B   B W W B   B W

63. L.

65. V.

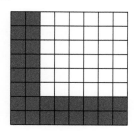   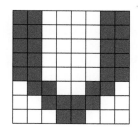

Section 10.2

1. $[1, 2, 4, 3, 5, 6, 7]$.

3. $[1, 2, 4, 6, 5, 3, 7]$.

5. $[1, 3, 4, 5, 2, 6]$.

7. $[1, 2, 3, 5, 4, 6]$.

9. $[5, 4, 3, 7, 6, 2, 1]$.

11. $[7, 6, 3, 5, 4, 2, 1]$.

13. $[4, 6, 5, 3, 2, 1]$.

15. $[6, 5, 3, 4, 2, 1]$.

17. Yes. K_3 with labels $1, 2, 3$ is the smallest example.

19. Yes. For C_4, input orderings $1, 2, 3, 4$ and $1, 4, 2, 3$ give different trees (from root 1).

21. No. $K_{1,3}$ with the degree-three vertex labeled 1 and the others $2,3,4$ yields the same list L as does P_4 with consecutive labels $2,3,1,4$.

23. $1,2,4$. No; $1,3,4$ is another.

25. True. From each vertex, the path in the breadth-first search tree to the root is a shortest such path.

27. Yes.

29. Let v be a vertex of $G = (V, E)$.
Perform Depth-First Search for G starting at v to obtain edge set F.
If $|F| = |V| - 1$, then G is connected.
Otherwise G is not.

31. Yes.

33.

35. (a) $S\ L\ -\ n\ 1\ -\ \div$.
(b) $\div\ -\ S\ L\ -\ n\ 1$.
(c) $S\ -\ L\ \div\ n\ -\ 1$.

37. 4.

39. 7.

41. 3.

43. -6.

45. Boston, New York, Toronto, Baltimore, Tampa Bay.

47. Hamiltonian cycle *adcbe* is found.

49. Depth-First Search completes finding no Hamiltonian cycle.

51. Colorings $\boxed{1\,|\,1\,|\,2\,|\,?\,|\ |\ }$ and $\boxed{1\,|\,2\,|\,1\,|\,?\,|\ |\ }$ are attempted, before realizing none can be found.

53. Colorings $\boxed{1\,|\,1\,|\,2\,|\,?\,|\ |\ }$ and $\boxed{1\,|\,2\,|\,?\,|\ |\ |\ }$ are attempted, before realizing none can be found.

55. Consider a graph H in which the vertices are paths of length at most k in K_n. Also add a trivial vertex to H that connects to each path of length 0. The graph H will be a tree with leaves corresponding to permutations of size k. So Depth-First Search will find each of these leaves.

Section 10.3

1. (a)

(b) $\$120,000$.

3. (a)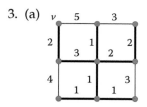

(b) $130,000.

5. (a)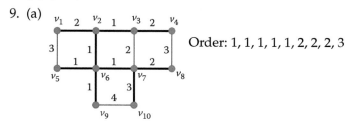

(b) $120,000.

7. (a)

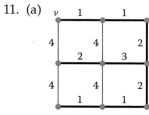

The answer depends on an edge ordering.

(b) $270,000.

9. (a)

Order: 1, 1, 1, 1, 1, 2, 2, 2, 3

(b) $140,000.

11. (a)

Order: 1, 1, 2, 2, 1, 1, 3, 2 Assuming the top left edge
of weight 1 is added first

(b) $130,000.

13. (a)

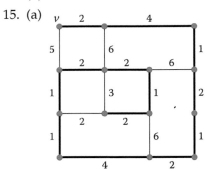

Order: 1, 2, 1, 4, 1, 1, 1, 2, 2, 2, 3 Assuming the top left edge of weight 1 is added first

(b) $200,000.

15. (a)

Order: 1, 1, 2, 2, 1, 2, 4, 2, 1, 2, 1, 4, 2 Assuming the bottom left edge of weight 1 is added first.

(b) $250,000.

17. Certainly not if it is a loop, but yes otherwise.

19. True.

21. $18,500.

23. Apply the result from Exercise 19.

25. (a)

(b) 8 PSI.

27. (a)

(b) 8 PSI.

29. (a)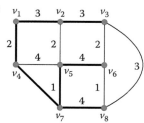

The answer is not unique since any of the three edges incident with the right middle vertex may be chosen.

(b) 8 PSI.

31. (a)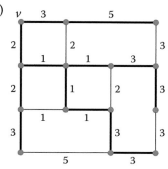

(b) 11 PSI.

33. False.

35. Take C_4 with three edges of weight 1 and one of weight 2, and let v be an endpoint of the weight-2 edge.

37. No.

39. True.

41. Adams 12, Johnson 7, Kennedy 5, Lincoln 7, Nixon 6, Polk 18.

43. Let $G = C_4$ with three edges of weight 1 and one of weight 2. Suppose v_1 is an endpoint of the weight-2 edge and v_2 is not.

45. Yes.

47. A shortest path tree.

Section 10.4

1. 3.

3. location $= 0$.
   low $= 1$, high $= 6$, mid $= 3$.
   low $= 4$, high $= 6$, mid $= 5$.
   low $= 4$, high $= 5$, mid $= 4$.
   low $= 4$, high $= 4$.
   location $= 4$.

5. location $= 0$.
   low $= 1$, high $= 7$, mid $= 4$.
   low $= 1$, high $= 4$, mid $= 2$.

low $= 3$, high $= 4$, mid $= 3$.
low $= 4$, high $= 4$.
location $= 0$. Value not found.

7. The first.

9. (a) 0.0000000105 seconds.
   (b) 500 seconds.

11. n.

13. *Proof.* Certainly, $\lceil \log_2(k+2) \rceil \geq \lceil \log_2(k+1) \rceil$. So suppose, to the contrary, that $\lceil \log_2(k+1) \rceil = n < \lceil \log_2(k+2) \rceil$. Hence, $\log_2(k+1) \leq n < \log_2(k+2)$. That is, $k+1 \leq 2^n < k+2$. Since $1 \leq 2^n - k < 2$, it follows that $2^n - k = 1$. However, $k = 2^n - 1$ is now odd, a contradiction. $\square$

15. **Maximum.**

   Let max $= 1$.
   For $i = 2$ to n,
      If $A[i] > A[\text{max}]$,
         then
            Let max $= i$.
   Return max.

17. $n - 1$.

19. $\forall\, x > 0,\ |g(x)| \leq |g(x)|$.

21. $\forall\, x > 0,\ |cg(x)| \leq |c||g(x)|$ and $|g(x)| \leq |\frac{1}{c}||cg(x)|$.

23. False.

25. True.

27. *Sketch.* Say, for $i = 1, 2$, that $\forall\, x > d_i,\ |f_i(x)| \geq C_i|g(x)|$. Let $C = C_1 + C_2$ and $d = \max\{d_1, d_2\}$. So $\forall\, x > d,\ |f_1(x) + f_2(x)| \leq |f_1(x)| + |f_2(x)| \leq C_1|g(x)| + C_2|g(x)| = C|g(x)|$. $\square$

29. *Sketch.* Suppose $f(x) \in O(g(x))$. By Exercise 19, $g(x) \in O(g(x))$. So Exercise 27 finishes the job. $\square$

31. *Sketch.* $c = c \cdot 1 \in O(g(x))$ by Exercise 21. So Exercise 27 finishes the job. $\square$

33. (a) Simple. (b) Compound. (c) 65 months.

35. *Sketch.* ($\subseteq$) $\forall\, n > 0,\ \log_2 n \leq 1 + \lceil \log_2 n \rceil$. ($\supseteq$) $\forall\, n > 0,\ 1 + \lceil \log_2 n \rceil \leq 2\lceil \log_2 n \rceil \leq 4\log_2 n$. $\square$

37. *Sketch.* ($\subseteq$) $\forall\, n > 0,\ \frac{n}{2} \log_2 \frac{n}{2} \leq n \log_2 n$. ($\supseteq$) $\forall\, n > 3$, $n \log_2 n \leq n \log_2 n + n(\log_2 n - 2) = 2n(\log_2 n - 1) = 4(\frac{n}{2} \log_2 \frac{n}{2})$. $\square$

39. *Sketch.* ($\rightarrow$) Say $\forall\, x > d_1,\ |f(x)| \leq C'_1|g(x)|$ and $\forall\, x > d_2,\ |g(x)| \leq C'_2|f(x)|$. Let $d = \max\{d_1, d_2\}$, $C_1 = \frac{1}{C'_2}$, and $C_2 = C'_1$. ($\leftarrow$) Let $d_1 = d_2 = d$, $C'_1 = C_2$, and $C'_2 = \frac{1}{C_1}$. $\square$

41. Apply Exercise 39.

43. False.

45. True.

47. (a) The second. (b) The first. (c) The first.

49. Apply Exercise 21.

51. *Sketch.* (i) For $2 \leq n$, we have $1 = \log_2 2 \leq \log_2 n$. (ii) Let C be given. For any choice of $n > 2^C$, we have $\log_2 n > C$. □

53. *Sketch.* This follows from Exercise 51 by multiplying by n. □

55. *Sketch.* (i) By induction, we see that $\forall\, n \geq 4$, $n^2 \leq 2^n$. (ii) Let C be given. It suffices to consider $C \in \mathbb{Z}$ with $C \geq 10$. By induction, we see that $\forall\, C \geq 10$, $2^C > C^3$. That is, for $n = C$, $2^n > Cn^2$. □

Section 10.5

1.

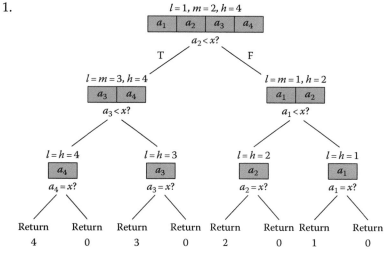

3.

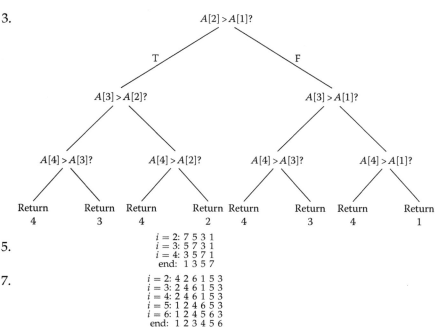

5.
$$
\begin{aligned}
i &= 2\text{: } 7\ 5\ 3\ 1 \\
i &= 3\text{: } 5\ 7\ 3\ 1 \\
i &= 4\text{: } 3\ 5\ 7\ 1 \\
&\text{end: } 1\ 3\ 5\ 7
\end{aligned}
$$

7.
$$
\begin{aligned}
i &= 2\text{: } 4\ 2\ 6\ 1\ 5\ 3 \\
i &= 3\text{: } 2\ 4\ 6\ 1\ 5\ 3 \\
i &= 4\text{: } 2\ 4\ 6\ 1\ 5\ 3 \\
i &= 5\text{: } 1\ 2\ 4\ 6\ 5\ 3 \\
i &= 6\text{: } 1\ 2\ 4\ 5\ 6\ 3 \\
&\text{end: } 1\ 2\ 3\ 4\ 5\ 6
\end{aligned}
$$

9.

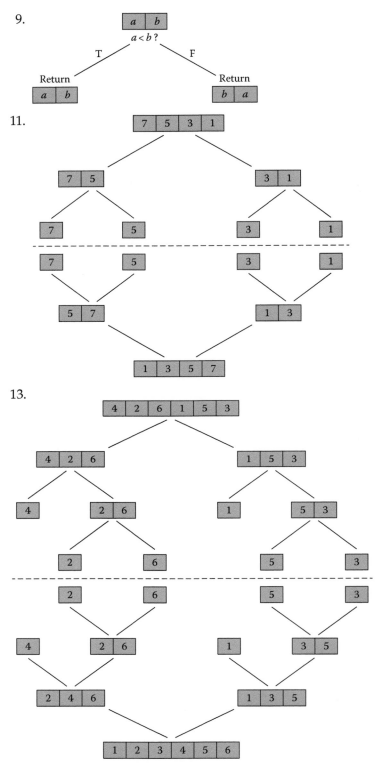

11.

13.

15.

$$\begin{array}{llll}
i = 6, j = 1: & 4\ 2\ 6\ 1\ 5\ 3 \\
j = 2: & 2\ 4\ 6\ 1\ 5\ 3 \\
j = 3: & 2\ 4\ 6\ 1\ 5\ 3 \\
j = 4: & 2\ 4\ 1\ 6\ 5\ 3 \\
j = 5: & 2\ 4\ 1\ 5\ 6\ 3 \\
i = 5, j = 1: & 2\ 4\ 1\ 5\ 3\ 6 \\
j = 2: & 2\ 4\ 1\ 5\ 3\ 6 \\
j = 3: & 2\ 1\ 4\ 5\ 3\ 6 \\
j = 4: & 2\ 1\ 4\ 5\ 3\ 6 \\
i = 4, j = 1: & 2\ 1\ 4\ 3\ 5\ 6 \\
j = 2: & 1\ 2\ 4\ 3\ 5\ 6 \\
j = 3: & 1\ 2\ 4\ 3\ 5\ 6 \\
i = 3, j = 1: & 1\ 2\ 3\ 4\ 5\ 6 \\
j = 2: & 1\ 2\ 3\ 4\ 5\ 6 \\
i = 2, j = 1: & 1\ 2\ 3\ 4\ 5\ 6 \\
\text{end:} & 1\ 2\ 3\ 4\ 5\ 6
\end{array}$$

17.

$$\begin{array}{llll}
i = 5, j = 1: & 7\ 5\ 4\ 2\ 1 \\
j = 2: & 5\ 7\ 4\ 2\ 1 \\
j = 3: & 5\ 4\ 7\ 2\ 1 \\
j = 4: & 5\ 4\ 2\ 7\ 1 \\
i = 4, j = 1: & 5\ 4\ 2\ 1\ 7 \\
j = 2: & 4\ 5\ 2\ 1\ 7 \\
j = 3: & 4\ 2\ 5\ 1\ 7 \\
i = 3, j = 1: & 4\ 2\ 1\ 5\ 7 \\
j = 2: & 2\ 4\ 1\ 5\ 7 \\
i = 2, j = 1: & 2\ 1\ 4\ 5\ 7 \\
\text{end:} & 1\ 2\ 4\ 5\ 7
\end{array}$$

19.

$$\begin{array}{ll}
i = 1: & 3\ 8\ 6\ 1\ 4 \\
i = 2: & 1\ 8\ 6\ 3\ 4 \\
i = 3: & 1\ 3\ 6\ 8\ 4 \\
i = 4: & 1\ 3\ 4\ 8\ 6 \\
\text{end:} & 1\ 3\ 4\ 6\ 8
\end{array}$$

21.

$$\begin{array}{ll}
i = 1: & 6\ 5\ 4\ 3\ 2\ 1 \\
i = 2: & 1\ 5\ 4\ 3\ 2\ 6 \\
i = 3: & 1\ 2\ 4\ 3\ 5\ 6 \\
i = 4: & 1\ 2\ 3\ 4\ 5\ 6 \\
i = 5: & 1\ 2\ 3\ 4\ 5\ 6 \\
\text{end:} & 1\ 2\ 3\ 4\ 5\ 6
\end{array}$$

23. $\frac{n(n-1)}{2}$.

25.

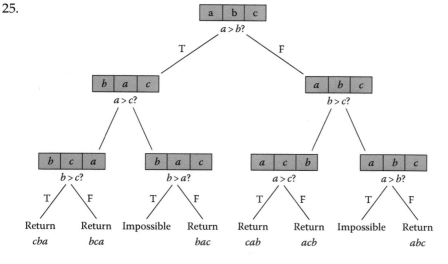

27.

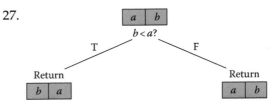

29. No.

31. Bubble Sort.

33. Order can be changed.

35. Same order.

37. *Sketch.* $2^{c_{k+1}} = 2^{c_{\lfloor \frac{k+1}{2} \rfloor} + c_{\lceil \frac{k+1}{2} \rceil} + k} = 2^{c_{\lfloor \frac{k+1}{2} \rfloor}} 2^{c_{\lceil \frac{k+1}{2} \rceil}} 2^k$

$\le 2^k \lfloor \tfrac{k+1}{2} \rfloor^{2\lfloor \frac{k+1}{2} \rfloor} \lceil \tfrac{k+1}{2} \rceil^{2\lceil \frac{k+1}{2} \rceil}$

$= \tfrac{1}{2^{k+2}} \cdot \begin{cases} (k+1)^{2(k+1)} & \text{if } k \text{ is odd} \\ k^k (k+2)^{k+2} & \text{if } k \text{ is even} \end{cases}$

$\le (k+1)^{2(k+1)}. \ \square$

39. The worst-case complexity is $n - 1$. Moreover, every input of size n uses $n - 1$ comparisons.

41.

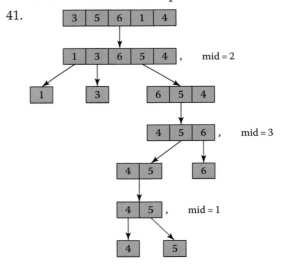

43.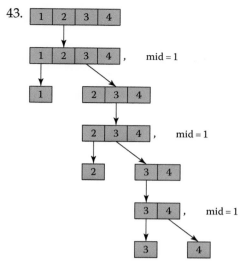

45. $\frac{n(n-1)}{2}$.

Review

1. *Sketch.* Suppose G is connected. So G has a spanning tree T. Hence, $|E_G| \geq |E_T| = |V| - 1$. $\square$

2. $4, 2, \ldots, 2, 1, 1, 1, 1$ or $3, 3, 2, \ldots, 2, 1, 1, 1, 1$.

3. 18.

4. 30.

5. $3n + 1$.

6. In a path with endpoints u and v, all vertices except u and v have degree 2.

7. (a) $4, 6, 7$. (b) 2. (c) 1. (d) 2. (e) No. (f) Yes. (g) Yes.

8. No. Two opposite edges on the square Q_2 form a subgraph that is a forest on the vertex set. However, it is not a spanning forest, which should be a tree in this case.

9. $i = n - l$ and $m = \frac{n-1}{n-l}$.

10. 32.

11. True. *Sketch.* Let v be the root. Let P be a path whose length is the diameter, and let u_1 and u_2 be its endpoints. There is a path Q_1 from u_1 to v and a path Q_2 from v to u_2. Note that the lengths of Q_1 and Q_2 cannot exceed the height of the tree. Since Q_1 followed by Q_2 is a walk from u_1 to u_2, its length is at least that of P. Thus, twice the height is at least the diameter. $\square$

12.

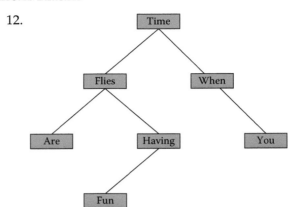

13.

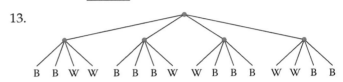

14. U.

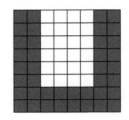

15. 1, 2, 3, 5, 4, 6, 7.

16. 1, 2, 4, 3, 5, 6.

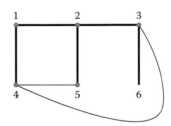

17. 1, 2, 3, 4, 6, 5.

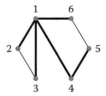

18. 3, 4, 6, 7, 5, 2, 1.

19. 5, 4, 6, 3, 2, 1.

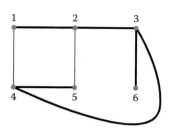

20. 3, 2, 6, 5, 4, 1.

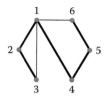

21.

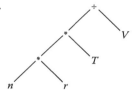

22. (a) b_1 b_2 + 2 ÷ h ∗ .
    (b) ∗ ÷ + b_1 b_2 2 h .
    (c) b_1 + b_2 ÷ 2 ∗ h .

23. 12.

24. 7.

25. Los Angeles, Houston, Chicago, New York, Boston.

26. vw, vwy.

27.

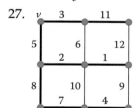

28. (a)

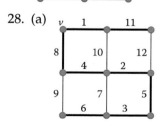

    (b) $400,000.

29.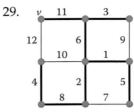

1, 2, 5, 6, 3, 8, 4, 11

30. (a)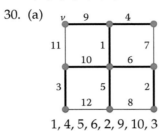

1, 4, 5, 6, 2, 9, 10, 3

(b) $400,000.

31.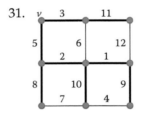

32. (a)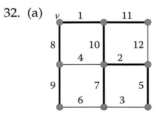

(b) 18 minutes.

33. (a) True. (b) True.

34. (a) $200 + 200 + 150 = \$550$. (b) A shortest path tree.

35. (a) $31,000. (b) A minimum spanning tree.

36.

$l=1$		$m=3$		$h=5$
1	2	4	6	8

$l=m=4$	$h=5$
6	8

$l=h=4$
6

Return location $= 0$.

37. 50 seconds.

38. $n - 1$.

39. True.

40. False.

41. False.

42. True.

43. False.

44. (a) The first. (b) The second. (c) 21 years.

45. $\exists\, C \in \mathbb{R}^+$ such that $\forall\, x > 0,\ |f(x)| \le C \cdot 1$.

46. $O(\log_2 n) \subset O(n)$ by Theorem 10.16. $O(n) \subseteq O(n^{\frac{3}{2}})$ by Lemma 10.11.

47. Yes, $O(n - 1) \subseteq O(n^2)$.

48. Yes. Minimum always uses $n - 1$ comparisons.

49.

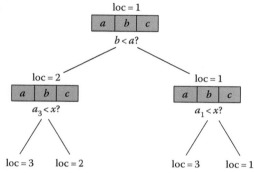

50.
$$
\begin{aligned}
i &= 2: \ 6\ 3\ 8\ 2\ 5 \\
i &= 3: \ 3\ 6\ 8\ 2\ 5 \\
i &= 4: \ 3\ 6\ 8\ 2\ 5 \\
i &= 5: \ 2\ 3\ 6\ 8\ 5 \\
\text{end} &: \ 2\ 3\ 5\ 6\ 8
\end{aligned}
$$

51.

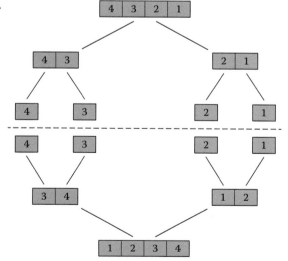

52.
$$i = 4, j = 1 : 4 \ 2 \ 1 \ 3$$
$$j = 2 : 2 \ 4 \ 1 \ 3$$
$$j = 3 : 2 \ 1 \ 4 \ 3$$
$$i = 3, j = 1 : 2 \ 1 \ 3 \ 4$$
$$j = 2 : 1 \ 2 \ 3 \ 4$$
$$\text{end} : 1 \ 2 \ 3 \ 4$$

53.
$$i = 1 : 2 \ 4 \ 3 \ 1$$
$$i = 2 : 1 \ 4 \ 3 \ 2$$
$$i = 3 : 1 \ 2 \ 3 \ 4$$
$$\text{end} : 1 \ 2 \ 3 \ 4$$

54. In order.

55. No.

56.
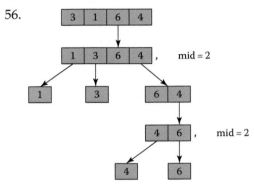

57. No.

Bibliography

Burton, D.M., *Elementary Number Theory*, 7th ed., McGraw-Hill, New York, 2011.

Burton, D.M., *The History of Mathematics: An Introduction*, 7th ed., McGraw-Hill, New York, 2011.

Cormen, T.H., Leiserson, C.E., Rivest, R.L., and Stein, C., *Introduction to Algorithms*, 3rd ed., MIT Press, Cambridge, MA, 2009.

Epp, S.S., *Discrete Mathematics with Applications*, 4th ed., Cengage Learning, Boston, MA, 2011.

Garfunkel, S., *For All Practical Purposes*, 9th ed., W. H. Freeman and Company, New York, 2013.

Harary, F., *Graph Theory*, 1st ed., Addison-Wesley, Reading, MA, 1969.

Ho, O., *Amazing Math Magic*, Sterling Publishing Company Inc., New York, 2001.

Pólya, G., Tarjan, R.E., and Woods, D.R., *Notes on Introductory Combinatorics*, 1st ed., Birkhäuser, Boston, MA, 1983.

Rosen, K.H., *Discrete Mathematics and Its Applications*, 7th ed., McGraw-Hill, New York, 2012.

Ross, S.M., *Introduction to Probability Models*, 10th ed., Academic Press, San Diego, CA, 2010.

Rudin, W., *Principles of Mathematical Analysis*, 3rd ed., McGraw-Hill, New York, 1976.

Sedgewick, R. and Wayne, K., *Algorithms*, 4th ed., Pearson Education, Boston, MA, 2011.

Tucker, A., *Applied Combinatorics*, 6th ed., John Wiley & Sons, Inc., Hoboken, NJ, 2011.

West, D.B., *Introduction to Graph Theory*, 2nd ed., Prentice Hall, Providence, RI, 2001.

Index